19500

World Mapping Today

R B Parry
Department of Geography, University of Reading, UK

C R Perkins
School of Geography, University of Manchester, UK

Graphic indexes prepared by Cartographic Unit,
Department of Geography, Portsmouth Polytechnic, UK

Butterworths
London Boston Durban Singapore Sydney Toronto Wellington

To Mary, Sarah, Matthew, Helen, Rose and Robin

First published 1987

© **Butterworth & Co. (Publishers) Ltd, 1987**

British Library Cataloguing in Publication Data

World mapping today.
 1. Maps
 I. Parry, R.B.   II. Perkins, C.R.
 III. Portsmouth Polytechnic. *Cartographic Unit*
 912   GA105

 ISBN 0–408–02850–5

Library of Congress-in-Publication Data

World mapping today.

 Bibliography: p.
 1. Acquisition of maps.   2. Libraries—Special
collections—Maps.   3. Maps—Publishing—Directories.
4. Map collections.   I. Parry, Robert B.
II. Perkins, C. R.   III. Portsmouth Polytechnic.
Cartography Dept.
Z692.M3W67 1987   025.2′86   87–25604
ISBN 0–408–02850–5

Photoset by Katerprint Typesetting Services, Oxford
Printed and bound by Anchor Brendon Ltd, Tiptree, Essex

Contents

World mapping

Acknowledgements

In preparing this book we have depended heavily on the goodwill and courteous responses of well over 1000 individuals and organizations. Our letters to map publishers and surveys prompted the return of catalogues, published material, long, carefully researched letters, map samples, and a great deal of encouraging comment. We warmly thank all these organizations and individuals, and hope that we have done justice within the constraints of this book to the information they supplied.

We also thank the many survey organizations that have allowed us to redraw graphic indexes of their map series for publication in the book.

Once the resources of our own libraries were exhausted, our researches took us on visits to several more, and into valuable contact with many people with more specialized knowledge than our own. In particular we thank the following:

- Miss Betty Fathers and staff of the Bodleian Library Map Room, Oxford,
- Miss M. B. McHugo and staff of the Overseas Surveys Directorate Library, Southampton,
- The Library of the School of Oriental and African Studies, University of London,
- Ms V. J. Galpin and staff of the Scott Polar Research Institute Library, Cambridge,
- The Library Association Library,
- The United Nations Library, London,
- Mr Philip Reilly and staff of the Library of the Land Resources Development Centre, Tolworth,
- The Inter Library Loans Departments of Reading University Library and the John Rylands University Library, Manchester.

Mrs Joy Liddell provided excellent cover at Reading during numerous sorties by RBP from the home base.

We specially thank Henderson McCartney of McCarta Ltd, London, for his encouragement in the early stages of this project.

We thank our respective heads of department, Mr Brian Goodall at Reading University and Dr Mike Pegg and Professor Brian Robson at Manchester University, for their encouragement and for the use of departmental facilities.

Finally, our wives and children deserve our greatest gratitude for tolerating an excessive burning of midnight oil through a project which sometimes seemed to have no end in sight.

List of graphic indexes

Senegal	140	1:200 000, 1:50 000 topographic
Sierra Leone	141	1:50 000 topographic
South Africa	144	1:250 000, 1:50 000 topographic; 1:250 000, 1:50 000 geological; 1:250 000 land use; 1:50 000 soils; 1:250 000 topo-cadastral
Sudan	146	1:100 000 topographic
Swaziland	148	1:50 000 topographic, geological
Tanzania	149	1:250 000 topographic
	150	1:50 000 topographic; 1:125 000 geological
Togo	151	1:200 000, 1:50 000 topographic; 1:200 000 soils
Uganda	154	1:250 000 geological
	155	1:50 000 topographic
Zambia	158	1:250 000 topographic, geological
	159	1:50 000 topographic; 1:100 000 geological
Zimbabwe	161	1:250 000 topographic
	162	1:50 000 topographic

Canada	174	1:250 000 topographic, Canada Land Inventory Maps
Mexico	182	1:250 000 topographic, thematic
	183	1:50 000 topographic, thematic
USA	194–5	1:250 000 topographic, land cover and associated maps
Alaska	197	1:250 000 topographic, land use and land cover

Belize	207	1:50 000 topographic
Costa Rica	209	1:50 000 topographic
Guatemala	212	1:50 000 topographic, geological, land use and land use potential
Honduras	214	1:50 000 topographic, geological
Panama	217	1:50 000 topographic

Bahamas	222	1:25 000 topographic
Jamaica	230	1:50 000 topographic (metric edition)
	231	1:50 000 geological
Puerto Rico	234	1:20 000 topographic, geological
Tobago	237	1:25 000 topographic
Trinidad	237	1:25 000 topographic
Turks and Caicos Islands	238	1:25 000 topographic

Argentina	245	1:500 000, 1:250 000 topographic
	246	1:200 000 geological
Bolivia	248	1:250 000, 1:50 000 topographic; 1:100 000 geological
Brazil	253	1:1 000 000 International Map of the World (IMW); 1:500 000, 1:250 000 topographic; 1:1 000 000 RADAMBRASIL
Chile	256	1:500 000 topographic
	257	1:250 000 topographic
Colombia	259	1:500 000 topographic
	260	1:200 000 topographic
	261	1:100 000 topographic, geological
Ecuador	263	1:100 000, 1:50 000 topographic
French Guiana	265	1:100 000, 1:50 000 topographic
Guyana	267	1:50 000 topographic
Paraguay	269	1:250 000 topographic
	270	1:50 000 topographic
Peru	272	1:250 000 topographic
	273	1:100 000, 1:50 000 topographic; 1:100 000 geological
Suriname	276	1:200 000, 1:100 000 topographic; 1:200 000, 1:100 000 soils
	277	1:50 000 topographic

Uruguay	279	1:200 000, 1:100 000 topographic; 1:100 000 geological
	280	1:50 000 topographic
Venezuela	282	1:500 000, 1:250 000 topographic
	283	1:100 000 topographic

Cyprus	301	1:25 000 topographic, soils, land classification
Hong Kong	303	1:20 000 topographic, geological
India	306	1:1 000 000 geological
	307	1:250 000, 1:50 000 topographic; 1:253 440/1:250 000 geological
Iran	312	1:250 000 geological, aeromagnetic
Israel	315	1:100 000 topographic; 1:50 000 geological
Japan	320	1:200 000, 1:50 000, 1:25 000 topographic; 1:200 000 geological
	321	1:500 000, 1:50 000 geological
Lebanon	328	1:20 000 topographic
Malaysia	331	1:63 360 geological
Papua New Guinea	338	1:250 000 topographic, geological, aeromagnetic
	339	1:100 000 topographic, geological
Philippines	341	1:250 000 topographic
	342	1:50 000 topographic
Qatar	344	1:50 000 topographic
Saudi Arabia	346	1:500 000 geographic, geological
Syria	350	1:200 000 topographic, geological
Thailand	354	1:250 000 topographic, geological
Yemen Arab Republic	361	1:50 000 topographic

Australia	372–3	1:100 000, 1:50 000 topographic; 1:100 000, 1:50 000 geological
	374	1:250 000 topographic, geological, aeromagnetic
New Zealand	381	1:250 000 topographic
	381	1:250 000 forest
	382	1:250 000 geological, erosion, magnetic, gravity; 1:63 360 topographic, geological
	383	1:50 000 topographic; 1:100 000 land inventory

Europe	390	1:1 500 000 International Geological Map of Europe, International Hydrogeological Map of Europe
Austria	395	1:200 000 topographic, 1:200 000 satellite image
	395	1:200 000, 1:50 000 topographic, 1:50 000 geological
Belgium	399	1:100 000 topographic
	399	1:50 000, 1:25 000 topographic; 1:25 000 geomorphological
	400	1:40 000, 1:25 000 geological; 1:20 000 soils, vegetation
Czechoslovakia	404	1:200 000 tourist series
	404	1:200 000 geological, vegetation
Denmark	407	1:100 000, 1:50 000, 1:25 000 topographic; 1:50 000 soils
Finland	410	1:200 000 road map, forests, population
	411	1:100 000, 1:50 000, 1:20 000 topographic; 1:100 000, 1:50 000/1:20 000 geological
France	417	1:100 000 topographic, soils
	418	1:50 000, 1:25 000 topographic; 1:50 000 geological, geomorphological, 'terres agricoles'
	419	1:250 000 geological, climatic
	419	1:200 000 gravimetric, vegetation
German Federal Republic	428	1:200 000 topographic, geological
	429	1:100 000 topographic, geological
	430	1:50 000 topographic, geological, soils
Great Britain	440	1:50 000 topographic
	441	1:250 000 geological UTM series
	442–3	1:25 000 topographic, soils, land use capability
	444	1:63 360, 1:50 000 geological; 1:63 360 soils
	445	1:63 360 agricultural land classification, England and Wales

1 Introduction

R B Parry and C R Perkins

You need a map? How do you know whether it's published, suitable, available, acquirable? Those who deal with cartographic information will know the inherent difficulties of these questions, although they will certainly also know how to begin the long-winded process of finding out. Those who are not familiar with maps will probably not know, and may rapidly give up trying.

This book addresses these problems by attempting to put together within one cover as much as possible of the various kinds of information needed for finding out about and acquiring modern topographic and thematic maps. Cognoscenti in the map library world will realize that this is an ambitious, even foolish task. The knowledge needed is vast and the goal, as we found, always over the horizon. Nevertheless, we hope that this book, for all its shortcomings, will be of some use both to those whose business is not normally cartographic, and to those who are more frequently involved in map acquisition. For the former we offer guidelines and basic information about what is available, for the latter we hope to offer some time saving in the usually laborious processes of search and acquisition.

Hitherto, there has been no general sourcebook covering world mapping and its availability on the grand scale. There are numerous specialized works, but none has attempted to bring it all together in the way that we do. Thus *GeoKatalog* (GeoCenter, current edition 1977–), the world's most comprehensive map retail catalogue, offers a far richer range of map citations than we have provided. Winch's *International Maps and Atlases in Print* (2nd edn, 1976) was, in its time, a very thorough listing but is now very dated: much has changed in the last 10 years. Lock's original *Modern Maps and Atlases* (1969) gave an impressive discursive account of world mapping in the 1960s, but this too has been left behind by the rapid developments in cartography since that time. There are also numerous publishers' and suppliers' address directories, some of them (e.g. Allin, 1982–) exclusively devoted to maps, though none is very satisfactory to use (see Chapter 3). For bibliographic information about cartography there is no better source than the annual *Bibliographia Cartographica* (München: K. G. Saur, 1974–). We make no claim to replace any of these individual sources, but have attempted to mix a distinctive and unique cocktail which both updates and brings together the various kinds of information found in these and many other sources. The book also offers an overview of the state of world mapping in the late 1980s, which we hope is of interest to a wide cartographic readership.

The nature and scope of the book

This book is about *current* maps, and we might first consider what is a current map. Our working definition is a map (or map series) which is currently available, useful in solving contemporary problems, and not superseded by a more recent map. Thus the book will not be much help for retrospective searching, but it does have the advantage that the cartographic material listed was, to the best of our knowledge, in print in the mid-1980s (the map listings were compiled largely from original sources during the period 1984–86).

The book is also about *available* maps, and this means that although we have discussed in our texts many restricted map series, the *map listings* and the accompanying *graphic indexes* which follow relate only to maps which we believe to be available on the international market. It does not necessarily follow that they are all easy to acquire, and the jungle of map acquisition is explored in Chapter 3.

We have limited the scope of the book in a number of ways (which are further discussed below), broadly concentrating on the kinds of material a good general map collection (with unlimited funding!) might wish to acquire in order to be able to meet a wide range of scientific enquiry.

Selection of material

Table 1.1 summarizes the kinds of mapping which have been included under the thematic headings used in the lists of maps which follow the text and address sections of each country. From this it will be apparent that the focus of attention has been on sheet maps, and particularly on series mapping in the topographic, geological and resource areas. We have not included many atlases in our listings, arguing that these really require separate treatment and are currently being catered for by Kister's series of atlas guides (Kister, 1984–). We have, however, included in-print national atlases or atlases of similar range and quality, and have also tried to include recent indigenous gazetteers for each country (often, however, falling back on the *US Board on Geographic Names* series). Meynen (1984) has provided a valuable retrospective listing of gazetteers, while Stams (1985) has produced a bibliographic survey of national and regional atlases up to 1978.

In the listings of general maps we have chosen small-scale maps of the government survey organizations where possible, but have added a selection of maps of commercial publishers as necessary. We have not attempted (as Winch

Table 1.1 Types of map included

Atlases and gazetteers	National atlases and atlases of similar quality Major gazetteers
General	Small-scale physical and road maps Satellite image maps
Topographic	Official contoured survey maps at scales generally between 1:20 000 and 1:1 000 000
Bathymetric	Maps of the configuration of the seabed
Geological and geophysical	Geological, tectonic, hydrogeological, geochemical, magnetic, gravimetric, geomorphological and mineral maps
Environmental	Maps of climate, soil, land capability, land cover, vegetation, forest and environmental hazards
Administrative	Maps of internal administrative and planning regions within countries Maps of census enumeration districts and other statistical mapping units
Human and economic	Maps of population, industry, and communications Maps of socioeconomic data
Town maps	Large-scale maps of major cities and street-finder maps

does) to list extensively the many small-scale road and tourist maps produced by commercial publishers. There is a slight prejudice in our choice towards English-language maps, but that is probably justified since English is also the language of the book. This section also includes a variety of small-scale satellite image maps, reflecting the many experiments over the last decade to adapt remotely sensed imagery to the geometry of map projections and to make viable substitutes for conventional maps. We have not included raw air photographs and satellite imagery in the book, however, since these are not amenable to the same kind of listing or systematic procurement as maps.

Detailed planimetric or cadastral mapping is not included in the listings, and in the topographic section we have restricted the scale range generally to between 1:20 000 and 1:1 000 000. (We have been flexible in drawing the line between topographic and general maps.) We do, however, discuss large-scale map series and cadastral mapping programmes in the texts.

In the geological and environmental sections we have tried to list all the major series mapping undertaken by government agencies. We have not listed monographic mapping covering only small areas of a country, however. This would have made the catalogue sections unduly long. We have tried to remedy this lack by discussing important regional environmental mapping programmes in the texts.

The human and economic section is something of a catch-all, and a great diversity of mapping will consequently be found here. But in particular we have included socioeconomic and demographic mapping based upon census statistics, and a number of census atlases thus also appear in this section.

The term 'town map' in the final section of the catalogues is used to cover both commercial street maps (for which the term 'plan' is often misapplied), and large-scale planimetric urban maps. This section, too, could have taken up many pages of catalogue space with very little benefit to the user. We have merely included a few citations of governmental and/or commercial urban maps of the capital city of each country (and sometimes a few other major cities have also been included). We hope we

have done justice to the many commercial publishers of street maps by giving due reference to their products in the texts.

A number of map categories have been almost totally excluded from our frame of reference. We have not discussed or listed any extraterrestrial maps, and most navigational (hydrographic and aeronautical) charts have been excluded. An exception has been made where such charts have more general value, such as the ICAO world charts. We have included bathymetric mapping, where the prime function of the map is to show the form of the ocean floor, rather than meet the needs of the navigator.

Early maps, reissued as reprints or facsimiles, have not been included, nor have historical atlases. The emphasis is always on the most recent mapping, and this has led to the omission of some in-print modern topographic series which have become fossilized through the arrival of a newer, active series.

We have not listed as separate items the digital map data that are becoming available from an increasing number of surveys. We do, however, include digital map availability as a footnote to map citations, where appropriate, and have discussed the development of digital mapping systems and products in the texts.

We believe that, although the number of map citations are fewer than in certain other sources, this is balanced by the texts which draw attention to the wider range of mapping produced by both governmental and commercial mapping establishments.

Collecting the data

The primary source of information used in compiling the book has been the catalogues and supplementary information provided by the survey organizations and map publishers themselves. A standard letter in one of several languages was sent to the majority of the addresses appearing in our address directories. The response rate was extremely good. For example, in the mailing to the individual state geological surveys of the USA, 47 of 49 surveys responded, most of them extremely promptly. Of some 40 addresses initially written to in the German Federal Republic, 36 replies were received. We have thus been able to validate a large number of the addresses listed in the book, a valuable exercise in itself, since there have been many recent changes and there are numerous inaccuracies in the many source lists we have used (see Chapter 3). The overall response rate to our mailings was about 70 per cent, but there were striking regional variations. Generally the pattern was predictable in that response was good from countries where mapping was known to be readily available, and poor from those where it was not!

We generally wrote a second or even third letter to major mapping organizations from whom we had not heard initially. There were also numerous follow-up letters written with specific queries. We are extremely grateful to the many people in mapping organizations around the world who responded so courteously and helpfully to our questions. Where we have not had replies from survey organizations (and there were a few countries from whom we received no returns at all), we have had to be cautious in our choice of material for the map listings, relying on many indirect sources including the catalogues of jobbers, acquisitions lists of major libraries, and publication announcements in cartographic serials such as *Cartinform* (Cartographia, 1971–).

For the texts we have used information supplied by survey organizations, but we have also delved deeply into all the literature accessible to us, both published and unpublished and in several languages. Needless to say, we have been 'lucky' in finding good material for some countries, and the texts are inevitably unequal in content as a result. Nevertheless, it is generally true that those countries which have active and accessible mapping programmes also offer the most accessible literature, and so it is likely that our texts do most justice to those countries whose mapping is available for acquisition by users of this book. We have not attempted to give full bibliographies of works consulted. Much of our source material was of the ephemeral kind, such as catalogues and unpublished conference papers; and it is not an aim of the book to be heavily bibliographic.

In the compilation of the regional sections, one of the authors (CRP) took primary responsibility for the African, Asian and Australasian sections, while the other (RBP) worked principally on the European and American sections. The polar areas and the Atlantic Ocean were also compiled by RBP, while CRP covered the Indian and Pacific Oceans. Nevertheless the contents are very much a joint responsibility, the authors sharing their knowledge and discoveries as the book progressed.

How the book is arranged

The main part of the book is arranged by country. The only geographical element is in the sorting of the world into continents and oceans: within these major units, countries are ordered alphabetically. A typical country unit has five elements:

• First there is a text which identifies and describes each of the principal mapping agencies, its policies and its products. The emphasis is on the current situation, although a little history is sometimes given to set the present organization and its maps in context.
• The second element is a section giving further information in the form of key references and publishers' catalogues and indexes to help the user who needs to know more.
• Third comes a directory of publishers' addresses. This includes addresses of indigenous publishing houses and other significant sources.
• Fourth there is a catalogue of current maps in print. This is selective, as has been explained above and in Table 1.1.
• Finally there are monochrome graphic indexes of major map series. Again these are necessarily selective. They have been specially redrawn to a common format and standard, and they are blank, i.e. they do not show the extent of published map cover. If this information had been included it would have dated very rapidly. The function of the graphic indexes is two-fold: to show the reader how the sheets of a series are arranged, and also to serve as base maps which may be used for plotting the holdings of a map library, or for entering the progress of a map series as new sheets are published. Many of the indexes serve several map series with different scales or themes. It is hard to find good blank indexes suitable for cataloguing and housekeeping purposes in a map collection, and we hope these will fill a need.

In addition to the core part of the book, there are seven introductory chapters. These offer an overview of the current state of mapping worldwide, describe the impact of new technologies on the map-making process, and explore the problems of map evaluation and acquisition.

In Chapter 2, Parry identifies the main groups of cartographic material in terms of theme and format, and discusses briefly their status in the late 1980s. In Chapter 3, Perkins considers the various methods of obtaining maps and the pitfalls which may be encountered along the way. Chapter 4 is a discussion by Farrell of the practical problem of selecting the right kinds of map for the uses they are expected to fulfil. Monmonier, in Chapter 5, explores the impact of remote sensing technology on production-line mapping, and offers a case study of the development of image maps at the United States Geological Survey. In Chapter 6, Blakemore and Rybaczuk discuss the nature of digital cartography in the 1980s and identify some key issues of the present time. Finally, in Chapter 7, Bickmore, one of the first people to link maps with computers back in the 1960s, considers why map libraries have been slow to become consumer outlets for the digital map, and offers his thoughts on the development of a world digital database of environmental information and its applications for the 1990s.

Using the book

It has been our aim to bring into close juxtaposition the various information needed for each country, so that the user can glean most of his needs within the span of a few pages: hence the integration of text, catalogue, addresses and graphics. We have made a compromise with the addresses, since to list every relevant address with each country would have led to excessive repetition of some addresses. We therefore list only significant or exclusive addresses with a country, and refer the reader to other parts of the book for the rest. There is also an index of publishers, to enable users to find the address of a specific publisher. We have included as many relevant addresses as possible but some have eluded us, and we have not always included addresses of well known book publishers, since these are easy to trace in several standard sources.

Because of their simple alphabetical arrangement within continents, countries can usually be found easily by flipping through the pages, but it should be noted that we have not adhered strictly to political definitions of each unit. Both location and the way mapping is organized have been taken into consideration. Thus for example Great Britain and Northern Ireland, although politically forming the United Kingdom, have been treated separately because of their distinct mapping systems. On the other hand, Martinique and Guadeloupe, although politically overseas departments of France with which their map series are integrated, have been placed with the other Caribbean islands for the sake of geographic continuity. The reader may therefore sometimes need to refer to the geographical index provided at the end of the book.

It is important that the text and catalogue of each country should be read and used together, since they complement each other. In some countries, the catalogue section is slim. A slim catalogue does not necessarily mean that few maps are published. There may be a substantial output of monographic maps but a dearth of series mapping, or mapping may be out of print, restricted or otherwise unavailable. We have tried to draw attention to these eventualities in the texts. The texts also give descriptions of the more significant mapping. We have aimed at objectivity in these descriptions, but they should help the user to evaluate the suitability of a map or map series for a particular requirement.

The texts are generally built around discussions of the main **mapping agencies**, but where possible the same thematic sequence is followed that occurs in the catalogue section. The addresses, too, are placed as nearly as possible in the order that they appear in the text (but with some exceptionally long address lists we have resorted to alphabetical order). Abbreviations are introduced in the text for many mapping agencies and publishers. They are repeated in the address sections and used again in the catalogues.

We have not adhered to library cataloguing rules in the construction of the map entries in the catalogue sections.

Table 1.2 Structure of the catalogue entries

Line 1 (a) *Map title* (b) Scale fraction (c) Series
        designation (d) Edition (e) Map author
Line 2 (a) Place of publication: (b) Publisher, (c) Date of
        publication
Line 3 (a) Number of sheets in series, (b) Number published.
        (c) ■ symbol indicates that a graphic index is provided
Line 4 (a) Additional notes.

Examples

An atlas

Tweede Atlas van België/Deuxième Atlas de Belgique
Bruxelles: Commission de l'Atlas National, 1983–

[Title given in the two official languages of Belgium, but rest of entry in French only; city of publication given in French form, as used on graphic indexes and in address list; dash after date of publication indicates that atlas is still in progress.]

A single sheet map

Konturkart over Norge med byer, jernbaner og riksveger 1 : 1 000 000
Hønefoss: SKV, 1976.
Outline map with towns, railways and main roads.
Also available showing only the drainage network.

[Title is in language of map: as this may be obscure to English readers, it is translated in the notes section; notes section also gives variant form of the map; full form of publisher abbreviation will be found in text and address list.]

A topographic series

Carta topografica d'Italia 1 : 50 000 Series M792
Firenze: IGMI, 1964–
636 sheets, *ca* 230 published. ■

[Title in language of map: meaning is obvious, so not translated; NATO series designation also given; date is of initiation of series, dash signifying that the series is still active; 636 sheets required to complete the series, of which about 230 are published; ■ indicates that a graphic index is provided.]

A thematic map

Carte tectonique de Bolivie 1 : 5 000 000
Bondy: ORSTOM, 1973.
Legend in French and Spanish.
Accompanies *Cahiers ORSTOM Sér. Géologie* IV (2), 1972.

[Title in principal language of map; notes section gives languages used in the map, and reference to accompanying literature.]

A thematic series

Land inventory maps 1 : 100 000 NZMS 290
Wellington: NZMS, 1978–
143 sheets in each of 7 series; *ca* 50 published. ■
Series issued on the same sheetlines to cover the following themes: soils, rock types and surface deposits, land slope, existing land use, indigenous forest, land tenure and holdings and wildlife.
Overlay editions available for agricultural and horticultural suitability, exotic forestry suitability and relief satellite image.

[Active series; seven different themes published on the same sheet lines; a total of 50 sheets so far published; notes give details of themes of the different series and overlays.]

The aim has been to identify simply but unequivocably each map or map series, so that it may be traced in a library or ordered from a supplier. Table 1.2 explains the components of a catalogue record and gives some examples. Table 1.3 gives further information on the definition of the catalogue elements for those who need it. We have tried to make the entries as consistent and helpful as possible, but in some cases have lacked the data to provide all the information we would have wished. Many mapping organizations, particularly in the developing world, give very minimal information in their catalogues, a point discussed in Chapter 3.

Table 1.3 Elements of the catalogue format

Title Given in the language of the map wherever possible. Non-roman scripts are transliterated. Multi-lingual titles are given in one or two of the principal languages.

Scale Given as a representative fraction.

Series designation An alphabetical/numerical code or name used by publisher or military authority to identify a map series.

Edition Given only when it is considered to be helpful in identifying the current map or map series.

Author Given only when a significant individual author is identified.

Place of publication During the life of a map series, this may change. The current, most recent or most usual place from which the maps are issued is given.

Publisher The names of map publishers frequently change. The current or most recent name is usually given.

Date The date of publication of a map or the date of initiation of a map series. If a series has gone through many editions, the date is that of the oldest sheet prepared to current specification. A dash not followed by a terminating date indicates that the series is still active (i.e. either not complete or under revision).

Sheets The number of sheets required to complete a series, or which are in the publisher's programme.

Sheets published The number or percentage of the series issued (generally to 1985–86). If a large number of sheets are out of print, this is indicated in notes or text.

Notes Used flexibly to give further information on map content, languages used, ancillary maps, supporting literature, and special formats. Maps are usually fully coloured unless otherwise stated.

The book and the future

This book has been constructed from a considerable database of information collected by the joint authors. We should like to keep this database up to date against the eventuality of a further edition of the book. We therefore welcome both constructive comments from users and further information from survey organizations and map publishers who do (or do not!) feature in the book. Information should be sent direct to the authors, whose addresses are given below. CRP would particularly welcome information on African, Asian, Australasian and Pacific and Indian Ocean mapping. RBP is primarily interested in European, North and South American, Polar and Atlantic Ocean mapping.

Mr R. B. Parry
Department of Geography, University of Reading, Whiteknights, READING RG6 2AB, UK.

Mr C. R. Perkins
School of Geography, University of Manchester, MANCHESTER M13 9PL, UK.

References

ALLIN, J. (1982–) *Map Sources Directory*. Downsview, Ontario: York University Libraries. 21 loose-leaf sections. Periodically updated

Bibliographia Cartographica (1974–) München: K.G. Saur. Annual

Cartinform (1971–) Budapest: Cartographia. Current information on cartographic publications. Supplied with the periodical *Cartactual*

GeoKatalog (1977–) Stuttgart: GeoCenter ILH. Two-volume loose-leaf retail catalogue of in-print mapping. Systematically updated

KISTER, K.F. (1984) *Kister's Atlas Buying Guide*. Phoenix, Arizona: Oryx Press. pp. 236. First of a planned series

LOCK, C.B.M. (1969) *Modern Maps and Atlases*. London: Clive Bingley. pp. 619

MEYNEN, E. (1984) *Gazetteers and Glossaries of Geographical Names*. Wiesbaden: Franz Steiner Verlag. pp. 518

STAMS, W. (1985) *National and Regional Atlases*. International Cartographic Association. pp. 249

WINCH, K.L. (1976) *International Maps and Atlases In Print*, 2nd edn. London and New York: Bowker. pp. 866

2 The state of world mapping

R B Parry

Since the eighteenth century, the preparation of a detailed basic reference map has been recognized by the governments of most countries as fundamental for the delimitation of their territory, for underpinning their national defence and for the management of their resources. The concept of the *topographic* map originated at that time. It is a term that is perhaps not easy to define (for a wide range of definitions, see Larsgaard, 1984), but it is easy to appreciate by example. When sorting a large number of maps by subject there is usually little problem in separating this major group of detailed, general-purpose maps from others of more specialist thematic content. Most literature on the development and progress of world mapping is largely preoccupied with topographic maps, which provide the spatial framework within which a society and its economy operate. Topographic maps at a scale usually not smaller than 1:100 000 (1 cm to 1 km) form the bread and butter of almost all government mapping programmes, and must form the most important element of any book devoted to world mapping.

However, the other major map group, which is covered by the broad collective term of *thematic* mapping, is also of considerable importance. Thematic maps are numerically far fewer than topographic and cover a great diversity of subjects. Of particular importance are those that have been designed to support the evaluation and management of the human environment.

The majority of maps, particularly at the larger end of the scale range, tend to fall into a few major clusters which reflect discrete areas of government interest in spatial data. Thus the governmental mapping of a country usually comprises a *basic* topographic map series (the basic scale is commonly about 1:25 000 but may vary from as large as 1:1000 to as small as 1:250 000), with various so-called *derived* series at smaller scales. These will be the responsibility of the principal mapping agency. Sometimes the same agency is also responsible for cadastral survey, i.e. the delimitation of property boundaries and the registration of land ownership. Such maps are necessarily of quite large scale. In many countries, cadastral survey pre-dates the establishment of a topographic survey, and it is then not unusual to find that the former continues to be carried out by a separate cadastral authority.

Geological and geophysical survey are usually the responsibility of a government department concerned with geology, mineral resources or mining. Soil, land capability and land cover mapping may be carried out by the national survey, or from within a department of agriculture or forestry, or by a quite separate soil survey institution. An increasing range of maps fall into this cluster which may collectively be termed *environmental* maps.

Other areas of official mapping are associated with highway or transportation departments, which have to maintain and publish up-to-date maps of the transportation infrastructure; with urban planning departments, which sometimes maintain large-scale mapping programmes to meet the needs of planners and of the public utility services; and with the national statistical and census authorities, which collect socioeconomic data which may be turned into distribution maps. Many countries also have aeronautical and hydrographic charting establishments, but this kind of mapping has not usually been included in this book, although *bathymetric* mapping is listed and discussed.

We have tended to group the mapping described in our texts and catalogues to reflect their provenance as well as their thematic content. In this chapter, we briefly review some of these main map clusters, together with national atlases and gazetteers, and discuss their characteristics and their status in the late 1980s.

Topographic mapping since World War II

In 1945, a world index of official topographic maps was published in the *Geographical Review* (Platt, 1945). The index was compiled by the American Geographical Society, which had at that time an active interest in mapping, and it summarized, with the help of an accompanying note, the state of world mapping at the outbreak of World War II. Even in the smallest of the three scale bands represented on the index (1:126 720 to 1:253 440) very substantial areas of the world appeared unmapped, including most of Africa, Central and South America, the USSR east of the Urals, and even Australia and New Zealand.

For much of the world therefore the history of topographic mapping is a story of the last 40 years. The war itself greatly stimulated both the production of maps and the technology behind their making. The postwar years saw concerted efforts by many nations to put their cartographic houses in order and to pursue new systematic surveys of their territories. At the same time much of the less developed world began to be mapped systematically for the first time, in many cases by the survey agencies of colonial powers or with the help of collaborative mapping agreements with western countries. In particular, the Inter-American Geodetic Survey, founded in 1946, signed mapping agreements with most Latin American countries (and continues to give support in this region); in Africa the French Institut Géographique National made a major impact on the mapping of substantial parts of the continent; while smaller areas were also mapped by the

Portuguese, Belgians and others. But the most extensive overseas mapping programme of all was achieved by the British Directorate of Overseas Surveys, like IAGS also founded in 1946 (originally as the Directorate of Colonial Surveys), which over a period of almost 40 years mapped, or in some cases remapped, more than 6.6 million square kilometres of the world's land areas. The history of the DOS programme has been described by McGrath (1983).

There have been several attempts to monitor postwar mapping progress on a continental or world basis. For example, UNESCO made a study of the mapping of Africa in 1959 (UNESCO, 1963), while more recently Hunting Surveys has charted the status of both topographic and resource surveys of the same continent (Hunting Geology and Geophysics Ltd, 1983). For the world as a whole, the United Nations Organization has conducted three surveys of topographic map cover, in 1968, 1974 and 1980, and the results have been tabulated in issues of *World Cartography* (Volumes 10, 14 and 17) and analysed and discussed by several authors (Meine, 1972; Brandenberger, 1976; De Henseler, 1982; Brandenberger and Ghosh, 1985; Carré *et al.*, 1985). The UN surveys are based primarily on the responses to a questionnaire survey sent to the national cartographic agencies. The results of the last survey look very unpromising, with about 70 per cent of the earth's land area mapped at 1:100 000 but only about 42 per cent at 1:50 000, and a mere 15 per cent at around 1:25 000. The UN survey results are, however, somewhat marred by the low response rates to the questionnaires (a rate which has also declined with each survey). Sixty-nine countries replied to the 1980 survey, covering only 50 per cent of the world's land area, although these data were supplemented from other sources (Brandenberger and Ghosh, 1985).

The state of world mapping

With such promising postwar developments, and with a succession of new surveying and mapping techniques becoming available (and described by Monmonier in Chapter 5), one might expect that by the late 1980s the world would have been mapped both thoroughly and well. Yet the UN surveys suggest that progress began seriously to stagnate during the 1970s. In fact the UN statistics even registered a *reduced* coverage in the 1:250 000 range, which is theoretically impossible.

This stagnation has occurred in the less developed world and is partly a reflection of the reduced survey activities of several western colonial nations. The overseas mapping efforts of the British Directorate of Overseas Surveys and the French Institut Géographique National, for example, have been substantially reduced (DOS essentially ceased to exist in 1984, when its much depleted staff and mapping hardware were transferred to the Ordnance Survey at Southampton to become the Overseas Surveys Directorate within the national mapping agency of the OS). Although both countries are continuing to give some survey and mapping support to developing countries as part of their aid programmes, the curtailment of external mapping effort has left many newly independent countries without sufficient survey infrastructure or financial resources of their own adequately to maintain, extend and improve their basic surveys. In parts of Africa, some of the first-time mapping is now more than 20 years old, stocks have become depleted, and in some cases the reprographic material has been lost or destroyed.

However, the situation is not everywhere so unpromising, and our own enquiries have revealed many examples of active and successful topographic mapping programmes in developing countries. Thus, in Africa, Burundi has a complete new 1:50 000 series produced by the French IGN as an aid project, while Tanzania has completed its 1:50 000 basic mapping by shopping around the aid-giving community. In Asia, there are ambitious basic mapping programmes in progress in Indonesia and Saudi Arabia, for example; while in South America, Brazil and Chile are instances of countries where very rapid progress has been made in the late 1970s and 1980s. Many developing countries continue to receive mapping aid from the developed nations through government mapping or development agencies, among them SwedSurvey of Sweden, the United States Geological Survey, the Geographical Survey Institute of Japan, Geokart of Poland, and the Canadian International Development Agency. Many private air survey companies in the western world have also carried out substantial mapping projects overseas, often using state-of-the-art mapping technologies.

While many developed nations have so far failed to bring order and integration into their diverse governmental mapping activities (Great Britain is a notable example), several developing nations have taken advantage of the previous weakness of their survey base by starting afresh and developing integrated topographic and thematic mapping programmes within a single government department. Mexico, through the foundation of CETENAL in 1968, has managed to make very rapid progress in a basic 1:50 000 topographic survey linked with resource mapping and evaluation at the same scale. Indonesia, using the high-altitude air photomapping methods developed in the USA, has made similar strides, while Brazil has used the potential of Side Looking Airborne Radar (SLAR) to map most of the country multi-thematically, albeit at the small scale of 1:1 000 000.

In the developed world, the emphasis since World War II has been on the renewal and densification of survey networks, and in a high proportion of cases a complete re-survey, generally using photogrammetric methods, has taken place. The end product is usually a multi-coloured basic map series typically at 1:25 000 or 1:50 000, together with a number of derived smaller scale series. The 1980s will have seen the completion of a large number of European postwar mapping programmes. So, while a cover index of topographic mapping such as we have supplied in Figure 2.1 gives a crude measure of the spatial progress of topographic mapping to the late 1980s, it fails to register the dynamism injected into the topographic mapping in the western world in the postwar years, and takes no account of the variety, changing specifications and historical development of mapping systems in individual countries throughout the world.

During the last decade, pressures brought by competing uses of the environment have stimulated increasing demands for larger scales of survey and for integration of the mapping requirements of, for example, the cadastral, census, public utility and agricultural services. Increasingly large scales have been adopted by Western European countries: West Germany, The Netherlands, France and Austria have all begun to produce relatively new basic series at a scale of 1:5000.

So if the world mapping statistics look rather unsatisfactory on a global scale, one finds, looking at nations individually, that there are pockets of considerable activity and innovation, and this is carried over into the development of thematic mapping programmes discussed below. More serious is perhaps the lack of availability to the public at large of topographic scale mapping in many countries. Sperling (1978) notes that even in the military

8

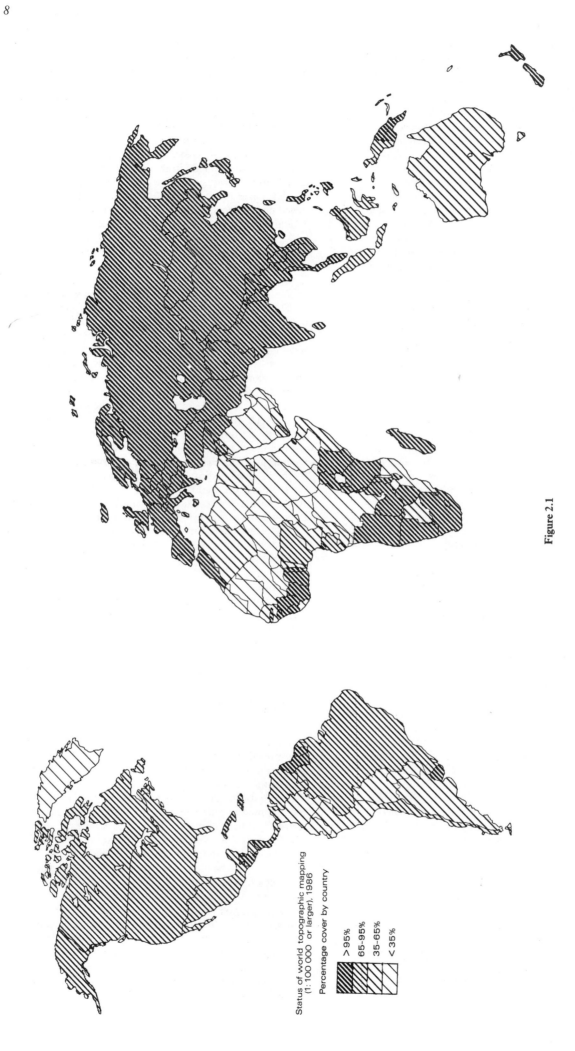

Status of world topographic mapping
(1:100 000 or larger), 1986

Percentage cover by country

>95%

65-95%

35-65%

<35%

Figure 2.1

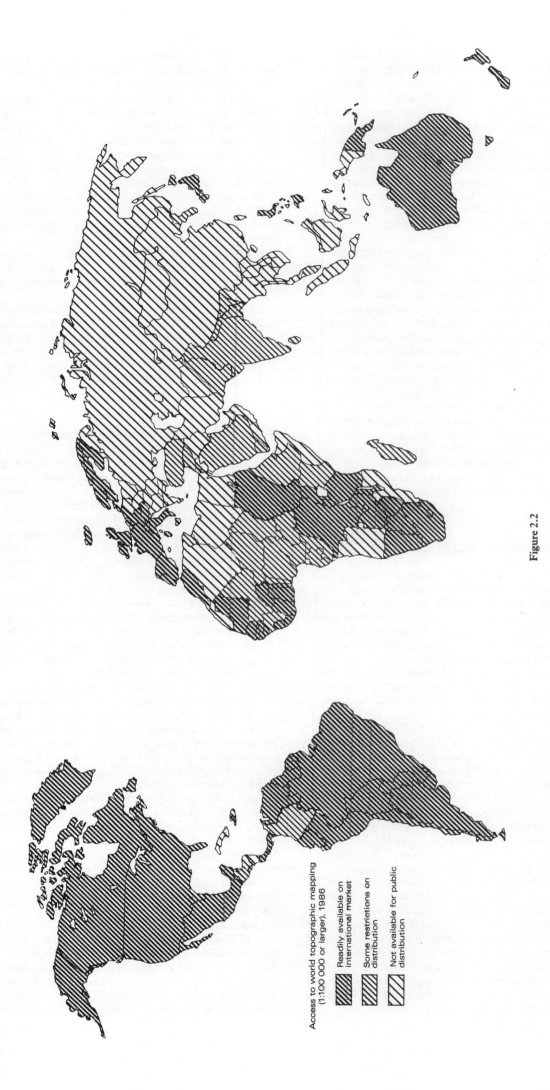

Access to world topographic mapping (1:100 000 or larger), 1986

Readily available on international market

Some restrictions on distribution

Not available for public distribution

Figure 2.2

mapping guide for East Germany the printed examples are fantasy maps and trenchantly observes that generations are not being educated in the nature or value of real maps. In view of the concern of this book with available mapping, we have balanced Figure 2.1 with a further map (Figure 2.2) showing countries where restrictions are imposed on the availability of topographic mapping. The question of availability is further analysed by Perkins in Chapter 3.

The use of proportional cover as a metric for assessing the state of world mapping is on the verge of redundancy. As Monmonier shows in Chapter 5, with present technology it is now possible very rapidly to cover large areas of land with scale-true image maps derived from satellite data, or indeed to use remotely sensed data as a map substitute for many purposes for which maps have been traditionally used. Colour image maps of the whole of Peru at 1:250 000 scale were produced in a single year (1984), and although the resolution of civilian satellite imaging systems during the 1970s was not good enough for basic topographic mapping scales in the 1:100 000 and larger-scale range, new mapping satellite systems, beginning with the Landsat Thematic Mapper in 1983 and now the French SPOT (which became operational in 1986) and the Japanese MOS (to be launched in 1987), promise to fulfil this requirement, while high-altitude air photography has been shown by the United States Geological Survey to meet national mapping standards for the production of 1:24 000 scale orthophotomaps.

Similarly the capability of global satellite positioning systems and of aerotriangulation methods has rendered unnecessary many of the tedious land-based methods of geodetic survey using fixed survey stations.

Geological and resource mapping

Traditionally, survey of a country's natural resources began with a basic geological survey in which rocks were classified mainly according to geochronological principles. In the developed world, many such surveys were begun in the 19th century, but they have often been allowed to lapse. Relatively few countries have doggedly pursued the goal of a complete and modern geological map cover at a scale of about 1:50 000. Some Western European countries have done so (France, for example, with a now almost complete modern 1:50 000 scale geological map; and Greece), but even in countries as generally well mapped as Great Britain or the German Federal Republic, there remain considerable gaps in geological map cover. In other parts of the world, the situation is very variable. Only about 25 per cent of the United States has been mapped geologically, but Australia boasts a complete cover at 1:250 000. Saudi Arabia has a substantial programme in progress at 1:100 000 and 1:250 000, which complements geographical and base map series at the same scales, and several African countries—Botswana, Zambia, Malawi—have well developed programmes. In Gabon, SLAR imagery has been used for geological mapping.

There have been few attempts to quantify the state of world geological mapping (some rather inexact maps are provided in Carré *et al.*, 1985), and it is not very meaningful to do so. Many countries do not have a policy of systematic cover, but concentrate survey activity in areas of known mineral potential. Furthermore, many geological and mineral surveys are issued only in semi-published form, the so-called 'open file' report. Namibia is an example of a country where almost all the geological information is disseminated in this way.

In recent years there has been a change of emphasis in geological mapping. Many new kinds of application-orientated maps have appeared. These include geotechnical, hydrogeological and mineral resource maps. Particularly significant are the geophysical (gravity and aeromagnetic) maps which can be produced very rapidly from aerial survey. Another growth area is in the collection and mapping of geochemical data from soil and stream sediments which can be used both for monitoring environmental pollution and for mineral prospecting. Thus both Britain and the German Federal Republic have developed digital data sets of grid-square geochemical information, from which geochemical atlases have been produced and published.

A number of these new geological map types are unique to the developed world, but some, such as those using airborne geophysical techniques have been exported to the developing world as well.

Geological mapping has also been extended to many offshore continental shelf areas in support of petroleum and mining interests in these areas. The North Sea is being mapped at 1:250 000 scale by Britain, Norway and the Netherlands, for example, in a suite of maps showing both seabed sediments and deeper structures.

New kinds of environmental maps

The term 'environmental map' covers a wide range of map types, some of which simply record the distribution of environmental attributes, while others are more synthetic, combining a number of attributes to form an interpretation of biomass potential or rangeland capacity, for example. The French have been particularly active in pursuing the concept of the environmental map and Journaux (1985) has suggested a three-tier classification. At the first level, *cartes d'analyse* map elements and simple processes, and this group includes maps of vegetation, environmental pollutants, urban structures and population (geological mapping is also included). The second level, *cartes de systèmes*, establishes relationships between environmental variables and the processes which result. This level includes environmental risk maps such as soil susceptibility to erosion. At level three are *cartes de synthèse*. These are complex maps which attempt to integrate environmental variables in a problem-solving context. Maps of this kind are exemplified by the French *Carte de l'environnement et de sa dynamique* which combines a template of environmental data with mapping of so-called 'environmental dynamics' which include soil degradation, water and air pollution, and human attempts to protect and improve the environment.

More traditional maps under the environmental heading include soil, land evaluation and land capability maps. Soil survey, in comparison with geological survey, is rather a young activity having developed in the 1930s. In the developing world more emphasis has been placed on land evaluation surveys, many of which have been carried out by western agencies or with the support of international organizations. Among the former, the acronyms LRDC, ORSTOM, INEAC and CSIRO recur, while internationally the United Nations FAO, UNEP, UNDP and the World Bank have all been active in promoting and financing such surveys. The FAO has provided a framework for land evaluation (Food and Agriculture Organization, 1976) which has been much used in the developing countries, while many soil surveys favour the taxonomy developed by the United States Department of

Agriculture or by the FAO for the *Soil map of the world*.

The problem with much original soil survey has been its rather introverted academic nature, and the difficulty of interpretation it presents for the people who manage the land and could benefit from it. But it is generally conceded that soil survey provides the essential basic data upon which more user-helpful evaluation and capability maps can be built. Increasingly soil surveys are making use of geographical information systems for the storage and manipulation of soil data (Burrough, 1986), and this gives an added flexibility to the kind of map output that can be offered to the client.

The French overseas mapping agency Institut Français de Recherche pour le Developpement en Cooperation, commonly known by the acronym ORSTOM, has produced a large number of environmental maps (Meunier and Hardy, 1986), and continues to innovate in this area.

Ocean mapping

It is not surprising that the oceans have been surveyed in much less detail than the land surfaces of the earth. Although about seven-tenths of the earth are covered by water, only about one-tenth of this area has been surveyed to an accuracy commensurate with modern topographic mapping. Charting of the oceans developed in relation to the need for safe navigation, and a large number of national hydrographic survey departments are engaged mainly in the preparation of navigational charts. A number of countries have, in addition to their charting organizations, specialist research institutes in oceanographic sciences, which have taken a lead in developing new marine mapping techniques.

There has been a great interest in the 1970s and 1980s in improving the mapping of the seabed, and a new generation of detailed bathymetric maps has begun to appear. The benchmark of bathymetric mapping in the 1980s has been the completion of the 5th edition of the world series of GEBCO charts, but already the achievement of this series has been overtaken by both the demand and the technical potential for better quality data on the morphology of the seabed.

The technical stimulation has come from several kinds of remote sensing operating from space and from the sea surface. The efficacy of radar altimetry for making precise measurements (to within a few centimetres) of variations in the surface level of the sea was demonstrated by the Seasat satellite launched in 1978, and equipped with Synthetic Aperture Radar (SAR). Initially the technique was used for studying wave patterns and ocean circulation, but many of the small distortions in sea level are due to variations in mass associated with the bottom topography, and so it has become possible indirectly to deduce gravity anomalies and ocean bottom morphology from altimeter measurements.

Seasat had a very short working life, but in the 1990s Canada will launch a new SAR imaging satellite (Radarsat).

In the deep ocean, the development of acoustic sensing techniques has produced further new and rapid methods of imaging and surveying the sea floor. Previously, depths were sounded using echo or line sounding techniques which measured only a linear profile of the floor. The most important new techniques are the American SEABEAM system (Ulrich, 1984), a multi-beam swathe sounding technique which has been used by the French, and GLORIA, a sidescan sonar technique which was pioneered by the British Institute of Oceanographic Sciences

(Laughton, 1981) and is currently mapping the United States Exclusive Economic Zone. Because GLORIA can survey swathes that are 30 km wide on either side (given a typical deep ocean water depth of 5 km), rapid progress can be made. The whole of the European continental shelf will have been surveyed at 1:250 000 by this method by 1990.

If progress in ocean mapping has partly been driven by new technology it has undoubtedly also been driven by political and economic motives. Coastal zone mapping is a growing area of interest, admirably investigated by Perrotte (1986). It is closely tied to national interests in the marine, seabed and subsurface resources of the continental shelf areas, and to the declaration of 200-mile Exclusive Economic Zones by numerous nations following the signing in 1982 of the United Nations Convention on the Law of the Sea by 119 nations. Large scale (>1:250 000) multi-thematic resources surveys of these offshore areas are in progress by, for example, the nations bordering the North Sea, the USA and Canada, and New Zealand. Computerized maritime resource information systems are also being developed by several nations.

The national atlas: survivor with a new future

National atlases are often overlooked as significant sources of spatial information about individual countries. This is unfortunate since for a number of countries where the larger-scale map series are restricted or otherwise unobtainable, they form one of the most readily accessible sources of such information, and for this reason we have listed such currently in-print atlases in our catalogue sections. Many national atlases include not only a wide range of fairly coarse resolution thematic material, but also a topographic cover of the country at scales in some cases as large as 1:250 000.

The now classic national atlas format was set by the first of the genre to be published, the *Atlas of Finland*, in 1899. Partly because of varying definitions of the term, there have been different estimates of the number of national atlases in existence at any one time. Nicholson (1973) quotes Ena Young's count of 26 in 1957 against Leszczycki's 'some dozen' in 1960. Whatever statistics one produces, there is no doubt that the national atlas is not a fading concept, for while many have gone out of print many more have appeared in new editions and formats. Stams, who has provided a very comprehensive listing of national and regional atlases up to 1978 (Stams, 1984), reported that about 50 nations then had a national atlas. Taking a rather conservative definition (see Glossary), we find that about 64 are actually in print in 1986. Table 2.1 shows the countries that have issued or begun to publish a new national atlas since 1979 (i.e. it updates Stams' compilation). There are no fewer than 37 new atlases (although a number of these are new editions rather than first-time atlases), and several more are known to be in preparation without a firm publication date.

Perhaps slightly surprising is the fact that, although often innovative in content, most of the new atlases are entirely conventional in format. But this will soon change. The electronic atlas has been much discussed, even if it has yet to appear. Canada, although currently publishing its fifth edition in a conventional loose-leaf format, plans an alternative electronic version, and France too is planning a new atlas of this kind. The Norwegian national atlas launched in 1983 is also being published

Table 2.1   Countries that have published, or begun to publish, national atlases (whether for the first time or as a new edition) since 1979

1979	Burundi, El Salvador, French Guiana, Ireland, Venezuela
1980	Australia, Tanzania
1981	Ethiopia, German Democratic Republic, Mexico, New Caledonia, Nigeria, Rwanda, Switzerland, Taiwan, Wales
1982	Guadeloupe, Qatar
1983	Belgium, Chile, Hawaii, Namibia, Norway, Sri Lanka, Swaziland, USSR
1984	China, Netherlands
1985	Afghanistan, Bolivia, Haiti, Israel, Malawi, Zambia
1986	
1987	Italy, Sudan
1988	Hungary

conventionally, but has been strongly underpinned by the digital storage and processing of the data. Monmonier (1983) has pointed out that while discussions have gone on in the USA for some years about a possible new edition of the atlas published in 1970, a *de facto* national atlas has appeared in the form of DIDS (the Digital Information Display System) developed by the National Aeronautics and Space Administration (NASA) and the Bureau of the Census, and used initially for presidential briefings. By the same token one could argue that Great Britain, which has never quite managed to produce a real national atlas of its own, now has a *de facto* atlas in the form of the BBC Domesday video disks, released in 1986 replete with topographical (all the OS 1:50 000 maps), census and other socioeconomic data, which are accessed interactively by means of a microcomputer, enabling the user virtually to create and compare thematic maps on a wide range of topics (Openshaw *et al.*, 1986).

Census mapping

The demographic censuses carried out decennially by most countries of the world provide a great wealth of mappable data within the socioeconomic field. Unfortunately many of these data sets remain unmapped. Nevertheless, censuses have been an important stimulus to national and regional atlas production and a number of national atlases have been timed to utilize mapping derived from a particular census. A world review of census mapping by Nag (1984), although marred by a very limited response (only 21 countries), reveals that many countries do not yet have an operative mapping system for their census data. But with the spread of cheap database management systems it is likely that census mapping, whether in purely electronic or hard copy form, will proliferate in the future. Several census atlases (as opposed to national atlases using census material) have been issued recently from countries as diverse as Sweden, Cuba and Britain.

The collection of census statistics also requires the preparation of large-scale maps of enumeration districts on which to record the data. The need for such maps stimulates basic survey effort in countries where good quality topographic survey does not already exist, and the maps themselves are often a valuable source of base mapping for other uses. In the USA a significant development in the preparations for the 1990 census has been the cooperation of the federal mapping agency

(USGS) with the Bureau of the Census in a joint digital mapping programme (see Chapter 6 by Blakemore and Rybaczuk).

Another by-product of censuses is the preparation of comprehensive gazetteers.

Toponyms and gazetteers

Geographical names—of settlements, administrative areas and natural features—form an important constituent of the information held on topographic maps. To retrieve these toponyms from the map, however, often requires the use of a separate list or gazetteer.

Some countries have official names authorities (in some countries there are also regional authorities) and in some cases the responsibility is invested in the national survey organization.

The gazetteers of such official sources have two practical values. First, they inform the user of the approved name and spelling for a given feature or settlement, and second, when combined with locational information, they enable the feature to be found on a suitable map. Many lists are in fact compiled from, or arranged to be used with, a particular map series. Thus at a small scale the Russian *Karta Mira* world map at 1:2 500 000 has an index of toponyms designed for use with it. At a large scale, the Ordnance Survey of Great Britain publishes a microfiche gazetteer containing all the toponyms appearing on its 1:50 000 scale series together with the sheet numbers of the appropriate maps and a grid reference enabling the feature to be located to a cell resolution of 100 m; and, similarly, the United States Geological Survey is issuing a series of state gazetteers in a choice of formats, which contain all the toponyms given on the 1:24 000 map series.

Gazetteers of toponyms to accompany large-scale maps are particularly useful, for it is the small and little known place which is most difficult to find and which most often needs a map for its identification.

Meynen (1984) has produced a very useful bibliography of *Gazetteers and Glossaries of Geographical Names* arranged nationally. In the introduction to this bibliography, he summarizes the various kinds of official and commercial gazetteers available, identifying six broad categories.

In this book we have tried to include the most up-to-date and useful gazetteer for each country, irrespective of whether it is an official government gazetteer or a commercial product. In many cases we have fallen back upon the extensive list of United States Board on Geographic Names gazetteers. Some commercially produced gazetteers have additional features not usually found in official lists, such as postcodes, population statistics, city street maps and descriptive information.

International interest in standardization of place names finds an outlet in the United Nations which has a Group of Experts on Geographical Names (UNGEGN) formed in 1960 (Breu, 1982), and there have been regular UN conferences on the standardization of geographical names. The most recent report on the work of UNGEGN constitutes a special volume of *World Cartography* entirely devoted to standardization of geographical names (World Cartography, 1986). This Group is charged with the task of fostering both national and international standardization of names and consideration of problems such as transliteration and the romanization of ncn-Roman alphabets. In 1967, the UN recommended that countries prepare standard gazetteers of their territories. Only a handful of countries have so far produced provisional gazetteers, prepared to the United Nations guidelines. An example is the Austrian gazetteer (Breu, 1975) which contains correct spelling, pronunciation, feature category, location, elevation and administrative region of each toponym. The ambitious future long-term object of the United Nations remains the preparation of a multi-volume standard *United Nations Gazetteer of the World*. A single-volume *Concise Gazetteer of the World* is in preparation. At present there is no satisfactory world gazetteer of substance, although a number are listed by Meynen (1984). The most celebrated is the *Colombia–Lippincott Gazetteer*, but the last supplement to this was published in 1961 and has been out of print for many years.

The United States has been particularly active in international geographical names programmes, and the United States Board on Geographic Names (USBGN) has published about 160 gazetteers of foreign countries or areas. USBGN also cooperates with the British Permanent Committee on Geographical Names for British Official Use (PCGN).

Commercial mapping

Commercial mapping is carried out by a large number of private companies most of which produce maps which are derived from government survey material and are adapted for the massed markets of traveller and tourist. Commercial mapping, as secondary material, therefore takes rather a low profile in this book, although in some instances (particularly in countries where government mapping is restricted) such mapping is the best that is readily available.

A number of commercial map publishers have recently begun to issue maps of the more exotic tourist areas, and these maps, albeit of small scale, often fill useful gaps in the pattern of availability. Two West German companies, Nelles Verlag and Karto+Grafik, may be cited as providing good quality small-scale maps of 'difficult' areas in the Caribbean and the Far East, for example.

Several of the larger commercial map publishers have adopted digital mapping systems for their own use, and are also beginning to offer digital products on the market. Thus Rand McNally, Chicago, launched its RANDMAP package in 1985. This includes digital boundary files and population data for the US states with mapping software for creating choropleth maps on an IBM PC or compatible. John Bartholomew, Edinburgh, has introduced the Scitex Response 280 system for its own in-house use, and has begun to sell digital products. There is no doubt that the digital map will become an increasingly marketable product for commercial map publishers, just as it has for government survey agencies.

Mapping and information

Traditionally, maps have been printed as 'hard copies' on paper and serve both to store and to communicate spatial information. In spite of the technological transition through which cartography is passing (Monmonier, 1985), the vast majority of maps are still available only in this form, and hence the catalogue sections of this book are overwhelmingly concerned with paper maps.

The paper map has proved its worth as an efficient and effective storage medium. How efficient is shown by the 0.3 Mbyte required to store an average 1:1250 Ordnance Survey sheet digitally. Nevertheless, there are many limitations to a spatial information system of this kind. For example, the resolution at which the information is stored is limited by the scale factor, and, even when the scale of the map permits fine detail to be shown, this is not always in the interests of the second function of the map, which is to communicate information effectively to the map user. Furthermore, map users rarely if ever need all the information stored in a map, so much of what it contains is redundant information so far as any given map user is concerned.

Digital cartography separates the store function from the display function, thus freeing the image so that it can more effectively serve the client. The potential of faster methods of data capture by sensors mounted in satellites, of digital image processing, together with developments in automated cartography, and the storage and manipulation of spatial data within sophisticated geographical information systems are giving us a new cartography. Chapters 5–7 explore these developments more fully.

References

BRANDENBERGER, A.-J. (1976) Study of the status of world mapping. *World Cartography*, **XIV**, 71–102

BRANDENBERGER, A.-J. and GHOSH, S.K. (1985) The world's topographic and cadastral mapping operation. *Photogrammetric Engineering and Remote Sensing*, **51**, 437–444

BREU, J. (1975) *Geographisches Namenbuch Österreichs*. Wien: Verlag der österreichischen Akademie der Wissenschaften. pp. 323

BREU, J. (1982) The standardization of geographical names within the framework of the United Nations. *International Yearbook of Cartography*, **22**, 42–47

BURROUGH, P.A. (1986) *Principles of Geographical Information Systems for Land Resources Assessment*. Oxford: Oxford University Press. pp. 193

CARRÉ, J. *et al.* (1985) La cartographie officielle dans le monde. *Bulletin du Comité Français de Cartographie*, **105**, 7–21

DE HENSELER, M.C. (1982) Mapping requirements in developing countries. *ITC Journal*, **1982-2**, 186–189

FOOD AND AGRICULTURE ORGANIZATION (1976) A framework for land evaluation. *Soils Bulletin*, No. 32. Rome: FAO

HUNTING GEOLOGY AND GEOPHYSICS LTD (1983) *African Resource Mapping and Development Chart*. Borehamwood, UK: Hunting Geology and Geophysics Ltd

JOURNAUX, A. (1985) Cartographie intégrée de l'environnement: un outil pour la recherche et pour l'aménagement. *Notes Techniques de MAB*, 16. Paris: UNESCO. pp. 55

LARSGAARD, M.L. (1984) *Topographic Mapping of the Americas, Australia and New Zealand*. Littleton, Colorado: Libraries Unlimited. pp. 180

LAUGHTON, A.S. (1981) The first decade of GLORIA. *Journal of Geophysical Research*, **86**, 11511–11534

McGRATH, G. (1983) Mapping for development. The contribution of the Directorate of Overseas Surveys. *Cartographica*, **20**(1&2), 264

MEINE, K.-H. (1972) Considerations on the state of development with regard to topographic maps of the different countries of the earth. *International Yearbook of Cartography*, **12**, 182–198

MEUNIER, F. and HARDY, B. (1986) *La Cartographie Thématique à l'ORSTOM*. Bondy: ORSTOM. pp. 8

MEYNEN, E. (1984) *Gazetteers and Glossaries of Geographical Names*. Wiesbaden: Franz Steiner Verlag. pp. 518

MONMONIER, M.S. (1983) DIDS: a *de facto* national atlas. *SLA Geography and Maps Division Bulletin*, **132**, 2–7

MONMONIER, M.S. (1985) *Technological Transition in Cartography*. Madison, Wisconsin: University of Wisconsin Press. pp. 282

NAG, P. (1984) Editor. *Census Mapping Survey*. New Delhi: Concept Publishing. pp. 299

NICHOLSON, N.L. (1973) The evolving nature of national and regional atlases. *SLA Geography and Maps Division Bulletin*, **94**, 20–25

OPENSHAW, S., RHIND, D. and GODDARD, J. (1986) Geography, geographers and the BBC Domesday project. *Area*, **18**, 9–13

PERROTTE, R. (1986) A review of coastal zone mapping. *Cartographica*, **23**(1&2), 3–72

PLATT, R.R. (1945) Official topographic maps, a world index. *Geographical Review*, **35**, 175–181

SPERLING, W. (1978) *Landeskunde DDR: eine annotierte Auswahlbibliographie*, Vol. 1. München: K.G. Saur

STAMS, W. (1984) *National and Regional Atlases, a Bibliographic Survey*. International Cartographic Association. pp. 249

ULRICH, J. (1984) Flächenhafte Kartierung des Meeresbodens. *Kartographische Nachrichten*, **34**, 41–47

UNESCO (1963) *Review of the Natural Resources of the African Continent*. Paris: UNESCO

World Cartography (1986) Volume XVIII. New York: United Nations. pp. 67

3 Map acquisition

C R Perkins

A bewildering variety of cartographic material is issued from all kinds of organizations in nearly every country in the world. It has been estimated that in 1985 there were 100 000 maps printed, and in one country alone (the German Federal Republic) there are nearly 1450 organizations involved in the compilation and publication of mapping. Anyone wishing to acquire some of this enormous output faces similar problems to those that confront any person concerned with the procurement of any published or semi-published format of information. It is first necessary to identify the item using some kind of bibliographic source, then one needs to find whether the item is available, and finally a decision must be made about how to acquire it. There are also other difficulties not usually faced by the bookbuyer, which affect each stage in the process. Thus in proportion to the number of maps in print there are more map than book publishers: map publication is dispersed and 90 per cent of map publishers are government organizations. The publisher, distributor, compiler and printer relationship is often very unclear for maps. Identifying published mapping is much more difficult than for book materials. There is a lack of adequate bibliographic control of the medium, very little published information about map publishers and few adequate buying aids. In part this reflects the nature of mapping: maps are ephemeral documents and go rapidly out of print; many are at best semi-published. Also a high percentage of published mapping is in series and therefore serial in nature, which presents problems in both identification and acquisition. Few bibliographic tools are well suited to describing sheets within map series: the graphic information needed to document the appearance of new editions of sheets is often not available. Even if information about published mapping is available the acquisition may be difficult. There are few distributors and wholesalers who deal in mapping and the unit costs of map acquisition are high in comparison to bookbuying. Also publication does not guarantee availability. Restrictions on mapping may mean that listed mapping cannot be procured.

This chapter discusses these problems and seeks to suggest some solutions. The emphasis is upon in-print mapping in the categories covered later in the book, but the discussion applies to both current and retrospective acquisition of this material. No attempt is made to cover ordering routines which are already well documented in works such as Farrell and Desbarats (1984) and Larsgaard (1986). Existing procurement policies of major collections are not discussed (see Rudd and Carver, 1981, for an attempt to quantify topographic map acquisition in American map libraries). The emphasis is upon the provision of a framework to guide those who need to acquire maps through the maze of cartobibliographic information and variety of possible acquisition methods towards a successful procurement. Familiarity with the need to evaluate different maps is assumed. This essential part of the acquisition process is discussed by Farrell in Chapter 4.

A literature search reveals that there are few very useful practical guides to map acquisition. Hodgkiss and Tatham (1986) list only eight works under this heading, and six of these are essentially source material rather than guides. Allison's (1983) annotated bibliography gives a more useful introduction to the range of published works about map acquisition. Much of the emphasis of the literature has been upon the need for a map librarian to develop a policy in which to set individual map purchase (Koerner, 1972; Schorr, 1974; Larsgaard, 1978; Prescott, 1978; Doehlert, 1984). Most articles are American, and most are set within an institutional context. Few emphasize practical ways for anyone to procure mapping. The best practical article was compiled in the late 1970s and was followed by several useful (though now dated) appendices (Wise, 1977–79). An earlier source book by Lock (1969) also provided some guidance. Larsgaard (1978) and Nichols (1982) both offer useful practical advice, the former in an American context the latter with the emphasis upon UK public libraries. Almost all of these articles and books are written with the needs of the specialist map librarian in mind and few consider the crucial factor of availability, or the mechanics of acquisition. Allison (1985) aims at a more general librarian audience but only covers the problems at a very superficial level. Other guides have tried to cover acquisition from an area or format based stance and have not been aimed at any particular audience. Parry (1983) provides a useful summary of acquisition sources for British mapping, Johnson (1974) discusses Latin American flat sheet acquisition, and Gardini (1979) considers Brazilian map sources. Acquisition of geological mapping is covered in an extensive literature review by Diment (1979), bathymetric charting by Hagen (1974). Many other articles and works can provide information about mapping of particular areas, but do not provide advice about acquisition techniques for procuring this material; see for instance Larsgaard (1984), Cobb (1983), Windheuser (1983), Williamson (1983), and country reports submitted to either International Cartographic Association Conferences, or to United Nations Regional Cartographic Conferences.

The published literature does not therefore provide very much practical assistance to anyone intending to procure mapping.

What influences the likely success of procurement? The rest of this chapter answers this question by examining the different influences upon the acquisition process: the publishers, the maps they issue, the information about these maps, and the possible acquisition techniques. Whenever possible examples are cited to illustrate the problems to be overcome. Only after such an exercise can conclusions be drawn about the overall availability of mapping.

Who issues maps: the nature of map publishing organizations

Map publisher listings

The large number of organizations that publish maps is mirrored by the large number of listings of map publishers. A map procurer needs different kinds of information about these bodies. He or she needs, first, to know which agencies publish maps, second to know which kinds of maps are issued from which agency, and third to verify addresses in order to request further information about mapping or to order the material. Larsgaard (1978) summarizes the problem of identifying addresses as being one of using any technique that works and is legal. The problem is to know which techniques work best, since there is no single and authoritative listing. It should be possible to evaluate listings in terms of their currency, accuracy, coverage (kinds of agencies included), content (the amount of descriptive information about the organizations' activities), comprehensiveness (are all relevant addresses identified) and ease of use. Table 3.1 summarizes such a comparative exercise on some of the more important lists. The overall conclusion to be derived from viewing this table and from our researches for this book is that most are in some way inadequate.

The accuracy of these listings varies very greatly: many addresses are wrong and some lists give both current and superseded addresses and names of the same organization. The coverage also varies. Few of the lists do more than indicate broad responsibilities. A national mapping agency may also have cadastral, soils, census, atlas, toponymic and even geological mapping functions, often not indicated in most listings. More general address source lists and reference works may be useful but are of no use in identifying map publishers and often omit many, especially government agencies. Their currency varies enormously from annual revision to one-off listings. Very few of these lists give any indication of the reliability of the information they contain (which is often drawn from inadequate secondary sources), and few indicate the sources of their information or attempt to verify the addresses they cite.

In addition to these published listings, addresses and information about organizations are to be found in a wide variety of published works, ranging from Lock (1969) to reports presented to United Nations Regional Cartographic Conferences and International Cartographic Association publications. Larsgaard (1978) exhorts the intending map procurer to use as many sources as possible—the journal literature, publishers' publicity, embassies all may produce publisher information.

The address lists and accompanying text in our work should provide a much better solution for the intending map procurer. Our addresses have where possible been verified, and the often disparate sources are brought together and explained in our work.

There are no authoritative published lists of map jobbers or of map distributors. Many western countries have a jobber who is prepared to procure mapping from a variety of publishers. We have included lists of a sample of their addresses in Table 3.2. It should, however, be stressed that GeoCenter is pre-eminent in the field, able to acquire mapping not otherwise available and the only very efficient jobber offering a service to all countries. The pros and cons of using a map jobber are considered below.

The nature of the map publishing agency

There are many different organizations involved in map publishing. In this work we have listed a total of 1173 agencies. The nature of these organizations can itself cause acquisition difficulties.

There is a continuum which ranges from the optimum (a public sector centralized official mapping agency in the western developed world, with civilian mapping responsibilities including the compilation of general and topographic mapping of its own state, which sells direct to the customer) to the worst possible organization for map acquisition (a public sector decentralized official mapping agency in the third world in a pro-Eastern block country, with military responsibilities including a very wide range of mapping tasks, whose maps are compiled through overseas aid programmes and which either operates only a very indirect sales service or does not sell maps).

Chapter 2 discusses something of the current variation in availability of different types of mapping by different organizations, topics and countries, and the country and continental sections of the book discuss the more detailed problems of procurement for specific countries. This section gives an overview of how these organizational factors work.

Civilian/military

Mapping is sometimes compiled by separate military and civilian authorities and sometimes by a national agency with both civilian and military responsibilities, which may either compile single editions of maps or issue both military and civil series. Military mapping is more difficult to obtain than maps designed for civilian uses. In some Western countries both civilian and military series are available, e.g. in Spain where both IGN and SGE publish mapping. Elsewhere only a subset of military mapping is available, e.g. in Singapore where the Defense Mapping Unit compiles topographic coverage and where some of these maps are sold by the Singapore Survey Department. As a generalization a military survey authority is much less likely to release mapping than a body with civilian responsibilities, though this situation depends very much on the nature of the society and the area of the world.

Geographical differences

Chapter 2 touches on the geographical variation in mapping availability. There are very big differences in the nature of organizations operating in different countries, which stem from the differing socioeconomic and political backgrounds. This in turn influences the chances of being able to procure mapping from these organizations. Thus Eastern bloc countries produce a great variety of official mapping which is very well documented in conference and periodical literature, but which cannot be obtained. The western developed world produces great quantities of mapping which is mostly obtainable. Here the problem is mainly one of publisher identification and map selection. In contrast, many third world countries do not produce much mapping, and what there is may not be well listed or available. Some third world countries with a long tradition

Table 3.1　Lists of map publishers

Listing	Subject coverage	No. of agencies listed	Date Currency Revision	Accuracy	Arrangement, ease of use	Descriptive information
Allin (1978–)	All map publishers, special emphasis on North America	~1750	1982– Information date varies with section Periodic sectional revision	Varied: many old and unverified addresses	By sections, alphabetic by organization or country in sections; useful index	None
USGS (Tinsley and Hollander, 1984)	Earth science agencies (NB includes map publishers and other agencies)	~900	1984 Information is as current as possible 3-year revision cycle	Varied	Alphabetic by country	Single letter to indicate responsibility
Wise (1977–79)	Wide range of map publishers	~500	1977–79 Varied information dates No revision planned	Varied: now very out of date	By section	None
DOS (Overseas Surveys Directorate, 1986)	National survey agencies	~300	1986 Indication of currency given Annual revision	Good: best for agencies with which DOS has contacts	Alphabetic by country	None
CFC (Carré, 1985b)	National, geological, hydrographic and selected commercial map publishers	~650	1985 Varied No revision planned	Good, best for francophone countries	Alphabetic by country within continent	Function responsibility shown
World Cartography (Brandenberger, 1984)	National survey agencies (selective)	73	1984 Varied information dates Periodic revision	Good where response to UN survey	Alphabetic by country	None
Hydrographic Department (1986)	Hydrographic agencies	46	1987 Varied information dates Annual revision	Good	Alphabetic by country	None
Statistics and Market Intelligence Library (1985)	National statistical agencies (includes census mapping bodies)	—	1985	Good	Alphabetic by country	None
Longman Earth Science (Fitch, 1984)	Earth science research agencies (NB includes map publishers and other agencies) Not all earth science map agencies listed	~3500	1984 Varied information dates No revision planned	Varied	Alphabetic by country	Brief functional and other information
Longman Agricultural Science (Harvey, 1983)	Agricultural science research agencies (NB includes map publishers and other agencies) Not all soil or land use surveys listed)	~11 000	1983 Varied information dates No revision planned		Alphabetic by country	Brief functional and other information
Publishers' International Directory (1984)	Comprehensive world listing of all publishers (NB the best source for commercial publishers)	153 878	1984 Varied information dates Annual revision	Good	Alphabetic by publisher within alphabetic country sections	None

of mapping and sufficiently established bureaucracies publish readily available series, e.g. Papua New Guinea, Malawi or Brazil. Others, such as Angola, Honduras, Afghanistan or Chad, have little current mapping available because of political affiliation or inadequate resources. In Kampuchea and other countries political anarchy restricts publication and availability. It is difficult to make generalizations about the chances of acquiring mapping of third world countries beyond the blanket statement that acquisition is likely to be more problematic, take longer and be more risky. It is also likely that mapping of third world countries will more quickly become unavailable: the window of availability is likely to swing shut rapidly.

Areas mapped

Some organizations map only their own territories, others map their own state and its colonies, while others in the developed world both map their own states and engage in aid mapping programmes of other countries. Some

Table 3.2   Map jobbers, in the order cited in the text

GeoCenter ILH
Postfach 80 08 30, D-7000, STUTTGART 80, German Federal
Republic

Edward Stanford Ltd
12–14 Long Acre, LONDON WC2E 9LP, UK

Bill Stewart
119 Grandview, ANN ARBOR, MN 48103, USA

Steve Mullin
456 Alcatraz Ave., OAKLAND, CA 94609, USA

Pacific Travellers Supply
529 State Street, SANTA BARBARA, CA 93101, USA

Geoscience Resources
2990 Anthony Rd, BURLINGTON, NC 27215, USA

Telberg Book Corporation
Box 920, SAG HARBOR, NY 11963, USA

Collets Holdings Ltd
Denington Estate, WELLINGBOROUGH, Northamptonshire NN8
2QT, UK

McCarta Ltd
122 Kings Cross Rd, LONDON WC1X 9DS, UK

Map Shop
15 High St, UPTON UPON SEVERN, Worcestershire WR8 0HJ, UK

Libraire l'Astrolabe
46 rue de Provence, 75009 PARIS, France

Esselte Kart Centrum
Vasagatan, STOCKHOLM, Sweden

agencies produce only aid mapping. Such mapping of overseas areas is often very difficult to procure even from the major western contracting agencies. This is because of the numbers of organizations involved and the difficulty of deciding which is involved in the distribution of map products. Also it is usually the client state which decides on the availability of any mapping produced. Thus for instance British OSD has produced mapping of the Yemen Arab Republic, of Kenya and of Ethiopia (to take just three examples) which is compiled and published in Southampton but which must be obtained on direct application to the client state survey organization.

Functional responsibilities

The exact functional responsibilities of a mapping organization vary: not only does this present problems in identifying the publisher of a map, but different bodies deal with map orders in rather different ways. It is difficult to make generalizations on the effect of an organization's function on acquisition. For instance, in some countries (e.g. Greece) geological series may be available where a comparable scale military topographic series is restricted. In others (e.g. in Qatar and other oil states) the opposite applies, where geological mapping is compiled by oil companies but is largely for commercial in-house use, whilst excellent topographic coverage is available to all. As a generalization it is usually the official topographic mapping which is restricted in a country: thematic series are usually at smaller scales, if available, and prepared for the civilian market. Cadastral mapping is also often available when topographic series are restricted.

Official/commercial

In the private sector, commercial agencies market their maps and must make a profit to survive. In theory this can be good for the map purchaser. However, in practice this

is often not the case. The information provided in commercial map catalogues is often inadequate—scale, title, specifications and especially date are often omitted. Ownership patterns may confuse intending purchasers and lead to further map identification problems in catalogues. In the UK, Reader's Digest, Geographia and Bartholomew are all now part of the same organization, whilst, in Australia, Universal now has a near monopoly of street plan production through ownership of the Robinson, Gregory, Broadbent and UBD imprints. Mergers and marketing arrangements between commercial map publishers can mean that the same map is marketed under two different publishers with only very minimal changes, e.g. the Recta Foldex and Geographia distribution of the Karto+Grafik Hildebrands Urlaubskarten, or the Ravenstein map of Africa which is a reprint of the Daily Telegraph map of Africa. Addresses and names of commercial agencies change with bewildering rapidity, e.g. Tamilnad Printers and Traders becoming TT maps, or Geobuch Verlag changing to Nelles Verlag. This problem affects the public sector as well, especially where administrative changes have meant that the affiliation of the National Mapping Agency to a Ministry shifts sometimes with every change of government. Perhaps the extreme case of address and responsibility confusion is in Mexico where there have been many changes in acronym of the official mapping body over the last 20 years. Sometimes even the agencies themselves are confused. For example, in the Philippines where requests to the Board of Coast and Geodetic Survey (BCGS) produced the reply that a new National Cartography Authority was now responsible for surveying and mapping. When contacted, the new authority claimed not yet to be established and passed the buck back to the BCGS! Functional shifts can also occur: for instance military organizations may assume civilian responsibilities should a change in regime bring about a liberalization of official attitudes. Not only are there problems over changing addresses and changing functions with official agencies. The same body often has several different branches at different locations which deal with different functions. For instance, IGN in France has a headquarters branch, a distribution branch and an air photo and map information branch at three different addresses.

Customer relations

The form of customer relationship varies too. Not all publishers are either equipped or prepared to sell direct. Many Eastern bloc agencies support their own marketing agencies (e.g. Cartographia in Hungary), which issue a subset of the officially produced mapping. In the USSR Mezhdunarodnaya Kniga distributes some Glavnoe Upravlenie Geodezii i Kartografii maps; in Taiwan Youth Cultural Services distributes official atlases. Other Eastern bloc mapping is available through a limited number of agents in the west, for instance geological mapping of the USSR is available through GeoCenter in West Germany. Some map publishers use a sole agent system but most employ a variety of agencies to distribute their mapping and acquisition may be more appropriate through one or other of these companies (see for instance Williamson (1983) for a short discussion of the relative merits of different agents for the procurement of UNESCO mapping). Not all of the publishers that are prepared to sell direct to the customer will sell direct to all customers. Restricted mapping may sometimes be obtained if permission is sought through a third official party, as in the case of IGN mapping of Niger, which may be obtained on application to the Niger Embassy in Paris. Others will sell in person over the counter but do not process postal applications. This is often not made explicit in publishers' information or catalogues, but is the case for many third world countries and especially for the Indian Subcontinent. The nature of agencies changes over time: even the longest established relationship can be broken. For instance, Edward Stanford no longer has the sole agency for OSD published mapping in the UK, following a change in OSD policy in 1986. Some types of agency are operated by mapping organizations for specific categories of customers. In the UK Ordnance Survey maps may be procured at a favourable discount by customers in educational establishments, provided they place orders with either the London Map Centre or with Thomas Nelson in Edinburgh. Other publishers operate discounts without the agents and sell direct, e.g. British Geological Survey educational customers can acquire mapping direct from BGS headquarters at Keyworth at a discount.

Centralized/decentralized

Large countries are often organized on a federal basis and the mapping of these states can usually also be expected to be divided between agencies at both national and provincial levels. This is significant for map acquisition because of the greater number of map-publishing bodies in federal countries. At one extreme is India which, though a federal country with decentralized government, has a largely centralized mapping establishment and therefore few publishers. At the opposite extreme are the USA and Canada which have both a well developed national mapping establishment and also mapping agencies at state, county and city level. Between these two extremes lie Australia, the German Federal Republic, Brazil, China, Nigeria, Malaysia and the USSR. A second problem for anyone acquiring mapping of such countries is to know where responsibilities lie, in terms of both the scale and functional responsibilities of national and provincial agencies. For instance, in Australia 1:100 000 and 1:250 000 mapping is chiefly a Commonwealth responsibility, whilst larger-scale and cadastral mapping is carried out by state surveys, whereas in Canada the 1:50 000 scale map is a national project, while provincial authorities have their own mapping programmes at both larger and smaller scales.

The published mapping

The nature of the map production process

It may be possible to acquire maps at different stages in the map publication process. First there is the fully published map: sheets often have limited print runs, and depending upon the agency may or may not be reprinted. Published mapping often goes out of print before it can be acquired, sometimes even before it can be listed. Acquisition may therefore be difficult.

Forms of semi-publication also occur. Particularly at larger scales it is becoming possible to obtain map information prior to publication. For instance the Ordnance Survey in Great Britain has operated a SIM and SUSI service for the last 15 years, which allows users to purchase amended copies of surveyors' Master Survey Drawings and so circumvent delays in revision programmes for sheets in large-scale series. Other forms of semi-publication are also becoming common. Maps held

on Open File are a common means by which earth science agencies in particular can allow access to compiled map information but not need to publish full editions of the mapping. Topographic and some geological agencies are also using printing on demand to produce quick semi-published (and often poor-quality) mapping where there is unlikely to be a high customer demand for the material. None of this semi-published mapping is likely to be well documented, except sometimes in the catalogues issued from the organization itself. It is likely that hard copy maps as a by-product of digital databases will increasingly be cheaply produced as on-demand and user-designed cartography. Such maps (particularly given the flexibility of the systems) are likely to be very difficult to identify as published mapping.

Mapping technology

Not only does the extent of the map publication process influence the chances of a successful acquisition, so too do the technology used and the form of map publication. Thus there are different levels of abstraction from reality ranging from air photographs and satellite images, through photomaps, image maps, controlled image maps, orthophotomaps, to line maps. The less processing of collected data is carried out the more difficult it often is to acquire that data, since map publication rather then spatial data provision is still the rationale of most surveying and mapping organizations. Many organizations offer alternative editions on identical sheet lines, e.g. the Australian State practice of publishing orthophoto, cadastral and topographic editions of the same quadrangle.

Identifying the published mapping

Anyone interested in the acquisition of cartographic materials needs to find out about individual maps. Few will be able to see the maps themselves before deciding whether or not to order: there are no 'map approval' schemes. It is therefore necessary to rely upon secondary cartobibliographic sources as an essential first stage in the acquisition process.

In an ideal world the source and the information describing mapping it contains would be ideally suited to the needs of the map procurer. These needs may be rather different if current acquisition is intended, rather than a retrospective 'catching up' exercise. What should the source do and what kinds of information ought it to include?

- *Frequency* It ought to be published at sufficient frequency to notify any intending purchaser of new mapping in time to take advantage of the window of availability.
- *Cost* The source should be inexpensive, or give a good return in terms of costs and benefits.
- *Currency* The gaps between compilation and publication should be as small as possible.
- *Ease of use* The source should be easy to use, indexed if necessary, in a suitable language and arranged in a convenient way.

The information in the source should be:

- *Current.*
- *Accurate*, both in terms of reliably describing the mapping and in fulfilling the stated aims of the tool; e.g. if

the tool is concerned with describing available mapping, are the maps it contains really available?

• *Appropriate* Map purchasers need to know, as a minimum, title, scale, area coverage, publisher, date (of publication and map information), sheet level information for series (including graphic indexes), specifications of mapping and ordering information, including cost.

• *Comprehensive*, in terms of area coverage (is the information representative of all published mapping of the world or is it biased to one area?), language coverage, subject coverage (are all themes included?) and time coverage (are all maps published in a given time period included?).

How do secondary sources measure up against these desiderata? The first obvious point is that few cartobibliographic sources are designed with map acquisition in mind. This is illustrated by checking the standard guides to bibliographic sources such as Sheehy (1976) or Walford (1973–80). More specific help is given by Lawrence (1985) and Hodgkiss and Tatham (1986).

Cartobibliographic sources can be divided into:

• bibliographies (national, subject, area, and by format),
• library catalogues,
• journal listings and reviews,
• acquisition lists,
• publishers' catalogues, and
• jobbers' lists.

Bibliographies

National bibliographies record the publications of a country. Not all countries issue such a publication, nor are maps always included in them. For instance, in the UK BNB includes atlases but not as yet other cartographic material.

Examples of national bibliographies that contain cartographic records in a separate section are Australia, Austria, France, the German Federal Republic, the German Democratic Republic, New Zealand, Poland, Switzerland, Turkey and, since 1983, the US. A comprehensive listing is given in Wise (1977–79). National bibliographies do not always identify newly published mapping and are usually dependent upon the reliability of the National Library Service in the country concerned. Even when cartographic materials are included the source is of course no indication of availability, and often copes poorly with the problems of series mapping. As retrospective records of the map publication of a country they may be authoritative and can be used to identify mapping, but they are not well suited to current acquisition. Also because of their comprehensive coverage they are usually very expensive and are unlikely to be readily available to all potential map procurers.

Most other types of bibliography also suffer from being rather inappropriate for map acquisition. Area bibliographies may include mapping, and we have cited several as sources of further information throughout this book, although they do not discuss availability. Subject bibliographies may also be useful but they too suffer from the retrospective listing problem and the lack of availability information. Some subject bibliographies have concentrated upon particular types of maps, for instance Küchler's *International Bibliography of Vegetation Maps*, the 11 *Listes des Cartes Géologiques Nationales et Internationales* prepared by the Commission for the Geological Map of the World, and the World Meteorological Organization *Bibliography of Climatic Atlases and Maps*. All of these

sources are difficult to use and none is well suited to map acquisition.

There is now no international single published bibliography of mapping. *Bibliographie Cartographique Internationale* used to fulfil this function but is now sadly defunct. Even when active it was much too out of date to be any use for current acquisition.

Some on-line databases have been set up which either include map records amongst other material, or are designed specifically for recording map information. An example of the former is *Georef*: its coverage is, however, very patchy and emphasizes American geological mapping. More specific on-line cartobibliographies have recently been created by the French **Bureau de Recherches Géologiques et Minières (BRGM)**. BRGM offers both a hard copy *Géocarte Information* service, providing regular regional information on new mapping, and an on-line database of mainly geological and earth science maps. Both are priced at a prohibitive level for most users.

There have been attempts to provide in-print bibliographies of mapping, e.g. Winch (1976). His *International Maps and Atlases in Print* grew from a sales catalogue maintained by the London map jobber Edward Stanford Ltd. It included a selection of graphic indexes and cartobibliographic descriptions of mapping but is now seriously out of date. *GeoKatalog* is essentially a jobber's listing and as such is not necessarily a guide to in-print mapping (see below), but it is the nearest to a current in-print listing that is available at present.

So bibliographies are not usually well suited to map acquisition.

Library catalogues suffer from similar deficiences to bibliographies. Many libraries issue their catalogues, and some are very useful in identifying mapping and in its description. Important examples are mainly American, e.g. the *Bibliographic Guide to Maps and Atlases*, issued annually since 1979, which lists cartographic material catalogued each year by the New York Public Library Map Division and the Library of Congress Geography and Map Division.

To complement catalogues and updates to catalogues issued at fairly infrequent intervals, there are *map acquisition lists* which are to current acquisition what library catalogues are to retrospective purchase. These may be very useful in identifying material newly published which would otherwise not be easily found in a single source elsewhere. A list of map libraries that produce accession listings is included in Larsgaard (1978: 248–250). The *Bodleian Library List* is perhaps the model of a useful accessions list and provides the right level of basic brief descriptive information, sufficient to identify mapping, but brief enough to allow easy use. Obviously all accession lists give no indication of availability of the items described, and like library catalogues simply reflect a library's acquisition policy.

Journal listings can fulfil a similar role to accession lists and are also useful in identifying recently published material. The currency of the information may be less useful than in the better accession lists. Larsgaard (1978) and Wise (1975) list journals containing map listings or reviews of mapping.

Journal reviews are of less value in terms of the currency of the information but give much more critical evaluation of the products. Akin to such works is the recent spate of acquisition guides for categories of map-related material, such as Kister's series of buying guides for atlases, Stams' bibliographic review of atlases, and Meynen's work on gazetteers (see Chapter 2 for more information about these).

In view of the lack of any adequate and updated in-print listing, map purchasers need to rely upon the commercial information given by map publishers, distributors and jobbers.

Jobbers' listings

GeoKatalog issued from ILH GeoCenter is by far the most comprehensive jobber's catalogue. It is available in two cumulative volumes which may be obtained on an annual subscription service. Volume 1 covers tourist and commercially published mapping, and Volume 2 is a loose-leaf reference work, describing the complete range of published topographic and geoscientific mapping. Information is arranged under the country names, and country sections are updated on about a 5-year cycle. To maintain the currency of the work, *Geokartenbrief* is issued as an updating service every three or four months. The full bibliographic description is not always available (publishers' names are left out, for instance) but specifications of mapping are given in full German language descriptions. For a consideration of the use of this tool see below.

Publishers' information

Individual map publishing organizations also issue information about the mapping they publish: this is usually more detailed than jobbers' listings and is usually more current. It may include annual reports of mapping activities, cumulative catalogues and sometimes, depending on the sophistication of the organization, regular notifications of newly published mapping. The frequency of update of the cumulative listing varies enormously. Western surveying and mapping bodies update on an annual basis; many third world countries produce catalogues only very sporadically. Sometimes different editions of the catalogue are published. In the west there is a trend towards more aggressive marketing of official mapping, and official surveys are gearing their publicity material to the needs of particular user groups. For instance, the Ordnance Survey issues both a very simple general guide to its maps and an annual trade catalogue, in addition to a monthly update service on new mapping. To cite another example the Queensland SUNMAP organization issues both a general catalogue and a more specialized and huge union catalogue of all mapping published in the State, and has recently begun to publish a newsletter.

Few publishers' catalogues contain all the information which would be needed for map procurement. Many do not provide sufficient information about the specifications of a series and do not always provide graphic indexes, which are essential for the procurement of series mapping. Many third world catalogues give just a title and scale for maps, and sometimes even for series. Date and edition information is often missing. Distributors' catalogues aimed at the wholesale bookshop and tourist market, e.g. McCarta and Lascelles listings, sometimes omit useful information but include curious 'star ratings' of the saleability or quality of mapping. The existence of a catalogue is of course not necessarily a guide to map availability: for instance, Ethiopian, Greek and Pakistani catalogues exist but the majority of the mapping they describe is impossible to obtain.

However, the publishers' catalogues are usually the best cartobibliographic source available, and despite their inadequacies ought to be used by anyone wishing to procure mapping.

Acquisition methods

Having identified and selected an appropriate map, what methods may be used to acquire the item? There are essentially three ways of acquiring mapping: by purchase, on deposit, or on a gift and exchange basis. Each method may be more appropriate in certain circumstances and in practice most map libraries will use a mixture of all three depending on their needs, resources and organizational background.

Purchase

A financial transaction is usually the safest way of acquiring the desired map. With whom should the order be placed? There are considerable debates in the literature about the pros and cons of alternative procurement sources (see Larsgaard, 1978, for a discussion of some of these).

Direct Purchase from the publisher has the advantage of cutting out the middle man and thereby reducing both costs and time, and in theory maximizing the chances of a successful acquisition. It may be possible to obtain maps at a considerable discount from some publishers, particularly if the customer is a library. The United States Geological Survey operates such a direct sales discount policy. In practice there are drawbacks. Most publishers require prepayment from organizations, and nearly all require it from individuals. Most require payment in local currency. Not all publishers are equipped for direct sale, but refer orders to an agent. In the case of many third world countries direct ordering is unlikely to succeed. In part this is because of some of the organizational factors discussed above, but it is also a reflection of the nature of many third world bureaucracies. Prepayment can in these cases be very risky, with little guarantee of success or returns on money committed. There are no published guides to give the map procurer any indication of the likely success of direct purchase attempts. The best advice is to absorb as much information as possible about the mapping agency, use common sense and build on experience by testing the market with sample and small orders. For instance many Western European, North American, South African and Australasian organizations are obvious targets for direct application. Less success is likely to be met with any attempts to order direct from the Indian Subcontinent, from South America, from much of Asia, from almost the whole of Africa and even from Italy!

Distributors are a very useful source for the acquisition of published mapping: they act as publishers' agents for the sale of maps in specific countries. Major wholesale distributors in the UK include Roger Lascelles, Bartholomew, and McCarta Ltd; an example from the US is the Star Map Service. Such indirect acquisition is often the cheapest way of acquiring mapping, and may enable the procurer to make use of attractive trade discounts, avoid currency charges and sometimes avoid prepayment. However, distributors usually sell only the products of a limited number of publishers and will not procure other mapping for the customer. Moreover, the distributor does not always handle the whole of the publisher's list. For instance, Geographia/Bartholomew distribute only a sample of Falk townplans in the UK.

Map jobbers may be the answer for many map acquisition programmes. They act as specialized middlemen in the map distribution network and are the obvious and easy way of acquiring mapping. The best jobber to use will depend on the individual, the map, the nature of the organization, and the country of purchaser, purchases and jobber. No prepayment is usually required from institutional customers, shipping documentation is not needed and some form of guaranteed supply, or at least a report on the progress of the order, is likely. The world's largest map jobber is **GeoCenter ILH** in Stuttgart. GeoCenter offers a fast and reliable service, advertises new mapping, and reports on the progress of orders. It is often able to procure material that would not otherwise be available, through a network of contacts. GeoCenter has taken over much of the business formerly held by **Edward Stanford Ltd** in London, which no longer attempts to advertise either available or new mapping on a worldwide basis and has retrenched into acting as the world's largest map retail shop holding stock of all tourist destinations and the more easily available published mapping. Stanford is still a useful jobber for certain categories of material. Other jobbers build up or claim to have built up expertise in particular areas: the efficiency of these organizations ought to be tested before one believes their claims about product availability, pricing, discounts or other services. Examples of regional specialization include in the US **Bill Stewart**, who makes regular map purchase trips to South American countries and who is able to procure mapping that would otherwise be unavailable, and **Steve Mullin** who is a Mexican specialist. **Pacific Travellers Supplies** concentrates upon mapping of the Pacific region, but has recently begun to issue an updated listing of available mapping on a worldwide scale. Certain subject areas are also covered by jobbers. Thus in the USA **Geoscience Resources** claims to have one of the largest inventories of geological maps in the world, and is also able to supply to order topographic and quadrangle maps of many countries. **Telberg Book Corporation** also offers considerable expertise in the procurement of worldwide geological mapping and publishes a priced catalogue. Telberg is useful for the procurement of otherwise difficult Eastern bloc mapping, as is **Collets** in the UK.

In the UK two smaller jobbers offer expertise in certain areas. **McCarta Ltd** grew from being the UK official agent for IGN mapping to providing a general service for map purchasers and has issued its own minicatalogues of available mapping. The **Map Shop**, Upton, stocks a surprising variety of official mapping and offers an excellent service. Other European countries with major map jobbers include France (e.g. **Astrolabe**) and Sweden (e.g. **Esselte**). However, all jobbers suffer in comparison with GeoCenter in terms of their experience in procuring 'difficult' items.

An alternative approach to using a map specialist dealer is to approach a *general agent* (often local to the area of acquisition) with expertise in procuring material from specific regions. An example in the UK is the use of Dick Phillips (who specializes in Icelandic holidays) for the acquisition of mapping of Iceland. Such a policy is often followed by the map sections of large libraries with active regional book acquisition programmes, and may procure items which would otherwise go rapidly out of print.

Deposit

Deposit is a particularly important source of map acquisitions for map libraries in many countries. It usually involves an arrangement for the officially published mapping in a country to be lodged at no charge in a certain number of collections, under certain terms and conditions. Many countries operate a deposit system through their national library, which is able to claim on copyright deposit all (or some) of the indigenous published mapping. For an example of the way such arrangements work see Lepine (1983), who considers Canada: few other reviews of practice exist.

In the US deposit is of rather wider significance and supports the majority of the major map collections across the country by freeing them of the need to acquire USGS, DMA and other federal mapping. AMS and DMA map series in American libraries are often the best available scales for many areas where the official series are restricted. Theoretically deposit enables map budgets to be spent on more exotic material, but in practice often results in map libraries receiving very limited funding. For the development of US map deposit programmes see Larsgaard (1978) and for recent changes in the operation and regulations of these programmes see regular discussion in the newsletter *Baseline*. Cobb (1986) lists the US libraries that benefit from deposit arrangements.

In the UK a rather more limited form of deposit is available. Only five copyright libraries have the right to claim UK published mapping. Other more limited deposit arrangements exist with specific publishers, for instance the British Geological Survey lodges geological mapping with UK university geology departments.

International deposit is less common, but some publishers deposit their mapping abroad. For instance NATMAP in Australia deposits copies of its maps for free in several major institutions throughout the world.

The problems with deposit arrangements may stem from the terms of the deposit and from its nature. Thus it may not be possible to limit what is acquired, and in the US in particular this may result in storage problems given the 5000 or more new maps a year which may be available to collections. Also the material is not the property of the collecting institution and can be reclaimed by the publisher if needed. For instance during the Falklands War the British Ministry of Defence asked for the return of deposited Falklands mapping from some collections in the UK. Also deposit arrangements are not permanent and may lull institutions into false senses of security. US map libraries would face major acquisition crises should deposit programmes be abandoned. In the UK the British DOS mapping used until 1978 to be deposited with some British university geography departments; many of these collections have been unable to continue to procure any DOS material after this date.

Gift and exchange can be used to support acquisition programmes for a wide variety of mapping. Many map collections hold stocks of duplicate mapping which they are prepared to exchange with other institutions. In the UK the source of much of this mapping is the British Mapping and Charting Establishment, the largest current map library in the country, which disposes of its duplicate and superseded stock to other UK academic and national collections. Collections are unable to specify in precise detail the maps required, but can ask for broad categories of mapping. Thus boxes of mapping enter the map library system and the material is recycled around collections throughout the country. Periodic help-yourself days have been held by collections with surplus stock in an attempt to move mapping to new homes.

In the US the Library of Congress disposes of its duplicate material in annual Summer Map Processing Projects, in which participants have a free choice of duplicate mapping and attend seminars, in return for work in the various LC divisions.

Map exchange columns are published in several of the map library journals, to notify potential homes of unwanted mapping.

The problem with all exchange programmes is in deciding whether staff time is adequate to sort through the volumes of free available mapping. Lists of mapping are rarely prepared in detail and collections risk being inundated with material no-one wants.

Free maps may also be obtained in a more specific way from their publishers. Wise (1981) discusses the likely sources for free city mapping in the US, and Tiberio (1979) considers the methods and potential sources for this kind of material.

Summary: the availability of mapping

There have been some attempts to clarify overall availability of map series on a worldwide scale (see, for instance, Carre *et al.*, 1985a, which includes a qualitative map of map availability), but all too often publication is assumed to equal availability. Yet published maps are not always available maps and this is often the largest problem to be overcome in any attempted map acquisition.

There are different kinds of unavailability, which can be seen from Table 3.3. Thus at one extreme is unavailability because of non-publication, whereas at the other extreme is freely available mapping. It is the categories between these two extremes which demand more attention. Restriction may be placed on all mapping, or upon some organizations' mapping, or upon some of an organization's mapping. A case study of India reveals some of the problems in assigning a simple indication of availability to a country. In India much thematic mapping is readily available in the west, including geological, census and small-scale thematic and atlas maps. The catalogues documenting their publication are of varied quality, but the published literature about the maps and mapping systems compensates for some of these drawbacks. However, certain categories of officially produced topographic mapping are more difficult to obtain. In practice it may be impossible for most people to obtain any scales larger than 1:1 000 000 coverage: Survey of India issues catalogues describing and pricing only their small-scale mapping. The larger-scale series are advertised as being available from GeoCenter. The jobber claims to be able to supply this material, probably by using a local agent who is able to procure on the spot, using bribes and generally oiling the wheels of acquisition. Obviously such an acquisition route lessens the chances of always being able to acquire the maps. Also, even GeoCenter is unable to acquire series mapping of border and coastal areas. So

there is both an overall export restriction (which may be circumvented), a scale restriction and an area restriction, imposed because of national security considerations.

Such a complicated description could be applied to many other countries whose maps fall into the 'grey areas' between the extremes of availability and non-availability. We have endeavoured to explain availability in the country sections of the book and discuss some of the reasons for mapping being restricted in our coverage of the organizational influences on acquisition. Table 3.4 attempts to synthesize the effects of some of these factors and illustrates the varying current patterns of availability of world mapping.

The publication and availability of mapping can be expected to change over time. This chapter and the individual acquisition problems covered in country sections of this work must therefore be seen as describing the situation in the late 1980s. Anyone wishing to keep abreast of current trends in map publication and availability must be prepared to update their information sources continuously. Only by a constant update of information can the chances of a successful map acquisition be improved.

Table 3.3 Degrees of map availability

1 *Available by postal application*
  (a) Many possible sources
  (b) Available only through jobber
  (c) Part series only available (area restriction)
  (d) Part of scale range only available (scale restriction)
  (e) Part of subject coverage only available (theme restriction)
  (f) Official permission required
  (g) Available only for approved projects: civilian or military

2 *Only available in country of publication*
  (a) Can be purchased over the counter and exported
  (b) Can be purchased over the counter but not for export
  (c) Can be purchased over the counter with bribe or personal contact with officials

3 *Not available on international market*
  (a) Published and documented in secondary literature
  (b) Published but not documented in secondary literature
  (c) Published but out of print

4 *Not published*

Table 3.4 Patterns of map availability

This table shows the overall pattern of in-print official map availability and reflects the broad headings used in Table 3.3. If material is available by post it will also be available in the country of publication: we have not repeated the categories. Unlike other tabulations and maps it attemps to assess both topographic and other official published mapping availability on a country-by-country basis. If not specified the availability criteria apply to all published official mapping of the country. Further explanations of the availability constraints are given for particular countries (e.g. for French African countries topographic mapping availability differs according to whether it is old colonial mapping or post-independence surveys). It should be noted that all too often the reasons for restriction are not made clear. This table therefore indicates the *likely* reasons for unavailability, or simply classifies the mapping within category 3.

Topo. = topographic; geol. = geological; env. = environmental.

Africa	
Algeria	Topo. 1b or 2c; geol. 1a
Angola	3b
Benin	Topo. (old stock 1b, new cover 2c); env. 1a
Botswana	1a
Burkina Faso	Topo. (old stock 1b, new cover 1a); geol. 1a; env. 1a
Burundi	1b
Cameroon	Topo. (old stock 1b, new cover 2c); geol. 1a (old); env. 1a
Central African Republic	Topo. (old stock 1b, new cover 2c); env. 1a
Chad	Topo. (old stock 1b but 1f); env. 1a
Congo	Topo. (old stock 1b, new cover 2c); env. 1a
Djibouti	Topo. (old stock 1b); geol. 1a
Egypt	Topo. 2c; geol. 1a but 1c, 1d and 1e
Equatorial Guinea	3a
Ethiopia	3a
Gabon	Topo. (old stock 1b, new cover 2c); env. 1a
Gambia	1a
Ghana	1a (also old stock 1b)
Guinea	Topo. (old stock 1b but 1f)
Guinea Bissau	3c
Ivory Coast	Topo. (old stock 1b, new cover 1a)

Kenya	1a, 1f for larger scale series
Lesotho	1a
Liberia	1a, through jobbers and aid mapping agencies
Libya	3a
Madagascar	1d (larger scales 2b)
Malawi	1a
Mali	Topo. (old stock 1b but 1f)
Mauritania	Topo. (old stock 1b)
Morocco	1a
Mozambique	Topo. 3b; geol. 1a
Namibia	1a
Niger	Topo. (old stock 1b but 1f)
Nigeria	1b, or 2b
Rwanda	1a
Senegal	Topo. (old stock 1b, new cover 1a); geol. and env. 1a
Sierra Leone	1a (old stock 1b)
Somalia	3b (but some old stock 1b)
South Africa	1a
Sudan	1a
Swaziland	1a
Tanzania	1a
Togo	Topo. (old stock 1b); env. 1a
Tunisia	2b, 2c
Uganda	1b
Zaire	1b
Zambia	1a
Zimbabwe	1a

North America

Canada	1a
Mexico	1a
United States	1a

Central America

Belize	1b
Costa Rica	1a
El Salvador	1d (larger scales 1g)
Guatemala	1d (larger scales 1f)
Honduras	1d (larger scales 1g)
Nicaragua	3a
Panama	1a

Caribbean

Anguilla (including Barbuda)	1a
Antigua	1a
Bahamas	1a
Barbados	1a
Cayman Islands	1a
Cuba	1d (larger scales 3a)
Dominica	1a
Dominican Republic	1b
Grenada	1a
Guadeloupe	1a
Haiti	1a
Jamaica	1a
Martinique	1a
Montserrat	1a
Netherlands Antilles	3c
Puerto Rico	1a
St Christopher-Nevis	1a
St Lucia	1a
St Vincent	1a
Trinidad and Tobago	1a
Turks and Caicos Islands	1a
Virgin Islands (British)	1a
Virgin Islands (US)	1a

South America

Argentina	1a
Bolivia	1a
Brazil	1a
Chile	1a
Colombia	1d (larger scales 1f, also 1c border areas)
Ecuador	1a
French Guiana	1a

Guyana	1a
Paraguay	1b
Peru	1a but 1c for some series
Suriname	1a
Uruguay	1a
Venezuela	1a

Asia

Afghanistan	3b
Bahrain	1a
Bangladesh	1d (larger scales 3a or 2c)
Bhutan	4
Brunei	1d (larger scales 1g or 2c)
Burma	3b
China	3a
Cyprus	1a
Hong Kong	1a
India	Topo. 1d (larger scales 2b or 2c, with borders 1c); geol. 1a; env. 1a
Indonesia	3a
Iran	Topo. 3a, geol. 1a
Iraq	3a
Israel	1a
Japan	1a
Jordan	3a
Kampuchea	3c
Korea (North)	3b
Korea (South)	1d (larger scales 1f); geol. 1a
Kuwait	3a
Laos	3c
Lebanon	1a but old
Macao	1a
Malaysia	1d (larger scales 1g or 2c)
Mongolia	3a
Nepal	1d (larger scales 2b or 2c)
Oman	3b
Pakistan	Topo. 1d (larger scales 2b or 2c, with 1c for border areas)
Papua New Guinea	1a
Philippines	1a
Qatar	1a
Saudi Arabia	1d and 1e (larger scales 2c)
Singapore	1a
South Yemen	3c, some old stock available from jobbers
Sri Lanka	1d (larger scales 2b or 2c or 3a)
Syria	Topo. 3a; geol. 1a
Taiwan	3b
Thailand	Topo. 1d (larger scales 1g); geol. 1a; env. 1a
Turkey	1d (larger scales 1g)
United Arab Emirates	3b
USSR	1d (larger scales 3a)
Vietnam	3a
Yemen Arab Republic	2b (1g or 3a)

Australasia

Australia	1a
New Zealand	1a

Europe

Albania	1d (larger scales 3b)
Andorra	1a
Austria	1a
Belgium	1a
Bulgaria	1d (larger scales 3a)
Czechoslovakia	Topo. 1d (larger scales 3a); geol. 1a
Denmark	1a
Finland	1a
France	1a
German Democratic Republic	1d (larger scales 3a)
German Federal Republic	1a
Gibraltar	1b
Great Britain	1a

Greece	Topo. 1d (larger scales 1g); geol. 1a		*Indian Ocean*	
Hungary	Topo. 1d (larger scales 3a); geol. 1a		Christmas Island	1a
			Cocos/Keeling Islands	1a
Iceland	1a		Comoros	1a
Ireland, Northern	1a		French Southern and Antarctic Territories	1a
Ireland, Republic	1a		Heard Island	1a
Italy	1a		Maldives	1a
Liechtenstein	1a		Mauritius	1a
Luxembourg	1a		Réunion	1a
Malta	1a		Seychelles	1a
Monaco	1a			
Netherlands	1a		*Pacific Ocean*	
Norway	1a			
Poland	Topo. 1d (larger scales 3a); geol. 1a		American Samoa	1a
			Cook Islands	1a
Portugal	1a		Easter Island	1a
Romania	Topo. 1d (larger scales 3a); geol. 1a		Fiji	1a
			French Polynesia	1a
San Marino	1a		Galapagos Islands	4
Spain	1a		Hawaii	1a
Sweden	1a		Kiribati	1a
Switzerland	1a		Nauru	1a
Yugoslavia	1d/3a		New Caledonia	1a
			New Zealand Dependencies	1a
			Norfolk Island	1a
			Pacific Trust Territories	1a
Atlantic Ocean			Solomon Islands	1a
			Tonga	1a
Ascension Island	1a		Tuvalu	1a
Azores	1a		Vanuatu	1a
Bermuda	1a		Wallis and Futuna Islands	1a
Bouvet Island	1a		Western Samoa	1a
Canary Islands	1a			
Cape Verde Islands	1d		*Polar Regions*	
Faeroe Islands	1a			
Falkland Islands	1d		*Arctic*	1a
Madeira	1a		Greenland	1a
St Helena	1a		Jan Mayen	1a
St Pierre and Miquelon	1a		Spitzbergen	1a
São Tomé e Principe	1a			
South Georgia	1a		*Antarctica*	1a (Russian areas 1d (larger scales 3a))
South Sandwich Islands	1a			
Tristan da Cunha	1a		Antarctic Islands	1a

References

ALLIN, J. (1978–) *Map Sources Directory*. Downsview, Ontario: York University Libraries. 23 loose-leaf sections

ALLISON, B. (1983) Map acquisition: an annotated bibliography. *Information Bulletin. Western Association of Map Libraries*, **15**(1), 16–25

ALLISON, B. (1985) Map acquisition: the major sources. *Wilson Library Bulletin*, **60**(2), 21–24

BRANDENBERGER, A.J. (1984) Names and addresses of cartographic agencies that answered the questionnaire on topographic mapping. *World Cartography*, **XVII**, 49–51

CARRÉ, J. *et al.* (1985a) La cartographie officielle dans le monde. *Bulletin du Comité Français de Cartographie*, 105

CARRÉ, J. *et al.* (1985b) Agences cartographiques éditeurs publics et privés dans le monde. *Bulletin du Comité Français de Cartographie*, **3**, 105, 29–49

COBB, D.A. (1983) Government mapping in the western world. *Government Publications Review*, **10**, 381–394

COBB, D.A. (1986) *Guide to US Map Resources*. Chicago, Illinois: American Library Association. pp. 196

DIMENT, J. (1979) Geological map acquisitions: a guide to the literature. *Geoscience Information Bulletin*, 111–144.

DOEHLERT, I.C. (1984) Basic considerations for the development of an effective selection and acquisition policy for an academic map library. *Information Bulletin. Western Association of Map Libraries*, **15**(2), 188–196

FARRELL, B. and DESBARATS, A. (1984) *Guide for a Small Map Collection*, 2nd edn. Ottawa: Association of Canadian Map Libraries

FITCH, J.M. (1984) Editor. *Earth and Astronomical Sciences Research Centres: A World Directory of Organizations*. Harlow: Longman. pp. 742

GARDINI, M.J. DE ALMEIDA (1979) Brazilian map sources. *Society of University Cartographers Bulletin*, **10**, 36–40

HAGEN, C. (1974) New series of international bathymetric charts. *Information Bulletin. Western Association of Map Libraries*, **5**(3), 3–8

HARVEY, N. (1983) Editor. *Agricultural Research Centres: A Worldwide Directory of Organizations and Programmes*, 2 vols. Harlow: Longman

HODGKISS, A.G. and TATHAM, A.F. (1986) *Keyguide to Information Sources in Cartography*. London: Mansell. pp. 253

HYDROGRAPHIC DEPARTMENT (1986) Countries with established hydrographic offices publishing charts of their national waters. In *Catalogue of Admiralty Charts and Other Hydrographic Publications*. Taunton: Hydrographic Department. pp. 5

JOHNSON, P.T. (1974) Sources and methods to Latin American map acquisition. *SLA Geography and Map Division Bulletin*, **95**, 40–46, 60

KOERNER, A.G. (1972) Acquisition philosophy and cataloguing priorities for university map libraries. *Special Libraries*, **63**, 511–516

LARSGAARD, M.L. (1978) *Map Librarianship*. Littleton, Colorado: Libraries Unlimited. pp. 330

LARSGAARD, M.L. (1984) *Topographic Mapping of the Americas, Australia and New Zealand*. Littleton, Colorado: Libraries Unlimited. pp. 180

LARSGAARD, M.L. (1986) *Map Librarianship*, 2nd edn. Littleton, Colorado: Libraries Unlimited

LAWRENCE, G.P. (1985) Maps, atlases and gazetteers. In *A Guide to Geographical Information Sources*, edited by S. Goddard, pp. 189–210. Beckenham: Croom Helm

LEPINE, P. (1983) Le dépot légal des documents cartographiques. *Bulletin de la Bibliothèque Nationale du Québec*, September–December, 17–18

LOCK, M.R. (1969) *Modern Maps and Atlases*. London: Bingley

OVERSEAS SURVEYS DIRECTORATE (1986) *Address List of Overseas Survey Organizations and Contacts*. Southampton: Ordnance Survey. pp. 14

PARRY, R.B. (1983) Selection and procurement of European maps for an American map library, with a case study of British maps. *SLA Geography and Map Division Bulletin*, **131**, 16–25

PRESCOTT, D.F. (1978) Building the map collection in an academic library. *The Globe*, **10**, 9–18

RUDD, J. and CARVER, L.G. (1981) Topographic map acquisition in US academic libraries. *Library Trends*, **29**(3), 375–390

SCHORR, A.E. (1974) Written map acquisition policies and procurement problems. *SLA Geography and Map Division Bulletin*, **98**, 28–30

SEAVEY, C.A. (1981) Collection development for government map collections. *Government Publications Review*, **8A**(1), 17–29

SHEEHY, E.P. (1976) *Guide to Reference Books*, and supplements issued 1980 and 1982. Chicago, Illinois: American Library Association

STATISTICS AND MARKET INTELLIGENCE LIBRARY (1985) *National Statistical Offices of Overseas Countries*. London: Department of Trade and Industry. pp. 47

TIBERIO, B. (1979) The acquisition of free cartographic materials: request and exchange. *Special Libraries*, May–June 1985, 233–238

TINSLEY, E.J. and HOLLANDER, J.P. (1984) *Worldwide Directory of National Earth Science Agencies and Related International Organizations*. Reston, Virginia: United States Geological Survey. pp. 102

WALFORD, A.J. (1973–80) *Guide to Reference Material*, 3 vols. London: Library Association

WILLIAMSON, L.E. (1983) A survey of cartographic contributions of international governmental organizations. *Government Publications Review*, **10**, 329–344

WINCH, K. (1976) Editor. *International Maps and Atlases in Print*, 2nd edn. New York: R.R. Bowker. pp. 866

WINDHEUSER, C.S. (1983) Government mapping in the developing countries. *Government Publications Review*, **10**, 405–409

WISE, D. (1975) Selected geographical and cartographical serials containing lists and/or reviews of current maps and atlases. *SLA Geography and Map Division Bulletin*, **102**, 42–45

WISE, D. (1977–79) Cartographic sources and procurement problems. *Special Libraries*, **68**, 198–205, and appendices issued in *SLA Geography and Map Division Bulletin*, June 1978, **112**, 19–26; September 1978, **113**, 65–68; December 1978, **114**, 40–45; March 1979, **115**, 35–40

WISE, D. (1981) Cartographic solicitation programmes for city maps. *Western Association of Map Libraries. Information Bulletin*, **12**(3), 284–287

Further reading

ARCHIER, E. and LAGARDE, L. (1980) Enquète sur les acquisitions de cartes. *LIBER Bulletin*, **15**, 7–18

COBB, D.A. (1979) The politics and economics of map librarianship. *SLA Geography and Map Division Bulletin*, **117**, 20–27

LOW, J.G.-M. (1976) *The Acquisition of Maps and Charts Published by the United States Government*. University of Illinois, Graduate School of Library Science, Occasional Paper No. 125

NICHOLS, H. (1982) *Map Librarianship*, 2nd edn. London: Bingley. pp. 272

STEVENS, S. (1985) Map librarianship: suggestions for improvement. *Wilson Library Bulletin*, **60**(2), 33–36

STRICKLAND, M. (1978) Library of Congress Summer Map Processing project acquisitions. *Western Association of Map Libraries. Information Bulletin*, **10**, 67–71

4 Map evaluation

Barbara Farrell

In the early days of the computer revolution predictions were made for the imminent demise of the paper map. Computer assistance in mapping during the last quarter of a century has indeed resulted in an increase in the number of electronic maps, but automation has also assisted vital map production and resulted in a vast increase in the number of printed maps produced throughout the world. Many of these maps achieve excellent scientific standards. Never before has there been such a profusion of good quality cartographic material from which the map selector must choose.

At the same time, as Chapter 3 so well demonstrates, the bibliographic control of the world's cartographic materials is, as yet, at such an elementary level that the problem of establishing which new maps have been published, by whom, and from whom they may be obtained, is a major burden to the potential purchaser. The complexity of procuring new maps is in danger of tempting the harassed map selector to order from the publisher's list without an adequate evaluation of the products so offered.

Whether one is selecting maps for a personal collection or as a public responsibility, for a small collection or a large one, for a limited or a broadly based purpose, the essential task of the map selector is to obtain the highest level of useful information within the budgetary resources at his or her disposal. This presupposes a search for high-quality mapping, since maps of poor quality will provide both erroneous and confusing information under the veneer of authenticity. It assumes a search for pertinent mapping, since maps extraneous to one's purpose squander precious resources for their processing and upkeep, as well as their purchase. It requires a search for informationally economical mapping, that is, for maps with an efficient density of information, such that the investment of time, space and financial resources is well rewarded in the intensity of use of each item purchased. In sum, the map selector needs the highest quality maps for the money invested, the most relevant maps for general and specific purposes, and as few maps as are necessary to serve those purposes.

How then does one set about verifying the quality of maps to be purchased and ensuring their suitability for likely needs? The answer is not simple, nor is it aided by the fact that all too often in practice one cannot examine a particular map sheet before ordering it. To say that the answer lies in an accumulation of experience and in a depth of knowledge about maps, whilst in many ways true, is in no way useful to the beginner. This chapter therefore uses ideas which have long been discussed by theoretical and educational cartographers (Keates, 1982; Robinson *et al.*, 1984) regarding map purpose and evaluative criteria, to suggest a conceptual framework to guide the selection process.

Map purpose and map type

Why can one not simply select maps for particular purposes on the basis of their declared map type? The terms 'topographic map', 'geological map', 'vegetation map' are surely clear enough! Map type can often be an adequate guide, but the argument is made here that map type is not always a sufficiently authoritative or consistent guide on which to take action. The meaning of 'topographic map' may indeed be widely understood, as long as one is not making an issue of the precise scale limits involved, but consider, for example, the terms 'road map', 'street map', 'city plan', 'urban plan', 'recreation map', 'tourist map'. Is it possible to tell from the terms themselves from which of these to expect:

- a large-scale plan with building outlines shown,
- a map with street names and an index,
- a topographic style map with vacational and recreational details added,
- a highly generalized sketch of the city centre replete with three-dimensional buildings and highly coloured pictorial symbols,
- a well classified road system with detailed tourism and recreational information,
- a poorly mapped and generalized road system and little else,
- a detailed analysis of the recreational potential of an area, but little practical advice for the holidaymaker?

The distinctions rapidly become difficult to evaluate. An almost endless list of map types can be identified. *The Canadian MARC Communications Format: Monographs*, the authoritative manual for map cataloguers in Canada, lists no fewer than 52 distinct terms characterizing map types. The terms, moreover, are not mutually exclusive: some distinguish map content, some map form, and yet others the production or presentation technique. The MARC code itself (National Library of Canada, 1979: 009 CM9-11) states: '. . .cartographic materials can occur in an indefinite variety of products and combinations of products collected by libraries. There is no agreement among cartographers (or librarians), national or international, about the exact meaning of such terms, nor does there exist any documentation which includes all the terms needed to produce an exhaustive (or complete) list . . . There is, in fact, no agreement as regards to the exact meaning of certain [map type] terms within the English language'.

It is important therefore to learn the meaning of terms denoting map types (Farrell and Desbarats, 1984; International Cartographic Association, 1972) but these terms provide a guide to map selection only when one already knows what to expect and which terms have a precise and widely recognized meaning.

Map use

A more helpful approach may be to look, in a more generalized way, at the uses to which maps are put. Map use has been dealt with in various ways over the years by many authors (Board, 1984; Keates, 1982; Guelke, 1981; Muehrcke, 1978). From this body of work it is possible to distinguish four characteristic classes of map use which, although not completely mutually exclusive, are valuable in categorizing maps for the purpose of selection and collection development.

Maps as sources of objective information

Undoubtedly the primary purpose for which maps are made is to provide factual information about locations and relationships in specific areas of the world. Map users ask such questions as:

- What is the feature at this   Locating
  location?
- What is this place called?   Naming
- How far is it between . . .   Measuring distance
  and . . .?
- Is . . . further south than   Establishing relationships
  . . .?
- How high/deep is . . .?   Comparative elevation
- Whose land/territory is   Establishing ownership
  this?
- Is this place mine or   Establishing boundaries
  yours?
- How large is this land?   Measuring areas
- How much larger is this   Comparing areas
  area?

All of these questions require measurement from a map based upon a precisely established and represented scale and coordinate system. They have specific implications, as will be seen below, for the properties a satisfactory map must possess. Maps that serve these purposes are those that have been characterized elsewhere as 'encyclopaedias of the mapping world' (Rudd and Carver, 1981: 376). Such maps are the *sine qua non* of cartography. They provide both generalized and detailed knowledge of the earth's surface. They are the backbone of any but the most specialized map collection.

In this group we have to include a wide range of published maps distinguished primarily by scale.

- *General reference maps* relating to the world, major areas and regions of the world, and individual countries, areas and oceans. A large number of these are small-scale reference and general-purpose maps up to scales of approximately 1 : 1 000 000. Larger-scale 'general reference maps' of smaller areas may be included. Such maps are produced by major national or international mapping agencies and commercial map producers.
- *Topographic maps*, defined here, for convenience, as those ranging in scale from 1 : 25 000 to 1 : 500 000, whose purpose is 'to portray and identify the features of the Earth's surface as faithfully as possible within the limitations imposed by scale' (ICA, 1972: 821.1).

Topographic maps are produced by the mapping agencies of most nations, each for their own territory. Depending on scale this means that territorial coverage may be complete, partial or anticipated.
- *Large-scale maps*, that is, maps at scales larger than 1 : 25 000, sometimes cover smaller countries but more often depict limited urban areas. These maps may include elevation (contours and/or spot heights), but are often planimetric in nature emphasizing the layout of land boundaries, streets, lots and buildings. They are produced in some cases by national mapping agencies, but more often by regional and municipal levels of government or commercial mapping firms. Cadastral and land value maps such as fire insurance maps may be included.

Maps for travel and navigation

A second major function which maps have served throughout their history is to facilitate movement from place to place. As defined here this group includes all maps designed specifically for the traveller—whether the travel is by land, water or air. The emphasis of these maps is upon connectivity and the relationships between places. They must enable people and vehicles to move from one place to another firstly safely, secondly as efficiently and quickly as is consistent with the mode of travel, and thirdly, optionally, with enrichment, that is, added experiential value derived from the phenomenological experience of place. Maps in this category include, for example, bicycle route or automobile association strip maps, tourist maps, road maps, rail routes, maps of canoe routes, maps for wilderness expeditions, inland waters navigation charts, hydrographic charts, and aeronautical charts at various scales.

There is some overlap here with maps in the first group, in that topographic maps are often used for travel, but this group is distinguished by the fact that the maps are designed specifically for travel, and therefore different properties are emphasized.

The properties needed vary according to the speed, surface and mode of travel. Larger-scale maps are required for slower speeds of travel and *vice versa*. For travel on land, maps will differ if the travel is on foot, by bicycle, by car or by train. For travel by water different maps are needed for journeys by canoe than for those by sailboat, small power boat or ocean going vessel. In the air, light planes operating under visual flight rules have very different mapping needs from jet planes which move beyond the scope of the printed map and require on-board automated navigation systems.

On water and in the air, directional navigation requires maps constructed on projections which show correct angular bearing and therefore allow the plotting of azimuths. This is not as important for maps for land travel where routes are provided by the transportation infrastructure of the country.

Whatever the variations, any map used for travel must allow the user to:

- measure distances;
- identify directions, normally with reference to an azimuthal system;
- select suitable routes;
- identify conditions and/or constraints associated with routes;
- identify legislative constraints, boundaries, or zones restricting routes.

Maps for interpreting spatial relationships: thematic maps

Maps in this category are used for recording, analysing, interpreting and understanding distributions and relationships between specific phenomena distributed over a given area of the earth's surface. Such phenomena may be physical, cultural, historical, economic, sociological or political in nature. They may be concrete physical objects, transient in nature, or purely conceptual. They may deal with distinctions of kind (qualitative maps) or with distinctions of quantity (quantitative maps). The group is commonly referred to by the name 'thematic maps' and is the most diverse and rapidly growing of all the categories described here.

This kind of map arises from human desire to understand the nature of, to control, improve or exploit the world's resources. It also serves as a means of exploring human social, cultural and other relationships. Only after the developed nations of the west had made substantial strides towards achieving topographic map coverage were they able to afford the luxury of, and to perceive the benefits to be obtained from, a visual expression and communication of thematic spatial information.

Thematic mapping found its earliest expressions in efforts to understand and map fundamental aspects of the physical nature of the earth: ocean currents, prevailing winds, climate, geology, soils and vegetation. The rapid progress made in this area of geoscientific mapping since World War II has been well documented (Seavey, 1983; Stegena, 1979). Thematic mapping has proceeded from here to ever more particular and ever more intangible, but no less important, themes.

This is the class with the widest variation in map type, the widest variation in scale, the widest variation in quality and the greatest number of new products. Many new maps are of exceptionally high quality and result from the research efforts of individuals, groups, regional and national bodies, and international cooperation. Much of the quality is due to the efforts of national cartographic organizations and the **International Cartographic Association**, the **International Geographical Union** and the **United Nations** to establish standards for the various stages of map design and production. Commercial cartography too has produced materials of high quality. Valuable products, but sometimes of a lesser technological standard, emanate from those areas where money and expensive equipment are not as readily available.

At the same time this group of maps contains numbers of maps of simply poor quality and both map selector and map user must be continually on guard. Some cartography of poor quality results from an emphasis on commercial viability, some from financial cutbacks and consequent corner-cutting even in the best cartographic organizations; some from 'contracting out' and lack of vigilance over standards; some from individuals and small companies acting in ignorance of current mapping standards. Lack of vigilance there may be and maps of poor quality are indeed to be found, but the assumption here is that maps included in this category are made with an intellectually honest purpose, with controlled methods of data collection, analysis and representation, and above all with a scientifically dispassionate objective.

Advocacy maps

Included in this group are maps produced for the formal educational system, for example wall maps, teaching maps and atlases; maps used in the publishing industry in books and magazines; maps used by the news media in newspapers and on television; maps used in advertising; and others created deliberately to propound a particular point of view, such as political propaganda maps.

All such maps are, by definition, selective. Most are visually simplified in order that their message may be readily grasped. Their objective is to demonstrate, explain, inform, convince or persuade. Many are intellectually honest products displayed in a simplified manner in order to meet their specific purpose. These may be high-quality cartographic products which achieve their aims by abstracting from a mass of detail and portraying only information essential to the purpose in a manner which has impact on the user. Yet the group as a whole includes the potential for distortion of facts resulting from the subjective perspective of the map creator; relevant facts may be subconsciously or even deliberately suppressed; in extreme cases, usually for political purposes, incorrect information is deliberately portrayed.

This group is the most difficult to evaluate because the highly simplified may be an authoritative and honest portrayal for an elementary level of understanding, as in maps created especially for children. At the other extreme, highly biased or devious content may be portrayed with cartographic precision, and unless carefully examined may seem to be of good quality.

Criteria for map evaluation

The criteria for evaluating maps and relevant questions to ask when considering purchase will be discussed with reference to the four groups identified above: information maps, travel maps, thematic maps and advocacy maps. Since not all criteria are of the same relevance for all groups, general concepts will be discussed under Information maps and specific expansions made as needed for the other groups.

The criteria considered are, in sequence: scale, media and format, currency of information, reference structure of the map, appropriateness of the cartographic symbolism, effectiveness of the graphic language, and identification and publication data.

Information maps: maps as sources of objective information

Scale

The scale of a map is the dominant criterion and is equally significant for all four map groups. It conditions the size of the area which can be covered on a map sheet of given size, and the amount of detail about that area which can be shown. The first question regarding scale which must be asked of a particular map is whether the level of detail provided by the map is appropriate for the stated scale. With some practice one can get to know what level of detail to expect from characteristic scales:

- 1:1000–1:2500 Building outlines, lot boundaries and road widths should be shown correctly to scale; maps may be planimetric, showing horizontal information only, or include spot heights and contours.
- 1:10 000–1:20 000 Individual buildings may be included, but are depicted by conventional symbols. For maps of cities, street names and indexes should be expected. In rural areas field and lot lines may be shown.

- 1:25 000–1:50 000   These are the characteristic topographic scales. All the elements of topographic maps should be included at the highest level of survey and cartography available for the specific area of the world. Relief is normally depicted by contours and spot heights, often supplemented by hill shading (Imhof, 1982). The disadvantages of shading for uses involving orientation should be noted. The contour interval is significant since the maximum accuracy of elevation is half the contour interval. Selectors should note whether or not the contour interval is constant over the entire surface of the map or series, as this influences the interpretation of the maps.
- 1:250 000–1:500 000   These are progressively more generalized maps of topographic type. The methods and implications of generalization at these scales is admirably explained in *Cartographic Generalisation* (Swiss Society of Cartography, 1977).
- 1:1 000 000 and smaller   Individual maps and reference series at these scales should show an appropriate level of detail in relation to the stated scale.

One might ideally like to have maps at large scales in a collection in order to provide detailed information. The practical implications of this are, however, significant. Each time the scale is doubled (i.e. the divisor is reduced by half, such as from 1:50 000 to 1:25 000) the number of similarly sized map sheets required to cover a given area is quadrupled (Farrell and Desbarats, 1984: 23–24). This has obvious budgetary and storage implications. The decision of the map selector must be based on an evaluation of whether, in a particular situation, one can expect a corresponding quadrupling of benefit in terms of the information provided.

Suitability of media and format

Media and format are considered next to scale, not because of their intrinsic importance, but because they are often interlinked with scale and the two factors must be evaluated in relation to one another.

Most information maps are produced on paper and fall within relatively standardized sheet sizes. One has no control over the sheet size selected by the producer, but the smaller the sheet size the more cut up will be the areas of interest and the more sheets will be required to complete the series.

The map selector should consider the following questions:

- Is the number of sheets in a map series economically reasonable when considered in relation to the size of sheet, the cost per sheet and the projected uses?
- What is the medium of the map and is it suitable for the anticipated map uses?
- Is the paper of good quality, resistant to aging and tearing?
- If a blueline is the only format available, will its relatively short life serve adequately the purposes for which the map is required?
- Are maps on fiche or aperture card, if available, an acceptable alternative to paper copies?
- Should electronic maps be considered?
- Is the size and layout of an individual sheet suitable for its intended use?

Currency of the information

We rely on maps for accurate and current information. The face of the earth itself may be slow to change, but day by day the infrastructure documenting human activity and

social, political, economic and cultural conditions change. Maps must be as current as possible, and users must have the means to know how up to date they actually are. Map selectors should ask themselves the following questions:

- Are the dates of publication, printing or copyright shown on the map? This is essential information.
- Is this the latest edition of the map?
- Is the edition indicated on the map?
- Does the literature on the series provide information about the revision cycle?
- Is the date of survey shown on the map?
- If the map is a derived map, is information about the date(s) of the original map provided?
- Is there a reliability diagram showing dates of survey and revisions for the various areas of the map?

It is necessary to bear in mind that maps in this group are often part of large topographic series which have major logistic and financial implications for the producer. Such maps are published in accordance with a production schedule based upon the total number of maps in the series, the length of time needed to complete an individual map, the available budget and the desirable revision schedule for rapidly changing areas. It is unlikely that such series will satisfy all needs with regard to date and it is important to realize that the latest edition, together with a few in-the-field modifications, can be quite acceptable for many purposes. So too, in a map collection, such maps should not be discarded merely because the edition is superseded. Many researchers using maps need to be able to trace the changes in the landscape from one time period to another. The collection of 'time series', that is, all the editions of maps for certain defined areas, is one of a map curator's more important tasks.

Reference structure

For all maps and especially those in the first three groups, expect only the highest standard in the reference structure of the map. Geographical or mathematical coordinates are your evidence for a scientifically constructed map from which precise locations may be identified and from which accurate distance, bearings and areas may be measured. Ask such questions as:

- On what reference structure is the map based: geographical coordinates? a national grid? an arbitrary (ABC/123) grid?
- What evidence does the map supply for the quality of this structure?
- Is the projection named?
- Is the north arrow shown, and the magnetic variation?
- Are they correct?

Appropriateness of the cartographic symbolism

On information maps one expects the greatest use of traditional cartographic symbolism. 'Conventional' symbols have long been used for representing the human and physical features of the earth's surface, such that, despite minor differences of symbol form or colour scheme, it is generally possible to interpret the symbols on an information map regardless of the language of the map. Differences which do occur, for example, in boundary symbols, or in colour schemes for hypsometric tints, do not detract from a fairly high level of standardization of symbols.

Questions which should be asked are:

- Are all the symbols understandable?
- Is the scale given?
- Are point symbols appropriate?
- Are classifications relating to size readily understandable?
- Are different line symbols distinctive?
- Are any obscure or purely local symbols used?
- Is colour used correctly as a symbol?
- Are all symbols used included in the legend?
- Do geographical names follow standard authorities?

Effectiveness of the graphic language

The potential for variability in the graphic quality of maps is enormous. Information maps generally achieve high standards in this regard, and are limited primarily by the technology and financial resources available for production and reproduction. There is sometimes a tendency to sacrifice graphic quality for expediency in the haste to use new, faster and cheaper methods of production. As a result the latest map is not always the best in graphic quality.

Questions which should be asked about graphic quality are:

- Are all the words and symbols large enough to read?
- Are the words and symbols sufficiently distinct from their backgrounds and from one another to be easily interpreted?
- Is the method by which the scale is shown suitable for the kind of map?
- What is the quality of the linework for the various line symbols?
- Are the overlays used in printing perfectly registered?
- Are the colours and tones visually pleasing?
- Are the typeface(s) used for lettering appropriate?

Identification and publication data

Map publishers have not yet reached the same level of sophistication in the provision of bibliographic information for maps as is current practice for books. Lack of consistency and standardization in titling, identification of authors and the provision of publication information, amongst other facets of information, impede the selector's work, the bibliographic control of cartographic materials and the map user.

The selector should ask the following questions:

- Is there a definite title on the face of the map?
- Are there different titles on a panel or accompanying material?
- Is the title carefully worded as to area and topic, if relevant?
- Is the series title, if applicable, clearly and consistently given?
- In a series is the sheet distinctly identified?
- Is the authorship of the map explained?
- Is information about the basic survey included?
- Is other relevant source information acknowledged?
- Are the publisher, place of publication and date of publication and/or copyright unequivocally given?
- Is the edition clearly specified and distinguished from a reprint?

Official government agencies producing information maps are fairly scrupulous in providing some elements of this information but there is still great room for improvement.

Maps for travel and navigation

What are the distinctive criteria of special importance for maps for travel and navigation?

Scale

Scale must allow an adequate level of detail to depict required information about the nature of the surface over which travel takes place. It must also take account of the speed and mode of transport. The faster the speed of travel the less local detail is required; the more remote from the surface, the less surface detail is required.

Format and media

Format and media are of critical importance for travel maps which must often be used in cramped and poorly lighted conditions, or outdoors and subject to the elements. Of significance for these purposes therefore are such considerations as waterproof coating of the map paper; plastic sleeves or pockets; whether or not the sheet lines overlap; whether the maps are printed on two sides of the sheet; the method of folding and unfolding the map; whether or not the map is sectionalized into a book format; whether the map is in strip format; and whether or not the map can be displayed on an on-board CRT screen.

Currency of information

Currency of the map information is vital for travel maps. Navigation charts used for their primary purpose *must* be up to date. Such maps are responsible for the safety of craft, vehicles and passengers alike. They denote routes which may be selected with safety, and they indicate legal and political as well as physical barriers to travel. In the case of land travel, maps must be as up to date as possible although here an out-of-date map may be more a cause of annoyance and frustration than a direct threat to safety.

An argument may be made for the use of obsolete air and sea navigation maps in a general collection when information maps of equivalent detail are not available: many general collections use superseded coastal and aeronautical charts for detailed depiction of areas of the world for which they would not otherwise have detailed coverage.

Reference structure

Reference structure is of the utmost importance for maps for air or ocean navigation. Because angular bearings are required for navigation these maps must be constructed on a projection which preserves angular bearing, that is, the lines of latitude and longitude must intersect at right angles as they do on the globe, and the scale of the map surrounding any small point must be the same in all directions. In practice this means that most navigation maps are constructed using an azimuthal or a version of the Mercator projection.

Appropriateness of the cartographic symbolism

Standardization of symbols is important for all travel maps: the meaning of symbols must be readily grasped by all potential users of the map.

Effectiveness of the graphic language

Graphic standards are equally important. Colour selection for difficult lighting situations is particularly important and symbols must be readily identifiable under conditions such as in car or cockpit lighting.

Identification and publication data

The same criteria apply as for information maps.

Thematic maps

Scale

In addition to general considerations it is necessary to evaluate whether or not the scale allows enough detail to portray the topic accurately. Is the scale sufficiently large that neither data nor geographic information is over-generalized? Consider for whom the map is intended: a prospector, for example, will require a larger-scale map than a geologist; an architect is more likely to require a large-scale urban plan than a street map.

Format and media

Thematic maps as a group have no special format requirements. Each must be considered on its own merits. Atlases often provide a convenient and popular format for thematic materials. Thematic maps are increasingly produced as electronic maps.

Currency of information

There are several dates which may be of significance for a thematic map and it is important to know which is which. It is of little use that the map was published in 1986 if the population distribution shown upon it is derived from the 1961 census, that is, unless the title clearly states that this is indeed what the map is intended to show. At the same time one must be aware that the date of thematic information is of varying significance for different kinds of phenomena. Some distributions do not change as rapidly as others. One would not be unduly disturbed by a geology map 5 years old; but if it is 30 years old then one might expect that methods of analysis and classification would have changed in the intervening period. It must be remembered too that many major series, for example those depicting land classification and evaluation, are the product of a one-time major commitment of funds and human resources for an extensive survey. It is quite unrealistic to expect such surveys to be repeated at frequent intervals. Such surveys, when out of date, become major sources of historical evidence.

Reference structure

There is a wide range in standards of reference structure for thematic maps, ranging from the authoritative large- and small-scale series and single maps produced under strict specifications to stringent standards of accuracy by some national and international bodies, to very low quality products. Use the evidence provided by information about the reference structure to determine how carefully the map has been constructed and whether or not any measurements you may need to make will be reliable. Treat with suspicion all maps which fail to provide sufficient evidence for the authenticity of their construction. This may be an indication of careless work in other areas.

Appropriateness of the cartographic symbolism

The nature of the data depicted on thematic maps determines the most appropriate symbolism to use. If there is a mismatch between data and symbolism the maps may easily misinform rather than communicate effectively.

- Is the map concerned primarily with point, line or area distributions?
- Is the data depicted qualitative or quantitative?
- If the latter, how many classes are used and how are the classes assigned . . .
- . . . or is the data shown by continuously variable symbols?
- Does the data refer to discrete information—that occurring separately at points—or does it relate to a continuous surface?
- Does the symbolism reflect this difference?
- Does area data refer to absolute figures (poor) or is it processed or related in some way to area?

Effectiveness of the graphic language

Graphic quality is very important for thematic maps: it can mean the difference between the success or failure of the map. But because thematic maps come from such widely varied sources and each must solve its own specific set of graphic problems there is a wide range in graphic quality. The same criteria apply as described for information maps.

Identification and publication data

The same criteria apply as for information maps. Confidence in the standards and authority of map author and publisher remains a significant guide to potential map quality, particularly in products issued by national and international bodies.

Advocacy maps

Scale

Maps in this group are often of small scale or are generalized and simplified versions of larger-scale maps. Even though precise measurements are often not required from this type of map, scale must be shown to allow comparisons to be made and relationships to be seen.

Format and media

When maps are intended for didactic purposes it is important that the media suit the purpose. Maps may need to be on film, slides or transparencies; they may need to be mounted and rolled; or hard-wearing plasticized surfaces may be required. For handling by individual students the size which can be accommodated at a classroom desk may need to be considered. Maps which are for publication must follow the strictures of the particular publication format: issues such as page layout, portrait or landscape format, and so forth, become important.

Currency of information

Many maps in this class neglect dates entirely. This is a reprehensible situation which causes untold problems for map use. When selecting maps use the presence or absence of date(s) as an indication of honesty, quality and authenticity in the map.

Reference structure

Measurement from these maps is not normally as important as for other categories, therefore the reference structure may be simplified. Nevertheless simplification is not an excuse for poor or careless cartography. Many of these maps represent a more abstract level of conceptualization. The more abstract the product the less is a real earth reference structure required, but be very suspicious, however, of two-dimensional shapes purporting to be maps but which contain no information as to how the map space was constructed. Map projection is often of significance in small-scale maps and should be indicated or stated.

Appropriateness of cartographic symbolism

Map symbols are designed for impact. In many cases pictorial symbols may be used. Look carefully at the classification schemes used in relation to the information portrayed. Maps in this category must often not be taken at their face value. In addition to deliberate or self-deluding sources of error there is potential for misinformation from the inherent limitations in the symbolism chosen.

In maps of this kind where data is classified and symbolized, the method of establishing class intervals should be carefully examined and understood.

Effectiveness of the graphic language

Errors in symbolism are often compounded by poor graphic representation. Graphic quality is important and colour gradations should be carefully examined, particularly when they represent classed data. The same criteria apply as for information maps.

Identification and publication data

Maps in this group are often not autonomous maps. Sources of information should be credited and authorship acknowledged.

Conclusion

This chapter has looked at four characteristic types of map use and has reviewed the range of criteria that should be considered when selecting maps for those uses. In conclusion it should be emphasized again that the spatial information which is expressed in cartographic form can range from the concrete to the abstract, from the particular to the general, from the objective to the subjective, and from long-term, lasting value to the ephemeral. The cartographic products that communicate this information therefore vary from those with a strong emphasis on accuracy, detail and precision to those where simplification, generalization, clarity and impact are dominant. Any one map or other cartographic product can be located anywhere along the continuum. The test of its quality is its suitability for the proposed use.

Sometimes, however, maps are used for several purposes, or are indeed planned to accommodate a range of uses. In this case a combination of characteristics will be required but essential qualities must not be compromised. The map selector must know when several purposes can be validly served by one product and when the distinctive qualities required are mutually exclusive. The greatest potential for misunderstanding lies when the purposes identified by the map producer at the map design stage, and those of the user in a given time and place, do not agree. It should be the aim of the map selector to reduce the frequency of such occurrences by diligent selection, forethought and the application of stringent standards to map purchase.

The challenge to the map selector is to understand the method of representing spatial data as well as to evaluate critically the quality of cartographic presentation, for the two aspects go hand in hand, and one cannot meaningfully judge the latter without considering the former. The minimum is to learn what to expect from given scale ranges, map types and publishers, but the authority of the map publisher, though a good starting point, cannot be accepted uncritically. A continuous process of study, use and evaluation of maps is essential, with the deliberate aim of developing a critical eye for such matters as symbolism, classification schemes and the use of colour. It is also necessary to bear in mind that although careless cartographic presentation may be an immediate clue to careless thought processes, the reverse is not necessarily true. One should not be deceived by the authoritative appearance of a map into accepting its message without further thought.

In sum the virtues of knowledge, integrity and judgement which J. K. Wright (1966) urged for map users as well as map makers, are required *ipso facto* by those who select maps on the user's behalf.

References

BOARD, C. (1984) Editor. *New Insights in Cartographic Communication*. Cartographica Monograph 31

FARRELL, B. and DESBARATS, A. (1984) *Guide for a Small Map Collection*, 2nd edn. Ottawa: Association of Canadian Map Libraries

GUELKE, L. (1981) Editor. *Maps in Modern Geography: Geographical Perspectives on the New Cartography*. Cartographica Monograph 27

IMHOF, E. (1982) *Cartographic Relief Presentation*. Berlin: Walter de Gruyter

INTERNATIONAL CARTOGRAPHIC ASSOCIATION (1972) *Multilingual Dictionary of Technical Terms in Cartography*. Wiesbaden: Franz Steiner Verlag

KEATES, J.S. (1982) *Understanding Maps*. London: Longman

MUEHRCKE, P.C. (1978) *Map Use: Reading, Analysis, and Interpretation*. Madison, Wisconsin: JP Publications

NATIONAL LIBRARY OF CANADA, CANADIAN MARC OFFICE (1979) *Canadian MARC Communication Format: Monographs*, 3rd edn. Ottawa: National Library of Canada

ROBINSON, A.H. et al. (1984) *Elements of Cartography*, 5th edn. New York: Wiley

RUDD, J.K. and CARVER, L.G. (1981) Topographic map acquisition in United States academic libraries. *Library Trends*, 29, 3

SEAVEY, C.A. (1983) Editor. Government mapping. *Government Publications Review*, 10, 4

STEGENA, L. (1979) Geoscientific world mapping: facts and representations. *World Cartography*, 15, 77–89

SWISS SOCIETY OF CARTOGRAPHY (1977) *Cartographic Generalisation*. Cartographic Publication Series, 2. Zurich: Swiss Society of Cartography

WRIGHT, J.K. (1966) *Human Nature in Geography*. Cambridge, Massachusetts: Harvard University Press

Further reading

BALOGUN, O.Y. (1982) Communicating through statistical maps. *International Yearbook of Cartography*, 22, 23–41

BLACKADAR, R.G. (1972) Compiler. *Guide for the Preparation of Geological Maps and Reports*, Rev. edn. Geological Survey of Canada: Miscellaneous Report 16

BOARD, C. (1980) Map design and evaluation: lessons for geographers. *Progress in Human Geography*, 4(3), 433–437

BRANDENBERGER, A.J. (1976) Study on the status of world topographic mapping. *World Cartography*, 14, 71–102

BREU, J. (1982) The standardization of geographical names within the framework of the United Nations. *International Yearbook of Cartography*, 22, 42–47

CANADIAN INSTITUTE OF SURVEYING (1985) *Report of the Task Force on the Surveying and Mapping Industry in Canada*. Ottawa: Department of Regional Industrial Expansion

CANADIAN PERMANENT COMMITTEE ON GEOGRAPHICAL NAMES (1983) *Geographical Names and the United Nations*. Ottawa: Department of Energy, Mines and Resources

DEMEK, J. (1972) Editor. *Manual of Detailed Geomorphological Mapping*. Prague: Academia

DEPARTMENT OF TECHNICAL COOPERATION FOR DEVELOPMENT (1983) *World Cartography*, **17**

DICKINSON, G.C. (1973) *Statistical Mapping and the Presentation of Statistics*. London: Edward Arnold

GODDARD, S. (1983) Editor. *A Guide to Information Sources in the Geographical Sciences*. London: Croom Helm

HAGEN, C.B. (1982) *Maps: An Overview of the Producer–User Interaction*. Paper presented to the 1982 Annual Convention of the American Congress on Surveying and Mapping (ACSM) and the American Society of Photogrammetry (ASP)

KEATES, J.S. (1973) *Cartographic Design and Production*. London: Longman

KERR, A.J. (1980) Editor. *The Dynamics of Oceanic Cartography*. Cartographica Monograph, 25

LARSGAARD, M.L. (1984) *Topographic Mapping of the Americas, Australia and New Zealand*. Littleton, Colorado: Libraries Unlimited

LARSGAARD, M.L. (1986) *Map Librarianship: An Introduction*, 2nd edn. Littleton, Colorado: Libraries Unlimited

LOW, J.G.-M. (1976) *The Acquisition of Maps and Charts Published by the United States Government*. Occasional Papers, 125. University of Illinois: Graduate School of Library Science

MCGRATH, G. (1983) *Mapping for Development: The Contribution of the Directorate of Overseas Surveys*. Cartographica, **20**, 264

MEINE, K.-H. (1972) Considerations of the state of development with regard to topographic maps of the different countries of the earth. *International Yearbook of Cartography*, **12**, 182–198

MONMONIER, M.S. (1985) *Technological Transition in Cartography*. Madison, Wisconsin: University of Wisconsin Press

MORRISON, J.L. (1976) The science of cartography and its essential process. *International Yearbook of Cartography*, **16**, 94–108

NICHOLSON, N.L. (1973) The evolving nature of national and regional atlases. *SLA Geography and Map Division Bulletin*, **94**, 20–25

NOYES, L. (1979) Are some maps better than others? *Geography*, **64**, 303–306

OLSEN, J.M. (1975) Experience and the improvement of cartographic communication. *Cartographic Journal*, **12**(2), 94–108

PERROTTE, R. (1986) A review of coastal zone mapping. *Cartographica*, **23**, 3–72

SHUPE, B. and O'CONNELL, C. (1983) *Mapping Your Business*. New York: Special Libraries Association

5 Maps and remote sensing

Mark Monmonier

By the 1980s the prevalence of satellite images of the earth had forced cartographers to recognize two principal types of map: the traditional *line map*, with crisply delineated boundaries and routes, and the more recent *photomap*, or *image map*, often with fuzzy boundaries and discontinuous routes. This distinction is similar to that between an artist's interpretative pencil sketch of a scene and a photographer's landscape portrait. The photograph captures most of what the eye can see, but parts of the image might be blurred or not readily identifiable, and objects closer to the camera can obscure more distant features of greater interest. In contrast, the sketch is more selective and possibly enhanced as a caricature so that its 'truth' reflects not only the landscape but the artist's skill, values and goals. The line map is also a caricature, but its symbols are uniform and its geometry rigid because its content and perspective must sacrifice aesthetics to promote systematic description for planning and way-finding.

Advances in imaging technology have improved both the line map and the photomap. Improvements to conventional base maps reflect the development of *photogrammetry*, the technical speciality concerned with making measurements from aerial photographs. In the 1920s and 1930s pioneer photogrammetrists developed stereoplotting techniques for making efficient contour maps and for accurately plotting roads, buildings, forest boundaries and other ground features. Because of photogrammetry, maps also became richer in content, as an allied speciality, *air photo interpretation*, emerged with procedures for using aerial imagery to compile land cover maps for urban land use, agricultural, and geological studies. Although used by the military for reconnaissance even before World War I, aerial photography in the 1940s became an important tool for terrain mapping, target identification and damage assessment. Military applications encouraged research on colour-infrared films, useful for distinguishing camouflage from live vegetation, and non-photographic sensors, useful on high-altitude reconnaissance aircraft flying over hostile territory at night or in cloudy weather. In the late 1950s the difficulties of retrieving film from orbiting satellites provided further impetus for sensing systems able to transmit non-photographic images to ground stations. Military applications and commercial telecommunications provided the most compelling incentives for developing the requisite launch vehicles and orbiting satellites, but an intriguingly wide range of potential civilian applications led scientists and engineers with a broad range of backgrounds to promote a new speciality called *remote sensing*. New sensor and image processing techniques increased the detail, enriched the content, and improved the timeliness of satellite and other remotely sensed imagery—and led

government mapping agencies and commercial cartographic publishers to adopt new map series and cartographic products.

This chapter addresses the impact of remote sensing on production-line cartography, topographic as well as thematic. It begins with an examination of the mechanics of collecting image data and how particular sensing systems induce various geometric distortions that must be removed before the image can be integrated with cartographically accurate grids and existing map data. It then discusses the electronic measurement of reflected or emitted energy and the trade-off between radiometric resolution and spatial resolution in the extraction of cartographic features from image data. A final section looks briefly at a number of current American mapping programmes and image products that illustrate the pervasiveness of the photomap in contemporary cartography.

Platforms, sensors and geometric distortions

Perhaps the most common, and surely the earliest, remotely sensed image is the low-altitude photograph. When taken from a balloon or a primitive biplane, the air photo was usually an *oblique* view, with the optic axis of the camera meeting the terrain at an angle. Yet the preferred orientation for mapping, shown schematically in Figure 5.1, is with the optic axis perpendicular or nearly so to a horizontal plane called the *datum*. Conventional large-scale line maps project surface features perpendicularly on to the map's *datum plane*, defined for the projection chosen for the map; these maps are called *planimetric* maps because distances between symbols on the map portray distances between the orthogonal projections of the corresponding objects on to this horizontal plane. An aerial photograph, however, is a perspective view, with lines of projection converging to a point on the optic axis in the centre of the lens. This perspective obviates a planimetrically correct image. In Figure 5.1, for instance, the image of point A appears at a, not at a', its perpendicular projection on to the datum. This phenomenon affects other positions in the terrain, and the aerial photograph is said to have *radial relief displacement* whereby objects above the datum are projected radially outward from the centre and those below the datum are projected radially inward. On a truly vertical aerial photograph, the centre of the photo, or the *principal point*, represents the *ground nadir*, at which the optic axis encounters the terrain. Relief displacement is radial from

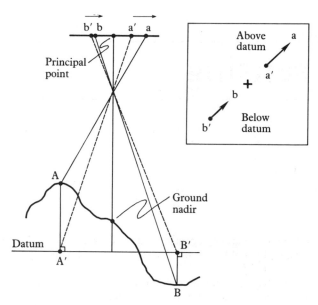

Figure 5.1   Perspective geometry of the vertical aerial photograph (*left*) displaces points above the datum radially outward and points below the *datum* radially inward toward the centre or *principal point*

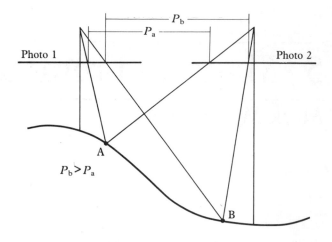

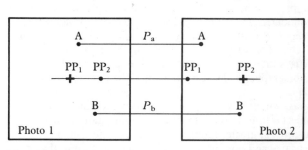

Figure 5.2   Geometry of a pair of overlapping vertical aerial photographs (*above*) produces a larger parallax p_b for point B, having a greater elevation than point A, with parallax p_a. When the photographs are oriented so that principal points and transposed principal points are collinear (*below*) the parallaxes can be measured and their difference used to compute the difference in elevation between A and B

the centre, and is directly proportional to both distance from the ground nadir and elevation above or below the datum. Distortion of shapes and distances can be particularly troublesome on the periphery of the photo. Users must avoid the temptation to measure distances on aerial photographs.

One air photo is not a planimetric map, yet two air photos can be used to make a planimetric contour map. The two photographs must overlap, and the ground nadir of each must be identified as the *transposed principal point* on the other. The two photos must have the proper relative orientation, with the principal points and transposed principal points aligned as in Figure 5.2, so that an instrument called a *stereoplotter* can measure the combined effect of each photograph's radial displacement. In a properly oriented *stereomodel*, the line between the two images of the same object will be parallel to the line through the original and transposed principal points; the length of the line between the images is called the object's *parallax*. If two objects are at the same elevation, they will have the same parallax. For two objects at different elevations, the *parallax difference* will be proportional to the elevation difference. A photogrammetrist can spend considerable time adjusting the stereoplotter to reflect the characteristics of the camera used to take the photographs, the relative orientation of the two photographs to each other, and the absolute orientation of the stereomodel to a set of *ground control points*, carefully marked on the ground for easy identification on the photos and precisely surveyed so that objects on the photo can be 'tied into' a system of plane coordinates (Moffitt and Mikhail, 1980). But once the stereomodel has been properly 'oriented', its operator can not only plot terrain features in their true planimetric positions but also trace contour lines through a series of points at the same elevation. Stereoplotters can also compensate for small amounts of *tilt*, or departure of the camera's optic axis from the true vertical. (A simpler device is the *stereoscope*, which focuses one eye on each photo so that the human eye–brain system can perceive the three-dimensional model inherent in the stereomodel.)

Computer technology has had a substantial impact on photogrammetry. Analytical plotters have reduced the time

required to orient a stereomodel, and, because fewer ground control points are required, have reduced survey costs as well (Friedman *et al.*, 1980). Contour lines, linear features such as roads and visible boundaries, and point locations can be entered directly into a digital cartographic database as a set of plane coordinates to describe geographic position and a set of feature codes to indicate the type of feature and its attributes. Thus a contour line might be described by its elevation, a road by its width and surface material, and a building by its ownership and number of storeys. This digital cartographic data can be enhanced further by the addition of feature and place names and by the generation of plotting symbols, which might require the displacement of close features to avoid overlap. Digital photogrammetric plotters are the modern approach to making large-scale line maps because these electronic images, once developed, can be used to generate film separations from which press plates are made to print multi-colour maps (Friedman *et al.*, 1980). The digital data also serves as an electronic database for modelling and information retrieval and for comparison with a new set of aerial photography when the database and map are revised.

Photogrammetric techniques for removing relief displacement from linear features extracted from aerial photography have also been used to remove relief displacement from the photographic image itself. Computers have made possible the efficient production of *orthophotos*, continuous-tone photographic images with all land cover elements in their true planimetric positions (Mullen *et al.*, 1980). Like a stereoplotter, machines for making orthophotos examine the overlap area between two photos, but each photo is divided into a large number of tiny picture elements, and electronic scanning and correlation are used to identify each object's corresponding images on the two photos. Parallax measurements are used to compute elevations relative to the datum, and these elevations then serve to estimate the relief displacement of individual picture elements so that a new photograph can be generated with planimetrically accurate greytones—making an orthophoto is conceptually akin to making a print with each small part of the negative brought back into its 'cartographically correct' location (Thrower and Jensen, 1976).

Production of orthophotos can be expedited by using photographs taken from a higher altitude than normally used in large-scale mapping—high-altitude photos cover larger areas, have less severe radial displacement, and are less likely to thwart electronic correlation with 'missing' images hidden on the other photo on the opposite side of a high, steep ridge. The US Geological Survey uses aerial photographs taken at 11–13 000 m over the centres of 7.5-min quadrangles to produce orthophotos that are published as *orthophotoquads*, quadrangle-format maps with grid lines, linear symbols for major boundaries and highways, and labels for important places and features. Orthophotos presented in a sheet-map format can be enhanced by line symbols to more precisely mark transportation routes, hydrographic features, and linear features not visible on the image, by colour to highlight differences in land cover or to simulate natural landscape colours, and by extensive labelling. An orthophoto might also be supplemented by its *stereomate*, an electronically produced companion photograph with landscape features given sufficient parallax displacement to yield a three-dimensional picture when viewed with the orthophoto under a stereoscope.

As altitude increases and the lines of projection from the terrain to the camera's film plane become almost parallel, radial displacement becomes a less prominent source of geometric distortion. At altitudes of roughly 800 km, the approximate distance above the surface of 'low-altitude' *earth resources satellites*, earth curvature and a wide field of view degrade the planimetric accuracy of a satellite photograph. And at much greater altitudes of 36 000 km or so, where some 'high-altitude' *environmental satellites* monitor atmospheric circulation, the effect of earth curvature is almost as blatant as that of the graticule on a small-scale global map projection. As with the cloud photographs distributed several times daily by various nations' meteorological services, a grid overlay is generated to provide a locational frame of reference. Small-scale maps of large areas must, after all, be presented on a projection of some sort—so what better approach than to accept the realistic 'earth-from-space' geometry of the satellite photomap and add appropriately distorted cartographic line symbols and labels?

An airborne platform allows aerial images to be taken over fixed points or in strips. Specifications might call for minimal cloud cover, and the photography can be 'flown' only once or at whatever frequency is appropriate. In contrast, a satellite platform commits the sponsor to a specific pattern of coverage. The principal distinction is between environmental satellites, which monitor large areas on a continual or frequent basis, and earth resources satellites, which provide greater detail less frequently at lower altitudes. Environmental satellites are generally in either a *geostationary orbit*, continually above the same point on the equator and making a complete rotation daily at 35 800 km, or in a *near-polar sun-synchronous orbit*, circling the earth every 12 or 24 h and passing over chosen areas at the same local times. A geostationary orbit is ideal for communications satellites as well as for the team of

equatorial weather satellites maintained by the European Space Agency, India, Japan and the United States, whereas a polar orbit is useful for monitoring extreme northern and southern latitudes and for providing frequent passes over a chosen mid-latitude region. For more detailed imagery, though, a low-altitude earth resources satellite can provide more focused coverage of smaller areas. For repeat detailed coverage of most of the earth, usually at an interval of 2 weeks or more, earth resources satellites have a precessing, sun-synchronous, near-polar orbit, like that described in Figure 5.3 for Landsats 1, 2 and 3. At an altitude of roughly 919 km, these satellites circled the earth every 103 min and always crossed the equator in a north-to-south direction at 9:30 a.m. local sun time, in an orbital path about 26° west of the previous crossing. These early Landsats circled the earth 14 times a day and repeated the complete cycle of orbital paths every 18 days. Landsats 4 and 5, at a lower altitude of 705 km, circled the earth every 99 min and completed the coverage cycle in 16 days (Southworth, 1985). France's SPOT-1 (Système Probatoire d'Observation de la Terre, or Probative System for the Observation of the Earth) satellite circled the earth every 101 min at 832 km in a 26-day coverage cycle (Courtois and Weill, 1985).

Except for intelligence satellites, which occasionally employ film drops, satellite imagery is sent electronically to

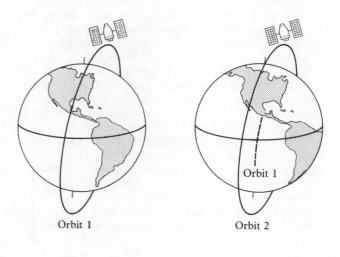

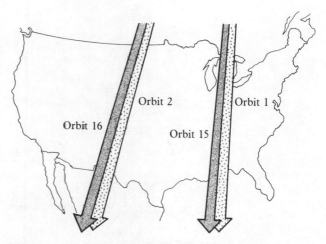

Figure 5.3 Earth resources satellites commonly have a sun-synchronous precessing near-polar orbit (*above*) yielding an orbital path somewhat west of the previous orbital path. The early Landsats circled the earth 14 times a day, with the ground swath of the fifteenth pass adjacent to that of the first pass (*below*). The entire coverage cycle was repeated every 18 days

a network of ground receiving stations, each receiving transmissions for a limited area called a *coverage circle*, for which line-of-sight signals from the satellite are not blocked by the earth. A typical coverage circle has a diameter of about 6000 km, depending upon the altitude of the satellite. Although altitudes as low as 300 km have minimal atmospheric drag and would provide more detailed coverage, somewhat higher altitudes are more efficient for the electronic recovery of image data because fewer ground stations are required. Satellite images can, of course, be recorded magnetically for later playback, but this practice is more common for intelligence satellites than for civilian satellites such as Landsat and SPOT.

A limited form of playback is used with the RBV (*return beam vidicon*) system, such as used in the late 1970s with Landsat 3. An RBV sensor recorded an instantaneous 'snap shot' image on a photosensitive surface for two adjacent, slightly overlapping, rectangular 98 × 98 km areas. These images were then scanned rapidly, line by line, and transmitted to the ground in a matter of minutes (Freden and Gordon, 1983). Because the entire *scene* was photographed instantaneously, with a camera pointing directly downward, RBV imagery is much like a very high-altitude aerial photograph, with earth curvature far more significant than relief displacement. But the image received at the ground station is a digital, not a photographic, image—with reflected light measured and recorded numerically for *pixels*, or picture elements, representing areas on the ground approximately 30 m wide. Because of earth curvature, the *ground spot* represented by the pixel tends to be slightly smaller for areas near the centre of the scene and slightly larger for areas near the edge. This ground resolution is akin to the graininess and level of detail of a photographic print, and sets a lower limit to the size of feature on the ground that can be recognized on the image. All or part of the digitally recorded scene can be viewed on a CRT monitor or used with a film-writer to generate a photographic film negative, which in turn might be used to make either positive photographic prints or a screened press plate for lithographic printing. The image can be enlarged or reduced but its sharpness is limited by the resolution of its ground spot.

More common, though, is the image scanned and transmitted continually as the earth resources satellite

follows its spiral-like *ground track* around the earth. The terrain below the satellite is examined in a *ground swath* about 185 km wide for the multi-spectral scanner system on Landsat 5 and 117 km wide for SPOT. Many scanner systems divide the ground swath into parallel *scan lines*, perpendicular to the ground track, and record reflected light for a series of ground spots spaced uniformly along each scan line to provide the gridded image of pixels shown schematically in Figure 5.4. The energy reflected or emitted from the circular ground spot is recorded for a rectangular pixel. Because the earth is rotating beneath the satellite, each row of ground spots is offset slightly from those in the previous scan line, and this 'raw' grid of pixels must be *resampled* to provide a regular grid of square pixels aligned in mutually perpendicular sets of parallel rows and parallel columns. Resampling can recover square pixels from measurements taken along the row at an interval slightly less than the spacing of the rows, as with Landsat 3's multi-spectral scanner, which recorded reflectance for ground spots 79 m in diameter with centres 56 m apart within rows 79 m wide. Resampling can also remove other geometric distortions, for example the slight, progressive decrease in scale for pixels closer to the edges of the ground swath and scale variations because of minor deviations from a circular orbit (Bernstein *et al.*, 1983). But even when reformatted to a square grid, continuous scanning yields a parallelogram-shaped scene, extending farther to the east at the north and farther to the west at the south. Because a rectangular array is more convenient for subsequent analysis and display, the grid is packed with wedges of 'dummy' pixels at the left and right. A special mathematical transformation, the Space Oblique Mercator projection, developed by US Geological Survey cartographer John Snyder, can be used to assign plane and spherical (latitude, longitude) coordinates to individual pixels and to generate graticules for satellite image maps (Snyder, 1977).

A somewhat similar scanning pattern is used at much lower altitudes with *side-looking airborne radar* (SLAR) sensors carried aboard aircraft. As illustrated in Figure 5.5, the sensor scans outward from one or both sides of the plane, along scan lines perpendicular to the flight path. Microwave radar is an *active sensing system*, which generates the energy reflected back to its antenna, in

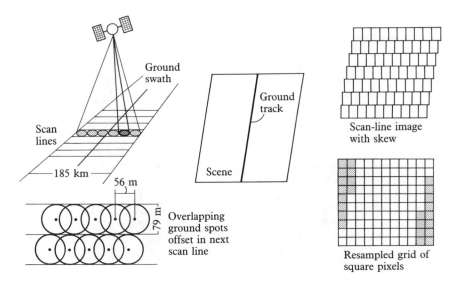

Figure 5.4   Landsat multispectral scanners examined a ground swath 185 km wide (*upper left*) divided into scan lines 79 m wide. A ground spot 79 m in diameter was measured every 56 m along the scan line (*lower left*). Because the earth was rotating beneath the satellite, the scene covered an area shaped like a parallelogram (*centre*) and the 'raw' grid of offset pixels (*upper right*) had to be resampled to yield a more convenient grid of square pixels (*lower right*)

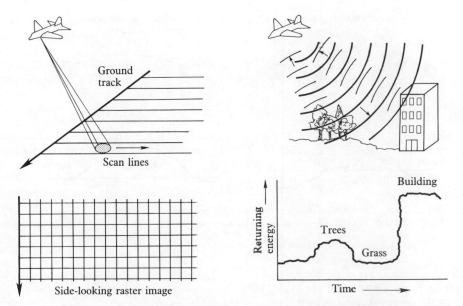

Figure 5.5 Side-looking airborne radar (SLAR) imagery based upon parallel scan lines perpendicular to the aircraft's ground track (*upper left*) can be formatted into a raster image of square pixels (*lower left*). Within each scan line radar pulses are emitted outward and backscattered by landforms and surface objects (*upper right*) to yield a profile of the strength of the backscattered energy as a function of time elapsed since the pulse was emitted (*lower right*). Substantial computational effort is required to convert the backscatter profiles to gridded data

contrast to the photographic and electronic systems discussed earlier, which are called *passive systems* because they rely upon reflected energy generated by another source, usually the sun. Radar pulses are reflected, or *backscattered*, from the land surface, with larger, more substantial objects such as office buildings returning stronger backscattered pulses than less substantial land covers such as an open field or golf course. For each scan outward from the ground track, the sensor records a two-dimensional profile of the strength and arrival time of the returning energy. Time is roughly equivalent to distance since energy backscattered by a distant object is received well after that from an object close to the flight path. SLAR imagery can be useful for detecting gross terrain features and for military reconnaissance.

SLAR's applications for production line cartography have here been limited in part by an incomplete understanding of the radar reflectivity of surface features and in part by the difficulty of registering imaged features on to a planimetric map (Moore *et al.*, 1983). Not only is the time–distance relationship affected by variations in the terrain, but slopes facing away from the aircraft do not receive—and thus cannot backscatter—radar pulses, leading to gaps across the scan line and black shadows on the image where the surface is hidden from the sensor. Moreover, returning wavefronts from the top of a steeply inclined feature might arrive before energy reflected from the base, thereby causing a severe type of relief displacement called *layover*. Although the geometric distortions of radar imagery are more challenging than the relief displacement on aerial photographs, microwave radar is a useful source of information about the land surface (Simonett and Davis, 1983).

Increasing the length of the antenna can improve the spatial resolution of an imaging radar system, but the fine resolution needed for detailed surface mapping would require a prohibitively long antenna were not a design available called *synthetic aperture radar* (SAR). SAR simulates the effect of a very long antenna, but requires considerable effort by a digital computer to convert backscattered pulses into an image map. Very high resolutions, of the order of several centimetres, are

possible, but economical commercial radar systems commonly offer a ground resolution of roughly 10 × 10 m. Another common acronym, at least until the Challenger disaster of early 1986, is SIR, for *shuttle imaging radar* sensors, carried aboard the National Aeronautics and Space Administration's Space Shuttle (Leberl *et al.*, 1985).

A satellite sensing technique potentially significant for the poorly mapped third world countries is the *Stereosat* concept—in essence the use of an earth resources satellite to acquire high-resolution imagery for stereocorrelation similar in principle to photogrammetric stereoplotting with aerial photographs. As with low-altitude photos taken from a plane, a pair of overlapping images taken from distinctly separated satellite positions, as shown in Figure 5.6, can provide parallax differences for measuring elevations. Features are not displaced radially, as with vertical aerial photos, but parallax can be measured either for overlapping *off-nadir* images sensed at different times from parallel orbital paths or for overlapping imagery obtained along the ground swath of a single orbital path with inclined, off-nadir sensors examining scan lines well ahead of and well behind the satellite (Welch and Marko, 1981). In order to recognize identical features on the two corresponding images, automated correlation techniques require high resolution, say, with a ground spot 10 m or less in diameter. SPOT's 10 m resolution has been judged suitable for producing 1 : 100 000 scale topographic maps and 1 : 25 000 scale photomaps (Welch, 1985). With little more than half the earth's surface mapped at the former, less detailed scale, satellite imaging and digital photogrammetry present a relatively rapid and comparatively cost-effective means for world mapping.

Sensor radiometry and feature extraction

Size of the ground spot is not the only kind of resolution important in remote sensing: of equal significance is the part or parts of the electromagnetic spectrum for which a sensor can measure and record energy. Most satellite

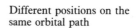

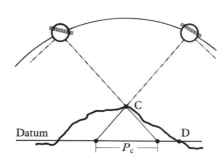

Adjoining orbital paths

Different positions on the same orbital path

Figure 5.6   The parallax needed to measure elevation differences on satellite imagery can be attained with overlapping off-nadir scenes recorded from adjoining orbital paths (*left*) or dual off-nadir scanning yielding overlapping images sensed from different positions on the same orbital path (*right*)

sensors and many airborne photographic systems are *multi-spectral*, recording separate images for several different, relatively narrow parts of the energy spectrum called *spectral bands*. A sensor and its bands can be tailored to detect specific kinds of feature under particular atmospheric and lighting conditions. Bands can be selected to accommodate not only differences in reflectivity or emissivity among important features but also haze, cloud cover and darkness. Because the atmosphere absorbs or otherwise interrupts the transmission upward of certain ranges of electromagnetic energy, the sensor's bands must be designed to fall within an *atmospheric window*—a range of wavelengths that can penetrate the atmosphere.

Perhaps the single most important concept in remote sensing is that of the *spectral signature*, that is, the pattern by which a feature or land cover's reflectivity varies with wavelength. Consider for a moment a lush maple tree in full foliage, viewed in daylight by the human eye. The maple receives solar energy over a wide range of wavelengths, some within the band from approximately 0.4 to 0.7 µm—the *visible band*, to which the eye is sensitive. Within and beyond the visible range the strength of the energy reflected by the maple varies with wavelength. But because its foliage absorbs comparatively more blue light (roughly 0.4–0.5 µm) and red light (0.6–0.7 µm), and thus reflects relatively more green light (0.5–0.6 µm), the tree appears green to the eye. This information is of little help, though, if visible light alone must be used to differentiate a green tree from, say, a green roof. Here, of course, differences in texture are diagnostic—but not so for defensive camouflage designed to mimic a vegetative land cover. Yet researchers working in military intelligence developed *colour-infrared film*, which extends the camera's vision beyond visible red to the *reflected-infrared* wavelengths (roughly 0.7–0.9 µm for infrared film, and to longer wavelengths for electronic sensors). In the reflected-infrared band the maple tree reflects far more light than either the green roof or the bogus vegetation. As illustrated in Figure 5.7, several important types of land cover have distinctive spectral signatures. And as shown in Figure 5.8, diagnostic differences in reflected energy can be detected by photographic and other sensors able to resolve separately the green, red and reflected-infrared bands.

To accommodate the human eye, *colour-infrared composite images* are displayed and reproduced with the *spectral shift* described schematically in Figure 5.9. Normal colour film has three layers of photosensitive emulsion to detect light in the three principal subdivisions of the visible spectrum—blue, green and red. Colour-infrared film also has three layers, sensitive to light in the

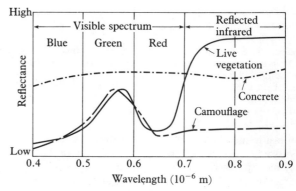

Figure 5.7   Spectral signatures of healthy vegetation and camouflage are nearly identical in the visible part of the spectrum but quite distinct in the reflected-infrared band. Concrete, which reflects visible and reflected-infrared wavelengths almost equally, has a distinctively different spectral signature

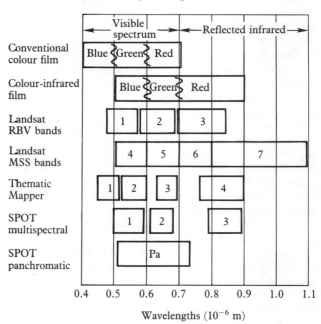

Figure 5.8   Ranges of sensitivity of colour and colour–infrared photographic film and spectral bands of five typical electronic sensing systems. The Thematic Mapper has three additional bands, described in Table 5.1

green, red and reflected-infrared bands. However, both photographic and printed images commonly assign blue to the green band, green to the red band, and red to the infrared band. Thus an area with normal, healthy vegetation will appear a vivid red, whereas an area covered

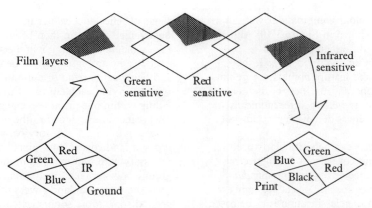

Figure 5.9 The spectral shift commonly used to produce a colour composite image for multispectral visible and reflected-infrared data yields the now traditional 'false-colour' red tones of healthy vegetation, with peak reflectance for infrared wavelengths

with green camouflage would be shown as blue. Multi-spectral sensors commonly ignore the blue band because blue light is scattered by haze and tends to make the image fuzzy—the reason why aerial photography taken with conventional colour film is seldom sharp and revealing. Clear blue water, which absorbs most of the sun's green, red and infrared rays, appears black on colour composite image maps. Concrete pavement, bare rocks, metal buildings and urban land cover in general reflect strongly and about equally in all three colour-infrared bands and produce image tones that are light grey or light greyish-blue.

Electronic sensors, airborne as well as satellite, collect colour imagery using several parallel sensors, one sensitive to each of the green, red and reflected-infrared bands but all examining the same ground spot simultaneously. In order to detect important diagnostic differences, many multi-spectral scanners have more than three bands. For example, the multi-spectral scanner on Landsat 1, launched in 1972, sensed not only separate green (0.50–0.60 μm) and red (0.60–0.70 μm) bands but two reflected-infrared bands (0.70–0.80 μm and 0.80–1.1 μm)—one for differentiating among vegetative covers and the other for gathering data on rocks and soil as well as for differentiating land from water. The Thematic Mapper scanner on Landsat 5, launched in 1984, had even greater radiometric resolution, with sensors for seven bands, including a thermal scanner (10.40–12.50 μm) recording emitted thermal-infrared energy with a 120 m ground spot for mapping temperature differences. As shown in Figure 5.8, the shorter of the other six bands were narrower than those on the early Landsats. Their spatial resolution was finer as well—30 m, in contrast to 79 m. Each band had at least one special function, as described in Table 5.1.

Additional, narrower bands and a smaller ground spot add to the amount of data that must be transmitted, processed and stored. A comparison of the Landsat Thematic Mapper and the *high-resolution visible-range* (HRV) sensors on France's SPOT-1 satellite, launched in 1986, demonstrate the trade-offs between radiometric resolution and spatial resolution. SPOT has a colour-infrared multi-spectral scanner with only three bands—green, red and reflected-infrared, as shown in Figure 5.8—but SPOT's 20 m ground resolution is significantly better than the Thematic Mapper's 30 m. Finer still, though, is the 10 m ground spot of the French satellite's panchromatic scanner, which covers a single, broad green–red band (0.51–0.73 μm). The 6-bit measurements transmitted for the panchromatic band are less precise than the 8-bit data for each multi-spectral band

Table 5.1 The seven bands of the Thematic Mapper

Band	Bandwidth (10⁻⁶ m)	Resolution (m)	Function
1	0.45–0.52	30	Water turbidity, soil, land use
2	0.52–0.60	30	Vegetative vigour
3	0.63–0.69	30	Vegetation mapping
4	0.76–0.90	30	Crop identification, vegetation/ water differentiation
5	1.55–1.75	30	Soil moisture, snow/cloud differentiation, crop moisture
6	10.4–12.5	120	Temperature differences, vegetation classification, land/ water differentiation, soil moisture
7	2.08–2.35	30	Rock formations, clay minerals, hydrothermal mapping

Source: Southworth (1985)

but each sensor generates the same amount of data for a given scene—the 24 bits needed for the three 8-bit colours measured for a 20 × 20 m pixel equal the 24 bits needed for four 10 × 10 m single-band pixels covering the same area with 6-bit precision (Courtois and Weill, 1985). But SPOT's ground swath of 117 km is narrower than the 185 km wide belt scanned by the Thematic Mapper, demonstrating an additional trade-off with the size of the region covered.

Colour-infrared film and three-band multi-spectral scanners leave little choice in the assignment of primary colours to spectral bands. With four-band scanners, usually the longer and wider of the two reflected-infrared bands is used to generate the red tones. More problematic is the assignment of three colour inks to the Thematic Mapper's six 30 m bands. For compatibility with 'traditional' multi-spectral scanner colour-infrared composite images, the National Oceanographic and Atmospheric Administration (NOAA) recommends assigning TM bands 2, 3 and 4 to the respective additive primaries blue, green and red (Clark and Johnson, 1985). But with printing inks an assignment of scanner bands to the subtractive primary colours is more convenient, and likely to become more conventional as image maps become more common. In an experimental 1 : 100 000 scale map of the Dyersburg, Tennessee, quadrangle the US Geological Survey (1983) used TM band 5 instead of band 4 because of the former's stronger response to roads, runways and urban areas. Reproducing the red band (TM band 3) in magenta and the reflected-infrared band (TM band 5) in cyan yielded the reddish fields and forests typical of colour-infrared imagery. Printed image maps displaying a multitude of subtly varied colour combinations warrant a relatively vague legend, as shown in Figure 5.10. Although cities, rivers, forest, agricultural land and major transportation corridors can be readily identified, the map's colours are mottled and feature boundaries are not sharply defined. But this inherent fuzziness of the image is both geographically realistic as well as cartographically appropriate, given the coarseness of a 30 m ground spot for a 1 : 100 000 scale map.

Techniques called *classifiers* can be particularly useful in extracting the pattern of land cover from multi-spectral data. Most classification strategies require *training data*, whereby the analyst identifies specific pixels belonging to land categories upon which the classification is to be based. The simplest approach is the two-band *parallelepiped classifier* illustrated in Figure 5.11, a two-dimensional plot of pixels according to their reflectances for red and reflected-infrared bands. The letters represent known land covers: deciduous forest (D), freshly ploughed field (F), cut meadow or turf (M), urban land (U) and water (W). A rectangle defined by the minimum and maximum reflectances for the two bands is fitted tightly around the pixels for each land cover type. Each rectangle represents a

TM wave bands used:
 #2, 0.52–0.60 μm, green, printed in yellow
 #3, 0.63–0.69 μm, red, printed in magenta
 #5, 1.55–1.75 μm, near infrared, printed in cyan

Color generally relates to typical features as follows:

 Blue-black . . . . **Deep clear water**
 Blue . . . . **Shallow or turbid water**
 Dark reddish brown . . . . **Deciduous forests**
 Light reddish brown . . . . **Crop and grasslands**
 Pink, gray, and white . . . . **Fallow or cleared areas, highways, runways, sand bars, etc.**
 Densely mixed patterns . . . . **Urban areas**

Figure 5.10 Legend on the US Geological Survey's (1983) Thematic Mapper image map of the Dyersburg, Tennessee, quadrangle shows the assignment of spectral bands to the subtractive primary colours (yellow, magenta and cyan), commonly used in colour printing

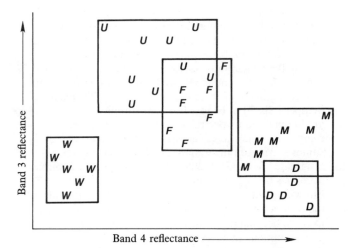

Figure 5.11 A simple, two-band parallelepiped classifier defines rectangular zones around pairs of reflectance values for training pixels, with known land covers. Letters in this example refer to deciduous forest (D), freshly ploughed fields (F), cut meadow or turf (M), urban land (U) and water (W). A pixel with a pair of reflectances falling within one of these zones is assigned the land cover category of the training pixels in the same rectangle

range of reflectance values in two dimensions. An unclassified pixel can then be assigned the land cover of the rectangle bracketing its reflectances. Where rectangles overlap, the ambiguity might be removed by consulting another spectral band or defining smaller, tighter stepped regions around each cluster of training pixels. Pixels not falling within one of these rectangles might remain unclassified, be assigned to the most similar category, or be placed in a 'mixed' category between two or more pure types of land cover. But 'mixed' categories are less necessary with high-resolution imagery because small ground spots are less likely to incorporate into a single pixel the variety of land covers mixed together by a large ground spot. More sophisticated classifiers rely upon a variety of statistical or heuristic principles (Fu and Yu, 1980; Estes *et al.*, 1983). Numerical classification techniques can also be used to extract specific themes, with well defined signatures such as the vegetation theme shown in Figure 5.12. Single-theme image maps can be presented in black and white, or as a colour overlay on an existing base map. Multi-category, multi-colour maps made from classified digital land cover data can be generalized and sharp black boundary lines added between adjoining land units.

Digital image analysis has become highly adept because of enhancement techniques for sharpening an image beyond its inherent ground resolution and artificial intelligence techniques for feature identification. These advances are fostered by improved computers called *image processors*, which can execute many more instructions per second than standard high-speed computers and are designed to examine arrays of pixels simultaneously. By looking at groups of neighbouring pixels, computer algorithms can more efficiently classify land cover as well as identify fault zones and other geological lineaments. Feature-tracking algorithms can follow roads and boundaries on high-resolution digital imagery. Current systems still require occasional consultation with a human operator, but rule-based expert systems that provide the computer with the logical framework of an experienced photo-interpreter should be able automatically to resolve most, if not all, apparent ambiguities. Future progress in machine vision, image processing and artificial intelligence appears destined to dilute the distinction between line map and image map by expediting the extraction of the former from the latter and promoting hybrid displays, in colour, with both line and image symbols. A plausible consequence of cartography's current electronic transition (Monmonier, 1985), in which the digital database is replacing the analogue map as the principal means for storing and analysing cartographic data, is that all map displays might eventually be image maps—but with line symbols readily available where appropriate.

Pervasiveness of the image map: examples from the United States

Aerial imagery has had a wide effect on mapping, but principally upon large-scale topographic and land cover mapping. Almost all large-scale land maps now are compiled from air photos, usually with only limited ground checking, and even property surveys and other largely unpublished maps requiring considerable fieldwork often rely on control data acquired in part through aerotriangulation or, more recently, satellite geodesy. Among small-scale thematic maps, in contrast, the

Figure 5.12   Black and white reproduction is sufficient to portray this reflected-infrared vegetation theme for the Chesapeake Bay region of eastern Maryland and Virginia (from Thompson, 1979)

contribution of remote sensing is less apparent. Most small-scale maps reporting news events or displaying census results or other administrative data have little, if any, tie to aerial or satellite imagery. Although small- and intermediate-scale image maps and atlases such as the National Geographic Society's recent *Atlas of North America* (Garrett, 1985) attract interest and acclaim, these products are still largely experimental—designed to explore, expound and exploit the newness of remote sensing. At neither National Geographic nor the World Bank, another enthusiastic publisher of occasional image maps, has the photomap replaced or become a standard supplement to the conventional line map. Although aerial photogrammetry is an established enterprise, satellite remote sensing—with an inherently greater potential for producing small-scale thematic maps—is an immature technology with an uncertain commercial future (Space Applications Board, 1985).

Despite these difficulties of judging the acceptance of the image map from commercial cartography, an examination of a nation's federal mapping programme provides a sound base for assessing the cartographic inroads of remotely sensed imagery. In the United States, for instance, the national mapping programme administered by the US Geological Survey uses photogrammetry heavily in the production of topographic maps but has produced comparatively few image maps, usually at scales of 1:100 000 and 1:250 000. These experimental image maps are frequently printed back to back with a conventional line map—a useful framework for comparison, to be sure, but also a reminder of the image map's inherent complexity. None the less, the Geological Survey maintains some programmes heavily reliant upon image displays or the interpretation of remotely sensed data, and

for a few of these plans to cover the entire middle 48, or 'conterminous', states.

One of these programmes provides comparatively raw high-altitude photographic images with no cartographic enhancements. The National High Altitude Photography (NHAP) programme consists of 1:80 000 scale monochrome and 1:56 000 scale colour-infrared photographs, taken from about 12 km during the leaf-off season and sold to the public as 23 × 23 cm (9 × 9 inch) prints or negatives. Not corrected for relief displacement, these photographs have a spatial resolution comparable to scanner imagery with a ground spot 2 m in diameter (Beetschen, 1984). Begun in 1980 as a joint effort of 13 federal agencies, under the administration of the Geological Survey, the NHAP programme reflects the wide range of uses of high-altitude aerial imagery and the need for government agencies to improve cost-benefit ratios through coordination. Initial coverage was to have been completed in 1986, and followed by a 'second cycle'.

Older but less enthusiastically pursued is the Land Use Data and Analysis (LUDA) programme, initiated in the early 1970s and originally planned for completion in the late 1970s (Place, 1977). Human photo-interpreters using high-altitude aerial imagery at scales of 1:60 000 or smaller classify land use and land cover into 37 categories. Data is available in digital form as well as on-line maps at scales of 1:100 000 or 1:250 000. The LUDA programme can be traced back to the Nixon administration, when Assistant to the President for Domestic Affairs John Ehrlichman, a former zoning lawyer, supported a national land use policy. GIRAS (Geographic Information Retrieval and Analysis System), developed by the Geological Survey to process the LUDA data, was an early successful step toward a national digital cartographic database. Although the planned national land use policy never crystallized, the LUDA programme survived budget cuts and limited support by Survey management, and was accorded a niche in the digital cartographic database that has become the focus of the United States national topographic mapping programme (McEwen *et al.*, 1983). In 1986 the series of published line maps was nearly complete for Hawaii and the 48 conterminous states.

Another facet of USGS mapping, as noted earlier, is the *orthophotoquads*, 1 : 24 000 scale orthophotos in a 7.5' quadrangle format similar to that of the nation's large-scale topographic map series. Prepared in one-third of the time required to produce a standard topographic map, and at a much lower cost, orthophotoquads are used as interim maps for areas without 1 : 24 000 scale mapping and as supplemental maps for other areas. Although there is no commitment to nationwide coverage, orthophotoquads are an important part of the national mapping programme; as of 1985 they were available for 57 per cent of the conterminous United States.

Of particular note is the role of remote sensing in the Alaska mapping programme (Brooks and O'Brien, 1986). In addition to the conventional use of aerial photography for photogrammetric compilation and map revision, the Geological Survey, in cooperation with various other federal and state agencies, is using computer classification techniques to produce land cover maps from Landsat data; a map series at 1 : 250 000 is underway, but with categories particularly appropriate to Alaska. For the oil-rich North Slope, USGS has produced 25 1 : 250 000 scale Landsat image maps: 14 based on RBV imagery and 11 based on multi-spectral scanner data. Photomaps at intermediate scales are particularly useful because of their relatively low cost, the vast area of the state, and the absence of cultural features, which provide most of the landmarks on line maps of more developed regions. Statewide photomapping in Alaska includes a series of 1 : 63 360 scale orthophotoquads and the Alaska High-altitude Aerial Photography (AHAP) programme, providing uniform black-and-white and colour-infrared imagery in a 23 × 23 cm (9 × 9 inch) format. Except for the Landsat image maps of the North Slope, these programmes are conceptually similar to mapping for the conterminous 48 states.

Another important similarity is the use in Alaska of SLAR imagery for special mapping projects on the North Slope and in the Aleutian Islands. As for scattered SLAR project areas elsewhere in the United States, products available include strip images at 1 : 400 000 and mosaic maps at 1 : 250 000, available on paper or film. (For some project areas in the 48 conterminous states, more detailed 1 : 250 000 scale strip images and 1 : 100 000 scale mosaics are available.) Microwave radar imagery is particularly valuable for geological interpretation, in contrast to Landsat multi-spectral data, which are better suited to mapping vegetation and land cover.

Although the improved spatial resolution of the Thematic Mapper and SPOT is yet to reveal its full impact on cartographic products, microwave radar imagery might well have the more striking future effect on the production of line maps. Indeed, radar's all-weather advantages have been demonstrated in the mapping of tropical rain forests, where frequent cloud cover mars much of the average Landsat scene. In the early 1970s, in fact, SLAR imagery was decisive in the reconnaissance mapping of that last cartographic frontier of sorts—the Amazon Basin (Koopmans, 1983). The short-lived Seasat, launched in 1978, was particularly prophetic in capturing radar altimeter data that could be used to construct a map of the ocean surface with 1 m contours! Continued improvement in the speed of digital computers promises the efficient generation of high-resolution SAR imagery, with resolution on the order of several centimetres. Radar's ability to penetrate sand cover in desert areas and to measure wave height over the oceans might promote a much broader, more detailed exploration of the earth than heretofore possible. As photogrammetry and the aerial camera have enhanced the precision of large-scale topographic mapping, imaging radar might expedite the mapping of surficial geology and soils—a tedious field exercise. But radar's unique contribution of important new information to a growing digital cartographic database should far outweigh whatever single-sensor image maps might be produced.

References

BEETSCHEN, C.W. (1984) The National High-Altitude Photography program. In *Yearbook of the US Geological Survey, 1983*, pp. 12–17. Washington, DC: USGPO

BERNSTEIN, R., COLBY, C., MURPHREY, S.W. and SNYDER, J.P. (1983) Image geometry and correction. In *Manual of Remote Sensing*, 2nd edn, edited by R.N. Colwell, pp. 873–922. Falls Church, Virginia: American Society of Photogrammetry

BROOKS, P.D. and O'BRIEN, T.J. (1986) The evolving Alaska mapping program. *Photogrammetric Engineering and Remote Sensing*, **52**, 769–777

CLARK, B.P. and JOHNSON, A.J. (1985) Color balance for Thematic Mapper: an improved image archive. *Landsat Data User Notes*, **33A** (supplemental issue), 1–9

COURTOIS, M. and WEILL, G. (1985) The SPOT satellite system. In *Monitoring Earth's Ocean, Land, and Atmosphere from Space—Sensors, Systems, and Applications*, edited by A. Schnapf, pp. 493–523. New York: American Institute of Aeronautics and Astronautics

ESTES, J.E., HAJIC, E.J. and TINNEY, L.R. (1983) Fundamentals of image analysis: analysis of visible and thermal infrared data. In *Manual of Remote Sensing*, 2nd edn, edited by R.N. Colwell, pp. 987–1124. Falls Church, Virginia: American Society of Photogrammetry

FREDEN, S.C. and GORDON, F., JR (1983) Landsat satellites. In *Manual of Remote Sensing*, 2nd edn, edited by R. N. Colwell, pp. 516–570. Falls Church, Virginia: American Society of Photogrammetry

FRIEDMAN, S.J., CASE, J.B., HELAVA, U.V., KONECNY, G. and ALLAM, H.M. (1980) Automation of the photogrammetric process. In *Manual of Photogrammetry*, 4th edn, edited by C.C. Slama, C. Theurer and S.W. Henriksen, pp. 699–722. Falls Church, Virginia: American Society of Photogrammetry

FU, K.S. and YU, T.S. (1980) *Statistical Pattern Classification Using Contextual Information*. New York: John Wiley

GARRETT, W.E. (1985) Editor. *Atlas of North America: Space Age Portrait of a Continent*. Washington, DC: National Geographic Society

KOOPMANS, B.N. (1983) Spaceborne imaging radars, present and future. *ITC Journal*, **1983-3**, 223–231

LEBERL, F.W., DOMIK, G. and KOBRICK, M. (1985) Mapping with aircraft and satellite radar images. *Photogrammetric Record*, **11**, 647–665

MCEWEN, R.B., CALKINS, H.W. and RAMEY, B.S. (1983) *USGS Digital Cartographic Data Standards: Overview and USGS Activities*. US Geological Survey circular No. 895-A. Reston, Virginia: USGS

MOFFITT, F.H. and MIKHAIL, E.M. (1980) *Photogrammetry*, 3rd edn, pp. 335–375. New York: Harper and Row

MONMONIER, M.S. (1985) *Technological Transition in Cartography*. Madison, Wisconsin: University of Wisconsin Press

MOORE, R.K., CHASTANT, L.J., PORCELLO, L. and STEVENSON, J. (1983) Imaging radar systems. In *Manual of Remote Sensing*, 2nd edn, edited by R.N. Colwell, pp. 429–474. Falls Church, Virginia: American Society of Photogrammetry

MULLEN, R., DANKO, J., WARREN, A., VAN WIJK, M. and BRUNSON, E. (1980) Aerial mosaics and orthophotomaps. In *Manual of Photogrammetry*, 4th edn, edited by C.C. Slama, C. Theurer and S.W. Henriksen, pp. 761–783. Falls Church, Virginia: American Society of Photogrammetry

PLACE, J.L. (1977) The land use and land cover map and data program of the US Geological Survey: an overview. *Remote Sensing of the Electromagnetic Spectrum*, **4**(4), 1–9

SIMONETT, D.S. and DAVIS, R.E. (1983) Image analysis—active microwave. In *Manual of Remote Sensing*, 2nd end, edited by R.N. Colwell, pp. 1125–1181. Falls Church, Virginia: American Society of Photogrammetry

SNYDER, J.P. (1977) Map projections for satellite tracking. *Photogrammetric Engineering and Remote Sensing*, **47**, 205–213

SOUTHWORTH, C.S. (1985) *Characteristics and Availability of Data from Earth-imaging Satellites*, US Geological Survey bulletin No. 1631. Washington, DC: USGPO

SPACE APPLICATIONS BOARD, COMMISSION ON ENGINEERING AND TECHNICAL SYSTEMS, NATIONAL RESEARCH COUNCIL (1985) *Remote Sensing of the Earth from Space: A Program in Crisis*. Washington, DC: National Academy Press

THOMPSON, M.M. (1979) *Maps for America*, p. 186. Washington, DC: USGPO

THROWER, N.J.W. and JENSEN, J.R. (1976) The orthophoto and the orthophotomap: characteristics, development and application. *American Cartographer*, **3**, 39–56

US GEOLOGICAL SURVEY (1983) *Dyersburg, Tennessee, Landsat Thematic Mapper 1 : 100 000 scale Image Map with Accompanying Line Map*. Reston, Virginia: US Geological Survey

WELCH, R. (1985) Cartographic potential of SPOT image data. *Photogrammetric Engineering and Remote Sensing*, **51**, 1085–1091

WELCH, R. and MARKO, M. (1981) Cartographic potential of a spacecraft line-array camera system: Stereosat. *Photogrammetric Engineering and Remote Sensing*, **47**, 1173–1185

6 Digital mapping

Michael Blakemore and Krysia Rybaczuk

Digital mapping no longer just concerns automating the production of a map sheet. Ten years ago this would have been the general case. Mapping systems such as GIMMS, CALFORM and others were still stand-alone suites of computer code that allowed thematic maps to be drawn by computer, and national surveying organizations were using the computer to automate their topographic mapping programmes. By the mid-1980s (see Carter, 1984) main attention was focused on what are termed Geographic Information Systems (GIS) which have map plotting as one part, but are crucially concerned with the management modelling and display of spatial information. The transition from mapping to GIS has been a complex technical, social and military process, documented by Monmonier (1985).

A mapping system allows a user to translate the lines on a map into digital coordinates, a process known as digitizing. Curved lines can be represented by closely spaced points which, when joined up (plotted) by straight lines (vectors) simulate the original line; hence the term 'vector-oriented mapping' (see Monmonier 1982 for a pioneering textbook on this area). Other raster methods exist, and are increasingly important. Here, rather than salient points being selected to represent the line, we may envisage a very fine-mesh graph paper being placed over the map, and coloured in where it overlies a line.

To these lists of coordinates (usually x and y dimensions on a rectangular—or cartesian—grid) can be added numbers or letters to represent their meaning. Hence there are feature-codes to note roads, rivers, building types, land use, etc. The digitized points, lines, areas or heights (dimensions 0,1,2,3) are taken usually from existing sheet maps, and this process is termed data capture. Of increasing importance are methods of direct data capture such as computerized surveying equipment, photogrammetric interpretation using digital workstations, and remote sensing. One aim of a computer map is to allow data digitized, say from a map of 1:50 000, to be plotted at another scale: the idealized goal would be a scale-free database. To plot the data at 1:100 000 would mean that only 25 per cent of the original paper size was available, forcing cartographers to omit certain details and levels of information; a process called generalization which is well provided in existing mapping systems. Conversely, to plot at 1:25 000 increases the paper area four-fold and, unless extra artistic detail is added to make lines seem smoother, the result is aesthetically poor. Techniques of enhancement are used by manual cartographers, but automated techniques are still under development even in the most sophisticated GIS. The accuracy of original maps, of digitizing and all the input or data capture process has a critical impact on the quality of eventual applications and this is discussed below.

Provided that information is digitized from a single map sheet, eventual plotting is easy. The problems start to arise when data are derived from more than one map sheet. Map sheets seldom match precisely at their edges. Paper deformation, or different dates of revision (particularly prevalent on topographic and cadastral map sheets) add to the imprecision of map sheet edges. Therefore systems are needed which have edge-matching facilities. Also, it is frequently the case that data may be derived from maps of different scales. These will have different generalization criteria, and indeed may be produced by different agencies for very different purposes. The shape of a river on one may differ subtly from another, and if superimposed will have minor deviations. The differences between lines are termed sliver-lines, and between closed shapes such as buildings or land-use polygons are sliver-polygons. The processes of cleaning up discrepancies, identifying and correcting digitizing errors, snapping together sheet edges, and building a unified database are a feature of the modern GIS.

We may wish to overlay and integrate layers of data (also termed coverages) from existing map sheets. Added to these can be survey results from the new generation of electronic equipment (total survey stations), interpreted output from aerial photographs using photogrammetric workstations, satellite imagery processed using image analysis, statistical data either on paper or computer-compatible media such as tapes and disks, and soft data such as written reports and newspaper articles. All of these need to be transformed and registered to a common map base, free from errors and inconsistencies and able to be updated at any time. The ability to do all of these is a feature of a GIS.

Once a unified map base is achieved questions can be asked of it, given an analytical capability in the system. A computer mapping system merely reproduces the data in a variety of graphical styles. By combining powerful mapping software with an (often relational) database system it is now possible to pose the questions in the form:

Display all areas where land use is type A *or* type B *and* where the relief is above 325 m *and* which are within 1 km of Class I roads *and* where the population density is greater than *y*.

The *if, and, or, not, equal to* conditions, known as relational operators, allow exploration and analysis to proceed naturally. In this way a GIS enables true interdisciplinary and inter-agency projects to be established. For example, medical researchers may wish to explore the relationship between disease and living environments. Data would come from the utilities agencies (water, power), communications and transport (road, rail), local administration (housing details), a census agency (population statistics), hospital and local medical services (disease and patient statistics), and many others.

It is indicative of the developments of machinery (hardware) that the programs (software) now run on smaller and smaller systems. In late 1986 it was possible to order GISs which before would have only run on large and powerful systems, but now operate on IBM Personal Computers.

Key issues in the mid-1980s

To make a statement on the actual activities in digital mapping throughout the world would be a daunting task, best left to those international organizations that have good information networks. For the International Cartographic Association (ICA), Starr (1984) presents a snapshot of key world activities. Another way of assessing current trends in system and applications is to attend international conferences, notably those run by organizations such as the ICA, the American Congress on Surveying and Mapping (ACSM), the Automated Mapping/Facilities Management (AM/FM), and the continuing 'Auto-Carto' conferences. Indeed it is becoming apparent that these are the *only* ways of keeping track of developments. To survey world developments would now take more than this entire book, though Opitz (1986) has arranged GIS developments under what are clearly accepted categories seen as being important. These include:

* microcomputer-based systems,
* specifying the requirements of users,
* military applications (many of which are secret),
* land use, planning and resource management,
* environmental assessment,
* training and system support,
* hardware and software developments.

(after Opitz, 1986)

What this presentation can do is to quote a series of case studies which typify current trends. For more background reading on developments in general, readers should consult Burrough (1986), Chrisman (1986c, and other papers in the ACSM Conference of that year), Marble *et al.* (1984), Opitz (1986), Rhind (1986), Taylor (1980) and Wiggins *et al.* (1986, and other papers of the Auto Carto London conference). It is now beyond debate that automated mapping and spatial database development is cost-effective in many ways. The fact that the United States Geological Survey is, like most North American agencies, heavily committed to automation, that the British Ordnance Survey is investing heavily like the French, West Germans, Australians and many others, all adds credibility to this viewpoint. What is more important is that the sheer weight and variety of data now presents a range of problems that must be answered if progress is to be maintained.

Two major issues surround digital cartographic data today. The first is that of data availability, and the second that of accuracy. The consumer needs data for specific areas. This data often needs to be of a defined quality level and to be available in sufficient quantity. There are several barriers to be overcome. Data may be physically not available, or may not exist in the required form. Linked to this there is the problem of cost, in terms of both data purchase and the outlay necessary to ensure adequate computing facilities for data analysis. Therefore although

data may physically exist, for those individuals or institutions unable to justify the expenditure the data remains unattainable. Finally there is a confidentiality barrier. If data is deemed to be of a sensitive nature, it remains within the confines of government archives in an effort to protect national and individual security. For example, military satellites are capable of resolutions of 1 m^2, but images from such systems are inaccessible to the general public. Additionally, much digital data now has a commercial intelligence value which means it stays within bureaus as proprietary information.

If, however, all the user requirements are met, a situation of abundant availability such as that described by Hardy (1985) and Peuquet (1984) comes into being and in the 1980s it is to this state that digital data is tending. Only by an increase in discriminating processes, the cardinal criterion of which is accuracy, can data be stratified. With the measurement of accuracy, comes the demand for cartographic standards to be enforced as an integral component of data capture. Aronoff (1985) suggests that qualitative statements relating to the accuracy of each map ought to be feasible, and should follow the form of a percentage accuracy and a percentage confidence level. The process could be carried out by verifying a sample of points on a map with ground data measurements. Chrisman (1983) goes a stage further by considering continuous accuracy statements as being of greater importance to digital cartographic standards than discrete ones. It is the lineage of a line that may prove the most value in dictating the accuracy of a map, since it holds a continuous record from the digitizing process to the hard-copy output, and notes the accuracy with which each task was performed.

With the onset of concern with standards the tendency has been to strive for an improvement in accuracy. Goodchild (1980) points out, however, that speedily generated, generalized images are usually as effective as their more accurate traditional counterparts. Whether such a statement is uniformly true depends again upon user needs. Certainly mass media maps are becoming increasingly more simplified, especially with televised graphics. Bearing in mind the standard of cartographic appreciation amongst the public at which these representations are being directed, such representations are effective and achieve their aim of rapid information transfer (Petchenik, 1985). Accuracy is an important issue, and ensuring the highest standards possible throughout the digital mapping process is still the aim of cartographic agencies. The first step towards fulfilling this aim is to ensure that input data is of the highest quality possible as 'eventual cartographic precision . . . is determined by the quality of input data and the types of usage' (Blakemore, 1984: 132).

Data capture

Digital data for cartographic purposes falls into two broad categories, that concerned with topographic or line data and that dealing with 'soft data', incorporating variables and attributes suitable for cartographic representation with or without topographical limits. Assimilating outline data into a computerized environment involves either digitizing or line scanning, depending on whether data is to be stored in vector or raster form, and also upon the nature of the collection process.

Orbital and suborbital platforms

The main sources of digital cartographic data are remotely sensed imagery from both orbital and suborbital platforms, surveyed data and derived data from existing maps. An increasingly utilized form of data is that obtained from remote sensing. Petrie (1985) argues that field survey methods are only cost-beneficial where the scale of map is 1 : 1000 or larger. In all other cases orbital and aerial imagery are better.

Suborbital platforms supply cartographic data mainly in the form of air photos. The technique of converting aerial photographs into maps or digital data is that of photogrammetry. Orbital imagery is a relatively new form of data capture, and has only been widely available for topographic mapping with the onset of the Landsat series in 1972. Satellite data stored in digital form may be incorporated into digital mapping in several ways. In most cases it is fed directly into raster image display and processing systems. Monmonier discusses the problems and processes of both suborbital and orbital data capture in Chapter 5.

Directly surveyed input

Surveyed data has been the traditional base for topographic mapping, and forms the foundation of most thematic maps. Although less important with the increasing influence of aerial and suborbital imagery, surveying techniques still are essential in the determination of control points for such imagery. It is the surveyed geodetic base that forms the metric against which map accuracy can be determined. While the actual methods of surveying have changed little over the centuries, the transfer mechanisms from field survey to map have. Increasingly electronic total survey stations have in-built computers which carry out calibrations, store observations and can download them on to a host computer for automated mapping. Positioning is being revolutionized, too, by the cluster of satellites forming the Global Positioning System (GPS), whereby portable equipment allows three or more satellites to be 'observed' and their geostationary positions used to assess position extremely accurately. Conventionally, however, surveyed data may be translated into cartographic images and then digitized. In fact surveyed maps form the basis for many digital mapping projects. The majority of Ordnance Survey data in the UK, for example, is the result of past surveys. Harley (1975) points out that many post-1945 maps in the UK are dependent on pre-1939 surveys for their accuracies as postwar maps were rarely re-surveyed, but merely updated.

Derived data capture (digitizing)

Digitizing is essentially a sampling process in which set machine parameters, or the human eye, select the most representative points on a line for storage. Labels or 'feature codes' are then added to denote what the points represent. Due to its very nature, the process involves an element of generalization and error. Digitizing may be carried out either automatically or manually. Automatic digitizing involves either time or distance as criteria for point selection. Using time as the criterion results in a point being selected every x seconds; as sinuous lines take longer to travel along, they become better represented with points. Using distance as a criterion will result in a point being registered every x mm; this has its difficulties as long straight lines have many redundant points. Thus both

cases often result in the need for thinning-out algorithms. If accuracy is to be maintained manual digitizing is more subjective in its choice of points, but is susceptible to handshake, poor concentration and the inability of the eye constantly to follow lines along their geographical centre, thus resulting in the created line 'bouncing' between the bands of the original line. Some of these problems can be reduced in semi-automated systems such as line followers, where a laser beam traces the path of the original line by perpendicular scanning. As soon as the machine reaches a point of indecision it switches to manual control allowing the operator subjectively to guide the tracing mechanism in the desired direction. Of increasing importance are raster scanning systems (Peuquet and Boyle, 1983). Here a map image is represented as a matrix of very fine cells (pixels) which take on a value of 1 if a line crosses them, or 0 if not. Problems of reconstructing the lines, thinning out redundant cells on thick lines, and other problems, have kept scanners in the top end of the market. Recently the reduction in hardware costs and increasing processing power have brought scanners into the purchasing realm of many more applications areas.

Soft data capture is less complicated than that concerned with the digital representation of graphical features. This data complements the hard numeric information and may take the form of descriptive and contextual items such as media reports, transcriptions of verbal interviews, or survey results. Often this can be categorized and added to a digital database as an extra 'layer' of data. The development of relational database management systems has freed digital cartographers from previous obsessions with strictly numerical information. A drawback with many such data sources is their considerable variation in reliability. Commercial market research bureaus and opinion polls have their own definitions of accuracy since they must justify their results to a critical audience, but their concern with good sampling technique may not extend to the myriad of small-scale surveys that generate data. In the UK the Data Archive of the Economic and Social Research Council, based at the University of Essex, at least asks academics depositing data there to fill in an exhaustive questionnaire on techniques and methodology. In late 1986 ESRC announced that it was setting up 'Regional Research Laboratories' in GIS, which would conform to an agreed coordinated plan. One stage further on in the US is the National Science Foundation's proposal to set up a National GIS Centre, for which proposals were being invited. In these ways standards may start to be enforced from above.

Problems relating to digital data

Data transfer

Data from such surveys may be entered and stored locally, or alternatively, if a suitable data set already exists, may be transferred from one system to another via a computer storage medium. The most conventional means for digital data transport between agencies is via magnetic tape or some other means of data storage. Recently, the option of networking has come into being for those who wish to transport data between two geographical locations. This involves sending data directly 'down the line' to be received in the machine at the other end. In the United Kingdom this is facilitated through British Telecom's PSS (Packet Switching Stream), IPSS (International PSS) and

Telecom Gold networks. Another option within academic circles allowing for similar transfers is the UK JANET (Joint Academic NETwork). This system allows fast file transfer, and permits users to access information on other machines and manipulate it as if the user were actually interacting with a computer in a different location. For example, users wishing to access the NOMIS database held at the University of Durham may do so in a matter of minutes (Nelson and Blakemore, 1986). Theoretically such innovations should result in fast and easy accessibility to data nationwide and even worldwide. Still, however, the standard 'urge' is for people to gather data around them, for use on their own systems, and the development of unified national or international databases is fraught with logistical and political problems (Wiggins *et al.*, 1986).

Data protection

Permeating all aspects of data availability are concerns with the rights of the individual. These vary between cultures and political systems. Confidentiality for individual concerns is guaranteed in the United Kingdom by the Data Protection Act and the Official Secrets Act. The former strives to ensure that no party can be directly identified and their personal details abstracted from any data set that holds data pertaining to that party without their knowledge. As the population census contains personal individual information in sparsely populated areas individuals have their anonymity maintained either by total suppression, or by the random addition of $+1$s, 0s and -1s to the data set pertaining to their enumeration districts. To counteract inaccuracies arising from such methods of suppression, and also to produce databases that related to homogeneous areas, rather than arbitrary administrative areas, Rhind (1985) suggested the possibility of storing data in postcode districts, from which it could be aggregated to suit user needs. Selected polygons which infringe upon individual anonymity would be refused access to information unless the data was suppressed first. The proposal to use postcodes was rejected in August 1986 on purely financial grounds.

Such stringent information restriction acts contrast markedly with the United States Freedom of Information Act. This allows data to be freely distributed in the interests of maximum information exchange. Not only this, but there has been an excellent development of inter-agency collaboration, coordinated by the US Geological Survey. They market extensive digital coverage at a range of standard scales, for topographic layers, land use, digital elevation models, and a 'Geographic Names Information System', and a $25 100 kbyte test area at 1:100 000 among other services. In collaboration with the US Bureau of the Census they have provided data that has allowed the Bureau to develop an integrated statistical and topographic database for the 1991 Census (Broome, 1986), called TIGER (Topologically Integrated Geographic Encoding and Referencing). USGS data is available at a cost which guarantees high-volume sales. This compares favourably with the UK Ordnance Survey who in 1986 announced the availability of 1:625 000 coverage of the UK at £7000, plus maintenance and royalty fees. TIGER (see Marx, 1986) is an exceptional example of inter-agency collaboration to produce a national database. 1:100 000 map sheets are scanned by the USGS and the data is checked and supplemented by the Census Bureau who add road names and classifications, to features such as water and transportation.

Data standards

As more digital cartographic data has become available, so its use and applications have increased. But even when machine protocols allow easy transfer of data at an acceptable cost, an increased movement of quality cartographic material does not necessarily follow. Data is often captured with specific standards relating to particular applications, which are not necessarily compatible with other uses. The user needs to be aware of these standards to ensure an accurate implementation of cartographic data. Failure to comply with such considerations leads to the misuse of data and erroneous assumptions as to the accuracy of the material being handled, and by implication to the material being produced (Chrisman, 1986a).

In the absence of adequate information about quality, judgements relative to data applicability are difficult to make. Crucial therefore to 'informed' digital cartography is a mechanism for communicating quality that is applicable not only to a particular piece of cartographic data, but also to some nationally and internationally recognized standard.

In the past, standards for traditional cartography have taken the form of a quality threshold which had to be equalled or exceeded for the standard to be met. Thus the responsibility for accuracy and quality control rested firmly on the shoulders of the producers. Proposed quality levels for digital cartographic data have witnessed a radical divergence from such traditionalist ideas. Rather than the producer striving to achieve a pre-defined standard of accuracy the responsibility should be transferred to the user, who must evaluate the data's 'fitness for use'. To enable the user to perform this task the producer must, however, provide as much information about the methods of data capture as possible, so that the user can make an informed decision as to whether the data is suited to his/her needs. To do so requires clearly understood criteria, terms and classificatory mechanisms (Sowton and Haywood, 1986).

Thus digital cartographic standards will relate to standards of information (rather than to cartographic entities), by applying the fitness of use concept in a method known as truth labelling. Using such an approach, 'the producer must disclose all the information needed to evaluate the data, and the user must perform the evaluation of fitness for use relative to the particular application. For such a system to operate a producer must have guidelines for the items which must be transmitted to permit evaluation' (Chrisman, 1986b: 53).

It was with this aim in mind that Working Group III of the US National Committee of Digital Cartographic Data Standards (NCDCDS) elected to proceed with its interim quality standards for digital cartographic data. Five categories of quality definition were selected as essential for the producer to provide lineage, positional accuracy, attribute accuracy, logical consistency and completeness (Moellering, 1986a). Although the initial complaint against such a system was that expenses incurred during map creation would increase, Chrisman (1986b) has shown that this is not the case. On the contrary, he points out that more mapping agencies would be able to comply with such a standard than with a traditional one, as the increment in cost is much lower. Therefore there is cause for optimism on all sides as producing agencies can comply with their targetted user demands and the user community will encounter a wide variety of quality coded cartographic material.

Aside from investigating methods to control quality, NCDCDS is concerned with other transfer problems such as defining an effective set of data exchange methods, defining a set of primitive and simple cartographic objects, and redefining a unified set of cartographic features (Moellering, 1986b). The work culminated in 1985 with an interim standard which has recently been tested by several American Federal agencies including USGS. The trial agencies provided a varied response, but in general they were approving. Brooks and Upperman (1986: 39), reporting on the comments made by USGS, The Defense Mapping Agency (DMA) and the National Ocean Survey (NOS), conclude that the standard 'requires considerable enhancement before two non-communicating implementations can exchange data'. Alternatively, Distefano (1986: 48), documenting the results of the Assignment Department of the City of Boston, painted a less gloomy picture by stating that generally the system provided an 'adequate format for encoding parcel lines and parcel centroids, graphics and non-graphic attribute data found in the city of Boston's database'.

In the United Kingdom similar moves are being made to achieve some standardization of cartographic material, following recommendation 25 of the House of Lords Select Committee on Remote Sensing and Digital Mapping (House of Lords, 1983). This is being carried out by the Ordnance Survey with the aim of creating a transfer standard which is versatile, flexible and has the ability to handle topological relationships (Sowton and Haywood, 1986). The philosophy behind such standards is similar to that pursued by the NCDCDS in that the transfer format must be able to handle all of the data generated by any cartographic system, that the techniques used be easily adapted to suit any system on the market, and that costs be kept to a minimum. As with NCDCDS the main areas of research are feature classification, data quality and transfer format, with an interim set of standards due to be brought out in early 1987.

Vahala (1986) illustrates the difficulties of exchanging data collected in Land Information Systems between separate administrative authorities in Finland, and outlines the efforts under way to create a set of classification standards that will allow a common reference system based on location to be built, and hopefully to overcome the problems of data movements. Once such standards are incorporated by mapping agencies in various countries, the eventual aim becomes a uniform international standard. Although such a proposition can only be foreseen for the very distant future, the need for internationally recognized standards can be justified with the rapidly increasing use and applications of Geographical Information Systems. Their range of facilities and applications areas are now warranting close attention to careful benchmarking criteria (Marble and Sen, 1986; Goodchild and Rizzo, 1986).

Geographic Information Systems

GISs have evolved as an important method of dealing with and integrating spatial data. They may be best described as systems able to 'accept large volumes of spatial data derived from a variety of sources including remote sensors and to effectively store, retrieve, manipulate, analyse and display these data according to user defined specifications' (Marble and Peuquet 1983: 923).

De Man (1984), however, insists that the 'prime function' of a GIS has to be its capacity for integration and the subsequent creation of new information. The major method for integrating spatial data is that of line and polygon overlay. This is a process whereby two images obtained from different sources are superimposed and the

discrepancies removed. The removal is determined by a set tolerance, which in most commercial systems takes the form of width specifications.

GISs have over the past decade become commonplace tools for the manipulation of spatial data, their exact specifications varying with respect to the mode of investigation for which they were designed. In the first instance there are the large-scale integrated GISs which run as turnkey systems on a small range of host machines and which are developed by established computer-graphics software companies. As these are non-specific in orientation they have a high versatility component and may be implemented by a wide range of institutions to perform a large array of tasks. One popular system is ARC/INFO marketed by ESRI (Environmental Systems Research Institute). Green *et al.* (1985: 301) regard this system as representing 'the most advanced GIS available at present', although this is by necessity a view not based on a full evaluation of all systems. For example, attractive GISs are now being offered by Intergraph, Siemens, Scitex, Sysscan, Geo Systems, McDonald Douglas, Synercom, Laser Scan, and many others. The main functions executed by ARC/INFO are those of data entry, data analysis and management, data display and both graphical and cartographic output. Green *et al.* (1985) consider two of the most powerful operators within the system to be those of overlay (of different layers of data) and storage, and the ability to build buffers. The overlay and storage option allows two or more coverages to be superimposed with their respective attributes, and the results to be stored as a new cartographic object. The buffer option enables a zone or buffer to be created around a predefined attribute to a predefined width specification. This can for example be utilized for the study of land use within a specified distance of a proposed road or aircraft flight path, and is therefore a useful planning aid.

Secondly there are the smaller more specific problem-orientated systems which have been developed as a result of user needs. These are developed either by software companies for institutions such as fiscal planning agencies or conservation groups, or as is more common they are conceived within the agencies themselves. IEMIS is one such a system, designed by the Federal Emergency Management Agency in Washington DC to assess its emergency preparedness. The system enables decision making to be aided by a suite of simulation models involving the diffusion of contaminants to air and water, evacuation dynamics and hydraulic flow. Combined with a resources database this system will help to maximize local government awareness and abilities to deal with emergencies.

Finally there are those GISs that revolve around one centrally located set of databases and are accessed either via networking links, such as NOMIS (Nelson and Blakemore, 1986) or via a consultancy as in the UK case with Pinpoint Analysis Ltd, or CACI's ACORN. ACORN has as its central database several socioeconomic surveys of the UK, the largest being the 1981 Census of Population. As it is a market research tool the services offered are often concerned with targetting 'ideal' populations. 'Direct Marketing', for example, seeks predefined consumer types and mails advertisements directly to them. 'Consumer types' can be selected by factors such as age group, type of housing or sex, and can be applied to specific areal units as defined by criteria such as postal districts, shopping centre catchment or administrative boundaries. The market research potential of postcoded data is being increasingly exploited. The *Yellow Pages* directories in the UK are now fully computerized with the Telecom 'Business Data Base' adding employment numbers to entries, as well as business classifications, address changes and other items.

Although most existing GISs incorporate some form of integration or overlay, they all rely on a crude predefined tolerance which tends not to embody any form of intelligence or qualitative discrimination. 'Intelligent' overlay has the ability to account for preselected influential factors other than size when assigning an area of overlap between two zones. These factors include concepts such as neighbourhood influence, time and culture. Neighbourhood influence assigns overlap areas (slivers) on the basis of the neighbouring zone with which it is most compatible, irrespective of size. Such a method is particularly important in urban cartography where small areas of land use are important in their own right, and cannot be classified with the closest, largest land use if they are to retain their significance. Temporal and cultural tolerancing are similar in that they consist of a system that gives adequate weighting to data created at particular points in both time and social environments.

Intelligent overlay should be achievable through the means of expert systems and artificial intelligence, collectively known as intelligent knowledge based systems (IKBS). 'At the most basic level an expert system is little more than a collection of rules built up into a structure from which deductions can be made along the lines of "if x is true and y is true then z must be true because it depends on the x and y conditions". The rules may be input as individual statements setting out the relationships between items, or deduced from examples and reasoning patterns set up by what is commonly described as the interface engine.' (McLening, 1984: 12).

These rules comprising the IKBS enable it to advise, help and solve real-world problems using a computer model of expert human reasoning, reaching the same conclusions as the expert would faced with a comparable problem. The initial problem is often deciding which decisions the rules are going to direct the user to, since large gaps exist in our understanding of human problem solving. The secondary problems as with most forms of digital research are the practical considerations of time, money and manpower. In this field these may be considerable as an IKBS can sometimes take between 10 and 25 man-years to produce, and cost as much as US$1–2 million (Robinson *et al.*, 1986b).

IKBS are not confined merely to the GIS overlay but cover other aspects such as map design, terrain/feature extraction and geographical database management/user interface. Robinson *et al.* (1986b) review some of the systems emerging at present relative to these areas of concern. MAP-AID and AUTOMAP deal with map design, and incorporate IKBS to utilize expert knowledge in deciding where titles, names and symbols ought to be positioned. MAP-AID is envisaged as an entire map design system, and is still in its very early stages. AUTOMAP, however, deals specifically with name placement and is probably the most successful system possessing this skill to date. It utilizes heuristic knowledge about name placement based upon existing conventions in the form of a knowledge base consisting of a small set of explicit rules. The approach is that features are annotated in the order of area, point and line so that the system progresses from the most constrained annotation task to the least. This allows for easy backtracking and replacement of names if it becomes impossible to locate a name.

FES and CEREBRUS are concerned with terrain/feature extraction from Landsat images. FES, for example, is a Forestry Expert System 'to analyse multi-temporal Landsat data for classification of land cover and land cover change

of interest to foresters' (Robinson *et al.*, 1986b). Two phases of rules are applied to this system; the first decides on the validity of classification change from one period to another, whilst the second generates a set of decision rules regarding the current state of the image. Following a test in Newfoundland, Canada over an area spanning 100 km^2 and using imagery obtained for a period of 6 years, the system is regarded as capable of tracking both subtle and apparent forest changes. OBRI and LOBSTER are examples of IKBS in the geographic database management/ user interface. Both are intelligent user interfaces to spatial database management systems. OBRI attempts to keep track of Portuguese environmental resources and provides a classification system for environmental data as well as a decision-support system for resource planning. Other features include graphic input and output of maps and a natural language parser for Portuguese.

Further research into this aspect of digital mapping is expected to enquire how expert systems rules will vary according to cartographic purpose, limitations, scale and cultural expectations. Although efforts are necessary to formulate concepts in such areas as map design, terrain analysis and the extraction of man-made features, Robinson (1987) suggests that these do not necessarily have to be concrete in every aspect but can be classified with permeable boundaries, so that definitions of entities remain fuzzy rather than precise.

Digital cartography is fast being subsumed into the wider facilities of Geographic Information Systems. While there are still systems which reproduce map images, and so mainly remove the need for human cartographic effort, they are becoming less important. Digital cartographic data is crucial, but not the only, data source in spatial decision making. Large-scale integration of data and the development of intelligent software systems are two key developmental areas of the late 1980s.

References

ARONOFF, S. (1985) The minimum accuracy value as an index of classification accuracy. *Photogrammetric Engineering and Remote Sensing*, **51**(1), 99–111

BLAKEMORE, M.J. (1984) Generalisation and error in spatial data bases. *Cartographica*, **21**(2–3), 131–139

BROOKS, A.A. and UPPERMAN, J.V. (1986) Interim NCDCDS testing report. In *Digital Cartographic Data Standards: A Report on Evaluation and Empirical Testing*, edited by H. Moellering, pp. 33–43. Columbus, Ohio: National Committee for Digital Cartographic Standards

BROOME, F. (1986) Mapping from a topologically encoded database: the US Bureau of the Census example. In *Proceedings, Auto Carto London*, Vol. 1, edited by M. Blakemore, pp. 402–411. London: Royal Institution of Chartered Surveyors

BURROUGH, P.A. (1986) *Principles of Geographical Information Systems for Land Resources Assessment*. Oxford: Oxford Scientific

CARTER, J.R. (1984) *Computer Mapping: Progress in the '80s*. Washington, DC: Association of American Geographers

CHRISMAN, N.R. (1983) The role of quality information in the long-term functioning of a geographic information system. In *Proceedings, Auto Carto Six*, Vol. 1, *Automated Cartography, International Perspectives on Achievements and Challenges*, edited by B.S. Wellar, pp. 303–312. Ottawa: Auto Carto Six

CHRISMAN, N.R. (1986a) Obtaining information on quality of digital data. In *Proceedings, Auto Carto London*, Vol. 1, edited by M. Blakemore, pp. 350–358. London: Royal Institution of Chartered Surveyors

CHRISMAN, N.R. (1986b) Testing the interim proposed standard for digital cartographic data quality. In *Digital Cartographic Data Standards: A Report on Evaluation and Empirical Testing*, edited by H. Moellering, pp. 53–55. Columbus, Ohio: National Committee for Digital Cartographic Standards

CHRISMAN, N.R. (1986c) Effective digitizing: advances in software and hardware. In *Technical Papers 1986 ACSM–ASPRS Annual Convention*, Vol. 1, pp. 162–171. Falls Church, Virginia: American Congress on Surveying and Mapping

DE MAN, W.H.E. (1984) Editor. *Conceptual Framework and Guidelines for Evaluating Geographic Information Systems*. Paris: UNESCO

DISTEFANO, J. (1986) Report to the National Committee for Cartographic Data Standards. In *Digital Cartographic Data Standards: A Report on Evaluation and Empirical Testing*, edited by H. Moellering, pp. 44–51. Columbus, Ohio: National Committee for Digital Cartographic Standards

GOODCHILD, M.F. (1980) The effects of generalisation in geographical data encoding. In *Map Data Processing*, edited by H. Freeman and G.G. Pieroni, pp. 191–205. New York: Academic Press

GOODCHILD, M.F. and RIZZO, B. (1986) Performance evaluation and workload estimation for geographic information systems. In *Proceedings, Second International Symposium on Spatial Data Handling*, pp. 497–509. Williamsville, New York: International Geographical Union

GREEN, N.P., FINCH, S. and WIGGINS, J. (1985) The 'state of the art' in Geographic Information Systems. *Area*, **17**(4), 295–301

HARDY, D.D. (1985) Today the earth is 100 000 000 000 000 bits. In *Advanced Technology for Monitoring and Processing Global Environmental Data*, pp. 1–4. London: Remote Sensing Society

HARLEY, J.B. (1975) *Ordnance Survey Maps: A Descriptive Manual*. Southampton: Ordnance Survey

HOUSE OF LORDS (1983) *Report of the Select Committee on Science and Technology: Remote Sensing and Digital Mapping*. London: HMSO

MARBLE, D. and PEUQUET, D. (1983) Editors. Geographic information systems and remote sensing. In *Manual of Remote Sensing*, edited by R.N. Colwell, pp. 923–958. Falls Church, Virginia: American Society of Photogrammetry

MARBLE, D.F. and SEN, L. (1986) The development of standardized benchmarks for spatial database systems. In *Proceedings, Second International Symposium on Spatial Data Handling*, pp. 488–496. Williamsville, New York: International Geographical Union

MARBLE, D.F., CALKINS, H.W. and PEUQUET, D.J. (1984) Editors. *Basic Readings in Geographic Information Systems*. Williamsville, New York: Spad Systems

MARX, R.W. (1986) The TIGER system: automating the geographic structure of the United States Census. *Government Publications Review*, **13**, 181–201

MCLENING, M. (1984) A solution looking for a problem. *Software*, June, 12–14

MOELLERING, H. (1986a) Developing digital cartographic data standards for the United States. In *Proceedings, Auto Carto London*, Vol. 1, edited by M. Blakemore, pp. 312–322. London: Royal Institution of Chartered Surveyors

MOELLERING, H. (1986b) Editor. *Digital Cartographic Data Standards: A Report on Evaluation and Empirical Testing*, No. 7 in Issues in Digital Cartographic Data Standards. Columbus, Ohio: Ohio State University/NCDCDS

MONMONIER, M.S. (1982) *Computer-Assisted Cartography: Principles and Prospects*. New York: Prentice Hall

MONMONIER, M.S. (1985) *Technological Transition in Cartography*. Madison, Wisconsin: University of Wisconsin Press

NELSON, R. and BLAKEMORE, M. (1986) NOMIS—a national geographic information system for the management and mapping of employment, unemployment and population data. In *Technical Papers 1986 ACSM–ASPRS Annual Convention*, Vol. 1: *Cartography and Education*, pp. 20–29. Falls Church, Virginia: American Congress on Surveying and Mapping

OPITZ, B.K. (1986) Editor. *Geographic Information Systems in Government*, 2 vols. Hampton, Virginia: A. Deepak

PETCHENIK, B.B. (1985) Maps, markets and money: a look at the economic underpinnings of cartography. *Cartographica*, **22**(3), 7–19

PETRIE, G. (1985) Remote sensing and topographic mapping. In *Remote Sensing in Civil Engineering*, edited by T.J.M. Kennie and M.C. Matthews, pp. 119–161. London: Surrey University Press

PEUQUET, D. (1984) A conceptual framework and comparison of spatial data models. *Cartographica*, **21**(4), 66–113

PEUQUET, D.J. and BOYLE, A.R. (1983) *Raster Scanning, Processing and Plotting of Cartographic Documents*. Williamsville, New York: Spad Systems

RHIND, D.W. (1985) Successors to the Census of Population. *Journal of Economic and Social Measurement*, **13**(1), 29–38

RHIND, D.W. (1986) Remote sensing, digital mapping, and geographic information systems: the creation of national policy in the United Kingdom. *Environment and Planning: Government and Policy*, **4**, 91–102

ROBINSON, V.B. (1987) Some implications of fuzzy set theory applied to geographic data bases. *Computers, Environment and Urban Systems*, forthcoming

ROBINSON, V.B., BLAZE, M. and THONG, D. (1986a) Representation and acquisition of a natural language relation for spatial information retrieval. In *Proceedings, Second International Symposium on Spatial Data Handling*, pp. 472–487. Williamsville, New York: International Geographical Union

ROBINSON, V.B., FRANK, A.U. and BLAZE, M. (1986b) Expert systems and geographic information systems: critical review and research needs. *Proceedings, GIS Workshop*, Springfield Virginia, December 1985

SOWTON, M. and HAYWOOD, P. (1986) National standards for the Transfer of digital map data. In *Proceedings, Auto Carto London*, Vol. 1, edited by M. Blakemore, pp. 298–311. London: Royal Institution of Chartered Surveyors

STARR, L. (1984) *Computer-Assisted Cartography. Research and Development Report July 1984*. London: International Cartographic Association

TAYLOR, D.R.F. (1980) Editor. *The Computer in Contemporary Cartography*. Chichester: Wiley

VAHALA, M. (1986) Data standardisation of the integrated LIS in Finland. In *Proceedings, Auto Carto London*, Vol. 1, edited by M. Blakemore, pp. 323–332. London: Royal Institution of Chartered Surveyors

WIGGINS, J.C., HARTLEY, R.P., HIGGINS, M.J. and WHITTAKER, R.J. (1986) Computing aspects of a large geographic information system for the European Community. In *Proceedings, Auto Carto London*, Vol. 2, edited by M. Blakemore, pp. 28–43. London: Royal Institution of Chartered Surveyors

7 Future trends in digital mapping

David Bickmore

Digital mapping implies formidable changes as much in the techniques as in the philosophy of cartography—perhaps more formidable than appeared at first sight. Inevitably these changes will affect map libraries, those who organize them and those who use them. A decade ago we referred to 'the new cartography', but computer techniques now seem likely to become a *sine qua non* of all cartography. This point was reinforced by Dr Joel Morrison in his opening address to Auto Carto London (September 1986) as President of the International Cartographic Association.

What is true for digital cartography seems equally true for its twin sister Remote Sensing. Preceding chapters in this volume have expounded on the techniques and on some of the objectives of both these developing fields. Many of the techniques were originated in the 1960s (and particularly in the UK and Canada). Maps—quite complicated and elegant ones—resulting from these techniques were being developed in the 1970s. Now in the 1980s we have rediscovered 'geographical information systems', though I suspect that logical correlations of this kind were notions that have been beloved by the makers of national atlases for most of this century. There are today many who accept that the database itself is the prime objective both for digital mapping and for remote sensing; individual maps become 'merely' one kind of derived graphic that can emerge from the golden cornucopia. I confess to having great sympathy for these views but at the same time I have vivid recollections of two decades of frustrations and obstacles that seem to have obfuscated the database strategy and delayed its progress into the map library world.

Assertions about the universality of computer cartography are all too commonplace: by contrast, a quick skim through the main cartobibliographic sections of this work emphasizes that in 1987 it is good straightforward printed maps that overwhelmingly dominate the marketplace. Magnetic tapes—let alone 'CD-ROMs' and their ilk—are such an invisibly small minority that one is bound to wonder what all the high-tech talk is about. It is evident that real geographic data, such as maps provide, are simply not available in computer form in the very places—map libraries—where you could reasonably expect to find them. What has gone wrong?

There is no doubt that all aspects of computing have an interest in this field simply because many different disciplines and professions are drawn to sets of promising and exciting ideas: this is certainly true in the geographic field. But there seems a strange gap that has separated actual data sets from theories of how they should be made, perhaps with too many of the latter and too few of the former. It may be useful to examine why the digital data sets just are not there—yet. Why is it that digital cartography still has problems in getting out of its ivory tower and getting massively involved in the real world of users and via map libraries?

During the 1960s and 1970s cognoscenti regarded computer techniques strictly as new methods of producing existing objects. The forms of thematic and topographic cartography had been made more or less sacrosanct by time. It was natural that computer maps were compared with classically produced ones and differences in standards, in production time, in cost, and in dependability were scrutinized. In many of these comparisons high technology as applied to cartography seemed expensive, slow and unreliable. Why pay more for a tape than a paper map if the tape called for additional capital investment on specialized equipment in order to read it?

Of course, maps and atlases based on the new techniques did emerge, many of them funded as experiments in what was feasible rather than as models for production processes. There was, for example, quite a plethora of black and white 'do-it-yourself' line printer maps, notably from the Laboratory for Computer Graphics at Harvard University. A smaller number of elegantly coloured oceanographic, geological and soil maps came from the Experimental Cartography Unit at the Royal College of Art, as did a world seismic atlas that displayed and quantified activity globally and decade by decade; furthermore, the maps in this atlas were derived directly from a seismic data bank that still lives on. Nearly all these and many other experimental productions emerged as printed maps ready to be used by the eye: they were not provided as digital data to a computer-using community. Indeed, the fact that such maps had actually been made by computers was explained in appropriately modest footnotes. In no way did these products rock the existing library system; indeed they assumed it. And at the time that was wise.

Today, however, we look to digital databases in this field that stand in their own right as coherent works of reference. Perhaps the computer-using community is becoming a reality.

It is possible to identify some factors which may have contributed to using computers just to make maps rather than to make databases. First of all, many of those concerned in the research and development had the considerable advantage of being professional cartographers and saw the conversion of their labours into elegant printed maps as a natural process and the appropriate one for their users. Second, one must record the deterrents of capital expenditure. Appropriate computer mapping or remote sensing equipments would be major new items of capital expenditure in any map library or in any business environment. They will be overheads of a specialized kind, and in addition to finding many tens of thousands of pounds or dollars they are also likely to demand acolytes with specialized technical skills not normally expected in a map library. Effective graphics, e.g. a graphic browse system, capable of replacing the equivalent of an hour's visit to the library at present seems an expenditure beyond the dreams of avarice and even of map librarians.

As if to clinch arguments about equipment, it is to be confessed that actual map data in digital form barely exists: medieval libraries came into existence because actual books did exist (and could indeed be read). The map librarian of today unhappily has to wait for his digital revolution and for the arrival of manipulable databases.

The third reason for the delays in development seems to lie in the difficult but not intractable nature of organizing GIS databases so as to provide the flexibility and versatility they need and to provide them for extremely large sets of data, e.g. a gigabyte at a time. Some neat solutions have been demonstrated in terms of geographical relationships for relatively small data sets but workability with much larger databases is still a somewhat uncertain matter. These uncertainties may be made more worrying by aspects of technical unreliability which, though greatly diminished during the last decade, do seem to occur. There *are* still tapes whose formats cannot be read; there are disks that crash; there is confusion about date stamping; there are new computer languages; there are, indeed, successive generations of computer systems. These complications may well be meat and drink to the experts but they are discouragements to the user, particularly to the user who *can* read a map but *cannot* read a tape. There are many such in the western world and many more in the developing countries.

Another problem area lies in the real difficulties and costs of creating effective map catalogue systems in digital form and of the labour involved in loading information about the assembled collections of maps into a system. Obviously the coding of sheet indexes is a simple problem of 'paging', but knowledge of different editions, revisions, etc., is not so obvious, and the sheet lines themselves need overlaying against a base map, for example to identify how much of sheet 71 falls on land rather than on sea. The base map to the catalogue should itself be a digital index in its own right, enabling the enquirer to identify which map sheets cover the boundaries of this province or can link this place name to that city, or can calculate a specific drainage catchment area by interpreting relief and river patterns. These catalogue indexing problems are even more complex in the case of air photographs and remote sensing with its well known problems of cloud cover and of variations in what may be observed by different wavelengths over time.

Feelings of despondency on the part of map librarians can readily be understood, e.g. data sets that do not exist and whose reading would require expensive equipment and staff who need as much experience in computing as in the sources of cartography are deterrents. By contrast printed maps exist, are (relatively) cheap, are directly readable by eye and have—by and large—been catalogued.

Despite these obstacles I am convinced that the need for basic map data in digital form will emerge and needs to be faced and accommodated by map librarians. It may well be that digital data may be called into existence less on its own cartographic account than in order to provide geography to other data sets, especially to statistical ones. There is no doubt that statistics are now more readily (and cheaply) available to the community of users in digital than in printed form. And indeed statistical data can be converted to machine-readable form much more readily

than is possible with the graphic subtleties implicit in map-based data. Census data is a particular example of statistics and is progressively emerging in digital form and in relation to provinces or nations or globally. Most censuses have to be based on spatial, if not geographical, units and sooner or later demand the existence of the shapes of these units as defined boundaries or polygons within the census system: here again is a need for base maps.

Of course there are innumerable statistical data sets unrelated to censuses in the field of environmental science which cry out for spatial analysis: examples abound, e.g. in oceanography, meteorology and climatology, in agriculture, in geology, and in geophysics. There is abundant need for a geography in a complementary computer form and this need really does seem logically inescapable. Modelling work has greatly increased as a direct consequence of available computing facilities and of the natural wish to predict courses of action resulting from past trends. In the environmental field these may relate to the spread of AIDS, or of desertification in Africa, or of the probability of North Sea surges.

Environmental problems are indeed complex and there is much scientific interest in examining them on a global rather than a local scale and on an interdisciplinary basis. The International Geosphere Biosphere Programme has very recently been set up by the International Council of Scientific Unions with massive backing from East and West and from an astonishing range of disciplines in the biological, chemical and physical fields. The programme is really focused on the sustainability of the environment of this planet. These global studies will take place during the 1990s. It is self-evident that much of the work in such a global programme makes demands on geography and, in particular, on map data. A digital base map does seem a necessity, if for no better reason than to provide basic stage scenery for more dramatic performing sciences.

This need to provide a World Digital Database for Environmental Science (WDDES) has been a concern of a Joint Working Group of the International Cartographic Association and the International Geographical Union. A brief resumé of the intentions of this project seems an appropriate epilogue to these introductory chapters. It also seems to go without much saying that such a database needs to be available in its digital form for use in the map libraries of the world. Is that too much to hope for the future?

The WDDES is intended as the computerized equivalent to a base map. It will exist *only* in digital form and will firmly assume that there are users enough with computer facilities to make it a product that will actually be used. The base map will be derived by digitizing maps that already exist. For those who only want maps, they can buy the sheets of the 1:1 000 000 Operational Navigation Charts or the sheets of the GEBCO maps of bathymetry at 1:10 000 000. Indeed, WDDES will not actually contain in digital form some of the details on those maps.

What, then, are the supposed advantages of putting existing maps into digital form? First of all, scale and projection can be changed at will. Segments of line will be represented by coordinates in geographicals which can readily be converted to any particular projection and can be windowed for the area required. Second, the user can select the elements required for a particular area of interest and can display them in whatever colours/symbols seem appropriate; furthermore, he/she can measure them, producing very rapidly calculations of area or length. Third, the user can 'model' the data, that is to say convert relief information into digital terrain models from which slope patterns can be derived—by combining these with river networks he/she can, for example, deduce catchment boundaries. There are, of course, many other modelling operations which are currently used by different scientists and operators for just these purposes. These techniques will unquestionably be used in the International Geosphere Biosphere Programme, and that is why there is urgency to develop WDDES by 1989/90.

Scale in a database is a far more flexible matter than on a map and these days is sometimes interpreted in terms of resolution on the ground. Evidence suggests that for a global database the aim should be to provide 1 km resolution, i.e. a scale of *ca* 1 : 1 000 000. For some parts of the globe this is a very small scale; for others, e.g. parts of the ocean and Antarctica, it is unreasonably large. The decision to use this resolution is partly related to the existence of a 1 : 1 000 000 world map series which is currently maintained for air navigation by the military: sheets of this series are available in most university map libraries and in many other business and academic institutions. At about this scale the ONC series is the only one regularly maintained at present. Of course, as with all map series, the ONC has deficiencies. It does not provide Antarctic cover, it is blank in the oceans, and it has some uncontoured gaps on the land (*ca* 10 per cent). Of course it is not perfect but digitizing it seems the first task: updating and improving it can follow. The same arguments also apply to the 1 : 10 000 000 GEBCO maps of the ocean.

Many readers will ask why such a database is not being derived directly from remote sensed data and we wish that such a strategy was indeed possible. Unfortunately, however, relief cannot be obtained from, for example, Landsat, and the use of Landsat or SPOT would probably increase the budget by two orders of magnitude. Both these sources can obviously provide updating material, hopefully when they have been translated into standard 1 : 250 000 mapping by the countries concerned.

WDDES will first concentrate on providing the physical geography of the world—by digitizing contours, height values, etc. It will also digitize the river patterns and coastlines and will relate these to the contour patterns (height values at contour intersections on river files). It will also include national and provincial boundaries so as to enable statistical data sets to be represented, and it will include the shapes of built-up areas. It will not include vegetation because topographical interpretations of vegetation are not highly regarded by botanists, and it will probably not include communications, for which no very meaningful classification is used on ONC. WDDES will include names of these features, e.g. rivers, which have been digitized and it will link such names to the line segments in the database and not, for example, to the point at which the lettering starts.

These brief notes reflect some of the structuring of the digital data which is intended as a significant feature of a database of this kind.

One of the purposes of a topographical database will be to enable fast (e.g. 24 hour) links to other databases of climatological information, or of river flows, or of desertification factors, or of seismic events. Other users may like to purchase the data in compact disk form, e.g. as CD-ROMs possibly for display on their own personal computers: perhaps this may be a route relevant to map libraries. The important point is that the database should be treated as a living and growing organism with many functions and capable of rapid access from anywhere in the world. These aspirations are particularly important if the International Geosphere Biosphere Programme hopes to have sets of standardized data in an agreed format and located in relation to the single index that WDDES aspires to represent.

WORLD MAPPING

THE WORLD

This section discusses mapping that provides worldwide monographic or series coverage. Many of the maps here included also provide the best available coverage of continents, or sometimes of countries, for particular themes. We have not usually repeated descriptions under the relevant country where this is the case. Much of the world mapping is published by international, non-governmental agencies, some by national survey authorities and some by commercial agencies.

The **United Nations** and its related agencies within the United Nations system have been responsible for the publication of much small-scale world mapping and for the coordination of cooperative series published by other mapping agencies, but the UN impact on world mapping is mostly indirect and has mainly stressed the encouragement and development of cartographic activities in UN member states. The **Cartographic Section** of its Department of Economic and Social Affairs in New York was established in 1950, and is the main agency in the UN Secretariat charged with carrying out UN cartographic policies. Since 1950 it has encouraged cooperation and the development of cartographic activity in member nations and has disseminated cartographic information. Regional Cartographic Conferences are sponsored by the UN: countries summarize their mapping activities and report on developments within the region. (See the continental and country section for further information about these conferences.) The journal *World Cartography* is issued from the Section, and training courses and aid programmes are organized.

Perhaps the most important map project in the twentieth century has been the publication of the *International Map of the World on the Millionth Scale* (IMW). This programme can be traced back to 1891, when a proposal was made at the fifth International Geographical Congress in Berne to compile a world map series at a scale of 1:1 000 000 to show both physical and human features. By 1913 agreements on specifications were reached, including the choice of a modified polyconic projection, and it was agreed that States would be responsible for the publication of mapping of their own areas. An office was established at the Ordnance Survey in Great Britain to oversee the huge cooperative task. Following World War II this Central Bureau was relocated to New York and the UN assumed responsibility for overseeing the publication programme. In 1962 it was agreed that specifications for relief depiction and symbology should be amended and that the Lambert conformal conic projection should be adopted to ensure conformity with the World Aeronautical Chart

series (see below). About half of the world's land area is still inadequately represented in this series, either because of very old and inaccurate compilations, or because areas have not yet been mapped. Because of the variability of publisher, source of the mapping and differing dates of the maps in the series it is difficult to generalize about specifications of IMW sheets beyond the facts that sheets usually cover 6° of longitude and 4° of latitude and are usually litho-printed colour editions. Relief depiction varies with area, date and publisher; some sheets are gridded, some not. In recent years the status of the IMW has been under much discussion. There has been criticism about the slow rate of progress of the series, about its fragmented nature and about the doubtful need for an IMW given complete world coverage in comparable scale aeronautical series, and the possibility of compiling photogrammetric mapping at 1:250 000 scale for large areas of the world. Examination of the IMW reports (see Further information) reveals that few new editions of IMW sheets have been published in the last 15 years: the British Directorate of Military Survey and the American Defense Mapping Agency are both no longer revising sheets in the series, and meetings have been held in 1986 to discuss the future of the project. The value of the IMW lies in its early stimulus towards cooperation in mapping standards and in its impact upon subsequent thematic series which have used IMW sheet lines. Such series have included the *International Map of World Vegetation* and the *International Map of the Roman Empire*, neither of which gives more than fragmentary coverage, but both of which use IMW sheet lines. Many national mapping agencies compile mapping of their territories that uses an IMW sheet breakdown. The UN has also compiled its own maps, through its **Cartographic Unit**. This office compiles mapping for use in UN official publications, but also publishes separate small-scale maps, obtainable through the **Publications Section**.

The UN is also involved in toponymic standardization and has sponsored five conferences on The Standardization of Geographical Names. Regional commissions of the UN have also been involved in mapping activities. We have described these and listed their products under the relevant continental sections. United Nations mapping (and that published by other International Organizations within the United Nations system) may be obtained through the distributor **UNIPUB** in New York.

The other major topographic world series is published in a collaborative project by the survey organizations of East Germany, Poland, Czechoslovakia, Hungary, Bulgaria,

Romania and the USSR. *Karta Mira* (World Map) at
1:2 500 000 scale gives complete coverage of the earth's
surface (including ocean areas) in 234 sheets. Each sheet
covers nine IMW quadrangles. Published in the 1960s and
1970s it uses an azimuthal projection for areas between the
poles and 60°, and a conical projection for all other areas.
Twenty-six overlap sheets are published for the areas in
the northern hemisphere where the projections meet. This
map uses contours and layer shading at intervals between
50 and 1000 m, depending upon terrain, to show relief.
Published in both English and Russian versions, the
English map is also printed with Russian character
explanations of symbols. This series is available through
various agencies, notably through the Hungarian
distributor **Cartographia**, though Pergamon Press in the
west was also distributing sheets until recently. The Russian
Glavnoe Upravlenie Geodezii i Kartografii (GUGK) has
recently begun to issue map sets of this series for certain
areas (see Latin America, USSR and Pacific).

The **International Civil Aviation Organization (ICAO)**,
established in 1947, acts as a clearing house for
aeronautical information, publishes a useful catalogue of
aerial navigation chart series and oversees conformity in
Standards and Recommended Practices for the publication
of maps in the *World Aeronautical Chart (WAC)* series.
IMW sheets can be used as part of the WAC series with
the addition of aeronautical information. Like the IMW
this is a cooperative project between the different national
aeronautical charting bodies and many countries have
published sheets in this series. International publication of
the WAC is no longer active but countries are still issuing
sheets that conform to WAC standards.

Current coverage of the world is available in aeronautical
chart series at scales of 1:5 000 000 (*Global Navigation and
Planning Chart, GNC*), 1:2 000 000 (*Jet Navigation Chart,
JNC*), 1:1 000 000 (*Operational Navigation Chart, ONC*)
and 1:500 000 (*Tactical Pilotage Chart, TPC*). These
series, like the IMW, are well known by their acronyms
and are published by the United States **Defense Mapping
Agency (DMA)**. They may be procured from the **DMA
Aeronautical Chart Distribution Center** attached to the
National Ocean Survey in Riverdale, Maryland. The
significance of the ONC and TPC series has increased with
the lack of progress in the IMW project, and though
intended primarily as navigational charts both of these
maps serve as world topographic series. The ONC series
gives complete world coverage in 271 large sheets, which
use the Lambert conical conformal projection (and a polar
stereographic projection for polar areas). Sheets are also
issued by British, Canadian, New Zealand and Australian
agencies. The graticule is overprinted at 1° intervals, relief
is shown by 500 foot contours and layer shading and
aeronautical information is overprinted. New editions of
these maps are regularly published. Selected information
captured from the ONC series is now available in digital
form as the Mundocart database from **Petroconsultants** in
Geneva and the ONC has become the standard source for
world topographic information. The TPC series is not
complete: coverage is at present available for about one-
third of the world. Sheets are similar in specification to the
ONC series, four TPCs comprise the area of one ONC,
and the series shows relief by 500 and 250 foot subsidiary
contours.

The Defense Mapping Agency is also involved in the
compilation of the most important source of placename
information. It issues the gazetteers compiled for the
United States Board on Geographic Names. These
volumes, which are available for nearly every country in
the world, were started in the 1950s and are periodically
revised. Volumes are listed under individual countries. A
recent change in the publication programme means that
new US Board gazetteers are only to be available as
microfiche editions.

Smaller-scale world coverage is available in several
different forms from various national mapping agencies.
We have listed only a sample of the best. The French
Institut Géographique National (IGN) publishes several
Cartes générales du monde at 1:5 000 000 (34 sheets using a
double Transverse Mercator projection); at 1:10 000 000
(12 sheets on the Mercator projection); at 1:15 000 000
(three sheets, Mercator projection) and at 1:33 000 000 (a
single sheet using the Aitoff–Wagner projection). The
American DMA issues a World outline series at
1:20 000 000 on the Miller cylindrical projection, but has
not updated sheets issued in *Series 1106 the world
1:5 000 000*. This map covered the world in 18 sheets and
was compiled by the US Army Map Service, the American
Geographical Society and later DMA in the 1960s and
1970s. Most sheets in this series are still available through
major jobbers but have not been revised. Smaller-scale
coverage has been compiled using many different
projections by a variety of commercial and official
agencies. We have listed a selection of these maps.

The UN agency that is most involved in thematic map
publication activities has been the **United Nations
Educational Scientific and Cultural Organization
(UNESCO)**. UNESCO publishes and sponsors mainly
small-scale thematic maps and series, often in association
with other international agencies. These projects involve
work from many scientists and institutions in individual
states. In the field of geological mapping UNESCO
cooperates with the **Commission for the Geological Map
of the World (CGMW)**. CGMW evolved in 1911 from the
Commission for the Geological Map of Europe, and had
the task of preparing a legend for, and drafting, a
geological map of the world, to give coverage in about 80
1:5 000 000 scale sheets. This project was never
completed, in part because of the lack of adequate
geographic base maps at a suitable scale. Initial effort was
therefore devoted to the publication of geological mapping
of smaller units, via the establishment of regional
subcommissions. Many of the maps that have resulted
from these programmes are listed under the relevant
continental sections. It was not until the 1970s that
CGMW and UNESCO began to publish an authoritative
world geological map series. Using the 1:5 000 000 scale
AGS mapping as a base, work began on the *Geological
World Atlas*. The entire world is now covered in 22 sheets:
16 continental sheets at 1:10 000 000 scale, five oceanic
and polar sheets at smaller scales between 1:36 000 000
and 1:20 000 000 and a legend sheet have been available
since completion of the project in 1984. The atlas also
includes explanatory information for each continent.
Amongst current worldwide CGMW and UNESCO
projects are the publication of a two-sheet 1:25 000 000
scale geological wall map of the world, and the continuing
publication of 1:5 000 000 scale continental geological
mapping (see continental sections). Other national
geological survey authorities such as the **United States
Geological Survey (USGS)** and the French **Bureau de
Recherches Géologiques et Minières (BRGM)** produce
small-scale geological and other earth science mapping at a
global scale. Commercial agencies are also active in this
field, including the **Geological Society of America (GSA)**,
the **American Association of Petroleum Geologists
(AAPG)** and **PennWell**. **W. H. Freeman** published a very
useful single-sheet bedrock geology map in 1985, which
shows symmetrical age belts on both ocean floor and land

surface. Other thematic earth resources mapping in the commercial area includes oil mapping, e.g. by **Oilfield Publications (OPL)**.

Subcommissions of the CGMW have been responsible for overseeing the compilation of more specialized earth science mapping, but their maps are published by CGMW in association with UNESCO. Thus the **Sub-Commission for the Tectonic Map of the World** was established in 1956 under CGMW, with a secretariat based in the Geological Institute in Moscow. This subcommission has been the most active, and 1:5 000 000 scale tectonic mapping has been issued for Africa, North America, South America, and South and East Asia, whilst Europe has been covered at 1:2 500 000 scale. These maps are described under the relevant continental section. Recent compilations at 1:15 000 000 and 1:45 000 000 give worldwide tectonic coverage.

The CGMW **Sub-Commission for the Cartography of Metamorphic Belts of the World** has been active since the 1970s, having been initiated following an International Union of Geological Sciences resolution of 1965. Less mapping has been published than by the Tectonic Sub-Commission: continental coverage is either published or on programme at a variety of scales and is listed under the relevant section.

The **Sub-Commission for the Metallogenic Map of the World**, based in Orléans-Cedex in the French Bureau de Recherches Géologiques et Minières, has published South and North American and Australian 1:5 000 000 metallogenic coverage and for Europe at 1:2 500 000 scale. Other continents are on programme. Other CGMW Sub-Commissions have as yet been less active. There is also an Environmental Sub-Commission, and since 1982 and reflecting current interest in the topic a Sub-Commission for the Geological Mapping of the Ocean Floor.

UNESCO has also been involved in cooperative projects concerned with the mapping of the world's oceans. We have discussed all oceanic mapping under Oceans or under one of the specific ocean sections.

UNESCO has published material in association with the **World Meteorological Organization (WMO)**. The *World Climatological Atlas* project was started in the mid-1950s, with the intention of publishing maps depicting temperature and precipitation in multi-lingual volumes at a continental scale. To date four volumes have been published, in conjunction with Cartographia, to cover Europe, South America, North and Central America, and Asia.

UNESCO has collaborated with the **Food and Agriculture Organization (FAO)** to publish the 1:5 000 000 scale *Soil Map of the World*. This project began in 1961 as an attempt to harmonize and synthesize knowledge about world soils. The 18 sheets were published with 10 accompanying texts and a legend sheet over a 9-year period in the 1970s. 106 different soil units are distinguished, the textural class of the dominant soil and the slope class are given for each association and 11 different phases are distinguished. Explanations of these units is given in English, French, Russian and Spanish. The FAO/UNESCO soil classification has been applied to many other soil mapping projects. The accompanying monographs are issued in two of the languages used in the legend, depending upon the area covered. The **International Society of Soil Scientists** is currently investigating the possibility of compiling a 1:1 000 000 scale global soils map, to be derived from a world soils database. This project was discussed at a seminar in January 1986 in the Netherlands but is not yet beyond the planning stage.

Other collaborative UNESCO/FAO projects include bioclimatic mapping of the Mediterranean. FAO itself is responsible for many thematic mapping projects. For instance a 1:25 000 000 scale map of world fuelwood distribution with explanatory brochure was published in 1981, in order to assess fuelwood shortages, whilst other ecological mapping projects in Africa have also been sponsored from Rome. The desertification map of the world at 1:25 000 000 was compiled as a joint project between FAO, UNESCO, WMO and the **United Nations Environment Program (UNEP)** for presentation to the world conference on desertification. UNEP in Nairobi prepared three other world desertification maps at 1:25 000 000 for this conference and has recently become involved in a major Geographic Information System Project. The *Global Resources Information Database* has been set up as part of UNEP's Global Environment Monitoring Programme. Information is being captured from a wide variety of spatial data sets, including many small-scale cartographic sources, both topographic and thematic. The descriptive literature on GRID provides a useful overview of what digital data is currently available at a worldwide scale. The aim is to use the ARC/INFO software to manipulate the data to provide environmental assessments at different levels of resolution. It is intended to present case studies of the application of this system and to follow up by the design of a worldwide GIS.

The GRID project is being carried out in the knowledge that it might be able to use the data collected in the potential International Cartographic Association/International Geographical Union *World Digital Database for Environmental Science*. At the end of 1986 this project was still at the design stage: it is intended (should funding be forthcoming) to provide complete worldwide relief, river pattern, bathymetric and statistical boundary information in a digital form, as far as possible at 1:1 000 000 scale, using data from ONC maps, supplemented by coastline and bathymetric data from GEBCO charts and the **World Data Centre** gridded bathymetric data set. Pilot studies on selected ONCs (in Kenya and Sumatra/Malaysia) have taken place. It is hoped to derive digital terrain models from the ONC contour information and to design data structures compatible with ARC/INFO database software.

Other UNESCO cooperative projects include the publication of Quaternary mapping in association with the **International Union for Quaternary Research (INQUA)** (see under Europe and Africa); hydrogeological maps in association with the **International Association of Hydrological Sciences** and **Bundesanstalt für Geowissenschaften und Rohstoffe (BfGR)** (see under Europe); and hydrological mapping in conjunction with the USSR National Committee the International Hydrological Decade (published as the *Atlas of World Water Balance*) in English and Russian language versions in 1978.

Vegetation mapping has also been published by UNESCO to cover the Mediterranean, South America and Africa and is listed under these sections in this book.

The **World Bank**, also known as the **International Bank for Reconstruction and Development (IBRD)**, publishes the annual *World Bank Atlas*, and its Cartographic Unit has also compiled both the *Landsat Index Atlas of the Developing Countries of the World*, issued by Johns Hopkins University Press, and a series of Image Maps issued by International Mapping Unlimited.

A very wide range of commercial publishers produce general small-scale world mapping or series of country maps, or publish general atlases of the world in different language editions. We have not attempted to list the complete variety of commercial world mapping available. Some useful thematic maps are published by Süddeutscher Verlag in the series *JRO Topic Maps*. Some official agencies also publish series of maps of individual countries in the world, some of which we have listed under the relevant country. The most useful series are published by **GUGK** in Russia, by the Chinese **Cartographic Publishing House (CPH)** and by the American **Central Intelligence Agency (CIA)**. The CIA also markets the digital databases World Data Bank I and II which give digital worldwide coastline and administrative outline data.

Further information

The best general review of world mapping carried out by international organizations is Williamson, L.E. (1983) A survey of cartographic contributions of international governmental organizations, *Government Publications Review*, **10**, 329–344.

The journal *World Cartography* includes many articles on global mapping, including Stegena, L. (1979) Geoscientific world mapping: facts and representations, *World Cartography*, **15**, 41–47.

Regular news of important new mapping projects and collaborative arrangements at an international level is reported in the *International Cartographic Association Newsletter*.

For further information about the IMW see status reports of the *International Map of the World on the Millionth Scale*, 1955–, New York: UN.

Articles about UNESCO-sponsored mapping are published in their journal, *Nature and Resources*, 1965–, Paris: UNESCO; including Dudel, R. and Batisse, M. (1979) Soil map of the world, *Nature and Resources*, **14**(1), 2–6.

Catalogues and information about most of the organizations cited in the text are readily available.

Addresses

United Nations, Cartography Section
Department of Technical Cooperation for Development, NEW YORK, NY 10017, USA.

United Nations, Cartographic Unit
United Nations Building, NEW YORK, NY 10017, USA.

United Nations, Publications Section
Room A-3315, NEW YORK, NY 10017, USA.

UNIPUB Inc.
PO Box 433, Murray Hill Station, NEW YORK, NY 10016, USA.

Cartographia
Bosnyack ter 5, POB 132, H-1443, BUDAPEST, Hungary.

Glavnoe Upravlenie Geodezii i Kartografii (GUGK)
Ul. Krzhizhanovskogo 14, Korp. 2, V218, MOSKVA, USSR.

International Civil Aviation Authority (ICAO)
International Aviation Building, 1080 University Street, MONTREAL, Québec, Canada.

Defense Mapping Agency (DMA)
WASHINGTON DC, 20305, USA.

Defense Mapping Agency Aeronautical Chart Distribution Center
National Ocean Service, RIVERDALE, Maryland 20737–1139, USA.

United Nations Educational, Scientific and Cultural Organization (UNESCO)
7 place de Fontenoy, F-75700 PARIS, France.

Commission for the Geological Map of the World (CGMW)
Maison de la Géologie, 77 rue Claude-Bernard, 75016 PARIS, France.

Sub-Commission for the Tectonic Map of the World
Geological Institute, USSR Academy of Science, Przhevsky per. 7, MOSKVA 109017, USSR.

Sub-Commission for the Cartography of Metamorphic Belts of the World
Instituut voor Aardwetenschappen-Utrecht, Budapestlaan 4, postbus 80021, 3508 TA UTRECHT, Netherlands.

Sub-Commission for the Metallogenic Map of the World
Bureau de Recherches Géologiques et Minières, BP 6009, 45060 ORLÉANS-CEDEX, France.

World Meteorological Organization (WMO)
41 Ave G. Motta, GENÈVE, Switzerland.

Food and Agriculture Organization (FAO)
via delle Terme di Caracalla, 1-00100 ROMA, Italy.

International Hydrographic Organization
7 Ave President J.F. Kennedy, BP 345, MONTÉ CARLO, Monaco.

United Nations Development Program (UNDP)
866 United Nations Place, Room CN – 300, NEW YORK, NY 10017, USA.

United Nations Environment Program (UNEP)
PO Box 30552, NAIROBI, Kenya.

International Union for Quaternary Research (INQUA)
Vrije Universiteit Brussel, Kwartaingeologie, Pleinlaan 2, B-1050 BRUXELLES, Belgium.

International Association of Hydrological Sciences
19 rue Eugene Carrière, 75018 PARIS, France.

Cartography Section and Publications Unit World Bank
1818 H Street NW, WASHINGTON DC, USA.

For AAPG, PennWell, NGS, AGS, USGS, CIA, World Data Center, NOAA, GSA, Office of Naval Research, Rand McNally, and W.H. Freeman, see United States; for Bartholomew, OPL, Lloyds, OSD and Oxford Cartographers see Great Britain; for Petroconsultants, Kümmerly + Frey see Switzerland; for IGN and BRGM see France; for DEMR, Inland Waterways Directorate and JRO see Canada; for CPH see China.

Catalogue

Atlases and gazetteers

Times Atlas of the World: Comprehensive Edition   7th edn
London: Times Publications, 1985.
pp. 520.

Times Concise Atlas of the World   4th edn
Edinburgh: Bartholomew, 1986.

New International Atlas
Chicago: Rand McNally, 1981.
pp. 568.

Webster's New Geographical Dictionary
Springfield: Webster, 1984.
pp. 1407.

Names of political entities of the world approved by the United States Board on Geographic Names
Washington, DC: DMA, 1985.

General

Welt   1 : 50 000 000
Bern: Kümmerly + Frey.
Also published at 1 : 32 000 000 and 1 : 23 000 000.
Available as physical or political edition.
Van der Grinten projection.

World map   1:40 000 000
Reston, VA: USGS, 1983.

The world   1:35 000 000   MCR46
Ottawa: DEMR, 1974.

Bartholomew: the world   1:30 000 000
Edinburgh: Bartholomew, 1982.
Times projection.

The world   1:25 000 000
Washington, DC: NGS, 1983.
With thematic insets.

The world   1:20 000 000   Series 1105
Washington, DC: DMA.
27 sheets, all published.
Outline coverage.

Carte générale du monde   1:15 000 000
Paris: IGN, 1980.
3 sheets, all published.
Mercator projection.

Orograficheskaja karta mira   1:15 000 000
Moskva: GUGK, 1983.
4 sheets, all published.
In Russian.
Orographic map.

The world   1:14 000 000
Washington, DC: DMA and NOAA, 1982.
6 sheets, all published.
Mercator projection.

Carte générale du monde   1:10 000 000
Paris: IGN, 1970.
12 sheets, all published.
Mercator projection.

World map in equal area presentation
Oxford: Oxford Cartographers, 1983.
Peters projection world map.
1 cm:63 360 km.

Topographic

Carte générale des continents   1:5 000 000
Paris: IGN.
34 sheets, all published. ■

Karta Mira World map   1:2 500 000
Various publishers, 1964–.
234 sheets, all published. ■
Complete world coverage including oceans.
Published in conjunction by the official surveys of the USSR,
German Democratic Republic, Hungary, Czechoslovakia,
Romania, Poland and Bulgaria.

International map of the world   1:1 000 000
Various publishers, 1912–.
2122 sheets, ca 750 published. ■
Series covers land areas only.

Operational navigational chart   1:1 000 000
St Louis, MO: DMA, 1961–.
271 sheets, all published. ■

Tactical pilotage chart   1:500 000
St Louis, MO: DMA, 1964–.
ca 1000 sheets, ca 300 published. ■

Geological and geophysical

World atlas of geology and mineral deposits   D.R. Derry
London: Mining Journal, 1980.
pp. 114.

Atlas of economic mineral deposits   C.L. Dixon
London: Chapman and Hall, 1979.
pp. 143.

Earth's dynamic crust   1:78 890 000
Washington, DC: NGS, 1985.
Double-sided.
With numerous block diagrams on sheet and back.

Seismicity of the earth 1960–1980   1:46 000 000   A.F. Espinosa,
W. Rinehart and M. Tharp
Washington, DC: Office of Naval Research, 1982.

Tektonicheskaja karta mira/Tectonic map of the world
1:45 000 000
Moskva: Sub-Commission for the Tectonic Map of the World,
1984.

The magnetic field of the earth, 1980   1:40 053 700
Reston, VA: USGS, 1983.
5 sheets, all published.
Sheets show magnetic declination; magnetic horizontal intensity;
magnetic inclination; magnetic total intensity; and magnetic
vertical intensity.

World map of gemstone deposits   1:40 000 000
Orléans: BRGM, 1969.

Map showing the world distribution of carbon dioxide springs and
major zones of seismicity   1:39 000 000   I. Barnes, W.P. Irwin
and D.E. White
Reston, VA: USGS, 1984.

World seismicity map   1:39 000 000   A.C. Tarr
Reston, VA: USGS, 1974.

Significant earthquakes 1900–1979   ca 1:32 000 000
Boulder, CO: GSA, 1980.

Tektonicheskaja karta mira   1:25 000 000
Moskva: Sub-Commission for the Tectonic Map of the World,
1982.
4 sheets, all published + 94 pp. text.
With English and Russian legends.

Bedrock geology of the world   1:23 230 000   R.L. Larson et al.
New York: W.H. Freeman, 1985.
Includes seabed floor geology.

Free-air gravity anomaly map of the world   1:22 000 000
Boulder, CO: GSA, 1982.

Carte minière du globe sur fond tectonique   1:20 000 000
Orleans: BRGM, 1965.
2 sheets, both published + 33 pp. text.

International tectonic map of the world   1:15 000 000
Moskva: Sub-Commission for the Tectonic Map of the World,
1984.
12 sheets, all published + 40 pp. text.

Geological world atlas/Atlas géologique du monde   1:10 000 000
Paris: CGMW and UNESCO, 1974–83.
20 sheets, all published.

Free air gravity anomaly atlas of the world   1:9 000 000
C. Brown, W. Warsi and J. Milligan
Boulder, CO: GSA, 1982.
84 maps.

Environmental

Climatic atlas of the world
Paris: UNESCO and Cartographia.
4 published vols.
Published in association with WMO and Cartographia.

Atlas of world water balance
Leningrad and Paris: Gidrometeoizdat and UNESCO, 1977.
Also published in Russian edition as Atlas mirovogo vodnogo
balansa.

World snow cover   1:50 000 000
Ottawa: Inland Waters Directorate, 1975.

Map of the world distribution of arid regions   1:25 000 000
Paris: UNESCO, 1977.
With MAB *Technical Notes* 7.

*Desertification map of the world/Carte mondiale de la
désertification*   1:25 000 000
Nairobi: UNEP, 1977.

*Map of the fuelwood situation in the developing provinces of the
world*   1:20 000 000
Rome: FAO, 1981.

*Soil map of the world/Carte mondial des sols/Mapa mundial de suelos/
Pochvennaja karta mira*   1:5 000 000
Paris: FAO and UNESCO, 1970–81.
18 sheets, all published. ■
With 10 accompanying explanatory texts and legend sheet.

*Carte internationale du tapis végétal et des conditions écologiques/
International map of world vegetation and environmental conditions/
Mapa internacional de la vegetación y de las condiciones
ecológicas*   1:1 000 000
Toulouse and Pondichery: ICITV, 1958–.
25 published sheets. ■

Administrative

The world   ca 1:70 000 000
Southampton: OSD, 1987.
Revised annually, shows Commonwealth neighbour countries.

Descriptive map of the United Nations   1:70 000 000
New York: UN, 1984.

Political map of the world   1:40 000 000
Reston, VA: USGS, 1983.

Le monde politique   1:33 000 000
Paris: IGN, 1984.
Available in French or international editions.

Human and economic

World Bank Atlas
Washington, DC: World Bank, 1987.
Revised annually.

Lloyds maritime atlas
London: Lloyds, 1983.

Sedimentary basins of the world   1:40 000 000   B. St John
Tulsa, OK: PennWell, 1980.

The world: nature, man and economy   1:32 000 000
Bern: Kümmerly + Frey, 1977.
In English, French, German and Italian.

World oil and gas activity map   1:30 000 000
Ledbury: OPL, 1984.

*Potential population supporting capacities of lands in the developing
world*   1:10 000 000
Rome: FAO, 1982.
15 sheets, all published + 169 pp. text.
FPA/INT/153 Land resources for populations of the future.

Produktion ausgewählte mineralisch und Rohstoffe in der Welt
Gotha: Haack, 1984.

AMERIQUE DU NORD

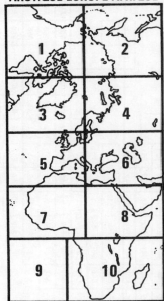

AMERIQUE DU SUD-ANTARCTIQUE

ANTARCTIQUE

ARCTIQUE-EUROPE-AFRIQUE

WORLD
1 : 5 000 000 topographic
Carte générale des continents

ASIE

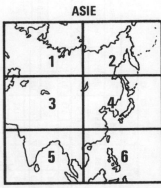

INDONESIE-AUSTRALIE

WORLD
1 : 2 500 000 Karta Mira

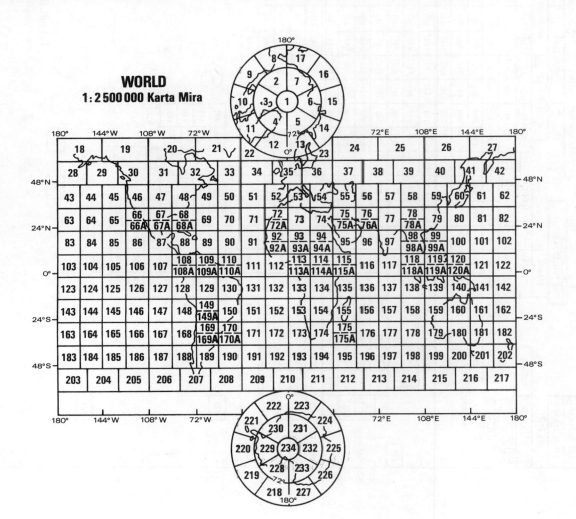

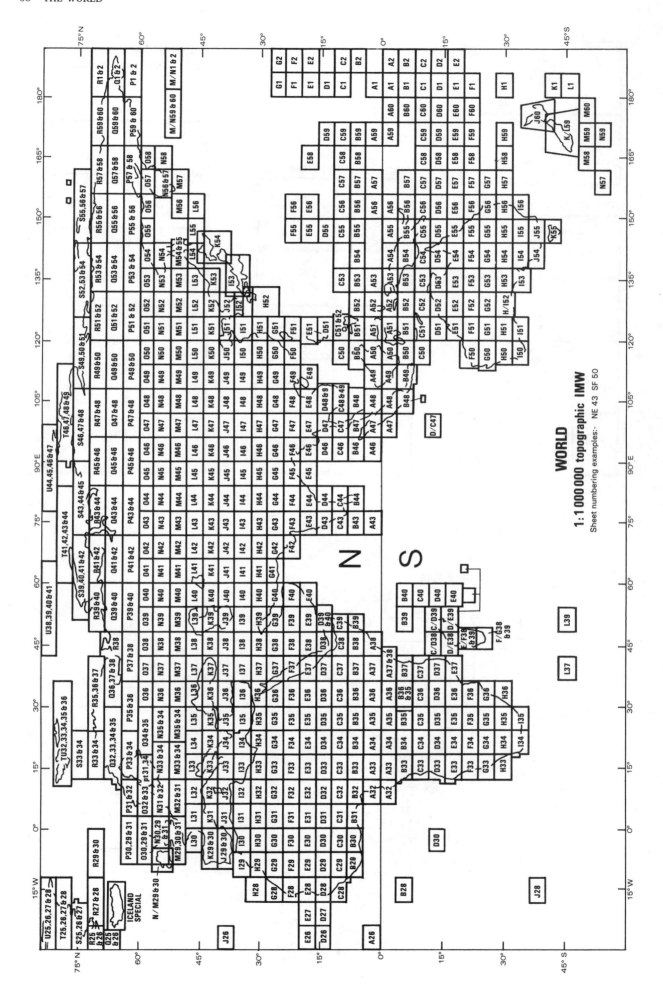

WORLD
1:1 000 000 topographic IMW

Sheet numbering examples:- NE 43 SF 50

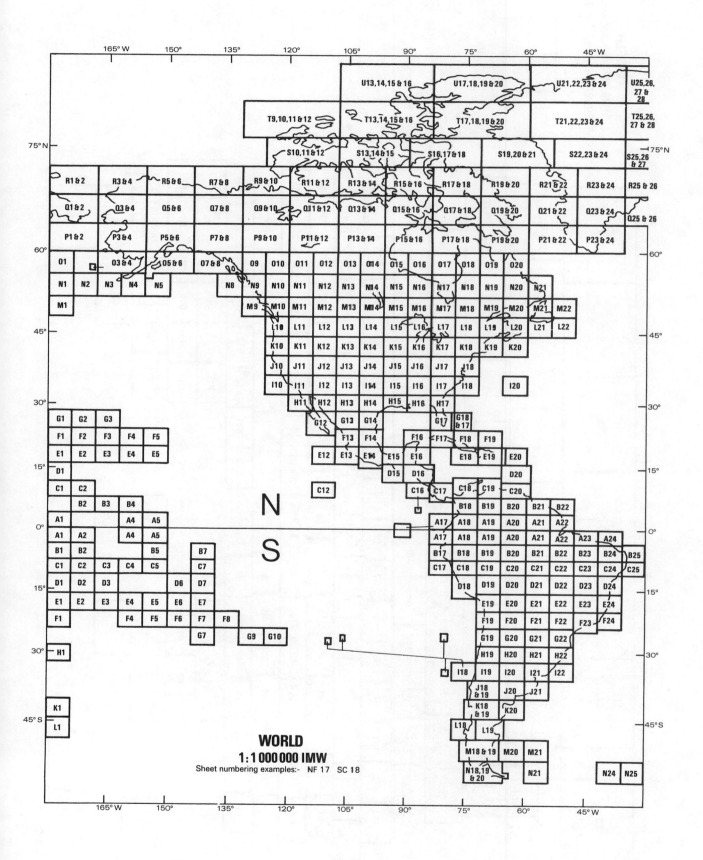

WORLD
1 : 1 000 000 IMW
Sheet numbering examples:- NF 17 SC 18

WORLD
1:1 000 000 aeronautical ONC
1:500 000 aeronautical TPC

A	B
	K6
D	C

Sheet numbering example:-
1:500 000 TPC K6 A

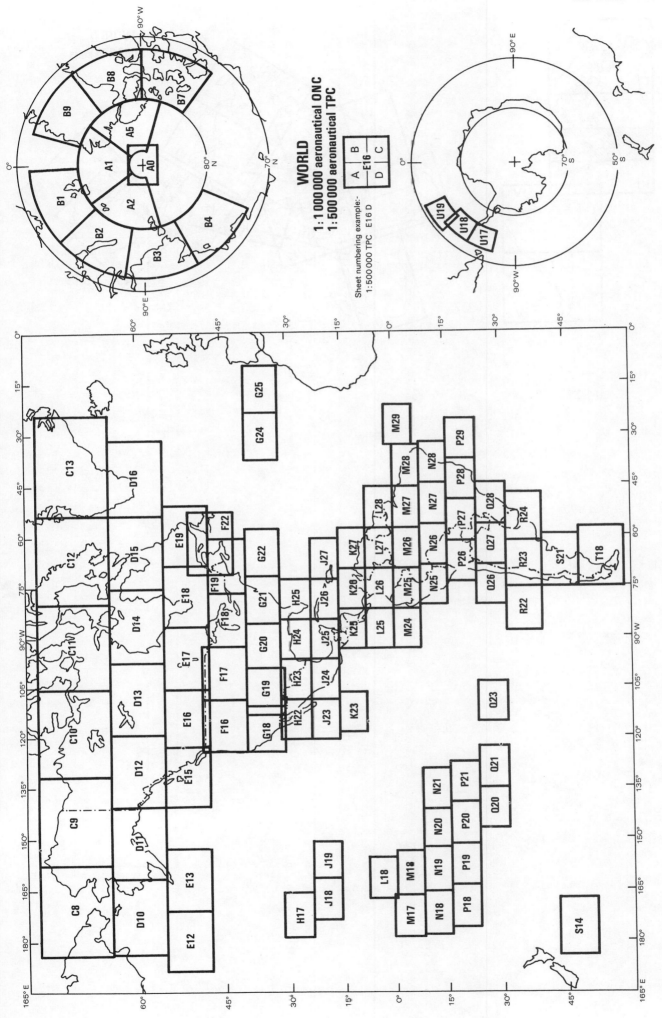

WORLD
1:1 000 000 aeronautical ONC
1:500 000 aeronautical TPC

Sheet numbering example:-
1:500 000 TPC   E16 D

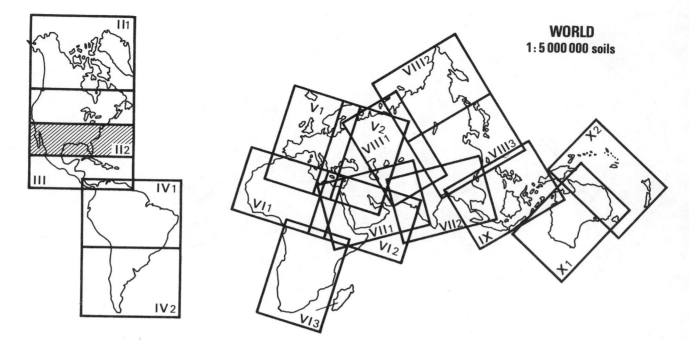

AFRICA

The mapping of Africa still reflects the colonial past of the continent: it is possible to identify groupings of countries whose map base was begun by one of the colonial powers: Britain, France, Belgium, and to a lesser extent Portugal, Spain, Italy and South Africa, have all influenced the mapping standards that now prevail (see the table). This influence has also affected the surveying and mapping infrastructure and has been continued after independence through aid projects, when other western and eastern bloc countries have also used their technical expertise to establish or develop mapping standards.

The **British Directorate of Overseas Surveys (DOS)** (now Overseas Surveys Directorate of the Ordnance Survey, OSD) was responsible for the creation of much of the modern map base in former British colonies. Common mapping standards were adopted, and series were compiled to similar specifications and using similar numbering systems, with some variation between countries or groups of countries according to local needs. Aid continues to be given to some of these countries and the legacy of 1:50 000 basic scale mapping is being built on. Individual surveys have adopted a variety of innovations in the post-colonial phase, but similar specifications are usually followed, e.g. the East African agreement to compile compatible mapping. Geological and earth resources mapping of many areas in Africa was also carried out by the DOS. Much of this mapping is still available from the former DOS main agents Edward Stanford, and in some cases is the best available coverage.

Other British mapping activities included the War Office 1:500 000 scale East African series, compiled from 1940 onwards and revised until the 1960s. This map is still available but is no longer being updated.

The French on the other hand started the map bases of their former colonies in West, Central and Northern Africa. Small-scale series were compiled by **Institut Géographique National (IGN)** to cover large areas, including the 1:500 000 scale five-colour *Carte de l'Afrique de l'Ouest*. This series covered the French West African territories in 81 sheets and used 100 m contours for relief: stocks of the more southern sheets are still available in Paris. In the north and in Madagascar IGN disengaged following independence in the 1950s or 1960s but has continued to influence map design. Morocco has for instance recently revised its old-style French specification mapping to conform to the new French *Série orange* 1:50 000 specification. The western and central African francophone states had until relatively recently little surveying and mapping infrastructure and the IGN

Involvement in mapping by state

French influence	British DOS influence
Benin	Botswana
Burkina Faso	Ethiopia
Cameroon	Gambia
Chad	Ghana
Central African Republic	Kenya
Congo	Lesotho
Djibouti	Malawi
Gabon	Nigeria
Guinea	Sierra Leone
Ivory Coast	Somalia (parts)
Mali	Sudan
Mauritania	Swaziland
Niger	Tanzania
Senegal	Uganda
Togo	Zambia
	Zimbabwe
Postwar and prewar French influence	
Algeria	*Other British influence*
Madagascar	Egypt
Morocco	
Tunisia	
Former Portuguese influence	*Former Spanish influence*
Angola	Equatorial Guinea
Guinea Bissau	Parts of Morocco
Mozambique	
Former Italian influence	*South African influence*
Libya	Namibia
Parts of Somalia	
Former Belgian influence	*US influence*
Burundi	Liberia
Rwanda	
Zaire	

1:50 000 and 1:200 000 series begun in the late 1950s and 1960s and carried through to the 1970s remain the basic scale coverage of many countries. Much of this mapping is still available from IGN in Paris, but the individual countries are now, through aid projects, beginning to update coverage, though mostly keeping to the specifications and design set by the French. IGN has in the last 10 years concentrated upon the production of small-scale tourist maps of individual countries and town mapping of capital cities. Geological mapping of the French African territories was carried out by the **Bureau de Recherches Géologiques et Minières (BRGM)** and its predecessors: some small-scale coverage is still available on a continental or sub-continental scale but no large- or medium-scale programmes comparable to the IGN topographic projects have been carried out. A variety of resources mapping of French African territories was carried out in the colonial period by **Office de la Recherche Scientifique et Technique Outre Mer (ORSTOM)**. This has been continued into the post-colonial era. Ecological mapping of rangelands in French countries in the Sahel zone was carried out in the 1960s and 1970s by **Institut d'Elevage et de Médecine Vétérinaire des Pays Tropicaux (IEMVT)**.

Belgium established map series for its central African colonies and the mapping continues to be available from **Musée Royale de l'Afrique Centrale (MRAC)** in Tervuren. There was no small-scale continental coverage and little current work is being carried out.

South Africa established a modern 1:50 000 and 1:250 000 map base for Namibia and also provided the necessary infrastructure to establish a geological survey of the country. It also compiled small-scale coverage of much of Southern Africa including series mapping at 1:500 000 and 1:1 000 000 scales.

Other colonial powers have had a less lasting impact on the current mapping: Spain has helped Equatorial Guinea to create a modern map base, but has had little lasting influence elsewhere in the continent. Likewise Italy, in Libya and parts of what is now Somaliland. Portugal, too, has completely disengaged from the surveying and mapping of its former territories, while German colonial involvement came too early to have any effect on the modern African map base.

There have been a variety of responses to post-colonial mapping needs. Some states have continued to rely upon former colonial powers or another current aid donor to survey, compile and publish their mapping. Some have developed a mixed response, using technical aid to set up their own mapping infrastructure, and surveying, compiling and printing their own mapping. Some have used foreign contractors for parts of the map compilation process; some, like Tanzania, have used a wide variety of aid donors to continue to publish maps to the same specification established in the colonial past. By and large there have been insufficient resources available in Sub-Saharan Africa radically to alter the map base established in the postwar years.

Other extra African influences on the mapping of Africa have begun in the 1970s and 1980s. These included Polish aid to Libya and eastern bloc assistance to Angola and Mozambique. The Americans continue to assist Liberian topographic and geological map compilation. Small-scale maps of individual states have been published by the Russian **Glavnoe Upravlenie Geodezii i Kartografii (GUGK)**, American **Central Intelligence Agency (CIA)** and **Chinese State Publishing House (CPH)**. These maps are compiled to a very similar format of a physical-based general map of the country with thematic insets. GUGK also publishes general physical, economic, political and climatic maps of the continent.

There has been a variety of continent-wide programmes, some the result of cooperation between different agencies, some the work of one body. Two organizations have been responsible for coordinating cartographic activities on the African continent. The **United Nations Economic Commission for Africa** is responsible for the *Regional Cartographic Conferences for Africa*, which are held every three or four years. These present essential background papers about the state of the art of cartography in the continent and promote cooperation between the different survey organizations. The other major body is the **African Association of Cartography**, based in Algiers, which has 29 member states and was established in 1975 through joint initiatives from within the Organization for African Unity and the United Nations. Amongst its recent projects have been the sponsoring with the UN of a *Basic Inventory of Cartographic Data for Africa* and the establishment of 1986 as a Year of Cartography for Africa. An international hydrogeological programme has been initiated, and the association is working towards the standardization of cartographical and technical specifications. Another recently created agency with continent-wide interests is the **Regional Centre for Services in Surveying and Mapping and Remote Sensing (RCSSMRS)** in Nairobi.

Few of these continental programmes have established series on a continental scale. The major exception to this is the American **Defense Mapping Agency Topographic Center (DMATC)** 1:2 000 000 scale Series 2201, which gives continental cover in 36 sheets. This was started after World War II: sheets are revised on a regular basis and portray relief with 200 m contours. The map uses the International Map of the World polyconic projection and is printed in six colours.

A recent project by the **Technische Fachhochschule** in Berlin has seen the starting of a satellite image series using International Map of the World sheet lines and specifications to cover parts of the north of the continent. The International Map of the World is no longer being published for Africa by French, UK or US agencies: some survey organizations continue to issue sheets for their own territories in this series and the French IGN still has stocks in either ground or air editions of the sheets it compiled to cover West Africa. See the World section for details of this and other world series mapping.

An important cooperative thematic map series was started in the late 1960s by the **German Research Society**. This *Afrika Kartenwerk* uses four sample 1:1 000 000 quads, in the north, west, east and south: areas were selected to cut across national boundaries and to illustrate the variation of natural and human phenomena across the continent. Data is presented on up to 16 themes for each quad, maps and associated explanatory monographs being compiled by experts in the field and area. The first map was published in 1976. It is planned that there will eventually be 64 map sheets, with 60 accompanying monographs, in German language but with English and French summaries.

The **United Nations Educational and Scientific Organization (UNESCO)** publishes a variety of thematic mapping of the African continent, in association with various UN or other scientific agencies. These include a nine-sheet 1:5 000 000 scale tectonic map, published in association with the **Association des Services Géologiques Africains (ASGA)** and 1:10 000 000 scale metamorphic and mineral maps. A third edition of the 1:5 000 000 scale *Geological map of Africa* is underway; two sheets have been published and the remainder should be issued by the end of the decade. This series is published under the aegis of the **Commission for the Geological Map of the World (CGMW)**. Another joint project of UNESCO and CGMW is the compilation of a new 1:5 000 000 scale metallogenic map of Africa, the first sheet of which is currently being prepared. A vegetation map of the whole continent was compiiled and published by UNESCO with the United Nations Sudano-Sahelian Office (UNSO) and the Association pour l'Etude taxonomique de la Flore de l'Afrique Tropicale (AEFAT) in 1983. This map uses the AGS sheet lines and is published at 1:5 000 000 scale with a large English language explanatory monograph.

Other UNESCO world series mapping also covers the

African continent, e.g. *Soil Map of the World* at
1 : 5 000 000. See the World section for details of these
maps.

The **Food and Agriculture Organization (FAO)** of the
United Nations is currently engaged with the **United
Nations Environment Programme (UNEP)** in the
compilation of a 1 : 5 000 000 scale map to illustrate world
desertification—a pilot 1 : 5 000 000 coverage of Africa was
compiled for the twelfth session of the UNEP Council in
1984 and coverage of the north of the continent was
published as a three-sheet map in 1979. UNEP has itself
been involved in mapping projects in several states.
Another UN agency active in cartography in Africa is the
United Nations Development Programme: it has been
involved in the publication of a variety of development
mapping in a number of countries. Other small-scale
thematic mapping of the continent has been published by
the Russians, including a 1 : 5 000 000 scale nine-sheet
mineral resources map published in the early 1980s.

A number of commercial map producers have produced
general small-scale maps of the continent. Jeune Afrique,
now part of **Editions Jaguar**, published a very useful series
of national atlases during the 1970s. Twelve francophone
countries have been covered in this series: for details, see
the relevant country sections. They also issue what is
perhaps the best available general atlas of the continent
and a political map with hill shading revised on a regular
basis, which also carries the flags of countries and basic
statistical information about each. Other commercial map
publishers active in Africa include **Bartholomew, Haack,
Michelin, Kümmerly + Frey**, and **VWK**.

Further information

The best sources of information about African mapping are the
papers presented at the *United Nations Regional Cartographic
Conferences for Africa*, organized by the UN Economic
Commission for Africa. We have cited papers presented at the
three most recent conferences—4th, 1979 (Abidian), 5th, 1983
(Cairo) and 6th, 1986 (Addis Ababa)—but unlike other Regional
Cartographic Conference papers these are not brought together at
later dates and published by the UN.

For a useful single-article overview of the state of the art of
surveying and mapping in Africa see Bos, E.S. (1982) Mapping in
Africa: aspects of manpower and development, *ITC Journal*, **2**,
191–199.

For a historical bibliographic treatment see Norwich, O.I.
(1983) *Maps of Africa: An Illustrated and Annotated
Cartobibliography*.

For background information to the development of Afrika
Kartenwerk see Freitag, U. *et al.* (1979) Afrika-Kartenwerk:
development, content and design of a thematic mapping
programme for Africa set up by the German Research Society,
paper presented to the *4th United Nations Regional Cartographic
Conference for Africa*, Addis Ababa: United Nations Economic
Commission for Africa.

Addresses

**United Nations Economic Commission for Africa, National
Resources Division**
PO Box 3005, ADDIS ABABA, Ethiopia.

African Association of Cartography
BP 102, Hussein Dey, ALGIER, Algeria.

**Regional Centre for Services in Surveying and Mapping and
Remote Sensing (RCSSMRS)**
Kenya Commercial Bank Building, PO Box 18118, Enterprise
Road, NAIROBI, Kenya.

**United Nations Educational Scientific and Cultural
Organization (UNESCO)**
Division of Earth Sciences, 7 Place de Fontenoy, 75007 PARIS,
France.

Association des Services Géologiques Africains (ASGA)
c/o CIFAG, 103 rue de Lille, 75007 PARIS, France.

Commission for the Geological Map of the World (CGMW)
Maison de la Géologie, 77 rue Claude-Bernard, 75005 PARIS,
France.

Food and Agriculture Organization (FAO)
Via delle Terme di Caracalla, 00100 ROMA, Italy.

United Nations Environment Programme (UNEP)
PO Box 30552, NAIROBI, Kenya.

United Nations Development Programme (UNDP)
One UN Plaza, NEW YORK, NY 10017, USA.

For addresses of DOS, Bartholomew, MCERE and OPL see Great
Britain; for IGN, CNRS, BRGM, ORSTOM, IEMVT, Michelin
and Editions Jaguar see France; for MRAC see Belgium; for
Technische Fachhochschule, Gebrüder Borntraeger and VWK see
German Federal Republic; for Haack see German Democratic
Republic; for Kümmerly + Frey and Hallwag see Switzerland; for
DMATC and CIA see United States; for GUGK and MGUSSR
see USSR; for CHP see China; for ONIG see Algeria; for DSM
see South Africa; for Freytag Berndt see Austria.

Catalogue

Atlases and gazetteers

Afrika – Kartenwerk   edited on behalf of the German Research
Society
Berlin: Gebrüder Borntraeger, 1976–.
64 maps, *ca* 50 published.
With accompanying monographs.

Grand atlas du continent africain
Paris: Editions Jaguar, 1973 (1986).
pp. 336.
Also available as English language edition *The Atlas of Africa*.

Atlas of African affairs   I.L. Griffiths.
London: Methuen, 1982.
pp. 208.

General

Africa   1 : 12 000 000
Bern: Kümmerly + Frey, 1986.

Africa   1 : 10 000 000
Edinburgh: Bartholomew, 1984.

Afrika   1 : 10 000 000
Moskva: GUGK.
In Russian.

Afrika   1 : 9 000 000
Bern: Hallwag, 1986–87.
With accompanying placename index.

Afrika: fizicheskaja karta   1 : 8 000 000
Moskva: GUGK.
2 sheets, both published.
In Russian.
Relief base.

Africa south of the Sahara/Afrika suid van die Sahara
1 : 7 500 000   3rd edn
Pretoria: DSM, 1984.
Topographic map.
Also available as political edition.

Europe–Afrique   1:6 000 000
Paris: IGN, 1981.
2 sheets, both published.
French and English legend.

Africa   1:5 000 000
Edinburgh: Bartholomew, 1985.
3 maps, all published.

Michelin Afrique/Africa   1:4 000 000
Paris: Michelin, 1986.

Political

Afrique: carte politique   1:10 000 000
Paris: IGN, 1984.
Legend in English and French.

Afrique et moyen orient   1:10 000 000   2nd edn
Paris: Jeune Afrique, 1983.

Afrika: politicheskaja karta   1:8 000 000
Moskva: GUGK.
2 sheets, both published.
In Russian.

Topographic

Africa   1:2 000 000   Series 2201
Washington, DC: DMATC, 1968–.
36 sheets, all published. ■

Geological and geophysical

Metamorphic map of Africa/Carte metamorphique de l'Afrique   1:10 000 000
Paris: Commission for the Geological Map of the World, 1978.

International tectonic map of Africa   1:5 000 000
Paris: UNESCO and Association of African Geological Surveys, 1968.
9 sheets, all published.

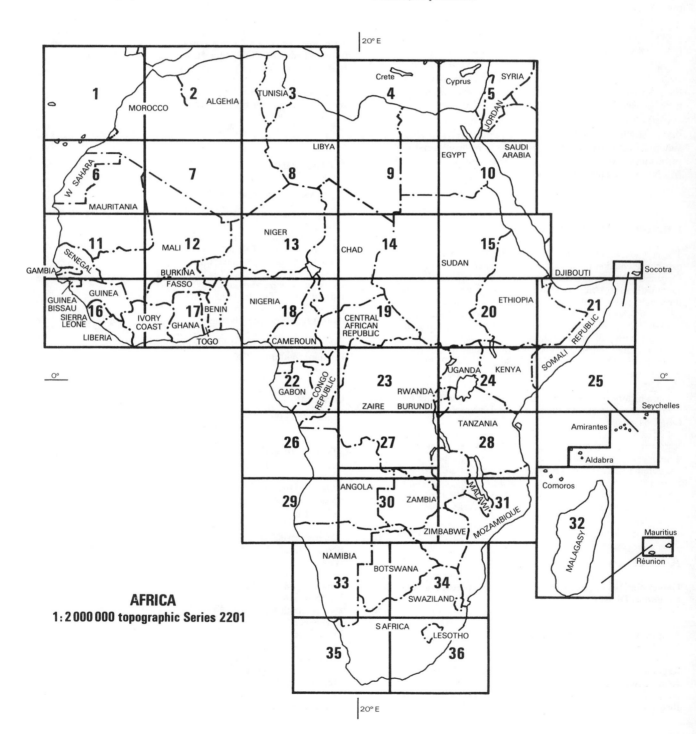

AFRICA
1:2 000 000 topographic Series 2201

International geological map of Africa   1:5 000 000   3rd edn
Paris: UNESCO and ASGA, 1985–.
6 sheets, 2 published.

Environmental

Klimaticheskaja karta Afriki   1:16 000 000
Moskva: GUGK.
In Russian.
Climatic map.

Mineral map of Africa   1:10 000 000
Paris: UNESCO and ASGA, 1969.
With accompanying explanatory brochure.

Vegetation map of Africa/Carte de végétation de l'Afrique
1:5 000 000
Paris: UNESCO, AEFAT and UNSO, 1983.
3 sheets, all published.
With accompanying explanatory text.

*Karta poleznych iskopaemych Afriki i Arabii/Map of the mineral
resources of Africa and Arabia*   1:5 000 000
Moskva: MGUSSR, 1982.
9 sheets, all published.
With accompanying Russian language explanatory text.

Metallogenic map of Africa   1:5 000 000
Paris: UNESCO and CGMW, 1986–.
6 sheets, 1 published.

Human and economic

Afrika: ekonomicheskaja karta   1:8 000 000
Moskva: GUGK.
2 sheets, both published.
In Russian.

CENTRAL AFRICA

Carte géologique de l'Afrique équatoriale et du Cameroun
1:2 000 000
Paris: BRGM, 1952.
3 sheets, all published.
With accompanying explanatory text.

*Carte géologique de reconnaissance des états d'Afrique équatoriale
française*   1:500 000
Paris: BRGM, 1950–60.
40 sheets, 11 published (see also Cameroon for sheets published
for that country).

EASTERN AFRICA

General

Ostafrika/East Africa/Afrique occidentale   1:2 000 000
Wien: Freytag und Berndt Artaria, 1979.

Topographic

East Africa   1:500 000
London: MCE RE, 1940–75.
130 sheets, all published.

NORTHERN AFRICA

Geological

Carte géologique du nord-ouest de l'Afrique   1:5 000 000
Alger: ONIG, 1978.

International Quaternary map of Africa   1:2 500 000
Paris: UNESCO, 1973.
3 sheets, all published.
Covers northern and western Africa to north of 20° N.

Carte géologique du Nord Ouest de l'Afrique   1:2 000 000
Paris: CNRS, 1962.
4 sheets, 2 available.

Environmental

Provisional methodology for soil assessment   1:5 000 000
Rome: FAO, 1979.
6 sheets, all published.
3 sheets in 2 series: present rate and state; and risks.

SOUTHERN AFRICA

General

Suidlike Afrika/Southern Africa   1:2 500 000
Pretoria: DSM, 1979.
General relief based map extending to 17° S.

Topographic

World aeronautical chart/Lugvaartkundige Wereldkaart   1:1 000 000
Pretoria: DSM, 1950–.
17 sheets, all published.

Suidlike Afrika/Southern Africa   1:500 000
Pretoria: DSM, 1977–.
33 sheets, 12 published. ■

WESTERN AFRICA

General

Zapadnaja Afrika   1:5 000 000
Moskva: GUGK, 1979.
With ancillary 1:15 000 000 scale thematic map and placename
index.

Nord und West Afrika/Africa north and west/Afrique nord et ouest
1:2 500 000
Obertshausen: VWK, 1983.
Road map with 6 language legend.

Topographic

Carte de l'Afrique de l'ouest   1:500 000
Paris: IGN, 1955–.
81 sheets, all published. ■

Geological and geophysical

Carte géologique de l'Afrique de l'Ouest   1:2 000 000
Paris: BRGM, 1960.
9 sheets, all published.
With a single accompanying explanatory text.

Carte géologique de reconnaissance de l'Afrique occidentale
1:500 000
Paris: BRGM, 1950–60.
26 sheets, 9 available.
With explanatory texts.

Environmental

*The West Africa oil and gas activity and concession
map*   1:2 000 000
Ledbury: OPL, 1982.
2 maps, both published.
Includes also Central Africa.

*Carte de planification pour l'exploitation des eaux souterrains de
l'Afrique soudana-sahelienne*   1:1 500 000
Orléans: BRGM, 1976.
3 sheets, all published.

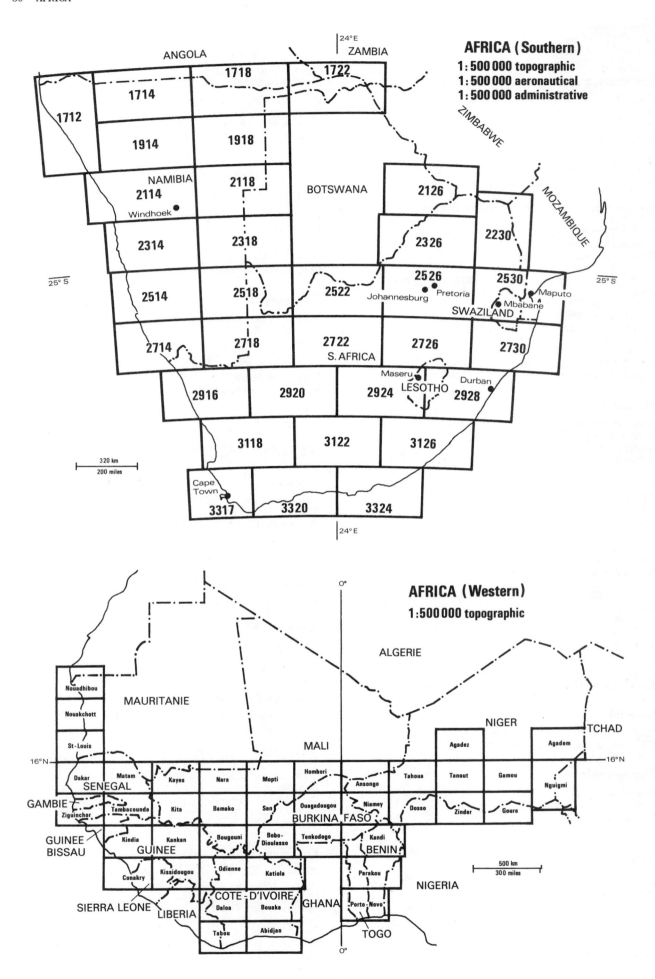

AFRICA (Southern)
1 : 500 000 topographic
1 : 500 000 aeronautical
1 : 500 000 administrative

AFRICA (Western)
1 : 500 000 topographic

Algeria

The **Institut National de Cartographie (INC)** has been the Algerian official survey organization since independence from France in 1962. There are three main topographic series, all of which have evolved from French series compiled by the **Institut Géographique National (IGN)** and its predecessors. The *1:25 000 Carte d'Algérie* is a four-colour map drawn on UTM projection, with 10 m contours. It is intended to cover the northern developed areas of the country in 1852 sheets. To date about 350 sheets have been published and the programme is continuing at a rate of about 20 new sheets a year. The *1:50 000 Carte d'Algérie* is more complete and will also cover areas north of 33° 30' N. Nearly 350 of the 524 sheets required are published in modern editions, five new sheets are issued a year and a further five are revised. It is planned to complete this series by 1995. Current sheets are drawn on Lambert conical conformal projection and use a 10 or 20 m contour interval, with hill shading. These two series are both now compiled from aerial coverage flown on a regular basis. The *1:200 000 Carte d'Algérie* gives nearly complete coverage of the country in a northern small-sheet series and a southern large degree sheet series. It is intended to complete the southern coverage by compiling the additional 24 sheets needed from satellite imagery. Both series use UTM grid and show relief by 50 m contour intervals and hill shading. Some 1:200 000 sheets have been partially revised in recent years using information from the 1:50 000 programme. A 1:500 000 map is also produced for the whole country based on subdivisions of the international numbering system.

Larger-scale programmes include 1:5000 and 1:1000 plans of reservoirs and development areas; many of these are produced as orthophotomaps. An urban mapping programme is also in operation involving the mapping at 1:2000 of more than 200 settlements: this has been carried out with Czechoslovak aid through the Algerian Ministry of Urbanization and Construction.

The availability of these officially produced topographic maps is currently in doubt. We have listed the three current series, as some stocks were advertised outside Algeria in the last 3 years, but these series and the larger-scale mapping are not likely to be available.

Geological and earth science mapping of Algeria is carried out by the **Office Nationale de la Géologie (ONIG)**, which was established in 1984 as part of the Ministère de l'Industrie Lourde. Prior to this date several different organizations within the Ministry had carried out earth science mapping activities, and French agencies were responsible for geological coverage prior to independence. A wide variety of different scales of geological mapping of the country has been produced. Series have been published at 1:500 000, 1:200 000, 1:100 000 and 1:50 000. None gives complete coverage. About 100 sheets in the 1:50 000 map are published, and these are accompanied by explanatory notes. Both 1:100 000 and 1:200 000 maps are published in blocks and cover large areas of the country. Neither is compiled on standard sheet lines; for instance, the 1:200 000 sheets are compiled on the new *Carte du Sahara* sheet lines for some areas, on old 1:200 000 sheet lines for other areas, and also as special sheets, unrelated to topographic map sheet lines. Only the 1:500 000 series gives nearly complete coverage; a French **Bureau de**

Recherches Géologiques et Minières (BRGM) produced map covers the southern Hoggar areas, whilst the rest of the country is covered in 16 sheets by ONIG maps.

The **Organisation Technique de Mise en Valeur du Sous Sol Saharien** carries out some soils mapping of the country.

Vegetation mapping of the country has been compiled by the French **Institut de la Carte Internationale du Tapis Végétal (ICITV)**. This includes 1:1 000 000 coverage on IMW sheet lines for part of Algeria (now available from CRBT in Algiers), together with 1:200 000 scale and 1:500 000 scale mapping of areas in the west of the country.

Amongst the other publishers of maps of Algeria are the American **Central Intelligence Agency (CIA)** and several western commercial houses such as **Hallwag** and **Michelin**. The best home-produced general map of the country is compiled with French aid by **SONATRACH**.

Further information

Institut National de Cartographie (1979, 1983 and 1986) Rapport national des activités. Presented to the United Nations Economic Commission for Africa *Regional Cartographic Conferences for Africa*, Addis Ababa: UNECA.
Catalogues are issued by ONIG.

Addresses

Institut National de Cartographie (INC)
BP 69, Hussein-Dey, ALGER.

Institut Géographique National (IGN)
136 bis rue de Grenelle, 75700 PARIS, France.

Office Nationale de la Géologie (ONIG)
18 Avenue Mustapha El Ouali, ALGER.

Bureau de Recherches Géologiques et Minières (BRGM)
BP 6009, 45060 ORLEANS-CEDEX, France.

Organisme Technique de Mise en Valeur du Sous Sol Saharien
ALGER.

Centre de Recherches Biologiques Tropicales (CRBT)
BP 812, Alger Gare, ALGER.

Société Nationale de Recherches et d'Exploitations Minières (SONATRACH)
10 rue du Sahara, Hydra, ALGER.

For ICITV and Michelin see France; for Hallwag see Switzerland; for CIA and DMA see United States.

Catalogue

Atlases and gazetteers

Algeria: official standard names approved by the United States Board on Geographic Names
Washington, DC: DMA, 1972.
pp. 754.

General

Carte touristique de l'Algérie   1 : 2 500 000
Alger: SONATRACH, 1976.

Algérie-Tunisie   1 : 1 000 000
Paris: Michelin, 1984.
French, English and Arabic legends.
Road map.
Includes insets at larger scales of Tunis, Algers, Tlemcen and
Oran regions and town sketches of Tunis and Alger.

Algeria-Tunisia road map   1 : 1 000 000
Bern: Hallwag.
In English, German, French and Italian.

Topographic

Carte d'Algérie   1 : 500 000
Alger: INC, 1961–.
43 sheets, 17 published.

Carte d'Algérie   1 : 200 000
Alger: INC, 1957–.
239 sheets, all published.

Carte d'Algérie   1 : 50 000
Alger: INC, 1960–.
463 sheets, *ca* 300 published. ■

Carte d'Algérie   1 : 25 000
Alger: INC, 1960–.
1852 sheets, *ca* 350 published. ■

Geological and geophysical

Carte géologique d'Algérie   1 : 500 000
Alger and Paris: ONIG and BRGM, 1951–.
25 sheets, all published.
16 northern sheets published by ONIG, 9 southern sheets
covering Hoggar by BRGM.

Carte géologique d'Algérie   1 : 200 000
Alger: ONIG, 1936–.
49 published sheets.

Carte géologique d'Algérie   1 : 50 000
Alger: ONIG, 1904–.
463 sheets, *ca* 100 published. ■
With accompanying explanations.

Environmental

Carte internationale du tapis végétal: Alger   1 : 1 000 000
Toulouse: ICITV, 1974.

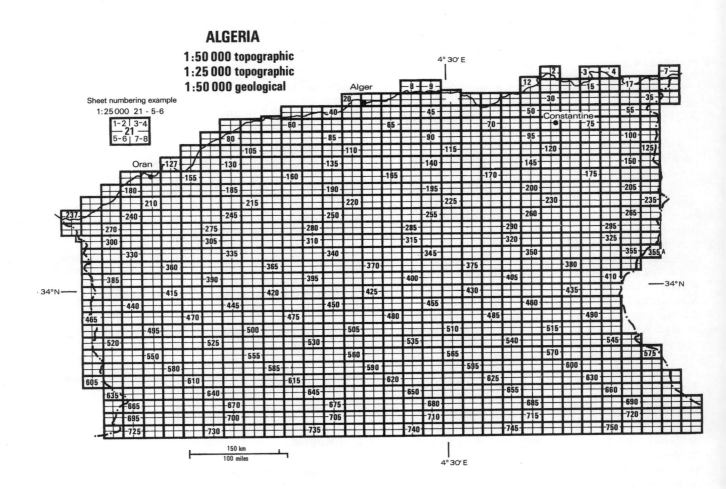

Angola

The national mapping agency in Angola is the **Direcção de Serviços de Geología e Minas (DSGM)**, which is responsible for the geodetic, photogrammetric, topographic and geological survey of the country. In the colonial period Portuguese mapping authorities produced a variety of map series of Angola, including a 1:100 000 scale topographic map: none are now available. There is little information about the post-independence topographic mapping of the country. A national atlas project was started in 1982 but is not available to the west. DSGM issued a catalogue in 1981 which describes the availability of geological mapping published at scales of 1:100 000 and 1:250 000. Coverage in these full-colour series is very patchy, and no sheets have been published since 1973. A recent general full-colour geological map of the country was published by the Portuguese **Instituto de Investigação Científica Tropical (IICT)** to cover Angola in four sheets at 1:1 000 000 scale. IICT has also compiled soils mapping at 1:500 000 scale in Cuenza Sul province.

Small-scale general maps of the country are available and published by the Russian **Glavnoe Upravlenie Geodezii i Kartografii (GUGK)**, the American **Central Intelligence Agency (CIA)**, the Hungarian official mapping distributor **Cartographia**, the Swedish commercial house **Esselte** and **Haack** in the German Democratic Republic.

Further information

Direcção Nacional de Geología e Industria Miniera (1981) *Tabela de Precos*.

Addresses

Direcção de Serviços de Geología e Minas (DSGM)
Caixo Postal 1260-C, LUANDA.

Instituto de Investigação Científica Tropical (IICT)
Rua da Junqueira 86, 1300 LISBOA, Portugal.

For GUGK see USSR; for CIA and DMA see United States; for Cartographia see Hungary; for Esselte see Sweden; for Haack see German Democratic Republic.

Catalogue

Atlases and gazetteers

Angola: official standard names approved by the United States Board on Geographic Names
Washington, DC: DMA, 1956.
pp. 234.

General

Angola 1:2 500 000
Moskva: GUGK, 1985.
With accompanying ancillary population, climatic and economic inset maps and name index.
Also available as Russian language edition.

Handy map of Angola 1:2 000 000
Budapest: Cartographia, 1986.

República Popular de Angola 1:1 400 000
Estocalmo: Ministerio de Educação and Esselte, 1982.
Rolled wall map.
Shows landscape types and administrative boundaries.

Geological and geophysical

Geológicade Angola 1:1 000 000
Lisboa: IICT, 1982.
4 sheets, all published.

Administrative

Angola 1:4 000 000
Gotha: Haack, 1984.
Administrative map with accompanying thematic insets and text.

Benin

The People's Republic of Benin (formerly Dahomey) was mapped by the French **Institut Géographique National (IGN)** which covered the country in its three West African series in the 1950s and 1960s. These maps use the usual French degree sheet lines, international sheet numbering system and UTM projection. The 1:200 000 scale map was completed in 1963 and shows relief with 40 m contours. This 19-sheet series was compiled from aerial coverage dating from the 1949–63 period. Some sheets were revised in the early 1970s, but there has been no systematic revision programme. The 1:50 000 scale map would require over 160 sheets for complete coverage. To date about 80 sheets have been published, mostly in the early

IGN period, to cover the southern areas. Relief is shown by 20 m contours. In 1978 the **Institut National de Cartographie (INC)** was created in Benin and assumed responsibility for geodetic, topographic and cadastral surveying and cartography. The major aim was to complete the 1:50 000 base map, and some new sheets have appeared as a result of a technical cooperation agreement with the Government of Nigeria. INC is, however, inadequately resourced to be able to reach this goal and relies upon foreign technical and financial aid. IGN produced a 1:500 000 two-sheet general map of the country in 1978 and a single-sheet 1:600 000 scale map in 1984. It has also been involved in a World Bank funded urban mapping programme to produce 1:2000, 1:5000 and 1:10 000 scale coverage of urban centres.

Geological surveying of Benin is the responsibility of the **Office Benois des Mines** but no new geological mapping of the country has been published and the best complete available coverage is still the 1:2 000 000 scale **Bureau de Recherches Géologiques et Minières (BRGM)** map issued in the 1960s as part of the *Carte géologique de l'Afrique Occidentale* (see West Africa).

Soil mapping of Benin has been carried out by the French **Office de la Recherche Scientifique et Technique Outre Mer (ORSTOM)**, including complete coverage with explanations at 1:200 000 scale, some larger-scale surveys of a few villages, and a single-sheet 1:1 000 000 scale map. ORSTOM has also compiled a bathymetric map of the continental shelf of Benin and Togo.

Further information

Moussedikou (Nadjim) (1979) Rapport sur la situation quant a l'équipement géographique en République Populaire du Benin, presented to the *4th United Nations Regional Cartographic Conference for Africa*, Addis Ababa: UN Economic Commission for Africa.

Addresses

Institut Géographique National (IGN)
136 rue de Grenelle, 75700 PARIS, France.

Institut National de Cartographie (INC)
Service Topographique et du Cadastre, BP 360, COTONOU.

Office Benois des Mines
Ministère de l'Equipement, BP 249, COTONOU.

Bureau de Recherches Géologiques et Minières (BRGM)
BP 6009, ORLEANS-CEDEX, France.

Office de la Recherche Scientifique et Technique Outre Mer (ORSTOM)
70–74 Route d'Aulnay, 93140 BONDY, France.

For GUGK see USSR; for DMA see United States.

Catalogue

Atlases and gazetteers

Dahomey: official standard names approved by the United States Board on Geographic Names
Washington, DC: DMA, 1965.
pp. 89.

General

Benin 1:1 000 000
Moskva: GUGK, 1986.
In Russian.
With ethnographic, climatic and economic insets and placename index.

République Populaire du Benin: carte générale 1:600 000
Paris: IGN, 1984.
With town plans of Porto Novo and Cotonou and administrative insets.

Topographic

République du Benin 1:200 000
Paris and Porto Novo: IGN and INC, 1945–.
19 sheets, all published. ■

République du Benin 1:50 000
Paris: IGN, 1950–.
160 sheets, *ca* 80 published. ■

Bathymetric

Fonds de pêche de long des côtes des républiques du Dahomey et du Togo 1:208 300 and 1:294 800   A. Crosnier and G.R. Berrit
Bondy: ORSTOM, 1966.
2 maps, both published.
Maps cover: bathymetric surveying tracks and bathymetry.

Environmental

Carte pédologique de la République du Dahomey 1:1 000 000
Bondy: ORSTOM, 1967.

République Populaire du Benin: carte pédologique de reconnaissance 1:200 000
Bondy: ORSTOM, 1975–78.
9 sheets, all published.
With 8 accompanying explanatory texts.

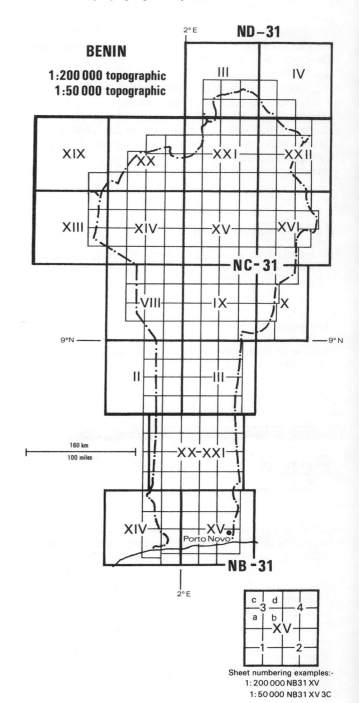

Botswana

There was very little mapping activity in Botswana before the end of World War II. The British **Directorate of Colonial Surveys** (now Overseas Surveys Directorate, Southampton, DOS/OSD) was responsible for the first planimetric surveys of Bechuanaland Territory in the early 1950s, which resulted in the publication of the 1:125 000 series still available for much of the Eastern part of Botswana. This series is described by the Botswana **Department of Surveys and Lands (DSL)** as 'first-stage mapping' and is almost entirely now superseded by more accurate and modern photogrammetrically compiled mapping. DOS established the national basic-scale series published at 1:50 000, on the Transverse Mercator projection and quarter degree sheet lines in the 1960s. Maps in this series use both UTM and South African Survey Grid. Conventional line maps in this series are published in five colours and include some cadastral information. Many were issued jointly by DSL and DOS. Areas of low relief and sparse settlement in the north have been published as photomaps, whilst DSL is issuing provisional monochrome sheets in an effort to speed the coverage of the series. The current state of the series is 27 per cent of the country covered in line maps, 17 per cent as photomaps and the remaining 56 per cent so far only taken as far as uncontrolled 1:70 000 photomosaics. It is intended that eventually the entire country will be covered in fully published mapping at this scale. A revision programme using 1982 flown aerial cover is underway at a rate of about 25 sheets a year. Full revision of the Gaborone and Lobatse areas is being undertaken with South African aid.

DOS has continued to aid DSL in mapping activities, but DSL is responsible for the publication of mapping and the geodetic, topographic and cadastral surveying of Botswana.

DSL and DOS have recently issued two large blocks of a photomap series at 1:100 000 for areas not covered by the 1:50 000 scale mapping.

The country is completely covered by maps published at 1:250 000. DOS issued a single sheet as a photomap, and eight further sheets have been published by DSL as conventional fully coloured maps with layer coloured relief. In the east of the country sheets have been published as Joint Operations Graphics and are available as either ground or air editions. Depending upon the reliability of the information, relief is shown by contours, form lines, hachures and colour tints. Information from Landsat images has been used in the production of the provisional monochrome sheets covering the sparsely settled Kalahari areas in the west and south-west of the country. A start was made in 1986–87 on the revision of all the maps in this series. A 1:500 000 scale series produced by DOS is currently under revision by DSL.

In addition to other smaller-scale topographic and general mapping, DSL is responsible for the compilation of a 1:500 000 administrative series showing the district divisions of the country.

DSL also publishes a wide variety of urban mapping; most settlements are mapped either as orthophoto plots or line maps at 1:2500, 1:5000 or 1:10 000 scale, and a special 1:25 000 sheet of Gaborone and surrounding area is under production.

Geological mapping of Botswana is carried out by the **Department of Geological Survey (DGS)**. The full-colour 1:125 000 series has been published for areas of the south and east of the country and new sheets are being issued to extend coverage into the desert. Sheets are issued with Bulletins or *District Memoirs* explaining the geology. The Survey has also issued gravity, hydrogeological and aeromagnetic series at smaller scales.

A soils mapping project has recently been started in Botswana. This is carried out by the **Department of Field Services (BMA)** of the Ministry of Agriculture and is intended to provide 1:250 000 soil coverage of the country. This project is funded by FAO and UNDP and was started in 1981. It is anticipated that 1:250 000 reconnaissance soil mapping and a 1:1 000 000 scale synthetic map will be completed by 1991. The mapping is available in limited colour editions, showing hydrology in blue with soil units separated by black lines.

Resource surveys were carried out by the British DOS for the **Land Resources Development Centre (LRDC)** in the 1960s, which provided 1:500 000 scale land facet, irrigation potential and vegetation mapping of the eastern, central and southern parts of the country. These maps are still available from LRDC or from OSD.

Other thematic mapping of Botswana has been carried out, but is no longer available. For instance, the **United Nations Development Programme** sponsored maps at small scales to cover land and water resource themes associated with special projects.

Small-scale general maps of the country have been compiled by both Russian **Glavnoe Upravlenie Geodezii i Kartografii (GUGK)** and American **Central Intelligence Agency (CIA)**.

Further information

Both DSL and DGS issue regular catalogues and annual reports.

Addresses

Overseas Surveys Directorate (DOS/OSD)
Ordnance Survey, SOUTHAMPTON SO9 4DH, UK.

Department of Surveys and Land (DSL)
Private Bag 0037, GABORONE.

Department of Geological Survey (DGS)
Private Bag 14, LOBATSE.

Department of Field Services, Ministry of Agriculture (BMA)
Private Bag 3, GABORONE.

For UNDP and FAO see World; for CIA and DMA see United States; for LRDC see Great Britain; for GUGK see USSR.

Catalogue

Atlases and Gazetteers

South Africa: official standard names approved by the United States Board on Geographic Names
Washington, DC: DMA, 1954.
2 vols.
Vol. 2 includes Bechuanaland (Botswana).

General

Botsvana 1 : 1 750 000
Moskva: GUGK, 1986.
In Russian.
With placename index and economic, climate and population insets.

The Republic of Botswana 1 : 1 500 000 5th edn
Gaborone: DSL, 1986.
Metric layer shaded relief.

World aeronautical chart ICAO 1 : 1 000 000
Causeway and Pretoria: Department of the Surveyor General
Zimbabwe and South Africa Government Printer, 1976.
6 sheets, all published.

Topographic

Republic of Botswana 1 : 500 000
Gaborone: DSL, 1965–69.
11 sheets, all published.
Diazo copies, relief by spot heights.

Botswana 1 : 250 000
Gaborone: DSL, 1974–.
41 sheets, all published. ■
Sheets published in 3 series: provisional monochrome; Joint
Operations Graphic available as either ground or air editions; and
conventional series.
Some only available in diazo form.

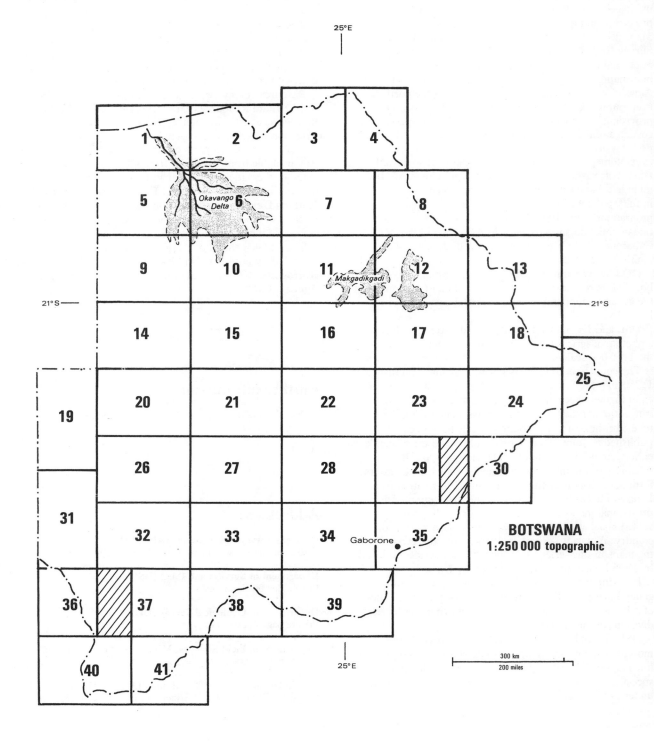

Botswana 1:125 000 DOS 621(Z 582)
Gaborone and Tolworth: DSL and DOS, 1955–66.
104 sheets. ■
Shows relief by formlines.

Republic of Botswana 1:100 000 DOS 547/1P
Tolworth and Gaborone: DOS and DSL, 1981–84.
39 sheets. ■
Photomap series covering parts of NE and NW of country.

Republic of Botswana 1:50 000 Z 761 (DOS 447)
Tolworth and Gaborone: DOS and DSL, 1967–.
856 sheets, *ca* 350 published. ■
Areas of low relief published as photomaps.

Geological and geophysical

Distribution of the Karoo system in Botswana 1:2 000 000
Lobatse: DGS, 1981.

Geological map of Botswana 1:1 000 000
Lobatse: DGS, 1984.
2 sheets, both published.

Photogeological map of Botswana 1:1 000 000
Lobatse: DGS, 1978.
2 sheets, both published.

Bouguer anomaly map 1:1 000 000
Lobatse: DGS, 1976.
2 sheets, both published.
Published with *The Gravity Survey of Botswana, 1972–3*, C.V.
Reeves and D.G. Hutchins.

*Mineral occurrences and metallogenic districts of Eastern
Botswana* 1:1 000 000 J. Baldock
Lobatse: DGS, 1977.

*Republic of Botswana: reconnaissance aeromagnetic
survey* 1:500 000
Lobatse: DGS, 1975–78.
35 sheets, all published.

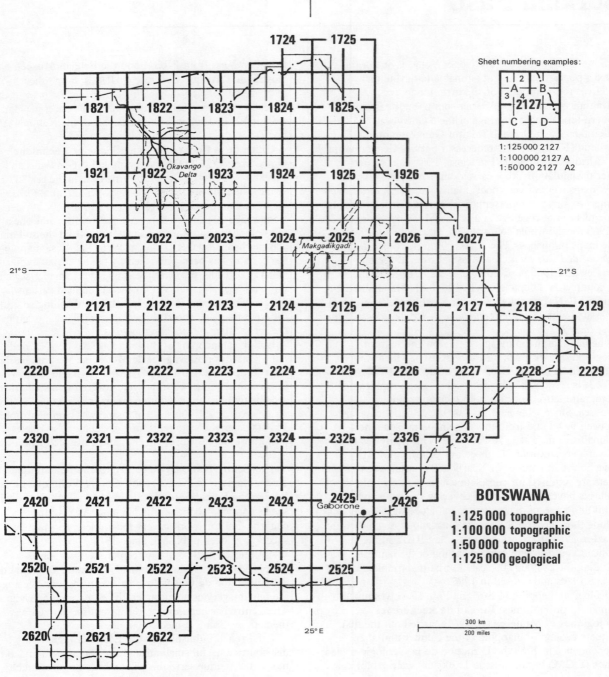

Sheet numbering examples:

1:125 000 2127
1:100 000 2127 A
1:50 000 2127 A2

BOTSWANA

1:125 000 topographic
1:100 000 topographic
1:50 000 topographic
1:125 000 geological

300 km
200 miles

Basement interpretation map of Botswana   1:500 000
Lobatse: DGS, 1978.
8 sheets, all published.

Superficial interpretation map of Botswana   1:500 000
Lobatse: DGS, 1978.
8 sheets, all published.

Quarter degree geological map series Botswana   1:125 000
Lobatse: DGS, 1963–.
104 sheets, *ca* 40 published. ∎
Recent sheets issued with explanatory texts.
Available either as multi-coloured or monochrome sheets.

Environmental

Botswana: soils map   1:250 000
Gaborone: BMA, 1985–.
41 sheets.

Administrative

Republic of Botswana: map of the districts
Gaborone: DSL, 1970–75.
8 sheets, all published.
Diazo prints.

Town maps

Gaborone: town centre   1:10 000
Gaborone: DSL, 1980.

Burkina Faso

Burkina Faso, until recently Upper Volta, is one of the
world's poorest nations and is unable to devote many
resources to the production of current mapping. Like
other francophone West African countries, the first
modern topographic coverage of the country was
undertaken by the French **Institut Géographique
National (IGN)** and its predecessor Service Géographique
de l'Afrique Occidentale Française. First- and second-
order triangulations were carried out from the 1930s but
early maps were of very poor quality and were unreliable
in their toponymic representation. Not until the provision
of regular aerial coverage at a scale of 1:50 000 from 1952
to 1960 for the whole country were any more adequate
base maps published. The 1:200 000 series *Carte de
l'Afrique de l'Ouest* which was issued from 1955 and
completed in 1960 is still the basic scale series for most of
the country, and depicts relief with 40 m contours in the
standard IGN four-colour format. Some sheets are revised
and issued with the title *République de Haute Volta/Burkina
Faso: carte 1:200 000*.

The national mapping agency is now the **Institut
Géographique de Burkina Faso (IGBF)**, which was
established as Institut Géographique de Haute Volta in
1976. It is responsible for geographic, cartographic and
photogrammetric work for both civil and military
purposes. Since 1979 new 1:50 000 scale aerial coverage
has been flown and from this is being compiled a new
national base map at 1:50 000 scale. This map uses IGN
style and specifications (four-colour printing, UTM
projection, 5 m contours). Nearly 400 quarter-degree
sheets are required for complete cover. To date only about
60 sheets have been compiled covering the main river
basins and areas of dense settlement. Published sheets
include those around the capital Ouagadougou which were
issued in 1984. IGN has been assisting IGBF in the
publication of this and other mapping of the country,
including the 1:1 000 000 scale tourist map published in
the late 1970s and revised in 1985.

Geological mapping of Burkina Faso also owes much to
French aid programmes. **Bureau de Recherches
Géologiques et Minières (BRGM)** carried out the first
adequate geological mapping of the country and after
independence in 1969 the **Direction de la Géologie et des
Mines (DGM)** began to issue 1:200 000 scale geological

maps, with aid from BRGM and the Belgian **Musée
Royale de l'Afrique Centrale (MRAC)**. Some sheets were
published as photogeological reconnaissance plots and the
series has not been completed. The single-sheet geological
map at 1:1 000 000 was issued as part of the National Atlas
in 1975.

**Office de la Recherche Scientifique et Technique
Outre Mer (ORSTOM)** has compiled various soils and
earth resources mapping of Burkina Faso in the last 30
years. The 1968–69 *Etude pédologique de la Haute Volta*
includes five soil maps published at 1:500 000, but derived
from 1:200 000 survey data. A series on the same sheet
lines, showing agronomic units derived from the soil map,
was published by ORSTOM in 1976. Land use mapping of
the country has been published by ORSTOM at 1:500 000
and 1:1 000 000, but only the single sheet smaller scale is
now still available. ORSTOM has also issued various larger
scale environmental mapping, usually publishing several
different thematic sheets for a single small area, often in
the series *Atlas des structures agraires*.

Ecological mapping of Burkina Faso has been carried
out by **Institut d'Elevage et de Médecine Vétérinaire des
Pays Tropicaux (IEMVT)**. The northern and north-
eastern zones of the country were covered in a set of maps
accompanying a three-volume report into grazing resources
in the area. 1:200 000 scale mapping was compiled from
1:50 000 aerial coverage to show vegetation (six sheets),
pastoral resources (three sheets), and cultivated land (six
sheets).

The **Centre National de la Recherche Scientifique et
Technologique (CNRST)** (now **Centre Voltaique de la
Recherche Scientifique**) began in 1968 to issue, as single
sheets with explanations, the National Atlas of the
country, but this work is very incomplete and no new
sheets have been published since 1975.

An international centre for remote sensing was
established in Ouagadougou in 1977 (**Centre Régional de
Télédetection de Ouagadougou**), with the aim of using
satellite imagery as an aid to earth resource monitoring.
This centre has begun to produce Landsat-derived
thematic mapping at medium scales.

The French publishing house **Editions Jaguar**
distributes a useful small atlas of the country, compiled in
the Jeune Afrique series in 1975, which contains much

thematic material not available elsewhere, and the American **Central Intelligence Agency (CIA)** publishes small-scale coverage of Burkina Faso.

Further information

Marchal, J.-Y. (1979) La cartographie et ses utilisateurs en pays africains à propos de la Haute Volta, *Cahiers ORSTOM, Série Sciences Humaines*, **XVI**(3), 261–272.

Addresses

Institut Géographique National (IGN)
136 bis rue de Grenelle, 75700 PARIS, France.

Institut Géographique de Burkina Faso (IGBF)
BP 7054, OUAGADOUGOU.

Bureau de Recherches Géologiques et Minières (BRGM)
BP 6009, 45060 ORLEANS-CEDEX, France.

Direction de la Géologie et des Mines (DGM)
BP 601, OUAGADOUGOU.

Musée Royale de l'Afrique Centrale (MRAC)
B 1980, TERVUREN, Belgium.

Office de la Recherche Scientifique et Technique Outre Mer (ORSTOM)
70–74 Route d'Aulnay, 93140 BONDY, France.

Institut d'Elevage et de Médecine Vétérinaire des Pays Tropicaux (IEMVT)
10 rue Pierre Curie, 94700 MAISONS-ALFORT, France.

Centre Voltaique de la Recherche Scientifique
BP 6, OUAGADOUGOU.

Centre Régional de Télédetection de Ouagadougou
BP 182, OUAGADOUGOU.

For Editions Jaguar see France; for CIA and DMA see United States.

Catalogue

Atlases and gazetteers

Atlas de Haute Volta
Ouagadougou: CNRST, 1968–.
Loose-leaf: includes both texts and maps; very incomplete.

Atlas de la Haute Volta   Y. Peron and V. Zalacain
Paris: Jeune Afrique, 1975.
pp. 48.

Upper Volta: official standard names approved by the US Board on Geographic Names
Washington, DC: DMA, 1965.
pp. 168.

General

Upper Volta   1:1 700 000
Washington, DC: CIA, 1968.
With population, economic and vegetation inset maps.

Haute Volta: carte touristique et routière   1:1 000 000
Ouagadougou: IGBF, 1985.
With administrative, population and Ouagadougou and Bobo-Dioulasso town plan insets.

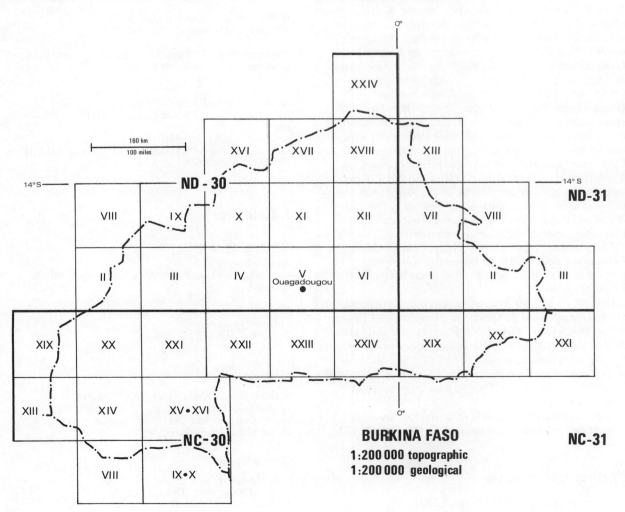

BURKINA FASO
1:200 000 topographic
1:200 000 geological

Topographic

République de Haute Volta: carte   1:200 000
Paris and Ouagadougou: IGN, 1949–.
35 sheets, 23 published. ■
Certain sheets being revised by IGBF.

Burkina Faso: carte   1:50 000
Paris and Ouagadougou: IGN, 1984–.
ca 400 sheets, 4 published.

Geological and geophysical

Carte géologique de la République de Haute Volta   1:1 000 000
Ouagadougou: DGM, 1976.
With text.

Carte géologique de Haute Volta   1:200 000
Paris and Ouagadougou: BRGM and DGM, 1967–.
35 sheets, 11 published. ■
More recent sheets produced by MRAC.

Environmental

L'occupation du sol en Haute Volta   1:1 000 000
Bondy: ORSTOM, 1970.
Single-colour map with text.

République de Haute Volta: carte pédologique de reconnaissance   1:500 000
Bondy: ORSTOM, 1968–69.
5 sheets, all published.

République de Haute Volta: resources en sol: carte des unités agronomiques deduites de la carte pédologique   1:500 000
Bondy: ORSTOM, 1976.
5 sheets, all published.

Burundi

The small central African state of Burundi was until independence a Belgian colony and was mapped in series similar in specification to Rwanda its near neighbour by Belgian survey organizations. A 52-sheet 1:50 000 map published by **Musée Royale de l'Afrique Centrale (MRAC)** and compiled in the 1930s was reprinted in the 1970s and was the basic scale series for most of this century. From it was derived a 1:100 000 scale map. Since independence responsibility for surveying and mapping activities in Burundi has been taken by the **Institut Géographique de Burundi (IGEBU)**. The first modern mapping of the country began recently under an aid project between IGEBU and the French **Institut Géographique National (IGN)**. This 42-sheet 1:50 000 series is compiled from 1980s aerial coverage, shows relief by 20 m contours, and has a bilingual French and Kirundi legend. Maps are drawn in IGN six-colour house style on a Gauss projection and each covers a quarter-degree square. The series is jointly published by IGN and IGEBU and was completed in the early 1980s. A 1:250 000 scale tourist and road map of the country was published in 1984, with 200 m contours, an inset town plan of the capital Bujumbura and an administrative inset map.

The National Atlas was published in 1981 by the **Association pour l'Atlas du Burundi** and shows the country in 30 thematic full-colour plates mostly at 1:750 000 scale.

Full-colour geological mapping of Burundi is carried out by the **Département de Géologie, Ministère de l'Energie et des Mines (MTPEM)**, in conjunction with MRAC in Belgium. A 1:100 000 scale series was begun in the 1970s. It is intended to cover the country in 13 sheets, each with an explanatory monograph: so far six have been published. MTPEM also compiles 1:500 000 scale geological and mineral maps.

Soils mapping of Burundi is the responsibility of the **Institut des Sciences Agronomique au Burundi (ISABU)**.

Further information

Catalogues are issued by MTPEM and IGEBU.

Addresses

Musée Royale de l'Afrique Centrale (MRAC)
Steenweg op Leuwen, B 1980, TERVUREN, Belgium.

Institut Géographique du Burundi (IGEBU)
BP 331, BUJUMBURA.

Institut Géographique National (IGN)
136 bis rue de Grenelle, 75700 PARIS, France.

Département de Géologie, Ministère des Travaux Publics de l'Energie et des Mines (MTPEM)
BP 745, BUJUMBURA.

Association pour l'Atlas du Burundi
3 Place des Chênes, La House Canejan, 33170, GRADIGNAN, France.

Institut des Sciences Agronomique au Burundi (ISABU)
BP 136, BUJUMBURA.

Catalogue

Atlases and gazetteers

Atlas du Burundi
Gradignan: Association pour l'Atlas du Burundi, 1979–81.
30 sheets.

Burundi: official standard names approved by the United States Board on Geographic Names
Washington, DC: DMA, 1964.
pp. 44.

General

Burundi: carte routière et touristique   1:250 000
Paris and Bujumbura: IGN and IGEBU, 1984.
Includes ancillary small administrative map, town plan of Bujumbura and name index.

Topographic

République du Burundi: carte au   1:50 000
Paris and Bujumbura: IGN and IGEBU, 1982.
42 sheets, all published. ■

Geological and geophysical

Carte géologique du Burundi 1 : 500 000
Bujumbura: MTPEM, 1981.

Carte métallogenique du Burundi 1 : 500 000
Bujumbura: MTPEM, 1981.

Carte géologique du Burundi 1 : 100 000
Bujumbura: MTPEM, 1975–.
13 sheets, 6 published. ■

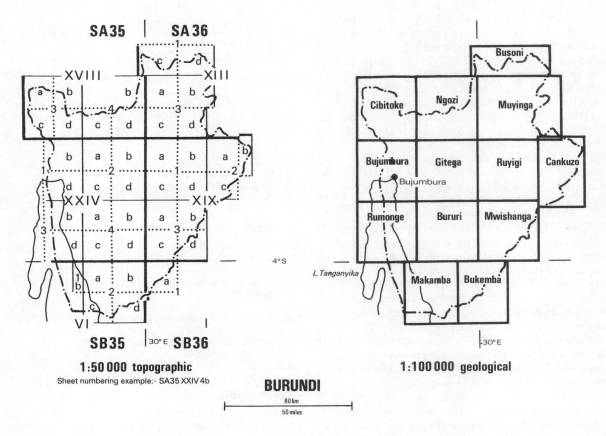

1:50 000 topographic
Sheet numbering example:- SA35 XXIV 4b

BURUNDI

80 km
50 miles

1:100 000 geological

Cameroon

The modern mapping of Cameroon was begun by the French **Institut Géographique National (IGN)** in the 1950s. The country was covered by three series entitled *Carte de l'Afrique Centrale*, which used UTM projection, were divided and numbered using the international sheet system, and were compiled from medium-scale aerial coverage of the country. The five-colour 1 : 200 000 scale series gave complete coverage in 55 sheets, and was completed in the early 1970s. Relief was shown using 40 m contours, though many sheets were first published as uncontoured Cartes Provisoires. Complete coverage was never reached in the 1 : 50 000 series. About 200 sheets out of a total of 667 were issued in this four-colour map and show relief by 20 m contours. Current responsibility for the surveying and mapping of the country falls to the **Centre Géographique National (ONAREST)** which has continued to produce maps in the same style as the IGN series, with the title *Carte de Cameroun*. The 1 : 200 000 map was regrouped in 1972 to avoid the largest border areas, into a 43-sheet series, with similar specifications to the IGN maps. Some sheets are therefore now larger than

degree squares. Revision of these maps is continuing. The 1 : 50 000 scale series has been extended, and many new sheets published. Amongst the other maps published by ONAREST is a 1 : 500 000 series covering the country in 10 sheets, which is revised on a regular basis and derived from the 1 : 200 000 series. Cooperation with IGN in Paris has resulted in the publication of smaller-scale general and road maps of the country and a variety of town plans.

Cadastral mapping of Cameroon is carried out by the **Direction du Cadastre**.

Geological mapping of the country is the responsibility of the **Direction des Mines et de la Géologie** but no current mapping is publicly available and the best complete geological coverage of the country is provided by the three-sheet 1 : 2 000 000 scale series published by the French **Bureau de Recherches Géologiques et Minières (BRGM)** in 1952. Some sheets are still available in the 1 : 500 000 scale series compiled as part of the *Carte Géologique de l'Afrique Centrale* by BRGM in the 1960s, and published under the title *Carte géologique de reconnaissance du Cameroun*.

The French **Office de la Recherche Scientifique et Technique Outre Mer (ORSTOM)** through its office in Yaoundé has produced many thematic maps of the country. Soils mapping at various scales is available, including a two-sheet 1:1 000 000 scale map giving complete coverage. The larger-scale maps on 1:200 000 and 1:50 000 topographic sheet lines are both very incomplete, but other significant local soil mapping projects have been carried out by ORSTOM, particularly in the north of the country, where a nine-sheet 1:100 000 series has been published. The National Atlas of Cameroon was compiled between 1960 and 1975 by the **Institut de la Recherche Scientifique et Technique** and is still available through ORSTOM. An orohydrographic map from the National Atlas is available as an individual sheet from ORSTOM. Another significant atlas project is the *Atlas Régional* series, which aimed to cover the country in 11 regional volumes of thematic maps and textual explanations. To date nine have been issued by ORSTOM, though two of these are already out of print. ORSTOM has also carried óut bathymetric mapping of the continental shelf and has published some geological coverage. The *Atlas des Structures Agraires* has been published by ORSTOM to cover four villages in different parts of the country and insect distribution mapping published in the 1950s and 1960s is still available.

A useful small atlas is published by the French commercial house **Editions Jaguar** in the series atlas Jeune Afrique, and the Russian **Glavnoe Upravlenie Geodezii i Kartografii (GUGK)** has published a general map of the country with thematic insets.

In the north of the country a significant mapping project was carried out in the 1970s for the **Commission du Bassin du Lac Tchad** by the British **Directorate of Overseas Surveys (DOS)** (now Overseas Surveys Directorate, OSD). This 1:50 000 photomap series was in part coordinated from offices in Maroua and covered the Chad Basin including parts of northern Cameroon. Sheets are still available from Edward Stanford or DOS/OSD and are listed in this book under Chad.

Addresses

Centre Géographique National (ONAREST)
BP 157, Ave. Monseigneur Vogt, YAOUNDÉ.

Institut Géographique National (IGN)
136 bis rue de Grenelle, 75700 PARIS, France.

Direction du Cadastre
Ministère de l'Urbanisme et de l'Habitat, BP 1614, YAOUNDÉ.

Direction des Mines et de la Géologie
Ministère des Mines et de l'Energie, BP 70, YAOUNDÉ.

Bureau de Recherches Géologiques et Minières (BRGM)
BP 6009, 45060 ORLEANS-CEDEX, France.

Office de la Recherche Scientifique et Technique Outre Mer (ORSTOM)
70–74 route d'Aulnay, 93140 BONDY, France.

Commission du Bassin du Lac Chad
BP 261, MAROUA.

For GUGK see USSR; for Editions Jaguar see France; for DMA see United States.

Catalogue

Atlases and gazetteers

Atlas du Cameroun
Yaoundé: Institut de la Recherche Scientifique et Technique, 1960–75.
16 parts.
Loose-leaf maps and explanations.

Atlas de la République Unie de Cameroun   G. Laclavère (Atlas Jeune Afrique)
Paris: Editions Jaguar, 1979.
pp. 72.
Also available as English language edition.

République Unie du Cameroun: Atlas Régionaux
Yaoundé: Institut de la Recherche Scientifique et Technique, 1965–.
11 vols, 9 published.

Cameroon: official standard names approved by the United States Board on Geographic Names
Washington, DC: DMA, 1962.
pp. 255.

General

Kamerun   1:2 000 000
Moskva: GUGK, 1985.
In Russian.
With placename index and climatic, economic and population inset maps.

République Unie de Cameroun/United Republic of Cameroon   1:1 000 000
Yaoundé and Paris: ONAREST and IGN, 1972.
2 sheets, both published.
Road map.

Topographic

Carte de Cameroun   1:500 000
Paris and Yaoundé: IGN and ONAREST, 1963–.
10 sheets, all published.

Carte du Cameroun   1:200 000
Yaoundé: ONAREST, 1971–.
47 sheets, all published. ■

Carte de Cameroun   1:50 000
Paris and Yaoundé: IGN and ONAREST, 1950–.
667 sheets, *ca* 250 published. ■

Bathymetric

Fond de pêches le long des côtes de la république fédérale du Cameroun   1:308 700
Bondy: ORSTOM, 1964.
2 sheets, both published.
Accompanying small-scale sheet shows location of dredging.

Geological and geophysical

Carte géologique de l'Afrique Équatorial et du Cameroun   1:2 000 000
Paris: BRGM, 1952.
3 sheets, all published.
With explanatory text.

Environmental

Schema de la répartition des Glossines   1:4 000 000
Bondy: ORSTOM, 1966.
Mosquito vector map.

Cartes des Glossines et Anophiles   1:2 000 000
Bondy: ORSTOM, 1953–55.
2 sheets, both published.
Mosquito distribution maps.

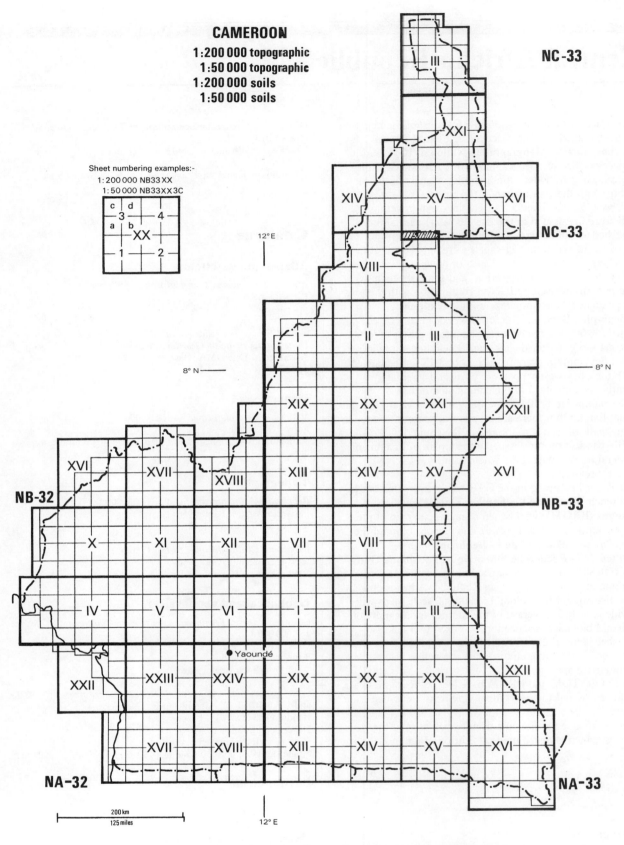

CAMEROON

1:200 000 topographic
1:50 000 topographic
1:200 000 soils
1:50 000 soils

Sheet numbering examples:-
1:200 000 NB33 XX
1:50 000 NB33 XX 3C

NC-33

NC-33

NB-32

NB-33

NA-32

NA-33

200 km
125 miles

Oro-hydrographie et réseau hydrometrique du Cameroun   1:2 000 000
Bondy: ORSTOM, 1984.

Carte pédologique du Cameroun   1:1 000 000
Bondy: ORSTOM, 1965–71.
2 sheets, both published.
With accompanying explanatory text.

Carte pédologique du Cameroun   1:200 000
Bondy: ORSTOM, 1974–.
47 sheets, 4 published. ■

Carte pédologique du Cameroun   1:50 000
Bondy: ORSTOM, 1968–.
667 sheets, 9 published. ■

Town maps

Yaoundé   1:10 000
Yaoundé: ONAREST, 1980.

Central African Republic

The French have been responsible for most of the mapping activities that have taken place in the Central African Republic. **Institut Géographique National (IGN)** and its predecessor Service Géographique de l'Afrique Centrale Française began modern topographic coverage of the country with the compilation of the *1 : 200 000 Carte de l'Afrique Centrale*. The 65 four-colour sheets covering the CAR were first issued in the late 1950s: 41 sheets were published as full editions showing relief with 40 m contours, the remaining 21 sheets were only issued at first as preliminary *fonds topographiques* or *fonds planimétriques*. This series uses UTM projection, and some sheets have been partially revised by IGN in conjunction with the post-independence survey organization the **National Topographic Service**. The series was originally based upon 1 : 50 000 aerial coverage. Some four-colour 1 : 50 000 scale maps on quarter-degree sheet lines showing relief with 20 m contours were published by IGN in the late 1950s for a few western parts of CAR and IGN has recently been involved in the compilation of new photogrammetric 1 : 50 000 mapping of the area around the capital Bangui, but there are no plans at present to extend coverage of this series to the rest of the country.

IGN has also recently issued a town plan of Bangui and a general tourist map of the whole country, which is regularly revised.

Geological mapping in the CAR has been carried out by the French **Bureau de Recherches Géologiques et Minières (BRGM)**, who issued a small-scale map of the whole country in 1952. It is now the responsibility of the **Direction des Mines et de la Géologie**.

Office de la Recherche Scientifique Technique Outre Mer (ORSTOM) has also carried out earth science mapping in the Central African Republic. A major current thematic series of full-colour 1 : 1 000 000 scale maps will provide two-sheet coverage of the country for a variety of earth and biological science topics. ORSTOM has also published some 1 : 200 000 scale soils mapping on the same sheet lines as the topographic series, as well as issuing gravimetric maps.

Both ORSTOM and BRGM have local offices in Bangui, but the maps and their accompanying texts are issued from France.

Jeune Afrique has issued a useful small atlas of the country, which is distributed by **Editions Jaguar**.

Addresses

Institut Géographique National (IGN)
136 bis rue de Grenelle, 75700 PARIS, France.

National Topographic Service
Base IGN de Bangui, BP 165, BANGUI.

Bureau de Recherches Géologiques et Minières (BRGM)
BP 6009, 45060 ORLEANS-CEDEX, France.

Direction des Mines et de la Géologie
BP 737, BANGUI.

Office de la Recherche Scientifique et Technique Outre Mer (ORSTOM)
70–74 Route d'Aulnay, 93140 BONDY, France.

For Editions Jaguar see France; for DMA see United States.

Catalogue

Atlases and gazetteers

Atlas de la République Centralafricaine   P. Vennetier
Paris: Jeune Afrique, 1984.
pp. 64.

Central African Republic: official standard names approved by the United States Board on Geographic Names
Washington, DC: DMA, 1962.
pp. 220.

General

République Centralafricaine   1 : 1 500 000
Paris: IGN, 1980.
With administrative map, town plan of Bangui and index to place names.

Topographic

République Centralafricaine: carte   1 : 200 000
Paris and Bangui: IGN, 1956–.
65 sheets, all published. ■

République Centralafricaine: carte   1 : 50 000
Paris and Bangui: IGN, 1950–.
1040 sheets, *ca* 80 published.

Geological and geophysical

Carte géologique de l'Afrique équatorial et du Cameroun
1 : 2 000 000
Paris: BRGM, 1952.
3 sheets, all published.
With French explanatory text.

Cartes gravimétriques de la République Centralafricaine   1 : 2 000 000
and 1 : 1 000 000   J. Albouy and R. Godivier
Bondy: ORSTOM, 1981.
3 sheets, all published.
With French explanatory text.
Single sheet 1 : 2 000 000 scale map, 2-sheet 1 : 1 000 000 scale map.

Environmental

Le milieu naturel physique Centralafricaine   1 : 1 000 000
Bondy: ORSTOM, 1983–.
10 sheets, 6 published.
Thematic series, 2 sheets for each theme with explanatory text in French.
Soils (1983), plant geography (1985), orohydrography (1985) so far issued. Forthcoming sheets to cover geomorphology and geology.

Carte pédologique de la République Centralafricaine   1 : 200 000
Bondy: ORSTOM, 1974–.
65 sheets, 3 published. ■

Town maps

Plan de Bangui   1 : 10 000
Paris: IGN, 1972.

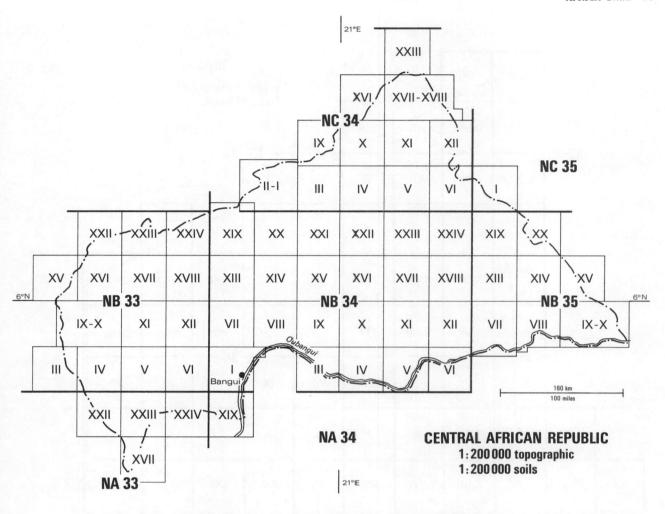

Chad

Chad has been mapped by the French **Institut Géographique National (IGN)** in the years since World War II. The country is completely covered by 128 sheets of the four-colour *Carte de l'Afrique Centrale* 1:200 000 which has sheet lines and numbering based on the international system, and uses a UTM projection. This series is derived from aerial coverage flown mainly in the early 1950s, but not all sheets are fully contoured editions. In the northern desert regions many of the published maps are provisional, either *fonds topographiques* or *fonds planimétriques*. Some revision of sheets has taken place in the 1970s to produce more *Cartes regulières* with 40 m contours. A few four-colour 1:50 000 sheets were published in the 1950s in the extreme south of the country and along the proposed line of the Chad–Cameroon railway. These used 20 m contour interval and UTM projection. The sale of all topographic mapping of Chad at scales greater than 1:1 000 000 is prohibited without prior permission from the Chadien embassy in Paris. Other maps produced by IGN and available in Paris include a 1:1 500 000 two-sheet general map, with partial revision to 1974. A town plan of the capital N'Djamena is no longer in print. The National Atlas, produced in association with the Institut National Tchadien in 1972, has recently once again become available. Current responsibility for mapping

and surveying activities in Chad is held by the **Direction du Cadastre et de la Topographie** in N'Djamena, but no new mapping is available and the activities of this organization are unknown.

There has been considerable cartographic work carried out on the Lake Chad Basin. The British **Directorate of Overseas Surveys (DOS)** produced a 1:50 000 photomap series for the **Lake Chad Basin Commission** in the late 1970s. These 59 sheets use the international numbering system and UTM projection. The French **Centre National de Recherche Scientifique (CNRS)** has compiled an important archaeological map of the basin and **ORSTOM** has published soil maps of the area.

Geological mapping in Chad has been carried out by the French **Bureau de Recherches Géologiques et Minières (BRGM)** whose 1:1 500 000 two-sheet full-colour geological map giving complete coverage of Chad is still available. Other more recent earth science mapping has concentrated upon the Tibesti region in the north, including an ongoing project investigating the geomorphology of the area, and recent interpretations of satellite imagery to produce geological mapping at 1:100 000 scale issued in the *Berliner Geowissenschaftlicher Abhandlungen* and published by **Dietrich Reimar**. No new geological mapping of the country has been carried out by

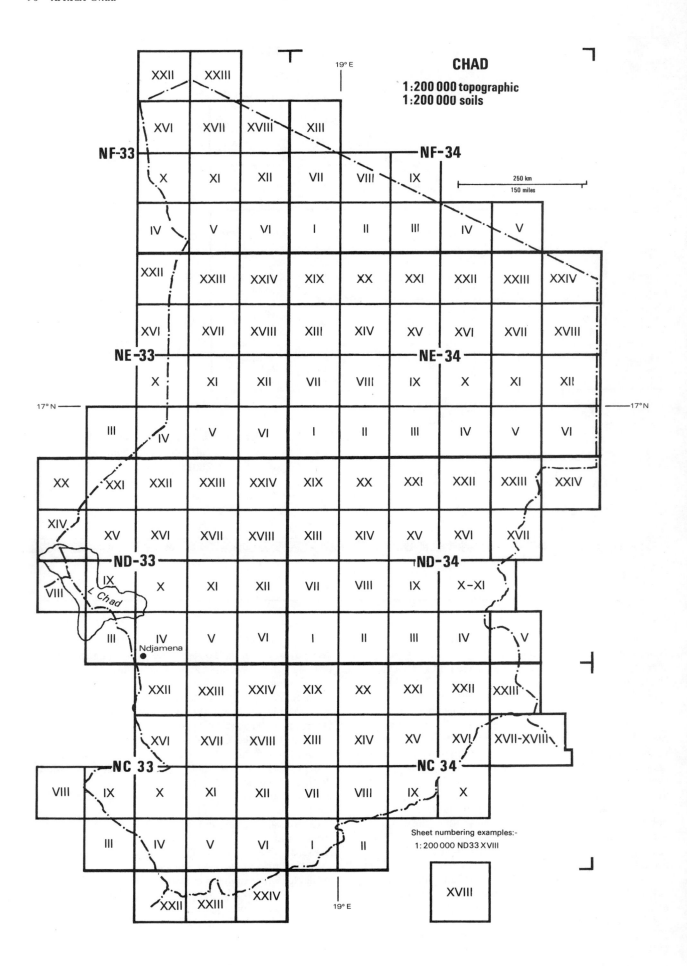

CHAD

1:200 000 topographic
1:200 000 soils

250 km
150 miles

Sheet numbering examples:-
1:200 000 ND33 XVIII

XVIII

Chadien organizations, but the responsible agency is the **Direction des Mines et de la Géologie** in N'Djamena.

The French **Office de la Recherche Scientifique et Technique Outre Mer (ORSTOM)** has undertaken a wide variety of thematic mapping of Chad. Soil maps at 1:1 000 000 give complete coverage and much of the south of the country is covered by 1:200 000 soil mapping published on topographic sheet lines. A six-sheet 1:1 000 000 scale gravimetric map was issued in 1968 and vegetation and hydrological maps covering southern areas have also been prepared by ORSTOM.

A useful small-scale Russian language reference map of the country is published by the Russian **Glavnoe Upravlenie Geodezii i Kartografii (GUGK)**.

Addresses

Institut Géographique National (IGN)
136 bis rue de Grenelle, 75700 PARIS, France.

Direction du Cadastre et de la Topographie (DCT)
N'DJAMENA.

Lake Chad Basin Commission (LCBC)
BP 261, MAROUA, Cameroon.

Overseas Surveys Directorate (DOS/OSD)
Ordnance Survey, Romsey Rd, Southampton SO9 4DH, UK.

Bureau de Recherches Géologiques et Minières (BRGM)
BP 6009, 45060 ORLEANS-CEDEX, France.

Direction des Mines et de la Géologie
BP 816, N'DJAMENA.

Office de la Recherche Scientifique et Technique Outre Mer (ORSTOM)
70–74 Route d'Aulnay, BONDY, France.

For GUGK see USSR; for Dietrich Reimar see German Federal Republic; for DMA see United States; for CNRS see France.

Catalogue

Atlases and gazetteers

Atlas pratique du Tchad
Paris: IGN, 1972.
pp. 77.

Chad: official standard names approved by the United States Board on Geographic Names
Washington, DC: DMA, 1962.
pp. 232.

General

République du Tchad: carte routière   1:1 500 000
Paris: IGN, 1974.
2 sheets, both published.

Chad   1:1 000 000
Moskva: GUGK, 1986.
In Russian.
With climatic, economic and ethnographic insets and placename index.

Topographic

République du Tchad   1:200 000
Paris and N'Djamena: IGN and DCT, 1957–.
130 sheets, all published. ■

République du Tchad   1:50 000
Paris and N'Djamena: IGN and DCT, 1950–.
1676 sheets, 33 published.

Lake Chad Basin/Bassin du Lac Tchad   1:50 000   LCBC 50P
Tolworth: DOS, 1978–79.
59 sheets, all published.

Geological and geophysical

Carte géologique de la République du Tchad   1:1 500 000
Paris: BRGM, 1962.
2 sheets, both published.

Carte gravimétrique de la République du Tchad   1:1 000 000
Bondy: ORSTOM, 1968.
6 sheets, all published.

Environmental

Carte pédologique du Tchad   1:1 000 000
Bondy: ORSTOM, 1970.
2 sheets, both published.
With accompanying explanatory text.

Carte pédologique de reconnaissance de la République du Tchad   1:200 000
Bondy: ORSTOM, 1964–.
130 sheets, 32 published. ■
With accompanying explanatory texts.

Congo

Prior to independence in 1960 Congo was part of French Equatorial Africa and was mapped by the French **Institut Géographique National (IGN)**. In January 1976 the **Institut Géographique National Congolaise (IGNC)** was established as the national mapping agency and continued the surveying and mapping activities initiated by IGN. The country is covered in 44 degree sheets in the 1:200 000 scale *carte de base*, which uses UTM projection and is published in either six- or four-colour editions. This map is derived from aerial coverage flown in the 1950s and 1960s and there is only a partial revision programme. The base map at 1:50 000 will require 570 sheets for complete

coverage—about 70 sheets were compiled by IGN in the colonial postwar period, and efforts are now being made to complete the series, with French and German aid, starting with areas of known natural resource potential. This map has 20 m contours, and uses four colours and UTM projection. IGNC has also compiled in joint projects with IGN town maps of the capital Brazzaville and of Pointe Noire: neither is currently in print.

Cadastral and large-scale mapping activities are carried out by the **Direction Topographique et du Cadastre (DCT)**.

Geological mapping is the responsibility of the **Direction**

des Mines et de la Géologie: no new mapping has been
published since independence. A 1:1 000 000 scale sheet
compiled by the French **Bureau de Recherches
Géologiques et Minières (BRGM)** covers the whole
country and is published with an explanatory text. Some
1:500 000 scale geological mapping covers Congo in the
equatorial African series compiled by the French in the
1950s (see Africa, Central).

The French **Office de la Recherche Scientifique et
Technique Outre Mer (ORSTOM)** has carried out a
variety of thematic mapping, including the production of
the National Atlas and atlas of Brazzaville, and some
incomplete 1:200 000 scale soils mapping in the extreme
south of the country. Sedimentological mapping of the
continental shelf has also been compiled.

Editions Jaguar distributes a useful small thematic atlas
published in the Jeune Afrique series.

Further information

Background information to the mapping of the Congo is contained
in articles prepared for the *4th UN Regional Cartographic
Conference for Africa* (1979): Progress report since the third
Conference in 1972: and Questions techniques relatives à la
cartographie de base. Addis Ababa: UN Economic Commission
for Africa.

Addresses

Institut Géographique National (IGN)
136 bis rue de Grenelle, 75700 PARIS, France.

Institut Géographique National Congolaise (IGNC)
BP 125, Route du Djove, BRAZZAVILLE.

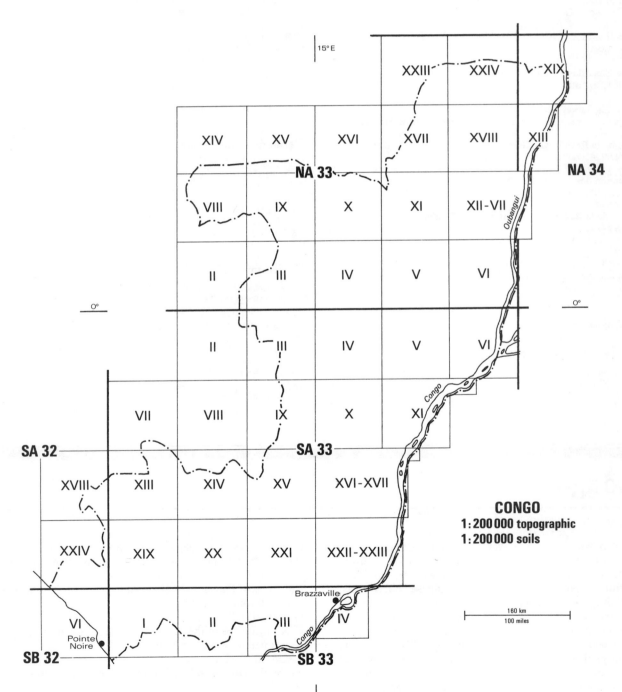

Direction du Cadastre et de la Topographie (DCT)
BP 125, Route du Djove, BRAZZAVILLE.

Direction des Mines et de la Géologie
BP 12, BRAZZAVILLE.

Bureau de Recherches Géologiques et Minières
BP 6009, 45060 ORLEANS-CEDEX, France.

Office de la Recherche Scientifique et Technique Outre Mer (ORSTOM)
70–74 Route d'Aulnay, 93140 BONDY, France.

For Editions Jaguar see France; for DMA see United States.

Catalogue

Atlases and gazetteers

Atlas du Congo
Bondy: ORSTOM, 1969.
10 sections, with accompanying explanatory texts.

Atlas de la République Populaire du Congo P. Vennetier,
G. Laclavère and G. Lasserre
Paris: Jeune Afrique, 1977.
pp. 64.

Congo: official standard names approved by the United States Board on Geographic Names
Washington, DC: DMA, 1962.
pp. 109.

Topographic

République du Congo: carte 1 : 200 000
Brazzaville: IGNC, 1950–.
44 sheets, all published. ■

République du Congo: carte 1 : 50 000
Brazzaville: IGNC, 1950–.
570 sheets, *ca* 100 published.

Bathymetric

Carte sedimentologique du plateau continental du Congo 1 : 200 000
Bondy: ORSTOM, 1980.
3 sheets, all published.
With accompanying explanatory text.

Geological and geophysical

Carte géologique du Congo-Brazzaville 1 : 1 000 000
Paris: BRGM, 1969.
With accompanying explanatory text.

Environmental

Carte pédologique du Congo 1 : 200 000
Bondy: ORSTOM, 1974–.
44 sheets, 5 published. ■
With accompanying explanatory texts.

Administrative

République populaire du Congo: organisation administrative
1 : 5 000 000
Brazzaville: IGNC, 1981.

Town maps

Atlas de Brazzaville R. Devauges
Paris: ORSTOM, 1984.
pp. 101.

Djibouti

The topographic mapping of Djibouti is currently largely obsolete. The French **Institut Géographique National (IGN)** carried out ground surveys in the 1940s and 1950s to produce the 12-sheet basic scale 1 : 100 000 map. This six-colour UTM gridded series showed relief by 25 m contour lines. A four-sheet 1 : 200 000 map was also derived from these surveys. There has been no updating of either series. After independence in 1977 the **Ministry of Public Works, Urban Planning and Housing (MPWUPH)** assumed mapping responsibilities in the country but little has been achieved and there are plans to establish a Central Topographic and Cartography Department, bringing together the different mapping interests in various Ministries. The French **Société Français de Travaux Topographiques et Photogrammétriques (SOFRATOP)** carried out mapping of six urban centres in the early 1970s. A photogrammetrically derived full-colour 1 : 10 000 scale plan of the capital Djibouti was published by IGN as a full-colour photomap in 1985 and a general map of the country was issued in 1983. For other general mapping of the country see Somalia and Ethiopia.

New mapping has been compiled notably by the **Institut Supérieur pour les Études et Recherches Scientifiques et Techniques (ISERST)** which is preparing a full-colour 1 : 100 000 scale 12-sheet geological map of the country, with French aid. Earlier sheets were published by **Centre d'Etudes Géologiques et de Développement, Bordeaux**, and distributed through **Bureau de Recherches Géologiques et Minières (BRGM)**. Later sheets are published by ISERST and distributed through **Office de la Recherche Scientifique et Technique Outre Mer (ORSTOM)**. ISERST is also making a 1 : 100 000 scale hydrogeological map of Djibouti with the cooperation of the German Federal Republic. A magnetic map was compiled by the **Institut du Globe de Paris** in 1983 and was distributed by BRGM.

Further information

For background information to the development of Djibouti mapping organizations and their current work see Republic of Djibouti (1983) Statement presented to the *5th United Nations Regional Cartographic Conference for Africa*, Addis Ababa: UN Economic Commission for Africa.

Addresses

Institut Géographique National (IGN)
136 bis rue de Grenelle, 75700 PARIS, France.

Topographic Section, Ministry of Public Works, Urban Planning and Housing (MPWUPH)
BP 11, DJIBOUTI.

Société Française de Travaux Topographiques et Photogrammétriques (SOFRATOP)
2 Ave. Pasteur, 94160 ST MANDE, France.

Institut Supérieur pour les Etudes et Recherches Scientifiques et Techniques (ISERST)
BP 486, DJIBOUTI.

Centre d'Etudes Géologiques et de Développement
Domaine Universitaire, 33415 TALENCE, France.

Office de la Recherche Scientifique et Technique Outre Mer (ORSTOM)
70–74 Route d'Aulnay, BONDY, France.

Bureau de Recherches Géologiques et Minières (BRGM)
BP 6009, 45060 ORLEANS-CEDEX, France.

For IGP see France.

Catalogue

General

République de Djibouti: carte touristique   1 : 385 000
Djibouti: MPWUPH, 1983.

Topographic

Térritoire française des Afars et Issas   1 : 100 000
Paris: IGN, 1948–61.
12 sheets, all published. ∎

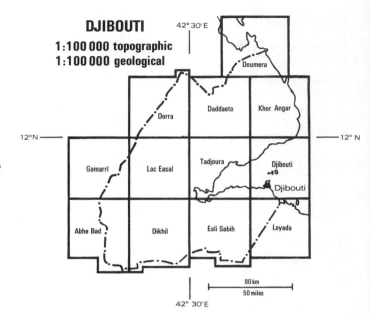

Geological and geophysical

Magnetic map of the Republic of Djibouti   1 : 250 000
Paris: Institut du Globe de Paris, 1983.

Carte géologique de la République de Djibouti   1 : 100 000
Djibouti: ISERST, 1974–.
12 sheets, 5 published. ∎

Town maps

Djibouti plan guide   1 : 10 000
Paris: IGN, 1985.
Photomap.
With 1 : 1 000 000 scale inset of whole country.

Egypt

The national mapping organization in Egypt was first brought together as long ago as 1889 and is currently the **Egyptian Survey Authority (ESA)**. It is now responsible for all civil topographic, geodetic and cadastral survey activities. Old medium- and small-scale series coverage of Egypt was never planned to cover the entire country; a 1 : 25 000 series had covered all cultivated land, and hachured 1 : 100 000 scale maps also covered the major settled areas, but these series were both inaccurate and outdated. Photogrammetric methods were introduced in 1954 and it was decided to compile five new series based on subdivisions of the International Map of the World and ranging from 1 : 25 000 to 1 : 1 000 000 in scale. There is now complete aerial coverage of the country in super-wide-angle photography in a scale range between 1 : 10 000 and 1 : 60 000, which is used to compile the new mapping. This programme includes a new series at 1 : 50 000 which now covers the agricultural areas of the country. This map is compiled at 1 : 25 000 and reduced for reproduction. A 1 : 100 000 scale map now covers about half the country, and there is complete coverage at 1 : 500 000 and 1 : 1 000 000 scales. The French Institut Géographique

National (IGN) has since 1975 been collaborating with ESA in a variety of photogrammetric and cartographic projects and there are plans for digital map production. Egyptian topographic mapping is probably not currently available for export, but a recently issued catalogue advertises 1 : 25 000 coverage, together with small-scale physical and administrative maps, geological coverage and town maps. We have not listed this material because of insufficient information and doubts about availability.

The **Military Survey Department** carries out military mapping of the country. The National Atlas of Egypt, compiled in the 1920s, is long out of print and there are no plans to publish a new edition.

The topographic survey of Sinai was carried out whilst the peninsula was under Israeli administration and maps at 1 : 250 000 and 1 : 100 000 scales were compiled to the same specifications as other **Survey of Israel** products. These maps are no longer available from Tel-Aviv.

Geological mapping of Egypt is the responsibility of the **Geological Survey and Mining Authority (EGSMA)**. Single-sheet 1 : 2 000 000 scale maps have been published to show geology, tectonics and mineral distribution.

Larger-scale earth science mapping has been compiled in the last decade by EGSMA in association with the **Egyptian Remote Sensing Center (ERSC)** in Cairo, which was set up in 1971 with American aid. EGSMA and ERSC have been particularly active in the production of multi-colour geological mapping derived from Landsat imagery. Series at 1:1 000 000 cover geology, structural lineaments and drainage and have been compiled in six sheets each on IMW sheet lines, from Landsat imagery and some field corroboration. 1:500 000 scale earth science series have also been compiled for much of the country. One-off thematic coverage of desertification, soil, land use, groundwater reserves, geomorphology, mineral resources, etc., has also been compiled from satellite imagery at a variety of scales between 1:1 000 000 and 1:250 000. Some of this mapping is available and distributed through EGSMA, including 1:500 000 and 1:250 000 scale quads.

The Sinai peninsula is covered in geological, geomorphological and aeromagnetic maps compiled by the **Geological Survey of Israel** and by the **Institute for Petroleum Research and Geophysics**. These maps are not available from Jerusalem or Holon.

Amongst the commercial mapping of the country has been town and tourist sheets issued by **Lehnart and Landrock**. Many western commercial cartographic houses have published useful small-scale maps of the country. These include **Kümmerly + Frey, Hallwag, Bartholomew, Mair, Ravenstein, Clyde Surveys, Geoprojects, Freytag-Berndt, Kompass** and **Karto+Grafik**.

Town plans of Cairo are also published by **Falk Verlag** and **Cartographia**, and planimetric urban coverage of Alexandria and Cairo in Arabic language editions is available from ESA.

Further information

El Shazly, E.M., Abdel Hady, M.A. and El Kassas, I.A. (1983) A review of space-borne imagery interpretation mapping in Egypt: paper presented to the *5th United Nations Regional Cartographic Conference for Africa, Cairo,* Addis Ababa: UN Economic Commission for Africa.

Egypt (1977) Egyptian national report: paper presented by Egypt *Technical papers presented to the 8th United Nations Regional Cartographic Conference on Asia and the Pacific,* New York: UN.

Egyptian Survey Authority (1986) *Catalogue,* Arabic language map listings.

Addresses

Egyptian Survey Authority (ESA)
308 El Haram St, Giza, Ormon, CAIRO.

Military Survey Department
Khalita El Mammoun St, Kobri El Kobba, CAIRO.

Cadastral Survey
18 Okasha St, Giza, ORMAN.

Geological Survey and Mining Authority (EGSMA)
3 Salah Salem St, Abbasia Post Office Bag, CAIRO.

Egyptian Remote Sensing Center (ERSC)
101 Kasr El Aini St, CAIRO.

Lehnart and Landrock
44 Sherif St, POB 1013/11511, CAIRO.

For Bartholomew, Clyde and Geoprojects see Great Britain; for Kümmerly + Frey and Hallwag see Switzerland; for Mair, Ravenstein, Kompass, Karto+Grafik see German Federal Republic; for Freytag-Berndt see Austria; for Geological Survey of Israel, Survey of Israel and Institute for Petroleum Research and Geophysics see Israel; for DMA see United States.

Catalogue

Atlases and gazetteers

Egypt and the Gaza strip: official standard names approved by the United States Board on Geographic Names
Washington, DC: DMA, 1959.
pp. 415

General

The Oxford map of Egypt 1:3 500 000
Beirut: Geoprojects, 1981.
Double-sided.
Includes 1:1 000 000 scale Nile Valley map and 1:1 500 000 Lake Nasser map; town plans of Cairo, and Alexandria; maps of Luxor and Giza Pyramids; businessman's guide and indexes.

Egypt: Hildebrands Urlaubskarte 1:1 500 000
Frankfurt AM: Karto+Grafik, 1985.
Covers areas to the north of Aswan.
Includes tourist notes and inset maps of Cairo area and of Egypt. Also available in English edition.

Egypt 1:1 000 000
Maidenhead: Clyde Surveys.
Double-sided.
Main map covers Nile valley to north of Aswan.
Includes indexed city maps of Cairo and Alexandria and inset maps of Luxor and Egypt.

Egypt 1:1 000 000
Wien: Freytag-Berndt.
Double-sided.
In English, French and German, with some Arabic placenames. Includes cultural guide on back.

Egypt 1:1 000 000
Edinburgh: Bartholomew.

Egypt 1:750 000
Bern: Kümmerly + Frey.
In English, French, German and Arabic.
Covers country to the north of Luxor.
Includes political inset map of whole country.

Geological and geophysical

Geologic map of Egypt 1:2 000 000
Cairo: EGSMA, 1981.

Tectonic map of Egypt 1:2 000 000
Cairo: EGSMA, 1967.
With accompanying Geological Survey explanatory paper.

Mineral map of Egypt 1:2 000 000
Cairo: EGSMA, 1979.
2 sheets, both published.
Single-colour maps.
Sheets cover: metallic mineral deposits and non-metallic mineral deposits.

Geological map 1:1 000 000
Cairo: EGSMA, 1979–.
6 sheets, 2 published.

Geologic map of Egypt 1 : 500 000
Cairo: EGSMA, 1978–.
15 sheets, 4 published.
Landsat image base.
Also available as metallogenic map.

Egypt: geological interpretation 1 : 500 000
Berlin: Technische Fachhochschule, 1979.
6 sheets published.

Town plans

Kairo 1 : 13 000
Hamburg: Falk.

Cairo 1 : 15 000
Cairo: Lehnart und Landrock.

Cairo
Budapest: Cartographia.

Equatorial Guinea

Equatorial Guinea comprises the former Spanish colonies
of Rio Muni and the islands of Fernando Po (now Bioko),
Pagalu and the Corisco group. They became fully
independent in 1968. Current mapping activities have been
carried out by the Spanish **Instituto Geográfico Nacional
(IGN)** for the **Servicio Geográfico** in Malabo, under an
aid agreement between the two countries. The programme
began in 1979 and involves the complete remapping of
mainland and island parts of the country. In the 1950s the
mainland had been mapped at 1 : 100 000 scale, but
representation of relief was very inadequate owing to
problems of obtaining aerial coverage. Using these maps as
a base, newly flown aerial cover supplemented by SLAR
radar coverage has been used to produce 1 : 100 000 scale
basic mapping of Rio Muni, and derived 1 : 200 000
coverage. A 1 : 50 000 scale photogrammetrically revised
map of Bioko has been published, and the basic scale for
Annoban is 1 : 10 000. There is no information about the
availability of these maps. A 1 : 750 000 scale four-colour
map of the whole country produced in 1979 and two other
single-colour maps are available from IGN.

Further information

Spain (Instituto Geográfico Nacional) (1983) Trabajos del Instituto
Geográfico Nacional de Madrid en la República de Guinea
Ecuatorial, *Bol. Inf. Servicio Geográfico del Ejército*, **56**, 11–18.

Addresses

Instituto Geográfico Nacional (IGN)
General Ibanez de Ibero 3, MADRID 3, Spain.

Servicio Geográfico
MALABO, Bioko.

Catalogue

General

República de Guinea Ecuatorial 1 : 750 000
Madrid: IGN, 1979.

República de Guinea Ecuatorial 1 : 400 000
Madrid: IGN, 1979.
Single-colour map.

*Mapa de la República de Guinea Ecuatorial (provincia de
Bioco)* 1 : 100 000
Madrid: IGN, 1979.

Ethiopia

The **Ethiopian Mapping Agency (EMA)** is the principal
national surveying and mapping organization. Its
responsibilities include aerial surveys, land survey,
mapping and research and development work. Prior to
1952 mapping of Ethiopia was carried out by various
colonial authorities, described in the latest available EMA
catalogue as 'unauthorized aliens'. Bilateral cooperation
with overseas surveys allowed the beginnings of national
medium-scale topographic coverage. After the revolution
of 1974 EMA began to produce modern photogrammetrically derived mapping with Russian aid. A
multi-colour *1 : 250 000 series* now gives complete coverage
of the country in 93 sheets. This series uses UTM
projection and shows relief by 100 m contours. It was

compiled in the 1970s from 1 : 50 000 wide-angle aerial
photography flown between 1963 and 1967. The central
area of the country has been mapped at 1 : 50 000 in a joint
project with the **British Directorate of Overseas Surveys
(DOS)** (now Overseas Surveys Directorate, Southampton,
OSD). This 1 : 50 000 series is being extended to the rest of
the country and has been adopted as the national standard
series. Over 200 sheets have now been published out of a
total of 1700 needed for complete coverage, and new
1 : 50 000 scale aerial coverage has been flown for high-
priority areas. 20 m contours are used. Both of these series
are only available to government-approved projects. Other
smaller-scale series produced by EMA include a
1 : 1 000 000 scale map and a multi-colour 1 : 2 000 000 scale

general map of the country (with 100 m contours and administrative overprint); both are revised on a regular basis. EMA also produces smaller-scale thematic mapping, including the preliminary black-and-white edition of the National Atlas which includes 91 pages of mostly 1:4 000 000 scale thematic maps of the country. A new multicolour edition of the National Atlas and a pilot regional atlas are currently being prepared. Administrative mapping in Amharic script editions and large-scale plans of 100 towns and 21 development projects are also published and a city guide map is issued for Addis Ababa.

All of these maps are listed in the 1983 Catalogue of Mapping but most are no longer available for distribution abroad.

Very little geological and earth science mapping is available for Ethiopia. The 1:250 000 scale geological survey of the country was begun in the early 1970s but is not now available. The **Ethiopian Institute of Geological Survey (EIGS)** is responsible for this work and other indigenous geological surveys, including recent mapping of the Omo river area. This was published in 1983 as 1:500 000 scale mineral and geological mapping, together with nine 1:250 000 scale regional mineral maps and an explanatory text. The Italian **Consiglio Nazionale delle Ricerche (CNR)** has produced a small-scale geological map of the country (distributed through **Pergamon Press**) and French research by **Centre National de la Recherche Scientifique (CNRS)** has resulted in some geological mapping of the rift valley and Afar areas.

The British DOS compiled resources mapping for the **Land Resources Development Centre (LRDC)** for several areas of the country. This has included medium-scale forest inventory mapping of areas in the south-west of the country, irrigation potential mapping of Lake Zwai and 1:1 000 000 scale mapping of the Southern rift valley to show development prospects. All these maps are still available from LRDC. The Swiss **University of Berne** has also compiled resources mapping of the Simien mountains.

Amongst the other mapping agencies in the country are the **United Nations Economic Commission for Africa (UNECA)**, whose offices are located in Addis.

Several foreign agencies have compiled small-scale general coverage: these include **Haack**, the Russian **Glavnoe Upravlenie Geodezii i Kartografii (GUGK)** and the American **Central Intelligence Agency (CIA)**.

Further information

Ethiopian Mapping Agency (1983) *Catalogue of mapping*, Addis Ababa: EMA.
Ethiopian Mapping Agency (1987) Cartographic activities in Ethiopia, paper submitted to the *11th United Nations Regional Cartographic Conference for Asia and the Pacific*.

Addresses

Ethiopian Mapping Agency (EMA)
PO Box 597, ADDIS ABABA.

Overseas Surveys Directorate (OSD/DOS)
Ordnance Survey, Romsey Rd, SOUTHAMPTON SO9 4DH, UK.

Ethiopian Institute of Geological Survey (EIGS)
Ministry of Mines, Energy, and Water Resources, PO Box 486, ADDIS ABABA.

Consiglio Nazionale delle Ricerche (CNR)
Piazzale Aldo Moro 7, 00185 ROMA, Italy.

United Nations Economic Commission for Africa (UNECA)
Africa Hall, PO Box 3001, ADDIS ABABA.

For LRDC and Pergamon Press see Great Britain; for CIA and DMA see United States; for GUGK see USSR; for Haack see German Democratic Republic; for CNRS see France; for University of Berne see Switzerland.

Catalogue

Atlases and gazetteers

Ethiopia: official standard names approved by the United States Board on Geographic Names
Washington, DC: DMA, 1982.
pp. 663.

General

Ethiopia 1:4 070 000
Washington, DC: CIA, 1976.
With population, economic and vegetation insets.

Äthiopien 1:4 000 000
Gotha: Haack, 1985.

Efiopia 1:2 500 000
Moskva: GUGK, 1986.
In Russian.
With population, economic and administrative insets.

Ethiopia 1:2 000 000
Addis Ababa: Ethiopian Tourist Commission and EMA, 1985.

Geological and geophysical

Geological map of Ethiopia and Somalia 1:2 000 000
Firenze: CNR, 1973.
Title also in Arabic and Amharic.
With explanatory text and accompanying 1:3 000 000 scale landform map.

Town Maps

Addis Ababa city guide map 1:20 000
Addis Ababa: EMA, 1980.

Gabon

Gabon was mapped by the French in their *Cartes d'Afrique Centrale* series, and the French influence on current mapping of the country is still high. After independence the Service Topographique et du Cadastre, now **Institut National de Cartographie (INCG)**, assumed responsibility

for topographic and planimetric mapping and cadastral survey work, which it carries out with French financial and technical aid. The 1:200 000 series compiled by the French **Institut Géographique National (IGN)** is still the basic scale: its 33 sheets are not all derived from aerial coverage, or based on accurate ground survey and levelling, because of atmospheric difficulties and dense vegetation. As with other French inspired series of Africa this map uses UTM projection and shows relief with 40 m contours where a full edition has been published. Sheets in this four-colour *carte de base* are now being published as full editions to replace the preliminary editions or *fonds planimétriques*; new maps are appearing at a rate of about a sheet a year. Since 1967 INCG has been involved in a 1:50 000 mapping programme. The aim is to cover the whole of Gabon at this scale; early sheets in this series were compiled in the 1950s by IGN, and effort is now concentrated upon the publication of new maps for areas of economic importance. To date 1:50 000 photogrammetric mapping of about two-thirds of Gabon is in progress and about 120 sheets have so far been published. This map is also four-colour and UTM based and shows relief by 20 m contours.

Other maps produced by INCG in collaboration with IGN include large-scale orthophotoplots of the capital area, town maps of Libreville and a general map of the country.

The National Atlas was prepared by the **Laboratoire National de Cartographie** and appeared in three instalments in the late 1970s. Most maps were published in this work at a scale of 1:2 000 000, but are not currently available. Another officially produced atlas was published in 1983 by the **Institut Pédagogique National**—we have listed this work but have not been able to assess its quality.

Geological mapping of Gabon is carried out by the **Service Géologique du Gabon (SGG)**, part of the Ministère des Mines et de la Géologie. Following a mineral resources assessment in 1971 it was decided to use SLAR radar techniques to produce national geological maps at 1:200 000 and 1:1 000 000. High-resolution imagery flown in 1981 to cover the whole country was interpreted and field checked, and supplementary geophysical surveys were carried out by the end of 1984. It is intended to issue 1:200 000 scale maps on the 1:200 000 *carte de base* sheet lines, with explanations, but the 1950s French *Carte géologique de l'Afrique Centrale 1:500 000* scale is still the best available geological coverage of the country. (For this map see Africa, Central.)

Soils mapping of Gabon is carried out by the French **Office de la Recherche Scientifique et Technique Outre Mer (ORSTOM)**. 1:2 000 000 maps give complete cover and are published with explanatory texts, and several sheets have been compiled in a 1:200 000 scale soil series. Earlier ethnic and entomological mapping is no longer available.

Further information

Various papers presented to UN Regional Cartographic Conferences for Africa provide useful background information about the mapping of Gabon: (1979) Cartographie du Gabon and Atlas du Gabon, *4th UN Regional Cartographic Conference for Africa*, Addis Ababa: UN Economic Commission for Africa; (1983) The status of 1:50 000 mapping of Gabon, *5th UN Regional Cartographic Conference for Africa*, Addis Ababa: UN Economic Commission for Africa.

Addresses

Institut National de la Cartographie (INCG)
BP 13600, LIBRÈVILLE.

Institut Géographique National (IGN)
136 bis rue de Grenelle, 75700 PARIS, France.

Laboratoire National Cartographique
Service de la Documentation et de la Cartographie au Ministère du Plan, Ermitage Ste Anne, BP 277, LIBREVILLE.

Institut Pédagogique National
BP 6, LIBREVILLE.

Service Géologique du Gabon (SGG)
Ministère des Mines et des Hydrocarbures, BP 576, LIBREVILLE.

Office de la Recherche Scientifique et Technique Outre Mer (ORSTOM)
70–74 Route d'Aulnay, 93140 BONDY, France.

For DMA see United States.

Catalogue

Atlases and gazetteers

Géographie et cartographie du Gabon: Atlas illustré
Libreville: Institut Pédagogique National, 1983.
pp. 135

Gabon: official standard names approved by the United States Board on Geographic Names
Washington, DC: DMA, 1962.
pp. 113.

General

Gabon   1:1 000 000   2nd edn
Paris: IGN, 1975.
Includes inset map of Libreville.

Topographic

République Gabonaise   1:200 000
Libreville: INCG, 1958–.
33 sheets, all published. ■

République Gabonaise   1:50 000
Libreville: INCG, 1951–.
360 sheets, 120 published. ■

Environmental

Les sols de Gabon cartes à 2 000 000
Bondy: ORSTOM, 1981.
2 sheets, both published.
Map titles: carte pédologique and carte des resources en sols.
With accompanying monograph.

Carte pédologique reconnaissance du Gabon   1:200 000
Bondy: ORSTOM, 1969–.
32 sheets, 6 published. ■

Town maps

Libreville et ses environs   1:20 000
Paris: IGN, 1983.

Plan de Libreville   1:10 000
Paris: IGN, 1977.

GABON

1:200 000 topographic
1:50 000 topographic
1:200 000 soils

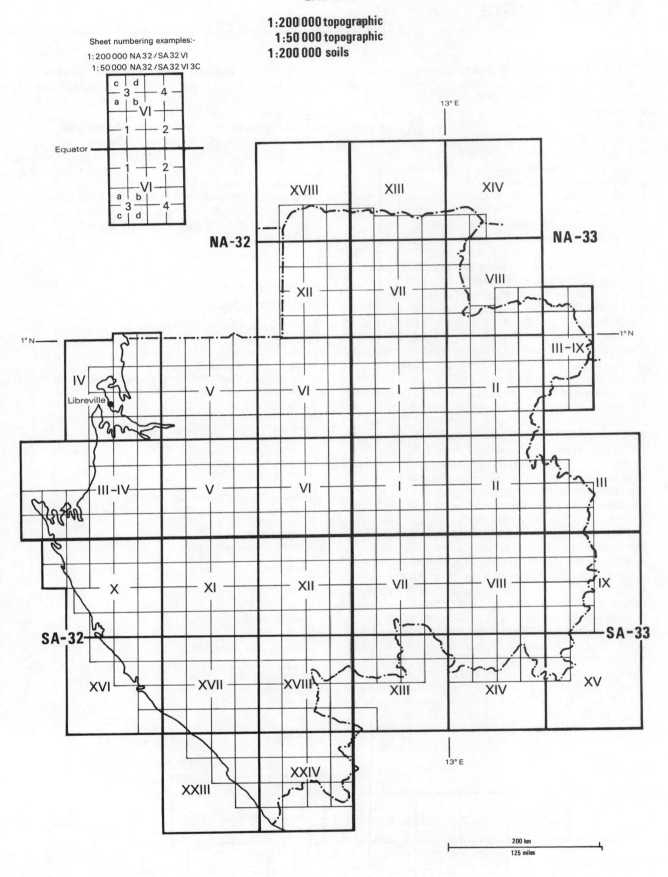

Sheet numbering examples:-
1:200 000 NA 32 / SA 32 VI
1:50 000 NA 32 / SA 32 VI 3C

The Gambia

The Gambia is a linear West African country extending 150 miles inland from the coast along the eponymous river. The **Survey Department** has responsibility for surveying and mapping activities including production of both topographic and cadastral maps. Current map series were compiled with technical aid from Britain and Germany: photomaps were adopted as the best means of portraying the flat topography of the Gambia and of establishing very rapidly a modern map base. The basic scale for the whole country is a 526-sheet diazo land use photomap produced by the German agency for Technical Cooperation with other German organizations in 1983 for the Gambian Government. Urban growth centres were mapped in 44 full topographic 1:10 000 editions, published in 1984. The British **Directorate of Overseas Surveys (DOS)** (now Overseas Surveys Directorate, Southampton, OSD) compiled a 20-sheet 1:50 000 scale photomap series DOS 415P in the 1970s. The programme was based on 1972 flown aerial photography, was completed in 1976 and all sheets in the series were published by the end of 1977. The multi-colour map uses the Transverse Mercator projection, and hills are shown by hachures and spot heights. Other DOS mapping in this aid programme included a 1:25 000 series of the Kombo peninsula (an eight-sheet topographic series showing relief by 10 m contours), which covered the most important economic area in the country. This series was in part derived from photogrammetric plotting and in part from a larger-scale project, the 1:10 000 coastal strip series, which showed relief by 5 m contours. Prior to this DOS mapping programme conventional survey methods had been used to compile large-scale cadastral surveys of the major settlements. These series were revised under the DOS programme using aerial coverage, and have been updated since then. Maps are available at 1:1250, 1:2500 and 1:5000 scales as diazo prints to cover all the main towns and villages in the Gambia. The Survey Department also compiles a diazo electoral constituency map at 1:250 000 and still has copies available of a DOS-produced general map of the country compiled in 1980.

Most thematic mapping of the country with the exception of the 1:10 000 land use photomap is now rather dated though still available. A planned 56-sheet land use series in the late 1950s was compiled by DOS, but only 35 sheets were published. Most are still available from Edward Stanford, or from the Survey Department. A four-sheet 1:125 000 scale soil association map was produced by DOS for the **Land Resources Development Centre (LRDC)** in the 1970s and is also still available from the Survey Department. Other DOS/LRDC mapping is either monographic or no longer available.

Geological mapping is the responsibility of the **Geological Unit**. A very poor reproduction of a 1925 monochrome Gold Coast Survey map is still available from the Survey Department and all the country is covered by geological mapping of Senegal at 1:500 000. A full programme is underway to produce modern geological mapping: data is being collected at 1:50 000 scale, with a view to a 1988 publication at either 1:125 000 or 1:250 000.

For other mapping of the country see Senegal.

Further information

Survey Department Gambia (1979) Progress report on mapping activities from 1972 to 1979, prepared for the *4th United Nations Regional Cartographic Conference on Africa*, Addis Ababa: UN Economic Commission for Africa.

Catalogues and information are available from both the Survey Department and the Geological Unit.

Addresses

Survey Department
Cotton St, BANJUL.

Overseas Surveys Directorate (OSD/DOS)
Ordnance Survey, Romsey Rd, SOUTHAMPTON SO9 4DH, UK.

Geological Unit
8 Marina Parade, BANJUL.

For LRDC see Great Britain; for DMA see United States.

Catalogue

Atlases and gazetteers

Gambia: official standard names approved by the United States Board on Geographic Names
Washington, DC: DMA, 1968.
pp. 35.

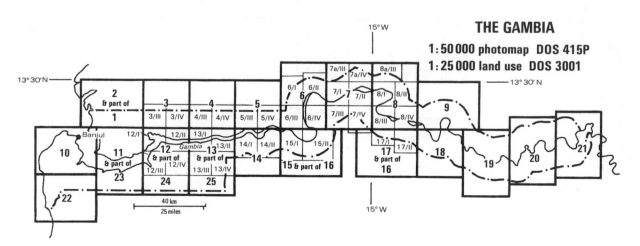

General

The Gambia 1:250 000 DOS 615
Tolworth: DOS, 1980.
Shows forest parks, administrative divisions and communications network.

Topographic

The Gambia 1:50 000 DOS 415P
Tolworth and Banjul: DOS and Survey Department, 1975–77.
20 sheets, all published. ■

Environmental

The Gambia: soil associations 1:125 000 DOS 3212A-D
Tolworth: DOS, 1976.
4 sheets, all published.

The Gambia: land use 1:25 000 DOS 3001
Tolworth: DOS, 1958.
56 sheets, 35 published. ■

Ghana

Ghana became independent in 1957 and has been mapped with British and more recently other foreign aid. The national mapping agency is the **Survey of Ghana**, which is responsible for geodetic, cadastral and topographic surveys, and the production of topographic and large-scale urban mapping. The current basic scale series is the 1:50 000 map which was begun in 1960 and is compiled entirely by photogrammetric methods to show relief by 50 foot contours. Though the country adopted the metric system in 1974 there are no plans for a 1:50 000 map drawn to full metric standards. The current 1:50 000 series will cover the country in 355 three-colour sheets using Transverse Mercator projection. The northern half of the country has been mapped under an aid project with the British **Directorate of Overseas Surveys (DOS)** (now Overseas Surveys Directorate, Southampton, OSD), whilst for areas south of 7° 30′ N the Canadians have flown 1:40 000 aerial coverage which is being used to compile the required 156 new sheets; mapping is now complete but many sheets remain unprinted. Other older series have been produced since 1947 using ground survey methods at scales of 1:62 500, 1:125 000 and 1:250 000. Sheets from these series are revised on an occasional basis but no new maps are being produced. Other small-scale maps are also published. Large-scale photogrammetric mapping of about 50 towns and cities in Ghana have been produced at a scale of 1:2500. A new programme of 1:10 000 mapping has been started, and Ghana is currently negotiating with Italy for an aid project to compile complete 1:25 000 scale aerial and map coverage of the country. Specifications of this project are not yet known.

We have listed the current series and some of the still available smaller scale SDG mapping.

The national atlas project was begun in 1969 with the aim of producing 150 maps at a scale of 1:1 500 000. This programme was begun as a joint project between the Survey Department and the **Council for Scientific and Industrial Research (CSIR)**. Since 1973 about 35 maps have appeared.

Geological mapping of Ghana is carried out by the **Geological Survey of Ghana**. Amongst the current projects is a 1:100 000 scale geological programme in conjunction with the **Bundesanstalt für Geowissenschaften und Rohstoffe (BfGR)** which will publish coverage in six sheets of south-western areas of the country. In the 1960s and 1970s the Geological Survey compiled a variety of small-scale earth science maps, and some coverage of a few areas on quarter-degree sheet lines, but the availability of these is in doubt and we have not listed them in our catalogue section.

Further information

Survey of Ghana (1979) Cartographic activities in Ghana 1972–79, paper presented to the *4th United Nations Regional Cartographic Conference for Africa*, Addis Ababa: UN Economic Commission for Africa.
Survey Department (1986) *Map catalogue*.

Addresses

Survey Department (SDG)
PO Box 191, Cantonments, ACCRA.

Overseas Surveys Directorate (DOS/OSD)
Ordnance Survey, Romsey Rd, SOUTHAMPTON SO9 4DH, UK.

Geological Survey of Ghana
PO Box M80, ACCRA.

Ghana National Atlas Project, Council for Scientific and Industrial Research (CSIR)
PO Box M32, ACCRA.

For GUGK see USSR; for BfGR see German Federal Republic; for DMA see United States.

Catalogue

Atlases and gazetteers

Ghana national atlas: folio of maps 1:1 500 000
Accra: CSIR, 1973–.
35 sheets.

Ghana: official standard names approved by the United States Board on Geographic Names
Washington, DC: DMA, 1967.
pp. 282.

General

Gana 1:1 250 000
Moskva: GUGK, 1980.
In Russian.
With placename index, climatic, population and economic inset maps.

Topographic

Ghana 1:50 000
Accra: SDG, 1960–.
355 sheets, *ca* 200 published. ■

Administrative

Ghana: administrative 1 : 2 000 000
Accra: SDG, 1977.

Town maps

Accra 1 : 10 000
Accra: SDG, 1975.
6 sheets, all published.

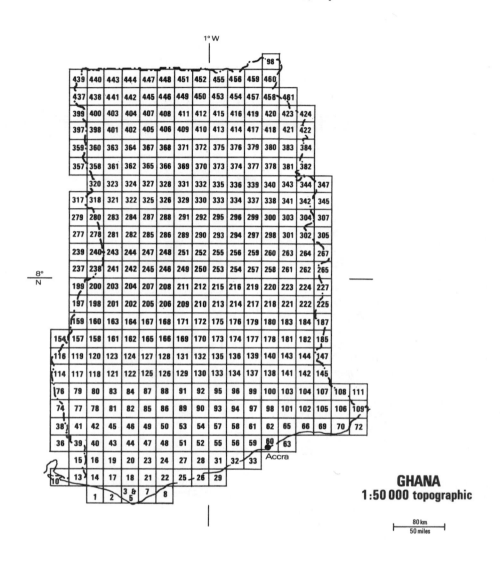

GHANA
1:50 000 topographic

80 km
50 miles

Guinea

Guinea was mapped by the French **Institut Géographique
National (IGN)** and its predecessors in the 1950s. Sheets
in the three *Cartes de l'Afrique Occidentale Française*
provided until recently the best published mapping. These
maps use UTM projection (see Africa, Western for
1 : 500 000 scale). Sheet lines of the 1 : 200 000 scale map
are arranged in degree quads and subdivided according to
the international sheet numbering system. This map was
never fully photogrammetrically produced—only the
extreme northern sheets and a block in the south-west of
Guinea are derived from aerial coverage and show relief
with 40 m contours. The central areas were only ever
issued as *cartes provisoires*. Complete coverage of the
country in 34 sheets was reached in 1957, but there has
been very little revision since then. The 1 : 50 000 series

would require about 360 sheets for complete coverage, but
only about 80 sheets were published by IGN to cover
coastal areas. These date from the 1950s, and show relief
by 20 m contours. IGN has also published a general map
of the country, which includes a list of place names, and a
five-colour orthophoto town map of the capital Conakry.

The national mapping agency in Guinea is now the
Institut Géographique National Guinéen (IGNG).
Under a technical cooperation agreement with Japan a new
basic scale series of 1 : 50 000 maps has been prepared.
This 350-sheet series was compiled as photomaps between
1977 and 1980, derived from 1 : 100 000 scale aerial
coverage. In addition 16 full-colour line maps have so far
been published, and a block of further conventional
mapping around Kerouane is also under compilation in an

aid project with France. These maps are not yet publicly available, and authorization is needed from IGNG for the purchase of any topographic mapping of the country.

Geological mapping of Guinea is the responsibility of the **Direction Général des Mines et Géologie**, but no earth science mapping of any kind has been published from Conakry and the best available geological coverage is still the *1:500 000 carte géologique de l'Afrique occidentale*, dating from the 1950s (see Africa, Western).

Addresses

Institut Géographique National (IGN)
136 bis rue de Grenelle, 75700 PARIS, France.

Institut Géographique National Guinéen (IGNG)
BP 159, CONAKRY.

Direction Général des Mines et de la Géologie
BP295, CONAKRY.

For BRGM see France.

Catalogue

Atlases and gazetteers

Guinea: official standard names approved by the United States Board on Geographic Names
Washington, DC: DMA, 1965.
pp. 175.

General

Guinée 1:1 000 000
Paris: IGN, 1980.
With ancillary thematic maps and placename list.

Topographic

République de Guinée 1:200 000
Paris: IGN, 1930–57.
34 sheets, all published. ■

République de Guinée 1:50 000
Paris and Conakry: IGN and IGNG, 1950–.
360 sheets, all published. ■
Complete photomap coverage; partial full-colour line map coverage.

Geological and geophysical

République de Guinée: géologie 1:700 000
Orléans: BRGM, 1984.
Double sided.
Includes soil map, with administrative map on back.

Town maps

Conakry 1:15 000
Paris: IGN, 1983.
Double-sided photomap.
Includes 1:10 000 scale plan of city centre.

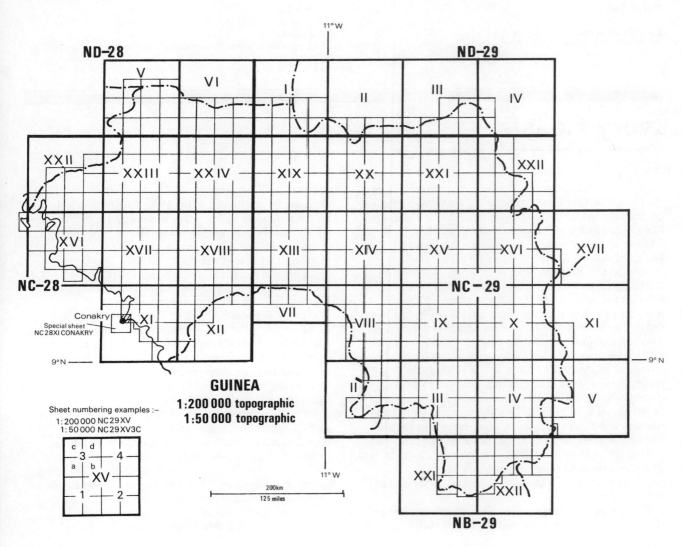

GUINEA
1:200 000 topographic
1:50 000 topographic

Sheet numbering examples:–
1:200 000 NC 29 XV
1:50 000 NC 29 XV 3 C

200km
125 miles

Guinea Bissau

The former colony of Guinea Bissau was mapped by the Portuguese **Instituto Geográfico e Cadastro**. In the 1950s and 1960 they produced a 72-sheet 1 : 50 000 series using UTM projection, with 10 m contours to cover the whole country. This map is no longer available. There is no information about current projects in Guinea Bissau, but the **Direcção de Topografia e Cadastro** in Bissau has since independence in 1974 been the national survey organization responsible for geodetic, topographic and cadastral surveying of the country. The French **Institut Géographique National (IGN)** recently produced a general map of the country with an inset town plan of the capital and has been involved in a project in the early 1980s to provide detailed accurate levelling of the entire country for the Ministère de l'Urbanisme in Bissau.

Geological mapping activities are the responsibility of the **Direcção Geral Geologia e Minas**, but again there is no information about current mapping activities, and no available geological mapping.

Another small-scale general map of the country was compiled by the Russian **Glavnoe Upravlenie Geodezii i Kartografii (GUGK)** in the early 1980s.

Addresses

Instituto Geográfico e Cadastro
Praça da Estrela, 1200, LISBOA 2, Portugal.

Direcçao de Topografia e Cadastro
CP 14, BISSAU.

Institut Géographique National (IGN)
136 bis rue de Grenelle, 75007 PARIS, France.

Direcção Geral Geologia e Minas
Comissariado dos Recursos Naturais, CP 37, BISSAU.

For GUGK see USSR; for DMA see United States.

Catalogue

Atlases and gazetteers

Portuguese Guinea: official standard names approved by the United States Board on Geographic Names
Washington, DC: DMA, 1968.
pp. 122.

General

Guinée-Bissau 1 : 500 000
Paris: IGN, 1981.
With small ancillary town plan of Bissau and administrative map.

Gvineja Bisau 1 : 500 000
Moskva: GUGK, 1982.
In Russian.
With placename index and economic and climatic inset maps.

Ivory Coast

Prior to 1961, mapping of Ivory Coast was undertaken by the French **Institut Géographique National (IGN)** and its predecessor Service Géographique de l'Afrique Occidentale Française, which covered the country in the three *Cartes de l'Afrique de l'Ouest*. After independence the **Institut Géographique de la Côte d'Ivoire (IGCI)** assumed responsibility for topographic, geodetic and other survey and mapping activities in the country. It became a fully autonomous public body in 1976 and since 1977 map compilation and printing has been carried out within Ivory Coast. The basic map is the 444 sheet 1 : 50 000 series. This four-colour map has 20 m contour interval, and uses UTM projection. Sheets cover quarter-degree quads. The series was begun by the French in the 1950s and is photogrammetrically compiled from medium-scale aerial coverage, which is now available for most of the country. Earlier sheets were published for the southern half of the country, and effort is now being concentrated upon the completion of the series by the compilation of new maps of northern areas at a rate of about a map every 2 months and upon the revision of some of the older maps. A 9 year programme of 247 new sheets to complete the series was agreed in 1976.

The 33-sheet 1 : 200 000 series gives complete coverage of the country, and is slowly being revised. Most sheets,

however, still date from the 1960s. A 1 : 500 000 series was compiled by the French in the 1950s and 1960s which has not been updated by IGCI. An administrative map of the country has been revised to a scale of 1 : 1 250 000 from an original 1 : 1 000 000 scale, and a 1 : 1 000 000 general map of the country is also available.

Larger-scale mapping of development areas has also been carried out and a two-sheet town plan of Abidjan has been published, again in conjunction with IGN.

There is at present a lack of current geological mapping of the country. The **Societé pour le Développement Minier de la Côte d'Ivoire (SODEMI)** has carried out some geological mapping including test 1 : 200 000 scale maps which are on programme for some areas, and a 1 : 2 000 000 single sheet of the whole country, published as part of the National Atlas. An earth science atlas of the country was also compiled by SODEMI in the 1960s, and a mineral distribution atlas in the 1970s. There is no information about the current availability of SODEMI published maps and we have excluded most from our catalogue section. The current official geological mapping agency in the country is **Direction de la Géologie et de la Prospection Minière** but there is no information about current mapping projects. The French-produced 1 : 500 000 and 1 : 2 000 000 scale geological mapping covering much of West Africa is still available and is described in the Africa, West section.

A wide variety of thematic mapping of the Ivory Coast has been carried out by the French **Office de la Recherche Scientifique et Technique Outre Mer (ORSTOM)**. Soils mapping has been carried out at scales of 1:200 000 and 1:500 000 for large areas of the country. The latter four-sheet series is no longer available but was generalized into a single-sheet 1:2 000 000 scale map published in the National Atlas. This atlas was compiled by ORSTOM in a joint project with Institut de Géographie Tropicale de l'Université d'Abidjan between 1970 and 1979, and contains nearly 50 maps. Amongst other ORSTOM mapping are bathymetric surveys of the continental shelf, insect distribution and vegetation maps, and some geomorphological mapping.

Amongst the other mapping of Ivory Coast is a road map revised on a regular basis by the French commercial house **Michelin**, and a useful small atlas in the Jeune Afrique series issued by **Editions Jaguar**.

Further information

The Fourth Regional Cartographic Conference for Africa was held in Abidjan in 1979 and several papers presented by the host country at this meeting present a good background to the mapping of Ivory Coast.

IGCI publishes a catalogue.

Addresses

Institut Géographique National (IGN)
136 bis rue de Grenelle, 75700 PARIS, France.

Institut Géographique de la Côte d'Ivoire (IGCI)
BP 3862, rue Jacques Aka, ABIDJAN 01.

Societé pour le Développement Minier de la Côte d'Ivoire (SODEMI)
BP 2816, ABIDJAN.

Direction de la Géologie et de la Prospection Minière
BP V 28,
ABIDJAN.

Office de la Recherche Scientifique et Technique Outre Mer (ORSTOM)
70–74 Route d'Aulnay, 93140 BONDY, France.

For Michelin and Editions Jaguar see France; for DMA see United States.

Catalogue

Atlases and gazetteers

Atlas de Côte d'Ivoire
Bondy and Abidjan: ORSTOM and Université d'Abidjan, 1979.
46 sheets, all published.

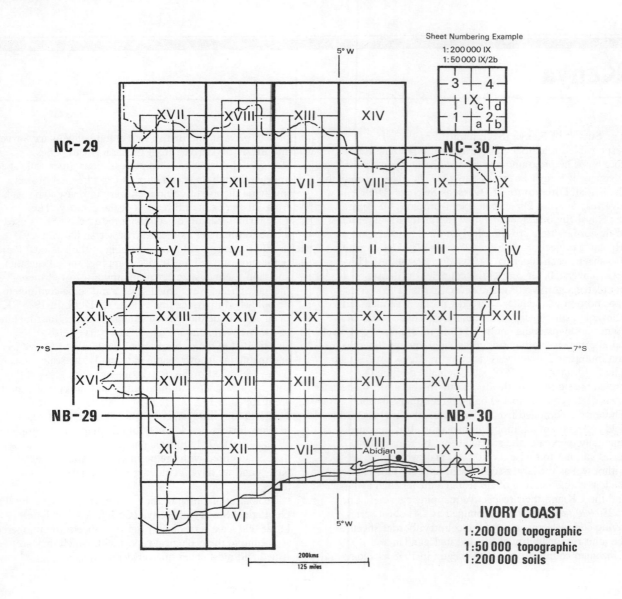

IVORY COAST
1:200 000 topographic
1:50 000 topographic
1:200 000 soils

Atlas de la Côte d'Ivoire   P. Vennetier and G. Laclavère
Paris: Editions Jaguar, 1978.
pp. 72.

*Ivory Coast: official standard names approved by the United States
Board on Geographic Names*
Washington, DC: DMA, 1965.
pp. 250.

General

République de la Côte d'Ivoire: carte générale et routière
1 : 1 000 000
Abidjan: IGCI, 1979.

Côte d'Ivoire/Ivory Coast   1 : 800 000
Paris: Michelin, 1980.

Topographic

République de Côte d'Ivoire: carte   1 : 200 000
Abidjan: IGCI, 1952–.
33 sheets, all published. ■

République de Côte d'Ivoire: carte   1 : 50 000
Abidjan: IGCI, 1952–.
444 sheets, *ca* 350 published. ■

Bathymetric

*Topographie générale du plateau continental de la Côte d'Ivoire et du
Libéria*   *ca* 1 : 294 000 and 1 : 945 000
Bondy: ORSTOM, 1968.

7 sheets, all published.
With accompanying explanatory text.
2 larger scale, and 1 smaller scale sheet cover Ivory Coast area.

Geological and geophysical

Côte d'Ivoire carte géologique   1 : 2 000 000
Abidjan: SODEMI, 1972.
With explanatory text.

Environmental

Carte de répartition des glossines   1 : 2 000 000
Bondy: ORSTOM, 1981.
2 sheets, both published.
Vegetation, isohyet and insect distribution maps.
With accompanying explanatory text.

Carte pédologique de la Côte d'Ivoire   1 : 200 000
Bondy: ORSTOM, 1975–.
33 sheets, 8 published. ■
2 maps (*paysages* and *unités*) and explanatory notice for each area.

Administrative

République de Côte d'Ivoire: carte administrative   1 : 1 250 000
Abidjan: IGCI, 1979.

Town maps

Plan d'Abidjan   1 : 10 000
Abidjan: IGCI, 1971.
2 sheets, both published.

Kenya

The **Survey of Kenya (SK)**, founded in 1906, is
responsible for all major surveying and mapping in the
country. The programme to provide basic topographical
coverage of Kenya was started in 1947, in cooperation with
the British **Directorate of Overseas Surveys (DOS)** (now
Overseas Surveys Directorate, Southampton, OSD) and
the Royal Engineers, and is now complete. All three series
in this programme are compiled on the Transverse
Mercator projection. *The 1:50 000 series Y731* comprises
435 sheets, each covering quarter-degree squares. This
series covers 50 per cent of the country, the denser
populated south-western and central parts. The current
specification is a standard agreed with the Ugandan and
Tanzanian survey authorities; contours are at 10, 20 or
40 m, the maps being compiled from medium-scale vertical
aerial coverage, with considerable field update. There is a
continuous and selective revision programme for this map.
The *1:100 000 series Y633* was originally designed to
provide basic cover of the dry northern and north-western
areas of the country in 130 half-degree square sheets.
These were compiled from aerial coverage with very little
field collection of additional information. It is intended
gradually to expand the areas covered by the 1:50 000
series and not to revise the 100 000 map which will be
withdrawn as the larger-scale sheets appear. This has been
made possible because of aid from countries such as Japan
and the UK and their respective mapping agencies. The
1:250 000 series Y503 is maintained as a 41-sheet series
giving complete coverage of the country; border sheets are
the joint responsibility of SK and its Ugandan and
Tanzanian counterparts. It was begun in 1958 as a joint

project between the British Directorate of Military Survey
and SK and is derived from the larger-scale series.
Revision of this series takes place as new larger-scale sheets
appear, though there is a current experimental project
involving the use of satellite imagery to revise two sheets
from this series in arid regions of the country. The
1 : 1 000 000 map of the country produced by SK is used as
a base for many of its thematic projects. The Survey
produces a range of tourist mapping of the National Parks,
population and administrative mapping, some vegetation
and land use maps, and road mapping. Most of these
thematic sheets are at small scales, and some continue to
be published in association with OSD/DOS. In 1985 SK
planned to issue the redesigned and enlarged fourth edition
of the *National Atlas of Kenya*. The Survey also
contributed to the publication of the US Board Gazetteer
and maintains its own name register. Large-scale cadastral
series are issued by SK for the major urban areas of the
country: townships, municipalities and the city of Nairobi
are mapped at 1 : 2500, 1 : 5000, 1 : 10 000, 1 : 14 000 and
1 : 20 000, but little large-scale aerial coverage is available
to allow further development of this mapping programme.
   The 1 : 50 000, 1 : 100 000 and 1 : 250 000 scale map series
may be acquired with the permission of the Survey of
Kenya and we have included graphic indexes to these
maps.
   Geological mapping of the country is carried out by the
Geological Survey of Kenya (GSK). A full-colour
1 : 125 000 scale series is in progress. Sheets are numbered
according to the geological report with which they are
issued and most cover quarter-degree quads.

Amongst the other Government bodies producing maps are the **Kenya Soil Survey** which recently issued 1 : 1 000 000 scale full-colour soil and agroclimatic maps of the whole country in a joint project with STIBOKA, the Dutch Soil Survey. These maps were issued with explanatory text and use the FAO/UNESCO classification. They are accompanied by two other 1 : 1 000 000 scale sheets, showing the information base and a combination of the two main maps designed for land evaluation purposes. KSS also publishes reconnaissance soil surveys at 1 : 100 000 and 1 : 250 000 scale, and some more detailed soil maps are issued for specific projects or important areas. Most of these maps cover parts of the more developed southern and western parts of Kenya.

Various other government agencies carry out mapping activities, including the **Kenya Rangeland Ecological Monitoring Unit (KREMU)**, which compiles ecological, land use and vegetation mapping of the country. KREMU was established in 1975 and has been particularly involved in the use of remotely sensed data for the production of land use/land cover mapping of Kenya. A national map has been published and effort is now being concentrated upon the production of more detailed land use maps for use at the district level.

The **Regional Centre for Services in Surveying and Mapping and Remote Sensing** in Nairobi is a major focus for research into cartography and remote sensing throughout the African continent. Another important international organization based in Nairobi is the **United Nations Environment Programme (UNEP)**. Within this organization is the Global Environment Monitoring Programme and the *Global Resources Information Database* (GRID) (see the World section for details of this database).

Amongst the other organizations who have published mapping of Kenya are **Bartholomew**, **Macmillan Kenya**, **Karto+Grafik**, **Nelles Verlag** and the American **Central Intelligence Agency (CIA)**. A recent town atlas of Nairobi was issued by **Kenway Publications**.

Further information

Kenya National Cartographic Committee (1984) *Kenya National Cartographic Report*, presented to the 12th International Conference and 7th General Assembly of the ICA, Perth: ICA.

Catalogues are issued by the Survey of Kenya, the Geological Survey, the Soil Survey and KREMU.

Addresses

Survey of Kenya (SK)
PO Box 30046, NAIROBI.

Overseas Surveys Directorate (DOS/OSD)
Ordnance Survey, Romsey Rd, SOUTHAMPTON SO9 4DH, UK.

Geological Survey of Kenya (GSK)
Madini House, Machakos Road, PO Box 30009, NAIROBI.

Kenya Soil Survey
PO Box 14733, NAIROBI.

Kenya Rangeland Ecological Monitoring Unit (KREMU)
Ministry of Finance and Planning, PO Box 47146, NAIROBI.

Regional Centre for Services in Surveying and Mapping and Remote Sensing (RCSSMRS)
Kenya Commercial Bank Building, PO Box 18118, Enterprise Rd, NAIROBI.

United Nations Environment Programme (UNEP)
PO Box 30552, NAIROBI.

Macmillan Kenya Ltd
Queensway House, Kaunda St, POB 30797, NAIROBI.

Kenway Publications
PO Box 18800, NAIROBI.

For Bartholomew see Great Britain; for Karto+Grafik and Nelles Verlag see German Federal Republic; for CIA and DMA see United States.

Catalogue

Atlases and gazetteers

National atlas of Kenya 4th edn
Nairobi: SK, 1985.

Kenya: official standard names approved by the United States Board on Geographic Names 2nd edn
Washington, DC: DMA, 1978.
pp. 470.

General

Kenya: tourist map 1 : 1 750 000
Nairobi: Macmillan Kenya, 1985.
Double-sided.
Includes gazetteer; The Kenya Coast 1 : 300 000; Central Nairobi 1 : 10 000 + index; Mombasa Island 1 : 22 000 + index; Nairobi National Park 1 : 48 000; Places of interest.

Kenya 1 : 1 250 000
Edinburgh: Bartholomew, 1981.

Kenya and Northern Tanzania: route map 1 : 1 000 000 SK 81
Nairobi: SK, 1978.
Also available as 5 different editions showing administrative boundaries, wildlife management, parliamentary constituencies, large/small-scale aerial coverage, and primary/secondary triangulation.

Topographic

East Africa 1 : 250 000 Series Y 503
Nairobi: SK, 1954–.
41 sheets, all published. ■

Kenya 1 : 100 000 Series Y633, SK 58
Nairobi: SK, 1958–70.
127 sheets, all published. ■

Kenya 1 : 50 000 Series Y731 (DOS 423)
Nairobi: SK, 1950–.
ca 500 sheets, 435 published. ■

Geological and geophysical

Geological map of Kenya 1 : 3 000 000
Nairobi: SK, 1985.
Single sheet from national atlas.

Kenya geological map 1 : 125 000
Nairobi: GSK, 1945–.
ca 250 sheets, ca 100 published.
Issued with accompanying texts.

Environmental

Exploratory soil map and agroclimatic zone map of Kenya 1 : 1 000 000
Nairobi: Kenya Soil Survey, 1982.
4 sheets.
Single sheets covering soils, agroclimatic zones, information base and land evaluation.
With explanatory report.

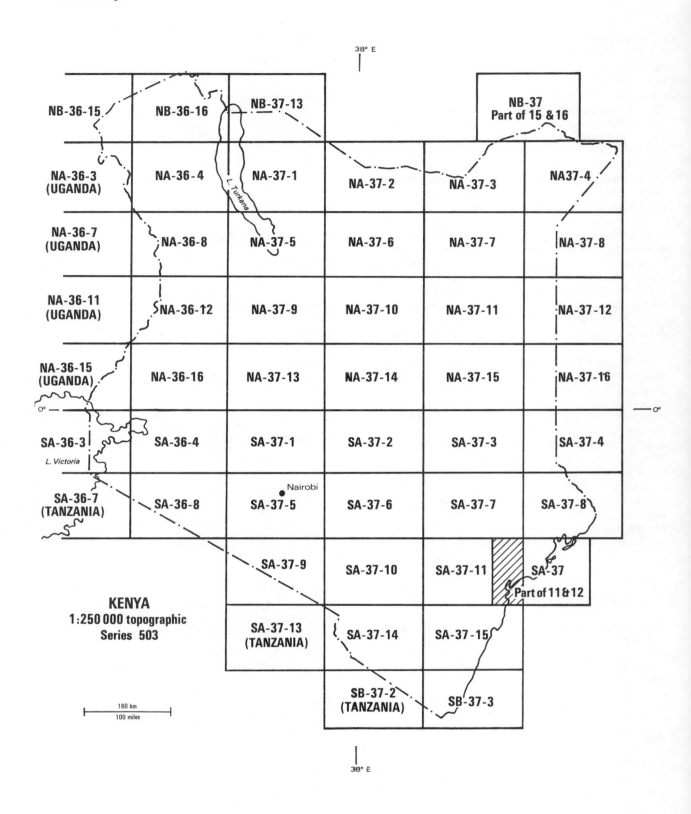

38° E

NB-36-15 | NB-36-16 | NB-37-13 | NB-37 Part of 15 & 16

NA-36-3 (UGANDA) | NA-36-4 | NA-37-1 | NA-37-2 | NA-37-3 | NA37-4

L. Turkana

NA-36-7 (UGANDA) | NA-36-8 | NA-37-5 | NA-37-6 | NA-37-7 | NA-37-8

NA-36-11 (UGANDA) | NA-36-12 | NA-37-9 | NA-37-10 | NA-37-11 | NA-37-12

NA-36-15 (UGANDA) | NA-36-16 | NA-37-13 | NA-37-14 | NA-37-15 | NA-37-16

0°

SA-36-3 | SA-36-4 | SA-37-1 | SA-37-2 | SA-37-3 | SA-37-4

L. Victoria

Nairobi

SA-36-7 (TANZANIA) | SA-36-8 | SA-37-5 | SA-37-6 | SA-37-7 | SA-37-8

SA-37-9 | SA-37-10 | SA-37-11 | SA-37 Part of 11 & 12

KENYA
1:250 000 topographic
Series 503

SA-37-13 (TANZANIA) | SA-37-14 | SA-37-15

SB-37-2 (TANZANIA) | SB-37-3

38° E

160 km
100 miles

Land cover/land use map of Kenya
Nairobi: KREMU.

Administrative

Kenya: administrative boundaries   1 : 1 000 000   SK 81A
Nairobi: SK, 1978.

Human and economic

Kenya: population   1 : 1 000 000
Nairobi: SK, 1969.
2 sheets, both published.

Town maps

A to Z: a pocket street atlas of Nairobi   1 : 22 000 and 1 : 11 000
R. Moss
Nairobi: Kenway Publications, 1981.
pp. 65.

City of Nairobi: map and guide   1 : 20 000   SK 46   7th edn
Nairobi: SK, 1978.

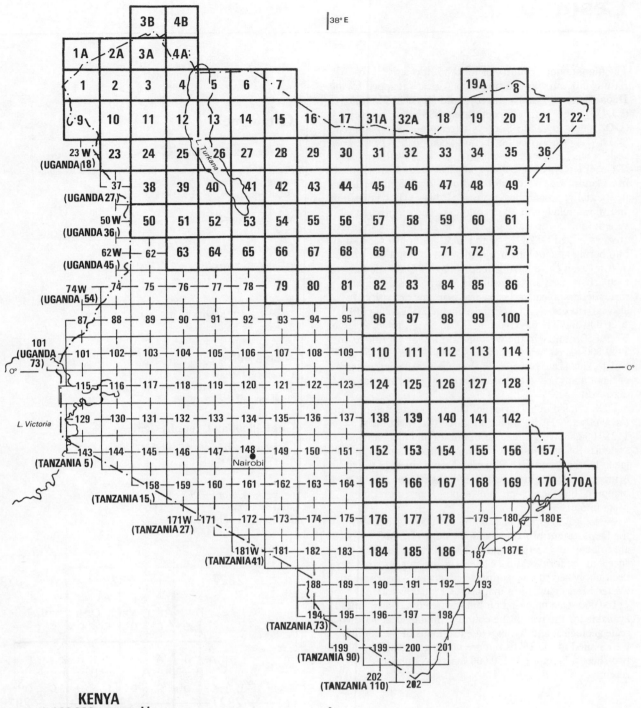

38° E

KENYA
1:100 000 topographic
Series Y633

1:50 000 topographic
Series Y731 (DOS 423)

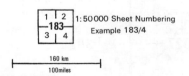

1:50000 Sheet Numbering

Example 183/4

160 km
100 miles

Lesotho

The mountainous country of Lesotho, formerly Basutoland, surrounded by South Africa is mapped by the **Department of Lands, Surveys and Physical Planning (LDLS)**. The British **Directorate of Overseas Surveys (DOS)** (now Overseas Surveys Directorate, Southampton, OSD) was responsible for the joint compilation of the basic scale series for the whole country, the *1:50 000 series L50 (DOS 421)*. This five-colour, 60-sheet series was completed in its latest (3rd) restructured edition in 1983, based on new aerial photographic coverage. It uses UTM projection and shows relief with 20 m contours. The numbering system follows the standard DOS breakdown but indexes are now issued by LDLS with a new additional sequential numbering system which it is planned to introduce in the next revision cycle for this map. A 1:250 000 scale map is regularly revised and is available as two sheets. Larger-scale topographical and cadastral mapping activities are also carried out by LDLS: these include full colour BASP 1:20 000 and Thaba-Thaka 1:25 000 series, to cover large blocks of country in the west and centre, and a full-colour 1:12 500 scale town plan of the capital Maseru. 1:5000 and 1:2500 scale plans cover the major settlements; some of these maps are issued as orthophoto plots, most are diazo editions.

DOS carried out a variety of land resources mapping in the 1960s and 1970s, including three two-sheet maps at 1:250 000 showing soils, agricultural potential and land systems. Responsibility for such resources mapping is now taken by the **Land Use Planning Department** of the Ministry of Agriculture, which still distributes the DOS mapping, but there is no information about current programmes.

Geological mapping of Lesotho has been carried out by the **Department of Mines and Geology (LDMG)**. There are full-colour 1:50 000 and 1:100 000 programmes; 15 sheets of the former and 12 sheets at the smaller scale have been published to give complete coverage. Northern and western areas have been mapped at 1:50 000 scale; 1:100 000 scale mapping on half-degree quad sheet lines is available for the rest of the country. Single-sheet 1:500 000 scale geological and hydrogeological maps of the country were issued in the 1970s but are no longer available. A two-sheet full-colour 1:250 000 map was published recently.

Further information

LDLS issues the useful *Lesotho map catalogue*.

Addresses

Department of Lands, Surveys and Physical Planning (LDLS)
Ministry of the Interior, PO Box 876, MASERU 100.

Overseas Surveys Directorate (OSD/DOS)
Ordnance Survey, Romsey Rd, SOUTHAMPTON SO9 4DH, UK.

Department of Mines and Geology (LDMG)
PO Box 750, MASERU 100.

Land Use Planning Department
Ministry of Agriculture, PO Box 24, MASERU 100.

For DMA see United States.

Catalogue

Atlases and gazetteers

South Africa: official standard names approved by the United States Board on Geographic Names
Washington, DC: DMA, 1954.
2 volumes.
Vol. 2 includes Lesotho.

General

Lesotho 1:750 000
Maseru: LDLS, 1977.

Map of Lesotho 1:250 000 DOS 621/1
Maseru and Tolworth: LDLS and DOS, 1978.
2 sheets, both published.

Topographic

Lesotho 1:50 000 L50 (DOS 421)
Maseru and Tolworth: LDLS and DOS, 1979–83.
60 sheets, all published. ■

Geological and geophysical

Geological map of Lesotho 1:250 000
Maseru: LDMG.
2 sheets, both published.

Geological map of Lesotho 1:100 000/1:50 000
Maseru: LDMG.
27 sheets, all published.

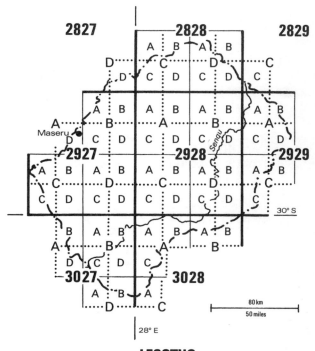

LESOTHO
1:50 000 topographic Series L50

Environmental

Lesotho land systems 1:250 000
Tolworth: DOS/LRDC, 1967.
2 sheets, both published.

Lesotho soils 1:250 000
Tolworth: DOS/LRDC, 1967.
2 sheets, both published.

Lesotho agricultural potential 1:250 000
Tolworth: DOS/LRDC, 1967.
2 sheets, both published.

Town maps

Tourist map of Maseru 1:12 500 DOS 221
Maseru: LDLS, 1975.

Liberia

The national mapping agency in Liberia is the **Liberian Cartographic Service (LCS)**. It was founded in 1951 with geodetic, cartographic, hydrographic and photogrammetric responsibilities and has carried out a variety of surveying and mapping activities in these fields, with the technical and financial assistance of the United States and United Kingdom. There is complete coverage of the country in a 10-sheet 1:250 000 scale series published by the **United States Geological Survey (USGS)**: this three-colour *Geographic map of Liberia* is not contoured. Twelve sheets in the six-colour series 1501 Joint Operations Graphic 1:250 000 were published in the early 1970s by the United States **Defense Mapping Agency Topographic Command (DMATC)**. This map has 50 m contours and uses UTM grid. It is, however, no longer available. The main priority at present in Liberia is the extension of 1:50 000 coverage. Thirty-seven sheets were published jointly with DMATC in the early 1970s to cover about one-quarter of the country, and the British **Directorate of Overseas Surveys (DOS)** (now Overseas Surveys Directorate, Southampton, OSD) is currently involved in the production of a further block of six-colour mapping. It is hoped to extend this scale to cover the remaining half of the country. None of these multi-colour maps is at present being revised.

Town mapping projects have produced coloured photomaps of Monróvia, Buchanan and Roberts Field and cadastral mapping at 1:1000 was compiled in 1974 with United Nations aid to cover the capital. Full-colour town maps at 1:5000 scale are currently being produced to cover the four major towns in the country, in conjunction with DOS.

The **Liberian Geological Survey (LGS)** was established in 1963. Like LCS it has little indigenous mapping capability and has relied upon aid projects, especially from the USGS. 1:250 000 scale coverage is available on the geographic map sheet lines and four different earth science editions of this 10-sheet series were published in the late 1970s: a full-colour geology series includes explanatory notes, and two-colour maps are published to show aeromagnetic intervals, total gamma count variation, and simple Bouguer gravity anomalies. A recent 1:1 000 000 scale single-sheet geological map has also been compiled by USGS.

Amongst the other organizations that have been involved in mapping in Liberia are the **International Institute for Aerospace Survey (ITC)** which has compiled a hypsometric tinted general map of the country and a town map of Monrovia. The French **Office de la Recherche Scientifique et Technique Outre Mer (ORSTOM)** has published bathymetric mapping of the Liberian and Ivory Coast continental shelves, which we have listed under Ivory Coast. The American **Central Intelligence Agency (CIA)** map compiled in the early 1970s is still available.

Further information

Liberia (1979) Status of surveying and mapping activities in Liberia, paper presented by Liberia in *Second United Nations Regional Cartographic Conference for the Americas, Technical Papers*, New York: UN, 1984.

Addresses

Liberian Cartographic Service (LCS)
Ministry of Lands, Mines and Energy, PO Box 9024, MONROVIA.

United States Geological Survey (USGS)
12201 Sunrise Valley Drive, RESTON, VA 22092, USA.

Defense Mapping Agency Topographic Command (DMATC)
WASHINGTON, DC 20315, USA.

Overseas Surveys Directorate (DOS/OSD)
Ordnance Survey, Romsey Rd, SOUTHAMPTON SO9 4DH, UK.

Liberian Geological Survey (LGS)
Ministry of Lands, Mines and Energy, PO Box 9024, MONROVIA.

For ITC see Netherlands; for ORSTOM see France; for CIA and DMA see United States.

Catalogue

Atlases and gazetteers

Liberia: official standard names approved by the United States Board on Geographic Names 2nd edn
Washington, DC: DMA, 1976.
pp. 167.

General

Liberia 1:1 350 000
Washington, DC: CIA, 1973.
With economic, vegetation and population insets.

Liberia 1:1 000 000
Enschede: ITC, 1982.

Topographic

Geographic map of Liberia 1:250 000
Reston: USGS, 1973.
10 sheets, all published.

Liberia 1:50 000
Monrovia: LCS, 1972–.
160 sheets, *ca* 60 published. ■
Jointly published with either DMATC or OSD.

Geological and geophysical

Geologic map of Liberia 1:1 000 000
Reston and Monrovia: USGS and LGS, 1983.

Geologic map of Liberia 1:250 000
Reston and Monrovia: USGS and LGS, 1977.
10 sheets, all published.
Also available as aeromagnetic, total gamma count radiation, and
Bouguer gravity editions.

Town maps

Tourist map of Monrovia 1:25 000
Enschede: ITC, 1980.
With city centre at 1:12 500 and index to places of interest.

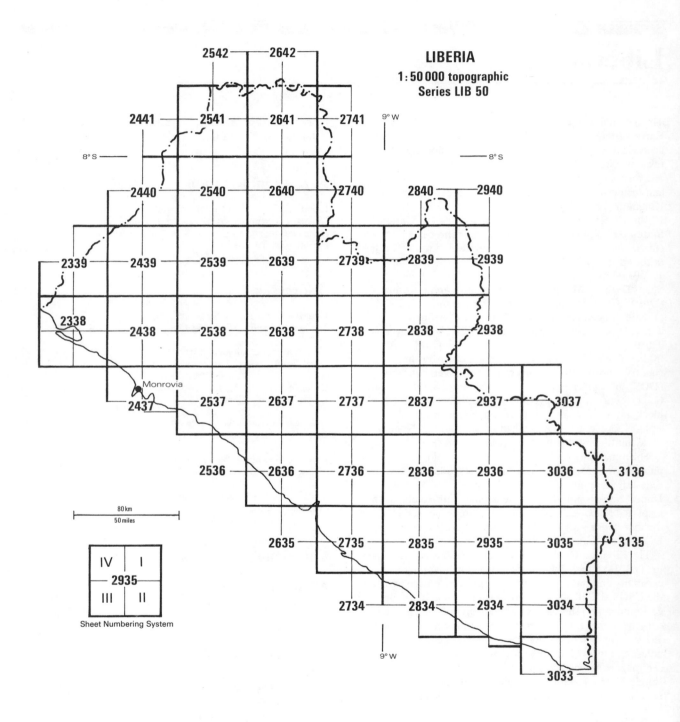

Libya

The national mapping agency in Libya is the **Survey Department of Libya (SDL)** which was established in February 1968 as an office within the Ministry of Planning. It became an independent department late in 1971 and is currently responsible for surveying and mapping activities throughout Libya, including geodetic, photogrammetric, cadastral and cartographic work within its remit. Mapping of Libya has reflected the very uneven distribution of population and economic activities: there has been almost no large- or medium-scale coverage of the southern desert areas of the country. An Italian reconnaissance series at 1:400 000 covered the whole country, and British and German wartime coverage was compiled at a variety of medium scales. Modern mapping at 1:250 000 compiled by the US Army Map Service (now **Defense Mapping Agency**) covered the area to the north of 28° N. This series was derived from 1:60 000 aerial photographs. Only the well populated coastal belt is mapped at medium scales: a 1:50 000 base map compiled in the 1960s from mid-1950s aerial coverage comprised 240 quarter-degree sheets. Both these series use a UTM grid. Following the rapid development of the 1970s decisions were made to revise this series and to compile a new basic scale map at a scale of 1:25 000 to cover the populated areas of the coastal belt. It was also decided to introduce a new Transverse Mercator 2° band-based projection, in order to minimize distortion in cadastral surveying activities. Some revised 1:50 000 sheets are being published as orthophotomaps with contour overprints. SDL is receiving aid from **Geokart** the Polish overseas survey authority in these topographic programmes. Amongst the other modern mapping compiled for the country have been satellite image maps produced by the American **Earthsat Corporation**. Urban mapping has been carried out, including recent contracts let to the Greek Doxiades Associates for photo interpretation of the eastern urban centres of the country and production of planimetric surveys.

A National Atlas was published in Arabic and English language editions in 1978.

None of these Libyan-produced maps is available outside the country and stocks of the earlier topographic series are no longer available from their publishers, but might be procured through the major map jobbers.

Geological mapping of Libya is the responsibility of the **Geological Research and Mining Department**. There is no information available about current mapping programmes, but a 1:250 000 scale 72-sheet series was begun in the mid-1970s under an aid project with the Czech agency Kartografie. This was completed in the early 1980s and sheets were issued between 1975 and 1983. Small-scale geological mapping of the whole country was compiled by the **United States Geological Survey (USGS)** in the 1970s: 1:2 000 000 scale single-sheet maps were prepared for a geology and mineral resource inventory, to show topography and mineral resources, tectonics and palaeogeography, water resources and geology. Only the geology sheet, compiled in the mid-1960s, was published as a full-colour edition.

Small-scale and tourist coverage of Libya has been produced by a variety of overseas agencies. USGS produced a 1:2 000 000 scale map in 1970 of the country

with joint English and Arabic language explanatory text, and 125 m contours. Both the American **Central Intelligence Agency (CIA)** and the Russian **Glavnoe Upravlenie Geodezii i Kartografii (GUGK)** have published useful small-scale coverage; the latter is now available as either Russian or English language editions. A tourist map has been produced for Malt International in the Arab Map Library series by **Geoprojects**, which also includes town maps of Tripoli and Benghazi, and **Cartographia** has compiled a 1986 map of Libya in its Handy Map series.

Further information

For background information to the development of Libyan series see Libyan Arab Jamahariya (1977) The state of cartographic services in the Libyan Arab Jamahiraya, paper presented to the *8th United Nations Regional Cartographic Conference for Asia and the Pacific*, New York: UN.

A useful historical cartobibliography is Allen, J.A. (1969) *A Select Map and Air Photo Bibliography of Libya*, London: Libyan University/London University Joint Research Project.

Addresses

Survey Department of Libya (SDL)
PO Box 600, TRIPOLI.

Geological Research and Mining Department
Industrial Research Centre, PO Box 3633, TRIPOLI.

For USGS, DMA and Earthsat see United States; for Geokart see Poland; for GUGK see USSR; for Geoprojects see Lebanon; for Cartographia see Hungary.

Catalogue

Atlases and gazetteers

Libya: official standard names approved by the United States Board on Geographic Names 2nd edn
Washington, DC: DMA, 1973.
pp. 746.

General

Map of the Socialist Peoples' Libyan Arab Republic 1:3 500 000
Beirut: Malt International, 1983.
Double-sided.
Includes city maps of Tripoli and Benghazi.

Libya 1:2 500 000
Moskva: GUGK, 1985.
With placename index and economic inset map.
Also available as Russian language edition.

Libia: handy map 1:2 000 000
Budapest: Cartographia, 1986.

Topographic map of the United Kingdom of Libya 1:2 000 000
Reston, VA: USGS, 1971.
English Arabic language explanation.

Geological and geophysical

Geology and mineral resources of Libya   1 : 2 000 000
Reston, VA: USGS, 1970.
4 sheets, all published.

Thematic set: sheets published to cover: topography and mineral resources; geology; tectonics and palaeogeography; and water resources.
With accompanying text.

Madagascar

The national mapping agency of the Malagasy Republic is **Foiben Taosarintanin' i Madagasikara (FTM)**. It is responsible for topographic, photogrammetric and geodetic surveying and for the production of a variety of maps, and took over responsibility from the French **Institut Géographique National (IGN)** following independence in 1960. Until recently IGN-produced maps of the island were available from Paris but all enquiries are now to be directed to Antananarivo. The basic scale mapping programme involves the compilation of maps at 1 : 50 000 and 1 : 100 000 scales. Complete coverage in 453 sheets has been attained in the seven- or eight-colour 1 : 100 000 scale series, which uses 25 or 50 m contours and is drawn on the Oblique Mercator projection. Since 1962, sheets have been published with violet administrative overprint. The 1 : 50 000 series is very incomplete: sheets have mainly appeared as seven-colour editions with 25 m contours, for eastern parts of the island. Many sheets in both these scale series are very old, but new sheets are being compiled to full photogrammetric standards at a rate of about 10 sheets in each series a year.

Basic scale coverage of the country may not be exported. The largest scale series still available is an 11-sheet 1 : 500 000 scale map using Lambert conformal conic projection, showing relief by 100 m contours, which was prepared in the 1960s, and which has been periodically revised since then. FTM also publishes smaller-scale mapping which is readily available. Sheets published include an eight-colour 1 : 2 000 000 road and tourist map, Malagasy language relief maps at 1 : 1 250 000 and 1 : 2 500 000 scales, and a 1 : 2 000 000 administrative and population map. Larger-scale topographic series are published for some urban areas, and a new 1 : 10 000 scale town map series was started in the early 1980s to cover the largest towns in Madagascar. Detailed tourist maps have been issued for resort areas. There are currently plans to revise the national atlas of the country, which was compiled in the early 1970s by the Association des Géographes at the Université de Madagasikara.

Geological mapping of Madagascar is carried out by the Service Géologique, in the **Ministère de l'Industrie et de Mines (MIEM)**. Some of the older geological mapping recently became available, but most has not been included in our listings because of doubt about supplies. The three-sheet 1 : 1 000 000 scale geology and mineral map compiled in 1960 is still available. More recent hydrogeological mapping of the island has been published by the Service Hydrogéologique in MIEM: a 1 : 500 000 scale map in eight sheets and a single-sheet 1 : 2 000 000 scale map both give complete coverage.

The **Ministère de la Recherche Scientifique et Technologique pour le Développement** is currently engaged in a major resource inventory mapping programme. This map is published at a scale of 1 : 250 000

and will be available in six different editions for each quad: soils, land use, agriculture, forests, pasture and population. There is no further information yet available about this project.

The French **Office de la Recherche Scientifique et Technique Outre Mer (ORSTOM)** has carried out considerable earth resources mapping of the island at a variety of scales. A three-sheet soil map gives complete coverage at 1 : 2 000 000 scale and ORSTOM has also issued about 60 soil maps at scales of 1 : 200 000 and greater of certain areas of the island. Small-scale gravimetric, bioclimatic and magnetic mapping has been published by ORSTOM with explanatory monographs, and a three-sheet geomorphological map of the island showing agricultural potential was recently published. Vegetation maps at 1 : 1 000 000 in the series *Carte Internationale du Tapis Végétal et des Conditions Ecologiques* were issued in the 1960s by the **Institut Français de Pondicherry (IFP)**.

Further information

FTM issued a catalogue in 1979, and a 1986 short map title listing. A country report was submitted by Madagascar to the *6th United Nations Regional Cartographic Conference for Africa*, Addis Ababa: UN Economic Commission for Africa (1986).

Addresses

Foiben-Taosarintanin' i Madagasikara (FTM) (Institut National de Géodesie et Cartographie)
3 Lalana Ravelomanantsoa, BP 323, ANTANANARIVO.

Institut Géographique National (IGN)
136 bis rue de Grenelle, 75700 PARIS, France.

Ministère de l'Industrie de l'Energie et de Mines (MIEM)
Service Géologique et Service Hydrogéologique, BP 280, ANTANANARIVO.

Ministère de la Recherche Scientifique et Technologique pour le Développement
ANTANANARIVO.

Office de la Recherche Scientifique et Technique Outre Mer (ORSTOM)
70–74 Route d'Aulnay, 93140 BONDY, France.

For IFP see India; for DMA see United States.

Catalogue

Atlases and gazetteers

Madagascar, Réunion and the Comores: official standard names approved by the United States Board on Geographic Names
Washington, DC: DMA, 1953.
pp. 498.

General

Madagasikara: sarintany fizika 1 : 2 500 000
Antananarivo: FTM, 1977.
Relief map with French and Malagasey legend.

Madagasikara: sarintanin-dalana 1 : 2 000 000 9th edn
Antananarivo: FTM, 1984.
Tourist and road map on relief base.

Madagasikara: sarintany fizika 1 : 1 250 000
Antananarivo: FTM, 1983.
2 sheets, both published.

Topographic

Madagasikara: carte routière 1 : 500 000
Antananarivo: FTM, 1964–.
11 sheets, all published.

Geological and geophysical

Carte de la déclinaison magnétique à Madagascar pour le 1er Janvier 1961 1 : 2 500 000
Bondy: ORSTOM, 1964.

Cartes gravimétriques de Madgascar . . . 1 : 2 000 000 and
1 : 1 000 000
Bondy: ORSTOM, 1978.
4 sheets, all published.
1 : 2 000 000 Bouguer and isostatic anomaly maps, 1 : 1 000 000 single-colour Bouguer anomaly maps.
With accompanying explanatory text.

Carte géologique scolaire 1 : 1 000 000
Antananarivo: MIEM, 1964.
3 sheets, all published.

Carte hydrogéologique 1 : 500 000
Antananarivo: MIEM.
8 sheets, all published.

Environmental

Carte bioclimatique de Madagascar 1 : 2 000 000
Bondy: ORSTOM, 1973.
With accompanying text.

Carte pédologique de Madagascar 1 : 1 000 000
Bondy: ORSTOM, 1968.
3 sheets, all published.

Carte internationale de tapis végétal et des conditions écologiques 1 : 1 000 000
Pondicherry: IFP, 1965.
3 sheets, all published.

Cartes des conditions géographiques de la mise en valeur agricole de Madagascar: potential des unités physiques 1 : 1 000 000
Bondy: ORSTOM, 1981.
3 sheets, all published.
With accompanying text.

Administrative

Carte administrative et de densité de population 1 : 2 000 000
Antananarivo: FTM, 1977.

Town maps

Sarintany fizahantany 1 : 10 000
Antananarivo: FTM, 1982–.
2 published sheets.
Town map series, includes map of surrounding areas at 1 : 500 000 scale on sheet.

Malawi

The **Department of Surveys (MSD)** is responsible for the compilation of topographic and some thematic mapping of Malawi. In its present form the Department was established as a separate office in 1950 but prior to independence in 1963 the country was a British colony and Nyasaland, as it was then, was mapped by the British **Directorate of Overseas Surveys (DOS)** (now Overseas Surveys Directorate, Southampton, OSD). Some sheets now available in both the standard series date from DOS days, but MSD is revising the earlier maps and there are plans for entirely new specifications. The basic scale map is the *1 : 50 000 Malawi National series*. This five-colour series uses standard DOS style and covers the country in 159 sheets, some of which data back to 1965. Contours are at 20 or 50 m intervals. The current revision programme began in 1972 and efforts are at present concentrated on revising sheets to the north of latitude 11° S. This revision should be complete by the end of 1987. This series is also available as monochrome reductions at 1 : 100 000 scale. It is planned to publish a new metric 1 : 50 000 series using the UTM projection, with sheet lines based on UTM grid zones; the first pilot sheet in this series was issued in 1985.

No further revision of the old series will be carried out when the current programme is complete but it is intended to issue the two series until complete UTM cover is attained. The *1 : 250 000 Malawi National series* covers the country in 10 sheets, with 100 m contours and relief shading. The current revision cycle for this map began in 1984 and will be completed in 1987. A new UTM-based map at this scale will be published from about 1988 onwards. A single-sheet 1 : 1 000 000 map is regularly revised and is also available as a single-colour base map edition. From 1987 onwards this map will be published to UTM specifications. A new 1986 tourist map of the country at 1 : 500 000 scale has also been published by MSD and is available in flat or folded editions. The National Atlas was published by MSD in 1985 in loose-leaf format and includes a national gazetteer; it is also published as a school edition. MSD also issues thematic mapping for other government departments such as the Development Division, including the 1 : 3 000 000 scale Development Report mapping which was last revised in 1983–84 but is presently out of print. Large-scale township orthophotomapping is published by MSD for the major urban centres as diazo prints, with the exception of a multi-colour 1:16 000 plan of Blantyre and a planned two-sheet 1:15 000 multi-colour map of Lilongwe, currently available as preliminary monochrome edition.

There are plans to introduce an automated cartography system and to digitize all future 1:2500 scale township maps.

Geological mapping activities in Malawi are the responsibility of the **Geological Survey Department (GSD)**. Maps have been issued with accompanying explanatory bulletins. A full-colour 1:100 000 scale series, on a topographic base, published jointly with the British DOS, gives complete coverage of the country in 40 sheets and a single-sheet geological map at 1:1 000 000 is still available. The 1:250 000 geological atlas project begun in the 1970s is not yet completed and only four sheets are published to date.

Further information

For the background to the development of current cartographic activities see Martin, C.G.C. (1980) *Maps and Surveys of Malawi*, Rotterdam: Balkema.

Both MSD and GSD issue current map catalogues.

Addresses

Survey Department (MSD)
PO Box 349, BLANTYRE.

Overseas Surveys Directorate (OSD/DOS)
Ordnance Survey, Romsey Rd, SOUTHAMPTON SO9 4DH, UK.

Geological Survey Department (GSD)
PO Box 27, ZOMBA.

Catalogue

Atlases and gazetteers

National atlas of Malawi
Blantyre: MSD, 1985.
pp. 85.
Loose-leaf.

General

Malawi report map 1:2 500 000 DOS 975
Blantyre: MSD, 1971.

Malawi 1:1 500 000
Blantyre: MSD, 1986.

Malawi 1:1 000 000 5th edn
Blantyre: MSD, 1983.
Relief map, also available as monochrome base map.

Topographic

Malawi 1:250 000
Blantyre: MSD, 1975–.
10 sheets, all published. ■

Malawi 1:50 000
Blantyre: MSD, 1965–.
159 sheets, all published. ■

Geological and geophysical

Geological map of Malawi 1:1 000 000
Zomba: GSD, 1973.
With accompanying monograph: *The geology and mineral resources of Malawi*.

Malawi: geological survey 1:100 000
Zomba and Tolworth: GSD and DOS, 1958–75.
40 sheets, all published. ■
With accompanying explanatory bulletins.

Human and economic

Republic of Malawi: maps illustrating development projects
1:3 000 000
Blantyre: MSD, 1983–84.
36 sheets, all published.
Available as individual sheets (some now out of print) or bound in atlas format.

Town maps

Street guide and gazetteer to the city of Blantyre 1:16 000
Blantyre: MSD, 1974.

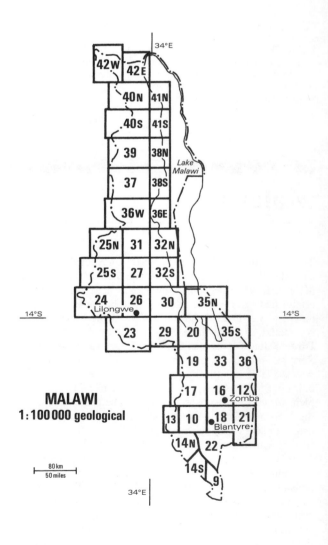

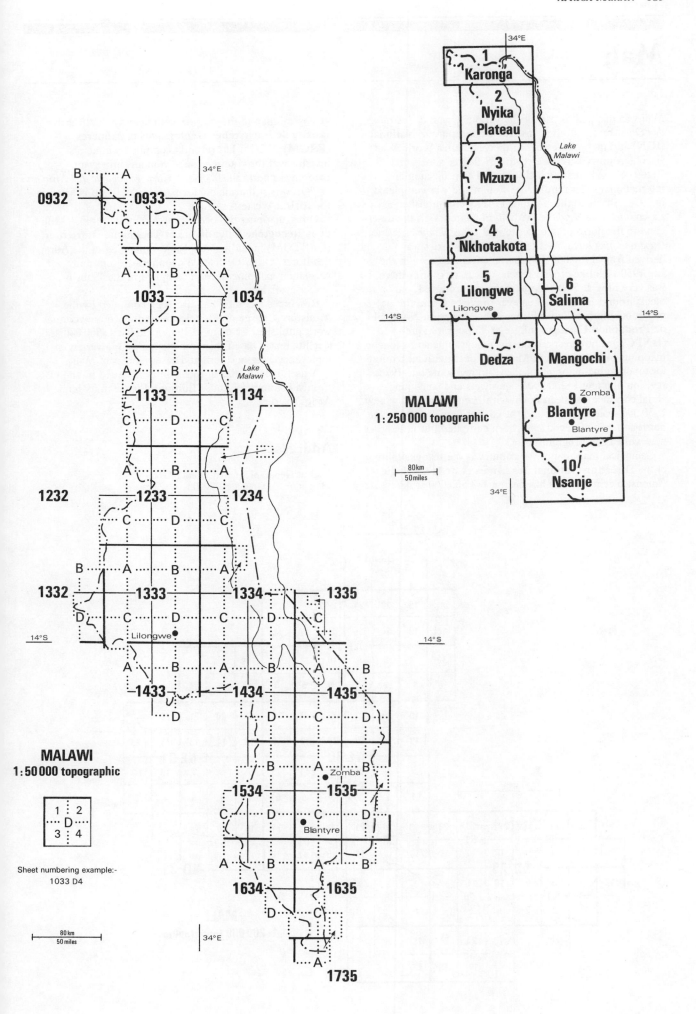

MALAWI
1:250 000 topographic

80km
50 miles

MALAWI
1:50 000 topographic

1	2
D
3	4

Sheet numbering example:-
1033 D4

80 km
50 miles

Mali

Mali was mapped as part of the various *Cartes de l'Afrique de l'Ouest* by the French **Institut Géographique National (IGN)** and its predecessors in the years after World War II. The country was covered in 133 degree sheets in a 1 : 200 000 scale map, which used UTM projection and international sheet numbering system and was completed in 1972. Sheets to the south of 17° N were published as six-colour *cartes regulières* with 40 m contours, in the desert areas to the north only uncontoured *fonds planimétriques* or *topographiques* have been published. This series was derived from 1 : 50 000 scale aerial coverage flown in the late 1950s. Unlike other French African territories there was very little 1 : 50 000 scale mapping produced, only a block around Bafoulabe was ever published, and this as provisional four-colour editions. The **Direction National de Production Cartographique et Topographique (DNPCT)** is now responsible for topographic and geodetic mapping and surveying of Mali, and deals with all requests for topographic mapping of the country. Some late-1970s revision of a few 1 : 200 000 scale sheets and some new aerial coverage has been flown with French aid. A few new 1 : 50 000 scale maps around the capital Bamako and to the north and east of Segou have been compiled but little other new work is on programme.

Geological mapping of the country is the responsibility of the **Direction National des Mines et de la Géologie**. Amonst recent projects has been a 1 : 1 500 000 scale geological map of the country compiled by the French **Bureau de Recherches Géologiques et Minières (BRGM)**. BRGM has recently compiled a minerals inventory of the country which contains important mapping. The 1 : 500 000 scale *Carte géologique de l'Afrique de l'Ouest* is still available for parts of Mali and is listed in the Africa, Western section.

Other mapping has been compiled by the French **Office de la Recherche Scientifique et Technique Outre Mer (ORSTOM)** including geophysical mapping of the Adrar des Ifores. Other ORSTOM geophysical, soils and vegetation mapping of Mali compiled in the 1950s is no longer available.

Resources mapping has also been compiled by the **Ministère Chargé du Développement**. A three-volume work published by them in 1983 included 1 : 500 000 scale satellite image based thematic mapping for a variety of topics, covering northern and eastern areas of Mali.

A useful small atlas of the country is issued by the French commercial house **Editions Jaguar** in its Jeune Afrique series.

Addresses

Institut Géographique National (IGN)
136 bis rue de Grenelle, 75700 PARIS, France.

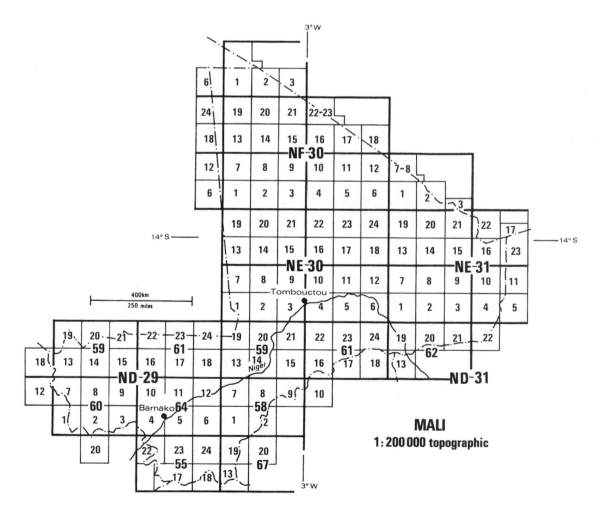

MALI
1 : 200 000 topographic

Direction National de Production Cartographique et
Topographique (DNPCT)
BP 240, BAMAKO.

Direction National des Mines et de la Géologie
BP 223, Koulouba, BAMAKO.

Bureau de Recherches Géologiques et Minières (BRGM)
BP 6009, 45060, ORLEANS-CEDEX, France.

Office de la Recherche Scientifique et Technique Outre Mer
(ORSTOM)
70–74 Route d'Aulnay, 93140 BONDY, France.

Ministère Chargé du Développement Rural
BAMAKO.

For Editions Jaguar see France; for DMA see United States.

Catalogue

Atlases and gazetteers

Atlas du Mali (Atlas Jeune Afrique)
Paris: Editions Jaguar, 1980.
pp. 64.

*Mali: official standard names approved by the United States Board on
Geographic Names*
Washington, DC: DMA, 1966.
pp. 263.

General

République du Mali   1 : 2 500 000
Paris: IGN, 1971.

Topographic

République du Mali   1 : 200 000
Paris and Bamako: IGN, 1945–.
133 sheets, all published. ■

République du Mali   1 : 50 000
Paris: IGN, 1950–.
1560 sheets, 20 published.

Geological and geophysical

Plan minéral de la République du Mali
Orléans: BRGM, 1978.
pp. 650.
Includes 11 accompanying map sheets.

République du Mali: carte géologique   1 : 1 500 000
Orléans: BRGM, 1981.
2 sheets, both published.
Includes tectonic inset.
With accompanying explanatory text.

Town maps

Bamako et ses environs   1 : 20 000
Paris: IGN, 1980.
4 sheets, all published.

Mauritania

The West African state of Mauritania was mapped by the
French **Institut Géographique National (IGN)** and its
predecessors in the various *Cartes de l'Afrique de l'Ouest*
series in the postwar period. There has been very little new
mapping carried out after independence and the 1 : 200 000
series begun by the French in the 1950s remains the basic
scale series for nearly all the country. This map uses UTM
projection and is derived from medium-scale aerial
coverage. Only in the more settled western areas are sheets
published as six-colour *cartes regulières*, with 40 m
contours. The remaining 50 sheets covering Saharan areas
have only ever been published as uncontoured four-colour
fonds topographiques or as *fonds planimétriques*, with no
indication of topography. After independence,
responsibility for national surveying and mapping passed
to the **Division de la Topographie et de la Cartographie
(MDTC)** in Nouakchott. There was some revision of some
1 : 200 000 scale sheets carried out by IGN in aid
programmes in the 1970s, and new maps appear with an
updated title but MDTC has no plans or resources to
update the series on any systematic basis. A few sheets in a
1 : 50 000 series were published in the 1950s to cover parts
of the Niger valley, but this was never extended to the rest
of the country.

IGN has also recently published a general map of
Mauritania.

Geological mapping of Mauritania is the responsibility of
the **Direction des Mines et de la Géologie**, but mapping
has been compiled by the French **Bureau de Recherches
Géologiques et Minières (BRGM)**. The six-sheet
1 : 1 000 000 map compiled in the 1960s is no longer
available and the best geological coverage of the country
that is currently available for sale is the 1 : 2 000 000 scale
Carte géologique de l'Afrique occidentale, which we have
listed under Africa, Western.

Other organizations involved in producing maps of
Mauritania include **Office de la Recherche Scientifique et
Technique Outre Mer (ORSTOM)** which has carried out
soils, sedimentological and geophysical mapping in the
country.

A useful small atlas in the Jeune Afrique series gives
thematic overviews of the country and is distributed by
Editions Jaguar, and the Russian **Glavnoe Upravlenie
Geodezii i Kartografii (GUGK)** recently published a
general Russan language map of Mauritania with thematic
insets. The **Food and Agriculture Organization (FAO)** of
the UN recently published an atlas to cover the
development of oases in the country.

Further information

Mauritania (1979) Cartographic activities in Mauritania, report
submitted by the Government of Mauritania to the *4th United
Nations Regional Cartographic Conference for Africa*, Addis Ababa:
UN Economic Commission for Africa.

Addresses

Institut Géographique National (IGN)
136 bis rue de Grenelle, 75700 PARIS, France.

Division de la Topographie et de la Cartographie (MDTC)
Ministère de l'Equipement, BP 237, NOUAKCHOTT.

Bureau de Recherches Géologiques et Minières (BRGM)
BP 6009, 45060 ORLEANS-CEDEX, France.

Direction des Mines et de la Géologie
Ministère de l'Industrialisation et des Mines, BP 199,
NOUAKCHOTT.

Office de la Recherche Scientifique et Technique Outre Mer (ORSTOM)
70–74 Route d'Aulnay, 93140 BONDY, France.

For GUGK see USSR; for FAO see World; for Editions Jaguar see France.

Catalogue

Atlases and gazetteers

Atlas de la République Islamique de Mauritanie (Atlas Jeune Afrique)
Paris: Editions Jaguar, 1977.
pp. 64.

Mauritania: official standard names approved by the United States Board on Geographic Names
Washington, DC: DMA, 1966.
pp. 149.

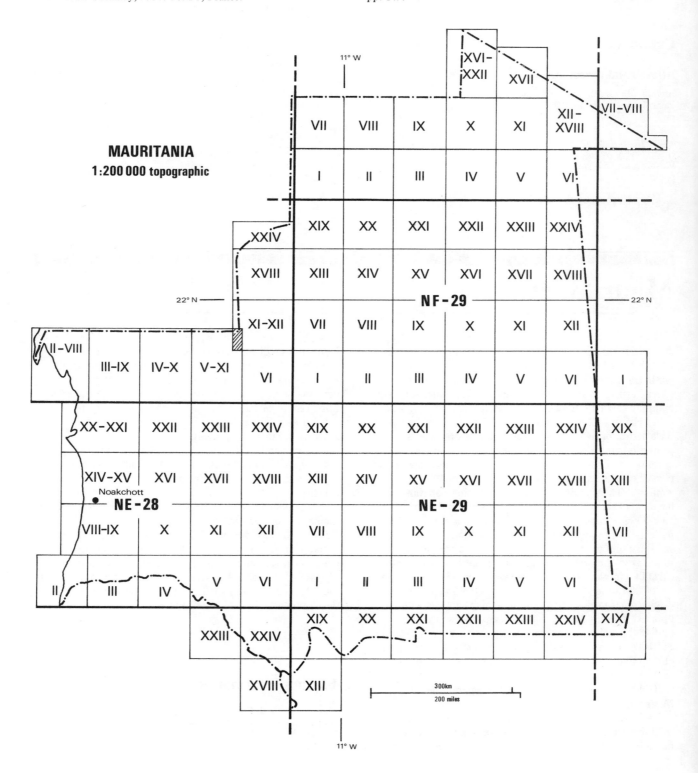

General

Mauritanie 1:2 500 000
Paris: IGN, 1980.

Mavritanija 1:2 500 000
Moskva: GUGK, 1986.
In Russian.
With economic and climate inset maps and placename index.

Topographic

République Islamique de Mauritanie 1:200 000
Paris and Nouakchott: IGN and MDTC, 1947–.
106 sheets, all published. ■

République Islamique de Mauritanie 1:50 000
Paris: IGN, 1950–.
1475 sheets, 27 published.

Bathymetric

Carte sédimentalogique du plateau continental mauritanien
1:200 000
Paris: ORSTOM, 1986.
With accompanying explanatory text.

Geological and geophysical

Cartes gravimétrique et magnétique du nord-Mauritanie 1:1 000 000
Paris: ORSTOM, 1971.
2 sheets, both published.
Single-colour magnetic map and gravimetric map with accompanying explanatory text.

Morocco

Morocco was mapped by French and Spanish colonial agencies prior to independence in 1956. 1:50 000 scale basic mapping was established for Spanish Morocco by Instituto Geográfico Nacional in Madrid and for French Morocco by **Institut Géographique National** in Paris. The French influence on Moroccan mapping has continued through technical and financial aid. The national mapping organization in Morocco since independence has been the **Division de la Cartographie (MDC)**. Current map series are all issued on the Lambert conical conformal projection and use the international hierarchical system for sheet numbering and division. Since 1976 Morocco has also been involved in the mapping of Western Sahara which is now designated as the 'Provinces du Sud'. The maps of this area are, however, not available for sale. The national basic scale series is published at 1:50 000 scale and will cover the country (including Western Sahara) in 1105 quarter-degree sheets. About 15 new sheets are being issued a year in the five-colour *Série orange*, older sheets in the four-colour *Série ancienne* are gradually being revised at about 20 sheets a year. The new specification sheets were first issued in 1977 with French and Arabic legends and show relief with 20 m contours. The *1:100 000 topo internationale géographique* series requires 300 sheets for complete coverage; sheets cover half-degree areas. Again the specification changed in 1977: *Série ancienne* sheets prior to this date were four-colour with 20 or 25 m contours, issued as Types 1922 or *cartes de reconnaissance*. The five-colour *Série orange* sheets after this date have 40 m contours and bilingual legends. Photomaps have been published to cover the whole of the area of former Spanish (Western) Sahara in 116 sheets at this scale. The 70-sheet 1:250 000 scale map is derived by generalizing 1:100 000 maps. To date only 18 sheets have been published, though another five are in the process of being compiled. An earlier 1:200 000 series compiled by the French is no longer available.

MDC is responsible for the compilation of map sheets and monographs in the National Atlas, which is published by the **Comité National de Géographie du Maroc**. This long project was begun in the 1950s and involved mapping at 1:1 000 000, 1:2 000 000 and 1:4 000 000 scales. New

sheets and accompanying *notices explicatives* are still being issued.

Large-scale urban mapping is also carried out by MDC. A five-colour series with Arabic and French legends is available as one- or two-sheet plans to cover about 50 of the major towns in the country. This series was begun in 1977 and is published at scales of 1:10 000 and 1:5000 (with the exception of Casablanca which is published at 1:12 500) and there is a revision programme, now producing seven revised plans a year. Photoplans are also available for a further 25 towns, mainly in Western Sahara.

Geological and earth science mapping of Morocco is carried out by the **Direction de la Géologie (MDG)** whose history can be traced back to the founding of the Service de la Carte Géologique in 1912. Prior to independence in 1956 both French authorities in French Morocco and Spanish in the former Spanish Morocco carried out geological surveys. The 1:50 000 Spanish sheets issued by the Instituto Geológico y Minera de España (IGME) were until recently still available. MDG now has an active publication programme of about eight new sheets a year covering geology, geotectonics, hydrogeology, minerals and other earth science themes. Small-scale general geological maps are issued to cover the whole country: important recent editions include a 1985 two-sheet 1:1 000 000 scale map, and a 1980 revision of the five-sheet 1:500 000 geological series. Thematic mapping at small scales is also issued, including aeromagnetic, structural and mineral maps of the country, and MDG has published two scales of satellite mapping of the country. Larger-scale series mapping is also available to give almost complete coverage of the country in nearly 200 sheets published in a mix of three scales; sheets are numbered according to the date of publication in chronological sequence. The oldest maps in these series date from 1928 and are the *cartes provisoires* published at 1:200 000 scale. Some modern postwar sheets are published at this scale but the current output is concentrating on 1:50 000 scale geological mapping of the northern areas (*carte géologique du Rif*) and on a 1:100 000 scale national geological series. Both of these larger-scale geological maps are published with accompanying explanations.

A variety of other governmental and foreign organizations are also responsible for the publication of earth science mapping in the country, including the **Division des Resources en Eau** and the **Service Cartographique des Sols et de l'Erosion**. The French **Office de la Recherche Scientifique et Technique Outre Mer (ORSTOM)** compiled a variety of larger-scale soils mapping, mainly of southern areas around Marrakech. Most of these 1 : 100 000 and 1 : 50 000 maps are no longer available. Amongst the commercial houses publishing mapping of Morocco are **Kümmerly + Frey, Editions Marcus, Reise und Verkehrsverlag, Ravenstein, Hallwag, Karto+Grafik** and **Michelin.**

Further information

Morocco (1979 and 1983) Country reports submitted to the *4th and 5th United Nations Regional Cartographic Conferences for Africa*, Addis Ababa: UN Economic Commission for Africa.
Catalogues are issued by both MDC and MDG.

Addresses

Institut Géographique National (IGN)
136 bis rue de Grenelle, 75700 PARIS, France.

Division de la Cartographie (MDC)
Direction de la Conservation Foncière et des Travaux Topographiques, 31 Ave Moulay Al Hassan, RABAT.

Comité National de Géographie du Maroc
Institut Scientifique, IURS, Chari Ma Al Ainine, AGDAL-RABAT.

Direction de la Géologie (MDG)
Ministère de l'Energie et des Mines, Nouveau Quartier Administratif, RABAT-CHELLAH.

Division de Resources en Eau
Direction de l'Hydralique, Ave. John Kennedy, BP 525, RABAT-CHELLAH.

Service Cartographique des Sols et de l'Erosion
Direction de la Recherche Agronomique, Ave. de la Victoire, BP 415, RABAT.

Office de la Recherche Scientifique et Technique Outre Mer (ORSTOM)
70–74 route d'Aulnay, 93140 BONDY, France.

For Editions Marcus see France; for Kümmerly + Frey and Hallwag see Switzerland; for Ravenstein, Karto+Grafik and Reise und Verkehrsverlag see German Federal Republic; for DMA see United States.

Catalogue

Atlases and gazetteers

Atlas du Maroc
Rabat: Comité National de Geographie du Maroc, 1954–.
24 published sheets and texts.

Morocco: official standard names approved by the United States Board on Geographic Names
Washington, DC: DMA, 1970.
pp. 923.

General

Carte générale du Maroc   1 : 5 000 000
Rabat: MDC, 1980.

Carte générale du Maroc   1 : 2 500 000
Rabat: MDC, 1979.

Photo satellite du Maroc   1 : 2 000 000
Rabat: MDG, 1982.

Maroc/Morocco   1 : 1 500 000
Rabat: MDC, 1980.
English, French and Arabic legend.

Maroc: carte routière   1 : 1 400 000   4th edn
Rabat: Marcus, 1986.
Road map with relief background.

Photo satellite du Maroc   1 : 1 000 000
Rabat: MDG, 1982.

Morocco: road map   1 : 1 000 000
Bern: Kümmerly + Frey, 1985.

Grosse Landkarte Marokko   1 : 800 000
München: Reise und Verkehrsverlag.
Includes Western Sahara at 1 : 2 500 000 scale and small town plans of Tangier, Rabat and Marrakech.

Marokko   1 : 750 000
Moskva: GUGK.
In Russian
With placename index and thematic inset maps.

Topographic

Carte du Maroc   1 : 250 000
Rabat: MDC, 1983–.
70 sheets, 18 published. ■

Carte du Maroc   1 : 100 000
Rabat: MDC, 1977–.
300 sheets, all published. ■
Sheets being revised to *Série orange* specifications.

Carte du Maroc   1 : 50 000
Rabat: MDC, 1977–.
1105 sheets, *ca* 400 published. ■
Sheets being revised to *Série orange* specifications.

Geological and geophysical

Carte structurale du Maroc   1 : 4 000 000
Rabat: MDG, 1982.

Carte structurale du Maroc   1 : 2 000 000
Rabat: MDG, 1982.

Carte métallogenique du Maroc   1 : 2 000 000
Rabat: MDG, 1965.
3 sheets, all published.
Single-theme sheets, with accompanying memoir.

Carte des gîtes minéraux du Maroc   1 : 2 000 000
Rabat: MDG.
7 sheets, all published.
Single-theme sheets.

Carte minière du Maroc   1 : 2 000 000
Rabat: MDG, 1982.

Carte géologique du Maroc   1 : 1 000 000
Rabat: MDG, 1985.
2 sheets, both published.

Carte aeromagnetique du Maroc   1 : 1 000 000
Rabat: MDG, 1980.
2 sheets, both published.

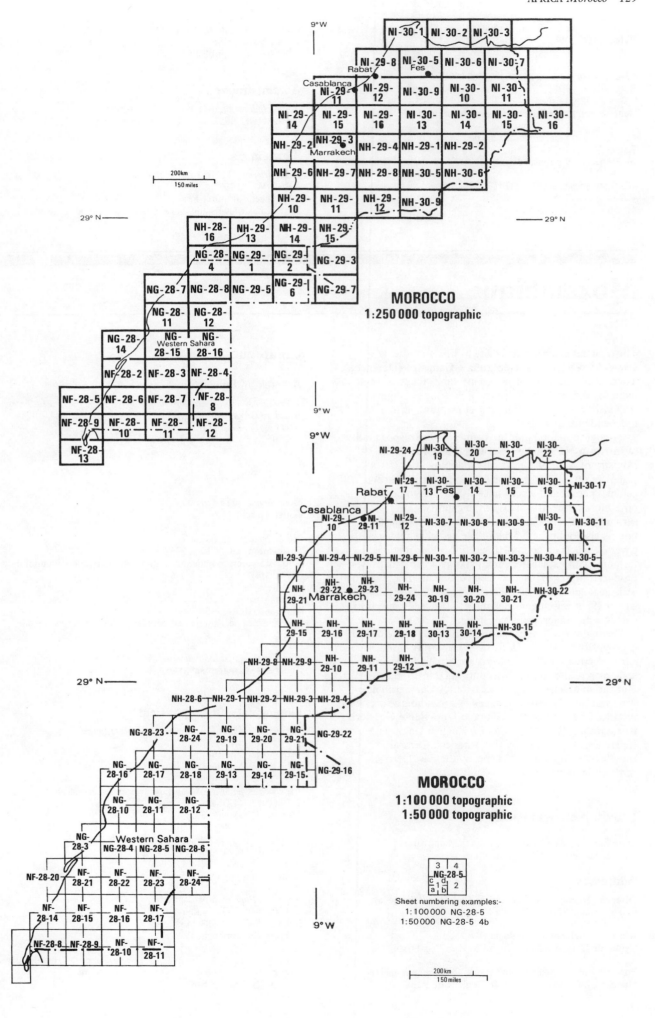

MOROCCO
1:250 000 topographic

MOROCCO
1:100 000 topographic
1:50 000 topographic

Sheet numbering examples:-
1:100000 NG-28-5
1:50000 NG-28-5 4b

Carte géologique du Maroc 1 : 500 000
Rabat: MDG, 1980.
5 sheets, all published.

Carte gravimétrique du Maroc 1 : 500 000
Rabat: MDG, 1971.
7 sheets, all published.
Bouguer anomaly editions with explanatory memoir.

Carte géologique du Maroc 1 : 200 000
Rabat: MDG, 1928–.
ca 90 sheets, 19 published.

Carte géologique du Maroc 1 : 100 000
Rabat: MDG, 1950–.
ca 160 sheets, 31 published.

Carte géologique du Rif 1 : 50 000
Rabat: MDG, 1966–.
ca 90 sheets, 25 published.

Administrative

Carte administrative du Maroc 1 : 2 500 000
Rabat: MDC, 1982.

Town maps

Plans urbains 1 : 5000 and 1 : 10 000
Rabat: MDC, 1977–.
ca 75 sheets, all published.
Covers major towns and cities in single or double sheets.

Mozambique

The national mapping authority in Mozambique is
Direcção National de Geografia e Cadastro (Dinageca).
There is very little information available about current
mapping activities in the country, but the Portuguese were
responsible for the compilation of mapping until
independence in 1975.

Geological mapping of the country is the responsibility
of **Instituto Nacional de Geologia (INGM)** in the
Ministry of Natural Resources. Mapping at 1 : 250 000
scale was established by Portuguese authorities in the
colonial period and stocks have recently become available
once more for central areas of the country. An
aeromagnetic series at this scale and on the same sheet
lines is also published. Small-scale geomorphological
coverage has been compiled and full-colour mineral,
tectonic (with explanatory text) and geological maps were
produced in the 1970s at 1 : 2 000 000 scale. A 1 : 1 000 000
scale geological map of the country was due for publication
late in 1986: this supersedes the 1 : 2 000 000 scale coverage
which is no longer available.

Thematic coverage of the southern areas of the country
is provided by the *Afrika Kartenwerk* southern series. (See
Africa section for details of this series.) The best small-
scale map of the country was published in 1986 by the
Hungarian official mapping distributor **Cartographia**.
Amongst other small-scale maps of Mozambique are sheets
produced by the Russian **Glavnoe Upravlenie Geodezii i
Kartografii (GUGK)** in the new English-language
Reference Map Series, by the American **Central
Intelligence Agency (CIA)** and by the East German
commercial house **Haack**.

Further information

A current Catalogue is available from the Centro de
Documentação of INGM in Maputo.

Addresses

Direcção Nacional de Geografia e Cadastro (Dinageca)
Caixa Postal 288, Aeroporto de Maputo, MAPUTO.

Instituto Nacional de Geologia (INGM)
Centro de Documentação, Caixa Postal 217, MAPUTO.

For CIA and DMA see United States; for Cartographia see
Hungary; for Haack see German Democratic Republic.

Catalogue

Atlases and gazetteers

*Mozambique: official standard names approved by the United States
Board on Geographic Names*
Washington, DC: DMA, 1969.
pp. 505.

General

Mozambique 1 : 4 400 000
Gotha: Haack, 1985.

Mozambique 1 : 3 860 000
Washington, DC: CIA, 1973.
With vegetation, population, mining, industry and agricultural
inset maps.

Mozambique 1 : 2 500 000
Moskva: GUGK, 1986.
With climatic, economic and population insets and place-name
index.

Handy map of Mozambique 1 : 2 000 000
Budapest: Cartographia, 1986.

Geological and geophysical

República Popular de Moçambique: carta tectônica
1 : 2 000 000   R.S. Afonso
Maputo: INGM, 1977.
With explanatory text.

República Popular de Moçambique: carta geomorphologica
1 : 2 000 000   I.G. Bondyerev
Maputo: INGM, 1983.

*República Popular de Moçambique: carta de jazigos e
ocorrencias* 1 : 2 000 000   J.A.C. Gouveia
Maputo: INGM, 1977.
With explanatory text.

República Popular de Moçambique: carta geomorfológica
1 : 1 000 000   I.G. Bondyrev
Maputo: INGM, 1983.
2 sheets, both published.

República Popular de Moçambique: carta geológica 1 : 1 000 000
Maputo: INGM, 1986.
2 sheets, both published.

República Popular de Moçambique: carta geologica 1:250 000
Maputo: INGM.
13 published sheets.

República Popular de Moçambique: carta aeromagnetometrica 1:250 000
Maputo: INGM.
20 published sheets.

Namibia

Until the 1960s there was no medium- or large-scale topographic mapping of Namibia. The German colonial authorities had carried out some plane table surveys at 1:100 000 and 1:200 000 prior to World War I, and a 17-sheet topocadastral series at 1:500 000 scale had been published as the national basic scale in 1925–27. Not until 1966 was a decision made by the South African **Directorate-General of Surveys** to begin a comprehensive modern basic scale map series. The scale chosen was 1:50 000 and the first sheets in this new series were issued in 1975, using a Gauss conformal projection, with a 10 m contour interval except in areas of low relief where 5 m were used. Like the South African 1:50 000 series, this map also portrays cadastral information. Orthophotomaps have been issued in this series to cover some of the unpopulated areas in the Namib desert. Sheets cover quarter-degree quads and use the same numbering system as the South African Survey. Much of the compilation work is carried out on contract in South Africa. Since 1973, the **South West Africa Surveyor General's Department (SGD)** has been responsible for the publication of this and other topographical series mapping of Namibia. The 1:50 000 series now gives complete coverage of the country and was completed in 1979. The full six-colour editions of all sheets are not yet available, but diazo copies of machine compilations are publicly available in Windhoek where the full edition has not yet been published. A limited revision programme, using high-altitude super-wide-angle aerial photography, has been started.

A 1:250 000 scale series was begun in 1960, compiled from 1:100 000 scale mapping. The current series at this scale is derived from the 1:50 000 mapping and uses a 50 m contour interval with 300 m interval hypsometric tints: it is as yet not complete, and some maps are out of print. It has recently become difficult to obtain certain published 1:50 000 and 1:250 000 sheets, particularly those near to the Angolan border and sheets to the north of 19° N are now officially restricted.

SGD also issues a 14-sheet 1:500 000 series using 100 m contour interval and 300 m hypsometric tints. This has the same specifications as the South African series at this scale and is issued in either topographic or aeronautical editions. The single sheet 1:1 000 000 wall map of Namibia compiled by SGD from recent 1:250 000 series maps uses Albers equal area projection, and presents cadastral information on a hypsometric tinted base. SGD has also published 1:10 000 orthophotomapping of the capital Windhoek and of Walvis Bay on the same specifications as the South African national series. A street plan of Windhoek is published at 1:20 000 scale by the **Town Clerk's Department**.

The *National Atlas of South West Africa/Namibia* issued by the **Director of Development Co-ordination** for SGD in 1983 was compiled in the Institute for Cartographic Analysis of the University of Stellenbosch in South Africa.

Geological mapping of Namibia has been carried out by the **Geological Survey of South West Africa**, a four-sheet 1:1 000 000 full-colour sheet gives complete coverage and there is a 1:250 000 programme. To date only four printed sheets on topographic sheet lines have been published in this series, but ammonia prints of maps at this scale are available from the Geological Survey. A wide variety of other provisional mapping is held as ammonia prints and available on demand, including aeromagnetic contour and aeroradiometric maps at 1:250 000 and 1:50 000.

A variety of other small-scale and general maps of the country are published by overseas agencies. The best of these is a two-sheet, six-colour, 1:1 000 000 scale **United Nations** map, which uses a combination of satellite imagery and thematic and topographic data to provide a general physical and human overview of the country. Among other publishers of maps of Namibia are the American **Central Intelligence Agency (CIA)**.

Further information

Martin, C. (1979) The survey of Namibia, *Chartered Land Surveyor/Chartered Minerals Surveyor*, **1**(3), 1–24.
Leser, H. (1982) Namibia-Südwestafrika: kartographische Probleme der neuen topographischen Karten 50 000 und 250 000 und ihre Perspektiven für die Landesentwicklung, *Mitteilungen der Basler Afrika-Bibliographien*, **26**, 1–56.
Both the Surveyor General's Department and the Geological Survey issue catalogues.

Addresses

Directorate of Surveys and Mapping
Private Bag, MOWBRAY, Republic of South Africa, 7705.

Surveyor General's Department
Private Bag 13182, WINDHOEK, 9000.

Director of Development Coordination
Private Bag 12025, WINDHOEK, 9000.

Town Clerk's Department
PO Box 59, WINDHOEK, 9000.

Geological Survey
PO Box 2168, WINDHOEK, 9000.

For United Nations see World; for CIA and DMA see United States.

Catalogue

Atlases and gazetteers

National atlas of South West Africa/Namibia compiled by the Institute for Cartographic Analysis, University of Stellenbosch Cape Town: National Book Printers, 1983.
pp. 92.

South Africa: official standard names approved by the US Board on
Geographic Names
Washington, DC: DMA, 1954.
2 volumes.
Vol. 2 covers South West Africa.

General

Namibia and Walvis Bay   1 : 4 000 000
Washington, DC: CIA, 1978.
Includes 4 small ancillary thematic maps.

Namibia   1 : 1 000 000
New York: United Nations, 1986.
2 sheets, both published.
Includes 3 inset maps at 1 : 4 000 000 showing land use, minerals,
fisheries and geology.

Topographic

Southern Africa   1 : 500 000
Windhoek: SGD, 1977–.
16 sheets.
Issued as topographical and aeronautical editions.

South West Africa   1 : 250 000
Windhoek: SGD, 1975–.
42 sheets, 29 published. ■

South West Africa   1 : 50 000
Windhoek: SGD, 1975–.
1218 sheets, all published. ■

Geological and geophysical

South West Africa/Namibia: geological map   1 : 1 000 000
Windhoek: Geological Survey, 1980.
4 sheets, all published.

South West Africa geological map series   1 : 250 000
Windhoek: Geological Survey, 1975–.
43 sheets, 4 published. ■

Administrative

South West Africa/Namibia   1 : 1 000 000
Windhoek: SGD, 1980.
Topocadastral wall map.

Town maps

Windhoek street plan   1 : 20 000
Windhoek: Town Clerk's Dept., 1985.

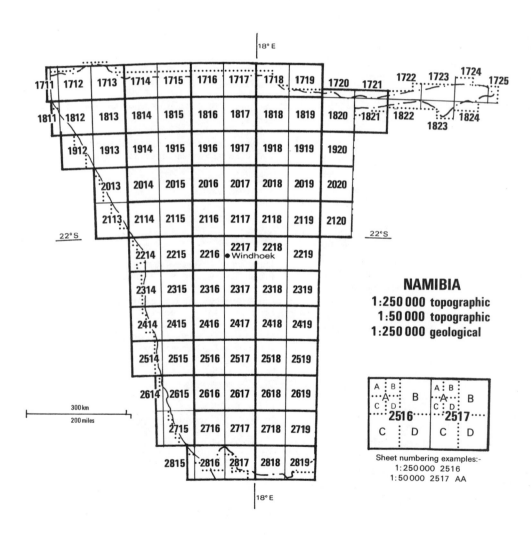

Niger

The French **Institut Géographique National (IGN)** and its agency in Niamey (which closed in September 1985) carried out the first modern mapping of Niger. Using the UTM projection and the international sheet numbering system the country was covered in 119 1° sheets of the 1 : 200 000 scale *Carte de l'Afrique de l'Ouest*, retitled after independence *République du Niger 1 : 200 000*. Sheets in the north-eastern part of the country were compiled as *fonds topographiques* or *planimétriques* without contours. Only the southern and western areas were published showing relief by 40 m contours. The four-colour 1 : 50 000 series was only ever compiled for the extreme southern border areas, mostly as provisional four-colour editions, and under 10 per cent of Niger falls within its coverage. IGN has also published a general map of the country and town plans of the major cities. The **Service Topographique et du Cadastre** has since independence been the government agency responsible for topographic and geodetic survey activities. It has set in train a limited revision programme, and new aerial coverage at 1 : 60 000 was flown in the mid-1970s for all the developed areas of the country. The 1 : 50 000 series is being revised for areas of economic interest and there are plans to extend the series to other southern areas. For permission to acquire mapping of Niger one should approach the **Ambassade du Niger** in Paris.

Geological survey activities are carried out by the **Direction des Mines et de la Géologie**, including several 1 : 200 000 full-colour geological sheets on 1 : 200 000 topographic sheet lines, compiled with assistance from the German **Bundesanstalt für Geowissenschaften und Rohstoffe (BfGR)** and the French **Bureau de Recherches Géologiques et Minières (BRGM)**. No current complete coverage of the country in geological mapping is available, other than the *Carte géologique de l'Afrique de l'Ouest* 1 : 2 000 000, published by BRGM in the 1960s, which we have listed under Africa, Western.

Other organizations involved in mapping Niger include **Office de la Recherche Scientifique et Technique Outre Mer (ORSTOM)**, which has compiled monographic soil mapping of parts of the country, especially of the south, and a 1 : 1 000 000 scale gravimetric map of the whole of Niger. A useful thematic atlas has been published in the Jeune Afrique series by **Editions Jaguar**, and the Russian **Glavnoe Upravlenie Geodezii i Kartografii (GUGK)** has compiled a Russian language general map of the country.

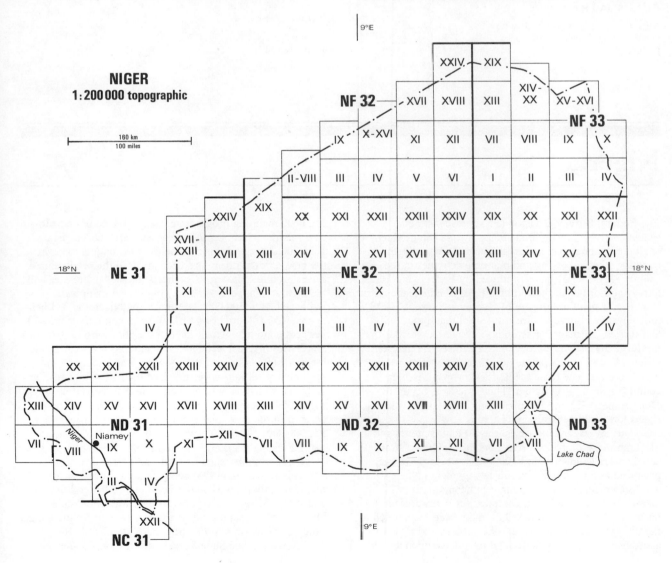

NIGER
1 : 200 000 topographic

160 km / 100 miles

Further information

Niger (1979) La cartographie du Niger: paper presented to the *4th United Nations Regional Cartographic Conference for Africa*, Addis Ababa: UN Economic Commission for Africa.

Addresses

Institut Géographique National (IGN)
136 bis rue de Grenelle, 75700 PARIS, France.

Service Topographique et du Cadastre
PO Box 250, NIAMEY.

Ambassade du Niger
154 rue de Longchamp, 75016 PARIS, France.

Direction des Mines et de la Géologie
BP 257, NIAMEY.

Office de la Recherche Scientifique et Technique Outre Mer (ORSTOM)
70–74 Route d'Aulnay, 93140 BONDY, France.

For Editions Jaguar and BRGM see France; for GUGK see USSR; for BfGR see German Federal Republic; for DMA see United States.

Catalogue

Atlases and gazetteers

Atlas du Niger (Atlas Jeune Afrique)
Paris: Editions Jaguar, 1980.
pp. 64.

Niger: official standard names approved by the United States Board on Geographic Names
Washington, DC: DMA, 1966.
pp. 207.

General

République du Niger   1 : 2 500 000
Paris: IGN, 1977.

Niger   1 : 2 500 000
Moskva: GUGK, 1982.
In Russian.
With economic, population and climatic inset maps and placename index.

Topographic

République du Niger   1 : 200 000
Paris and Niamey: IGN, 1955–.
119 sheets, all published. ∎

République du Niger   1 : 50 000
Paris and Niamey: IGN, 1952–.
1810 sheets, 134 published.

Geological and geophysical

Cartes gravimétriques du Niger   1 : 1 000 000
Bondy: ORSTOM, 1969.
5 sheets, all published.
With accompanying explanatory text.

Town maps

Niamey et ses environs   1 : 20 000
Paris: IGN, 1978.

Nigeria

The **Federal Survey Division (FS)** is responsible for geodetic surveys, topographic mapping and the production of all the national map series, and acts as a coordinating body for other surveying and mapping activities in the country. It prints all maps produced by Federal and State Survey authorities in the country. Prior to independence in 1960 the British **Directorate of Overseas Surveys (DOS)** (now Overseas Surveys Directorate, Southampton, OSD) had carried out mapping activities in Nigeria, which produced a variety of partial coverage in several different series. Most of the map base now dates from after independence. The basic scale series for much of the country is the 1 : 50 000 map. This series was begun by DOS in the 1950s and requires about 1200 quarter-degree sheets for complete coverage. There is now almost complete coverage. This series uses 50 foot contours and the Transverse Mercator projection and is derived from aerial coverage flown at 1 : 40 000 scale. Sheets have been produced as a result of joint projects involving Nigerian and foreign surveyors. Preliminary uncontoured planimetric maps had been compiled for many areas in the north-east by the DOS; these have now been recompiled to the full specification as a result of Canadian and British assistance. In 1981 it was decided to embark upon a

1 : 25 000 mapping programme using 10 m contours. Most of the country has been flown in new aerial coverage at 1 : 25 000 scale, and this is to be used to compile maps in the new series. Priority has been given to the production of sheets around State capitals: sheets for the Sokoto area will be the first to appear. A variety of derived mapping is published. A 1 : 100 000 series covering Nigeria in 343 half-degree, five-colour sheets with 100-foot contours is now 60 per cent complete, whilst the full specification 1 : 250 000 map series (five-colour, 100 foot contours) is also available for about 60 per cent of the country. Uncontoured preliminary editions cover much of the rest. Other smaller-scale maps include 16-sheet complete coverage in old series 1 : 500 000 maps, and about one-quarter of the country is covered in the new smaller format 1 : 500 000 series. 1 : 1 000 000, 1 : 2 000 000 and 1 : 3 000 000 physical and administrative maps are also published. The *National Atlas of Nigeria* was published in 1981 and a gazetteer of place names is at present (1986) being printed. Larger-scale mapping activities are shared between the Federal Survey and the 19 State Survey Offices. Sixty towns and cities including all 19 state capitals are in the township mapping programme which is producing maps at either 1 : 1000 or 1 : 2000 scales. The State Surveys are responsible for the

production of cadastral mapping, and the joint compilation of township maps, administrative and road maps of their respective States.

The availability of Nigerian FS mapping must at present be regarded as problematic. Major jobbers record material as being available and FS issued a 1984 map catalogue, but no replies were received from FS in our questionnaire survey. We have nevertheless listed the major maps and series.

Geological mapping activities are the responsibility of the **Nigerian Geological Survey Department**. A 1:250 000 scale series using the topographic base will cover the country in 85 sheets and there is 80 per cent coverage in a 1:100 000 scale series. Smaller-scale geological coverage is also compiled and unpublished geophysical mapping is maintained and is available as diazo prints. Efforts are being concentrated on those areas with rich mineral resources such as the Niger Delta oilfields and the potential water resource developments of the Sokoto-Rima and Chad basins.

An interesting application of SLAR imagery resulting in the publication of a full-colour 69-sheet vegetation map of the country took place in 1976–78. This use of radar imaging overcame cloud cover problems and allowed accurate and very detailed depiction of vegetation formations and sub-formations on to 1:250 000 JOG base maps. The maps were compiled for the **Federal Department of Agriculture** in a joint project between the American Motorola Company and the British Hunting Surveys, and were eventually published in 1983.

Other Nigerian official mapping agencies include the **Land Resources Department** (soil mapping), the **Hydrographic Surveys Department** (marine and nautical charting), **Department of Civil Aviation** (aeronautical charting) and the **Inland Waterways Department**. None of their maps is available at present.

Other thematic mapping of Nigeria has been carried out by the British DOS for the **Land Resources Development Department (LRDC)**. This includes soil and land resources mapping of the northern and central areas of the country, which was published throughout the 1970s at scales of 1:250 000, 1:500 000 and 1:1 000 000 to accompany land resource studies. Smaller-scale overviews of the resources of Central Nigeria were issued in atlas format by LRDC in 1981. Most of this mapping is still available from Tolworth, and we have listed the atlas in the catalogue section.

Amongst the commerical maps published which provide general coverage of Nigeria are a Russian language map issued by **Glavnoe Upravlenie Geodezii i Kartografii (GUGK)**, a **Bartholomew** world travel series sheet, and a road map of the country issued by **Macmillan Nigeria Publishers**. There is no single commercial company in Nigeria whose products are solely maps: road maps and street guides of major cities are produced and some book publishing houses like Macmillan market atlases and wall maps.

Further information

The best short summary of current mapping activities in Nigeria is Barbour, K.M. (1982) Surveys and mapping, in *Nigeria in Maps*, edited by K.M. Barbour *et al.*, London: Hodder and Stoughton.

A rather fuller and more recent treatment is Balogen O.Y. (1985) Surveying and mapping in Nigeria, *Surveying and Mapping*, **45**(4) 347–355.

Background information is given in Nigerian reports to UN Regional Cartographic Conferences, Nigeria (1979 and 1983) Reports on Cartographic Activities in Nigeria, presented to the *4th and 5th United Nations Regional Cartographic Conferences for Africa*, Addis Ababa: UN Economic Commission for Africa.

For further information about the Hunting Surveys compiled radar vegetation mapping see Parry, D.E. and Trevett, J.W. (1979) Mapping Nigeria's vegetation from radar, *Geographical Journal*, **145**(2), 265–281.

Addresses

Federal Survey Division (FS)
Federal Ministry of Works, 5 Tafewa Balewa Square, PMB 12596, LAGOS.

Overseas Surveys Directorate (DOS/OSD)
Ordnance Survey, Romsey Rd, SOUTHAMPTON SO9 4DH, UK.

Nigerian Geological Survey Department
Ministry of Mines and Power, PMB 2007, KADUNA SOUTH.

Federal Department of Forestry
Ministry of Agriculture and Rural Development, PMB 12613, LAGOS.

Land Resources Department
Ministry of Agriculture and Rural Development, PMB 12613, LAGOS.

Hydrographic Surveys Department
Nigerian Ports Authority, 26–28 Marina, PMB 12588, LAGOS.

Department of Civil Aviation
Ministry of Aviation, New Federal Secretariat Building, Bedwell Rd, LAGOS.

Inland Waterways Department
Federal Ministry of Transport, Old Secretariat Buildings, Marina, LAGOS.

Macmillan Nigeria Publishers Ltd
Scheme 2, Oluyole Industrial Estate, Lagos–Ibadan Expressway, PO Box 1463, IBADAN.

For Bartholomew and LRDC see Great Britain; for GUGK see USSR; for DMA see United States.

Catalogue

Atlases and gazetteers

National atlas of Nigeria
Lagos: FS, 1981.
pp. 136.

Nigeria in maps K.M. Barbour *et al.*
London: Hodder and Stoughton, 1982.
pp. 148.

Nigeria: official standard names approved by the United States Board on Geographic Names
Washington, DC: DMA, 1971.
pp. 641.

General

Nigerija 1:2 000 000
Moskva: GUGK, 1977.
In Russian.
With climatic, ethnographic and economic inset maps and placename index.

Nigeria 1:1 500 000
Edinburgh: Bartholomew, 1980.

Road map of Nigeria 1:1 500 000
Ibadan: Macmillan, 1982.

Nigeria   1:500 000
Lagos: FS, 1972–.
33 sheets, 10 published.

Topographic

Nigeria   1:250 000
Lagos: FS, 1958–.
85 sheets, *ca* 50 published.

Nigeria   1:100 000
Lagos: FS, 1950–.
343 sheets, *ca* 200 published. ■

Nigeria   1:50 000
Lagos: FS, 1954–.
1200 sheets, all published. ■

Geological and geophysical

Geology and minerals map of Nigeria   1:3 000 000
Lagos: FS, 1981.

Geological survey of Nigeria   1:250 000
Lagos: FS, 1957–.
85 sheets, *ca* 30 published.

Environmental

Land resources of Central Nigeria: an atlas of resources maps   1:1 500 000
Tolworth: LRDC, 1981.
19 maps.

Nigeria: vegetation and land use   1:250 000
Lagos: Federal Forestry Dept., 1983.
69 sheets, all published.

Administrative

Administrative map of Nigeria   1:1 000 000
Lagos: FS, 1977.
4 sheets, all published.

Town maps

Lagos: 1:20 000
Lagos: FS.

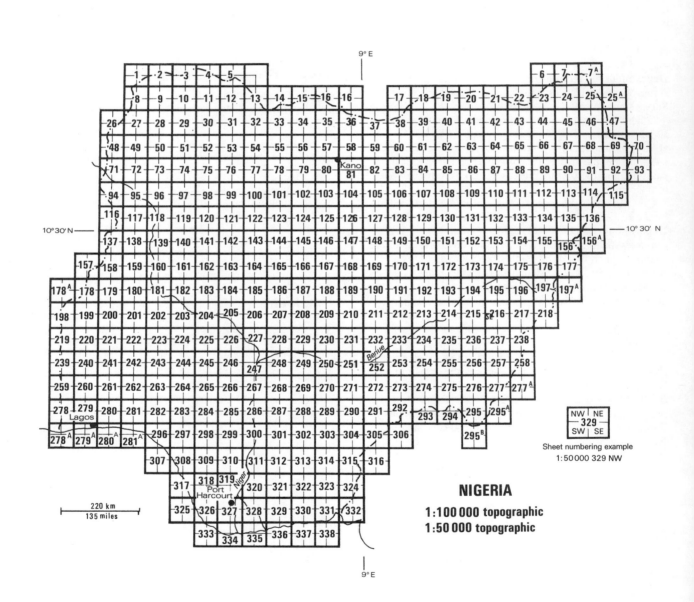

NIGERIA

1:100 000 topographic
1:50 000 topographic

Rwanda

Rwanda has been mapped by Belgian colonial authorities and, after independence in 1962, with Belgian assistance by the **Service Cartographique du Rwanda (SCR)**. Plane table surveys carried out in the 1930s were used in the production of a monochrome 1:100 000 scale series produced by the Belgian Ministry of Colonies. This map is still available from the **Musée Royale de l'Afrique Centrale** (MRAC) and shows relief with 50 m contours. The basic scale map is a 1:50 000 series, using UTM projection. The old 1:50 000 series was published by MRAC and is available as one-colour ozalid prints. This planimetric map is derived by aerotriangulation from the first aerial coverage flown in 1960 and gives complete coverage of the country in 48 sheets. From these surveys a 1:100 000 scale map in 13 sheets was also compiled (only six sheets of this smaller-scale planimetric map were issued), as was a single-sheet 1:250 000 scale map.

An aid project was set up to equip a new cartographic service within Rwanda, one of its aims being to set up a new multi-colour 1:50 000 base map, to be photogrammetrically derived, but only eight sheets reached publication when the Belgian aid ran out in 1980. We have included the graphic index for the new series, but have provided catalogue entries for the MRAC mapping as well. Aerial coverage has been flown at 1:20 000 scale, with urban areas flown at 1:10 000. SCR is producing 1:20 000 and 1:10 000 maps of some areas and urban mapping programmes are being carried out at 1:5000 and 1:2000 scales. A range of administrative mapping of the country is available including 143 sheets published in the monochrome 1:25 000 scale *carte communale*. A range of other thematic mapping is also available at 1:500 000 and 1:250 000 scales. Many of these sheets were prepared for the National Atlas, which was published in 1981 by **Université National de Rwanda**, but they are available as individual sheets from SCR and some are listed in the catalogue section.

Other mapping produced as a result of bilateral agreements includes a 1:10 000 town plan of the capital Kigali and a new administrative and road map produced by the Belgian company Geotechnique for **Editions Delroisses**. Geological mapping of Rwanda has been carried out by MRAC, including a multi-colour 13-sheet 1:100 000 series and a 250 000 scale lithological map. The former was a joint project with the Rwandan **Service Géologique**, and is still in progress. Published 1:100 000 scale sheets cover the eastern areas of the country.

Further information

For background information to surveying and mapping activities in Rwanda see Rwanda (1979 and 1983) Rapport d'activités du Service Cartographique du Rwanda, papers presented to the *4th and 5th United Nations Regional Cartographic Conferences for Africa*, Addis Ababa: UN Economic Commission for Africa.
SCR publishes a useful *Catalogue des Cartes*.

Addresses

Service Cartographique du Rwanda (SCR)
Ministère des Travaux Publics et d l'Energie, BP 413, KIGALI.

Musée Royale de l'Afrique Centrale (MRAC)
BP 1980, TERVUREN, Belgium.

Service Géologique
Ministère du Commerce, des Mines et de l'Industrie, BP 15, RUHENGIRI.

Université Nationale de Rwanda
BP 117, BUTARE.

For Editions Delroisses see France; for INEAC see Zaire.

Catalogue

Atlases and gazetteers

Atlas du Rwanda
Kigali: National University, 1981.

Rwanda: official standard names approved by the United States Board on Geographic Names
Washington, DC: DMA, 1964.
pp. 44.

General

Carte hypsométrique du Rwanda   1:500 000
Kigali: SCR, 1980.

République Rwandaise: carte administrative et routière   1:250 000
Boulogne-Billancourt: Editions Delroisses, 1980.
Includes street plan of Kigali.

Topographic

Carte hypsométrique du Rwanda   1:200 000
Kigali: SCR.
4 sheets, all published.

République Rwandaise   1:100 000
Tervuren: MRAC, 1965–71.
13 sheets, 6 published. ■

République Rwandaise   1:50 000
Tervuren and Kigali: MRAC and SCR, 1962–68.
48 sheets, all published. ■

République Rwandaise   1:50 000
Kigali: SCR, 1980–.
43 sheets, 8 published. ■

Geological and geophysical

Carte lithologique du Rwanda   1:250 000
Tervuren and Ruhengeri: MRAC and Service Géologique, 1981.

Carte géologique du Rwanda   1:100 000
Tervuren and Ruhengeri: MRAC and Service Géologique, 1964–.
13 sheets, 6 published. ■

Environmental

Carte de la végétation du Rwanda et du Burundi   1:1 000 000
Bruxelles: INEAC, 1963.
Soils map.
With explanatory text.

Carte des régions naturelles du Rwanda   1:400 000
Kigali: SCR.

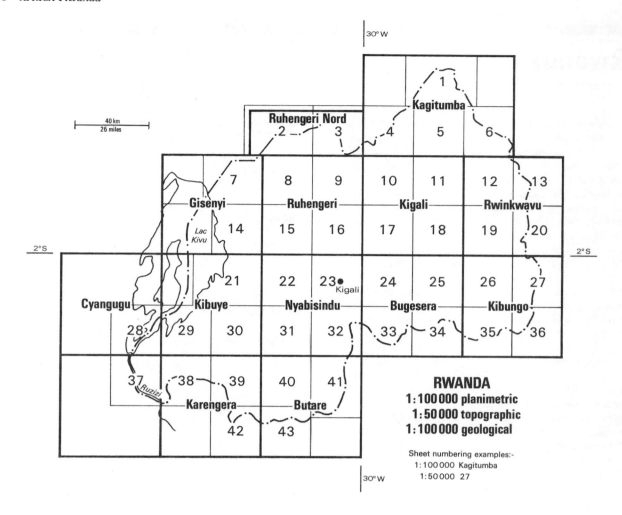

Sheet numbering examples:-
1:100000 Kagitumba
1:50000 27

Carte climatologique du Rwanda   1:250 000
Kigali: SCR.

Administrative

Carte administrative du Rwana   1:500 000
Kigali: SCR, 1985.

Human and economic

Carte de la santé publique   1:500 000
Kigali: SCR, 1973.

Town maps

Kigali   1:10 000
Kigali: SCR, 1980.

Senegal

Since 1972 the major responsibility for mapping and surveying of Senegal has been taken by the **Service Géographique National (SSGN)**. Prior to that date the French **Institut Géographique National (IGN)** and its predecessors produced maps of the country in the three *Cartes de l'Afrique Occidentale* series, through it agency in Dakar. IGN continues to influence mapping of Senegal and to publish some maps for the tourist. Senegal was covered in basic scale series of 1:200 000 and 1:50 000. These maps were compiled from complete 1:50 000 aerial coverage flown in the 1950s. There is complete coverage in 27 sheets at 1:200 000 scale in the four-colour, UTM based *Carte du Senegal,* which shows relief with 40 m contours. This series was begun by IGN in the 1950s and was completed in the early 1970s. New editions of some

sheets appeared in the early 1980s, after new 1:60 000 aerial coverage was flown to facilitate revision, but it is not intended to up-date this series as the basic scale map. From this 1:200 000 series were derived 1:1 000 000 and 1:500 000 scale mapping. SSGN intends to devote resources to the completion of the 1:50 000 series begun by the French in the 1950s and now published as *Carte du Sénégal 1:50 000.* This four-colour map will cover the country in 290 sheets on UTM projection, showing relief by 20 m contours. Blocks of nearly 90 sheets were published in the 1950s and 1960s to cover the Senegal valley, the coastal areas to the south of the River Gambia and the region around the capital Dakar. It is intended to up-date this coverage and to extend the series to the rest of the country. Recently, however, SSGN has concentrated on mapping for development schemes in the two major valley systems in the country. SSGN has flown new aerial cover at a variety of scales of the whole Senegal river basin and has produced a 1 : 10 000 scale orthophotomap series, whilst similar schemes in the Gambia valley in the south have seen the production of a 1 : 25 000 orthophotomap of the whole basin, with selective 1 : 10 000 coverage. Large-scale urban mapping projects are also in hand, with the intention of producing coverage of the regional capitals and urban centres. The different mapping projects are now integrated into a single policy to update the whole map base, with priorities allocated according to the economic development potential of the area concerned. Included in this policy are plans to cover low-priority areas in a new 43-sheet 1 : 100 000 scale map. The National Atlas maps the country at 1 : 1 500 000 scale, was published in 1977 and is still available through IGN. Other recent French published mapping includes a town plan of Dakar at 1 : 10 000 scale and the 1 : 1 000 000 scale tourist map published in 1980. Both were compiled by IGN.

Geological mapping of Senegal is now the responsibility of the **Direction des Mines et de la Géologie (SDMG)**, but much of the mapping was compiled during the years of French involvement by **Bureau de Recherches Géologiques et Minières (BRGM)**. A 1 : 200 000 scale full-colour geological series on the topographic sheet lines gives coverage of half of the country in 14 sheets, some of which are issued with explanatory texts. At 1 : 500 000 scale SDMG publishes or distributes four-sheet geological, mineral and geotectonic maps of the whole country. A recent project has involved the 1 : 20 000 scale mapping of the Cap Vert peninsula.

Soils mapping of Senegal is carried out by the French **Office de la Recherche Scientifique et Technique Outre Mer (ORSTOM)**. 1 : 200 000 scale soils maps have been compiled for large areas of the south of the country, and a single sheet covering the whole country at 1 : 1 000 000 scale was published in the 1960s. ORSTOM has also produced population and agricultural coverage at 1 : 1 000 000 scale of the Senegal river valley. Bathymetric and sedimentological mapping of the Seno-Gambian continental shelf was compiled by ORSTOM in the mid-1970s and is still available, as is the 1 : 2 000 000 scale insect distribution map of the country. Vegetation mapping of the country is available on IMW sheet lines: the single Dakar sheet at 1 : 1 000 000 gives almost complete coverage and was compiled in 1962. We have listed the vegetation mapping under Africa, Western. Another recent thematic mapping project has been funded through USAID, with the aim of producing a complete small-scale resources inventory of Senegal. Sheets are issued at scales of 1 : 500 000 and 1 : 1 000 000 for 10 different themes. There is no information about the availability of these recent compilations.

A useful small atlas of the country was published in the Jeune Afrique series and is distributed by **Editions Jaguar**.

Further information

Senegal (1979, 1983 and 1986) Communications de Sénégal to the *4th, 5th and 6th United Nations Regional Cartographic Conferences for Africa*, Addis Ababa: UN Economic Commission for Africa.

Addresses

Institut Géographique National (IGN)
136 bis rue de Grenelle, 75700 PARIS, France.

Service Géographique National (SSGN)
14 rue Victor Hugo, BP 740, DAKAR.

Direction des Mines et de la Géologie (SDMG)
Route de Ouakam, BP 1238, DAKAR.

Bureau de Recherches Géologiques et Minières (BRGM)
BP 6009, 45060 ORLEANS-CEDEX, France.

Office de la Recherche Scientifique et Technique Outre Mer (ORSTOM)
70–74 Route d'Aulnay, 93140 BONDY, France.

For Editions Jaguar see France; for DMA see United States.

Catalogue

Atlases and gazetteers

Atlas National du Sénégal
Dakar: Institut Fondamental d'Afrique Noire, 1977.
pp. 147.

Atlas du Sénégal   P. Pelissier and G. Laclavère (Atlas Jeune Afrique)
Paris: Editions Jaguar, 1980.
pp. 72.

Senegal: official standard names approved by the United States Board on Geographic Names
Washington, DC: DMA, 1965.
pp. 194.

General

Sénégal   1 : 1 000 000   2nd edn
Paris: IGN, 1980.

Sénégal   1 : 500 000
Paris: IGN, 1969.
4 sheets, all published.

Topographic

République du Sénégal   1 : 200 000
Dakar: SSGN, 1955–.
27 sheets, all published. ■

République du Sénégal   1 : 50 000
Dakar: SSGN, 1950–.
290 sheets, *ca* 100 published. ■

Bathymetric

Cartes des fonds du plateau continental sénégambien et sénégelo–mauritanien   *ca* 1 : 300 000
Bondy: ORSTOM, 1975–76.
2 sheets, both published.

Carte sédimentologique du plateau continentale sénégambien
1 : 200 000
Bondy: ORSTOM, 1977.
3 sheets, all published.
With accompanying explanatory text.

Geological and geophysical

Carte géologique de la République du Sénégal et de la Gambie
1 : 500 000
Dakar: SDMG, 1962.
4 sheets, all published.

Carte géotechnique de la République du Sénégal   1 : 500 000
Dakar: SDMG, 1964.
4 sheets, all published.

Carte des gîtes minéraux de la République du Sénégal   1 : 500 000
Orléans: BRGM, 1966.
4 sheets, all published.

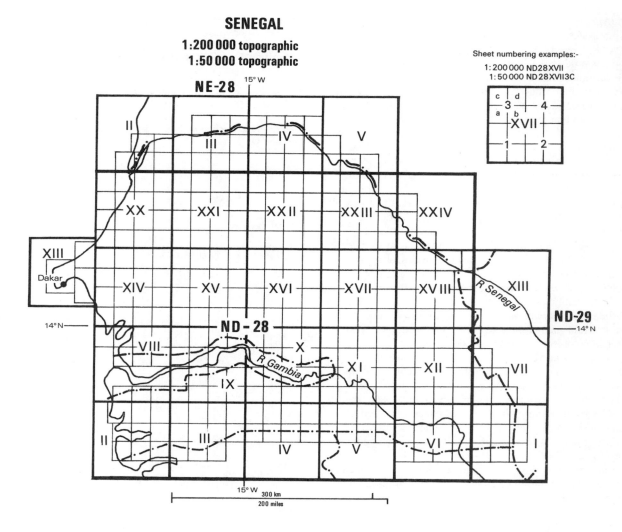

SENEGAL

1:200 000 topographic
1:50 000 topographic

Sheet numbering examples:-
1:200 000 ND28XVII
1:50 000 ND28XVII3C

Carte géologique du Sénégal   1:200 000
Dakar: SDMG, 1963–.
27 sheets, 14 published. ■

Carte pédologique du Sénégal   1:1 000 000
Bondy: ORSTOM, 1965.
With accompanying explanatory text.

Environmental

Carte des répartitions des glossines au Sénégal   1:2 000 000
Bondy: ORSTOM, 1981.
With accompanying explanatory text.

Town maps

Dakar: plan   1:10 000
Paris: IGN, 1983.

Sierra Leone

Sierra Leone was well mapped under British aid projects in the 1960s but little current information about mapping of the country is available. Topographic mapping of the country was carried out by the British **Directorate of Overseas Surveys (DOS)** (now Overseas Surveys Directorate, Southampton, OSD) which compiled basic scale coverage of the whole country in the 114-sheet *1:50 000 series DOS 419 (G 742)*. This five-colour map used Transverse Mercator projection, with UTM grid, and showed relief with 50 foot contours. Sheets are still available from Edward Stanford in London or from OSD, and some revised editions have been published. A full-colour four-sheet 1:250 000 scale map, and single-sheet

1:500 000 were published by DOS in the early 1970s and are still available in diazo print form. Large-scale projects included 1:2500 scale township maps of the major settlements, most published as two-colour photogrammetric plots.

Current responsibility for the topographic and cadastral survey of Sierra Leone falls to the **Survey and Land Division** of the Ministry of Lands, Housing and Country Planning. It maintains the topographic base and has published 1:500 000 scale chiefdom and administrative boundary maps. There is no information about current mapping availability from this source.

Geological and resources mapping is carried out by

Geological Surveys, in the Ministry of Mines. 1:1 000 000 and 1:250 000 scale mapping has been carried out under British aid projects, and DOS printed some 1:50 000 scale geological maps in the 1960s. Only some of the larger-scale mapping is still available.

Land use mapping of the country is the responsibility of **Land Resources Surveys**. Some large-scale full-colour soil and resources sheets were published for the British **Land Resources Development Centre** in the Overseas Development Administration by DOS during the period of British involvement.

Addresses

Overseas Surveys Directorate (DOS/OSD)
Ordnance Survey, Romsey Rd, SOUTHAMPTON SO9 4DH, UK.

Survey and Lands Division (SLD)
Ministry of Lands, Housing and Country Planning, New England, FREETOWN.

Geological Surveys
Ministry of Mines, New England, FREETOWN.

Land Resources Surveys
Ministry of Agriculture and Natural Resources, New England, FREETOWN.

For LRDC see Great Britain; for DMA see United States.

Catalogue

Atlases and gazetteers

Sierra Leone: official standard names approved by the United States Board on Geographic Names
Washington, DC: DMA, 1966.
pp. 125.

Topographic

Sierra Leone 1:50 000 DOS 419 (G742)
Tolworth and Freetown: DOS and SLD, 1960–.
111 sheets, all published. ∎

Administrative

Sierra Leone: chiefdom boundaries map 1:500 000 DOS 2618
Tolworth and Freetown: DOS and SLD, 1970.

SIERRA LEONE

1:50 000 topographic Series DOS 419

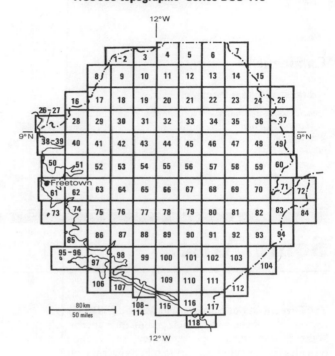

Somalia

Somalia comprises the former territories of British and Italian Somaliland, which became an independent state in 1960. The national mapping agency is the **National Cartographic Directorate**. Current mapping includes a 106-sheet 1:200 000 scale topographic series but no further information is available about current specifications of this or other topographic or thematic mapping produced in the country. No Somali-produced mapping is currently available for purchase.

The British **Directorate of Overseas Surveys (DOS)** (now Overseas Surveys Directorate, Southampton, OSD) compiled 68 single-colour uncontoured preliminary plots in 1:125 000 series DOS 539 in the 1950s. This series covered the whole of former British Somaliland: sheets may still be obtained from OSD and from the large map jobbers. In view of the lack of other mapping of the country we have listed this series despite the poor quality and dated nature of the information. On the same sheet lines DOS published a 1:125 000 scale full-colour geological map. Most sheets were issued in the early 1970s, but this series was never intended to give complete

coverage: 22 sheets for coastal areas were published, of which 17 are still available from Edward Stanford in London or from OSD. Mapping of former Italian Somaliland is no longer available, but included 1:200 000 and 1:50 000 scale preliminary coverage of settled areas, compiled in the 1930s. Small-scale geological coverage of Somalia and Ethiopia produced by the Italian **Consiglio Nazionale delle Ricerche (CNR)** and distributed through Pergamon Press is still available. The **Geological Survey** in the Ministry of Minerals and Water Resources is currently responsible for geological mapping activities. Some recent thematic mapping at 1:500 000 scale was compiled by OSD for the Land Resources Development Centre. These maps cover southern parts of the country and show vegetation, land use, land capability and livestock distribution. Russian language small-scale mapping of the country was produced in the early 1970s by **Glavnoe Upravlenie Geodezii i Kartografii (GUGK)**, and the American **Central Intelligence Agency (CIA)** compiled a similar sheet covering Somalia and Djibouti in 1977.

Addresses

Overseas Surveys Directorate (OSD/DOS)
Ordnance Survey, Romsey Rd, SOUTHAMPTON SO9 4DH, UK.

National Cartographic Directorate
PO Box 1188, MOGADISHU.

Geological Survey Department
Ministry of Minerals and Water Resources, PO Box 744,
MOGADISHU.

For GUGK see USSR; for CIA see United States; for CNR see
Italy.

Catalogue

Atlases and gazetteers

*Somalia: official standard names approved by the United States Board
on Geographic Names*
Washington, DC: DMA, 1982.
pp. 231.

General

Somali   1 : 2 500 000
Moskva: GUGK, 1983.
In Russian.
With ancillary thematic maps on sheet and accompanying
placename index.

Somalia and Djibouti   ca 1 : 3 500 000
Washington, DC: CIA, 1977.
With population, ethnographic and economic insets.

Topographic

British Somaliland   1 : 125 000   DOS 539
Tolworth: DOS, 1952–55.
68 sheets, all published.

Geological and geophysical

Geological map of Ethiopia and Somalia   1 : 2 000 000
Firenze: CNR, 1973.
Title also in Arabic and Amharic.
With explanatory text and accompanying 1 : 3 000 000 scale
landform map.

Somali Democratic Republic geological survey   1 : 125 000
Tolworth: DOS, 1960–.
68 sheets, 22 published.

South Africa

The **Directorate of Surveys and Mapping (DSM)** is
responsible for the compilation and maintenance of all
South African official topographic map series, and also
provides cartographic backup to other governmental
agencies. It carries out geodetic and cadastral surveying
and has been responsible for the compilation of the map
base of Namibia, as well as that of the Republic and its
quasi-independent homelands. The Mapping Division of
DSM compiles sheets which are printed by the
Government Printer in Pretoria. DSM is still responsible
through technical cooperation agreements for the mapping
of the newly declared black homeland areas, though
Transkei now distributes its own mapping through the
Surveyor General of Transkei.

Maps of South Africa may also be obtained through the
Government Printer. We have listed here only coverage of
the Republic of South Africa: for smaller-scale series
mapping, which includes South Africa itself but extends
over much of the southern part of the continent, see
Africa, South.

The basic topographic map for most of the country is
the 1916-sheet *South Africa 1 : 50 000 map series*. This series
was completed in 1978, but most sheets were issued as
non-metric editions with 50 foot contours prior to
metrication in 1970. Since completion of the series DSM
has been concentrating upon issuing revised sheets with 20
m contour intervals. In 1986 70 per cent of the series was
available in full metric editions and a further 20 per cent
was under revision: a maximum interval of 20 years
between revisions has been set for this series and sheets are
regularly updated depending upon the development of the
terrain covered. In the revision process orthophotomaps
derived from 1 : 150 000 scale aerial photography are used.
It is expected that full metric coverage of the country will

be attained by 1990. DSM has been considering altering
the specification of this series and a trial 1 : 50 000 sheet has
been prepared to the new design to test map user reaction.

The *South Africa 1 : 250 000 series* is available in two
versions: a topographical map without cadastral
information and a topocadastral map using the same base
and contours but without the layer shaded relief and with a
cadastral overprint. The series is fully metricated and 10 of
its 70 sheets are currently being revised. This map is used
by various other Departments as a base for their own
thematic mapping activities.

The *Southern Africa 1 : 500 000 series* is derived from
larger-scale mapping and is published in three versions:
aeronautical, topographical and administrative. Its 23
sheets replace the earlier non-metric 1 : 500 000 series.

DSM is responsible for the production of other smaller-
scale maps, including 16-sheet ICAO 1 : 1 000 000 series
and two smaller-scale wall maps which have regularly
appeared in new editions. There are plans to produce a
new four-sheet 1 : 1 000 000 scale topographical series.

Urban and peri-urban areas, and rural areas with growth
potential, are mapped by the *South Africa 1 : 10 000 series*,
an orthophotomap series on the Gauss Conformal
Projection, with 5 m contour interval, and some
cartographic enhancement, including placenames. This
map was started in 1971 and to date nearly 7500 3' × 3'
sheets have been published as diazo prints. There is an
average annual production of about 1000 new sheets, and
many earlier sheets are already being revised.

The **Geological Survey (GSSA)** produces geological,
geophysical, mineral and geochemical maps of South
Africa. A wide variety of old scales of mapping of the
country were published; some are still available, including
1 : 125 000 scale geological coverage, which was compiled

until the mid-1970s. The main current series are issued at scales of 1:250 000 and 1:50 000 using national map series sheet lines. Twenty-six sheets in the 250 000 geological series have now been produced and the series is in progress at a rate of about two new sheets a year. Explanatory notes are sometimes available printed on sheets, but some maps in this series are already out of print. It is intended to publish this map for the whole of the country. The 1:50 000 series is to be published for areas around growth points and around important mineral deposits. To date mapping has been published with accompanying monographs for areas around Richards Bay, Pietermaritzburg, Durban, Pretoria, Barburton and Cape Town. Smaller-scale earth science mapping of South Africa is also issued by GSSA: a new edition of a multi-colour geological map of South Africa at 1:1 000 000 has recently been published and a mineral map, using the same legend as the metallogenic map of Africa, was issued at the same scale in 1984. There are plans to publish structural and metamorphic maps of South Africa, also at 1:1 000 000 scale, as well as a gravity overprint to the geological sheet. All these maps may be acquired either through GSSA or through the Government Printer.

Other earth science series mapping is only available as open file publications: the mineral map series at 1:250 000 is now nearly complete, as is the aeromagnetic map at this scale. Seventy-five per cent coverage is available in an airborne radiometric series at 1:250 000, and a geochemical series at 1:50 000 is also available for fewer than 100 quads.

The **Soils and Irrigation Research Unit (SIRU)** is responsible for mapping land and soil types. Since 1977 SIRU has been issuing 1:250 000 scale land type maps with information on soils, terrain form, and climate on a 1:250 000 topocadastral base. Almost one-quarter of the country has been covered in this series. The other major project concerning SIRU is the production of a 1:50 000 soil map of South Africa. Two blocks of 57 sheets have so far been surveyed and are due to be published.

The **Department of Forestry** has produced tourist maps at 1:50 000 of various national hiking trails, together with waterproof recreation maps at the same scale of areas in the Cederberg and Drakensberg mountains. They have also compiled terrain inventory maps at 1:250 000 overprinted on the standard 1:250 000 sheets in the national series.

The **Hydrographic Office of the South African Navy** is responsible for the compilation and maintenance of all charting of the South African coast and waters.

The **Institute for Cartographic Analysis of the University of Stellenbosch** has compiled numerous thematic works covering a variety of topics. Notable amongst these have been an economic atlas based upon computerized census data, and a population atlas with more than 200 full-colour maps presenting results of the 1980 Census.

Three organizations in the private sector produce significant available mapping. The **Automobile Association of South Africa** produces a variety of road maps for distribution to members and for sale. The **Reader's Digest Association** has compiled, in association with DSM, the first modern national atlas, which was published in 1984. **Map Studio**, which produces a wide variety of tourist maps, town plans and other commercial products, has recently installed a fully automated cartographic system. Another tourist and town map publisher is the **Swan Publishing Company**, which publishes mapping of Natal and Transkei.

Nearly all South African mapping is issued with bilingual English and Afrikaans script: we have listed maps under their English language titles.

Further information

South African National Committee for IGU/ICA (1984) *Report on cartography in the Republic of South Africa*, presented to the International Cartographic Association Seventh General Assembly Perth Australia 1984, Perth: ICA.

Chief Director of Surveys and Mapping: Map Catalogues and Annual Reports.

Thomas, P.W. (1982–83) The topographical 1:50 000 map series of South Africa, Parts 1–3. *South African Journal of Photogrammetry, Remote Sensing, and Cartography*, **13**(2), 77–88; **13**(3), 171–184; **13**(5), 331–343.

Addresses

Directorate of Surveys and Mapping (DSM)
Mapping Division, Private Bag, MOWBRAY, 7705.

Surveyor-General's Office
Private Bag X5031, UMTATA 5100, Transkei.

Geological Survey (GSSA)
Private Bag X112, PRETORIA, 0001.

Government Printing Works
Private Bag X85, PRETORIA, 0001.

Soils and Irrigation Research Unit (SIRU)
Private Bag X79, PRETORIA, 0001.

Department of Environment Affairs
Directorate of Forestry, PRETORIA, 0001.

Hydrographic Office
South African Navy, Private Bag, TOKAI, 7966.

Institute for Cartographic Analysis, University of Stellenbosch
STELLENBOSCH, 7600.

Automobile Association of South Africa (AA)
PO Box 91257, AUCKLAND PARK, 2006.

Reader's Digest Association for South Africa (Pty) Ltd
Reader's Digest House, 130 Strand St, CAPE TOWN, 8001.

Map Studio
PO Box 76199, WENDYWOOD, 2144.

Swan Publishing Company (Intratex)
PO Box 37125, OVERPORT, 4067.

For DMA see United States; for Quail see Great Britain.

Catalogue

Atlases and gazetteers

Reader's Digest Atlas of Southern Africa
Cape Town: Reader's Digest Association South Africa, 1984.
pp. 140.

Official placenames in the Republic of South Africa and in South West Africa, approved to 1 April 1977
Mowbray: DSM, 1978.
pp. 450.

South Africa: official standard names approved by the United States Board on Geographic Names
Washington, DC: DMA, 1954.
2 volumes.

General

Southern Africa 1:2 500 000 5th edn
Mowbray: DSM, 1984.

South Africa 1:2 400 000
Wendywood: Map Studio, 1985.
Road map.

Topographic

Southern Africa topographical series 1:250 000
Mowbray: DSM, 1977–.
70 sheets, all published. ■

South Africa 1:50 000
Mowbray: DSM, 1950–.
1916 sheets, all published. ■
Some sheets available as coloured orthophotomaps.

Geological and geophysical

*Geological map of the Republics of South Africa, Transkei,
Bophuthatswana, Venda, and Ciskei, and the Kingdoms of Lesotho
and Swaziland* 1:1 000 000 3rd edn
Pretoria: GSSA, 1984.
4 sheets, all published.
Also available with a gravity overprint.

*Mineral map of the Republics of South Africa, Transkei,
Bophuthatswana, Venda, and Ciskei, and the Kingdoms of Lesotho
and Swaziland* 1:1 000 000
Pretoria: GSSA, 1983.
4 sheets, all published.

Republic of South Africa: geological series 1:250 000
Pretoria: GSSA, 1957–.
70 sheets, 26 published. ■

Republic of South Africa: geological series 1:50 000
Pretoria: GSSA, 1973–.
1916 sheets, 25 published. ■

Environmental

*Terrain morphological map of the Republic of South
Africa* 1:2 500 000
Pretoria: SIRU, 1983.
Includes Lesotho and Swaziland.

Veldtypes of South Africa 1:1 500 000
Pretoria: SIRU, 1975.
2 sheets, both published.

Land type map 1:500 000
Pretoria: SIRU, 1977–.
23 sheets, 1 published.

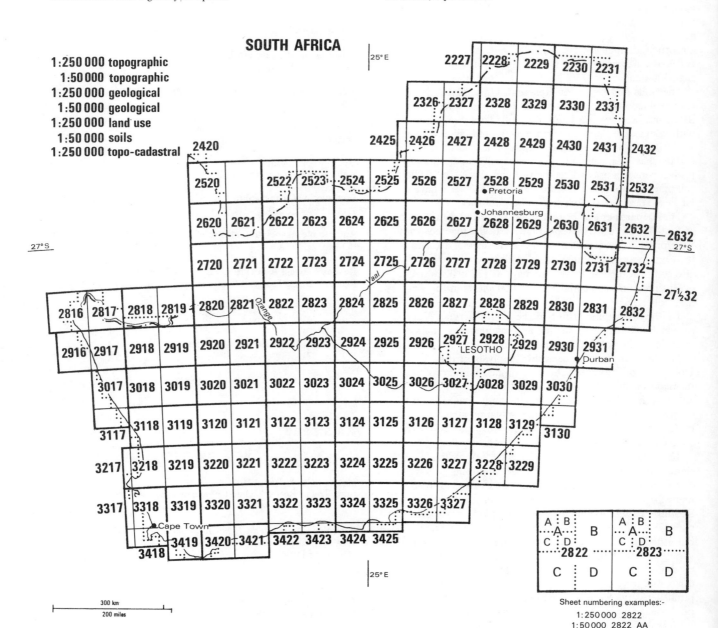

Land type map 1 : 250 000
Pretoria: SIRU, 1978–.
70 sheets, 16 published. ■

Administrative

Republic of South Africa: magisterial districts map 1 : 3 000 000
Mowbray: DSM, 1984.

Republic of South Africa 1 : 1 124 000
Wendywood: Map Studio, 1983.
Economic and administrative wall map.

South Africa: topocadastral series 1 : 250 000
Mowbray: DSM, 1977–.
70 sheets, all published. ■

Human and economic

Population census atlas of South Africa
Stellenbosch: Institute for Cartographic Analysis, 1986.
pp. 200.

Economic atlas of South Africa
Stellenbosch: Institute for Cartographic Analysis, 1981.
pp. 160.

South Africa: railway map ca 1 : 3 200 000
Exeter: Quail Map Co., 1980.

Town maps

South Africa: street plans Various scales
Wendywood: Map Studio, 1986.
10 sheets, all published.
Covers Johannesburg, Pretoria, Verwoerdburg/Midrand,
Vereeniging/Meyerton/Vanderbijlpark/Sasolburg; Bloemfontein;
Port Elizabeth; Durban; Cape Town; Soweto/Lenasia/Eldorado
Park.

Sudan

The national mapping agency in the Sudan is the **Sudan
Survey Department (SSD)**, which was originally
established under the Anglo-Egyptian Corps of Engineers
in 1899. SSD is now responsible for topographic, cadastral
and geodetic activities in the country. Standard map series
are published at scales of 1 : 250 000 and 1 : 100 000. The
1 : 250 000 scale series gives complete coverage of the
country in 170 sheets, with 50 m or 200 foot contours. It is
drawn on a modified polyconic projection, and dates from
prewar surveys; sheets are one-, three- or four-colour
editions. Railway and boundary information was up-dated
for most of the country in 1975–76, but the series remains
very outdated. There are plans to revise the series using
satellite imagery but no sheets have yet appeared. Foreign
aid projects have ensured blocks of more modern
1 : 100 000 mapping. The British **Directorate of Overseas
Surveys (DOS)** has mapped a large area of the Red Sea
Hills at this scale and other blocks have been published in
the Jebel Marra area, in the Nuba Hills and around
Khartoum. These new maps were photogrammetrically
compiled, and use UTM projection. Sheets cover 30′
quads, some show relief using 40 m contours, and most
have English and Arabic script. Very little of the country
is on programme for 1 : 100 000 mapping; 920 sheets would
be required for complete coverage, and only about 200
have been issued.

SSD is also involved in the production of smaller-scale
maps at 1 : 1 000 000, 1 : 2 000 000, 1 : 4 000 000 and
1 : 8 000 000. Some up-dating of these maps has been
carried out and the latter three scales are used as bases for
a wide variety of thematic mapping. OSD has been
collaborating with SSD in the production of a new
National Atlas and Gazetteer which is scheduled for
publication in June 1987. It has also recently been involved
in an aid project to provide base mapping in the form of
photomosaics and preliminary monochrome maps showing
settlement, drainage and communications for areas of the
country coping with large numbers of refugees displaced
by the drought conditions of the mid-1980s. This project
was undertaken from August 1984 in collaboration with

the United Nations High Commission for Refugees.
Amongst other SSD projects are plans to revise the map
catalogue.

Geological mapping of Sudan is carried out by the
Geological Survey Department, but SSD distributes their
maps. No large-scale programme is carried out, though
some sheets and memoirs on the same sheet lines as the
1 : 250 000 topographic series have been issued. The
Egyptian Remote Sensing Center (ERSC) has recently
produced Landsat-derived geological and earth science
mapping of southern areas of the country and the **Freien
Universität Berlin** has compiled six-sheet 1 : 500 000 scale
Landsat-based geological mapping of parts of the north-
west of Sudan.

A map in the *Arab World Map series* provides the best
small-scale tourist coverage of the country and is compiled
by **Geoprojects**. The Russian **Glavnoe Upravlenie
Geodezii i Kartografii (GUGK)** has compiled a useful
small-scale map of the country with placename index and
thematic inset maps. An atlas of the Khartoum
conurbation was compiled recently and issued by
Khartoum University Press.

Further information

Sudan (1977 and 1986) Cartographic activities in Sudan: papers
presented by Sudan to the *8th United Nations Regional
Cartographic Conference for Asia and the Far East* and to the *6th
United Nations Regional Cartographic Conference for Africa*, New
York: UN.

Addresses

Sudan Survey Department (SSD)
PO Box 306, KHARTOUM.

Overseas Surveys Directorate (DOS/OSD)
Ordnance Survey, Romsey Rd, SOUTHAMPTON SO9 4DH, UK.

Geological Survey Department
Ministry of Mining and Energy, PO Box 410, KHARTOUM.

Egyptian Remote Sensing Center (ERSC)
101 Kasr El Aini Street, CAIRO, Egypt.

Khartoum University Press
POB 321, KHARTOUM.

For Freien Universität Berlin see German Federal Republic; for
Geoprojects see Lebanon; for GUGK see USSR; for DMA see
United States.

Catalogue

Atlases and gazetteers

*Sudan: official standard names approved by the United States Board
on Geographic Names*
Washington, DC: DMA, 1962.
pp. 358.

National atlas of the Sudan
Khartoum: SSD, 1987.

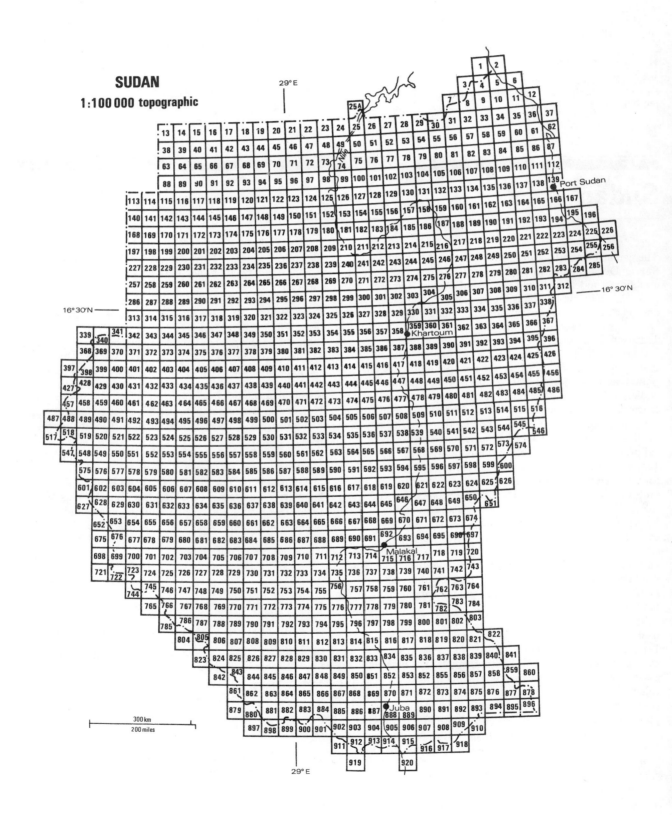

SUDAN
1:100 000 topographic

General

The Oxford map of the Sudan 1:4 000 000
Beirut: Geoprojects, 1980.
Double-sided.
Includes tourist information and city maps of Khartoum and
Omdurman.

Sudan 1:4 000 000
Khartoum: SSD, 1983.
Available in English or Arabic editions.
Also available as roads edition.

Sudan 1:4 000 000
Moskva: GUGK, 1985.
In Russian.
With placename index and climatic, economic and population
inset maps.

Sudan 1:2 000 000
Khartoum: SSD, 1984.
3 sheets, all published.

Topographic

Sudan 1:250 000
Khartoum: SSD, 1932–76.
171 sheets, 167 published.
Railways and provincial boundaries revised to 1975–76.

Sudan 1:100 000
Khartoum: SSD, 1963–.
920 sheets, *ca* 200 published. ■
Specifications vary.

Geological and geophysical

Sudan geological map 1:4 000 000
Khartoum: SSD, 1981.
Published as English or Arabic edition.
Also available as sheet showing superficial deposits.

Geological map of the Democratic Republic of Sudan 1:2 000 000
Khartoum: SSD, 1984.
With accompanying explanatory bulletin.

Environmental

Sudan 1:8 000 000
Khartoum: SSD, 1976–.
28 sheets, all published.
Single-sheet thematic maps covering 17 topics.

Sudan 1:4 000 000
Khartoum: SSD, 1976–.
14 sheets, all published.
Single-sheet thematic maps covering a variety of topics including
vegetation, animal density (4 versions).

Sudan 1:2 000 000
Khartoum: SSD, 1976–.
16 sheets, all published.
Single-topic thematic maps, 3 sheets per map.

Town maps

Atlas of the Khartoum conurbation
Khartoum: Khartoum University Press, 1981.
pp. 97.

Khartoum 1:20 000
Khartoum: SSD.

Swaziland

The national mapping organization in Swaziland is the
Department of Surveys (SDS) which has produced series
mapping with the assistance of the British **Directorate of
Overseas Surveys (DOS)** (now Overseas Surveys
Directorate, Southampton, OSD). The country is also
covered on many of the maps produced by the South
African survey organizations. The basic map scale is
1:50 000. The six-colour *DOS series 435 (Z771)* covers the
country in 31 sheets, and conforms to usual DOS
specifications. Relief is shown by 50 foot contours, and
SDS has a revision programme in operation. A diazo
orthophotomap series at 1:5000 scale was compiled by
SDS and dates from 1972–79 aerial coverage: this map
covers the whole country and is now the largest scale
available to give national coverage. A 1:250 000 full-colour
topographic map was jointly published by SDS and DOS
with metric 125 m contours and relief shading. SDS has
also issued a road map at this scale. A false colour
composite satellite image map of the country was
published for SDS, by DOS with the Regional Centre for
Remote Sensing in Nairobi, in 1984. Other responsibilities
of SDS include cadastral township mapping: urban areas
are covered in 1:2500 and 1:5000 scale topocadastral diazo
or monochrome maps.

Geological coverage of the country is also provided in
jointly published series. The **Geological Survey and
Mines Department (GSMD)** publishes the *1:50 000*

geological series DOS 1216 with DOS, which gives
complete coverage of the country on the same sheet lines
as the topographic map. This series is under revision; 16
sheets are 1979–80 editions, and a further 13 sheets are
published with explanatory notes printed on the sheets.
GSMD also publishes a 1:250 000 scale multi-colour
single-sheet geological map of the country, and a four-
sheet 1:25 000 scale geological series of the north-west.

Agricultural mapping activities are the responsibility of
the **Land Use Planning Department (SLUPD)**. 1:125 000
scale soil and land use capability maps of the country were
prepared by G. Murdoch in the 1960s, and published by
the Ministry of Agriculture. Each covers the country in
two sheets and both are still available from SLUPD.

The National Atlas was issued by the **Swaziland
National Trust Commission** in 1983, and contains many
black-and-white maps of economic, cultural and
environmental characteristics of the country.

Amongst the commercial concerns producing maps are
the **Swan Publishing Company** from South Africa which
publishes town maps.

Further information

A national map catalogue published by SDS gives details of most
of Swaziland's mapping.

Addresses

Department of Surveys (SDS)
PO Box 58, MBABANE.

Overseas Surveys Directorate (OSD/DOS)
Ordnance Survey, Romsey Rd, SOUTHAMPTON SO9 4DH, UK.

Geological Survey and Mines Department (GSMD)
PO Box 9, MBABANE.

Swaziland National Trust Commission
PO Box 100, LOBAMBA.

Land Use Planning Department (SLUPD)
Ministry of Agriculture, PO Box 57, MBABANE.

For Swan Publishing Company, see South Africa.

Catalogue

Atlases and gazetteers

The Atlas of Swaziland
Lobamba: Swaziland National Trust Commission, 1983.
pp. 98.

General

Swaziland   1 : 250 000   DOS 635   3rd edn
Mbabane: SDS and DOS, 1982.

Swaziland satellite image map   1 : 250 000
Mbabane: SDS and DOS, 1984.

Swaziland road map   1 : 250 000
Mbabane: SDS.

Topographic

Swaziland   1 : 50 000   DOS 435 (Z771)
Mbabane and Tolworth: SDS and DOS, 1968–.
31 sheets, all published. ■

Geological and geophysical

Geological map of Swaziland 1 : 250 000
Mbabane: GSMD, 1982.

Swaziland: geological series   1 : 50 000
Mbabane and Tolworth: GSMD and DOS, 1961–.
31 sheets, all published. ■
Some sheets published with textual explanations on sheets.

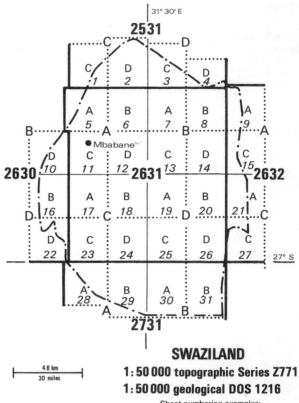

SWAZILAND
1 : 50 000 topographic Series Z771
1 : 50 000 geological DOS 1216
Sheet numbering examples:-
1 : 50 000 topographic   2631 DA
1 : 50 000 geological   *5*

Environmental

Soil map of Swaziland   1 : 125 000   G. Murdoch
Mbabane: Ministry of Agriculture, 1968.
2 sheets, both published.

Land capability map of Swaziland   1 : 125 000   G. Murdoch
Mbabane: Ministry of Agriculture, 1968.
2 sheets, both published.

Town maps

Mbabane   1 : 14 500
Overport: Swan Publishing Company, 1982.

Tanzania

The first accurate planimetric maps of Tanzania were compiled by the **Directorate of Colonial Surveys (DCS)** in the late 1940s and early 1950s. Later topographic mapping of Tanzania has been carried out by DCS's successor from 1957, the **Directorate of Overseas Surveys (DOS)** (now Overseas Surveys Directorate, Southampton, OSD) and following independence by the **Surveys and Mapping Division (SMD)** of the Ministry of Lands, Housing and Urban Development of Tanzania. SMD is currently responsible for topographic, geodetic and cadastral surveying and mapping activities. The basic scale

series is published at 1 : 50 000 and current sheets are issued with 20 m contour interval and are published in five colours in standard British DOS style. Some earlier sheets still use a 50 foot contour interval and some are planimetric preliminary plots dating from DCS days: the earlier sheets are being withdrawn as new full specification sheets are issued. This series (Y742) is compiled by photogrammetric methods from aerial photographs and is used to derive the other smaller-scale maps published by SMD. This series ought to give complete cover by the end of 1986. Foreign aid continues to be important in the production of the 1:50 000 map: Japanese, Polish, British and Finnish projects have all enabled SMD to print sheets, and the massive task of completing coverage is currently being facilitated by Canadian participation in the production of at least 600 sheets.

Series Y503 is the standard 1:250 000 map for East Africa, and is derived from current 1:50 000 mapping where this is available. Relief is shown by layer colouring: where contour information is not available sheets are published as first editions in unlayered format.

Smaller-scale maps are published by the SMD: for instance, six sheets are issued according to the IMW specifications and sheet lines at 1:1 000 000 scale. OSD is currently involved in the trial production of a Landsat-based 1:1 000 000 scale map of Tanzania. District and region mapping is issued at scales between 1:100 000 and 1:500 000 depending on the size of the unit to be mapped. These series show roads, settlements and administrative information with an overprint of forest information in green, and farm and estates in red.

Larger-scale urban and township maps are published at scales of 1:1250, 1:2500 and 1:5000 for about 60 settlements throughout the country. SMD also issues tourist mapping of National Parks, and was responsible for the compilation of a National Atlas, the second edition of which was published recently. Individual sheets at a scale of 1:3 000 000 from this atlas are still available but the bound volume of the work is no longer for sale.

The island of Zanzibar has its own survey authority, the **Department of Lands and Surveys Zanzibar**, which has recently issued high-quality 1:10 000 mapping produced by the British DOS. This seven-colour series DOS 208 covers the island in 57 sheets, the more southern sheets appearing first.

Geological mapping of Tanzania is the responsibility of the **Ministry of Energy and Minerals (MEM)**, whose Geology Division has been responsible for the compilation of the basic scale geology series at 1:125 000. These quarter-degree sheets have been published for almost a half of the country: some were compiled with technical assistance from the West German Bundesanstalt für Bodenforschung (now BfGR). Areas of important mineral resources are also covered at different scales by single sheets. A new edition of MEM's small-scale geological map of the country has recently been issued.

The **Central Bureau of Statistics** has participated with SMD in the production of mapping derived from census results, showing census divisions in regions and districts, together with settlement patterns.

Amongst the other map publishers who have issued mapping of Tanzania are **Freytag Berndt**, the American **Central Intelligence Agency (CIA)** and the Russian **Glavnoe Upravlenie Geodezii i Kartografii (GUGK)**.

Further information

Surveys and Mapping Division, Ministry of Lands, Housing and Urban Development (1983) *Catalogue of maps*, Dar es Salaam: SMD.

Ministry of Energy and Minerals (1981) *Geological and mining publications*, Dodoma: MEM.

The best overall study (though now rather dated) is Kaduma, J. (1974) Maps and air photographs of Tanzania, in *Tanzania in Maps*, London: Macmillan.

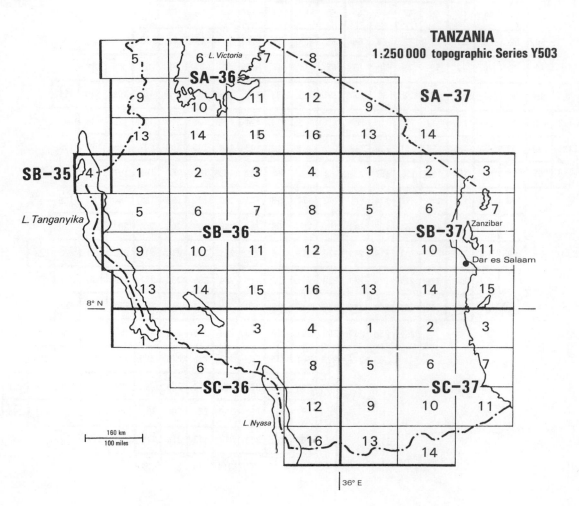

Addresses

Overseas Surveys Directorate (OSD/DOS)
Ordnance Survey, Romsey Rd, SOUTHAMPTON, UK.

Surveys and Mapping Division (SMD)
Ministry of Lands, Housing and Urban Development, PO Box 9201, DAR ES SALAAM.

Department of Lands and Surveys
PO Box 811, ZANZIBAR.

Geology Division, Ministry of Energy and Minerals (MEM)
PO Box 903, DODOMA; and PO Box 3060, DAR-ES-SALAAM.

Central Bureau of Statistics
POB 796, DAR-ES-SALAAM.

For Freytag Berndt see Austria; for GUGK see USSR; for CIA and DMA see United States; for ITC see Netherlands.

Catalogue

Atlases and gazetteers

Atlas of Tanzania 2nd edn
Dar-es-Salaam: SMD.
38 single sheets available.

Tanzania: official standard names approved by the United States Board on Geographic Names
Washington, DC: DMA, 1965.
pp. 236.

General

Tanzania 1 : 3 220 000
Washington, DC: CIA, 1970.
With 4 thematic ancillary maps on sheet.

Tanzanija 1 : 2 000 000
Moskva: GUGK, 1986.
In Russian.
With thematic insets and placename index.

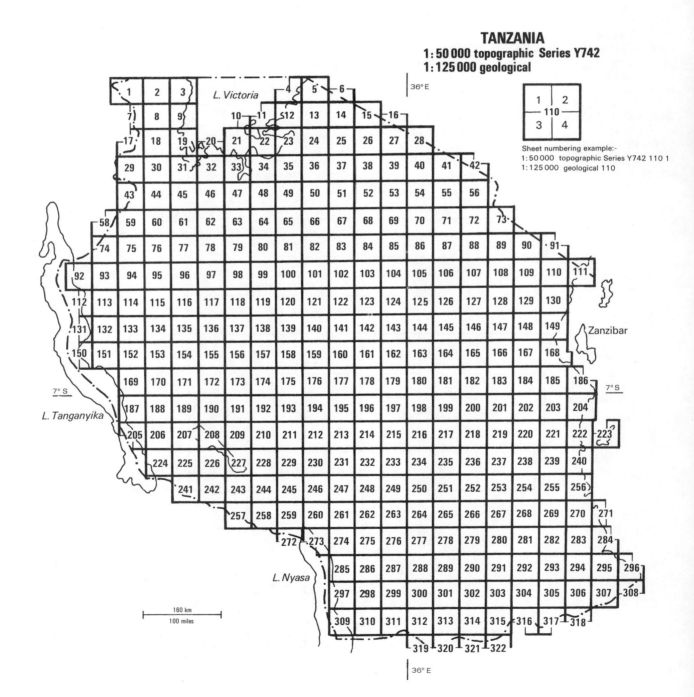

TANZANIA
1 : 50 000 topographic Series Y742
1 : 125 000 geological

Sheet numbering example:-
1 : 50 000 topographic Series Y742 110 1
1 : 125 000 geological 110

Tanzania 1:2 000 000
Dar-es-Salaam: SMD, 1968.
Available as layer coloured or contoured edition.

Tanzania: land of Kilimanjairo 1:2 000 000
Dar-es-Salaam: Tanzania Tourist Corporation and Freytag
Berndt, 1978.

IMW series 1:1 000 000
Dar-es-Salaam: SMD.
6 sheets, all published.

Topographic

East Africa 1:250 000   Series Y503
Dar-es-Salaam: SMD, 1956–.
63 sheets, 47 published. ■

Tanzania 1:50 000   Series Y742
Dar-es-Salaam: SMD, 1947–.
1250 sheets, all published. ■

Geological and geophysical

Geological map of Tanzania 1:2 000 000
Dodoma: MEM, 1985.

Tanzania: geological map series 1:125 000
Dodoma: MEM, 1954–.
330 sheets, *ca* 140 sheets published. ■
14 sheets published by the German Geological Mission in
Tanzania.

Environmental

Tanzania: vegetation and mean annual rainfall 1:5 000 000
Enschede: ITC, 1979.

Administrative

District and Region series 1:500 000–1:100 000
Dar-es-Salaam: SMD.
53 sheets published.

Togo

The West African State of Togo was mapped by the
French **Institut Géographique National (IGN)** and its
predecessors in the 1950s and 1960s as part of the *Cartes de
l'Afrique de l'Ouest*. The country was covered by 90 four-
colour, 1:50 000 sheets showing relief with 20 m contours
and by 12 four- or five-colour, 1:200 000 scale sheets with
40 m contours. Both series used UTM projection and were
compiled from aerial coverage. Responsibility for
topographic and cadastral surveying and mapping of Togo
is now vested in the **Service Topographique** which
operates with French aid. Little recent topographic
cartography has been carried out because of lack of
resources, although some revision of a few 1:200 000
sheets has taken place (including recasting of sheet lines in
the south), and a general map of the country and town
plan of Lomé have been published in conjunction with
IGN. The last two are currently not available. The **Bureau
National de Recherches Minières** is responsible for
geological surveying, but no recent maps have been
published and the best available coverage remains the
1:500 000 and 1:2 000 000 scale French West African
series published in the 1960s by **Bureau de Recherches
Géologiques et Minières (BRGM)** and listed in this work
under Africa, Western. Soils mapping of Togo at
1:200 000 and 1:1 000 000, with accompanying
monographs, has been carried out by **Office de la
Recherche Scientifique et Technique Outre Mer
(ORSTOM)**, which has also compiled bathymetric
coverage of the continental shelf.

Editions Jaguar issued a useful atlas in its Jeune Afrique
series in 1981, which gives a good overall thematic
coverage of the country.

Further information

Togo (1979 and 1986) Report on cartographic developments in
Togo . . ., papers presented to the *4th and 6th United Nations
Regional Cartographic Conferences for Africa*, Addis Ababa: UN,
Economic Commission for Africa.

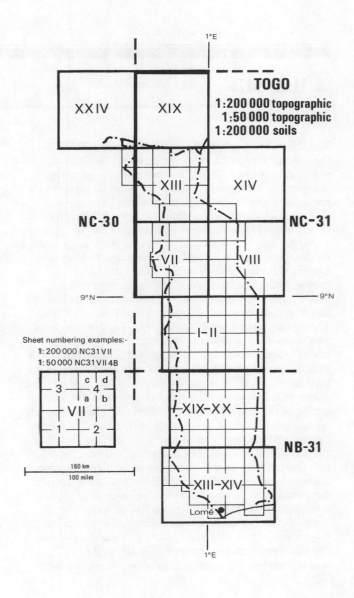

Addresses

Institut Géographique National (IGN)
136 bis rue de Grenelle, 75700 PARIS, France.

Service Topographique
BP 508, LOMÉ.

Bureau National de Recherches Minières
BP 356, LOMÉ.

Bureau de Recherches Géologiques et Minières (BRGM)
BP 6009, 45060, ORLEANS-CEDEX, France.

Office de la Recherche Scientifique et Technique Outre Mer (ORSTOM)
70–74 Route d'Aulnay, 93140 BONDY, France.

For Editions Jaguar see France; for DMA see United States.

Catalogue

Atlases and gazetteers

Atlas du Togo Y. E. Gukonu and G. Laclavère (Atlas Jeune Afrique)
Paris: Editions Jaguar, 1981.
pp. 64.

Togo: official standard names approved by the United States Board on Geographic Names
Washington, DC: DMA, 1966.
pp. 100.

Topographic

République du Togo 1 : 200 000
Paris and Lomé: IGN, 1955–.
12 sheets, all published. ■

République du Togo 1 : 50 000
Paris and Lomé: IGN, 1955–.
90 sheets, all published. ■

Bathymetric

Fonds de pêche le long des côtes des républiques du Dahomey et du Togo 1 : 208 300 and 1 : 294 800
Bondy: ORSTOM, 1966.
2 sheets, both published.
Maps cover bathymetric surveying tracts and bathymetry.

Environmental

Carte pédologique du Togo 1 : 1 000 000
Bondy: ORSTOM, 1962.

Carte pédologique du Togo 1 : 200 000
Bondy: ORSTOM, 1979.
3 sheets, all published. ■
With accompanying explanatory text.

Tunisia

The **Office for Topography and Cartography (OTC)** is the national mapping and surveying agency in Tunisia. It was established in 1974 and is responsible for geodetic, topographic, aerial photographic and cadastral surveys in the country and for the publication of topographic, thematic and marine maps. Prior to the establishment of OTC topographic series were available at 1 : 50 000, 1 : 100 000 and 1 : 200 000 scales. Most of these maps were produced by French colonial authorities and published by **Institut Géographique National (IGN)** and its predecessors. They were derived from surveys of very varied quality and date, some 1 : 50 000 sheets were still based on early plane table surveys dating from the 1890s. These older series used Bonne's projection. New mapping programmes using modern photogrammetric techniques and UTM projection were set up by OTC in the mid-1970s with the aim of providing complete basic scale coverage of the country in various scales by the year 2000. A four-colour 1 : 25 000 series is intended to cover the northern section of the country. Some 227 sheets are planned, which are being published at a rate of about 20 a year. The seven-colour 1 : 50 000 map covers areas to the north of the Chott and coastal areas further east in about 170 sheets, some 82 new sheets are in the OTC programme and are appearing at about three a year. A new 1 : 200 000 scale map has been completed by OTC to give basic scale coverage of the remaining areas of the country south of latitude 34° in 13 sheets. OTC also carries out an extensive town mapping programme, which involves 1 : 2000 scale mapping of over 100 settlements; this operation was established in 1975 and is now complete. Cadastral plans are now produced from

the topographic data bank using digital methods, having been compiled manually since 1964. OTC edits and prints geological and administrative maps for other government organizations and is also responsible for the publication of the National Atlas.

None of the current topographic products prepared by OTC is available outside of Tunisia.

A small-scale tourist map of the country is regularly updated by OTC and is available.

Earth science mapping of the country is the responsibility of the **Service Géologique**. Geological mapping is compiled on 1 : 50 000 topographic sheet lines and base: about 30 sheets have been published, mainly covering northern areas. Small-scale mineral and geological coverage of the country was also compiled in the 1960s but is no longer available.

The French **Office de la Recherche Scientifique et Technique Outre Mer (ORSTOM)** has published soils mapping in the 1960s. Ecological mapping of the country was prepared by the **Institut National de la Recherche Agronomique de Tunisie (INRAT)** in the 1960s but is no longer available. The **Institut de la Carte Internationale du Tapis Végétale (ICITV)** still distributes the vegetation and climatic maps compiled in the 1940s and 1950s.

Perhaps the most useful thematic coverage of the country is available in the *Afrika Kartenwerk* northern series, published by the German Research Society. This provides 1 : 1 000 000 scale coverage of the whole country for 16 themes, published with accompanying monographs. The sheet lines chosen make this series almost a Tunisian national atlas, and maps are readily available. The other

recent publication covering a variety of thematic mapping
is the atlas published in French language by **Editions
Jaguar** in its Jeune Afrique series. Various small-scale
maps of the country have been prepared for the tourist
market by the following commercial houses: **Hallwag** (see
under Algeria), **Karto+Grafik**, **Kümmerly + Frey**,
Michelin (see under Algeria), and **Ravenstein**. An indexed
street map of Tunis is published by **Editions de
l'Enterprise**.

Further information

Tunisia Office for Topography and Cartography (1979 and 1983)
Tunisia country reports presented to the *4th* and *5th United
Nations Regional Cartographic Conferences for Africa*, Addis Ababa:
Economic Commission for Africa.

Addresses

Office de la Topographie et de la Cartographie (OTC)
Cité Olympique, TUNIS.

Institut Géographique National (IGN)
136 bis rue de Grenelle, 75700 PARIS, France.

Service Géologique
Office National des Mines, 95 ave. Mohamed V, TUNIS.

**Office de la Recherche Scientifique et Technique Outre Mer
(ORSTOM)**
70–74 route d'Aulnay, 93140, BONDY, France.

**Institute National de la Recherche Agronomique de Tunisie
(INRAT)**
Ave de l'Indépendance, Ariane, TUNIS.

Editions de l'Enterprise
16 ave. de Carthage, TUNIS.

For Karto+Grafik, Ravenstein and the Gebrüder Borntraeger see
German Federal Republic; for Editions Jaguar, ICITV and
Michelin see France; for Kümmerly + Frey and Hallwag see
Switzerland; for DMA see United States.

Catalogue

Atlases and gazetteers

*Afrika Kartenwerk: Northern folio edited on behalf of the German
Research Society*
Berlin: Gebrüder Borntraeger, 1976–.
16 sheets.
With accompanying monographs.

Atlas de Tunisie (Atlas Jeune Afrique)
Paris: Editions Jaguar, 1985.
pp. 72.
Available as French or Arabic editions.

*Tunisia: official standard names approved by the United States Board
on Geographic Names*
Washington, DC: DMA, 1964.
pp. 399.

General

Tunisia: Hildebrands Urlaubskarte 1 : 1 900 000
Frankfurt AM: Karto+Grafik, 1985.

Tunisia 1 : 1 000 000
Bern: Kümmerly + Frey, 1986.

Tunisien: Strassenkarte 1 : 1 000 000
Frankfurt AM: Ravenstein, 1985.

Tunisie: carte touristique et routière 1 : 500 000
Tunis: OTC, 1984.
English and French legends.
Includes insets of Tunis area and southern Tunisia.

Environmental

Carte international du tapis végétal: Tunis Sfax 1 : 1 000 000
Toulouse: ICITV, 1958.
Covers northern and central parts of Tunisia.

Carte des précipitations 1 : 500 000 H. Gaussen and A. Vernet
Toulouse: ICITV, 1948.
2 sheets, both published.
With accompanying explanatory text.

Town maps

Plan de Tunis
Tunis: Editions de l'Enterprise.
Index on reverse.

Uganda

The **Department of Lands and Surveys (ULSD)** is
responsible for surveying and mapping activities in
Uganda. The basic scale series is the 1 : 50 000 map
prepared by the British **Directorate of Overseas Surveys
(DOS)** (now Overseas Surveys Directorate, Southampton,
OSD) which gives complete coverage of Uganda in 310
five-colour sheets. This map was completed in 1974 and is
drawn on the Transverse Mercator projection with a UTM
grid. Older sheets in the series use a contour interval of 50
feet, whilst more recent revisions use 20 m contour
intervals. This series has been compiled
photogrammetrically and sheets are being revised by
ground survey and more recently from aerial photographs.

From this map was derived a 1 : 250 000 scale series
covering the country in 20 sheets. Again more recent
revisions have been produced with metric relief
information. Other smaller-scale series have also been
published and a town plan of Kampala was issued at a
scale of 1 : 25 000. A 1 : 2500 scale series of town maps has
been compiled from aerial photographs, which has been
used to derive 1 : 10 000 and 1 : 12 500 scale maps. A
National Atlas and gazetteer were produced in the 1970s.

Thematic mapping activities are also coordinated by
ULSD but data is compiled by other departments. The
1 : 250 000 scale soils and vegetation maps compiled in the
1960s and 1970s both use the topographic series sheet lines

and base and are overprinted with four-colour thematic information. Both maps were prepared by the Department of Agriculture. The availability of all this Ugandan published mapping of the country is currently uncertain.

The **Geological Survey and Mines Department (UGSMD)** carries out geological and other earth science mapping activities and recently advertised that its mapping compiled in the 1960s and 1970s is still available. It has published maps at scales of 1:100 000 and 1:250 000. We have indexed the 1:250 000 scale geological series, which is also available as a Bouguer contours and gravity stations edition. This map is compiled on the topographic map sheet lines. Coverage at 1:100 000 scale is in half-degree quads and is rather more patchy. Smaller-scale maps were compiled for the National Atlas and are available as separate sheets at 1:1 250 000 scale to show geology and at 1:1 500 000 scale to show geophysics, minerals and geomorphology. A geochemical atlas plots the distribution of individual mineral deposits.

Small-scale general coverage of Uganda is published by the Russian **Glavnoe Upravlenie Geodezii i Kartografii (GUGK)** and by the American **Central Intelligence Agency (CIA)**.

Other maps covering the country are described under Africa, East.

Further information

Uganda (1979) Report on cartographic activities in Uganda for the period 1972 to 1979, presented to the *4th United Nations Regional Cartographic Conference for Africa*, Addis Ababa: UN Economic Commission for Africa.

A current catalogue is issued by the Department of Geological Survey and Mines.

Addresses

Lands and Surveys Department (ULSD)
PO Box 7061, 15 Obote Avenue, KAMPALA.

Overseas Surveys Directorate (DOS/OSD)
Ordnance Survey, Romsey Rd, SOUTHAMPTON SO9 4DH, UK.

Geological Survey and Mines Department (UGSMD)
PO Box 9, ENTEBBE.

For DMA and CIA see United States; for GUGK see USSR.

Catalogue

Atlases and gazetteers

Uganda: official standard names approved by the United States Board on Geographic Names
Washington, DC: DMA, 1964.
pp. 187.

General

Uganda 1:1 500 000
Moskva: GUGK, 1986.
In Russian.
With economic, climatic and ethnographic insets, and placename index.

Topographic

Uganda 1:50 000 DOS 426 Y732
Kampala and Tolworth: DOS and ULSD, 1950–74.
310 sheets, all published. ∎

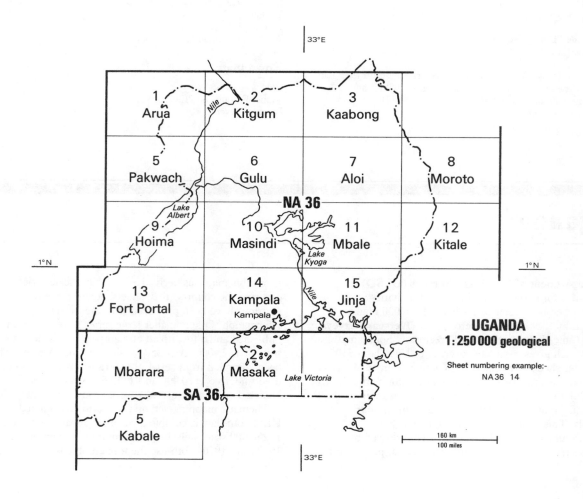

Geological and geophysical

Geochemical atlas of Uganda 1 : 2 000 000 and 1 : 4 000 000
Entebbe: UGSMD, 1973.
pp. 32.

Uganda geology 1 : 1 500 000
Entebbe: UGSMD.

Uganda geophysics and seismology 1 : 1 500 000
Entebbe: UGSMD.
With explanatory text.

Uganda geomorphology 1 : 1 500 000
Entebbe: UGSMD.
With explanatory text.

Uganda: geology 1 : 1 250 000
Entebbe: UGSMD, 1960.
Also available with Bouguer contours and gravity stations
overprint.

Uganda geological survey 1 : 250 000
Entebbe: UGSMD, 1961–.
17 sheets, 12 published. ■
Also available with Bouguer contour and gravity station overprint:
unpublished areas available as topographic edition with gravity
information overprint.

Uganda: geological survey 1 : 100 000
Entebbe: UGSMD, 1959–.
ca 75 sheets, 24 published. ■

Human and economic

Uganda industries and mining 1 : 1 500 000
Entebbe: UGSMD.
With explanatory text.

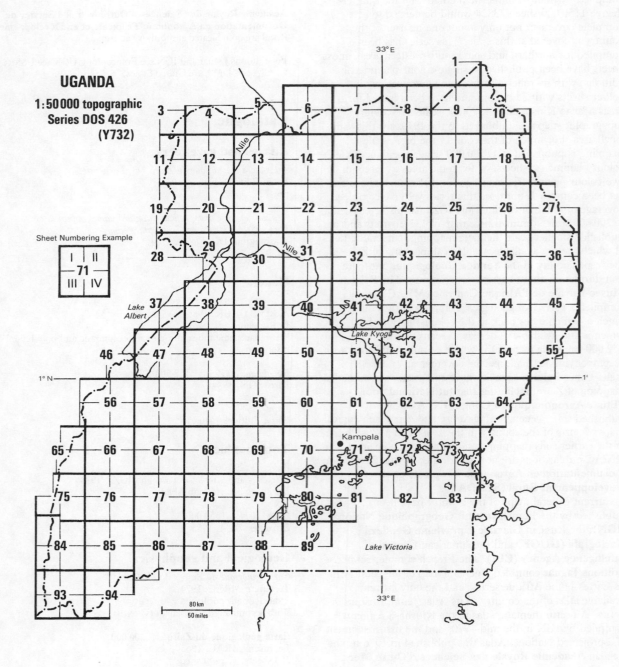

Zaire

The national mapping agency is the **Institut Géographique du Zaire (IGZa)** which is responsible for photogrammetric, geodetic and topographic surveys of the country and for the publication of Zaire's national mapping. The mapping base was established by Belgian colonial authorities, and the older colonial mapping in the 1:50 000, 1:100 000 and 1:200 000 series may still be acquired. Current series are, however, unavailable outside Zaire; we have listed the 1:200 000 and 1:50 000 maps in our catalogue section but have not included graphic indexes to these series because of the availability problems. All series use UTM specifications; recent sheets are being compiled by photogrammetric methods. Of the quarter-degree 1:50 000 sheets, 3824 would be needed to give complete coverage, but only about one-quarter of the country is covered at this scale. This series has been compiled for northern and some south-western areas: most sheets have been published as single-colour planimetric editions, with no relief, though some more recent six-colour sheets with 25 m contours have been issued for the areas near to Kinshasa. A 1:25 000 scale five-colour topographic map was published in the 1970s for this same small area. For much of the country the basic scale mapping is published at 1:200 000 scale, as either single-colour planimetric sheets, or for some areas as three- or two-colour planimetric maps. Full-relief editions have not yet been compiled, nor does this series yet give complete coverage of much of the interior. IGZa has also compiled a 1:200 000 scale administrative map of 132 districts and has published smaller-scale thematic mapping, current editions of which are not available. Geological mapping of Zaire is the responsibility of the **Service Géologique**, but there is no information about current programmes. The Belgian **Musée Royale de l'Afrique Centrale (MRAC)** has compiled a variety of geological mapping of the country, much of which dates back to the colonial period, but which is still available. There is a recent two-sheet 1:2 000 000 scale map of the whole country published with accompanying text, and partial coverage at 1:200 000 scale. Other Belgian agencies have compiled environmental mapping of Zaire, including **Institut National pour l'Etude Agronomique du Congo (INEAC)**, which published soils, vegetation, land use and geomorphological maps of parts of the country in the 1960s and 1970s. Some of the earlier soils mapping is no longer available and INEAC publications are now distributed by the **Service de Documentation en Agronomie Tropicale et en Développement Rural (SERDAT)**.

Current general maps of the country have been published by the French **Institut Géographique National (IGN)**, the Russian **Glavnoe Upravlenie Geodezii i Kartografii (GUGK)** and the American **Central Intelligence Agency (CIA)**. The French commercial house **Editions Jaguar** compiled a general map of the country in its Cartes Jeune Afrique series, and the only current available atlas of the country in its Atlas Jeune Afrique Series. A useful thematic atlas of the Kinshasa region was compiled by IGN in the mid-1970s and the first instalment of a projected National Atlas was published in 1976 by the Belgian **Académie Royale des Sciences d'Outre Mer**.

Addresses

Institut Géographique du Zaire (IGZa)
106 Boulevard du 30 Juin, BP 3086, KINSHASA 1.

Musée Royale de l'Afrique Centrale (MRAC)
13 Steenweg op Leuwen, B1980, TERVUREN, Belgium.

Service Géologique du Zaire
BP 898, 44 Avenue des Huileries, KINSHASA.

Institut Agronomique pour l'Etude Agronomique du Congo (INEAC)
Rue Defacqz 1, 1050, BRUXELLES, Belgium.

Academie Royale des Sciences d'Outre Mer and **Service de Documentation en Agronomie Tropicale et en Développement Rural** are both located at the above address.

For Editions Jaguar and IGN see France: for GUGK see USSR; for CIA and DMA see United States.

Catalogue

Atlases and gazetteers

Atlas de la République du Zaire   G. Laclavère
Paris: Editions Jaguar, 1978.
pp. 72.

Republic of the Congo: official standard names approved by the United States Board on Geographic Names
Washington, DC: DMA, 1964.
pp. 426.

General

Zair   1:2 500 000
Moskva: GUGK, 1985.
In Russian.
With placename index and climatic, economic and population inset maps.

République du Zaire   1:2 500 000
Paris: IGN, 1984.

Zaire
Paris: Editions Jaguar.

Topographic

Zaire   1:200 000
Bruxelles and Kinshasa: IGM and IGZa, 1934–.
230 sheets, *ca* 100 published.

Congo Belge   1:50 000
Bruxelles: IGM, 1955–.

Geological and geophysical

Carte géologique du Zaire   1:200 000
Tervuren: MRAC, 1974.
2 sheets, both published.
With accompanying explanatory text.

Carte géologique du Zaire   1:200 000
Tervuren: MRAC, 1963–.
230 sheets, 17 published.
Some sheets published with title *Congo-Belge – Katanga*   1:200 000, or as part of *Atlas du Katanga*.

Environmental

Carte des sols du Congo-Belge et du Ruanda-Urundi   1 : 5 000 000
Bruxelles: INEAC, 1960.
With accompanying explanatory text.

Administrative

Carte politique et administrative   1 : 3 000 000
Kinshasa: IGZa, 1980.

Town maps

Atlas de Kinshasa
Paris: IGN, 1975.
44 map plates.

Zambia

The national mapping organization in Zambia is the
Survey Department (ZS) which is responsible for the
production of topographic mapping and the publication of
smaller-scale thematic coverage, as well as the compilation
of large-scale street maps and township plans. The national
basic scale is the *1 : 50 000 ZS 51 series*. This map was
started by the British **Directorate of Colonial Surveys**
(now Overseas Surveys Directorate, Southampton, DOS/
OSD) after World War II. Early sheets were issued with
25 or 50 foot contours, or in some areas with no contour
information, as preliminary plots. DOS continued to
provide aid in the production of new maps in this series
throughout the 1970s: many sheets still carry the DOS
series designation DOS 424 (Z741). Since 1970 five-colour
metric editions have been published by ZS with 20 m
contour intervals, and on a UTM grid. Details are revised
from the latest available aerial photography. Complete
coverage of the country would require over 1000 sheets,
but at present there are no plans to extend the series into
sparsely populated south-western areas of Western Prairie,
which are to be covered only by 1 : 100 000 scale mapping.
This series ZS 41 has been in progress since 1980 and will
be completed by 1991. In the early 1980s SwedSurvey, the
Overseas Agency of the National Land Survey of Sweden,
provided aid in order to allow improved production
methods for the 1 : 50 000 map; changes included the
introduction of process colours. The other national series is
the *1 : 250 000 scale ZS 31* map which covers the country in
54 sheets. New sheets in this series are published by ZS
with 50 m contours and on a UTM grid; some older sheets
are issued with layered relief as non-metric versions. These
are being replaced in a revision programme. There are no
plans to revise the 1960s-produced provisional 1 : 500 000
scale series which covered the country in 17 sheets. Some
sheets in this series are still available, however.

Township mapping is produced by ZS at scales of
1 : 2500, 1 : 5000, and 1 : 10 000. All major centres of
population are covered; sheets are available as dyeline
copies. Full-colour street maps have been issued at
1 : 20 000 to cover the main cities. Six sheets have been
published, including a new four-colour street map of
Lusaka, produced with aid from SwedSurvey.

ZS also issues smaller-scale and thematic mapping. At
1 : 1 500 000 scale there is a full-colour metric general map
of the country; outline editions at this scale are available to
show medical facilities, roads, national parks and
administrative boundaries, whilst full-colour maps of
forests and tsetse fly distribution are also published.
Administrative mapping is available at 1 : 500 000, as
outline maps of each province, and at 1 : 250 000 in a
District Series of outline maps with red overprints of ward

boundaries. A four-sheet 1 : 750 000 map is issued as a
relief or land use edition. Another important example of
ZS's role in thematic mapping is the full-colour nine-sheet
1 : 500 000 scale vegetation map of the country. This map
was compiled with the assistance of the German Survey
Authority **Institut für Angewändte Geodäsie (IfAG)** for
the Zambian Forest Department and maps 17 different
vegetation types.

The *Republic of Zambia Atlas* was begun in 1966; 22
sheets are still available, most at 1 : 2 500 000. A new
edition of the National Atlas is under compilation which
will when complete give 1 : 3 000 000 scale thematic
coverage of the physical environment, population,
agriculture, industry and socioeconomic activities. This,
too, was conceived with Swedish aid. A National Gazetteer
is currently under revision.

The **Cartographic Location Analysis Research Unit** of
the National Council for Scientific Research is involved in
the production of the demographic, social and economic
maps in the new National Atlas. It is also producing
1 : 20 000 land use maps of major towns derived from
1 : 5000 orthophotomapping and produced the *Atlas of the
Population of Zambia*. This demographic atlas is being
updated using data from the 1980 Census of Population
and Housing, and will be published at 1 : 3 000 000 scale to
conform to the new National Atlas scale. The first two
sheets in the new edition became available in 1986 and a
further four maps will be published by the end of 1986.
The **Central Statistical Office** provides the census data for
this exercise and has also carried out pre-census
enumeration district mapping.

Geological mapping of Zambia has been carried out by
the **Geological Survey Department (ZGS)**. A quarter-
degree sheet series has been issued since 1960 at 1 : 100 000
scale (each sheet covers four 1 : 50 000 topographic maps).
This series is available either with explanatory monographs
or as separate sheets. Only a small part of the country has
been mapped in this series; coverage is best around areas
of economic importance, that is the Copperbelt and around
Lusaka. A new full-colour 1 : 250 000 geological series on
topographic sheet lines has recently been initiated—five
sheets were printed by the UK Overseas Surveys
Directorate in 1986, and it is planned to extend this series
to give national coverage. Smaller-scale mineral and
geology mapping of the country is also available, including
a 1 : 1 000 000 scale full-colour national map.

Soil mapping of Zambia is carried out by the **Soil
Survey Unit**, part of the Department of Agriculture. The
Unit was set up in 1969 and has mostly to date compiled
large-scale mapping of small areas, published with
accompanying *Soil Survey Reports*. In recent years

emphasis has shifted towards the publication of exploratory mapping at a district level. This programme includes a 1:500 000 scale soil map of the north-west region, derived from Landsat imagery and compiled in 1984 with the aid of the Regional Centre for Remote Sensing in Nairobi. In 1986 the first systematic quarter-degree soil map at 1:100 000 was published. A national soil map at 1:2 500 000 has been compiled and is available as a diazo print, as are maps at the same scale covering agro-ecology, soil acidity and crop suitability. There are plans to publish a 1:1 000 000 scale national soil map in 1988.

Further information

Catalogues are issued on a regular basis by both ZS and ZGS. Other organizations provide information about their mapping and organizational set-up.

For information about the role of SwedSurvey in the mapping of Zambia see Palm, C. (1985) Cartographic education in Zambia, in Fraser-Taylor D.R., Editor, *Education in Contemporary Cartography*, Chichester: Wiley.

For information about the Soil Survey Unit see Dalal-Clayton (1984) The development of the Soil Survey in Zambia, *Soil Survey and Land Evaluation*, **4**(1), 18–22.

Addresses

Survey Department (ZS)
Mulungushi House, PO Box 50397, LUSAKA.

Overseas Surveys Directorate (DOS/OSD)
Ordnance Survey, Romsey Rd, SOUTHAMPTON SO9 4DH, UK.

Cartographic and Location Analysis Research Unit
National Council for Scientific Research, PO Box CH.158, Chelston, LUSAKA.

Central Statistical Office
PO Box 1908, LUSAKA.

Geological Survey Department (ZGS)
PO Box 50135, Ridgeway, LUSAKA.

Soil Survey Unit
Department of Agriculture, Mount Makalu Central Research Station, Private Bag 7, CHILANGA.

For IfAG see German Federal Republic; for DMA see United States.

Catalogue

Atlases and gazetteers

National atlas of Zambia
Lusaka: ZS, 1985–.
Single sheets from earlier atlas still available.

Atlas of the population of Zambia
Lusaka: National Council for Scientific Research, 1977.
15 maps, all published.
Explanatory text accompanies maps.

Zambia: official standard names approved by the United States Board on Geographic Names
Washington, DC: DMA, 1972.
pp. 585.

General

Republic of Zambia   1:1 500 000
Lusaka: ZS, 1981.
Metric contoured relief.

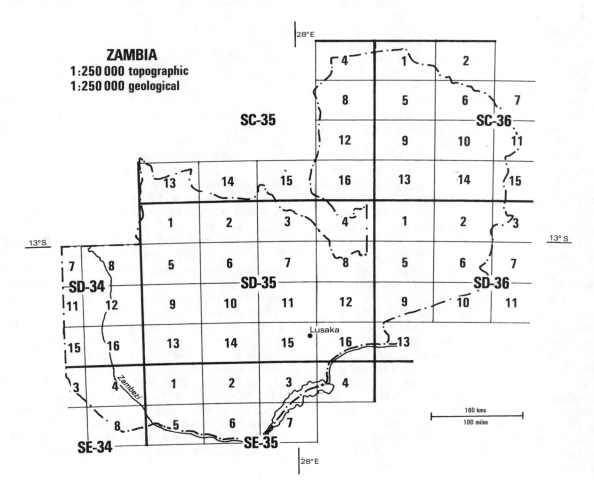

ZAMBIA
1:250 000 topographic
1:250 000 geological

Republic of Zambia: metric road map   1 : 1 500 000
Lusaka: ZS and Roads Dept, 1979.
Outline map showing road distances.

Republic of Zambia: designated roads   1 : 1 000 000
Lusaka: ZS and Roads Dept, 1977.
2 sheets, both published.

Republic of Zambia   1 : 750 000
Lusaka: ZS, 1975.
4 sheets, all published.

Topographic

Republic of Zambia   1 : 250 000   ZS 31
Lusaka: ZS, 1971–.
54 sheets, all published. ■

Republic of Zambia   1 : 100 000   ZS 41
Lusaka: ZS.
71 sheets, 13 published. ■
Uncontoured, some sheets only available as diazo prints.

Republic of Zambia   1 : 50 000   ZS 51
Lusaka: ZS, 1970–.
810 sheets, *ca* 750 published. ■
Many sheets available as DOS 424 (Z741) editions.

Geological and geophysical

Provisional mineral map of the Republic of Zambia   1 : 1 500 000
Lusaka: ZGS, 1973.

Geological map of the Republic of Zambia   1 : 1 000 000
Lusaka: ZGS, 1977.
4 sheets, all published.
Reprinted in 1981 with some modifications.

Geological map of the Republic of Zambia   1 : 250 000
Lusaka: ZGS, 1986–.
54 sheets, 5 published. ■

*Republic of Zambia: geological survey quarter degree
sheets*   1 : 100 000
Lusaka: ZGS, 1960–.
250 sheets, *ca* 30 published. ■
Available with geological reports.

Environmental

Map of Zambia: Forest Estate 1975   1 : 1 500 000
Lusaka: ZS, 1977.

Map of Zambia showing tsetse fly distribution   1 : 1 500 000
Lusaka: ZS, 1974.

The Republic of Zambia land use map   1 : 750 000
Lusaka: ZS, 1976.
4 sheets, all published.

The Republic of Zambia: vegetation map   1 : 500 000
Lusaka: ZS, 1976.
9 sheets, all published.

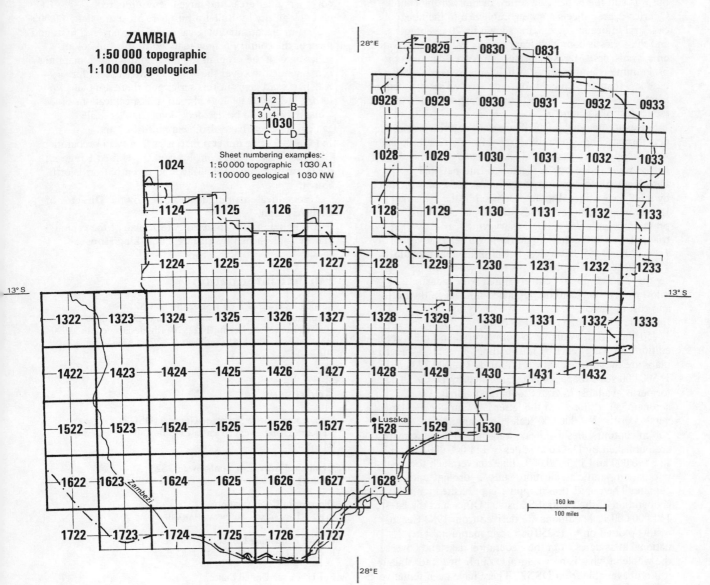

Administrative

Zambia 1 : 1 500 000 SDT No. 324/1
Lusaka: ZS.
Outline map showing province and district boundaries.

Town maps

Republic of Zambia: street maps 1 : 20 000
Lusaka: ZS, 1985.
7 sheets, all published.
Series covers the major cities in the country.

Zimbabwe

The national surveying and mapping agency in Zimbabwe is the **Department of the Surveyor General (DSGZ)** which was established in the 1890s with mainly land registration responsibilities. Since 1933 DSGZ has been responsible for topographic, geodetic and cadastral surveying of the country and publishes topographic, thematic and cadastral mapping. The map base of Zimbabwe was established by British mapping agencies when the country was part of the Federation of Rhodesia and Nyasaland. A 1 : 50 000 series was begun after World War II, and complete cover of 561 sheets was available by 1970. Until the 1970s this map showed relief with 50 foot contours. The specification of this series has since changed but it is still the basic scale series for most of the country. Quarter-degree sheets are now published to include cadastral information; relief is shown by 20 m contours, and the series is compiled from aerial coverage and plotted on a Transverse Mercator projection. A regular revision programme is in operation, which aims to keep the map base less than 10 years old for most areas. This programme is intended to revise all sheets to the new five-colour metric topocadastral standard.

Sheets to the north of 17° S are not yet published in the new format. They are to be derived from a new 1 : 25 000 scale mapping programme. This *Zimbabwe Air Survey Project* was funded by the Canadian International Development Agency and technical expertise was provided by the Surveys and Mapping Division of the Canadian Department of Energy Mines and Resources. This programme involved geodetic surveying, aerial triangulation and complete newly flown aerial coverage at 1 : 80 000 or 1 : 65 000 scales. It will ultimately provide 1 : 25 000 scale orthophotomaps of the whole of the country, which are to be overprinted with cadastral information. Blocks to be published first are in the north and east of the country.

The 1 : 50 000 map is used to compile a 33-sheet 1 : 250 000 series, available in both layered and unlayered editions showing relief by hill shading. These topocadastral maps are overprinted with UTM grid. 1 : 1 000 000 and 1 : 500 000 scale maps are also published by DSGZ. The former is available in either topographic or ICAO aeronautical editions and the latter is a four-sheet enlarged version of the 1 : 1 000 000 scale topographic map, with 300 m contours, designed as a wall map. Thematic maps are published by DSGZ at scales of 1 : 1 000 000, 1 : 2 500 000 and 1 : 3 000 000. These cover such themes as population, land classification, soils, hydrology and climate. The population mapping is published with the cooperation of the Central Statistical Office and the latest 1 : 1 000 000 scale editions are derived from 1982 Census results plotted on to 1 : 250 000 scale mapping. The national atlas dates from the Federation days and covered the whole of what is now Zambia, Zimbabwe and Malawi: it is still available from DSGZ. There have been tentative discussions for the production of a new National Zimbabwe Atlas but no firm timetable or definite programme has been established. Larger-scale programmes are carried out by DSGZ. 1 : 5000 scale topocadastral maps based on UTM sheet falls, derived from aerial coverage with a 4 m (or 10 foot for older sheets) contour interval are published for the major settlements in the country. Harare is also mapped in a 1 : 2500 series, and DSGZ publishes town plans of Harare and Bulawayo. Tourist maps are available to cover certain areas of the country.

Geological mapping is carried out by the **Geological Survey of Zimbabwe** (GSZ), which was founded in 1910. Maps have been published with accompanying bulletins, cover irregular areas and have been numbered chronologically by Bulletin number. Most are at 1 : 100 000 scale, important mineralogical areas were covered first, and most of the country is now mapped in this full-colour series or is in the programme for reconnaissance mapping. It is intended to cover the whole country except for Kalahari sand areas in this series and there are plans for a future 1 : 250 000 geological map, the specifications of which have yet to be decided. Some smaller-scale geological sheets have also been published and a 1 : 1 000 000 scale mineral map is currently in preparation. GSZ has received aid from France, the United Kingdom, the German Federal Republic, and recently from North Korea.

Aeronautical mapping is published by the **Director of Civil Aviation**.

A useful and fully indexed A to Z plan of the capital Harare is published by **Harare Publishing House**.

Further information

Catalogues are issued by DSGZ and by GSZ.

For further information about the new specifications of the 1 : 50 000 series see Urban, F. (1970) Rhodesia's new 1 : 50 000 map series, in *South African Journal of Photogrammetry*, **3**(4), 301–310.

Addresses

Department of the Surveyor General (DSGZ)
Samora Machel Avenue Central, PO Box 8099, Causeway, HARARE.

Geological Survey of Zimbabwe (GSZ)
Ministry of Mines and Lands, PO Box 8039, Causeway, HARARE.

Director of Civil Aviation
Private Bag 7716, Causeway, HARARE.

Harare Publishing House
Robinson House, Ringwa St, HARARE.

For DMA see United States.

Catalogue

Atlases and gazetteers

Federal atlas Federation of Rhodesia and Nyasaland 1:2 500 000
Causeway: DSGZ, 1960–63.
24 sheets, all published.
Published sheets include Malawi and Zambia.

Southern Rhodesia: official standard names approved by the United States Board on Geographic Names
Washington, DC: DMA, 1973.
pp. 362.

General

Zimbabwe: general relief 1:2 500 000
Causeway: DSGZ, 1983.
Layered or unlayered editions.

Rhodesia: relief 1:1 000 000
Causeway: DSGZ, 1984.
Available as layered or unlayered editions or with 1:50 000 sheet lines overprint.

Zimbabwe 1:500 000
Causeway: DSGZ, 1975–.
4 sheets, all published.
Layered or unlayered editions.

Topographic

Zimbabwe 1:250 000
Causeway: DSGZ, 1968–.
33 sheets, all published. ■

Zimbabwe: 1:50 000
Causeway: DSGZ, 1954–.
561 sheets, all published. ■

Geological and geophysical

Provisional; geological map of Rhodesia (Zimbabwe) 1:3 000 000
Causeway: GSZ, 1977.

Provisional geological map of Rhodesia 1:1 000 000
Causeway: GSZ, 1978.
With accompanying explanatory monograph.

Zimbabwe: geological map series 1:100 000 (mostly)
Causeway: GSZ, 1925–.
78 sheets published.

Environmental

Zimbabwe–Rhodesia climatic comfort/discomfort belts and building design 1:2 500 000
Causeway: DSGZ, 1979.

Rhodesia: rainfall maps 1:2 500 000
Causeway: DSGZ, 1968.
3 sheets.
Maps cover mean annual and monthly rainfall and highest 24 h totals on record.

Provisional soil map of Zimbabwe–Rhodesia 1:1 000 000
Causeway: DSGZ, 1979.

Rhodesia: hydrological zones 1:1 000 000
Causeway: DSGZ, 1970.

Zimbabwe: natural regions and farming areas 1:1 000 000
Causeway: DSGZ, 1980.

Zimbabwe–Rhodesia: land classification 1:1 000 000
Causeway: DSGZ, 1979.
Also available with 1:250 000 and 1:50 000 sheet line and UTM grid overprint.

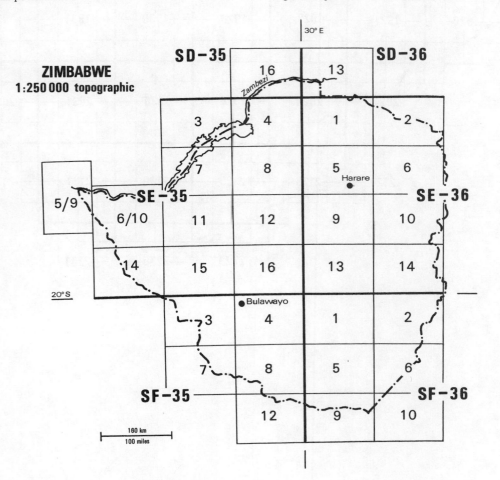

Zimbabwe: population map 1 : 1 000 000
Causeway: DSGZ, 1984.
3 sheets, all published.
Sheets cover distribution, density and administrative areas.

Administrative

Zimbabwe: administrative provinces and districts 1 : 1 000 000
Causeway: DSGZ, 1984.

Town maps

The A to Z of Harare
Harare: Harare Publishing House, 1981.
pp. 100.

Street map of Harare 1 : 33 333
Causeway: DSGZ, 1979.

Street map of Greater Bulawayo 1 : 33 333
Causeway: DSGZ, 1980.

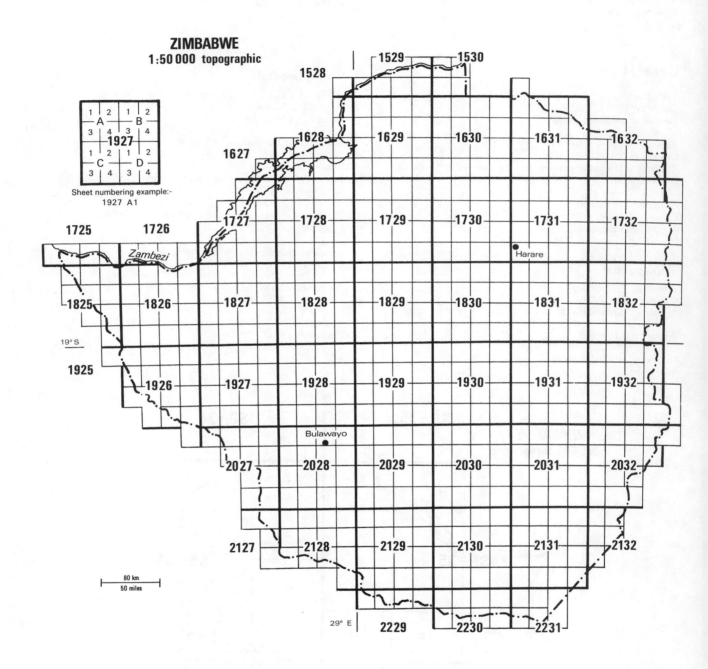

THE AMERICAS

NORTH AMERICA

Mapping in North America is dominated by the three great nations of Canada, the United States and Mexico, each with its distinctive mapping systems and programmes. The best complete topographic coverage of the continent is currently at 1:250 000. Basic scale mapping differs between the US (1:24 000 or 25 000) and Canada and Mexico (1:50 000), and although none of these basic series is yet complete all are nearing that state.

A number of small-scale general maps of the whole continent are published by the major federal mapping agencies in the US and Canada, and by numerous commercial publishers both on and off the continent.

The **United States Geological Survey (USGS)** has published several thematic maps of the whole continent on earth science themes, and has cooperated with the **Association of American Petroleum Geologists (AAPG)** and the **Geological Society of America (GSA)** in the preparation of some of these.

The new edition of *The Tectonic Map of North America* scheduled for publication in 1987 is the result of a project of the AAPG to produce a map in a plate tectonics framework to replace the map by P. B. King first published in 1969. The map is in four sheets on a Transverse Mercator projection. The base map is a new one and is also used for three new maps (*Geologic map of North America*, *Gravity map of North America*, and *Magnetic map of North America*) which are being produced or marketed through GSA in its Decade of North American Geology (DNAG) programme.

Addresses

For USGS, GSA, NGS, AAPG and PennWell see United States; for DEMR, GSC and Queen's University, see Canada; for Bartholomew, see Great Britain; for Kartographische Verlag Wagner, see German Federal Republic; for UNESCO, see World.

Catalogue

Atlases and gazetteers

Atlas of North America
Washington, DC: NGS, 1985.
pp. 264.

General

The Americas   1:20 000 000
Washington, DC: NGS, 1979.
Inset map: *Physical map of the Americas*.

North America   1:13 825 000
Washington, DC: NGS, 1981.

North America (base)   1:10 000 000
Reston, VA: USGS, 1982.
Map compiled from World Data Bank II and other sources. Transverse Mercator projection.

North America   1:10 000 000   MCR 31
Ottawa: DEMR, 1971.

North America   1:10 000 000
Edinburgh: Bartholomew, 1983.

Nordamerika   1:10 000 000   Wagnerkarte
Berlin: Kartographischer Verlag Wagner, 1982.

Geological and geophysical

Geological map of North America   1:5 000 000
Reston, VA: USGS/Boulder: GSA, 1965.
2 sheets, both published.

The tectonic map of North America   1:5 000 000
Tulsa, OK: AAPG, 1987.
4 sheets, all published.

Generalized tectonic map of North America   1:5 000 000
Reston, VA: USGS, 1972.
2 sheets, both published.

Preliminary metallogenic map of North America   1:5 000 000
Reston, VA: USGS, 1981.
4 sheets, all published.

Mineral deposits map of North America   1:5 000 000
Reston, VA: USGS, 1982.
4 sheets, all published.

Geothermal gradient map of North America   1:5 000 000
Reston, VA: USGS, 1976.
2 sheets, both published.

Subsurface temperature map of North America   1:5 000 000
Reston, VA: USGS, 1976.
2 sheets, both published.

Retreat of Wisconsin and Recent ice in North America   1:5 000 000
Ottawa: GSC, 1965.

Environmental

Climatic atlas of North and Central America
Paris: UNESCO, 1979.

Les regions pluviométriques de l'Amerique Nord 1:13 750 000
Paris: Institut de Géographie, 1974.

Human and economic

Indians of North America 1:10 610 000
Washington, DC: NGS, 1982.

Isodemographic map of North America 1980–81
Kingston, Ontario: Department of Geography, Queens
University, 1983.

Isodemographic map of North America 1975–76
Kingston, Ontario: Department of Geography, Queens
University, 1978.

Natural gas pipelines of the United States and Canada 1:3 600 000
Tulsa, OK: PennWell, 1985.

Crude oil pipelines of the United States and Canada 1:3 600 000
Tulsa, OK: PennWell, 1984.

Canada

In Canada today, mapping is carried out by agencies of
both the federal and provincial governments. In
topographic surveying and mapping, the federal
government is responsible for the first-order geodetic
network and for mapping at scales of 1:50 000 and
smaller, while provincial governments and the
administrations of the larger cities take on the tasks of
more detailed mapping, usually at 1:20 000 or larger as
required for resource development, cadastral purposes and
for urban planning. Similarly, much thematic mapping,
including geological, soil, land use and forest surveys, is
carried out by provincial agencies sometimes cooperatively
with, and sometimes independently of, the federal ones.

Most federal mapping in Canada falls within the control
of the **Department of Energy, Mines and Resources
(DEMR)**, and within this department there are three
branches with significant mapping interests. Topographic
mapping is carried out by the **Surveys and Mapping
Branch**, geological mapping by the **Geological Survey of
Canada (GSC)**, and gravity surveys by the **Earth Physics
Branch (EPB)**, which maintains a National Gravity Data
Base. There are also important federal thematic
programmes within the **Lands Directorate** and the **Inland
Waters Directorate** of Fisheries and Environment Canada,
and at the **Land Resource Research Centre** of Agriculture
Canada. All these federal organizations have their
headquarters at Ottawa.

Modern topographic mapping of Canada dates from
1922 when a Board of Topographic Surveys and Maps was
set up to coordinate the rather haphazard situation that
had resulted from the activities of three major topographic
mapping authorities which then existed. The Board set out
to ensure the adoption of uniform mapping standards and
to devise a National Topographic System of scales and
sheet lines. The projection eventually adopted (in 1946) for
national mapping was the Universal Transverse Mercator
(UTM). The sheet lines for the topographic series are
graticule based, and are broken down from primary
quadrangles each covering 8° of longitude and 4° of
latitude in the pattern shown in the following diagram.
This sheet system is officially known as the National
Topographic System (NTS). It is important to note that it
differs from the International Map of the World system,
used by many countries, which is based on a 6° division of
longitude. IMW sheets lines are used in Canada only for
the 1:1 000 000 IMW maps themselves, for some thematic

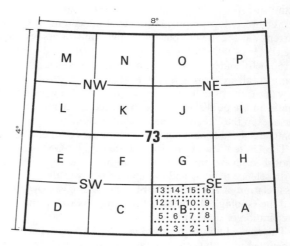

Sheet numbering examples: –

Primary 1:1 M quadrangle	73
1:500 000	73 NW
1:250 000	73 G
1:50 000	73 B/3

maps at this scale, and for the 1:250 000 Joint Operations
Graphic (JOG) series.

The National Topographic System set up in 1922
originally envisaged a series of scales based on imperial
units of measurement, and sheets were published at one,
one-half and one-quarter inch to the mile, but these were
slowly phased out as the rational scales of 1:50 000,
1:125 000 (now also discontinued), 1:250 000 and
1:500 000 were introduced to replace them. The principal
nationwide series being produced today are therefore at
1:50 000, 1:250 000, 1:500 000 and 1:1 000 000 (IMW).
Federal maps at 1:25 000 and 1:125 000 for some areas
are still available, but both these series have been
discontinued. Some of the provincial authorities maintain
series at these and other intermediate scales, however.

As in the US, the largest scale at which complete
topographic cover is presently available is 1:250 000, and
this series was completed in 1970. The 1:250 000 series
began in 1923 as a 4 mile to the inch (1:253 440) map, the
present scale ratio being introduced in 1950. The series
progressed most rapidly after 1947 when it benefited from
an intensive programme aimed at carrying out a complete
trilateration of the country and the acquisition of

comprehensive air photo cover from which maps could be compiled by photogrammetry. The mapping was shared at this time between the Surveys and Mapping Branch of DEMR (then called the Department of Mines and Technical Surveys) and the Army Survey Establishment, but sheets were produced to a uniform specification.

As is common with a long-lived series, there have been a number of specification changes to the 1:250 000 series, and because some sheets have been revised more frequently than others there are dissimilarities between sheets of different areas. The modern sheets are printed in six colours with contours ranging in interval from 20 to 200 m according to the prevailing terrain. A red stipple is used for built-up areas. Roads are shown in colour with a simple classification based on suitability for traffic. The UTM grid is shown in light blue at 10 000 m intervals. Several revision cycles have been established to satisfy different rates of human-induced change. For the most settled areas the cycle is 10 years.

The most important federal topographic map, although incomplete, is the 1:50 000 series which forms the basic scale mapping for the country as a whole. The series requires more than 13 000 sheets, and, initially as a 1 inch to the mile (1:63 360) map, has been in progress since 1904. About 10 000 sheets have so far been produced, and the aim is to complete cover of the national territory in both hard copy and digital form by the end of the century. Most of the remaining work is in the Canadian north, beyond the provincial boundaries.

The 1:50 000 scale was adopted in 1950, with new maps produced at this scale and others progressively converted by photographic enlargement or rescribing. Other elements of metrication—contours and distances—were not introduced until 1975, and although all new maps since that date have been metric there remain many sheets which still have imperial measurements. Since 1977, new maps have also been fully bilingual.

Since the new scale was introduced in 1950, there have been many changes in style and format. Currently the main distinction is between the six-colour specification (1967 Standard) used for southern Canada and the 1972 monochrome specification used for the emptier northern areas defined by the so-called Wilderness Line, beyond which it was decided that full-colour mapping was unnecessary. The 1:50 000 sheets are now all of a common 15' × 30' format (until 1967, a half-sheet format had been used in the south), and there is no longer a separate military edition, so that all sheets have the UTM grid printed in blue or purple. Information content includes much attention to road classification (on the coloured sheets), while the northern sheets take account of the variety of swamp types, and of special relief forms such as eskers and pingos, which have their own symbols. Contour intervals vary with the type of relief from 25 foot (or 5 m) to 100 foot (or 40 m). The inclusion of various types of administrative and cadastral boundary varies from province to province.

In 1962, photomaps were introduced into the 1:50 000 production line. Photographic imagery would seem well suited to the texture of permafrost landscapes, whose surface features are difficult to classify by conventional symbology. A number of contoured and colour enhanced photomaps of high arctic settlements were also produced by the **Mapping and Charting Establishment (MCE)**, Department of National Defense, and are still available. Current interest, however, is in the production of orthophotomaps by a fully automated system.

The revision cycle for the 1:50 000 map varies regionally according to the rate of change in the landscape, from 5 years for cities and suburban areas to 30 years for slow-change wilderness areas.

The 1:500 000 scale federal series has its origins in the 8 mile to the inch (1:506 880) series initiated in 1929, and the new scale was adopted in 1958. The series is issued in two versions, topographic and aeronautical, the latter being an overprint on the topographic base. The maps have layer coloured relief, and graticule lines are printed on the map face. However, as a topographic map this series is rather generalized.

Canada is one of few countries to have produced a recent and complete cover of its territory to the specification of the International Map of the World. Most of the 74 sheets were published in first editions during the period 1969–80. Sheets conform to the standard IMW system which is shown in the world index (see pages 68-69).

In addition to the regular topographic series described above, the Surveys and Mapping Branch also publishes a number of small-scale maps of Canada, maps of most of the provinces (some of which are rather dated) and some fine National Park maps.

The **Geological Survey of Canada (GSC)** was founded in 1842, achieving permanent status in 1877. It has always had a primary interest in mapping, and was initially involved in topographic as well as geological survey, retaining a topographic division until 1936. It has published a number of small-scale, general maps on earth science topics (mainly dating from the 1960s) and geological maps of regions and provinces, and has also been responsible for series mapping (incomplete) of the country at 1:1 000 000, 1:500 000 (surficial) and 1:250 000, as well as much detailed mapping. The 1:1 000 000 Geological Atlas was previously published using NTS sheet lines, but future sheets will use the IMW base. GSC has also published more than 10 000 aeromagnetic maps, and a complete magnetic anomaly cover at 1:1 000 000 is in progress. Much modern mapping carried out by GSC, however, is attached to specific projects, often in association with the provincial geological surveys, and the range and variety of this material can only be fully appreciated by browsing through the catalogues issued by the various provincial agencies.

The **Earth Physics Branch (EPB)** was so designated in 1970. It is concerned *inter alia* with geophysical survey and observations, and has published a number of small-scale magnetic and gravity maps.

Canada has a long-established tradition of inventorying, mapping and monitoring land use and land cover. In 1963 the **Canada Land Inventory (CLI)** was set up under the Agriculture and Rural Development Act as a cooperative federal–provincial programme to make a comprehensive reconnaissance-level survey of land use and capability for the purposes of land use and resource planning. It is now administered by the Lands Directorate. About 2.6 million km² have been mapped, covering the Maritime Provinces and the southern agricultural areas of Quebec, Ontario, the Prairie Provinces and British Columbia.

The system uses a seven-category rating of land capability for each of four sectors: agriculture, forestry, recreation and wildlife. Three main scales are used. Compilation maps are at 1:50 000 and are available as diazo copies from provincial sources. The main published series is at 1:250 000 (1:125 000 for agriculture and forestry in British Columbia). These are colour maps with written reports on the reverse, and are available from the Canada Map Office or from Supply and Services Canada. Finally, there is a series of generalized maps at 1:1 000 000 issued on a provincial basis for most provinces. The maps are published for each of five themes, namely soil capability for agriculture, and land capability for forestry, recreation, wildlife-ungulates and wildlife-waterfowl. The 1:1 000 000 maps have the same themes with additional maps of land capability for sport fish, and of critical capability areas. A few general land use maps (*ca* 1967) at 1:250 000 have also been published. The CLI cover is substantially complete, and a current status index is available from the Lands Directorate.

From the very beginning of the programme there was interest in computerizing the Canada Land Inventory, and in 1968 the Canada Geographic Information System, one of the first such systems in the world, was set up to do this. Since then some 10 000 map sheets have been input and at the same time the system has continuously evolved to keep pace with developing computer and database technology. Currently a group of systems is in operation at the Lands Directorate offices at Hull, Ottawa, collectively known as the **Canada Land Data Systems (CLDS)**. Besides the Canada Land Inventory maps, CLDS currently holds digitized administrative and shoreline boundaries, CLUMP information (see below) and Federal Lands and Parks data.

A series of land use maps to cover the Yukon and Northwest Territories was initiated in 1971. Designated the **Northern Land Use Information Series (NLUIS)**, this is a federal project funded by Environment Canada and the Department of Indian and Northern Affairs. Thematic maps on such topics as wildlife and fish resources, native land uses and ecological land classification are issued at a scale of 1:250 000, and by 1986 some 410 sheets had been published covering about one-half of the two territories. Maps are issued as regional sets, the whole of northern Canada including the arctic islands being organized into 16 study areas for the purpose of the project. The Lands Directorate supplies a current status index of this series.

In 1978, the Lands Directorate of Environment Canada established the **Canada Land Use Monitoring Program (CLUMP)** to help planners make informed decisions about the management of the resource base. This system uses a new classification which takes account of both formal (land cover) and functional (activity) aspects of land use, and is intended to complement the long-established Canada Land Inventory. Information on land use change is acquired from many sources including air photographs and Landsat imagery. The mapping is at a resolution of 1:50 000 or 1:250 000 and is stored digitally in computer databases including the CLDS. Rate, extent, location and nature of land use change can all be evaluated over time and can be compared with other data sets such as those provided by the Census of Canada. Output can be in a variety of formats including maps and other graphics.

Maps are a potentially important component of the work of the **Canada Committee on Ecological Land Classification (CCELC)**, formed in 1976 to develop a uniform ecological approach to the classification of Canada's land resource base. Within this committee, an Ecoregion Working Group is involved in producing a national map of the ecoclimatic regions of Canada. The map and its accompanying report is expected to be published in 1987 and may be incorporated into the *National Atlas of Canada*, 5th edn. Other working groups are preparing standards for the classification of vegetation and of wetlands in Canada (maps of the latter to be published in the national atlas) and for the integration of wildlife resource assessment into ecological land survey. The terrestrial ecozones of Canada have recently been described in *Ecological Land Classification Series* No. 19, published in 1986 by the Lands Directorate. As with other resource surveys in Canada, ecological land classification is

increasingly moving away from conventional cartography to a GIS environment.

Soil mapping in Canada has been carried out from a number of provincial centres, in cooperation with the **Land Resource Research Centre** of Agriculture Canada. Maps, usually accompanied by monographs, have been published for only limited areas and usually at the old imperial scales of 1:63 360 or 1:126 720. For the eastern provinces they are usually published on a county basis rather than as grid maps. An important two-volume work on the *Soils of Canada* has been prepared jointly by the Canada Soil Survey Committee and the Land Resource Research Centre and includes a 1:5 000 000 Soil Map of Canada and a 1:10 000 000 map of Soil Climates of Canada, both in colour. The work is available from **Supply and Services Canada.**

A series of agroclimatic maps has been prepared by the Agrometeorology Branch of the Land Resource Research Centre and together these comprise the *Agroclimatic Atlas of Canada*. The maps, at a scale of 1:5 000 000, show distribution of climatic parameters affecting crop production such as potential evapotranspiration, seasonal water deficits, soil water reserves, summer soil temperatures and frost risk.

The **Inland Waters Directorate** has published maps of flood risk for settlements located on flood plains, and numerous glacier surveys in the western provinces.

The **Canadian Hydrographic Service (CHS)** is responsible for the production of nautical charts (which fall outside the frame of this book), but also prepares a range of *Natural Resource Maps* which includes 1:250 000 scale bathymetric and geophysical mapping of coastal waters, some 1:2 000 000 bathymetric mapping, and a number of special-purpose seabed maps which are published within its *Marine Science Paper* series. A 1:1 000 000 series will form an offshore component of the *National Earth Science Series* on IMW sheet lines. Several earth science themes are planned in cooperation with GSC and EPB.

Canada's first national atlas was produced in 1906, with second and third editions published in 1915 and 1958 respectively. A fourth edition, based on the 1961 census and other data, was issued in loose-leaf format from 1968 and finally published in a bound volume in 1974. Some individual sheets from this edition, in English or French, are still available from the Canada Map Office, but currently the fifth edition is in progress.

This edition, begun in 1978, is expected to have at least 200 sheets covering some 40 themes and issued over a 10-year time period. This atlas is intentionally of a more flexible format than its predecessors. It employs larger scales (principally 1:7 500 000, but a number of 1:5 000 000 maps have also been published) and although the sheets are compiled by the National Geographical Services Directorate within the Surveys and Mapping Branch of DEMR, there is opportunity for a wide variety of other agencies to contribute. Maps using time-dependent data, such as census information, will and have been up-dated as required. So far the atlas has been issued as a series of individual, conventional printed sheets each with an individual MCR number and available from the Canada Map Office; a cased selection of these maps has also been offered. There are plans to develop alternative formats including an electronic atlas probably with interactive capabilities. Although the maps are available individually we have not attempted to list them all in the catalogue.

A number of provincial, regional and monothematic atlases have also been published, and a selection of the major ones in print have been listed in the catalogue.

Canada has 350 000 official names, and these are recorded and approved by the Canadian Permanent Committee on Geographical Names. In addition, some provinces have their own name boards. All the names appearing on federal topographic maps have been approved by the Permanent Committee and they are listed in a series of 12 gazetteers collectively forming the *Gazetteer of Canada*. Eleven of these are published by DEMR but distributed through **Supply and Services Canada** (Canadian Government Publishing Centre) while the twelfth, *Repertoire toponymique du Québec* is available from **L'Editeur Officiel du Québec**.

Statistics Canada is responsible for the collection of statistical information for the government. Within the organization a Geography Division supplies outline maps showing the boundaries of the census geostatistical units at various levels of the hierarchy. Small-scale reference maps of these areas are published in the form of two Reference Bulletins, while the more detailed divisions are shown on sheet maps. Eight series of the latter, involving several thousand sheets, at various levels of detail are available for all the census geostatistical areas of the 1981 census.

Thematic mapping of statistics data is carried out by the **Geocartographics (GCG) Subdivision** whose role is to apply computer techniques to handling of geographical data, and the production of graphical output in the form of maps and statistical charts and diagrams. It serves a range of clients, both inside and outside the federal government, and has produced, among other things, a number of maps and atlases for sale to the public. Recent hard-copy products include the *Metropolitan Atlas Series* based on the 1981 Census of Canada, the *1981 Census of Agriculture*, the *Mortality Atlases of Canada*, the *1981 Forest Inventory* and maps of 1981 Census subdivision boundaries. The principal mapping systems in use are GIMMS and PILLAR, and recently the ARC/INFO GIS has also been acquired.

The *Metropolitan Atlas Series* combines text with maps and other graphics to present 1981 census data for the census metropolitan areas of St Johns, Halifax, Quebec, Montreal, Ottawa–Hull, Toronto, Hamilton, Winnipeg, Regina, Calgary, Edmonton and Vancouver. Computer mapping was provided by GIMMS Release 4 software and produced on a GERBER 4442 drum plotter.

Provincial mapping

Although substantial mapping programmes are carried out by federal agencies, as described above, all the 10 provinces also have their own mapping organizations. In particular many of the provinces have initiated new large-scale provincial topographic and/or cadastral series in recent years, and there are also many provincial resource surveys of various kinds in progress. There is no set pattern; these programmes are unique to each province. In the next few paragraphs we have therefore described some of the more important provincial series. In view of the great detail as well as the monographic (as opposed to series) character of many of the maps, we can only give a sampling. Most provincial authorities issue very comprehensive catalogues of maps and related products and these may be obtained from the addresses given in the address section.

In the catalogue section, we have adopted the same approach as for the United States, namely to list mainly the smaller-scale general and thematic maps available for each province.

Alberta

Provincial mapping in Alberta is carried out by the Alberta Departments of Forestry, Environment, Municipal Affairs and the **Alberta Research Council** (the last named being responsible for geological mapping). Fortunately for the map user, the distribution of the products of all these organizations was reorganized in the early 1980s, and they may all be obtained from Maps Alberta at the **Alberta Bureau of Surveying and Mapping**, Edmonton. A very comprehensive catalogue of provincial mapping (including also federally published maps) is issued by this agency.

Scales range from 1:5000 and 1:1000 cadastral/orthophotomaps of municipal areas, and planimetric city maps at 1:20 000/1:40 000 to provincial topographic series with or without contours at 1:250 000 and 1:100 000. Thematic mapping includes a wide range of forestry mapping, a ' eries of maps associated with the Alberta Oil Sands Environmental Research Program, and incomplete provincial series of bedrock geology, surficial geology and hydrogeology at 1:250 000.

British Columbia

British Columbia has a **Surveys and Resource Mapping Branch (SRMB)**, which formerly collaborated with the federal agency in producing the standard topographic map series of the province, and has continued to publish the 1:125 000 series (discontinued elsewhere), converting it progressively to 1:100 000. This series is on a UTM projection with UTM grid and has metric contours.

As well as a number of small-scale general, thematic and administrative boundary maps, there are two special maps of *Victoria Island* (1:380 160) and *Vancouver–Kamloops* (1:400 000), and a variety of large-scale cadastral, orthophoto and planimetric mapping, the latter mainly at 1:15 800 or 1:20 000 and covering large parts of the province.

Geological mapping is carried out by the **Ministry of Energy, Mines and Petroleum Resources (MEMPR)**, and forestry mapping by the **Ministry of Forests**, while floodplain mapping is undertaken by the **Inventory and Engineering Branch** of the Ministry of Environment.

The company of **Macdonald Dettwiler** has developed MERIDIAN, a computer-based system for making maps from digital imagery, and in a cooperative project with SRMB has produced experimental ortho-image maps from Landsat Thematic Mapper data.

Manitoba

A 1:20 000 topographic series (with 5 or 10 m contour interval) has been started by the **Surveys and Mapping Branch (SMB)**, Department of Natural Resources, Winnipeg. Sheets are issued as diazos (described as 'white prints') or contoured photomaps. Cover so far is very limited. Surveys and Mapping Branch also publish small-scale base maps of the province, administrative area maps, and for outdoor recreation there are canoe route maps, angling maps and lake depth charts.

Geological mapping is undertaken by the **Mineral Resources Division (MRD)** of the Manitoba Department of Energy and Mines, Winnipeg. There is a 1:1 000 000 multi-coloured 'geo-scientific' series which gives an overview of specific geology-related subjects. There are currently five maps in the series (see catalogue below), with further structural, tectonic and metallogenic maps in preparation. A 57-sheet 1:250 000 *Bedrock geology compilation map series* has been started. Mineral Resources also produce a wide range of detailed special project mapping, most of it published in monochrome or two colours.

Soil maps are published by the **Manitoba Department of Agriculture** and by the Manitoba Soil Survey, University of Manitoba.

Maritime Provinces

The three Maritime Provinces (New Brunswick, Nova Scotia and Prince Edward Island) are being mapped in detail under a programme organized by the Surveys and Mapping Division of the **Land Registration and Information Service (LRIS)**. The mapping is designed to provide a base for cadastral purposes and for planning for economic development in these provinces. There are three distinct series in this programme, designated resource, urban and property maps. Resource maps are orthophotomaps generally at 1:10 000 in Nova Scotia and New Brunswick (1:20 000 in little-developed areas) and 1:5000 for the whole of Prince Edward Island. They have added contours and spot heights. The urban maps are line maps at scales ranging from 1:1000 to 1:4800, and are planned for all communities with populations greater than 300. Property maps are overlays showing property boundaries and lot numbers. LRIS has its headquarters at Fredericton, New Brunswick.

Provincial geological mapping of New Brunswick is carried out by the **Geological Surveys Branch** of the **Department of Forests, Mines and Energy**, Fredericton, and includes mineral occurrence maps and a large amount of geophysical and geochemical mapping. In Prince Edward Island and Nova Scotia, most geological mapping has been by the GSC. Based at Amhurst, Nova Scotia is the **Maritime Resources Management Service (MRMS)**, which produces thematic maps of various kinds, including some geological mapping. It also publishes special topographic series of Nova Scotia at 1:250 000 and 1:125 000, and a base map at 1:20 000.

Soil mapping has been carried out in all three provinces, and maps are available from the provincial Departments of Agriculture, whose addresses are given below.

Newfoundland–Labrador

Large-scale mapping in Newfoundland–Labrador is prepared by the Lands Branch of the **Department of Forest Resources and Lands** at St Johns. The Island of Newfoundland is partially covered by orthophotomaps at 1:12 500 while the communities have topographic maps mostly at 1:2500. Labrador communities are also mapped at this scale. The Department of Forest Resources and Lands has a computerized Geographic Information System which holds information about forests throughout the province.

Geological mapping is carried out by the **Department of Mines and Energy**, St Johns. *Aggregate resource maps* and *Mineral occurrence maps* at 1:250 000 are published, as well as numerous geological maps usually at 1:50 000.

Ontario

The Ontario **Ministry of Natural Resources** has its own Ontario Basic Mapping Program (OBM), begun in 1978. The maps in this series provide simple, generally unannotated, base maps at scales of 1:20 000 or, in Southern Ontario, 1:10 000. The projection and grid are UTM, and sheet format is a standard 50 × 50 cm. Most maps are conventional line maps with a contour interval of 10 m (1:20 000 sheets) or 5 m (1:10 000 sheets), but for the Hudson Bay Lowlands photomaps are used. Conventional maps of 35 per cent of the province are so far available, and a project to test the production of digital data and graphics for the remaining area has begun. It is also planned to produce urban area mapping at 1:2000 with a 1 m contour interval.

In Ontario a cooperative research project began in 1984 to develop and implement a digital database for land-related information systems. The goal is to provide a common base of digitized land-related information which can be built on by users in both public and private sectors.

Provincial geological mapping is carried out by the **Ontario Geological Survey (OGS)** within the Ministry of Natural Resources, and by the Ontario Ministry of Northern Development and Mines. There is a very wide range of detailed mapping, including aeromagnetic, geochemical, Quaternary, and bedrock maps, and maps of industrial mineral resources.

The **Ontario Centre for Remote Sensing (OCRS)** produces a range of thematic mapping derived from satellite data using digital image processing techniques. OCRS is a provincial government organization and carries out special projects for the Ontario government and for other provinces. A peatland resource mapping project in cooperation with OGS was completed in 1985 and a wetlands mapping programme in the Hudson Bay–James Bay lowlands is in progress. Surficial geological mapping of northern Ontario is also being carried out.

Soil mapping is available from the **Ontario Ministry of Agriculture and Food**.

The **Ministry of Municipal Affairs** has published *Maps: A Map Index for Community Planning in Ontario* (1986), a comprehensive guide to the wide range of provincial and federal maps available for the province.

Quebec

Quebec's **Ministère de l'Energie et des Ressources (MER)** also publishes its own 1:20 000 series of base maps. This is an ongoing series which covers much of the south of the Province and is extending into more northerly areas as well. The map is prepared in several variants. There is thus a planimetric version, a cadastral version, and a *carte topographique* with 10 m contours. Since 1982, sheets have been made available to the public from the **Photocartothèque québecoise**. They are generally available in lithographic or diazo form, although some sheets are also available in digital form on magnetic tape. All sheets are monochrome except for 19 sheets covering the metropolitan areas of Montreal and Quebec, which are printed in five colours.

The **Direction Générale de l'Exploration Géologique et Minerale** publishes a vast range of geological, geochemical and geophysical maps, usually issued with accompanying reports, available from MER's Centre de diffusion de la géoinformation.

Soils maps, some of them included with monographs, are published by the **Ministère de l'Agriculture, des Pêcheries et de l'Alimentation**, Quebec.

The **Ministère des Transports**, Quebec is also a significant publisher of maps (including a series of *Cartes du reseau routier* at 1:50 000) many of them in colour, and some regional tourist maps. This ministry also publishes plans of all the small municipalities at scales of 1:50 000, 1:20 000 or 1:10 000. Maps are available from the **Editeur Officiel du Québec**.

Saskatchewan

Large-scale mapping in Saskatchewan is the responsibility of the **Central Surveys and Mapping Agency**, Regina.

Geological mapping is carried out by **Saskatchewan Energy and Mines** at Regina. There is a recent small-scale geological map of the whole province, and a large number of monographic maps published at a range of scales and usually included in Saskatchewan Geological Survey Reports, although they are invariably also available as separates. Many maps are associated with oil and gas exploration and are therefore concerned with subsurface structure, aeromagnetic and gravity anomalies, and well locations.

Geological maps are also published by the **Saskatchewan Research Council**, while forest inventory is carried out by the **Forestry Branch, Department of Tourism and Renewable Resources.**

Private mapping

There are rather few commercial mapping companies in Canada. Among those producing mainly regional road maps and city street maps are **Allmaps Canada Ltd** (formerly Rolph-McNally), Markham, Ontario (provincial maps and street maps of about 24 major cities), **Pathfinder Air Survey Ltd**, **MapArt** and **Shaw Photogrammetric Services**. Provincial road and city maps are also published by the provincial and municipal tourist and transport authorities.

Some interesting experimental mapping emanates from cartographic units in the universities, including the **University of Waterloo Cartographic Centre** (recreational mapping), and the **Department of Geography, Queen's University** (isodemographic maps).

Further information

A thorough description of the evolution of the modern mapping of Canada, and which also includes information about thematic maps and provincial mapping programmes, is given in Nicholson, N.L. and Sebert, L.M. (1981) *The Maps of Canada*. Hamden, Connecticut: Archon Books.

DEMR publishes annually a set of three indexes showing the current status and availability of all the federal topographic map series, and of general and national atlas maps.

The Geological Survey of Canada does not publish a comprehensive map listing, but there is a *Monthly Information Circular* which includes notification of new maps, and also an annual *Index to Publications*. The best source of information about geological mapping of specific areas is, however, the catalogues issued by the provincial geological and natural resource agencies.

The Lands Directorate issues brochures showing the publication status of the Canada Land Inventory maps, NLUIS, etc. The CLI programme is described in *Canada Land Inventory Report* No. 1 (2nd edn, 1970), while the evolution of the Canada Land Data Systems is discussed by Crain, I.K. and Macdonald, C.L. (1984) From land inventory to land management, *Cartographica*, **21**(2 & 3), 40–46.

The Land Resource Research Centre of Agriculture Canada publishes a collection of indexes showing soil survey cover in each province and giving the addresses of the relevant provincial sources.

Statistics Canada has issued a *Guide to reference maps available for the 1981 Census*.

Most provincial mapping authorities publish catalogues and announcements about new maps and associated publications. Some catalogues are very voluminous and comprehensive. Of special note are the *Natural Resources Information Directory* published in 1984 by Alberta Energy and Natural Resources, with

about 150 pages devoted to maps, and the *Guide de la Géoinformation québecoise 1984* published by the Ministère de l'Energie et des Ressources, Québec, which has a particularly useful listing of geoscientific mapping arranged by NTS sheet numbers at the 1:250 000 scale level. The New Brunswick Mineral Resources Division also has an excellent indexed catalogue of both provincial and GSC geological publications and maps, and in 1986 the Ontario Ministry of Municipal Affairs published *Maps: a map index for community planning in Ontario*, which gives a very good overview of the various topographic and thematic series available for that province.

Addresses

Federal mapping organizations

Canada Map Office, Department of Energy, Mines and Resources (DEMR)
615 Booth Street, OTTAWA, Ontario K1A 0E9.

Geological Survey of Canada (GSC)
Department of Energy, Mines and Resources, 601 Booth Street, OTTAWA, Ontario K1A 0E8.

Earth Physics Branch (EPB)
Department of Energy, Mines and Resources, 1 Observatory Crescent, OTTAWA, Ontario K1A 0Y3.

Mapping and Charting Establishment (MCE)
Department of National Defence, 615 Booth Street, OTTAWA, Ontario K1A 0K9.

Land Resources and Data Systems Branch (CLI, NLUIS, CLUMP)
Lands Directorate, Environment Canada, OTTAWA, Ontario K1A 0E7.

Canada Committee on Ecological Land Classification (CCELC)
Lands Directorate, Environment Canada, 20th Floor, Place Vincent Massey, OTTAWA, Ontario K1A 0E7.

Land Resource Research Centre
Agriculture Canada, K.W. Neatby Building, Central Experimental Farm, OTTAWA, Ontario K1A 0C6.

Inland Waters Directorate
Water Resources Branch, Department of Fisheries and Environment, OTTAWA, Ontario K1A 0H3.

Canadian Hydrographic Service (CHS)
Department of Fisheries and Oceans, Box 8080, 1675 Russell Road, OTTAWA, Ontario K1G 3H6.

Supply and Services Canada
Canadian Government Publishing Centre, OTTAWA, Ontario K1A 0S9.

Statistics Canada, Geocartographics Subdivision (GCG)
Jean Talon Building, 2nd Floor, Section A1, OTTAWA, Ontario K1A 0T6.

Statistics Canada (orders)
Publication Sales and Services, Coats Building, OTTAWA, Ontario K1A 0T6.

Atmospheric Environment Service
Environment Canada, 4905 Dufferin Street, DOWNSVIEW, Ontario M3H 5TA.

Provincial mapping organizations

Alberta

Maps Alberta, Alberta Bureau of Surveying and Mapping (ABSM)
2 Floor, North Tower, Petroleum Plaza, 9945–108 Street, EDMONTON, Alberta T5K 2G6.

Alberta Research Council
Alberta Geological Survey, 3rd Floor, Terrace Plaza, 4445 Calgary
Trail South, EDMONTON, Alberta T6H 5R7.

Energy Resources Conservation Board
640 Fifth Avenue SW, CALGARY, Alberta T2P 3G4.

British Columbia

Maps—BC, Surveys and Resource Mapping Branch (SRMB)
Ministry of Environment, Parliament Buildings, VICTORIA, British
Columbia V8V 1X5.

The Map Library—distribution
Assessment and Planning Division, Ministry of Environment,
VICTORIA, British Columbia V8V 1X4.

Ministry of Energy, Mines and Petroleum Resources (MEMPR)
Geological Branch, Mineral Resources Division, Parliament
Buildings, 617 Government Street, VICTORIA, British Columbia
V8V 1X4.

Inventory and Engineering Branch, Ministry of Environment
Parliament Buildings, VICTORIA, British Columbia V8W 2Z2.

Ministry of Forests
Parliament Buildings, VICTORIA, British Columbia V8V 1X5.

University of British Columbia Press
2075 Wesbrook Mall, VANCOUVER, British Columbia V6T 1W5.

Macdonald Dettwiler
3751 Shell Road, RICHMOND, British Columbia V6X 2Z9.

Manitoba

Map Sales Office, Surveys and Mapping Branch (SMB)
1007 Century Street, Attn. Dept 989, WINNIPEG, Manitoba R3H
0W4.

Mineral Resources Division (MRD)
Manitoba Energy and Mines, 555–330 Graham Avenue,
WINNIPEG, Manitoba R3C 4E3.

Publications Branch, Manitoba Department of Agriculture
Manitoba Archives Building, 200 Vaughan Street, WINNIPEG,
Manitoba R3C 1T5.

New Brunswick

Land Registration and Information Service (LRIS)
PO Box 6000, FREDERICTON, New Brunswick E3B 5H1.

**Geological Surveys Branch, Department of Forests, Mines and
Energy**
PO Box 6000, FREDERICTON, New Brunswick E3B 5H1.

Department of Agriculture and Rural Development
Communications Branch, PO Box 6000, FREDERICTON, New
Brunswick E3B 5H1.

Newfoundland

Lands Branch, Department of Forest Resources and Lands
Howley Building, Higgins Lane, ST JOHNS, Newfoundland A1C
5T7.

Newfoundland Department of Mines and Energy
PO Box 4750, 95 Bonaventure Avenue, ST JOHNS, Newfoundland
A1C 5T7.

Canada Soil Survey
Research Branch, Agriculture Canada
PO Box 7098, ST JOHNS, Newfoundland A1E 3Y3.

Northwest Territories

Department of Indian Affairs and Northern Development
Box 1500, YELLOWKNIFE, Northwest Territories O4E 1H0.

Nova Scotia

Department of Mines and Energy (DME)
Box 1087, HALIFAX, Nova Scotia 83J 2X1.

Maritime Resource Management Service (MRMS)
PO Box 310, AMHERST, Nova Scotia B4H 3Z5.

Atlantic Soil Survey Unit
NSAC, PO Box 550, TRURO, Nova Scotia B2N 5E3.

Ontario

**Surveys and Mapping Branch (OMNR), Ontario Ministry of
Natural Resources**
Whitney Block, 99 Wellesley Street West, TORONTO, Ontario M7A
1W3.

Ontario Geological Survey (OGS)
Ontario Ministry of Natural Resources, Whitney Block, 99
Wellesley Street West, TORONTO, Ontario M7A 1W3.

**Cartography Section, Ontario Ministry of Transportation and
Communications (OMTC)**
1201 Wilson Avenue, East Building, DOWNSVIEW, Ontario M3M
1J8.

Ontario Ministry of Municipal Affairs
Plans Administration Branch, 14th Floor, 777 Bay Street,
TORONTO, Ontario M5G 2E5.

Ontario Ministry of Agriculture and Food
Soils and Water Management Branch, 801 Bay Street, TORONTO,
Ontario M7A 2B2.

Ontario Centre for Remote Sensing (OCRS)
Ontario Ministry of Natural Resources, 880 Bay Street, 3rd Floor,
TORONTO, Ontario M5S 1Z8.

Ontario Ministry of the Environment
135 St Clair Avenue West, Suite 100, TORONTO, Ontario M4V
1P5.

Prince Edward Island

Surveys and Mapping Division
120 Water Street, SUMMERSIDE, Prince Edward Island C1N 1A9.

Department of Agriculture
Plant Industry Branch, PO Box 1600, CHARLOTTETOWN, Prince
Edward Island C1A 7N3.

Quebec

Ministère de l'Energie et des Ressources (MER)
Centre de diffusion de la géoinformation, 1620 blvd de l'Entente,
local 1,04, QUEBEC, Québec G1S 4N6.

Photocartothèque québecoise
Ministère de l'Energie et des Ressources, 1995 boul. Charest ouest,
SAINTEFOY, Québec G1N 4H9.

Ministère de l'Agriculture, des Pêcheries et de l'Alimentation
Renseignements, publications et documentations, 200a Chemin
Ste-Foy, QUEBEC, Québec G1R 4X6.

Ministère des Transports
Division de la cartographie, Service des relevés techniques, 200
Dorchester sud, 2e étage, QUEBEC, Québec G1K 5Z1.

Ministère des Transports
Division de la cartographie, 201 Cremazie est, MONTREAL, Québec
H2M 1L2.

Editeur officiel du Québec
Service commercial, 1283 boul. Charest ouest, QUEBEC, Québec
G1N 2C9.

Office de Planification et Développement du Québec
Edifice E, 3e étage, Hôtel du gouvernement, QUEBEC, Québec
G1A 1A5.

Société pour Vaincre la Pollution
445 rue Saint-François-Xavier, MONTREAL, Québec H2Y 2T1.

Saskatchewan

Central Surveys and Mapping Agency
Department of Highways and Transportation, REGINA,
Saskatchewan S4S 0B1.

Saskatchewan Energy and Mines
Toronto-Dominion Bank Building, 1914 Hamilton Street, REGINA,
Saskatchewan S4P 4V4.

Saskatchewan Research Council
Geology Division, 15 Innovation Blvd, SASKATOON, Saskatchewan
S7N 2X8.

Saskatchewan Institute of Pedology
Room 143, John Mitchell Building, University of Saskatchewan,
SASKATOON, Saskatchewan S7N 0W0.

**Forestry Branch, Department of Tourism and Renewable
Resources**
1840 Lorne Street, REGINA, Saskatchewan S4P 2L7.

Yukon

Department of Indian Affairs and Northern Development
Geology Section, 200 Range Road, WHITEHORSE, Yukon Territory
Y1A 3V1.

Yukon Soil Survey Unit
Agriculture Canada, Box 2703, WHITEHORSE, Yukon Y1A 2C6.

Private and commercial map publishers

Canadian Society of Petroleum Geologists
612 Lougheed Building, CALGARY, Alberta.

Allmaps Canada Ltd
390 Steelcase Road E., MARKHAM, Ontario L3R 1G2.

Pathfinder Air Survey Ltd
3363 Carling Avenue, OTTAWA, Ontario K2H 7V6.

MapArt
25 Athlone Avenue, BRAMPTON, Ontario.

Shaw Photogrammetric Services Ltd
6 Bexley Place, OTTAWA, Ontario.

University of Waterloo Cartographic Centre
Faculty of Environmental Studies, University of Waterloo,
WATERLOO, Ontario N2L 3G1.

Department of Geography, Queen's University
KINGSTON, Ontario K7L 3N6.

National Film Board of Canada
1 Lombard Street, TORONTO, Ontario M5C 1J6.

Reader's Digest Association (Canada) Ltd
215 Redfern Avenue, MONTREAL, Québec H3Z 2V9.

Catalogue

Atlases and gazetteers

The National Atlas of Canada   5th edn
Ottawa: DEMR, 1978–.
Issued as series of separate sheets, *ca* 40 published.

Canada Gazetteer Atlas
Ottawa: Canadian Government Publishing Centre, 1980.
pp. 104.
In English or French.

Atlas of Canada
Reader's Digest Association (Canada) Ltd/Canadian Automobile
Association, 1981.
pp. 220.

Gazetteer of undersea feature names 1983
Ottawa: Department of Fisheries and Oceans, 1983.
pp. 191.
Names approved by CPCGN for Canadian waters.

General

National Film Board map of Canada
Ottawa: National Film Board of Canada, 1984.
Very large oblique view wall map computer generated from
700 000 reference points and then hand-painted.

Canada   1:15 840 000   MCR 18
Ottawa: DEMR, 1973.
English or French (MCR 19) versions.
Outline and 2-colour editions also available.

Canada   1:8 870 000   MCR 15
Ottawa: DEMR, 1970.
English or French (MCR 16) versions.

Canada (base map)   1:7 500 000   MCR 128
Ottawa: DEMR, 1985.
Included in *National Atlas* 5th edn.
English or French (MCR 128F) versions.

Canada road map   1:5 300 000
Bern: Kümmerly + Frey, 1984.

Canada (base map)   1:5 000 000   MCR 126
Ottawa: DEMR, 1986.
English or French (MCR 126F) versions.

Canada   1:2 000 000   MCR 5
Ottawa: DEMR, 1972.
6 sheets, all published.
English or French (MCR 82) versions.

Eastern Canada and adjacent areas. Physiography   1:2 000 000
Ottawa: GSC, 1976.
Map 1399A.
4 sheets, all published.

Atlantic Provinces   1:2 000 000   MCR 77
Ottawa: DEMR, 1973.

Prairie Provinces   1:2 000 000   MCR 27
Ottawa: DEMR, 1973.

Maritime Provinces   1:633 360   MCR 38
Ottawa: DEMR, 1968.

Topographic

Canada   1:500 000
Ottawa: DEMR, 1958–.
218 sheets, all published.

Canada   1:250 000
Ottawa: DEMR, 1923–.
918 sheets, all published. ■

Canada   1:50 000
Ottawa: DEMR, 1950–.
13 150 sheets, *ca* 80% published.

Bathymetric

Natural resource maps   1:2 000 000
Ottawa: CHS.
8 sheets, 2 published.
Bathymetric maps, other themes planned.

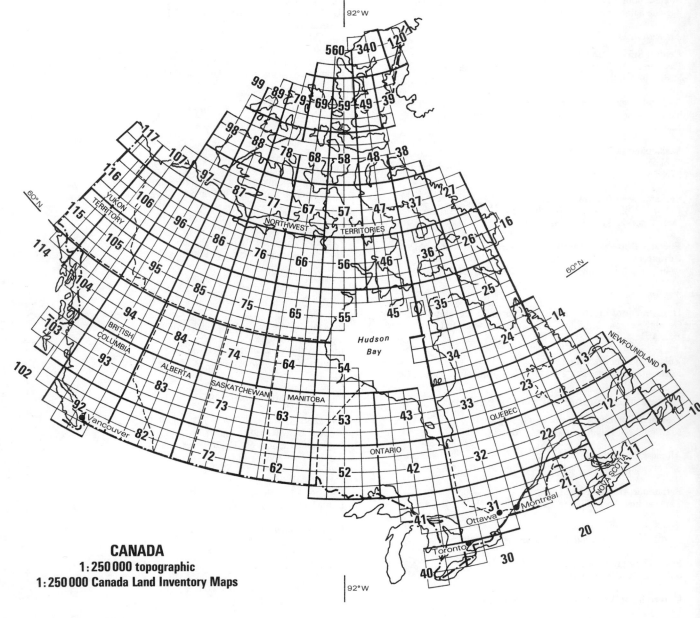

CANADA
1 : 250 000 topographic
1 : 250 000 Canada Land Inventory Maps

Sheet numbering examples:-
31 K   57 F

Natural resource maps   1 : 250 000
Ottawa: CHS.
Series of bathymetric and geophysical maps of Canadian waters.
ca 125 sheets published.

Geological and geophysical

Canada total intensity chart 1985.0   1 : 10 000 000   MCR 702
Ottawa: DEMR, 1985.

Canada horizontal intensity chart 1985.0   1 : 10 000 000   MCR 703
Ottawa: DEMR, 1985.

Canada magnetic inclination chart 1985.0   1 : 10 000 000   MCR 704
Ottawa: DEMR, 1985.

Canada vertical intensity chart 1985.0   1 : 10 000 000   MCR 705
Ottawa: DEMR, 1985.

Principal mineral areas of Canada   1 : 7 603 200   Map 900A
34th edn
Ottawa: GSC, 1984.

Geological map of Canada   1 : 5 000 000   Map 1250A
Ottawa: GSC, 1969.

Tectonic map of Canada   1 : 5 000 000   Map 1251A
Ottawa: GSC, 1965.

Mineral deposits of Canada   1 : 5 000 000   Map 1252A
Ottawa: GSC, 1968.

Glacial map of Canada   1 : 5 000 000   Map 1253A
Ottawa: GSC, 1968.

Physiographic regions of Canada   1:5 000 000   Map 1254A
Ottawa: GSC, 1970.

Magnetic anomaly map of Canada   1:5 000 000   Map 1255A
Ottawa: GSC, 1984.

Isotopic age map of Canada   1:5 000 000   Map 1256A
Ottawa: GSC.

Bouguer gravity anomaly map of Canada   1:5 000 000   Gravity map series 74-1
Ottawa: GSC, 1974.

Radioactivity map of Canada   1:5 000 000   Map 1600A
Ottawa: GSC, 1986.
Shows radioactivity of upper 30 cm of ground.

Magnetic anomaly map of Arctic Canada   1:3 500 000   Map 1512A
Ottawa: GSC, 1982.

Tectonic assemblage map of the Canadian Cordillera and adjacent parts of the USA   1:2 000 000   Map 1505A
Ottawa: GSC, 1981.

Mineral deposits and principal mineral occurrences of the Canadian Cordillera and adjacent parts of the USA   1:2 000 000   Map 1513A
Ottawa: GSC, 1984.

Geological atlas   1:1 000 000
Ottawa: GSC, 1965–.
65 sheets, 9 published. ■

Aeromagnetic anomaly maps   1:1 000 000
Ottawa: GSC, 1982–.
65 sheets, 39 published. ■

Environmental

Agroclimatic atlas of Canada
Ottawa: Land Resource Research Centre, 1976–.
Series of 1:5 000 000 scale maps.
37 published.

Climatic atlas Canada: a series of maps portraying Canada's climate
Ottawa: Canadian Government Publishing Centre, 1984.

Hydrological atlas of Canada
Ottawa: Inland Waters Directorate, 1977.
34 coloured plates.
Bilingual.

Great Lakes Climatological Atlas
Ottawa: Canadian Government Publishing Centre, 1986.
pp. 145.
Bilingual.

Lakes, rivers and glaciers in Canada   1:7 500 000   MCR 50
Ottawa: DEMR, 1969.

Soils of Canada   1:5 000 000
Ottawa: Agriculture Canada, 1981.
The map forms part of a two-volume report *Soils of Canada* which includes a soil inventory and maps and diagrams of soil climates.

Wetlands of Canada   1:5 000 000
Ottawa: CCELC, 1981.
Two sheets: Distributions of wetlands, Wetland regions (provisional 1981).

Canada Land Inventory   1:1 000 000
Ottawa: Lands Directorate, 1963–.
Series of maps covering southern Canada and with the themes of soil capability for agriculture, and land capability for forestry, wildlife-ungulates, wildlife-waterfowl, sport fish, and critical capability areas.

Canada Land Inventory   1:250 000
Ottawa: Lands Directorate, 1963–.
Series of land capability maps covering southern Canada.
Themes same as in previous entry.
British Columbia published at 1:125 000.

Northern Land Use Information Series   1:250 000
Ottawa: Lands Directorate, 1972–.
410 sheets published.

Administrative

Canada's federal lands   1:5 000 000
Ottawa: Lands Directorate, 1986.

Canada. Census divisions and subdivisions 1971   1:7 500 000   MCR 4000 Bil.
Ottawa: DEMR, 1978.

Human and economic

The Metropolitan Atlas Series
Ottawa: Statistics Canada, 1984.
Series of 12 census atlases of major metropolitan areas.

Mortality Atlas of Canada
3 volumes: 1, Cancer; 2, General mortality; 3, Urban mortality.
Ottawa: Statistics Canada, 1980–84.

A profile of Canadian Agriculture, 1981 Census of Canada
Ottawa: Statistics Canada, 1984.

Canada—total population by census divisions 1981
Ottawa: Statistics Canada, 1983.
'Pillar map'.

Canada—population change 1976–1981 by subdivisions/Variations de la population par divisions de rencensement de 1981
Ottawa: Statistics Canada, 1984.

Isodemographic map of Canada
Ottawa: Lands Directorate, 1977.

Town maps

Military city maps   1:25 000   Series A 902
Ottawa: MCE, 1970–.
Maps of about 46 Canadian cities.

(Commercial street maps of cities are available from the companies mentioned in the text.)

ALBERTA

Atlases and gazetteers

The atlas of Alberta   general editor Ted Byfield
Edmonton: Interwest Publications, 1984.
pp. 160.

Gazetteer of Canada. Alberta   2nd edn
Ottawa: DEMR, 1974.
pp. 153.

General

Provincial general base   1:2 500 000
Edmonton: ABSM, 1982–.
Outline map available in six additional versions showing judicial districts, electoral divisions, municipalities, regional planning areas, census divisions and transportation regions.

Alberta Landsat mosaic   1:1 500 000
Edmonton: Department of Transportation, 1981.

Alberta relief map   1:1 000 000
Edmonton: ABSM.

Provincial base map 1:1 000 000
Edmonton: ABSM, 1986.
Also available in 10 further versions.

Alberta 1:750 000 MCR 83
Ottawa: DEMR, 1973.

Provincial general base map 1:750 000
Edmonton: ABSM, 1976–.
Choice of thematic overlays projected on provincial base map.

Provincial base map 1:500 000
Edmonton: ABSM, 1982–.
4 sheets, all published.
Available in 4 versions.

Topographic

Provincial access series 1:250 000
Edmonton: ABSM, 1950–.
49 sheets, all published.
Available in topographic or planimetric versions.

Alberta topographic series 1:100 000
Edmonton: ABSM, 1970–.
192 sheets, 8 published.

Geological and geophysical

Geological map of Alberta 1:1 167 000
Edmonton: Alberta Research Council, 1972.

Geological highway map of Alberta 1:800 000 2nd edn
Calgary: Canadian Society of Petroleum Geologists, 1981.

Major coal-bearing formations in Alberta 1:3 000 000
Calgary: Energy Resources Conservation Board, 1983.

Alberta bedrock topography 1:250 000
Edmonton: Alberta Research Council.
13 sheets published.

Alberta surficial geology 1:250 000
Edmonton: Alberta Research Council.
9 sheets published.

Environmental

Alberta resource maps 1:5 000 000
Edmonton: ABSM, 1977–.
Series of 82 small thematic maps.

Agroclimatic zones of Alberta 1:2 534 400
Ottawa: Agriculture Canada, 1967.

BRITISH COLUMBIA

Atlases and gazetteers

Atlas of British Colombia
Vancouver: University of Columbia Press, 1979.
pp. 144.

Gazetteer of Canada. British Columbia 3rd edn
Ottawa: DEMR, 1985.
pp. 281.

General

British Columbia 1:2 000 000 Map 1J
Victoria: SRMB, 1980.

British Columbia relief map 1:2 000 000 Map 1JR
Victoria: SRMB, 1980.
Inset map of precipitation.

British Columbia physical—landforms and roads 1:1 900 800 Map 1JP
Victoria: SRMB, 1964.
Inset map of geology.

British Columbia 1:600 000
Victoria: SRMB, 1976.
6 regional sheets, all published.
Available in planimetric or landform versions.

Topographic

Province of British Columbia 1:100 000 Edition 3
Victoria: SRMB, 1971–.
ca 30% of the province covered.

Geological and geophysical

British Columbia geological highway map 1:800 000
Victoria: MEMPR, 1983.

Geothermal potential map of British Columbia 1:2 000 000
Victoria: MEMPR, 1983.

Energy resources of British Columbia 1:2 000 000
Victoria: MEMPR, 1982.

Environmental

Biogeoclimatic units: Victoria–Vancouver 1:500 000
Victoria: Ministry of Forests, 1981.

Administrative

British Columbia. Electoral districts 1:2 000 000 1JF
Victoria: SRMB, 1978.

MANITOBA

Atlases and gazetteers

Manitoba atlas
Winnipeg: SMB, 1983.
pp. 150.

Gazetteer of Canada. Manitoba 3rd edn
Ottawa: DEMR, 1981.
pp. 113.

General

Manitoba Landsat mosaic 1:2 000 000
Winnipeg: SMB, 1980.
Colour mosaic.

Manitoba relief 1:1 000 000
Winnipeg: SMB, 1979.

Manitoba base map 1:1 000 000
Winnipeg: SMB, 1979.

Landsat 1 mosaic 1:1 000 000
Winnipeg: SMB, 1975.
Black and white mosaic.

Manitoba 1:760 320 MCR 26
Ottawa: DEMR, 1964.
2 sheets, both published.

Geological and geophysical

Geological map of Manitoba 1:1 000 000 Map 79–2
Winnipeg: MRD, 1979.

Surficial geological map of Manitoba 1:1 000 000 Map 81-1
Winnipeg: MRD, 1981.

Mineral map of Manitoba 1:1 000 000 Map 80-1
Winnipeg: MRD, 1980.

Metallogenic map of Manitoba 1:1 000 000 Map 66-2
Winnipeg: MRD, 1966.

Residual magnetic anomaly map of Manitoba 1:1 000 000 Map 75–3
Winnipeg: MRD, 1975.

Surficial geology, Northeastern Manitoba 1:500 000 Map 1617A.
Ottawa: GSC, 1986.

Bedrock geology compilation map series 1:250 000
Winnipeg: MRD.
57 sheets, 1 published.

Administrative

Manitoba. Provincial Electoral Divisions 1:1 000 000
Winnipeg: SMB, 1979.

Manitoba: Municipalities and LGD 1:1 000 000
Winnipeg: SMB, 1977.

Manitoba. Municipalities and LGD 1:500 000
Winnipeg: SMB, 1978.

NEW BRUNSWICK

Atlases and gazetteers

Gazetteer of Canada. New Brunswick 2nd edn
Ottawa: DEMR, 1972.
pp. 213.

Geological and geophysical

Geological map of New Brunswick 1:500 000 NR-1 2nd edn
Fredericton: Geological Surveys Branch, 1979.

Surficial geology map of New Brunswick 1:500 000 Map 1594A
Ottawa: GSC, 1984.
Available also with text GSM 416.

Mineral occurrence map of New Brunswick 1:500 000
Fredericton: Geological Surveys Branch, 1979.

NEWFOUNDLAND–LABRADOR

Atlases and gazetteers

Gazetteer of Canada. Newfoundland and Labrador
Ottawa: DEMR, 1983.

General

Newfoundland 1:500 000 MCR 30
Ottawa: DEMR, 1975.

Geological and geophysical

Geological map of Newfoundland 1:1 000 000 Map 1231A
Ottawa: GSC, 1967.

Geological map of Labrador 1:1 000 000
St. Johns: Department of Mines and Energy, 1970.

Newfoundland mineral occurrence maps 1:250 000
St Johns: Department of Mines and Energy.
23 sheets published.

Labrador mineral occurrence maps 1:250 000/1:500 000
St Johns: Department of Mines and Energy.
ca 25 sheets published.

Environmental

Ecological land classification of Labrador 1:1 000 000
Amhurst: MRMS, 1977.

NORTHWEST TERRITORIES

Atlases and gazetteers

Gazetteer of Canada. Northwest Territories 1st edn
Ottawa: DEMR, 1980.
pp. 184.

General

Northwest Territories and Yukon Territory 1:4 000 000 MCR 36
Ottawa: DEMR, 1974.

Geological and geophysical

Northwest Territories economic geology series
Ottawa: Department of Indian Affairs and Northern
Development, 1975–.

NOVA SCOTIA

Atlases and gazetteers

Gazetteer of Canada. Nova Scotia 2nd edn
Ottawa: DEMR, 1977.
pp. 477.

General

Nova Scotia and Prince Edward Island 1:500 000 MCR 37
Ottawa: DEMR, 1974.

A map of the province of Nova Scotia 1:250 000
Halifax: Surveys and Mapping Branch, 1979.
Atlas of 46 map sheets.

Geological and geophysical

Geologic highway map of Nova Scotia 1:633 600
Halifax: DME, 1980.

Geologic and geophysical map of the province of Nova Scotia 1:500 000
Halifax: DME, 1979.

Tectonic map of the Province of Nova Scotia 1:500 000
Halifax: DME, 1982.

Metallogenic map of the Province of Nova Scotia 1:500 000
Halifax: DME, 1983.

ONTARIO

Atlases and gazetteers

North of 50. An atlas of far northern Ontario
Toronto: University of Toronto Press for Royal Commission on
the Northern Environment, 1985.
pp. 119.

Gazetteer of Canada. Ontario
Ottawa: DEMR, 1975.
pp. 822.

General

Ontario 1:2 000 000 MCR 39
Ottawa: DEMR, 1973.

Official road map of Ontario 1:1 600 000/1:800 000
Downsview: OMTC.

Ontario transportation map series 1:250 000
Downsview: OMTC.

Geological and geophysical

Ontario mineral map 1:1 584 000
Toronto: OGS, 1985.

Industrial minerals of Ontario   1 : 1 500 000   P 2591
Toronto: OGS, 1983.

Ontario, magnetic anomaly map: Ontario and Quebec
1 : 1 000 000   Map NL-18-M
Ottawa: GSC, 1984.

Geological highway map. Southern Ontario   1 : 800 000
Toronto: OGS.

Administrative

*Reference map of Ontario counties, districts and regional
municipalities*   1 : 3 200 000/1 : 1 600 000
Downsview: OMTC.

PRINCE EDWARD ISLAND

Atlases and gazetteers

Gazetteer of Canada. Prince Edward Island   2nd edn
Ottawa: DEMR, 1973.
pp. 35.

General

Prince Edward Island   1 : 250 000   MCR 41
Ottawa: DEMR, 1974.

Geological and geophysical

Surficial deposits of Prince Edward Island   1 : 126 720   Map 1366A
Ottawa: GSC, 1973.

QUEBEC

Atlases and gazetteers

Le Nord du Québec: profil régional
Québec: Office de planification et de développement du Québec,
1983.
pp. 184.

Repertoire toponymique du Québec
Québec: Editeur Oficiel du Québec.

General

Le Québec vue par satellite   1 : 2 500 000   2nd edn
Québec: MER, 1980.

Québec   1 : 2 000 000   MCR 42
Ottawa: DEMR, 1973.
English and French (MCR 42F) versions.

Le Québec en relief   1 : 2 000 000
Québec: MER, 1984.

Le Québec meridional   1 : 1 250 000
Québec: MER, 1984.

La carte routière du Québec   1 : 1 000 000
Montreal: Ministère des Transports, annual.

Cartes touristiques et routières   1 : 250 000
Québec: MER, 1981–84.
Series of 18 regional maps.

Geological and geophysical

Carte géologique du Québec   1 : 1 500 000
Québec: MER, 1985.
2 sheets, both published.

Carte minière du Québec   1 : 1 500 000
Québec: MER, 1985.
Covers only south part of Quebec.

*Quebec, magnetic anomaly map: Quebec, New Brunswick and Nova
Scotia*   1 : 1 000 000   Map NL-19-M
Ottawa: GSC, 1984.

Environmental

*Les régions écologiques du Québec meridional: deuxième
approximation*   1 : 1 250 000
Québec: MER, 1985.

Carte acide du Québec   1 : 1 400 000
Montreal: Société pour Vaincre la Pollution, 1984.
+ 15 pp. text.

*Les zones de conservation et de récréation sur les terres
publiques*   1 : 1 250 000
Québec: MER, 1983.

Administrative

Les conscriptions électorales   1 : 1 250 000
Québec: MER, 1980.
Several other administrative area maps also available at this scale.

SASKATCHEWAN

Atlases and gazetteers

Atlas of Saskatchewan   J.H. Richards and K.I. Fung
Saskatoon: University of Saskatchewan, 1969.
pp. 236.

Gazetteer of Canada. Saskatchewan   3rd edn
Ottawa: DEMR, 1985.

General

Saskatchewan   1 : 760 320   MCR 45
Ottawa: DEMR, 1963.
2 sheets, both published.

Geological and geophysical

Geological map of Saskatchewan   1 : 1 000 000
Regina: Saskatchewan Geological Survey, 1980.
2 sheets, both published.

*Quaternary geology of the Precambrian shield,
Saskatchewan*   1 : 1 000 000
Regina: Saskatchewan Geological Survey, 1984.

Mineral map of Saskatchewan   1 : 1 225 000
Regina: Saskatchewan Geological Survey, 1980.

Geology and groundwater maps   1 : 250 000
Saskatoon: Saskatchewan Research Council, 1967–.
20 sheets published.

Surficial geology of Saskatchewan: preliminary maps   1 : 250 000
Saskatoon: Saskatchewan Research Council, 1984–.
11 sheets published.

YUKON

Atlases and gazetteers

Gazetteer of Canada. Yukon Territory   4th edn
Ottawa: DEMR, 1981.
pp. 70.

General

Yukon Territory   1 : 2 000 000   MCR 47
Ottawa: DEMR, 1976.

Yukon Territory   1 : 1 000 000   MCR 25
Ottawa: DEMR, 1972.

Mexico

Mexico has a long and complex history of survey and mapping, but in 1968, by a decision of the Ministry of the Presidency, both topographic and thematic mapping were largely centralized into one organization, known initially as the Comisión de Estudios del Territorio Nacional (CETENAL). Since that date there have been several name changes, and a confusing variety of acronyms—DETENAL, DGGTN, CGSNEGI, INEGI, DGIAI, DGG, etc.—have come and gone. The present title for the coordinating body for statistical and geographic information is **Instituto Nacional de Estadística, Geografía e Informática (INEGI)**. Operating within this body are a number of *Direcciónes* including the **Dirección General de Integración y Análisis de la Información (DGIAI)** and the **Dirección General de Geografía (DGG)**, the latter being immediately concerned with map production.

When CETENAL was set up it was given the task of preparing an integrated series of topographic and resource maps covering the whole country. There was to be a basic publication scale of 1:50 000, with the production of derived maps at smaller scales. The system, designed in terms of late-1960s technology, was to provide a comprehensive set of planning maps over a 10 year period, produced by the systematic application of photogrammetric and air photo interpretation techniques (backed up with field observations) and providing a fundamental inventory of the country's resources. This programme was accompanied by a considerable public-awareness campaign designed to promote the use of the maps in planning and development.

Considerable progress has been made towards the achievement of this goal. By 1985 about 88 per cent of the country was covered by the definitive edition of the 1:50 000 topographic map, while the remainder was available in a provisional edition. In addition the whole country had been covered by 1:250 000 and 1:1 000 000 scale maps, and there was also a 60 per cent cover at 1:100 000. Less rapid progress has been made with the thematic series, although their production rate is nevertheless very impressive.

The mapping system includes about 11 000 map sheets in the basic 1:50 000 combined topographic and thematic series. The topographic maps are produced by photogrammetry using Wild A7 Autographs for aerial triangulation and Stereosimplex plotters for the compilation. This work is supplemented by field classification of roads and other cultural features, and the collection of toponyms. A UTM projection is used and the sheets each cover an area of 15′ latitude by 20′ longitude. Maps are printed in four colours with a kilometric grid and contours at 10 or 20 m.

The suite of thematic maps at the 1:50 000 scale comprises a geology, a land use, a soil and a land capability map, and these are produced from the field collection of relevant data, coupled usually with interpretation of 1:25 000 scale colour air photographs. An applied emphasis is given to all this mapping. Geological maps show superficial geology, classified genetically. The soils maps use an adaptation of the International Soils Classification of FAO/UNESCO, while for the potential land use maps the land capability classification of the US Department of Agriculture has been adopted.

The 1:250 000 series covers the whole country in 122 sheets and is derived from the 1:50 000 basic map. Sheets each cover an area of 1° by 2°, and have contours at 50 or 100 m and UTM grid at 10 km intervals. There is also a further group of thematic maps at this scale, including hydrological, geological, soils and land use. These themes have recently been extended to include separate maps of land use potential for forestry, arable and pastoral agriculture, and also of seasonal climatic variables. The 1:250 000 topographic sheets are also available as a bound volume forming the *Atlas carta de México topográfica*.

Complete cover of Mexico at 1:250 000 may also be acquired (on different sheet lines) under the PAIGH Hemispheric Mapping Programme (more information about this programme is given in the introduction to Central America).

The country is covered at 1:1 000 000 in eight sheets, and these too are used as a base for an even wider range of thematic mapping. A 1:1 000 000 tourist map of Mexico is also published in six regional sheets (plus an extra sheet of the Mexico City area at 1:250 000). The original map is in Spanish, but four sheets are also available in English.

The *Atlas Nacional del Medio Físico* may be regarded as the National Atlas of Mexico. It is a collection of 1:1 000 000 six-colour maps brought together from the thematic mapping programme of the DGG and covering the subjects of topography, relief, climate, geology, soils, land cover, geostatistics and tourism. It also includes a Landsat image map and a general map of the country at 1:5 000 000.

INEGI is responsible for a broad range of geographical information, not all of it cartographic. In 1981, it began a series of descriptive geographies of each Mexican state. Each is in three parts: a text, a map folio and a gazetteer. By 1986, 17 of these had been published.

In 1973, a programme of urban mapping at 1:5000 was started. Thirteen cities have been mapped at this scale, but due to slow progress of the series a new programme at 1:10 000 was substituted, and some 52 cities with populations greater than 15 000 have so far been mapped at this scale using either line or photomapping techniques. A number of urban land use and land use potential maps have also been published at the 1:10 000 scale. An *Atlas urbana de la República Mexicana* has been issued with 39 maps covering the urban areas at various scales and including eight of the metropolitan area of Mexico City at 1:25 000.

The 1:500 000 climatic map in 45 sheets was produced by CETENAL in collaboration with the **Instituto de Geografía** of the **Universidad Nacional Autónoma de México (UNAM)**. It uses a modification of the Köppen classification and is based on 15 years of observations at 2400 climatic stations. This series does not form a part of the integrated thematic programme, and as sheets go out of print they are not being reissued. The 1:1 000 000 climate series, also based on Köppen, is being maintained however.

In 1979, work began on the production of enumeration district mapping for the 1980 population census and for

the economic and agricultural censuses. More than 27 000
maps at a variety of scales were required. The results of
the 1980 census have been published by state, and each
has a cartographic annexe (*Cartografía geoestadística de la
República Mexicana*).

While the original concept embodied in the creation of
CETENAL in 1968 was to use the technologies associated
with air photography to produce an integrated series of
conventional colour printed maps and associated
descriptive monographs, more recent technology has
revealed the advantages of automated cartography and of
computer-based geographical information systems. Thus,
in the 1980s, a new policy of automating the cartographic
processes has been introduced, beginning with automation
of the 1:250 000 scale products. Research is also being
carried out on the digital image processing of satellite data
for the production of land use maps.

The **Dirección General de Oceanografía (DGO)**
publishes nautical charts and has also produced a set of
1:1 000 000 bathymetric maps of Mexican waters with
100 m isobaths. Recent work has included the production
of prototype maps showing variables concerned with
physical oceanography, and the development of a Centro
de Datos Oceanográficos.

Before the establishment of CETENAL, the **Instituto
de Geología** of the **Universidad Nacional Autónoma de
México** had started to produce a series of geological maps
at 1:100 000, and new sheets have continued to appear
intermittently into the 1980s. About 13 sheets were
available in 1986, several of them as heliographic copies,
but the more recent ones in colour. UNAM has also
published a series of geological maps of each state at
various scales (1965–78), which are also available as
heliographic prints, and two 1:2 000 000 scale series of the
distribution respectively of metallic and non-metallic
mineral deposits.

There are a number of private publishers producing
tourist and road maps and guides of Mexico. Thus **Guia
Roji** publishes a series of up-to-date road maps of all the
states (*Mapas de los estados*) at scales varying with the size
of each state. Other products include a guide to postal
codes, a series of *Guías turísticas* (also by state), street maps
of Mexico City and of other towns, and a 366 page *Guía
Turística República Mexicana*.

Patria also publishes a series of maps of individual states
at varying scales and with explanatory texts in Spanish and
English. Road maps are also published by the **Asociación
Mexicana Automovilística (AMA)**.

Further information

INEGI publishes an index of the current status of all the
topographic and thematic series, *Inventario de Información
Geográfica*, which is up-dated every 3 months. A comprehensive
Catálogo de publicaciones is also published.

A *Guía de información cartográfica para investigadores* (40 pp.) in
the PAIGH series was published in 1980.

There is a prolific literature on the mapping policies and
progress of Mexico since the establishment of CETENEL in 1968.
Mexico is a frequent contributor to international conferences on
cartography, and papers will be found in the volumes of technical
papers of the UN Cartographic Conferences for the Americas and
the ACSM Fall Technical Meetings.

A very useful discussion of the impact of the foundation of
CETENAL upon the production and distribution of Mexican
mapping is provided by Hagen, C. (1979) The new mapping of
Mexico, *Western Association of Map Libraries Information Bulletin*,
10, 108–115.

A useful, though now dated summary of map-producing
agencies in Mexico is given by Jensen, J.G. (1977) Some
important cartographic agencies and available maps of Mexico,
Bulletin, Western Association of Map Libraries, **8**, 102–114.

Thematic cartography is described in Puig de la Parra, J.B.
(1979) Thematic cartography in Mexico, *World Cartography*, **15**,
45–51.

Addresses

**Instituto Nacional de Estadística, Geografía e Informática
(INEGI—DGIAI—DGG)**
Col. San Juan Mixcoac, Delegación Benito Juárez, 03730 MEXICO
DF.

Universidad Nacional Autónoma de México (UNAM)
Instituto de Geografía, Ciudad Universitaria, MEXICO 20 DF.

Universidad Nacional Autónoma de México (UNAM)
Instituto de Geología, Apartado Postal 70-296, Ciudad
Universitaria, 04510 MEXICO DF.

Dirección General de Oceanografía (DGO)
Secretaria de Marina, MEXICO 1 DF.

Guia Roji
República de Colombia No 23, Col. Centro, Delegacion
Cuahtemoc, 06020 MEXICO DF.

Editorial Patria
Av. Uruguay 25, Periférico Sur, Apartado 784, 06000 MEXICO DF.

Editorial Porrúa SA
Av. República Argentina 15, MEXICO 1 DF.

Asociación Mexicana Automovilística (AMA)
Avenida Orizaba 7, Colonia Roma, 06700 MEXICO DF.

For Bureau of Business Research, see United States; for
Bartholomew, see Great Britain; for Karto+Grafik, see German
Federal Republic.

Catalogue

Atlases and gazetteers

Atlas Nacional del Medio Físico
México: DGG, 1981.
pp. 224.

Nuevo atlas porrúa de la República Mexicana E. García de
Miranda and Z. Falcón
México: Patria, 1984.
pp. 219.

Atlas of Mexico S.A. Arbingast *et al.*
Austin, Texas: Bureau of Business Research, 1975.
pp. 164.

General

República Mexicana. Carta geográfica 1:5 000 000
México: DGG, 1979.
Available in colour or black and white.

República Mexicana. Carta geográfica 1:4 000 000
México: DGG, 1981.
Available also with shaded relief or shaded relief and layer
colouring.

Mexico 1:3 000 000
Edinburgh: Bartholomew, 1983.
Inset map of Mexico City.

Mexico. Hildebrands Urlaubskarte 1:3 000 000
Frankfurt AM: Karto+Grafik, 1985.

Carta geográfica de México 1 : 2 500 000
México: AMA, 1980.
+ 64 pp. index of places.

República Mexicana. Carta topográfica 1 : 2 000 000
México: DGG, 1978.
Also available in hypsometric version.

Carta turística Mexico 1 : 1 000 000 2nd edn
México: DGG, 1980.
6 sheets, all published.
4 sheets, also available in English edition.

Topographic

Estados Unidos Mexicanos. Carta topográfica 1 : 1 000 000 2nd
edn
México: DGG, 1983.
8 sheets, all published.
Also available in hypsometric version.

Estados Unidos Mexicanos. Fotomapa 1 : 1 000 000
México: DGG, 1980.
8 sheets, all published.
Based on Landsat imagery.

Estados Unidos Mexicanos. Carta topográfica 1 : 250 000
México: DGG, 1976–.
121 sheets, all published. ■

Estados Unidos Mexicanos. Carta topográfica 1 : 50 000
México: DGG, 1971–.
2348 sheets, all published. ■
Some sheets in a provisional edition.

Bathymetric

Carta de la zona económica exclusiva 1 : 5 000 000
México: DGG, 1976.

Carta batimétrica 1 : 1 000 000
México: DGG.
8 sheets, all published.
Covers seas adjacent to Mexico.

*Atlas geofísica de la margen continental oeste de México, Sección 1
Gravimetría-Magnetometría*
México: DGO, 1980.
pp. 37.

Geological and geophysical

Carta geológica de la República Mexicana 1 : 2 000 000 4th
edn E.L. Ramos
México: UNAM, Instituto de Geológia, 1976.

República Mexicana. Yacimientos minerales metalicos 1 : 2 000 000
México: UNAM, Instituto de Geológia, 1966.
13 sheets published.
Heliographic copies.

República Mexicana. Yacimientos minerales no metalicos
1 : 2 000 000
México: UNAM, Instituto de Geológia, 1966.
17 sheets published.
Heliographic copies.

Estados Unidos Mexicanos. Carta geológica 1 : 1 000 000
México: DGG, 1980.
8 sheets, all published.

Estados Unidos Mexicanos. Carta geológica 1 : 250 000
México: DGG, 1978–.
122 sheets, 78 published. ■

Cartas geológicas de México 1 : 100 000
México: UNAM, Instituto de Geológia, 1962–.
13 sheets published.

Estados Unidos Mexicanos. Carta geológica 1 : 50 000
Mexico: DGG, 1971–.
2348 sheets, *ca* 40% published. ■

Environmental

Estados Unidos Mexicanos. Carta de uso de suelo 1 : 4 000 000
México: DGG.

Estados Unidos Mexicanos. Carta edafológica 1 : 4 000 000
Provisional edition
México: DGG.

Estados Unidos Mexicanos. Carta de uso del suelo 1 : 2 000 000
México: DGG.

Estados Unidos Mexicanos. Carta climática fisiográfica
1 : 2 000 000
México: DGG, 1978.

Estados Unidos Mexicanos. Carta uso del suelo y vegetacion
1 : 1 000 000
México: DGG, 1978–.
8 sheets, all published.

Estados Unidos Mexicanos. Carta fisiográfica 1 : 1 000 000
México: DGG.
8 sheets, all published.

Estados Unidos Mexicanos. Carta de uso potencial agricultura
1 : 1 000 000
México: DGG.
8 sheets, all published.

Estados Unidos Mexicanos. Carta de uso potencial ganaderia
1 : 1 000 000
México: DGG.
8 sheets, all published.

Estados Unidos Mexicanos. Carta edafológica 1 : 1 000 000
México: DGG, 1981–.
8 sheets, all published.

Estados Unidos Mexicanos. Carta de climas 1 : 1 000 000
México: DGG, 1980–.
8 sheets, all published.

*Estados Unidos Mexicanos. Carta de evapo-transpiración y deficit de
agua* 1 : 1 000 000
México: DGG.
8 sheets, all published.

*Estados Unidos Mexicanos. Carta de temperaturas medias
anuales* 1 : 1 000 000
México: DGG, 1980–.
8 sheets, all published.

*Estados Unidos Mexicanos. Carta de precipitación total
anual* 1 : 1 000 000
México: DGG, 1980–.
8 sheets, all published.

*Estados Unidos Mexicanos. Carta de humedad en el
suelo* 1 : 1 000 000
México: DGG, 1980–.
8 sheets, all published.

*Estados Unidos Mexicanos. Carta de hydrología aguas
superficiales* 1 : 1 000 000
México: DGG, 1981–.
8 sheets, all published.

*Estados Unidos Mexicanos. Carta de hydrología aguas
subterráneas* 1 : 1 000 000
México: DGG, 1981–.
8 sheets, all published.

México. Carta de climas 1:500 000
México: DGG/UNAM, 1970.
45 sheets, all published.
Some sheets out of print.

*Estados Unidos Mexicanos. Carta uso del suelo y
vegetación* 1:250 000
México: DGG, 1980–.
122 sheets, 75 published. ■

*Estados Unidos Mexicanos. Carta uso potential
agricultura* 1:250 000
México: DGG.
122 sheets, 18 published. ■

*Estados Unidos Mexicanos. Carta uso potential
ganaderia* 1:250 000
México: DGG.
122 sheets, 18 published. ■

Estados Unidos Mexicanos. Carta uso potential forestería 1:250 000
México: DGG.
122 sheets, 18 published. ■

Estados Unidos Mexicanos. Carta edafológica 1:250 000
México: DGG, 1982–.
122 sheets, 63 published. ■

*Estados Unidos Mexicanos. Carta de efectos climáticos
regionales* 1:250 000
México: DGG.
122 sheets, 68 published. ■
In two versions: periods May–October and November–April.

*Estados Unidos Mexicanos. Carta de hidrología de aguas
superficiales* 1:250 000
México: DGG, 1978–.
122 sheets, 91 published. ■

*Estados Unidos Mexicanos. Carta de hidrología de aguas
subterraneas* 1:250 000
México: DGG, 1978–.
122 sheets, 91 published. ■

Estados Unidos Mexicanos. Carta uso del suelo 1:50 000
México: DGG, 1971–.
2348 sheets, *ca* 35% published. ■

Estados Unidos Mexicanos. Carta edafológica 1:50 000
México: DGG, 1971–.
2348 sheets, *ca* 35% published. ■

*Estados Unidos Mexicanos. Carta de uso potencial del
suelo* 1:50 000
México: DGG, 1971–.
2348 sheets, *ca* 25% published. ■

Town maps

Plano de bolsillo Cd. de México 1:30 000
México: Guia Roji.

Ciudad de México Seccionado 1:10 000
México: Guia Roji.
61 sheets, all published.

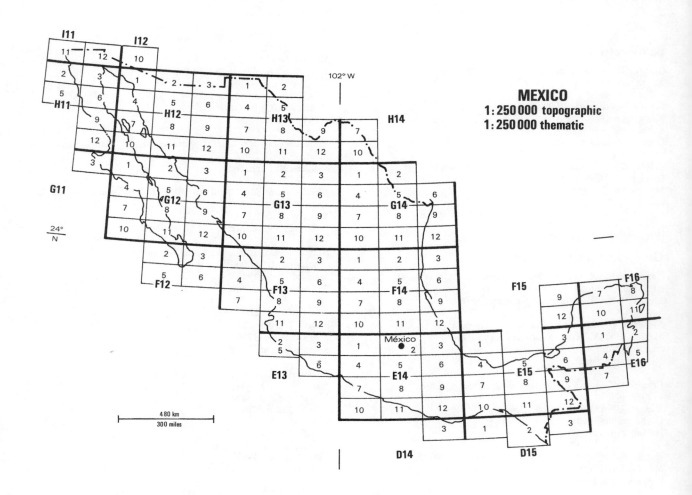

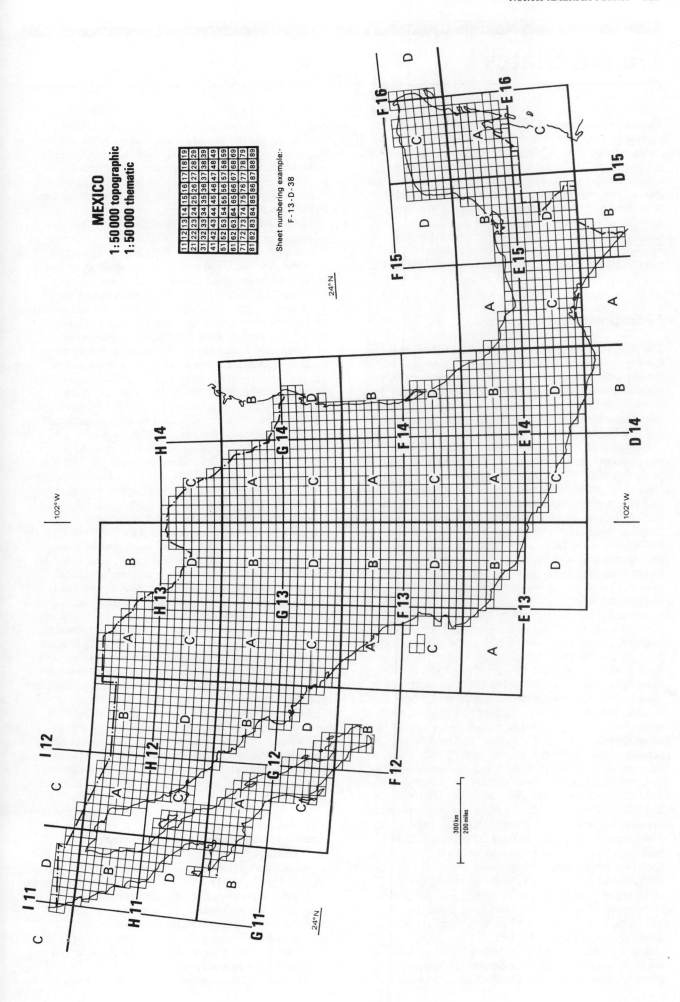

MEXICO
1:50 000 topographic
1:50 000 thematic

11	12	13	14	15	16	17	18	19
21	22	23	24	25	26	27	28	29
31	32	33	34	35	36	37	38	39
41	42	43	44	45	46	47	48	49
51	52	53	54	55	56	57	58	59
61	62	63	64	65	66	67	68	69
71	72	73	74	75	76	77	78	79
81	82	83	84	85	86	87	88	89

Sheet numbering example:-
F-13-D-38

United States

The mapping of the United States is highly complex because it is carried out at both federal and state level. In addition to the 39 or so federal agencies which are involved in some kind of mapping activity, there are also mapping agencies in each of the 50 states, which carry out their own mapping projects as well as cooperative programmes with the federal government. There are of course also very many private mapping companies, of which only a sampling can be given here. Although its mapping is somewhat distinctive, Alaska has been included in this section. Hawaii, however, will be found under the Pacific.

Federal mapping agencies

The principal federal mapping agency is the **United States Geological Survey (USGS)**. USGS was founded in 1879 initially with the task of classifying (i.e. surveying) public lands as well as surveying the geological and mineral resources of the country, and today it continues to be responsible for topographic mapping as well as geological and resource mapping.

The mapping itself is conducted from four regional mapping centres, and funding is partly from state and partly from federal sources. Although there is some variation in priorities between states, the topographic map series are prepared to a common specification and form part of a unified National Mapping Program whose present form was determined in 1975 under a directive from the Department of the Interior. In 1980, a restructuring took place within USGS and a new national Mapping Division was created, which may mark the beginning of a trend towards a more far-reaching rationalization of federal mapping activities.

Systematic topographic survey was initiated in 1882. The scales adopted at that time were 1 : 125 000 and 1 : 62 500 (a little more than an inch to the mile), and in the early days progress was quite rapid, 20 per cent cover being achieved in the first 10 years. Since 1885, the topographic survey has proceeded through cooperative agreements between federation and state, with 50 per cent funding from each. Most of the country's topographic archive is a product of this cooperation, but it has led to unequal progress between the states, and by 1950 only 50 per cent of the land area had been covered at the basic mapping scale. Until about 1950, this basic scale was 1 : 62 500, but from this time on there was a shift to 1 : 24 000 (2000 feet to the inch), and since the 1960s, the 1 : 62 500 has become a 'dead' series, with sheets remaining available but unrevised. Currently the main thrust of the programme is completion of the 1 : 24 000 (or 1 : 25 000) basic mapping and in order to achieve this by 1991, some sheets are being issued in provisional editions. In Alaska the basic mapping scale is 1 : 63 360 (1 inch to the mile), a series begun in 1948, with some limited 1 : 25 000 scale mapping of urban and other developing areas. By 1985, basic scale cover was about 85 per cent complete in both Alaska and the conterminous US.

Although the 1 : 24 000 basic scale adopted by USGS is not large compared with other developed countries, it nevertheless requires about 57 000 sheets to provide full cover of the conterminous states. Following the Metric Conversion Act of 1975, there was a move to convert the mapping to a 1 : 25 000 scale with metric contours, and a new 'double quad' format map sheet was designed. However, most states opted to complete mapping at 1 : 24 000 first, and since that time metrication plans seem to have lost impetus.

The 1 : 24 000 series is commonly known as the *7.5 minute topographic quadrangle series*, since each sheet covers 7.5′ of latitude and longitude. Projection is Transverse Mercator, and the UTM grid is shown on the map. Sheets are printed in five or six colours, with brown for relief features, black for cultural information and place names and blue for water. Woodland is in green, and red is used for land division boundaries, built-up areas and major roads. The contours are generally in feet (metres on the 1 : 25 000 sheets), and the interval varies according to the ruggedness of the terrain on any given sheet. A rather complicated sheet numbering system has been devised for the series, but the quads are most usually referred to by name. The series is accessed on a state-by-state basis, and since the number of sheets needed to cover a single state is normally counted in thousands it is impossible to include indexes in this book, and the reader is referred to the excellent state catalogues and indexes issued by USGS (see Further information).

A particular boost was given to 1 : 24 000 scale mapping with the development of high-altitude photography (photos taken by aircraft from altitudes of 40 000 feet or more). Experiments conducted in 1969 showed that it was possible to acquire such photographs to provide stereo cover of each map quadrangle without the need to mosaic them. One photo is accurately centred over the middle of the quad while the adjacent photos, centred over the edges, provide two stereo models which enable the scene to be converted into an orthophoto. Having established that such orthophotos meet National Map Accuracy Standards, this new product entered the National Mapping Program in 1973. Orthophotoquads have been used in two ways: to provide rapid revision of existing line maps by photo-inspection (the revision data is overprinted on the maps in purple), and as substitutes for conventional maps. Fulfilling the latter function are two versions: the *orthophotoquad*, which is a black-and-white scale-true photo-image combined with a grid and a few place names and highway numbers, and the *orthophotomap* which is a colour-enhanced photo image combined with traditional line map detail. Orthophotoquads are mostly available only in diazo form.

The success of high-altitude photography and its numerous potential applications led to agreement in 1980 between 12 federal agencies to fund a complete photographic cover of the 48 states in black and white and colour-infrared. Photography for the first National High Altitude Program (NHAP I) has been flown during seasons of dormant vegetation, and was due to be complete in 1986. A second programme is in progress with photographs being taken during the growing season to facilitate study of vegetation, and the conducting of agricultural resource surveys. This programme should be complete by the end of 1990.

The largest scale topographic map to cover the whole of

the US at present is the 1 : 250 000 series. This began as an AMS (Army Map Service) series, but since 1958 has been maintained by USGS as part of the civil topographic mapping programme. Many sheets of this series are rather out of date, although revisions are continuously in progress. A total of 472 sheets are needed to cover the conterminous United States, while a further 153 cover Alaska. The projection is Transverse Mercator, and the sheet system is based on that of the International Map of the World, each sheet covering 1° × 2° (12 sheets cover the area of one 1 : 1 000 000 sheet). Although sheets are numbered they are most usually identified by name (see index map). Sheets east of the Mississippi are entitled *Eastern United States* and those west of the great river *Western United States*. This series has a UTM grid but also shows township and range lines on the western sheets. The contour interval varies with the nature of the terrain, between 25 and 200 feet on older maps, while some newer sheets have metric contours. Terrain data digitized from the contours on these maps are available for purchase on computer tapes.

USGS also publishes maps of each state at a scale of 1 : 500 000 (Alaska is at the smaller scale of 1 : 1 584 000, while some of the smaller states share a sheet).
There are usually three versions of the state map: base (printed from the black-and-blue separates), topographic and shaded relief. Dates of publication may vary between these versions and the date given in the catalogue is of the latest topographic edition.

Although the progress towards a satisfactory complete topographic map of the US may appear to have been surprisingly slow, new mapping introduced in 1975 to plug the scale gap between the 1 : 24 000 and 1 : 250 000 series has been equally remarkable for the rapidity of its

progress. This is the *intermediate scale series* at 1 : 100 000 and 1 : 50 000 issued in quadrangle or county formats. More than 75 per cent of the conterminous United States is already available at one or other of these scales, and the 1 : 100 000 quadrangle series should be complete by the end of the 1980s. The choice of county format mapping was made by individual States (Colorado, Kansas, Connecticut, Virginia, California and Pennsylvania) under cooperative mapping agreements, and Colorado already has a complete 1 : 50 000 county series. Built into the design of the intermediate-scale maps was a high degree of feature separation (to facilitate digitization) and a fully metric specification. The 100 000 series is being offered in a digital version known as DLGs (Digital Line Graphs). A DLG sampler, containing digital planimetric data for hydrography and transportation separates of a 30′ × 60′ block is available from the **National Cartographic Information Center (NCIC)** computer to allow potential users to experiment with the data on their own systems. Standard DLGs began to be marketed in 1986.

Since 1972, USGS has published a number of mainly experimental satellite image maps based upon Landsat imagery, and printed in colour or black and white. Many of these cover a whole state and are included (when still in print) in the catalogue section. With the advances in digital image processing techniques it has become possible to produce a great variety of image maps geometrically adjusted to map projection geometry and combined if desired with conventional map detail. USGS has been in the forefront of research with these techniques, and a number of experimental maps have been published in recent years. Thematic Mapper imagery has been successfully applied to 1 : 100 000 scale mapping, while Return Beam Vidicon and MSS image mapping have

USA
This map serves as an index to State base maps,
State geological and other resource maps

found practical application in Alaska. In 1984, satellite image mapping became formally part of the National Mapping Program and it is planned to produce 20–25 image maps per year.

An important thematic series published by USGS is the two-level land use and land classification map devised for use with remote sensor data and in progress since 1975. This mapping now covers more than 90 per cent of the conterminous states. The scale is 1 : 250 000, with some limited mapping at 1 : 100 000, and sheets are keyed to the national topographic maps. Maps are published in two (or sometimes more) colours, or are on open file (in which case they are available as photographic copies). The principal map is of land use and cover, and there are five further outline maps showing political units, hydrological units, census county subdivisions and federal land ownership. Currently a level III land classification system is being developed for use with larger-scale mapping. The units shown on the land classification and associated maps are being digitized in a polygon format. They can be converted to grid cells for the calculation and analysis of statistical data, and the aim is to register the cells with Landsat digital data in order periodically to map land use changes.

The US National Academy of Sciences recommended in the early 1960s that USGS should prepare a National Atlas, and this was completed in 1970 under the editorship of A. C. Gerlach. It was issued as a bound volume in a 15 000 print run and also selected sheets were published in separate sales editions. Some additional sheets were also published which did not appear in the original volume. The atlas as a whole is out of print. There are plans for a new edition, although at the time of writing no schedule has been fixed (see Chapter 2).

USGS has developed a computerized *Geographical Names Information System* (GNIS) which contains 2 million official names with their feature category, location (state, county and geographic coordinates), 1 : 24 000 map sheet name, and elevation. GNIS is available by direct on-line access and also in various other formats including computer tape, microfiche and bound alphabetical listings. It also forms the basis of a series of bound state gazetteers which began to be issued in 1982 as chapters of the USGS *Professional Paper* 1200.

Geological and mineral resource mapping is carried out by the Geologic Division of USGS. Maps are published in a wide range of scales from 1 : 24 000 quadrangle maps to 1 : 2 500 000 scale maps overviewing aspects of the geology or geophysics of the whole country. Quadrangle maps show bedrock, surficial or engineering geology. *Miscellaneous investigations maps* cover a wider variety of subject matter in a range of formats. About half of the US is covered by geological maps at 1 : 250 000 or larger. Other map products include coal, oil and gas investigation maps, and geochemical and geophysical maps. However, because most geological mapping projects are carried out cooperatively with individual states, a full appraisal of map availability is best acquired from state rather than federal sources (see below).

The USGS also has a Water Resources Division, and publishes a range of water resource mapping including *Hydrologic investigations atlases* showing geohydrological data in multi-colours or black and white on a topographic base map at 1 : 24 000 scale or smaller, and *Hydrologic unit maps* of each state. Other map products include water availability maps and flood-prone area maps.

Automated methods of data capture, editing and map production are all in use in the USGS mapping system. Furthermore, a national digital cartographic database (DCDB) is being developed which holds the digital data

captured from the various topographic and land use series, and from the 1 : 2 000 000 sectional maps, together with digital elevation data generated from the production of orthophotoquads. The database files are managed by the SYSTEM 2000 DBMS. A number of digital products are already available on tape, and some have already been referred to. US GeoData tapes of large-scale data are sold by 7.5′ and 15′ blocks corresponding to the topographic quadrangle maps and are available in four thematic layers, namely boundaries, transportation, hydrography and US Public Land Survey System.

Other federal mapping agencies

The **National Geophysical Data Center (NGDC)**, under the National Oceanic and Atmospheric Administration (NOAA), collects and disseminates data in the subject areas of solid earth geophysics (seismology, gravity, geothermics, magnetics, tsunami, volcanology) and marine geology and geophysics. Much of this data is in digital or computer printout form, but a number of conventional maps are published from time to time. In particular, NGDC has worked in conjunction with the state agencies in the production of maps summarizing the geothermal energy resources of a number of states. The maps are mainly intended for interpretation by the general public, and in some cases a second 'technical' map of the same area has also been produced for specialist users. The 'public usage' maps are included in our catalogue. The data collected for these maps has been entered into a geothermal database being constructed by the State Resource Assessment Program, funded by the Department of Energy. As the home of World Data Center-A, NGDC also provides an international geophysical data service.

The **National Ocean Service (NOS)** primarily carries out geodetic and hydrographic surveys, but since 1975 has also worked cooperatively with USGS in producing topobathymetric maps of the coast showing both land and seafloor detail. These maps form part of the standard 1 : 250 000 and 1 : 100 000 land-based series. Some 1 : 24 000 maps are also being published through cooperative mapping agreements with individual states. The topobathymetric sheets continue offshore as pure bathymetric maps, and most areas of the continental shelf and slope are now covered. NOS also produces *bathy fishing maps* (1 : 100 000) showing seabed sediment type and bottom obstructions. Geophysical maps are published in two series: 1 : 250 000 covering the continental shelf and slope (with sheets of base bathymetry, magnetic and gravity anomalies) and 1 : 1 000 000 (*Seamap Series*) giving geophysical information on the deep sea area. 1 : 1 000 000 *Regional Maps* are compiled from the 1 : 250 000 bathymetric data.

Soil mapping is carried out mainly by the federal **Soil Conservation Service (SCS)** under the Department of Agriculture, although a few state authorities also publish soil maps. The US Department of Agriculture has been making soil surveys since 1899 and so a very large number of increasingly sophisticated maps have been published, many of which have however gone out of print, although they may often be consulted in libraries. The more recent soils maps are frequently overprinted on the USGS orthophotoquads. By early 1986 a total of 3509 surveys had been published of which 1679 were out of print. About 17 per cent of the country has been covered by soil maps.

A series of maps with the title *Important farmlands* has also been published by SCS at 1 : 100 000 or 1 : 50 000 in association with the USGS. These sheets, printed in colour, distinguish land of prime quality for agriculture, as well as agricultural land of 'statewide' and of 'local' importance.

The **Bureau of Land Management (BLM)** is responsible for the administration of public lands in the Western States and Alaska. Maps published by BLM include 1:100 000 quadrangle maps of surface management status and surface minerals management status. Base maps are provided by USGS to which BLM adds the thematic overlays. BLM Land Management and Surface Minerals maps are distributed by USGS at Denver. BLM also provides USGS with the public land survey grid of the Western States. Regional offices of BLM issue a number of recreational maps.

The **Fish and Wildlife Service (FWS)** has worked with USGS in the production of 83 *Coast Ecological Inventory* maps at 1:250 000 scale. These five-colour maps, published in 1981–82, assign major land use designations to the entire coastal zone, and indicate all important fish and wildlife species and their habitats. The FWS is also preparing a series of *National Wetlands Inventory* maps based on USGS series mapping.

The **Forest Service (FS)** maintains a primary base map (1:24 000) for management purposes covering all the National Forest areas, and distributes *Forest visitor maps* from its regional offices.

The **National Parks Service (NPS)** is not a major map producer: a few simple maps are issued of individual parks. The USGS, however, issues a number of specially formatted Parks maps based on the national topographic series.

The **United States Bureau of the Census (USBC)**, Washington, DC, has for many years produced a variety of choropleth maps based on the US census statistics, and has also been responsible for the production of outline maps of census enumeration districts. Recent mapping activity has been concerned with the 1980 Decennial Census, and the maps produced may be divided into three kinds:

● Five series of outline maps defining the various census tracts, blocks and enumeration districts.
● Summary reference maps, at smaller scales, designed to accompany the written report series.
● Special-purpose maps, which are mainly thematic maps showing the distributional aspects of specific census statistics. For previous censuses these included the GE50, GE70 and GE80 map series, but it is unlikely that a similar fulsome range of colour choropleth maps will be produced from the 1980 census data, although a few have been published and are included in our catalogue.

In 1981, the USGS and the USBC formed an inter-agency task force to review the future needs of both organizations for digital map products, and in the case of the USBC with regard to its needs for the 1990 Decennial Census. As a result, the need for the 1:100 000 scale database was identified, and following a pilot study for the State of Florida, carried out in 1983, an agreement was signed committing the two agencies to production of this national Cartographic Data Base (code-named TIGER—see Chapter 6) by 1987.

The **Central Intelligence Agency (CIA)** has been involved in mapmaking since its foundation in 1947, but it is only since 1968 that unclassified maps produced by the agency have become widely distributed, beginning with a *China map folio* issued in that year. Since then, a number of useful atlases and sheet maps of overseas areas have become available from the **US Government Printing Office (USGPO)**. These are usually of smallish format, and are mainly general maps of whole countries supported by

several thematic inset maps. The CIA has also produced the world digital databases known as *World Data Bank* I and II, and described in the World section.

The **Defense Mapping Agency (DMA)**, formed in 1972 from a combination of other military agencies, is responsible for defence mapping, but a number of its products are also available to the public. These include small-scale world series, and the US Board on Geographic Names country gazetteers, which are processed and published by DMA. The DMA also has a new 1:50 000 mapping programme of the USA which is available to the public. This is a metric map on Transverse Mercator projection and is now prepared by USGS for the DMA, and is sold by USGS. Only a small part of the country has so far been covered in this series, however.

State mapping

Each state government carries out its own local mapping, and mapping activities are usually spread through several agencies. The main agencies, however, are usually the State Transportation Departments and the State Geological Surveys. The former issue small-scale general-purpose road maps of the whole state, and planimetric base maps usually in a county format, which are used to display various thematic data such as road quality and traffic density. A very few Transportation Departments have a well developed map information and retail unit, the best example being that of New York State, which issues an impressive catalogue and publishes an excellent 1:250 000 scale state map in sheet or atlas format.

Each state has its own geological survey division whose activities are funded partly by federal and partly by state sources. Thus mapping is carried out in close liaison with USGS. Much state mapping in the earth science area is available from USGS, but there is also a proportion which is issued exclusively by the state survey and is not obtainable from the USGS Distribution Center at Denver. For this reason we have included below addresses of all the state surveys of the 48 conterminous states and Alaska (for Hawaii, see the Pacific Ocean section). Maps figure prominently among the numerous publications of the surveys, and cover a wide range of themes: from bedrock and surficial geology through hydrogeology and economic geology. Delaware has flood-prone area maps, Louisiana maps of oil basins, California of slope stability and seismic hazards. Many coastal states also have coastal and marine mapping programmes. In addition to small-scale maps on such various themes, and special-purpose local studies, states also have regular series of geological quadrangle maps based on the standard topographic sheet lines. Indexes to geological mapping for each state are issued by USGS, while the state organizations issue their own catalogues as well. Clearly it is beyond the scope of this book to list systematically all this wealth of mapping. Instead, we have selected a few small-scale thematic maps for each state, generally of the one-sheet/one-state kind. We have tried to include the most recent maps published and have omitted older superseded maps, but where recent basic geological maps have not been published older material is cited. A number of new sheets are in process, but have only been listed if a firm publication date is known. For more information about individual state mapping, the reader is advised to contact the addresses given or to consult the cumulative catalogues mentioned under Further information.

Many individual states also have their own atlas of near National Atlas standard, published by universities with state support or by private publishers, and many of these have been included in our catalogue section under the individual states.

A number of state organizations have introduced digital mapping programmes and in many cases are operating complete geographical information systems. In 1983, it was estimated that 24 states had some form of GIS. NCIC is collecting and cataloguing information about spatially related data files and GIS software held by federal, state and other agencies.

Private mapping organizations

A number of private companies, academic and research organizations are producers of maps within the geological and geophysical subject area. Thus the **PennWell Publishing Company (PennWell)** of Tulsa, Oklahoma includes a number of maps among its publications. These are not confined to the USA but are mostly concerned with oil- and gas-producing areas and include maps of oil and gas fields and pipelines. Structural maps of several US states are included in the catalogue section. The **American Association of Petroleum Geologists (AAPG)** has published a number of geological and geophysical maps, but of particular note are the *Geological highway maps* which cover the whole country in a set of 12 regional sheets each comprising a geological map supplemented by cross-sections, physiographic maps and descriptive texts, and the *Circum-Pacific map series* (described under Pacific Ocean).

The **Geological Society of America (GSA)**, Boulder, Colorado, is celebrating its 1988 centenary with the publication of a 28-volume *Geology of North America*. Associated with this work, seven continental-scale (1:5 000 000) maps are scheduled for publication in 1987 covering the themes of geology, gravity, magnetic anomaly, seismicity and neotectonics, and an accompanying volume of text. There will also be a transect package including 23 transects in colour at the scale of 1:500 000.

Other publishers within this area include the **Society of Exploration Geophysicists** who produce gravity maps of the United States.

The **American Geographical Society** was formerly an important publisher of both general and thematic maps, and notable achievements were the 1:1 000 000 *Hispanic America* series, subsequently absorbed into the military Series 1301 by the Army Map Service, and the 1:5 000 000 world map in 18 sheets.

The **National Geographic Society** issues maps as supplements to its *National Geographic* magazine. They are also available as separates and are listed in the NGS *Educational Services Catalog*. A number of unusual and innovative maps have been published from time to time, including the ocean floor maps (1974–), a Landsat mosaic *Portrait USA* (1976) and *Heart of the Grand Canyon* (1978). A particularly fine *Atlas of North America* was published in 1985. NGS has introduced a Scitex computer mapping system, and is implementing a computerized Geographic Names Data Base.

Rand McNally, founded 1856, is a major producer of educational maps and atlases and of city street maps and tourist maps and guides. The annual *Rand McNally road atlas of USA/Canada/Mexico* is one of its most familiar products, and the massive *Commercial Atlas and Marketing Guide*, also revised annually, is particularly valuable for its detailed indexes (cities, towns, counties, transportation lines, banks, post offices), presented state by state. There are also a number of thematic maps, mostly designed for business users, but including such themes as manufacturing, zip codes, metropolitan statistical areas, and population.

The HM Gousha Company, founded 1926 and now forming the combination **Gousha/Chek-Chart**, publishes road atlases, individual maps of all the states, and street maps of 63 US cities. The state and city maps may each be purchased as a complete 'library'. Gousha has also specialized in the production of 'Natural Color Relief Maps', and provide customized trip routing for automobile clubs.

There are very many other companies specializing in road and city maps, including **Arrow Publishing**, **Champion Map Corporation**, **Dolph Map Company**, **Hearne Brothers** and **Thomas Brothers**. **EXXON Touring Service** distributes a complete set of state road maps under their imprint and also specializes in customized mapping for travellers, as does the **American Automobile Association**, who provide 'Triptik' and more conventional maps for their members. Among the more unusual approaches to city mapping are the axonometric maps of city centres produced by **David A. Fox** and **Pierson Graphics**, the map guides of **Flashmaps Publications**, which have a clever integration of service directories with locator maps, and the maps of **Perspecto Map Co**, who publish maps for car rental companies and the like but whose list includes bird's-eye views, and university campus and airport maps.

There are many private publishers with a mainly regional specialism, and only a few will be mentioned. The **International Aerial Mapping Company** produces city and recreational maps of the southern states (as well as carrying out aerial photography and photogrammetric services), while **Hagstrom** produces pocket maps of cities in the north-east. The **Western Economic Research Company** specializes in census mapping of California. The **Sidwell Company** carries out aerial mapping in the midwest and publishes tax and cadastral maps in that region. **Rockford Map Publishers** also specialize in land ownership mapping, producing a series of *County plat books* for several midwestern and eastern states.

The major private publisher of large-scale city maps, which have evolved from the early fire insurance maps, is the **Sanborn Map Company**, New York.

Several well known companies—**Nystrom, Cram, Dennoyer-Geppert, American Map Corporation**—have an output mainly of educational maps and atlases. **Kistler Graphics** produce raised relief maps, and **Panoramic Studios** earth curved relief maps, while **Hubbard** markets the raised relief 1:250 000 topographic maps of the western states and the Appalachian region originally prepared by DMA.

Ryder Geosystems have made a speciality of publishing 'display-quality' satellite imagery, and have also published a *Satellite Photo-Atlas of the United States of America*.

Recreational mapping is carried out locally by many organizations. On the west coast, **Wilderness Press** adapts USGS mapping to the needs of the hiker, and similar work is done by the **Telecote Press, Potomac Appalachian Trail Club, Geo-Graphics** and others.

Acquisition of American maps

The distribution of USGS products (maps and texts) was centralized in 1986 at the Federal Center, Denver, Colorado. (Previously, distribution had been from two centres at Arlington, Virginia and Denver, Colorado.)

As described above, the State Geological Surveys are funded partly from federal and partly from state sources. Thus some state geological and earth resource mapping is available only from the state offices and not from USGS.

Further information

The principal source of information about published topographic mapping at the larger scales is provided by indexes for individual states issued by USGS. In recent years the single, folded index sheets have been replaced by comprehensive state catalogues in booklet form. These are issued in pairs, one giving the permanent indexes and quadrangle names, the other listing actual maps available. They were due to be available for all states by the end of 1986, and may be obtained from NCIC.

New USGS maps are notified in the monthly *New Publications of the US Geological Survey* and in annual cumulative listings. The *USGS Yearbooks* also provide valuable summaries of mapping progress as well as feature articles on various developments.

The USGS has its own National Cartographic Information Center at Reston, Virginia, established in 1974, which gives information not only about USGS products but on US mapping in general, and especially on the various other federal agencies which produce maps. There are also five additional regional NCIC offices and a growing number of state affiliates where users can obtain local map information and browse through quick-look satellite image files, etc. There is a trend for US maps to be better publicized and marketed. Even the DMA has opened a customer assistance office.

The *Guide to Obtaining USGS Information* (USGS Circular 900), 1986, gives information about how to acquire USGS information of all kinds. USGS also issues a number of general map indexes showing the current status of topographic mapping, orthophotoquad mapping, NHAP photography, land use and land cover mapping, and Digital Line Graphs, and there is a special index to USGS topographic map cover of the National Parks. NCIC has a number of leaflets giving information about state atlases, commercial map publishers and state geological maps, for example, and issues a *Newsletter* from time to time. There is also a *Map Data Catalog* which colourfully illustrates and explains the less conventional map-related products.

The individual state geological surveys also issue their own catalogues, and these are essential to give a complete picture of the geological mapping pertaining to a particular state. Some include local topographic map information as well.

Several other federal agencies issue catalogues or map indexes including the National Ocean Service, the Bureau of the Census and the Soil Conservation Service. The DMA has a *Public Sales Catalog* listing small-scale world and overseas series mapping and giving a full list of the US Board on Geographic Names gazetteers, while the CIA issues a catalogue, *CIA maps and publications released to the public*. Most commercial publishers also issue publication lists or catalogues.

A thorough description of American mapping is provided by Thompson, M.M., *Maps for America* (2nd edn, 1981), published by the USGS and distributed by the Superintendent of Documents, US Government Printing Office (USGPO). This volume is mainly devoted to USGS mapping of all kinds, but also includes a chapter on maps from other agencies. A good summary of federal mapping and policy is given by North, G.W. (1983) Maps for the nation: the current federal mapping establishment, *Government Publications Review*, **10**, 345–360.

An excellent overview of recent developments in US mapping in all areas will be found in the *American Cartographer Special Issue: US National Report to ICA*, 1984 (Supplement to Volume 11). This also includes (pp. 73–77) a useful guide to 'Sources of US maps and map information' by M. Galneder.

A further useful introduction to American maps aimed at the business user is Shupe, B. and O'Connell, C., *Mapping your Business* (1983), published by the American Special Libraries Association.

The classification system developed for the USGS land use maps is described by Anderson, J.A. *et al.* (1976) *A Land Use and Land Cover Classification System for Use with Remote Sensor Data*, Geological Survey Professional Paper 964.

The mapping of Alaska (including some of the innovative image mapping projects) has been described in a Special Alaska Issue of the journal *Photogrammetric Engineering and Remote Sensing*, **52**(6), 1986.

Addresses

The addresses given here are necessarily selective. They include principal federal mapping agencies, all state geological mapping agencies and a list of private mapping companies which publish significant mapping usually on a nationwide basis, or which are otherwise cited in the text or catalogue. More comprehensive listings will be found in Shupe and O'Connell (cited above) and in Allin (see Chapter 3), and NCIC issues a list of *State information sources*, giving addresses of all government map-producing offices in each state. The newsletter *Baseline* published by the American Library Association's Map and Geography Round Table (MAGERT) is a useful source of information about new (and less familiar) map publishers.

Federal mapping organizations

US Geological Survey (USGS)—map distribution
Map Distribution, Federal Center, Bldg 41, Box 25286, DENVER, Colorado 80225.

National Cartographic Information Center (NCIC)—information
508 National Center, US Geological Survey, RESTON, Virginia 22092.

US Government Printing Office (USGPO)
Superintendent of Documents, WASHINGTON, DC 20402.

National Geophysical Data Center (NGDC)
NOAA, Code E/GC4, Dept ORD, BOULDER, Colorado 80303.

National Ocean Service (NOS)
Distribution Division N/CG33, 6501 Lafayette Avenue, RIVERDALE, Maryland 20737.

Soil Conservation Service (SCS)
US Department of Agriculture, PO Box 2890, WASHINGTON, DC 20013.

US Bureau of Land Management (BLM)
Office of Public Affairs, Eighteenth & East Street NW, WASHINGTON, DC 20240.

Fish and Wildlife Service (FWS)
Division of Realty, WASHINGTON, DC 20240.

US Forest Service (FS)
Information Office, PO Box 2417, WASHINGTON, DC 20013.

National Park Service (NPS)
Office of Public Enquiries, 1013 Interior Building, Eighteenth & C Street NW, WASHINGTON, DC 20240.

US Bureau of the Census (USBC)
United States Department of Commerce, WASHINGTON, DC 20233.

Central Intelligence Agency (CIA)
Public Affairs, WASHINGTON, DC 20505.

Defense Mapping Agency (DMA)
Office of Distribution Services, Attn. DOA. WASHINGTON, DC 20315-0010.

Woods Hole Oceanographic Institution
WOODS HOLE, Massachusetts 02543.

State mapping organizations

Alabama

Geological Survey of Alabama (GSA)
420 Hackberry Lane, PO Box 0, University Station, TUSCALOOSA, Alabama 35486.

Cartographic Research Laboratory
Department of Geography, 324 Farrah Hall, Box 1982, University of Alabama, UNIVERSITY, Alabama 35486.

Alaska

Division of Geological and Geophysical Surveys (DGGS)
3601 C Street, Pouch 7-028, ANCHORAGE, Alaska 99510.

Arizona

Arizona Bureau of Geology and Mineral Technology (ABGMT)
845 North Park Avenue, TUCSON, Arizona 85719.

University of Arizona Soils Club
410 Agricultural Sciences Building 38, University of Arizona, TUCSON, Arizona 85721.

Arkansas

Arkansas Geological Commission (AGC)
3815 West Roosevelt, LITTLE ROCK, Arkansas 72204.

California

Division of Mines and Geology (CDMG)
California Department of Conservation, 1516 Ninth Street, Fourth Floor, SACRAMENTO, California 95814.

Colorado

Colorado Geological Survey (CGS)
Department of Natural Resources, 1313 Sherman Street, DENVER, Colorado 80203.

Natural Hazards Information Center
University of Colorado, IBS No. 6, Campus Box 482, BOULDER, Colorado 80309.

Connecticut

Department of Environmental Protection (CGNHS)
Natural Resources Center and Geological and Natural History Survey, 165 Capitol Avenue, Room 555, HARTFORD, Connecticut 06106.

Delaware

Delaware Geological Survey (DGS)
University of Delaware, 101 Penny Hall, NEWARK, Delaware 19716.

Florida

Florida Geological Survey (FGS)
Department of Natural Resources, 903 West Tennessee Street, TALLAHASSEE, Florida 32304.

Georgia

Georgia Geologic Survey (GGS)
Department of Natural Resources, 19 Martin Luther King Jr Drive, SW, ATLANTA, Georgia 30334.

Idaho

Idaho Geological Survey (IGS)
University of Idaho, Morrill Hall, Room 332, MOSCOW, Idaho 83843.

Illinois

Illinois State Geological Survey (ISGS)
Natural Resources Building, 615 East Peabody Drive, CHAMPAIGN, Illinois 61820.

Illinois Department of Conservation
Lincoln Tower Plaza, 524 South Second Street, SPRINGFIELD, Illinois 62706.

Indiana

Geological Survey (IGS)
Department of Natural Resources, 611 North Walnut Grove, BLOOMINGTON, Indiana 47405.

Iowa

Iowa Geological Survey (IGS)
123 North Capitol Street, IOWA CITY, Iowa 52242.

Kansas

Kansas Geological Survey (KGS)
1930 Constant Avenue, Campus West, The University of Kansas, LAWRENCE, Kansas 66046-2598.

Kentucky

Kentucky Geological Survey (KGS)
University of Kentucky, 311 Breckinridge Hall, LEXINGTON, Kentucky 40506-0056.

Louisiana

Louisiana Geological Survey (LGS)
University Station, Box G, BATON ROUGE, Louisiana 70893.

Maine

Maine Geological Survey (MGS)
Department of Conservation, State House Station 22, AUGUSTA, Maine 04333.

Maryland

Maryland Geological Survey (MGS)
2300 St Paul Street, BALTIMORE, Maryland 21218.

Michigan

Geological Survey Division (MGSD)
Department of Natural Resources, Box 30028, LANSING, Michigan 48909.

Minnesota

Minnesota Geological Survey (MGS)
2642 University Avenue, ST PAUL, Minnesota 55114-1057.

Mississippi

Bureau of Geology (MBG)
Department of Natural Resources, PO Box 5348, JACKSON, Mississippi 39216.

Missouri

Division of Geology and Land Survey (MDGLS)
Department of Natural Resources, 111 Fairgrounds Road, PO Box 250, ROLLA, Missouri 65401.

Montana

Montana Bureau of Mines and Geology (MBMG)
Main Hall, Montana Tech, BUTTE, Montana 59701.

Nebraska

Conservation and Survey Division (NCSD)
The University of Nebraska, 113 Nebraska Hall, LINCOLN, Nebraska 68588-0517.

Nevada

Nevada Bureau of Mines and Geology (NBMG)
University of Nevada-Reno, RENO, Nevada 89557-0088.

New Hampshire

Department of Resources and Economic Development
117 James Hall, University of New Hampshire, DURHAM, New Hampshire 03824.

New Jersey

New Jersey Geological Survey (NJGS)
1474 Prospect Street, TRENTON, New Jersey 08625.

New Mexico

New Mexico Bureau of Mines and Mineral Resources (NMBM&MR)
Campus Station, SOCORRO, New Mexico 87801.

New York

Geological Survey Office (NYSGS)
New York State Education Department, Empire State Plaza, ALBANY, New York 12230.

New York State Department of Transportation (NYSDT)
Map Information Unit, State Campus, Bldg 4, ALBANY, New York 12232.

North Carolina

North Carolina Geological Survey (NCGS)
Department of Natural Resources and Land Resources, PO Box 27687, RALEIGH, North Carolina 27611.

North Dakota

North Dakota Geological Survey (NDGS)
University of North Dakota, GRAND FORKS, North Dakota 58201.

Ohio

Division of Geological Survey (ODGS)
Ohio Department of Natural Resources, Building B, Fountain Square, COLUMBUS, Ohio 43224.

Oklahoma

Oklahoma Geological Survey (OGS)
The University of Oklahoma, 830 Van Vleet Oval, Room 163, NORMAN, Oklahoma 73019.

Oregon

State Department of Minerals and Mineral Industries (ODGMI)
910 State Office Building, 1400 SW Fifth Avenue, PORTLAND, Oregon 97201-5528.

Pennsylvania

Pennsylvania Geological Survey (PBTGS)
Department of Environmental Resources, PO Box 2357, HARRISBURG, Pennsylvania 17129.

Department of General Services (for map orders)
State Book Store, PO Box 1365, HARRISBURG, Pennsylvania 17105.

South Carolina

South Carolina Geological Survey (SCGS)
Harbison Forest Road, COLUMBIA, South Carolina 29210.

South Dakota

South Dakota Geological Survey (SDGS)
University of South Dakota, Science Center, VERMILLION, South Dakota 57069.

Tennessee

Tennessee Department of Conservation (TDC)
701 Broadway, NASHVILLE, Tennessee 37219-5237.

Tennessee Valley Authority (TVA)
Mapping Services Branch, 200 Haney Building, CHATTANOOGA, Tennessee 37401.

Texas

Bureau of Economic Geology (BEG)
University Station, Box X, AUSTIN, Texas 78713-7508.

Utah

Utah Geological and Mineral Survey (UGMS)
606 Black Hawk Way, SALT LAKE CITY, Utah 84108-1280.

Vermont

Vermont Geological Survey (VGS)
Agency of Environmental Conservation, 103 South Main Street, WATERBURY, Vermont 05676.

Vermont Department of Libraries (map orders)
Geological Publications, c/o State Office Building Post Office, MONTPELIER, Vermont 05602.

Virginia

Virginia Division of Mineral Resources (VDMR)
Box 3667, CHARLOTTESVILLE, Virginia 22903.

Washington

Division of Geology and Earth Resources (WDGER)
Department of Natural Resources, MS: PY-12, OLYMPIA, Washington 98504.

West Virginia

West Virginia Geological and Economic Survey (WVGES)
PO Box 879, MORGANTOWN, West Virginia 26507-0879.

Wisconsin

Wisconsin Geological and Natural History Survey (WGNHS)
University of Wisconsin-Extension, 3817 Mineral Point Road, MADISON, Wisconsin 53705.

Wyoming

Geological Survey of Wyoming (GSW)
Box 3008, University Station, University of Wyoming, LARAMIE, Wyoming 82071.

Private mapping agencies—in alphabetical order

Alexandria Drafting Company
6440 General Green Way, ALEXANDRIA, Virginia 22312.

American Association of Petroleum Geologists (AAPG)
PO Box 979, 1444 S Boulder, TULSA, Oklahoma 74101.

American Automobile Association (AAA)
8111 Gatehouse Road, Room 335, FALLS CHURCH, Virginia 22047.

American Geographical Society (AGS)
29 West 39th Street, Suite 1501, NEW YORK, New York 10018.

American Map Corporation
46–35 54th Road, MASPATH, New York 11378.

Arrow Publishing Co. Inc.
1020 Turnpike Street, CANTON, Massachusetts 02021.

Audits and Surveys
1 Park Avenue, NEW YORK, New York 10016.

Champion Map Corporation
PO Box 5545, CHARLOTTE, North Carolina 28225.

George R. Cram Co. Inc.
301 South La Salle Street, PO Box 426, INDIANAPOLIS, Indiana 46206.

Danger Zones
Drawer 2606, WASHINGTON, DC 20013.

DeLorme Publishing Co
PO Box 298, FREEPORT, Maine 04032.

Denoyer-Geppert
5235 N Ravenswood Avenue, CHICAGO, Illinois 60640.

Dolph Map Co.
430 North Federal Highway, FORT LAUDERDALE, Florida 33301.

Earth Satellite Corporation
7222 47th Street, CHEVY CHASE, Maryland 20815.

ENMAP Corporation
PO Box 4430, BOULDER, Colorado 80306.

EXXON Touring Service
PO Box 10210, HOUSTON, Texas 77206.

Flashmaps Publications Inc.
Box 13, Chappaqua, NEW YORK, New York 10515.

David A. Fox Studios
203 Lantwyn Lane, NARBERTH, Pennsylvania 19072.

Geo-Graphics
519 SW 3rd Street, Suite 418, PORTLAND, Oregon 97204.

Geological Society of America (GSA)
3300 Penrose Place, PO Box 9140, BOULDER, Colorado 80301.

Gousha/Chek-Chart
2001 The Alameda, PO Box 6227, SAN JOSE, California 95126.

Hagstrom Company
450 West 33rd Street, NEW YORK, New York 10001.

Hammond Inc.
515 Valley Street, MAPLEWOOD, New Jersey 07040.

Hearne Brothers
First National Building, Twenty-fifth Floor, DETROIT, Michigan 48226.

Hubbard
1946 Raymond, NORTHBROOK, Illinois 60062.

Infomap
3300 Arapahoe, Suite 207, BOULDER, Colorado 80303.

International Aerial Mapping Company
8927 International Drive, SAN ANTONIO, Texas 78216.

International Mapping Unlimited (distribution of World Bank maps)
4343 39th Street, Northwest, WASHINGTON, DC 20016.

Kistler Graphics Inc.
PO Box 5467, 4000 Dahlia Street, DENVER, Colorado 80217-5467.

Macmillan Publishing Co.
866 Third Avenue, NEW YORK, New York 10022.

Middle East Information Company
160 Hawley Lane, TRUMBULL, Connecticut 06611.

National Geographic Society (NGS)
Department 80, WASHINGTON, DC 20036.

Nystrom
3333 Elston Avenue, CHICAGO, Illinois 60618.

Panoramic Studios
2243 W Allegeny Avenue, PHILADELPHIA, Pennsylvania 19132.

PennWell Books
PO Box 21288, TULSA, Oklahoma 74121.

Perspecto Map Co.
5702 George Street, RICHMOND, Illinois 60071.

Petroleum Information Corp.
4150 Westheimer Road, HOUSTON, Texas 77027.

Petrophysics Inc.
2141 West Governors Circle, Suite A, HOUSTON, Texas 77902.

Pierson Graphics Corp.
820 16 St No. 718, DENVER, Colorado 80202.

Potomac Appalachian Trail Club
1718 North Street NW, WASHINGTON, DC 20036.

Ptolemy Press
PO Box 243, GROVE CITY, Pennsylvania 16127.

Rand McNally & Co.
PO Box 7600, CHICAGO, Illinois 60680.

Rhode Island Publications Society
150 Benefit Street, PROVIDENCE, Rhode Island 02906.

Rockford Map Publishers Inc.
PO Box 6126, 4525 Forest View Avenue, ROCKFORD, Illinois 61125.

Ryder Geosystems
445 Union, Suite 304, DENVER, Colorado 80228.

Sanborn Map Company
659 Fifth Street, PELHAM, New York 10803.

Sidwell Company
28W240 North Avenue, CHICAGO, Illinois 60185.

Society of Exploration Geophysicists
PO Box 3098, TULSA, Oklahoma 74101.

Stone Mountain Map Studio
Box 1851, STONE MOUNTAIN, Georgia 30083.

Telecote Press Inc.
PO Box 188, GLENWOOD, New Mexico 88039.

Thomas Brothers
550 Jackson Street, SAN FRANCISCO, California 94133.

Weber State College
3750 Harrison Blvd, OGDEN, Utah 84408.

Western Economic Research Co.
13437 Ventura Blvd, SHERMAN OAKS, California 91423.

Western Geographics
PO Box 2204, CANON CITY, Colorado 81212.

Wilderness Press
2440 Bancroft Way, BERKELEY, California 94704.

Catalogue

Atlases and gazetteers

The atlas of the United States
New York, NY: Macmillan, 1986.
pp. 128.

Rand McNally Commercial Atlas and Marketing Guide
Chicago, IL: Rand McNally, annual.

Satellite photo-atlas of the United States
Ryder's standard geographic reference.
Denver, CO: Ryder Geosystems, 1983.
pp. 223.

National gazetteer of the United States
Reston, VA: USGS, 1983–.
Issued state by state as bound volumes (3 published).
Many states also available as bound alphabetical listings or on microfiche or computer tape.

General

Digital terrain map of the United States   1:7 500 000
I-1318
Reston, VA: USGS, 1981.

United States base map   1:7 000 000   7-A
Reston, VA: USGS, 1916.
In two colours.
Also available with contours (7-B) and with physical divisions
(7-C).

United States   1:6 000 000   6-A
Reston, VA: USGS, 1975.

Outline map of the United States   1:5 000 000   5-D
Reston, VA: USGS, 1940.
Black and white map showing state boundaries and names.

United States of America   1:5 000 000
Edinburgh: Bartholomew, 1986.

The United States   1:4 560 000
Washington, DC: NGS, 1986.

Portrait USA   1:4 560 000
Washington, DC: NGS, 1976.
Landsat simulated natural colour mosaic.

United States   1:3 168 000   3-A
Reston, VA: USGS, 1965.
State boundaries and names, state capitals and principal cities.
Available with historical boundaries and public land surveys as
Map 3-B.

United States   1:2 500 000   2-A
Reston, VA: USGS, 1972.
2 sheets, both published.
State and county boundaries and names, state capitals and county
seats.
Available without land tint as Map 2-B.

United States of America, Eastern   1:2 500 000
Edinburgh: Bartholomew, 1984.
Inset maps of major cities.
Placenames index on the reverse.

United States of America, Western   1:2 500 000
Edinburgh: Bartholomew, 1984.
Inset maps of major cities.
Placenames index on the reverse.

United States sectional maps   1:2 000 000
Reston, VA: USGS.
20 sheets, all published.
Digital data derived from these maps is available.

Topographic

National topographic series   1:250 000
Reston, VA: USGS.
607 sheets, all published. ■
Digital data also available.

Intermediate scale topographic quadrangle series   1:100 000
Reston, VA: USGS, 1975–.
Digital data also available.

Intermediate scale topographic county/quadrangle series   1:50 000
Reston, VA: USGS, 1975–.

USGS/DMA 15 minute series   1:50 000
Reston, VA: USGS/Washington, DC: DMA, 1982.

7.5 minute topographic quadrangle series   1:24 000/25 000
Reston, VA: USGS.
ca 75 000 sheets, *ca* 85% complete.

Geological and geophysical

*Preliminary overview map of volcanic hazards in the 48
States*   1:7 500 000   MF-0786
Reston, VA: USGS, 1976.

*Preliminary map showing Quaternary deposits and their dating
potential in the conterminous United States*   1:7 500 000   MF-1052
Reston, VA: USGS, 1979.

*Digital colored residual and regional Bouguer gravity maps of the
United States*   1:7 500 000   GP-0953A
Reston, VA: USGS, 1982.
2 sheets, both published.

*Magnetic total intensity in United States, Epoch
1980*   1:5 000 000   I-1370
Reston, VA: USGS, 1981.

Oil and gas fields of the United States   1:3 530 000
Tulsa, OK: PennWell, 1985.

Geologic map of the United States   1:2 500 000
Reston, VA: USGS, 1974.
3 sheets, all published.

Basement rock map of the United States   1:2 500 000
Reston, VA: USGS, 1968.
2 sheets, both published.

*Composite magnetic anomaly map of the United States, part A:
conterminous United States*   1:2 500 000   GP-0954A
Reston, VA: USGS, 1982.
2 sheets, both published + 59 pp. text.

*Black and white composite magnetic anomaly map of the United
States Part A. Conterminous United States*   1:2 500 000   GP-
0906A
Reston, VA: USGS, 1983.
2 sheets, both published.

Gravity anomaly map of the United States   1:2 500 000
Tulsa, OK: Society of Exploration Geophysicists.
2 sheets, both published.

Geothermal gradient map of the United States   1:2 500 000
Boulder, CO: NGDC, 1982.
2 sheets, both published.

Geothermal energy in the Western United States   1:2 500 000
Boulder, CO: NGDC, 1979.

Geological highway maps   1:1 875 000
Tulsa, OK: AAPG, 1966–.
12 sheets, all published.

Environmental

America's federal lands   1:15 889 000
Washington, DC: NGS, 1982.
Main map: federal property.
Inset map: natural resources.

A national floodplain map   1:8 500 000
Boulder, CO: University of Colorado Natural Hazards
Information Center, 1985.

Landslide overview map of the conterminous United States
1:7 500 000
Reston, VA: USGS, 1983.

*Land resource regions and major land resource areas of the United
States*   1:7 500 000
Washington, DC: SCS, 1978.
Accompanies *Agricultural Handbook* 296.

*Hydrologic unit map of the United States/Hydrologic unit map of
Alaska and Hawaii*   1:5 000 000
Washington, DC: FWS, 1982.

Ecoregions and land-surface forms of the United States   1:5 000 000
Washington, DC: FWS, 1982.
On reverse: ecoregions and land-surface forms of Alaska and
Hawaii.

110° W

40° N

Cape Flattery

Victoria

Concrete · Okanogan · Sandpoint · Kalispell · Cut Bank · Shelby · Havre · Glasgow · Wolf Point · Williston · Minot · Devils Lake · Thief River Falls

Copalis Beach

Seattle · Wenatchee · Ritzville · Spokane · Wallace · Choteau · Great Falls · Lewistown · Jordan · Glendive · Watford City · McClusky · New Rockford · Grand Forks

Cape Disappointment

Hoquiam · Yakima · Walla Walla · Pullman · Hamilton · Butte · White Sulphur Springs · Roundup · Forsyth · Miles City · Dickinson · Bismarck · Jamestown · Fargo

Vancouver

The Dalles · Pendleton · Grangeville · Elk City · Dillon · Bozeman · Billings · Hardin · Ekalaka · Lemmon · McIntosh · Aberdeen · Milbank

Salem

Bend · Canyon City · Baker · Challis · Dubois · Ashton · Cody · Sheridan · Gillette · Rapid City · Pierre · Huron · Watertown

Coos Bay

Roseburg · Crescent · Burns · Boise · Hailey · Idaho Falls · Driggs · Thermopolis · Arminto · New Castle · Hot Springs · Martin · Mitchell · Sioux Falls

Medford

Klamath Falls · Adel · Jordan Valley · Twin Falls · Pocatello · Preston · Lander · Casper · Torrington · Alliance · Valentine · O'Neill · Sioux City

Crescent City

Weed · Alturas · Vya · McDermitt · Wells · Brigham City · Ogden · Rock Springs · Rawlins · Cheyenne · Scottsbluff · North Platte · Broken Bow · Fremont

Eureka

Redding · Susanville · Lovelock · Winnemucca · Elko · Tooele · Salt Lake City · Vernal · Craig · Greeley · Sterling · McCook · Grand Island · Lincoln

Ukiah

Chico · Reno · Millett · Ely · Delta · Price · Grand Junction · Leadville · Denver · Limon · Goodland · Beloit · Manhattan

Santa Rosa

Sacramento · Walker Lake · Tonopah · Lund · Richfield · Salina · Moab · Montrose · Pueblo · Lamar · Scott City · Great Bend · Hutchinson

San Francisco

San Jose · Mariposa · Goldfield · Caliente · Cedar City · Escalante · Cortez · Durango · Trinidad · La Junta · Dodge City · Pratt · Wichita

Monterey

Fresno · Death Valley · Las Vegas · Grand Canyon · Marble Canyon · Shiprock · Aztec · Raton · Dalhart · Perryton · Woodward · Enid

San Luis Obispo

Bakersfield · Trona · Kingman · Williams · Flagstaff · Gallup · Albuquerque · Santa Fe · Tucumcari · Amarillo · Clinton · Oklahoma City

Santa Maria

Los Angeles · San Bernardino · Needles · Prescott · Holbrook · St Johns · Socorro · Ft Sumner · Clovis · Plainview · Lawton · Ardmore

Long Beach · Santa Ana · Salton Sea · Phoenix · Mesa · Clifton · Tularosa · Roswell · Brownfield · Lubbock · Wichita Falls · Sherman

San Clemente Is

San Diego · El Centro · Ajo · Tucson · Silver City · Las Cruces · Carlsbad · Hobbs · Big Spring · Abilene · Dallas

Lukeville · Nogales · Douglas · El Paso · Van Horn · Pecos · San Angelo · Brownwood · Waco

Marfa · Ft Stockton · Sonora · Llano · Austin

Presidio · Emory Peak · Del Rio · San Antonio · Seguin

Eagle Pass · Crystal City · Beeville

Laredo · Corpus Christi

McAllen

Brownsville

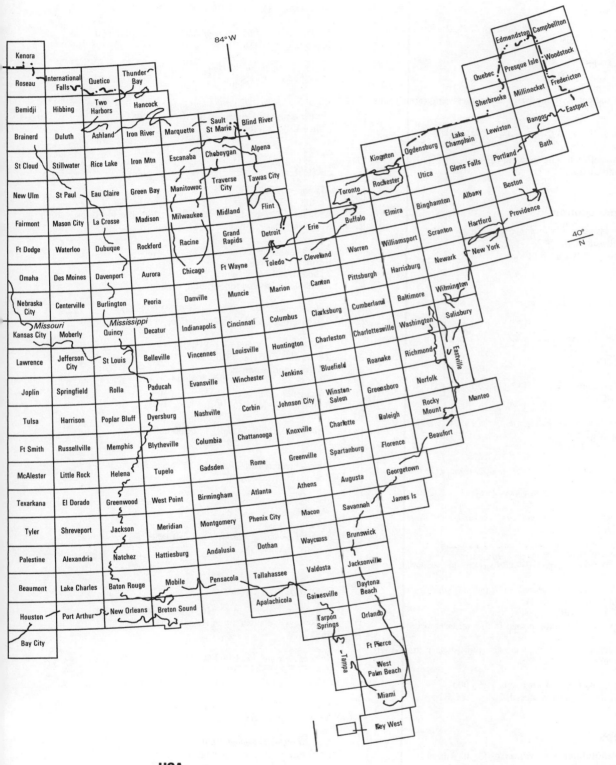

84° W

40° N

Kenora
Roseau
International Falls
Quetico
Thunder Bay
Bemidji
Hibbing
Two Harbors
Hancock
Brainerd
Duluth
Ashland
Iron River
Marquette
Sault St Marie
Blind River
St Cloud
Stillwater
Rice Lake
Iron Mtn
Escanaba
Cheboygan
Alpena
New Ulm
St Paul
Eau Claire
Green Bay
Manitowoc
Traverse City
Tawas City
Fairmont
Mason City
La Crosse
Madison
Milwaukee
Midland
Flint
Ft Dodge
Waterloo
Dubuque
Rockford
Racine
Grand Rapids
Detroit
Erie
Omaha
Des Moines
Davenport
Aurora
Chicago
Ft Wayne
Toledo
Cleveland
Nebraska City
Centerville
Burlington
Peoria
Danville
Muncie
Marion
Canton
Missouri
Mississippi
Kansas City
Moberly
Quincy
Decatur
Indianapolis
Cincinnati
Columbus
Lawrence
Jefferson City
St Louis
Belleville
Vincennes
Louisville
Huntington
Charleston
Joplin
Springfield
Rolla
Paducah
Evansville
Winchester
Jenkins
Bluefield
Tulsa
Harrison
Poplar Bluff
Dyersburg
Nashville
Corbin
Johnson City
Winston-Salem
Ft Smith
Russellville
Memphis
Blytheville
Columbia
Chattanooga
Knoxville
Charlotte
McAlester
Little Rock
Helena
Tupelo
Gadsden
Rome
Greenville
Spartanburg
Texarkana
El Dorado
Greenwood
West Point
Birmingham
Atlanta
Athens
Augusta
Tyler
Shreveport
Jackson
Meridian
Montgomery
Phenix City
Macon
Savannah
Palestine
Alexandria
Natchez
Hattiesburg
Andalusia
Dothan
Waycross
Brunswick
Beaumont
Lake Charles
Baton Rouge
Mobile
Pensacola
Tallahassee
Valdosta
Jacksonville
Houston
Port Arthur
New Orleans
Breton Sound
Apalachicola
Gainesville
Daytona Beach
Bay City
Tarpon Springs
Orlando
Tampa
Ft Pierce
West Palm Beach
Miami
Key West

Toronto
Kingston
Ogdensburg
Lake Champlain
Lewiston
Bangor
Eastport
Rochester
Utica
Glens Falls
Portland
Bath
Buffalo
Elmira
Binghamton
Albany
Boston
Warren
Williamsport
Scranton
Hartford
Providence
Pittsburgh
Harrisburg
Newark
New York
Clarksburg
Cumberland
Baltimore
Wilmington
Charlottesville
Washington
Salisbury
Roanoke
Richmond
Eastville
Greensboro
Norfolk
Winston-Salem
Rocky Mount
Manteo
Raleigh
Beaufort
Florence
Georgetown
James Is

Edmundston
Campbellton
Quebec
Presque Isle
Woodstock
Sherbrooke
Millinocket
Fredericton
Eastport

USA
1:250 000 topographic
1:250 000 land cover and associated maps

Solar energy in the United States and southern Canada 1 : 5 000 000
Boulder, CO: ENMAP, 1980.

*Potential natural vegetation of the conterminous United
States* 1 : 3 168 000
New York, NY: AGS, 1975.

Administrative

United States county outline map: January 1, 1980 1 : 5 000 000
Washington, DC: USBC.

United States 1 : 2 500 000 2-A
Reston, VA: USGS, 1972.
2 sheets, both published.
Shows state and county boundaries and names, state capitals and
county seats.

1980 state and county subdivision maps Scales vary
Washington, DC: USBC.
Series of maps of each state showing census county divisions,
townships and districts.

Human and economic

Atlas of Demographics: US by County
Boulder, CO: Infomap, 1982.
16 colour census maps + 38 pp. tabular data.

The United States Energy Atlas 2nd edn D.J. Cuff and W.J.
Young
New York, NY: Macmillan, 1986.
pp. 415.

*Distribution of older Americans in 1970 related to year of maximum
population* GE-70, No. 2
Washington, DC: USBC.

Employment in manufacturing, 1982 GE-70, No. 5
Washington, DC: USBC.

*Primary home heating fuel by counties of the United States, 1950,
1960, 1970* GE-70, No. 3
Washington, DC: USBC.

*1985 retail map of the United States: a graphic profile of the nation's
marketplace*
New York, NY: Audits and Surveys, 1985.

*Population distribution, urban and rural, in the United States,
1980* GE-70, No. 4
Washington, DC: USBC.
'Night time view' map.
A similar 1970 census map (GE-70, No. 1) is also available.

*Population distribution, urban and rural, in the United States,
1970* 1 : 5 000 000 GE-50, No. 45
Washington, DC: USBC, 1973.
A 1980 population distribution map in this series is in
preparation.
Many other census-related maps are also published in this GE-50
series: most relate to 1970 census.

Nuclear targets and high risk areas 1 : 3 168 000
Washington, DC: Danger Zones, 1982.

Town maps

There is a plentiful range of town maps for all American cities.
For publishers, see text.

State maps

Alabama

Uncontrolled ERTS-1 mosaic of the State of Alabama IM 2
Tuscaloosa, AL: GSA, 1974.

State of Alabama 1 : 500 000
Reston, VA: USGS, 1966.
Available in base or topographic versions.

Minerals map of Alabama SM 193
Tuscaloosa, AL: GSA, 1984.

Energy resources of Alabama SM 196
Tuscaloosa, AL: GSA, 1983.

Seismicity map of the State of Alabama 1 : 1 000 000
Reston, VA: USGS, 1979.

Oil and gas fields in Alabama 1 : 500 000
Tuscaloosa, AL: GSA.

State of Alabama. Land use and land cover 1 : 500 000
University, AL: University of Alabama, 1985.
2 sheets, both published.

Alaska

Alaska 1 : 2 500 000
Reston, VA: USGS.
Available in base, shaded relief or version showing Conservation
System areas.

Alaska 1 : 1 584 000
Reston, VA: USGS, 1973.
2 sheets, both published.
Available in base or topographic version.

Alaska region Forest Service map 1 : 2 000 000
Washington, DC: US Forest Service, 1981.

Geologic map of Alaska 1 : 2 500 000
Reston, VA: USGS, 1980.
2 sheets, both published.

Geothermal resources of Alaska 1 : 2 500 000
Boulder, CO: NGDC, 1984.

Permafrost map of Alaska 1 : 2 500 000 I-445
Reston, VA: USGS, 1965.

Bouguer gravity map of Alaska 1 : 2 500 000 GP-913
Reston, VA: USGS, 1977.

Oil and gas basins map of Alaska 1 : 2 000 000
Anchorage, AK: DGGS, 1983.

Surficial geology of Alaska 1 : 1 584 000 I-357
Reston, VA: USGS, 1964.

Arizona

The Arizona Atlas M.E. Hecht and R.W. Reeves
Tucson, AZ: University of Arizona Press, 1981.
pp. 174.

State of Arizona 1 : 1 000 000
Reston, VA: USGS, 1981.
Available in base, topographic or relief version.

NASA ERTS 1 Satellite image map. Arizona 1972–73 1 : 500 000
Reston, VA: USGS, 1976.

Arizona highway geologic map 1 : 1 000 000
Tucson, AZ: Arizona Geological Society, 1967.

Geological map of Arizona 1 : 500 000
Reston, VA: USGS, 1969.

Geothermal resources of Arizona 1 : 500 000
Boulder, CO: NGDC, 1982.

Arizona general soil map 1 : 1 000 000
Tucson, AZ: University of Arizona Soils Club, 1983.

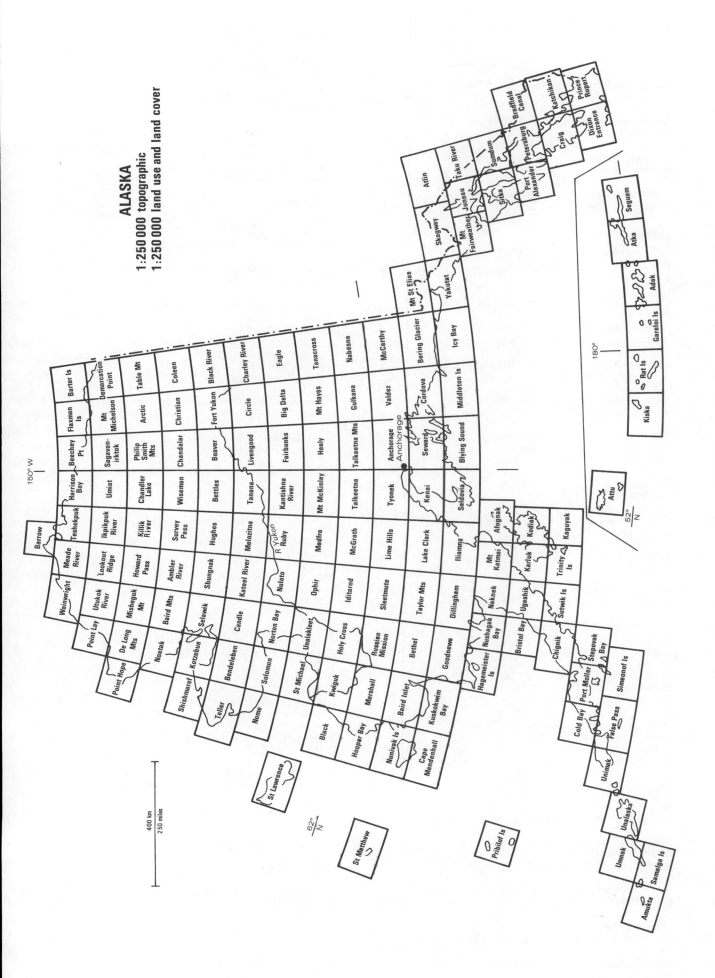

ALASKA
1:250 000 topographic
1:250 000 land use and land cover

Arkansas

State of Arkansas 1:500 000
Reston, VA: USGS, 1967.
Available in base, topographic or relief version.

Bouguer gravity map of Arkansas 1:500 000 GP-944
Reston, VA: USGS, 1981.

Geologic map of Arkansas 1:500 000
Little Rock, AK: AGC, 1976.

Arkansas seismicity map 1:1 000 000 MF-1154
Reston, VA: USGS, 1979.

Arkansas population distribution 1:1 000 000
Stone Mountain, GA: Stone Mountain Map Studio, 1984.

California

Atlas of California M.W. Donley *et al.*
Culver City, CA: Pacific Book Center, 1979.
pp. 191.

State of California 1:1 000 000
Reston, VA: USGS, 1982.
2 sheets, both published.
Available in base, topographic or relief version.

Geologic map of California 1:2 500 000 I-512
Reston, VA: USGS, 1966.

Geomorphic map of California 1:2 000 000
Sacramento, CA: CDMG, 1938.

*Preliminary mineral resource assessment map of
California* 1:1 000 000 MR-0088
Reston, VA: USGS, 1985.

Geologic map of California 1:750 000
Sacramento, CA: CDMG, 1977.

Gravity map of California and its continental margin 1:750 000
Sacramento, CA: CDMG, 1980.

Fault map of California 1:750 000
Sacramento, CA: CDMG, 1975.

Geothermal resources of California 1:750 000
Boulder, CO: NGDC/Sacramento: CDMG, 1981.

California oil and gas 1:500 000
Tulsa, OK: PennWell, 1983.
2 sheets, both published.

Geologic atlas of California 1:250 000
Sacramento, CA: CDMG, 1958–69.
27 sheets + legend, 14 available.

Bouguer gravity atlas of California 1:250 000
Sacramento, CA: CDMG, 1971–.
27 sheets, 23 published.
With explanatory notes.

Colorado

Atlas of Colorado K.A. Erickson and A.W. Smith
Boulder, CO: Colorado Associated University Press, 1985.

State of Colorado 1:500 000
Reston, VA: USGS, 1980.
Available in base, topographic or relief version.

State of Colorado Landsat satellite image mosaic 1:500 000
Fort Collins, CO: Colorado State University, 1981.

Colorado geologic highway map 1:1 000 000
Canon City, CO: Western Geographics, 1985.

Colorado seismicity map 1:1 000 000
Reston, VA: USGS, 1984.

Geologic map of Colorado 1:500 000
Reston, VA: USGS, 1979.

Geothermal resources of Colorado 1:
Boulder, CO: NGDC, 1980.

Colorado oil and gas fields 1:500 0
Denver, CO: CGS, 1983.

Coal resources and development map of Coloraao 1:
Denver, CO: CGS, 1978.

Bouguer gravity map of Colorado 1:500 000
Reston, VA: USGS, 1974.

Connecticut

State of Connecticut 1:125 000
Reston, VA: USGS, 1974.
Available in base or topographic version.

Natural drainage basins in Connecticut 1:125 000
Hartford, CT: CGNHS, 1981.

Connecticut seismicity map 1:1 000 000
Reston, VA: USGS, 1981.
Includes Rhode Island.

Bedrock geological map of Connecticut 1:125 000 J. Rodgers
Hartford, CT: CGNHS, 1985.
2 sheets, both published.

Aeromagnetic map of Connecticut 1:125 000
Reston, VA: USGS, 1975.

Community water systems in Connecticut 1:125 000
Hartford, CT: CGNHS, 1987.

Delaware

Maryland and Delaware 1:500 000
Reston, VA: USGS, 1973.
Available in base or topographic version.

Generalized geologic map of Delaware 1:576 000
Newark, DE: DGS, 1976.

Seismicity map of Delaware 1:1 000 000
Reston, VA: USGS, 1981.
Includes Maryland.

Florida

Atlas of Florida Edited by E.A. Fernald
Tallahassee, FL: Florida State University Foundation, 1981.
pp. 276.

State of Florida 1:500 000
Reston, VA: USGS, 1967.
Available in base or topographic version.

NASA ERTS 1 satellite image mosaic. Florida 1973 1:500 000
Reston, VA: USGS, 1975.

Physiographic divisions of Florida 1:500 000
Gainesville, FL: IFAS, University of Florida, 1982.

Geologic map of Florida 1:500 000
Gainesville, FL: IFAS, University of Florida, 1982.

Environmental geology series 1:250 000
Tallahassee, FL: FGS, 1978–.
13 sheets published.

Wetlands in Florida 1:2 000 000
Reston, VA: USGS.

Georgia

The Atlas of Georgia
Athens, GA: The University of Georgia, 1985.
pp. 288.

State of Georgia   1:500 000
Reston, VA: USGS, 1970.
Available in base or topographic version.

*NASA LANDSAT 1 satellite image mosaic. State of Georgia
1973–74*   1:500 000
Reston, VA: USGS, 1976.

Physiographic map of Georgia   1:2 000 000
Atlanta, GA: GGS, 1976.

Aeromagnetic map of Georgia   1:1 000 000
Reston, VA: USGS, 1981.

Seismicity map of Georgia   1:1 000 000
Reston, VA: USGS, 1979.

Geologic map of Georgia   1:500 000   GM-7
Atlanta, GA: GGS, 1976.

Slope map of Georgia   1:500 000
Reston, VA: USGS, 1976.

Hawaii

See under Pacific Ocean.

Idaho

The compact atlas of Idaho   Edited by A.A. DeLucia
Moscow, ID: Center for Business Development and Research/
Cart-o-Graphics, 1983.
pp. 105.

State of Idaho   1:500 000
Reston, VA: USGS, 1976.
Available in base, topographic or relief versions.

Geologic map of Idaho   1:500 000
Moscow, ID: IGS, 1978.

Complete Bouguer gravity anomaly map of Idaho
1:500 000   MF-1773
Reston, VA: USGS, 1985.

Aeromagnetic map of Idaho   1:1 000 000
Reston, VA: USGS, 1979.

Energy resources of Idaho   1:1 000 000   Map 3
Moscow, ID: IGS, 1980.

Oil and gas exploration in Idaho
Moscow, ID: IGS, 1982 (with 1983 update).

Geothermal resources of Idaho   1:500 000
Boulder, CO: NGDC, 1980.

Illinois

State of Illinois   1:500 000
Reston, VA: USGS, 1972.
Available in base or topographic version.

Satellite image map of Illinois   1:500 000
Champaign, IL: ISGS, 1986.
Colour map constructed from Landsat TM imagery.

Landforms of Illinois   1:1 000 000
Champaign, IL: ISGS, 1980.

Seismicity map of Illinois   1:1 000 000
Reston, VA: USGS, 1979.

Geologic map of Illinois   1:500 000
Champaign, IL: ISGS, 1967.

Quaternary deposits of Illinois   1:500 000
Champaign, IL: ISGS, 1979.

Coal resources of Illinois   1:500 000
Champaign, IL: ISGS, 1984.
5 sheets.

The forests of Illinois   1:750 000
Springfield, IL: Department of Conservation, 1982.

Satellite image map of Northeastern Illinois   1:200 000
Champaign, IL: ISGS, 1986.
Covers Chicago area.
On reverse: land use and land cover map of Illinois.

Indiana

State of Indiana   1:500 000
Reston, VA: USGS, 1973.
Available in base or topographic version.

Geologic map of Indiana   1:1 000 000
Bloomington, IN: IGS, 1956.

Glacial geology of Indiana   1:1 000 000
Bloomington, IN: IGS, 1958.

Bedrock geology map of Indiana   1:500 000
Bloomington, IN: IGS, forthcoming.

Quaternary geology map of Indiana   1:500 000
Bloomington IN: IGS, forthcoming.

Iowa

State of Iowa   1:500 000
Reston, VA: USGS, 1984.
Available in base or topographic version.

Seismicity map of Iowa   1:1 000 000
Reston, VA: USGS, 1981.

Geological map of Iowa   1:500 000
Iowa City, IA: IGS, 1969.

Aeromagnetic map of Iowa   1:500 000
Reston, VA: USGS, 1977.

Kansas

State of Kansas   1:500 000
Reston, VA: USGS, 1984.
Available in base or topographic version.

Geologic map of Kansas   1:500 000
Lawrence, KS: KGS, 1964.

Principal structural features of Kansas   1:500 000
Tulsa, OK: PennWell, 1984.

Geothermal resources of Kansas   1:500 000
Boulder, CO: NGDC, 1982.

Oil and gas fields in Kansas   1:500 000
Lawrence, KS: KGS, 1983.

Aeromagnetic map of Kansas   1:500 000
Lawrence, KS: KGS, 1981.

Kentucky

Atlas of Kentucky   Edited by P.P. Karan and C. Mather
Lexington, KY: University Press of Kentucky, 1977.
pp. 188.

State of Kentucky   1:500 000
Reston, VA: USGS, 1973.
Available in base, topographic or relief version.

Satellite image map of Kentucky   1:1 000 000
Lexington, KY: KGS, 1982.

Generalized geologic map of Kentucky   1:1 000 000
Lexington, KY: KGS, 1979.

Aeromagnetic map of Tennessee and Kentucky   1:1 000 000
Lexington, KY: KGS, 1984.

Seismicity map of the State of Kentucky 1:1 000 000 MF-1144
Reston, VA: USGS, 1979.

Simple Bouguer gravity map of Kentucky 1:500 000
Reston, VA: USGS, 1963.

Geologic map of Kentucky 1:250 000
Reston, VA: USGS, 1981.
4 sheets, all published.

Residual total intensity aeromagnetic map of Kentucky 1:250 000
Lexington, KY: KGS, 1978–80.
3 sheets, all published.

Louisiana

State of Louisiana 1:500 000
Reston, VA: USGS, 1968.
Available in base or topographic version.

State of Louisiana. A computer-enhanced Landsat image mosaic 1:500 000
Tulsa, OK: PennWell, 1985.

Geologic map of Louisiana 1:500 000 J.I. Snead and R.R. McCulloh
Baton Rouge, LA: LGS, 1984.

Principal structural features of Louisiana 1:500 000
Tulsa, OK: PennWell, 1984.

Oil and gas map of Louisiana 1:500 000
Houston, TX: Petroleum Information Corporation, 1983.

Oil and gas map of Louisiana/Offshore Louisiana oil and gas map 1:380 160
Baton Rouge, LA: LGS, 1981.

Maine

Maine atlas and gazetteer 9th edn
Freeport, ME: Delorme Publishing Co., 1984.

State of Maine 1:500 000
Reston, VA: USGS, 1973.
Available in base, topographic or relief version.

Bedrock geologic map of Maine 1:500 000
Augusta, ME: MGS, 1984.

Surficial geologic map of Maine 1:500 000
Augusta, ME: MGS, 1984.
2 sheets, both published.

Maryland

States of Maryland and Delaware 1:500 000
Reston, VA: USGS, 1973.
Available in base or topographic version.

Geologic map of Maryland 1:250 000
Baltimore, MD: MGS, 1968.

Aeromagnetic map of Maryland 1:250 000 GP-923
Reston, VA: USGS, 1978.

Massachusetts

Massachusetts, Rhode Island and Connecticut 1:500 000
Reston, VA: USGS, 1948.
Available in base, topographic or relief version.

Bedrock geologic map of Massachusetts 1:250 000
Reston, VA: USGS, 1983.
3 sheets, all published.

Michigan

Atlas of Michigan Edited by L.M. Sommers
East Lansing, MI: Michigan State University Press, 1977.
pp. 242.

State of Michigan 1:500 000
Reston, VA: USGS, 1970.
2 sheets.
Available in base or topographic version.

Quaternary geology of Michigan 1:500 000
Lansing, MI: MGSD, 1982.
2 sheets, both published.

Minnesota

Atlas of Minnesota: resources and settlement 3rd edn
J.R. Borchert and N.C. Gustafson
Minneapolis, MN: University of Minnesota/Minnesota State Planning Agency, 1980.
pp. 309.

State of Minnesota 1:500 000
Reston, VA: USGS, 1965.
Available in base or topographic version.

Geologic map of Minnesota 1:1 000 000
St Paul, MN: MGS, 1970.

Geological map of Minnesota: bedrock topography 1:1 000 000 S-15
St Paul, MN: MGS, 1982.

Geologic map of Minnesota: Quaternary geology 1:500 000 S-1
St Paul, MN: MGS, 1982.

Bedrock geologic map of Minnesota 1:250 000
St Paul, MN: MGS, 1966–.
11 sheets, 7 published.

Aeromagnetic map of Minnesota 1:250 000
St Paul, MN: MGS, 1983–.
11 sheets.

Mississippi

Elevation map of Mississippi 1:1 000 000
Jackson, MS: MBG, 1975.

State of Mississippi 1:500 000
Reston, VA: USGS, 1971.
Available in base or topographic version.

Geologic map of Mississippi 1:500 000
Jackson, MS: MBG, 1969.

Economic minerals map of Mississippi 1:500 000
Jackson, MS: MBG, 1983.

Missouri

State of Missouri 1:500 000
Reston, VA: USGS, 1973.
Available in base, topographic or relief version.

Surficial materials map of Missouri 1:1 000 000
Rolla, MO: MDGLS, 1982.

Magnetic anomaly map of Missouri 1:1 000 000
Rolla, MO: MDGLS, 1984.
+ additional map giving interpretation of Precambrian structure.

Geologic map of Missouri 1:500 000
Rolla, MO: MDGLS, 1979.

Mineral resources map of Missouri 1:500 000
Rolla, MO: MDGLS, 1987.

Energy resources and facilities map of Missouri 1:500 000
Rolla, MO: MDGLS, 1983.

Montana

State of Montana 1:500 000
Reston, VA: USGS, 1983.
2 sheets.
Available in base, topographic or relief version.

Satellite photomap of the northern Rocky Mountains and Great Plains, Montana, Wyoming and adjacent regions
Butte, MT: MBMG, 1983.

Geologic map of Montana 1:500 000
Reston, VA: 1955.
2 sheets, both published.

Aeromagnetic map of Montana 1:500 000
Reston, VA: USGS, 1982.

Geothermal resources of Montana 1:1 000 000
Boulder, CO: NGDC, 1981.

Nebraska

Satellite image of Nebraska GRM-9
Lincoln, NE: NCSD, 1977.

State of Nebraska 1:500 000
Reston, VA: USGS, 1972.
Available in base, topographic or relief version.

Geologic bedrock map of Nebraska 1:1 000 000 GMC-1
Lincoln, NE: NCSD, 1969.

Bouguer gravity map of Nebraska 1:1 000 000 GP-1
Lincoln, NE: NCSD, 1982.

Mineral resources of Nebraska 1:1 000 000 RM-6
Lincoln, NE: NCSD, 1980.

Geothermal resources of Nebraska 1:500 000
Boulder, CO: NGDC, 1982.

General land use in Nebraska 1:1 000 000 LUM-1
Lincoln, NE: NCSD, 1973.

Vegetation map of Nebraska 1:1 000 000 GRM-8
Lincoln, NE: NCSD, 1975.

Generalized soil map of Nebraska 1:500 000 SM-1
Lincoln, NE: NCSD.

Nevada

Synthetic natural color Landsat color photomosaic of Nevada 1:1 000 000 I-1219
Reston, VA: USGS, 1981.

Satellite photomap of Nevada 1:1 000 000 M51
Reno, NV: NBMG, 1976.

Topographic map of Nevada 1:1 000 000 M43
Reno, NV: NBMG, 1972.

Shaded relief map of Nevada 1:1 000 000 M71
Reno, NV: NBMG, 1981.

State of Nevada 1:500 000
Reston, VA: USGS, 1984.
Available in base, topographic or relief version.

Rockhound's map of Nevada 1:1 000 000 SP1
Reno, NV: NBMG, 1975.

Million-scale geologic map of Nevada 1:1 000 000 M57
Reno, NV: NBMG, 1977.

Cenozoic rocks of Nevada 1:1 000 000 M52
Reno, NV: NBMG, 1976.
4 sheets + text.

Aeromagnetic map of Nevada 1:1 000 000 GP-992
Reston, VA: USGS, 1978.

Geologic map of Nevada 1:500 000 MF-930
Reston, VA: USGS, 1978.
2 sheets, both published.
Discussion paper (SP4) to accompany map is published by NBMG, Reno.

Geothermal resources of Nevada 1:500 000
Boulder, CO: NGDC, 1984.

New Hampshire

New Hampshire atlas and gazetteer
Freeport, MA: DeLorme Publishing Co., 1983.
pp. 88.

States of New Hampshire and Vermont 1:500 000
Reston, VA: USGS, 1972.
Available in base, topographic and relief version.

Bedrock Geology of New Hampshire 1:250 000
Reston, VA: USGS, 1955.

New Jersey

State of New Jersey 1:500 000
Reston, VA: USGS, 1974.
Available in base, topographic or relief version.

New Jersey 1972. NASA ERTS 1 Satellite image mosaic 1:500 000
Reston, VA: USGS, 1973.

Geologic map of New Jersey 1:1 000 000
Trenton, NJ: NJGS, 1984.

Magnetic declination map of New Jersey for 1980 Epoch 1:500 000
Trenton, NJ: NJGS, 1980.

Geologic map of New Jersey 1:250 000
Trenton, NJ: NJGS, 1950.

New Mexico

New Mexico in maps J.L. Williams and P.E. McAlister
Albuquerque, NM: University of New Mexico Press, 1981.
pp. 177.

Base map of New Mexico 1:1 000 000 RM-11
Socorro, NM: NMBM, 1983.

State of New Mexico 1:500 000
Reston, VA: USGS, 1985.
Available in base, topographic or relief version.

Satellite photomap of New Mexico 1:1 000 000 RM-12
Socorro, NM: NMBM, 1981.

Geologic highway map of New Mexico 1:1 000 000
Socorro, NM: New Mexico Geological Society.

Geologic map of New Mexico 1:500 000
Socorro, NM: NMBM, 1965.
2 sheets, both published.

Surficial geology of New Mexico 1:500 000 GM 40-43
Socorro, NM: NMBM, 1977–78.
4 maps.

Seismicity map of the State of New Mexico 1:1 000 000 MF-1660
Reston, VA: USGS, 1983.

Hydrothermal anomalies in New Mexico 1:1 000 000 RM-1
Socorro, NM: NMBM, 1979.

Geothermal resources of New Mexico 1:500 000
Boulder, CO: NGDC, 1980.

Energy resources of New Mexico 1:500 000 I-1327
Reston, VA: USGS/Socorro, NM: NMBM, 1981.

Vegetation and land use in New Mexico 1:1 000 000 RM-8
Socorro, NM: NMBM, 1977.

New York

New York state: a socio-economic atlas   R. Beach and
M. Fairweather
Plattsburg, NY: State University of New York, 1983.
pp. 141.

State of New York   1:500 000
Reston, VA: USGS, 1974.
Available in base, topographic or relief version.

New York State Map   1:250 000
Albany: NYSDT, 1983.
4 sheets, all published.

Landforms and bedrock geology of New York State   1:1 000 000
Albany, NY: NYSGS, 1966.

*Seismic activity and geologic structure in New York and adjacent
areas*   1:500 000   Map & Chart 27
Albany, NY: NYSGS, 1977.

Geologic map of New York   1:250 000   Map & Chart 15
Albany, NY: NYSGS, 1970.
5 sheets + legend, all published.

Surficial geologic map of New York   1:250 000   Map & Chart 40
Albany, NY: NYSGS, 1987–.
5 sheets, 3 published.

Simple Bouguer gravity anomaly map of New York   1:250 000
Map & Chart 17
Albany, NY: NYSGS, 1971–73.
5 sheets, all published.

North Carolina

North Carolina Atlas: portrait of a changing southern state   Edited
by J.W. Clay, D.M. Orr and A.W. Stuart
Chapel Hill, NC: University of North Carolina Press, 1975.
pp. 331.

State of North Carolina   1:500 000
Reston, VA: USGS, 1972.
Available in base, topographic or relief version.

Geologic map of North Carolina   1:500 000
Raleigh, NC: NCGS, 1985.

Aeromagnetic map of North Carolina   1:1 000 000   GP-958
Reston, VA: USGS, 1984.

North Dakota

State of North Dakota   1:500 000
Reston, VA: USGS, 1983.
Available in base, topographic or relief version.

Relief map of North Dakota   MM-14   Karol-lyn Knudson.
Grand Forks, ND: NDGS, 1973.

Bedrock geologic map of North Dakota   1:1 000 000   MM-21
Grand Forks, ND: NDGS, 1982.

Geologic highway map of North Dakota   1:1 000 000
Grand Forks, ND: NDGS, 1976.

Geologic map of North Dakota   1:500 000
Reston, VA: USGS, 1980.

Surface geology of North Dakota   MM-15
Grand Forks, ND: NDGS, 1973.

Geothermal resources of North Dakota   1:500 000
Boulder, CO: NGDC, 1981.

Ohio

State of Ohio   1:500 000
Reston, VA: USGS, 1971.
Available in base, topographic or relief version.

Ohio landform map   1:1 000 000   J.A. Bier
Columbus, OH: ODGS, 1967.

Residual total intensity magnetic map of Ohio
1:1 000 000   GP-961
Reston, VA: USGS, 1984.

Gravity anomaly maps of Ohio   1:1 000 000   GP-0963
Reston, VA: USGS, 1986.
Colour Bouguer anomaly map + 6 filtered gravity anomaly maps.

Geologic map of Ohio   1:500 000
Columbus, OH: ODGS, 1947.

Glacial map of Ohio   1:500 000   I-316
Reston, VA: USGS, 1961.

Mineral industries map of Ohio   1:500 000
Columbus, OH: ODGS, 1977.

Oil and gas fields of Ohio   1:500 000
Columbus, OH: ODGS, 1974.

Oil and gas pipelines in Ohio   1:500 000
Columbus, OH: ODGS, 1973.

Complete Bouguer gravity anomaly map of Ohio
1:500 000   GP-962
Reston, VA: USGS, 1984.

Aerial radiometric contour maps of Ohio   1:500 000   GP-968
Reston, VA: USGS, 1985.
3 sheets, all published.

Oklahoma

State of Oklahoma   1:500 000
Reston, VA: USGS, 1972.
Available as base or topographic version.

Tectonic map of Oklahoma   1:750 000   GM-3
Norman, OK: OGS, 1955.

Vertical intensity magnetic map of Oklahoma   1:750 000   GM-6
Bouguer gravity-anomaly map of Oklahoma   1:750 000   GM-7
Norman, OK: OGS, 1964.
Available as a set + text.

Mineral map of Oklahoma (exclusive of oil and gas fields)   1:750 000
GM-15
Norman, OK: OGS, 1970.

Earthquake map of Oklahoma   1:750 000   GM-19
Norman, OK: OGS, 1979.
+ 15 pp. text.

Geologic map of Oklahoma   1:500 000
Reston, VA: USGS, 1954.
2 sheets, both published.

Principal structural features of Oklahoma   1:500 000
Tulsa, OK: PennWell, 1983.

Map of Oklahoma oil and gas fields   1:500 000   GM-28
Norman, OK: OGS, 1985.

*Geothermal resources and temperature gradient of
Oklahoma*   1:500 000   Map GM-27
Boulder, CO: NGDC/Norman, OK: OGS, 1984.

Hydrologic atlas of Oklahoma   1:250 000
Norman, OK: OGS, 1969–.
9 sets of 4 sheets, 8 sets published.
Includes geology sheets.

Oregon

Atlas of Oregon   W.G. Loy
Eugene, OR: University of Oregon Books, 1976.
pp. 215.

State of Oregon 1:500 000
Reston, VA: USGS, 1982.
Available in base, topographic or relief version.

Oregon Landsat mosaic 1:500 000
Corvallis, OR: ERSAL, 1982.
2 sheets.
Composite of Landsat 3 RBV images.

*Geologic map of Oregon east of the 121st
meridian* 1:500 000 I-902
Reston, VA: USGS, 1977.

*Geologic map of Oregon west of the 121st
meridian* 1:500 000 I-325
Reston, VA: USGS, 1961.

Geothermal resources of Oregon 1:500 000
Boulder, CO: NGDC, 1982.

Gravity maps of Oregon 1:500 000 GMS-4
Portland, OR: ODGMI, 1967.
3 sheets, including onshore and offshore areas.

Mineral resources map, offshore Oregon 1:500 000
Portland, OR: ODGMI, 1985.

Pennsylvania

Pennsylvania atlas: a thematic atlas of the Keystone State
P.F. Rizza
Grove City, PA: Ptolemy Press, 1982.

State of Pennsylvania 1:500 000
Reston, VA: USGS, 1975.
Available in base, topographic or relief version.

Rock types of Pennsylvania 1:500 000
Harrisburg, PA: PBTGS, 1984.

Geologic map of Pennsylvania 1:250 000
Harrisburg, PA: PBTGS, 1980.
3 sheets, all published.

Aeromagnetic map of Pennsylvania 1:250 000
Reston, VA: USGS, 1978.
2 sheets, both published.

Rhode Island

The Rhode Island atlas M.I. Wright and R.J. Sullivan
Providence, RI: Rhode Island Publications Society, 1982.
pp. 240.

Massachusetts, Rhode Island and Connecticut 1:500 000
Reston, VA: USGS, 1948.
Available in base, topographic or relief version.

Bedrock geologic map of Rhode Island 1:125 000
Reston, VA: USGS, 1971.

South Carolina

State of South Carolina 1:500 000
Reston, VA: USGS, 1970.
Available in base, topographic or relief version.

Aeromagnetic map of South Carolina 1:250 000
Reston, VA: USGS, 1982.
2 sheets, both published.

South Dakota

State of South Dakota 1:500 000
Reston, VA: USGS, 1984.
Available in base, topographic or relief version.

Magnetometer map of South Dakota ca 1:735 000
Vermillion, SD: SDGS, 1967.

Geologic map of South Dakota 1:500 000
Reston, VA: USGS, 1951.

Tennessee

State of Tennessee 1:500 000
Reston, VA: USGS, 1973.
Available in base, topographic or relief version.

Poster for Homecoming '86
Murfreesboro: Department of Geography and Geology, Middle
Tennessee State University, 1986.
Landsat mosaic of the state.

Aeromagnetic map of Tennessee and Kentucky 1:1 000 000
GP-0959
Reston, VA: USGS, 1984.

Bouguer gravity anomaly map of Tennessee 1:500 000
Nashville, TN: TDC, 1967.

Mineral resources and mineral industries of Tennessee 1:500 000
Nashville, TN: TDC, 1959.

Geologic map of Tennessee 1:250 000
Nashville, TN: TDC, 1966.
4 sheets, all published.

Magnetic maps of Tennessee 1:250 000
Nashville, TN: TDC.
4 sheets, all published.

Geologic hazards map of Tennessee
Nashville, TN: TDC, 1978.

Texas

Atlas of Texas S.A. Arbingast *et al.*
Austin, TX: Bureau of Business Research, University of Texas,
1976.
pp. 179.

State of Texas 1:500 000
Reston, VA: USGS, 1982.
4 sheets, all published.
Available in base, topographic or relief version.

Texas false-color Landsat map J.H. Smith
Houston, TX: Petrophysics Inc., 1985.

Geological highway map of Texas 1:1 875 000
Tulsa, OK: AAPG, 1973.

Geothermal resources of Texas 1:1 000 000
Boulder, CO: NGDC, 1983.

Energy resources of Texas 1:1 000 000
Austin, TX: BEG, 1976.

Mineral resources of Texas 1:1 000 000
Austin, TX: BEG, 1979.

Geologic atlas of Texas 1:250 000
Austin, TX: BEG, 1965–.
38 sheets, 37 published.

Bouguer gravity atlas of Texas 1:250 000
Austin, TX: BEG.
38 sheets, 1 published.

Utah

Atlas of Utah D.C. Grer *et al.*
Ogden, UT: Weber State College, 1981.

State of Utah 1:500 000
Reston, VA: USGS, 1976.
Available in base, topographic or relief version.

Physiographic subdivisions of Utah 1:2 500 000
Salt Lake City, UT: UGMS, 1977.

Aeromagnetic map of Utah 1:1 000 000 GP-907
Reston, VA: USGS, 1976.

Simple Bouguer gravity anomaly map of Utah Map 37
Salt Lake City, UT: UGMS, 1975.

Oil and gas fields and pipelines of Utah 1:750 000 Map 61
Salt Lake City, UT: UGMS, 1982.

Utah mining district areas and principal metal occurrences 1:750 000 Map 70
Salt Lake City, UT: UGMS, 1983.

Non-metallic mineral resources of Utah 1:750 000 Map 71
Salt Lake City, UT: UGMS, 1983.

Geologic map of Utah 1:500 000
Salt Lake City, UT: UGMS, 1981.
2 sheets, both published.

Geological highway map of Utah 1:500 000
Provo, UT: Brigham Young University Geology Department, 1975.

Energy resources map of Utah 1:500 000
Salt Lake City, UT: UGMS, 1983.

Vermont

The Vermont atlas and gazetteer
Freeport, ME: DeLorme Publishing Co., 1983.
pp. 88.

New Hampshire and Vermont 1:500 000
Reston, VA: USGS, 1972.
Available in base, topographic or relief version.

Topographic map of Vermont 1:250 000
Montpelier, VT: VGS, 1970.

Centennial geologic map of Vermont 1:250 000
Montpelier, VT: VGS, 1961.

Surficial geologic map of Vermont 1:250 000
Montpelier, VT: VGS, 1970.

Virginia

State of Virginia 1:500 000
Reston, VA: USGS, 1973.
Available in base, topographic or relief version.

Virginia. Gravity map of the simple Bouguer anomaly 1:500 000
Charlottesville, VA: VDMR, 1977.

Geologic map of Virginia 1:500 000
Charlottesville, VA: VDMR, 1963.
Black-and-white ozalid copy.

Seismicity map of the State of Virginia 1:1 000 000
Reston, VA: USGS, 1983.

Aeromagnetic map of Virginia 1:500 000
Reston, VA: USGS, 1978.

Virginia. Map of mineral industries and resources 1:500 000
Charlottesville, VA: VDMR, 1983.

Washington

State of Washington 1:500 000
Reston, VA: USGS, 1982.
Available in base, topographic or relief version.

Geologic map of Washington 1:2 000 000 I-583
Reston, VA: USGS, 1969.

Complete Bouguer gravity anomaly map of Washington 1:500 000 GM-11
Olympia, WA: WDGER, 1974.

West Virginia

State of West Virginia 1:500 000
Reston, VA: USGS, 1984.
Available in base, topographic or relief version.

Landsat image of eastern West Virginia 1:1 000 000 MAP-WV5
Landsat image of western West Virginia 1:1 000 000 MAP-WV6
Morgantown, WV: WVGES, 1979.
2 maps.
Colour-infrared images.

Landsat linear features of West Virginia 1:250 000 MAP-WV7
Morgantown, WV: WVGES, 1979.
2 sheets, both published.

Seismicity map of the State of West Virginia 1:1 000 000 MF-1226
Reston, VA: USGS, 1980.

Geologic map of West Virginia 1:500 000
Morgantown, WV: WVGS, 1932.

Geologic map of West Virginia 1:250 000
Morgantown, WV: WVGS, 1968.
2 sheets, both published.

Aeromagnetic map of West Virginia 1:250 000
Reston, VA: USGS, 1978.

Gravity map of West Virginia 1:250 000 MAP-WV25
Morgantown, WV: WVGS, 1985.

Oil and gas fields map of West Virginia 1:250 000 MRS-7
Morgantown, WV: WVGS, 1982.
2 sheets + 171 pp. text.

Transportation map of West Virginia 1:500 000
Morgantown, WV: WVGS/WV Railroad Maintenance Authority, 1986.

Wisconsin

State of Wisconsin 1:500 000
Reston, VA: USGS, 1984.
Available in base, topographic or relief version.

Wisconsin bedrock geology map 1:1 000 000
Madison, WI: WGNHS, 1982.

Glacial deposits of Wisconsin 1:500 000
Madison, WI: WGNHS, 1976.

Wyoming

State of Wyoming 1:500 000
Reston, VA: USGS, 1980.
Available in base, topographic or relief version.

Landsat image mosaic of Wyoming 1:500 000 MS-11
Laramie, WY: GSW, 1982.

Geologic highway map of Wyoming 1:1 000 000
Laramie, WY: GSW, 1986.

Geologic map of Wyoming 1:500 000 J.D. Love and A.C. Christiansen
Laramie, WY: GSW, 1985.
3 sheets, all published.

Oil and gas map of Wyoming 1:500 000 MS-12
Laramie, WY: GSW, 1984.

Geothermal resources of Wyoming 1:500 000
Boulder, CO: NGDC, 1984.

Seismicity map of the State of Wyoming 1:1 000 000 MF-1798
Reston, VA: USGS, 1985.

Metallic and industrial minerals map of Wyoming 1:500 000 MS-14
Laramie, WY: GSW, 1985.

Construction materials map of Wyoming 1:500 000 MS-21
Laramie, WY: GSW, 1986.

CENTRAL AMERICA

Detailed topographic mapping of the seven republics of Central America has largely taken place since World War II.

With the exception of Belize (formerly British Honduras), which was mapped by the British Government's Directorate of Overseas Surveys, all the Central American states were party to collaborative mapping agreements made with the Inter-American Geodetic Survey (IAGS) in the 1940s. The IAGS was set up by US President Truman in 1946 to help Latin America undertake modern topographic, hydrographic and geodetic survey and to tie the mapping systems together in a common geodetic network. It is therefore no surprise that the basic topographic mapping of these countries has a certain uniformity of style and format. Many of the map series begun in the 1940s were completed quite rapidly, notwithstanding the problems of acquiring satisfactory air photo cover in the persistently cloud-covered and forested lowland areas of the inter-tropical zone. The acquisition of cloud-penetrating radar imagery in the 1960s and early 1970s partly eased this problem, and for example allowed Panama to complete its 1:250 000 mapping. The Side Looking Airborne Radar (SLAR) imagery obtained for the Province of Darien in 1967 was the earliest major civilian application of this technology.

The initiative for the creation of the IAGS came from the **Pan American Institute of Geography and History (PAIGH)**, which was itself founded in 1928, becoming in 1949 a specialist body within the Organization of American States (OAS). Its headquarters are in Mexico City. PAIGH also promoted the preparation of a series of research guides to the individual Latin American republics (the series *Guías para Investigadores*, some of which are published in an English as well as a Spanish edition). These list and index the contemporary geographic and cartographic materials relating to each country. Most of the guides appeared in the mid- to late 1970s and, although the series is not complete, all the six Central American PAIGH members have a guide. They provide useful and comprehensive, though somewhat dated information about these countries' cartographic resources. There is of course no guarantee that items listed in them remain in print or are currently available to the public.

A more recent and important initiative by PAIGH has been to sponsor the production and publication of a *Unified Hemispheric Map Series* at 1:250 000. As originally conceived, the programme was intended to yield, over a period of eight or so years beginning in 1980, a complete and substantially uniform topographic map cover of North, Central and South America. The sheets are being produced by the various national surveys and are full-colour topographic editions on a UTM projection, generally conforming to a specification approved by PAIGH and with marginal information in four languages. At the time of writing, not all countries have joined the programme, nor has the original schedule been kept. Nevertheless, good progress has been made, and the series offers a unique opportunity to acquire topographic maps of some countries (not only in Central America) from which map acquisition has sometimes been difficult. Of the Central American countries, Costa Rica, Panama and Guatemala are currently in the programme.

These maps are available worldwide on subscription, and the address of the **PAIGH Program Coordinator** is given below.

The United Nations has held a series of *Regional Cartographic Conferences for the Americas*, the third and most recent having been held in 1985 in New York, and these provide a forum for Latin American and other countries to share knowledge about their progress and problems in developing modern cartographic information systems.

Relatively few of the Central American countries have extensive, country-wide programmes of resource mapping. Guatemala is a notable exception with its integrated programme of 1:50 000 scale geological, land use and land capability mapping. In most countries, there have been specific surveys for cadastral, land reform or resource exploitation purposes, usually covering limited areas. Some of this resource mapping has been carried out with the aid of foreign expertise.

Small-scale maps, whether general or thematic, of the Central American area as a whole are rather few in number. Some of these regional maps are listed in the catalogue below, but other maps in this scale range will be found in world series such as the *International Geological Map of the World* 1:10 000 000 and the *Soil Map of the World* 1:5 000 000, which are included in the World section of this book. In addition, some of the small-scale mapping of North America and of the Caribbean also covers Central America.

Addresses

Instituto Panamericano de Geografía e Historia (PAIGH)
Secretaria General, Ex-Arzobispado 29, Col. Observatorio, 11860 MEXICO, DF, Mexico.

PAIGH Program Coordinator
Robert L. Senter, PAIGH Hemispheric Map Series,
DMAIAGS, Bldg 144, FORT SAM HOUSTON, TX 78234, USA

Bureau of Business Research
PO Box 7459, University Station, AUSTIN, Texas 78712, USA.

For Academía de Ciencias de Cuba, see Cuba; for Haack, see
German Democratic Republic; for Universität Hamburg, see
German Federal Republic; for Kümmerly + Frey, see Switzerland;
for NGS, CIA, PennWell and NGDC, see United States; for IGN,
see Guatemala; for GUGK, see USSR.

Catalogue

Atlases and gazetteers

Atlas of Central America and the Caribbean
New York: Macmillan, 1985.
pp. 144.

Atlas of Central America
Austin, TX: Bureau of Business Research, 1979.
pp. 62.

Atlas regional del Caribe
La Habana: Academía de Ciencias de Cuba, 1979.

*Zentralamerika: Karten zur Bevölkerungs- und
Wirtschaftsstruktur* H.Nuhn, P. Krieg and W. Schlick
Hamburg: Universität Hamburg, 1975.
pp. 180, including 10 maps.

General

Mittelamerika 1:6 000 000
Gotha: Haack, 1978.

Central America 1:5 000 000
Bern: Kümmerly + Frey, 1986.
Includes Mexico and Caribbean.

Central America 1:2 534 000
Washington, DC: NGS, 1986.
Map of Mexico 1:3 803 000 on reverse.

Central America 1:2 500 000
Washington, DC: CIA, 1974.

Tsentral'naja Amerika i Anti'skie Ostrova 1:2 500 000
Moskva: GUGK.

Mapa de America Central 1:2 000 000
Guatemala: IGN, 1972.
2 sheets, both published.

Geological and geophysical

Seismicity map of Middle America 1:8 000 000
Boulder, CO: NGDC, 1982.

Oil and gas map of Mexico/Central America 1:3 600 000
Tulsa, OK: PennWell, 1982.

Environmental

Atlas climatológico e hidrológico del istmo Centro-americano
1:2 000 000
Mexico City: PAIGH, 1976.
13 sheets.

Belize

Belize became an independent nation in 1981, after 17
years of full internal self-government under the British
Crown.

Modern topographic mapping of Belize has been
undertaken by the British **Directorate of Overseas
Surveys (DOS)** (now Overseas Survey Directorate, OSD).
Early detailed mapping was a series of 'sketch maps' at
1:50 000 produced by the Directorate of Colonial Surveys
(as it was then called) in the 1940s and 1950s. These
monochrome maps lacked contours but were annotated
with land cover and other surface feature information. In
1963 a new series of contoured maps at 1:50 000 was
initiated in collaboration with the Directorate of Military
Surveys, British War Office. The 1970s saw the production
of a further series based on air photography flown in 1969
and 1972. This entire series was published between 1973
and 1977. There has been revision of some sheets since
1979 in the military (GSGS) edition, incorporating chiefly a
new vegetation specification and an updating of road
networks. Sheets are printed in four colours with contours
at 20 or 40 m intervals, and with a nine-category
classification of vegetation. The projection is Transverse
Mercator with UTM grid. Sheets in the series are
numbered 1 to 44, but due to some pairing in coastal
areas, cover is complete in 39 sheets.

The 1:250 000 series in two sheets supersedes an earlier
three-sheet series. The current (1st) edition, published in
1980, shows relief by layer tinting at 200 m intervals, and

has been compiled from the 1:50 000 series. On the
reverse of the two sheets are maps of all the principal
towns (scale approximately 1:7500) compiled from air
photography flown in 1977. The projection is Transverse
Mercator and the UTM grid is shown by marginal ticks.

Earth resource mapping has been carried out by the
Land Resource Division of DOS (now **Land Resources
Development Centre**, Surbiton), and includes assessments
of the soils, land use and land capability of the Belize
River Valley at 1:100 000 (DOS 3144) and several regional
forest resource studies. The 1:250 000 scale maps of soils,
vegetation and land use published in 1958, which cover the
whole country, had been out of print but were reprinted in
1981. Since Belize is claimed as a province of Guatemala
by the Guatemala Government, the country is also covered
by some of the smaller-scale thematic mapping of the
Instituto Geográfico Militar of Guatemala (see entries
under Guatemala).

A *Geological map of the Maya Mountains, Belize*
1:130 000 was prepared by the Institute of Geological
Surveys of Great Britain and published by DOS in 1975. It
accompanies an IGS *Overseas Memoir* No. 3.

The *Atlas of Belize* published by Cubola Productions,
and periodically revised, has simple coloured thematic
maps of Belize as well as numerous coloured photographs.
The **Belize Tourist Board** publishes a *Facilities map of
Belize* (also produced by Cubola), which includes small
street maps of major settlements.

Addresses

Lands and Surveys Department
Ministry of Natural Resources, BELMOPAN.

The Belize Tourist Board
53 Regent Street, PO Box 325, BELIZE CITY.

Overseas Surveys Directorate (DOS/OSD)
Ordnance Survey, Romsey Rd, Maybush, SOUTHAMPTON
SO9 4DH, UK.

Land Resources Development Centre (LRDC)
Tolworth Tower, SURBITON KT6 7DY, UK.

Catalogue

Atlases and gazetteers

Atlas of Belize   7th edn
Belize City: Cubola Productions, 1982.

General

Belize   1:750 000   DOS 958   Edition 4
Tolworth: DOS, 1981.

Belize   1:250 000   DOS 649/1
Tolworth: DOS, 1980–81.
2 sheets, both published.
Maps of major towns on the reverse.

Topographic

Belize   1:50 000   DOS 4499 (Series E755)
Tolworth: DOS, 1973–.
39 sheets, all published. ■

Ambergris Cay   1:50 000   DOS 4499P (Series E755)
Tolworth: DOS, 1980.
Photomap.
Special tourist edition of sheet 7 in the standard series with inset
map of San Pedro.

Environmental

British Honduras: provisional soil map   1:250 000   DOS
(Misc)241A
Tolworth: DOS, 1958.
2 sheets, both published.

British Honduras: natural vegetation map   1:250 000   DOS
(Misc)241B
Tolworth: DOS, 1958.
2 sheets, both published.

British Honduras: potential land use map   1:250 000   DOS
(Misc)241C
Tolworth: DOS, 1958.
2 sheets, both published.

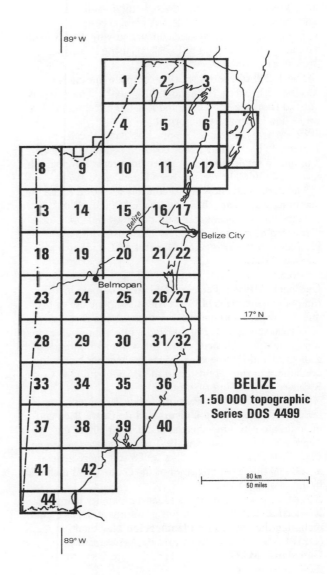

BELIZE
1:50 000 topographic
Series DOS 4499

80 km
50 miles

Town maps

Belize town plans   1:5000   Belize City Series E953
Edition 4   GSGS
Tolworth: DMS, 1983.
Colour map with indexes of streets and principal buildings.
UTM grid.

Belize town plans   1:5000 (approx.)   Belmopan Series
E953   Edition 1   GSGS
Tolworth: DMS, 1979.
Photomap with arbitrary grid.

Costa Rica

The Instituto Geográfico, San José, was founded in 1944
and was given its present name of **Instituto Geográfico
Nacional (IGN)** in 1968. Systematic large-scale
topographic mapping began in the mid-1940s in
collaboration with the IAGS. Initially this was a 1:25 000
scale photogrammetric map, issued from 1953, but the
series was terminated in 1962 with only 99 sheets

published. Although sheets from this series are still
available, they have not been revised and cover only part
of Guanacaste province and the Central Valley.

The present basic topographic cover is the complete
1:50 000 series (E762) in 133 sheets. This is a five-colour
map with 20 m contour interval (and 10 m auxiliary
contours). The projection is Lambert conformal conic.

This series was published between 1955 and 1971, with some sheets appearing in a revised edition subsequently. The sheet format is 10' by 15'. A parallel series of 1:50 000 maps was prepared concurrently by the US Army Topographic Command in 137 sheets using a Transverse Mercator projection.

The 1:200 000 series (E561), complete in nine sheets, was issued between 1955 and 1971 by IGN. This series is derived from the 1:50 000, and has 100 m contours. The projection is Lambert conformal.

Urban areas have been mapped by IGN at scales of 1:5000, 1:10 000 and 1:12 500, with contours at 10 or 5 m, and a small number of sheets in a general topographic series at 1:10 000 have also been issued since 1974.

The horizontal and vertical controls for Costa Rican mapping were established with the aid of the IAGS.

Maps for census enumeration purposes and small-scale demographic mapping based on census statistics are produced by the **Dirección General de Estadística y Censos (DGEC)**.

Geological maps have been issued by the **Dirección de Geología, Minas y Petrólio (DGMP)**, founded in 1951. Many thematic maps of Costa Rica have gone out of print due to a paper shortage in the country, but a new series of geological sheets at 1:200 000 and a mineral resources map were recently published. Other maps, including regional geological and hydrogeological maps, may not be readily available, but they are listed in the PAIGH guide to Costa Rica (see Further information).

Some geological and geomorphological mapping has been carried out by the **Universidad de Costa Rica** and includes a recent nine-sheet series at 1:50 000 of the Central Valley published in 1981 (*Carta geomorfológica de Valle Central de Costa Rica*), as well as the more general 1:500 000 map covering most of the country and included in the catalogue.

A range of soil, land use and land capability mapping has been undertaken by various government departments, including the **Oficina de Planificación Nacional (OFIPLAN)** and the **Ministerio de Agricultura y Ganadería (MAG)**. These maps, too, are likely to be difficult to obtain.

The **Centro de Ciencia Tropical (CCT)** is a private organization which carries out ecological surveys and has published a number of small-scale ecological maps (not restricted to Costa Rica) stemming from the so-called 'life zone system' first elaborated by L. R. Holdridge.

Further information

A comprehensive listing of map resources up to 1977 is found in *Research Guide to Costa Rica* prepared by the IGN for the Pan American Institute of Geography and History (PAIGH, 1977, 137 pp.).

The IGN, San José, issues a leaflet with indexes and sheet names of the 1:50 000 and 1:200 000 topographic series.

Addresses

Instituto Geográfico Nacional (IGN)
Apartado Postal 2272,1000 SAN JOSÉ.

Instituto Geográfico Nacional (orders)
Apartado Postal 148, 1001 Plaza González Víquez, SAN JOSÉ.

Dirección General de Estadística y Censos (DGEC)
Sección de Cartografía Censal, Ave. 6, Calle Ctl. y 2, SAN JOSÉ.

Dirección de Geología, Minas y Petrólio (DGMP)
Apartado Postal 2549, SAN JOSÉ.

Oficina de Planificación Nacional (OFIPLAN)
División de Planificación Regional y Urbana, Ave. 3 y 5, Calle 4, SAN JOSÉ.

Ministerio de Agricultura y Ganadería (MAG)
Ave. Ctl. y 1, Calle 1, SAN JOSÉ.

Universidad de Costa Rica
Ciudad Universitaria Rodrigo Facio, San Pedro de Montes de Oca, SAN JOSÉ.

Centro de Ciencia Tropical (CCT)
Apartado Postal 8-3870, SAN JOSÉ.

For DMA, see United States.

Catalogue

Atlases and gazetteers

Gazetteer of Costa Rica. Names approved by the US Board on Geographic Names 2nd edn
Washington, DC: DMA, 1983.

General

Costa Rica. Mapa físico-político 1:500 000
San José: IGN, 1983.

Topographic

Mapa de Costa Rica 1:200 000 Series E561
San José: IGN, 1969–.
9 sheets, all published.

Costa Rica. Mapa topográfico 1:50 000 Series E762
San José: IGN, 1955–.
133 sheets, all published. ∎

Geological and geophysical

Mapa geomorfológico de Costa Rica 1:1 000 000
San José: IGN, 1978.

Mapa de recursos minerales de Costa Rica 1:750 000
San José: DGMP, 1982.

Geomorfología de algunos sectores de Costa Rica 1:500 000
J.-P. Bergoeing
San José: IGN/Universidad de Costa Rica, 1982.
4 sheets + 16 pp. text.

Mapa geológico de Costa Rica 1:200 000
San José: DGMP, 1982.
9 sheets, all published.
19 pp. text in Spanish.

Environmental

República de Costa Rica. Mapa ecológica 1:750 000
J.A. Tosi Jr
San José: CCT, 1969.

Costa Rica. Uso potencial de la tierra 1:750 000 J. Coto and J. Torres
San José: MAG, 1970.

Administrative

Mapa politico administrativo de Costa Rica 1:1 000 000
San José: IGN, 1980.
Shows provinces in different colours.

Town maps

San José centro 1:12 500
San José: IGN, 1982.

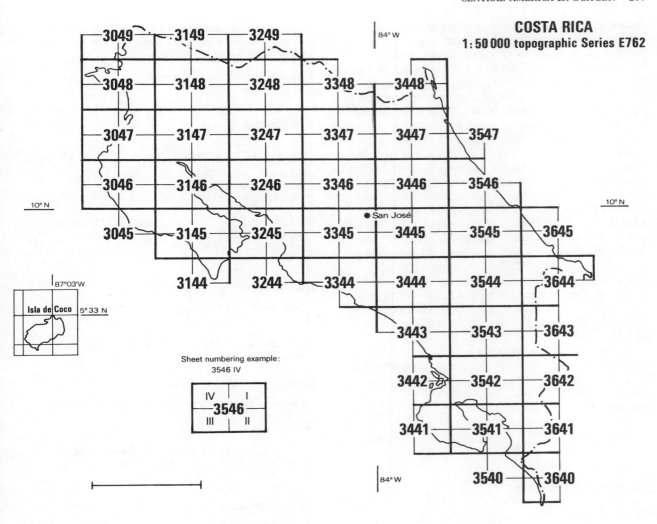

Sheet numbering example:
3546 IV

IV	I
III	II

3546

El Salvador

Topographic mapping in El Salvador is the responsibility of the **Instituto Geográfico Nacional 'Ingeniero Pablo Guzman' (IGN)**, located at Ciudad Delgado. The organization was founded, as the *Oficina de Mapa*, in 1946. Modern topographic mapping was carried out with IAGS collaboration from the early 1950s, following acquisition of a complete air photo cover in 1954 at scales of 1:30 000 and 1:60 000, and at larger scales subsequently. Basic mapping has been compiled at 1:20 000 and sheets issued as monochrome heliographic copies with 10 m contours. These maps have been used as base maps for thematic mapping. The main printed series, however, is the 1:50 000 topographic map in 54 sheets, issued since 1955. This five-colour map is on a Lambert conformal conic projection and contours are at 20 m. Sheets each cover an area of 10' by 15'. There is complete cover of the country with most sheets now in a second edition. A series of 1:100 000 scale departmental maps was derived from the 1:50 000, and subsequently a six-sheet quadrangle cover of the country at 1:100 000 was also published (1974–77).

IGN has also undertaken an extensive programme of cadastral mapping, with a rural cadaster at 1:5000/ 1:10 000, and urban at 1:1000, to cover the whole country. The rural cadaster is produced partly as

orthophotomaps and partly as photogrammetrically produced line mapping with 5 m contours.

Other publications of IGN include a four-volume *Diccionario geográfico de El Salvador*, published in 1970, a number of pictomaps of cities, and a variety of special project mapping at large scales.

At the time of writing, almost all the IGN mapping is restricted, with the main exception of the general map at 1:300 000, and we have not been able to ascertain whether this restriction is likely to be lifted. We have therefore not been able to include most of it in our catalogue of available mapping.

Geological mapping has been carried out through a collaborative programme by the West German Geological Mission to El Salvador with the **Centro de Investigaciones Geotécnicas (CIG)**, San Salvador, and a series of six 1:100 000 geology sheets has been published and is available from the **Bundesanstalt für Geowissenschaften und Rohstoffe (BfGR)**, Hannover. A 1:500 000 scale general geological map was also published, but is not now available.

Coloured soil maps at 1:50 000 have been published by the **Centro Nacional de Tecnología Agropecuaria (CENTA)**—29 sheets covering the central south of the country are available—and both soil and land capability mapping by the Departamento de Inventario y Clasificación de Suelos which is part of the **Centro de Recursos Naturales (CENREN)**. Maps published by the latter include several 1 : 50 000 soil maps, Departmental land capability maps at 1 : 100 000, and a number of land capability and socioeconomic maps in black and white at 1 : 20 000 (the latter issued as heliographic prints).

Resource studies have been carried out with the aid of several overseas agencies. Thus a study of soil conservation, river stabilization and water pollution control in the Acelhuate River catchment was undertaken by the British Overseas Development Agency for the El Salvador Government, and the report (*Land Resource Study* 30) published in 1981 includes 1 : 100 000 scale maps of soils, land use and land capability. A cooperative study between CENREN and the Organization of American States (OEA-CONAPLAN) in the early 1970s, to evaluate the productive capacity of the land and to outline priorities for agricultural development, also yielded 1 : 250 000 maps of land capability and of agricultural development zones, which form the second part of the two-volume report published in 1977.

In 1979, the third edition of the *Atlas Nacional de El Salvador* was published. It has 61 pages of maps, mostly at 1 : 750 000, and provides a comprehensive review of the country's resource base and its economic and social geography.

Further information

The map resources of El Salvador up to 1977 are listed and indexed in *Guía para Investigadores República de El Salvador* prepared by the IGN for the Instituto Panamericano de Geografía e Historia (81 pp.). A list of soils mapping is issued by CENREN.

Addresses

Instituto Geografico Nacional 'Ingeniero Pablo Arnoldo Guzman' (IGN)
Avenida Juan Bertis 79, Apartado Postal 247, CIUDAD DELGANO.

Centro de Investigaciones Geotécnicas (CIG)
Apartado 109, Final Avenida Peralta, SAN SALVADOR.

Centro Nacional de Tecnología Agropecuaria (CENTA)
Final la Avenida Norte, SAN SALVADOR.

Centro de Recursos Naturales (CENREN)
Ministerio de Agricultura y Ganaderia, Apartado Postal 2265, Cantón El Matasano, SOYAPANGO.

Bundesanstalt für Geowissenschaften und Rohstoffe (BfGR)
Stilleweg 2, Postfach 510153, D-3000 HANNOVER 51, German Federal Republic.

For CIA and DMA see United States.

Catalogue

Atlases and gazetteers

Atlas de El Salvador 3rd edn
Ciudad Delgano: IGN, 1979.
88 pp. plates and text.

Gazetteer of El Salvador: names approved by the US Board of Geographic Names E.S. Szmanski and G. Quinting
Washington, DC: DMA, 1982.
pp. 180.

General

El Salvador 1 : 750 000
Washington, DC: CIA, 1980.

Mapa oficial de la República de El Salvador 1 : 300 000
Ciudad Delgano: IGN, 1978.

Geological and geophysical

Carta gravimétrica provisional. Mapa de anomalias Bouguer simple 1 : 300 000
Ciudad Delgano: IGN, 1976.

Mapa geológico de la República de El Salvador 1 : 100 000
San Salvador: CIG/Hannover: BfGR, 1978.
6 sheets, all published.

Environmental

Uso actual predominante de la tierra 1 : 500 000
Washington, DC: OEA-CONAPLAN, 1974.

Sistemas predominantes de tenencia de la tierra 1 : 500 000
Washington, DC: OEA-CONAPLAN, 1974.

Capacidad productiva de la tierra 1 : 250 000
Washington, DC: OEA-CONAPLAN, 1974.

Zonas de desarrollo agrícola prioritário 1 : 250 000
Washington, DC: OEA-CONAPLAN, 1974.

Mapas de capacidad de uso de las tierras (agrológico) 1 : 100 000
Soyapango: CENREN.
14 sheets, 3 published.
Maps of Departments.

Mapas de suelos (pedológico) 1 : 50 000
Soyapango: CENREN.
5 sheets published.

Cuadrantes pedológicos 1 : 50 000
San Salvador: CENTA.
29 sheets published.

Guatemala

Topographic mapping in Guatemala is now the responsibility of the **Instituto Geográfico Militar (IGM)**, Guatemala City. Modern topographic survey began in earnest only after 1945 with the formation of a Department of Mapping and Cartography, and the initiation (in 1948) of IAGS involvement in the mapping of this, one of the largest of the Central American republics. First-order triangulation was completed about 1954 and tied to Mexico, El Salvador and Honduras. Also in 1954 the Department was renamed the Dirección General de Cartografía, and the first air photography sorties were flown upon which the main topographic series were subsequently based. There are two main series, at 1:50 000 (E754) and 1:250 000 (E503), both prepared in collaboration with IAGS and the Army Map Service. Both series are complete, except that in the case of the E754 series, Belice, claimed as one of Guatemala's provinces, is not yet covered (hence the catalogue entry suggests only partial cover). Belice (Belize) is however covered by the mapping of the British Directorate of Overseas Surveys. Some sheets in the 1:250 000 series have been revised for minor cultural changes in the late 1970s and early 1980s.

The 1:50 000 series is on the Transverse Mercator projection and is in five colours with contours at 20 m intervals and supplementary contours at 10 m. Sheets have the usual 10' by 15' format. The 1:250 000 series was completed in 1969. It is in eight colours with a 100 m contour interval and 50 m supplementary contours. Marginal information is in English as well as Spanish.

Large-scale mapping has been undertaken in urban areas and the populous areas south of 15°. There is a programme of cadastral mapping at scales of 1:2000 (photomosaics) or 1:1000 (line maps) for urban areas and 1:10 000 for rural areas.

IGM also undertakes the geological mapping of Guatemala, and two main series have been initiated, at 1:250 000 and 1:50 000 scale, and on the same sheet lines as the topographic series. The larger-scale series, printed in eight colours, was undertaken in collaboration with a number of universities in the US, through a programme initiated in 1962. Thus each published sheet is accompanied by a monograph in English representing an American doctoral thesis. It is intended as a partial series, concentrating on areas of potential economic value. The German Technical Mission has also been active in Guatemala, carrying out investigations from 1967 to 1969 in part of the Guatemalan Cordillera, and a map of *Baja Verapaz* at 1:125 000 was published in 1971 by the **Bundesanstalt für Geowissenschaften und Rohstoffe (BfGR)**. There has been little progress on the two main series since the late 1960s.

The *Atlas Nacional de Guatemala* was published in 1972 by the then civilian Instituto Geográfico Nacional under the Ministério de Comunicaciones y Obras Publicas. It includes 130 maps and diagrams in seven colours as well as colour and black-and-white photographs, and is organized into eight chapters covering physical and human resources, economic activities, service industries and tourism.

The 1:50 000 land use series covers only a small part of the south of the country, and about half the sheets are available as eight-colour litho-printed colour productions, the others being available in diazo form. The land capability maps cover a rather larger area of southern Guatemala and new sheets are being issued. They are available only as diazo prints, and use an eight-category capability classification, based on soil, slope and drainage criteria.

The **Instituto Nacional de Estadística (INE)** maintains mapping for census enumeration purposes at a range of scales, including 1:1000 and 1:2000 for municipal areas.

All the maps listed in the catalogue below are distributed by IGM. There may currently be some restriction on the availability of the 1:50 000 scale maps.

Further information

IGM issues a list of published topographic and thematic mapping, *Guía Geográfica de Guatemala para investigadores* (Publicacion IPGH No 319—Abril 1978). This is one of the research guides initiated by PAIGH. It includes details of maps and other resource publications with parallel texts in Spanish and English.

Addresses

Instituto Geográfico Militar (IGM)
Avenida Las Americas 5–76, Zona 13, GUATEMALA CA.

Instituto Nacional de Estadística (INE)
Edificio America, 8a Calle 9–55, Zona 1, GUATEMALA CA.

For CIA and DMA, see United States; for BfGR, see German Federal Republic.

Catalogue

Atlases and gazetteers

Atlas nacional de Guatemala
Guatemala, CA: IGM, 1972.
pp. 104.

Gazetteer of Guatemala: names approved by the United States Board on Geographic Names 2nd edn
Washington, DC: DMA, 1984.

General

Guatemala 1:1 500 000
Washington, DC: CIA, 1983.

Mapa de la República de Guatemala 1:1 000 000
Guatemala, CA: IGM, 1981.

Guatemala: mapa vial turístico 1980 1:1 000 000 (main map)
Guatemala, CA: IGM, 1980.
Maps of Guatemala City and other major towns on reverse.

Mapa hipsométrico de la República de Guatemala 1:500 000
Guatemala, CA: IGM, 1979.
4 sheets, all published.

Topographic

Mapa basico de la República de Guatemala 1:250 000 Series E503
Guatemala, CA: IGM, 1961–.
13 sheets, all published.

Mapa topográfico de la República de Guatemala 1 : 50 000 Series
E754
Guatemala, CA: IGM/DMA 1960–.
386 sheets, 259 published. ■

Geological and geophysical

Mapa de la República de formes de la tierra 1 : 1 000 000
Guatemala, CA: IGM, 1984.

Mapa geológico de la República de Guatemala 1 : 500 000
Guatemala, CA: IGM, 1970.
4 sheets, all published.
Legend in Spanish and English.

Carta isogónica de la República de Guatemala 1 : 500 000
Guatemala: Instituto Nacional de Sismología, 1981.
4 sheets, all published.

Mapa geológico de la República de Guatemala 1 : 250 000
Guatemala, CA: IGM, 1966–.
13 sheets, 2 published.

Mapa geológico de la República de Guatemala 1 : 50 000
Guatemala, CA: IGM, 1966–.
386 sheets, 35 published. ■

Environmental

*Mapa climatológico de la República de Guatemala (segun el sistema
Thornthwaite)* 1 : 1 000 000
Guatemala, CA: IGM, 1975.

Mapa de cuencas de la República de Guatemala 1 : 500 000
Guatemala, CA: IGM, 1978.
4 sheets, all published.

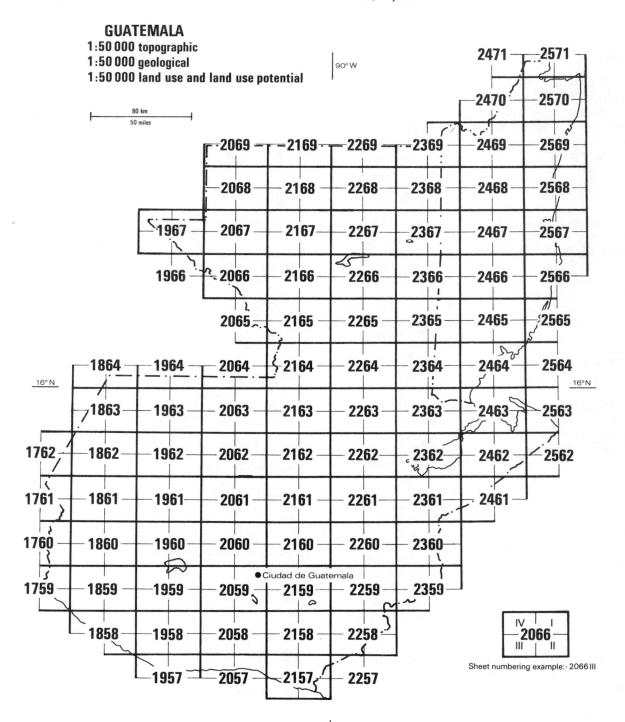

GUATEMALA
1 : 50 000 topographic
1 : 50 000 geological
1 : 50 000 land use and land use potential

80 km
50 miles

Sheet numbering example:- 2066 III

*Mapa de cobertura y uso actual de la tierra de la República de
Guatemala* 1:500 000
Guatemala, CA: IGM, 1982.
4 sheets, all published.

Mapa de capacidad productiva de la tierra 1:500 000
Guatemala, CA: IGM, 1980.
4 sheets, all published.

Mapa usó de la tierra de la República de Guatemala 1:50 000
Guatemala, CA: IGM, 1966–.
386 sheets, 16 published. ■

*Mapa uso potencial de la tierra de la República de
Guatemala* 1:50 000
Guatemala, CA: IGM, 1978–.
386 sheets, 51 published. ■

Town maps

Ciudad de Guatemala 1:25 000
Guatemala, CA: IGM, 1981.

Plano de la ciudad de Guatemala a escala 1:12 500
Guatemala, CA: IGM.
Edition in English–Spanish.

Honduras

The official mapping agency in Honduras is the **Instituto
Geográfico Nacional (IGN),** Comayagüela (the twin town
of Tegucigalpa).

Honduras was little mapped before the 1950s, when a
1:50 000 series was started by the Honduran Comisión
Geográfica Especial in collaboration with the IAGS and
AMS (Series E752). A 1:60 000 air photo cover was
acquired in 1954–56, and horizontal and vertical control
for the mapping was established by IAGS in the same
period. The 1:50 000 series is in about 275 sheets but is
not quite complete. Some sheets (in border areas) are
restricted, while others are out of print. The projection is
Transverse Mercator, and sheets are in the usual 10' by
15' Army Map Service format and printed in five colours
with a 20 m contour interval. Some areas have been
mapped at 1:25 000 and a derived series at 1:100 000 was
also started, but we have obtained no information on the
progress of this series. The only maps at 1:250 000 form
part of the Inter-American JOG Charts (Series 1501), and
only three of 12 sheets have been published, all in 1971.

Semi-controlled photomosaics and pictomaps at
1:50 000 cover rural areas in the eastern half of the
country.

Urban planimetric mapping of Tegucigalpa and environs
has been carried out at scales of 1:5000 and 1:10 000, and
pictomaps have been produced of some of the smaller
cities. 1:5000 cadastral maps are also made by the IGN,
with support from the municipal authorities.

Some limited geological mapping was done in the 1950s
in collaboration with the universities and the Peace Corps
of the United States, and a few 1:50 000 maps covering
areas near the capital have been published. These sheets
are apparently issued by the IGN, although there is a
Departamento de Minas (DM), which has responsibilities
for geological and mineral resource mapping.

There seems to have been little recent cartographic
activity in Honduras. In the 1980s, the country received
United Nations technical cooperation in the execution of a
regional planning project, and a proposal was made for the
strengthening of IGN to meet present mapping needs.

Further information

Research Guide to Honduras (45 pp.), a PAIGH guide, was
prepared by the Director of IGN and published in 1977. It
includes lists and indexes of topographic and other maps
published up to that time.

Addresses

Instituto Geográfico Nacional (IGN)
Apartado Postal 758, Barrio la Bolsa, Comayagüela, TEGUCIGALPA
DC.

Departamento de Minas (DM)
Dirección General de Recursos Naturales, Boulevard a la Colonia
Kennedy, TEGUCIGALPA DC.

For CIA and Rand McNally, see United States.

Catalogue

Atlases and gazetteers

*Gazetteer of Honduras. Names approved by the US Board on
geographic names* 2nd edn
Washington, DC: DMA, 1983.
pp. 478.

General

Honduras 1:1 500 000
Washington, DC: CIA.

Honduras. mapa turístico
Tegucigalpa: Instituto Hondureño de Turismo/IGN, 1978.
In Spanish and English.
Street maps of Tegucigalpa, San Pedro Sula and La Ceiba on
reverse.

Honduras 1:1 100 000
Chicago: Rand McNally/Texaco, 1979.
Road map.

República de Honduras. Mapa general 1:1 000 000 8th edn
Tegucigalpa: IGN, 1980.
Map of Central America on reverse.

Topographic

Honduras 1:50 000 Series E752
Tegucigalpa: IGN, 1961–.
275 sheets, *ca* 237 published. ■

Geological and geophysical

Mapa geológico de la República de Honduras 1:500 000
R.E. Aceituno
Tegucigalpa: DM, 1974.

Mapa geológico de Honduras 1:50 000
Tegucigalpa: DM, 1957–.
275 sheets, 12 published. ■

Town maps

Ciudad Tegucigalpa 1:12 500 2nd edn
Tegucigalpa: IGN, 1978.

Tegucigalpa 1:10 000 2nd edn
Tegucigalpa: IGN, 1971.

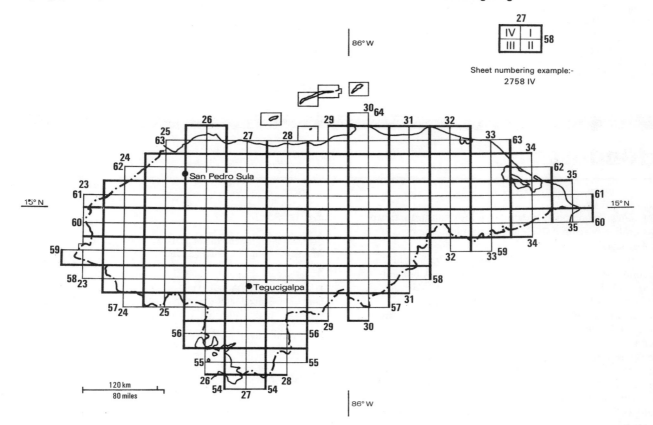

HONDURAS
1:50 000 topographic Series E752
1:50 000 geological

Sheet numbering example:-
2758 IV

Nicaragua

Detailed mapping of Nicaragua, the largest of the Central American countries, was initiated in 1946 with the signing of an agreement with IAGS to establish geodetic control for a national survey, and with the foundation of an Office of Geodesy, now the **Instituto Geográfico Nacional (IGN)**, Managua.

In 1954, complete air photo cover of the Pacific western part of the country was flown by the USAAF (nominal scale 1:64 000), and this formed the basis of the first photogrammetric mapping. The basic series is at 1:50 000 (Series E751), covering the country in 300 sheets. The first sheets were issued in 1956 and the series is now almost complete. The projection is UTM and the map is published in five colours with contours at 20 m. Sheets have a format of 10' by 15'. There are also series at 1:100 000 (E651) and 1:250 000 (E503). Only 27 of the 90 or so 1:100 000 sheets needed to cover the country have been published. These are planimetric maps, mostly monochrome, covering mainly the Pacific west. The 1:250 000 series is complete in 12 sheets. There is also a fragmentary series at 1:25 000 (E851) but only a handful of sheets have been published at this scale.

IGN has also carried out cadastral surveying, and several hundred contoured sheets at 1:10 000 have been produced, as well as a series of photomaps of the most westerly departments of the country.

Major cities are covered by a series of pictomaps, mostly at scales of 1:4000 or 1:5000.

Since the Revolution in 1979, these map series have all been restricted for reasons of national security, and we have received no notification that this ban may be lifted. We have therefore included only small-scale maps in the catalogue section.

In addition to conventional mapping, a series of radar (SLAR) image maps has been constructed. The radar was flown in 1971 by Westinghouse Electric Corporation at 1:250 000 and was converted into 1:100 000 scale mosaics by Hunting Surveys, Toronto. These have been used to create interpretative maps of land cover, geomorphology and geology.

The radar imagery also contributed to the preparation of a preliminary 1:1 000 000 scale geological map by IGN in collaboration with the **Servicio Geológico Nacional (SGN)** and the Dirección Ejecutiva de Catastro y Recursos

Naturales. A number of 1:250 000 and 1:50 000 geological sheets of the western part of the country have also been published.

Systematic soil and land use surveys were initiated in 1968 by the Departamento de Suelos of the Ministry of Agriculture as an integral part of the cadastral surveying programme. Soil classification is based on the USDA system. This work, in common with the cadastral programme, has been concentrated in the western area.

Further information

Cartographic publications up until about 1977 are included in *Guia de recursos basicos contemporaneos para estudios de desarrollo en Nicaragua*, one of the PAIGH guides, which was prepared by the IGN and published in 1977.

Addresses

Instituto Geográfico Nacional (IGN)
Km 6, Carretera Norte, Apartado Postal 2110, MANAGUA.

Servicio Geológico Nacional (SGN)
Apartado Postal 1347, MANAGUA.

Instituto Nicaragüense de Turismo
Apartado Postal 122, MANAGUA.

For DMA and CIA, see United States; for Haack, see German Democratic Republic; for Cartographia, see Hungary.

Catalogue

Atlases and gazetteers

Nicaragua. Official standard names approved by the US Board on Geographic Names   2nd edn
Washington, DC: DMA, 1976.
pp. 129.

General

Nicaragua   1:2 500 000
Gotha: Haack, 1986.

Nicaragua   1:2 500 000
Budapest: Cartographia, 1987.

Nicaragua   1:1 500 000
Washington, DC: CIA, 1979.

República de Nicaragua   1:1 000 000
Managua: Instituto Nicaragüense de Turismo/Wuppertal: Edition Nuevo Hombre, 1986.
On reverse: Spanish, German and English texts.

Nicaragua. Mapa hipsografica   1:1 000 000
Managua: IGN, 1978.

Geological and geophysical

Nicaragua. carta isogónica para 1975.0   1:1 000 000
Managua: IGN, 1978.

República de Nicaragua. Mapa geológico preliminar   1:1 000 000
Managua: SGN, 1974.

Nicaragua. Mapa geológico   1:250 000
Managua: SGN, 1972–.
5 sheets published.

Panama

Detailed mapping of Panama began in the 19th century, receiving a stimulus from the need for topographic surveys to precede the construction of the Panama Railway and the Panama Canal. Notable maps of that century were those of H. Tiedermann (1851) and Agustin Codazzi (1855). These maps covered only limited areas, however, and accurate general maps of the whole country did not appear until the early decades of the present century.

A regular programme of topographic surveying began in 1946 when a Cartography Section was established for the Pan-American Highway. The modern topographic survey was established in 1967 under the title Instituto Cartográfico 'Tommy Guardia' (Tomas Guardia having been the first director of its antecedent, the Dirección de Cartografia). In 1969 the name was changed to **Instituto Geográfico Nacional 'Tommy Guardia' (IGNTG)**, and this institute now has responsibilities for topographic and geodetic survey, cadastral and hydrographic survey, and various development projects requiring cartographic work.

Much of the initial basic scale mapping at 1:50 000 (Series E762) was carried out after 1946 in cooperation with the IAGS. But since 1962, most work has been undertaken independently by the Institute, which now has facilities for air photo survey and photogrammetry. Air photo cover exists for almost all the country, while radar

imagery of the Province of Darien was acquired in 1967–72, an area where persistent cloud cover made conventional photography impossible to obtain.

The 1:50 000 map is in five colours with a 20 m contour interval, and is drawn on the Transverse Mercator projection with a UTM grid. The legend is in Spanish and English. The sheet format is 10' by 15'. The general map at 1:250 000 is also on the Transverse Mercator projection and has 200 m contours. This latter map is frequently revised.

A project to map the country at 1:25 000 (Series E866) was abandoned in 1960, but 46 sheets had been printed to that date and cover the Panama Canal Zone. A 1:10 000 series is being prepared by IGNTG for cadastral purposes. This is a UTM map with 5 m contours.

The **Dirección General de Recursos Minerales (DGRM)** was founded in 1970 and carries out various mapping activities in relation to the exploitation of mineral resources, but systematic geological mapping is provided by the regional maps, printed in colour, at 1:250 000 and the 1:1 000 000 map of the whole country.

In 1964 a Rural Land and Water Cadastral Programme was set up to make an extensive inventory of natural resources. The work involved air photo interpretation in association with field and laboratory work and the project ran for four years, covering the 53 per cent of the country where most of the population and agricultural activity is concentrated. A large number of special-purpose maps were produced, mostly at 1:50 000, and covering the themes of geology, geomorphology, soils, drainage, ecology and land use. These maps are held by the Instituto de Investigaciones Agropecuarias.

Several other government departments make maps. These include the **Dirección de Estadística y Censos (DEC)** (mapping of census enumeration districts, and certain sheets in the National Atlas), the Departamento de Suelo (various local maps of soils, geology, land use, climate and ecology), the Servicio Forestal (forest inventories and forest classification) and the Dirección Nacional de Mantenimiento of the Ministry of Public Works (highway maps and maps of transport infrastructure). Planimetric maps of all the main cities are prepared by the Ministerio de Vivienda.

The first National Atlas of Panama was published in 1965, the inspiration of the geographer Angel Rubio, who died however 2 years before its publication. The current edition was published in 1975 and contains 161 maps with textual descriptions. The maps cover the following major themes: general regional maps of the provinces, natural resources, population and human resources, agricultural development, industry, commerce and fishing, public services and tourism.

The **Instituto Panameño de Turismo** publishes a simple 1:1 000 000 tourist map of the country which includes a street map of Panama City on the reverse.

Further information

IGNTG produces an index of topographic and other mapping (*Indices de material cartográfico*).

The following two publications give accounts of the development of Panamanian mapping: Actividades cartográficas en Panamá, in *Technical Papers, First United Nations Regional Cartographic Conference for the Americas*, Panama, 8–19 March 1976, Vol. 2, pp. 20–41, New York: United Nations, 1979; and Barahona, J. (1972) *El desarrollo de la cartografía en Panamá*, IGNTG, 1972.

A thorough listing of maps produced up to about 1978 is included in *Guide for Research Workers of Panama*, one of the PAIGH guides, produced in collaboration with IGNTG (PAIGH, 1978: 48).

Addresses

Instituto Geográfico Nacional 'Tommy Guardia' (IGNTG)
Apartado 5267, PANAMA 5.

Dirección General de Recursos Minerales (DGRM)
Apartado 8515, PANAMA 5.

Dirección de Estadística y Censos (DEC)
Contraloría General de la República, Apartado 5213, PANAMA 5.

Universidad de Panama
Urbanización El Cangrejo, PO Box Estafeta Universitaria, PANAMA.

Instituto Panameño de Turismo
Apartado Postal 4421, PANAMA 5.

Catalogue

Atlases and gazetteers

Atlas nacional de Panamá
Panama: IGNTG, 1975.
pp. 116.

Diccionario Geográfico de Panamá
Panama: Universidad de Panama, 1972–.
3 volumes.

Panama. Official names approved by the US Board on Geographic Names
Washington, DC: DMA, 1969.

General

Panama 1:1 500 000
Washington, DC: CIA, 1970.
Inset maps of proposed canal routes, Canal Zone, population, vegetation and economic activity.

Panamá. Mapa turístico 1:1 000 000
Panamá: IGNTG.
Double-sided map which includes plans of Panama City, Colon and David.

Panamá. Mapa físico 1:500 000
Panamá: IGNTG, 1980.
2 sheets, both published.

Topographic

Panamá 1:250 000 (Especial) 10th edn
Panamá: IGNTG, 1985.
12 sheets, all published.

Mapa topográfica de Panamá 1:50 000 Series E702
Panamá: IGNTG, 1967–.
ca 200 sheets, 160 published. ■

Geological and geophysical

Mapa geológico de la República de Panamá 1:1 000 000
Panamá: DGRM, 1976.

[Panamá]. Mapa geológico 1:250 000
Panamá: DRGM, 1976–.
7 sheets (1–2, 3–4, 6–7 are paired), all published.

Administrative

Panamá. Mapa político 1:500 000
Panamá: IGNTG, 1980.
2 sheets, both published.

Town maps

Mapa de la Ciudad de Panamá y Alrededores 1:5000
Panamá: IGNTG, 1975–76.
65 sheets, all published.

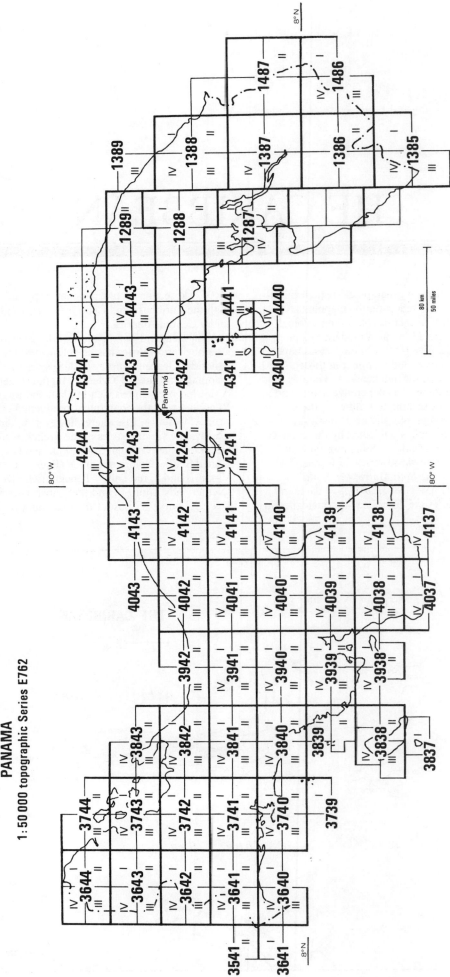

PANAMA

1:50 000 topographic Series E762

THE CARIBBEAN

The Caribbean forms a natural geographical region, but its cartography is quite diverse, reflecting the mapping styles of the several nations with former or continuing colonial interests in the area. These all initiated modern mapping programmes after World War II, and in most cases have continued to help with survey and mapping in those Caribbean states which are now independent. There is thus rather little regional mapping of a cooperative nature and a dearth of small-scale maps seeking to synthesize these mapping efforts either topographically or thematically. Apart from a few thematic maps published by the United States Geological Survey and other US organizations, it is necessary to resort mainly to world series for a general overview, as for example the *Soils Map of the World* 1:5 000 000 and the CGMW *Geological Map of the World* 1:10 000 000. These series are described and catalogued in the World section. A few regional maps and atlases cover both the Caribbean and Central America and are included under Central America.

As in mainland Latin America, the geodetic network of a number of countries has been linked to the Inter-American Geodetic Survey (IAGS) system, and cooperative mapping with the US Defense Mapping Agency (formerly AMS) has been undertaken in Haiti and the Dominican Republic, with the consequent production of 1:50 000 scale AMS series. Elsewhere, however, the mapping displays the varying styles of the French Institut Géographique National (Martinique and Guadeloupe), the United States Geological Survey (Puerto Rico and US Virgin Islands), the Dutch Topografische Dienst (Netherlands Antilles), and the British Directorate of Overseas Surveys (most of the remaining islands). In

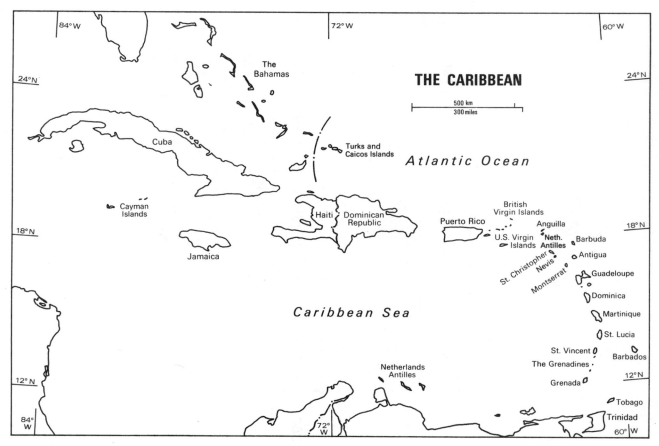

Cuba, a post-revolutionary national survey came into existence in 1967, but this too has received outside help, in this case from the USSR and Eastern Europe.

Of all the Caribbean territories, a majority have been mapped by Britain. Thus 15 out of the 23 islands or island groups identified below were originally mapped by the Directorate of Overseas Surveys (DOS), and most continue to rely on British support for their mapping programmes. Systematic survey began in the period immediately following World War II. The DOS (originally called the Directorate of Colonial Surveys, and now—since 1984—the Overseas Surveys Directorate of the Ordnance Survey, OSD) was formed in 1946 and began its work in Antigua, Barbados and Jamaica, initially using US Army Air Force (USAAF) photography and control, but subsequently contracting new photographic sorties to commercial air survey companies and establishing its own mapping control (the history of DOS mapping is thoroughly chronicled by McGrath, 1983).

Users of DOS maps of the Caribbean will generally find that there are three scale 'levels' to choose from. The standard topographic scale is normally 1:25 000 (sometimes 1:10 000), and this provides a detailed gridded and contoured map of prime value for scientific users. More general maps are usually represented by a scale of 1:50 000 (1:250 000 for a large island like Jamaica). Many of these have been adapted in recent editions for tourist use, with the addition of text, indexes and insets of the capital towns and popular tourist locations. There are also small-format 'country maps' of some islands at scales usually of 1:100 000 or smaller. These latter serve the purpose of general orientation and are sometimes included in our catalogue, especially where they put an entire island state (such as St Vincent and the Grenadines) on to one sheet.

The third scale level is the large-scale planimetric mapping most commonly at 1:2500 or 1:5000. Maps at these large scales were initially prepared for urban or other densely populated areas, but there has been a tendency to extend and revise these large-scale surveys in recent years, usually in support of land registration requirements. The DOS has run a Regional Cadastral Survey and Registration Project in several of the islands. Most of the large-scale maps are held in transparency form, from which diazo copies can be made as required.

Since Guadeloupe and Martinique are overseas departments of France, French mapping in the Caribbean is simply an extension of the French metropolitan series. Their basic scale mapping has recently been converted to the specifications of the French 1:25 000 *série bleue*.

It is not only colonial powers, however, who have been active in mapping Caribbean America. The USSR and Czechoslovakia have been involved in the mapping of Cuba and the former was responsible for the publication of the national *Atlas de Cuba*. The German Federal Republic has been involved, under its aid programmes, in mapping activities in Haiti, while topographic mapping initiated by the DOS in Trinidad was completed with Canadian help. The United Nations Development Programme has also contributed to mapping in the region, in supporting the basic mapping programmes of Jamaica and of Trinidad and Tobago, for example, as has the Food and Agriculture Organization of the United Nations.

The result of all this foreign interest and activity has been the production of relatively complete and generally high-quality mapping for the whole island chain. Most of it is available on the international market, although some difficulties will be encountered in politically sensitive areas. In preparing this section we had little success in eliciting a *direct* response from Cuba, the Dominican Republic or the Netherlands Antilles, and maps of these countries may prove difficult to obtain. DOS mapping may be readily obtained either from the mapping office or agency in the country concerned or, in most cases, from the Overseas Surveys Directorate, Southampton or the OSD agents Edward Stanford in London.

Further information

The mapping by DOS in the Caribbean is particularly well documented. The DOS annual reports give information on survey and mapping progress country by country, and these include a survey status map for the Caribbean. The French IGN also issues an annually revised status map for its overseas mapping.

For more information on DOS mapping in the Caribbean, see also: McGrath, G. (1983) Mapping for development: the contribution of the Directorate of Overseas Surveys, *Cartographica*, **20**(1&2), 264; and Howells, L.J. (1985) Systematic mapping programmes in the Commonwealth countries of the Caribbean: a review, paper presented to the *Third United Nations Regional Conference for the Americas* (E/CONF.77/L.45).

Addresses

For ACC, see Cuba; for NGS, AAPG, NOS and USGS see United States; for GUGK see USSR; for Bartholomew see Great Britain.

Catalogue

Atlases and gazetteers

Atlas regional del Caribe
La Habana: Instituto de Geografía, ACC, 1979.
pp. 69.
In Spanish with contents in English.

General

West Indies and Central America   1:4 500 000
Washington, DC: NGS, 1981.
On reverse: tourist islands of the West Indies.

West Indies and the Caribbean   1:3 250 000
Edinburgh: Bartholomew, 1985.

Tsentral'naja antil'skie ostrova   1:2 500 000
Moskva: GUGK.

Bathymetric

Bathymetry of the Gulf of Mexico and the Caribbean Sea
1:3 289 263
Tulsa, OK: AAPG, 1984.

Caribbean region—bathymetric map   1:2 500 000
Washington, DC: NOS, 1981.

Geological and geophysical

Geologic-tectonic map of the Caribbean region   1:2 500 000   MIS Map I-1100
Reston, VA: USGS, 1980.
3 sheets, all published.

Anguilla

Anguilla is a self-governing colony of Britain, but it was briefly linked in an independent union with St Christopher (St Kitts) and Nevis. Following its succession in 1967, it returned to British rule in 1971 acquiring self-governing status in 1976.

The original mapping by the **Directorate of Overseas Surveys (DOS)** (now Overseas Surveys Directorate, Southampton, OSD) was a three-sheet series at 1:25 000 which included St Kitts and Nevis (DOS 343). This was published in 1961 and based on air photography flown in 1946 by the USAAF and in 1956 by Hunting. New editions of the Anguilla sheet were published in 1986 (Edition 5 with BWI grid and Edition 6 with UTM grid). Contour interval is 25 feet. There is also a 1:50 000 map (DOS 443), compiled from the earlier 1:25 000 map with revision from 1968 air photography and 1971 ground survey. This sheet is in five colours with 25 foot contours. The projection is Transverse Mercator.

In the early 1980s, densification of the survey control network was undertaken, and a 54 sheet series covering the whole island at 1:2500 and based on new air photography has been published.

Addresses

Lands and Surveys Department
The Valley, ANGUILLA.

Overseas Surveys Directorate (OSD/DOS)
Ordnance Survey, Romsey Rd, Maybush, SOUTHAMPTON
SO9 4DH, UK.

Catalogue

General

Anguilla   1:80 000   OSD 943
Southampton: OSD for Anguilla Government, 1986.

Topographic

Lesser Antilles   1:50 000   DOS 443   Edition 1   Anguilla
Tolworth: DOS, 1973.

Lesser Antilles   1:25 000   DOS 343 (E803)   Editions 5/6   Anguilla
Southampton: OSD for Anguilla Government, 1986.

Antigua (including Barbuda)

Antigua, together with neighbouring islands Barbuda and Redonda, formed a self-governing state in association with the UK in 1967. In 1981 the islands became a fully independent nation.

Antigua was one of the first Caribbean islands to be mapped by the Directorate of Colonial Surveys (later **Directorate of Overseas Surveys (DOS)**, now Overseas Surveys Directorate, Southampton, OSD) with publication of a 1:25 000 scale map (DCS 6) in 1946 based on US Army Air Corps photography and military ground control. A second edition of this map (DOS 306/E843) was published in 1962 but was not subsequently revised. DCS also published an early 1:10 000 contoured map of St John's, the capital (DCS 11, 1947). New air photography was flown by Fairey Surveys in 1968, and 1:2500 scale contour mapping of the north part of the island including St John's was completed in 1970. This urban series was revised in the late 1970s, based on photography flown in 1975. The whole island was also mapped at 1:500, with litho-printed contoured sheets being issued from 1971 to 1973.

The 1:50 000 map of Antigua (DOS 406/E703) was derived from the 1:500 series and was first published, as a tourist map, in 1973. The most recent editions were issued in 1980, and are printed in process colours with BWI grid (as Edition 7) or UTM grid (as Edition 8). Contours are at 50 foot intervals, and the entire land area is printed in shades of green (distinguishing cultivation or grass from bush and trees) or blue (for marsh and mangrove swamp). The projection is Transverse Mercator.

New DOS mapping of Antigua at 1:2500 and 1:5000 (the latter scale for the less developed areas) was completed in the early 1980s.

A map of *Barbuda* in nine sheets was published at 1:10 000 in 1970 (DOS 257) based on 1966 air photography, and a derived 1:25 000 map in four colours with 25 foot contours was issued in 1970–71 (DOS 357/ E803). The projection is Transverse Mercator and the grid UTM. *Redonda* appears on the DOS map of Montserrat.

Addresses

Lands and Survey Officer
Survey Division, Ministry of Agriculture, Lands and Fisheries, ST JOHN'S.

Overseas Surveys Directorate (OSD/DOS)
Ordnance Survey, Romsey Rd, Maybush, SOUTHAMPTON
SO9 4DH, UK.

Catalogue

General

Antigua   1:100 000   DOS 993   Edition 3
Tolworth: DOS, 1980.

Topographic

Lesser Antilles 1:50 000 DOS 406 (E703) Edition 7/8
Antigua
Tolworth: DOS, 1980.
Insets: St John's 1:10 000, English Harbour 1:10 000.

Lesser Antilles 1:25 000 DOS 357 (E803) Edition 1 Barbuda
Tolworth: DOS, 1970–71.
2 sheets, both published.

Bahamas

The Commonwealth of the Bahamas has been an independent nation since 1973. It consists of about 700 islands (29 of them inhabited) spread over about 750 miles (1200 km) of ocean.

Topographic survey was initiated by the British Directorate of Overseas Surveys, which began issuing a 1:25 000 series (DOS 358) from 1962. These map sheets were issued in multi-sheet sets covering each island group, and the following were completed:

- Cat Island (Series E818), 6 sheets
- Eleuthera (Series E817), 8 sheets
- Grand Bahama and Abaco (Series E811), 33 sheets
- New Providence (Series E815), 2 sheets

These sheets were all on a Transverse Mercator projection with UTM grid and, when appropriate, contours at 25 foot intervals. DOS also published 11 sheets of urban mapping at 1:2500 scale covering Nassau in 1961–63 (DOS 158/E819).

Since 1972, topographic surveying has been in the hands of the **Department of Lands and Surveys (BLS)**, Nassau, and all the remaining island groups have been mapped at 1:25 000 in new BLS series, with the exception of the Bimini Island group, for which 1:10 000 scale maps are available. A total of 182 sheets at 1:25 000 and 10 at 1:10 000 are needed to give complete cover of the Bahamas. BLS has also issued 1:10 000 maps of most islands and some 1:2500 mapping (e.g. New Providence in 265 sheets), mostly available as diazo prints.

The BLS maps are listed in our catalogue in the island group series by which they are published. Although some stocks of the DOS maps were still available in the mid-1980s, they have all been superseded by BLS editions published in the 1970s, and are therefore not included in the catalogue.

A number of resource surveys have been carried out for the Bahamas Government by the Overseas Development Administration through the **Land Resources Development Centre (LRDC)**, Tolworth. These include an island-by-island survey, published between 1971 and 1976, of vegetation, land use/land capability and land ownership with associated maps at 1:50 000 or 1:25 000 (not now available at LRDC) and summarized as *Land Resources of the Bahamas: a summary* (LRS 27) published in 1976 with several 1:1 000 000 maps of the island group. This report is still available.

In 1971, the Directorate of Military Surveys, UK, conducted gravity and magnetic surveys for the Bahamas Government, and these were published at 1:250 000.

Further information

An up-dated list of available topographic maps is issued by BLS from time to time. A *Land Resource Bibliography of the Bahamas* by N.W. Posnett and P.M. Reilly was published by LRDC in 1971.

Addresses

Department of Lands and Surveys (BLS)
PO Box N-592, NASSAU.

Land Resources Development Centre (LRDC)
Tolworth Tower, SURBITON KT6 7DY, UK

For USGS, see United States; for GUGK, see USSR.

Catalogue

Atlases and gazetteers

Atlas of the Commonwealth of the Bahamas
Nassau: Ministry of Education, 1985.
pp. 48.

General

The Commonwealth of the Bahamas 1:2 300 000 BLS 000
Nassau: BLS, 1978.

Bagamskiye ostrova 1:1 500 000
Moskva: GUGK, 1985.
Map of Bahamas with placename index.

Bahamas map 1:1 000 000
Nassau: BLS, 1980.
Available laminated or folded.

Berry Islands: The Bahamas 1:500 000
Reston: USGS, 1980.
Satellite image map.

Topographic

Abaco 1:25 000 BLS 311 Edition 1
Nassau: BLS, 1975.
27 sheets, all published. ■

Acklins Island 1:25 000 BLS 326 Edition 1
Nassau: BLS, 1972.
13 sheets, all published. ■

Andros Island 1:25 000 BLS 315 Edition 1
Nassau: BLS, 1970.
42 sheets, all published. ■

Berry Islands 1:25 000 BLS 312 Edition 1
Nassau: BLS, 1969.
4 sheets, all published. ■

Bimini Island 1:10 000 BLS 213 Edition 1
Nassau: BLS, 1969.
10 sheets, all published. ■

Cat Island 1:25 000 BLS 318 Edition 1
Nassau: BLS, 1972–75.
8 sheets, all published. ■

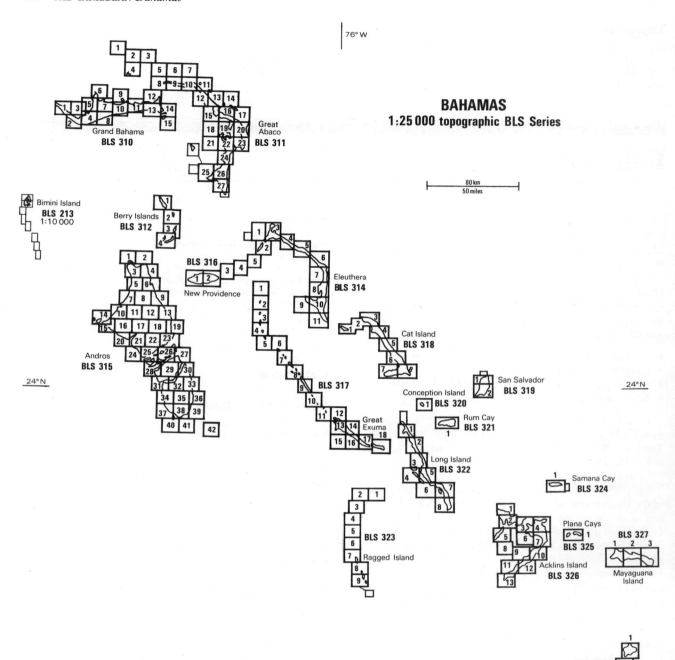

BAHAMAS
1:25 000 topographic BLS Series

Conception Island 1:25 000 BLS 320 Edition 1
Nassau: BLS, 1972.
1 sheet. ■

Eleuthera 1:25 000 BLS 314 Edition 1
Nassau: BLS, 1969–75.
11 sheets, all published. ■

Exuma Island 1:25 000 BLS 317 Edition 1
Nassau: BLS, 1969.
18 sheets, all published. ■

Grand Bahama 1:25 000 BLS 310 Edition 1
Nassau: BLS, 1968–75.
16 sheets, all published. ■

Inagua Island 1:25 000 BLS 328 Edition 1
Nassau: BLS, 1972.
14 sheets, all published. ■

Long Island 1:25 000 BLS 322 Edition 1
Nassau: BLS, 1972.
8 sheets, all published. ■

Mayaguana Island 1:25 000 BLS 327 Edition 1
Nassau: BLS, 1972.
3 sheets, all published. ■

New Providence 1:25 000 BLS 316 Edition 1
Nassau: BLS, 1972–75.
5 sheets, all published. ■
Includes 'Cays to Eleuthera'.

Plana Cay 1:25 000 BLS 325 Edition 1
Nassau: BLS, 1972.
1 sheet. ■

Ragged Island 1:25 000 BLS 323 Edition 1
Nassau: BLS, 1972.
9 sheets, all published. ∎

Rum Cay 1:25 000 BLS 321 Edition 1
Nassau: BLS, 1972.
1 sheet. ∎

Samana Cay 1:25 000 BLS 324 Edition 1
Nassau: BLS, 1972.
1 sheet. ∎

San Salvador 1:25 000 BLS 319 Edition 1
Nassau: BLS, 1972.
2 sheets, both published. ∎

Town maps

Street map of Nassau, New Providence 1:5000 BLS 319
Edition 1
Nassau: BLS, 1973.
Bound volume with gazetteer.

Barbados

Barbados became an independent sovereign state within
the British Commonwealth in 1966, and now has its own
Lands and Surveys Department at St Michael,
Bridgetown.

The basic mapping of the island was carried out by the
Directorate of Colonial Surveys (late DOS and now
Overseas Surveys Directorate, Southampton, OSD) at
1:10 000 scale in the early 1950s, and this was one of the
earliest Caribbean series to be completed. The 1:50 000
map was compiled from this basic scale series, and the
latest edition was published in 1974 by DOS for the
Barbados Government, revised from air photography flown
in 1972–73. The 1:50 000 has the BWI Grid as well as a
local alphanumeric grid. Contours are at 100 foot vertical
interval, and sugar cane plantations are shown in green.

The 1:50 000 geology map was prepared from a
1:10 000 scale diazo series produced in 1980–81 (DOS
1227). It is a colour map with text printed in the margin.
A special sheet *Geology of the Scotland Area* 1:20 000
(DOS 1228) was published in 1982.

Large-scale contour mapping of Bridgetown, and of
Speightstown and the Seawell Airport area, was
undertaken in the 1970s, and the 1:5000 tourist map of
Bridgetown (1975) is derived from this mapping.

In the early 1980s work began on remapping the country
at 1:10 000, and new derived tourist maps at 1:50 000 and
1:100 000 will be produced from this. The 1:10 000 map
will be in 12 sheets and Canadian aerial photography and
aerial triangulation data is being used.

Addresses

Lands and Surveys Department
Culloden Road, St Michael, BRIDGETOWN.

Energy and Natural Resources Division
Ministry of Finance and Planning, Bay Street, St Michael,
BRIDGETOWN.

Overseas Surveys Directorate (OSD/DOS)
Ordnance Survey, Romsey Road, Maybush, SOUTHAMPTON
SO9 4DH, UK.

Catalogue

Topographic

Barbados 1:50 000 Series DOS 418 Edition 2
Tolworth: DOS, 1974.

Geological and geophysical

Geology of Barbados 1:50 000 Series DOS 1229
Tolworth: DOS, 1983.

Town map

Bridgetown and environs, Barbados 1:5000 Series DOS 118
Tolworth: DOS, 1975.

Cayman Islands

The Cayman Islands are a British dependency, formerly
under the administration of Jamaica. There are three
islands in the group: Grand Cayman, Little Cayman and
Cayman Brac.

The **Directorate of Overseas Surveys (DOS)** (now
Overseas Surveys Directorate, Southampton, OSD) first
published a 1:25 000 map cover (DOS 328) in 1965–66
based on air photography flown in 1958. Complete air
photo cover was again flown in 1970–71, and the 1:25 000
sheets were revised and published in a second edition in
1978 with planimetry derived from large-scale cadastral

plans, and swamp vegetation from an ODM survey carried
out in 1976. A road classification was supplied for this
edition by the Survey Department of the Cayman Islands.
This map series has 20 foot contours, and is on a
Transverse Mercator projection with UTM grid. In 1979, a
single-sheet 1:50 000 visitors' map combining both line
and photomapping techniques (DOS 428P) was issued.
This map has an inset of the capital, Georgetown, at
1:10 000.

Large-scale mapping of the Caymans includes 1:2500
maps of Georgetown (1965), and 1:2500 cadastral mapping

carried out in the 1970s. In 1982 the **Land Resources Development Centre (LRDC)** published three 1:25 000 swamp maps based on work by the Mosquito Control and Research Unit in the Caymans.

Addresses

Lands and Survey Department
PO Box 1085, GRAND CAYMAN.

Overseas Surveys Directorate (OSD/DOS)
Ordnance Survey, Romsey Road, Maybush, SOUTHAMPTON SO9 4DH, UK.

Land Resources Development Centre (LRDC)
Tolworth Tower, SURBITON KT6 7DY, UK.

Catalogue

General

Cayman Islands 1:150 000 DOS 928 Edition 3
Tolworth: DOS, 1975.

Topographic

Cayman Islands 1:50 000 DOS 428P Edition 1
Tolworth: DOS, 1979.
Inset of Georgetown 1:10 000.

Cayman Islands 1:25 000 DOS 328/E821 Edition 2
Tolworth: DOS, 1978.
4 sheets, all published.

Cuba

The **Instituto Cubano de Geodesia y Cartografía (ICGC)**, Havana, was founded in 1967 and assigned the tasks of executing a geodetic survey and preparing topographic maps, as well as creating a national cadastre and maintaining records of land tenure and use. During the 1970s, first, second and third order triangulation networks were completed. At the same time, a basic series of 1:10 000 scale maps was initiated, and this was 80 per cent complete by 1980. Series at 1:50 000, 1:100 000, 1:250 000 and 1:500 000 have also been completed. In spite of this very active programme, topographic maps do not appear to be in circulation to the public, at least outside Cuba, and so only more readily obtainable general maps are listed in our catalogue.

A major geographical synthesis of the country was provided by the excellent *Atlas Nacional de Cuba*, prepared by the **Academia de Ciencias de Cuba (ACC)** in collaboration with the Academy of Sciences of the USSR and published in 1970. This includes 143 pages of full-colour maps covering physical, economic, social and historical themes, and has an index of place names. This atlas was published in both Spanish and Russian editions, but the Spanish edition is now out of print. The *Atlas de Cuba*, published in 1978, is however akin to a smaller version of the National Atlas, offering 93 thematic maps (at 1:1 750 000 scale) and a set of general maps of the island at 1:300 000 together with descriptive text.

Geological mapping was formerly undertaken by the now defunct Instituto Cubano de Recursos Minerales (ICRM), and several small-scale maps on geology-related themes were published in the 1960s. The 1:1 000 000 geological map of 1962 may still be available.

From 1972 to 1981 a 1:250 000 scale geological map of Cuba was prepared by the **Instituto de Geología y Paleontología** of the Cuban Academy of Sciences. Scientists from Eastern Europe and the USSR collaborated in its production. This map has not yet been published.

Soil mapping has been carried out by the **Instituto de Suelos** of the Cuban Academy of Sciences, and a set of 19 coloured 1:250 000 scale sheets distinguishing 14 soil types was published in 1971. There is a text (*Informe sobre el mapa genetico de los suelos de Cuba* by A. Hernandez *et al.*) describing these soils.

Further information

A record of soil mapping in the 1960s is given in *Suelos de Cuba. Tomo I* (La Habana: Editorial Orbe, 1975, pp. 352).

The development of geological mapping up to the present period is outlined by Franco, G.L. (1984) A historical outline of the geological cartography of Cuba, pp. 205–218, in *Contributions to the History of Geological Mapping*, edited by E. Dudich, Budapest: Akademiai Kiado.

Addresses

Instituto Cubano de Geodesia y Cartografía (ICGC)
Loma y 39, Nuevo Vedado, LA HABANA.

Instituto de Geografía, Academia de Ciencias de Cuba (ACC)
Calle 11, No 514 e/D y E, Vedado, LA HABANA.

Instituto de Geología y Paleontologia
Academia de Ciencias de Cuba, Calzado No 851 esq. 4 Vedado, LA HABANA 4.

Instituto de Suelos, Academia de Ciencias de Cuba (ACC)
Apartado 8022, LA HABANA 8.

Comité Estatal de Estadísticas (CEE)
Calle 46 No 307, Mirimar, Mun. Playa, LA HABANA.

Editorial Orbe
Calle 17, No. 903 entre 6 y 8, redado, LA HABANA.

For Karto+Grafik, see German Federal Republic; for GUGK, see USSR; for CIA, see United States.

Catalogue

Atlases and gazetteers

Atlas de Cuba
La Habana: ICGC, 1978.

Atlas Nacional de Cuba/Natsional'ny atlas Kuby
La Habana: ICGC/Moskva: GUGK, 1970.
Only Russian language version still in print.

Miniatlas de Cuba
La Habana: ICGC.
pp. 60.

General

Cuba 1:2 450 000
Washington, DC: CIA, 1984.

Kuba 1:1 500 000
Moskva: GUGK, 1982.

Mapa turístico de Cuba ca 1:1 450 000
La Habana: ICGC, 1983.
On reverse: street map of La Habana and Varadero.
Legend in Spanish and English.

Cuba 1:1 100 000 Hildebrands Urlaubskarte
Frankfurt AM: Karto+Grafik, 1985.
Tourist map with text in German.
Small inset maps of La Habana and Santiago de Cuba.

Geological and geophysical

Mapa geológico de Cuba 1:1 000 000
La Habana: ICRM, 1962.

Environmental

Genetic map of soils of Cuba 1:250 000
La Habana: Instituto de Suelos, ACC/ICGC, 1971.
19 sheets, all published.

Administrative

Cuban administrative units 1:1 000 000
Washington, DC: CIA, 1986.

Cuba político-administrativo 1:900 000
La Habana: ICGC, 1982.

Human and economic

Atlas demográfico nacional
La Habana: CEE/ICGC, 1985.
96 maps + 17 pp. text.
Atlas of 1981 population census.

Town maps

Mapa turístico de la Habana
La Habana: ICGC, 1982.

La Habana. Guía general de la ciudad ca 1:17 500 6th edn
La Habana: ICGC, 1978.

Dominica

The Commonwealth of Dominica gained full independence from Britain in 1978, and is now a republic and a member of the British Commonwealth. The island has been mapped by the **Directorate of Overseas Surveys (DOS)** (now Overseas Surveys Directorate, Southampton, OSD) with the principal map being the 1:25 000 DOS 351. This was first issued in 1961, and was plotted from air photography flown in 1956 with local information supplied by the Crown Surveyor, Dominica. Originally a six-colour production, the fourth edition (1978) was printed in process colours. The projection is Transverse Mercator, the grid BWI and there is a 50 foot contour interval.

The 1:50 000 map (DOS 451), derived from the 1:25 000, was first issued in 1963. It was conceived as a tourist map, and the first edition incorporated shaded relief. The third edition (1982) was considerably redesigned, with the adoption of a more practical road classification and the addition of layer tints as well as shading to the 250 foot contoured relief.

The 1:5000 map of Roseau is a full-colour production and has an index of public buildings referenced by an alphanumeric grid. The second edition also has 50 foot contours printed in green.

Contoured 1:2500 scale mapping has been produced for selected areas, and new mapping at this scale and at 1:5000 was in progress in the 1980s.

Resource mapping has been carried out in Dominica by the Overseas Development Administration of the British Government through the **Land Resources Development Centre (LRDC)**. 1:25 000 soil mapping was completed in 1972, and in the 1980s a project was undertaken to produce land suitability maps for 20 major crops and a 'best solution' crop development map. The final report, *LRDC Project Report* 127, was issued in 1983 and included six 1:30 000 scale land suitability maps.

Further information

A *Land Resource Bibliography of Dominica* by N.W. Posnett and P.M. Reilly was published by LRDC in 1978.

Addresses

Lands and Surveys Division
Ministry of Agriculture, Lands and Fisheries, ROSEAU.

Overseas Surveys Directorate (OSD/DOS)
Ordnance Survey, Romsey Rd, Maybush, SOUTHAMPTON SO9 4DH, UK.

Land Resources Development Centre (LRDC)
Tolworth Tower, SURBITON KT6 7DY, UK.

Catalogue

General

Dominica 1:125 000 DOS 998 Edition 2
Tolworth: DOS for Dominica Government, 1982.

Topographic

Dominica 1:50 000 DOS 451 (E703) Edition 3
Tolworth: DOS for Dominica Government, 1982.
Inset: Roseau 1:10 000.

Dominica 1:25 000 DOS 351 (E803) Edition 4
Tolworth: DOS for Dominica Government, 1978.
3 sheets, all published.

Environmental

Land suitability for major crops in Dominica 1:30 000
Surbiton: LRDC, 1983.
Set of six maps accompanying REP 127.

Dominica. Land use 1:25 000 DOS 3140B
Tolworth: DOS, 1972.
3 sheets, all published.

Town map

Roseau and environs 1:5000 DOS 151 Edition 2
Tolworth: DOS, 1974.

Dominican Republic

An Instituto Cartográfico Militar was established in Santo
Domingo in 1955, and this was combined in 1960 with the
Comisión de Limites Geográficos Nacionales under the
authority of the University of Santa Domingo to form the
Instituto Cartográfico Universitario. This is now called the
Instituto Geográfico Universitario (IGU), and is
responsible for all the survey and mapping programmes in
the Dominican Republic.

The Inter-American Geodetic Survey (IAGS) was active
in the Dominican Republic from about 1946, when it
helped to establish a survey control network for
topographic mapping, and subsequently a topographic
series was prepared jointly by the US Army Map Service
(now DMA) and the IGU. Today IGU continues to
cooperate closely with the IAGS.

The principal topographic series is the 1:50 000 joint
DMA/IGU series (E733), issued from about 1960 and
complete in 122 sheets. This is a five-colour map with
20 m contours on a Transverse Mercator projection. There
is also a 1:250 000 series in five sheets based on the JOG
1501 specification, and a series of 1:5000 city maps (E931)
dating from the 1960s. The 1:50 000 also forms the basis
of a series of planimetric province maps, prepared by IGU
in collaboration with the Secretaría de Estado de
Agricultura, which are issued as monochrome ozalid copies
for use in planning and development.

Some geological mapping has been carried out by the
Dirección General de Minería (DGM), and a 1:250 000
geological atlas was published in 1969, but does not appear
to be still available.

A set of 1:250 000 resource maps of the north-east was
published in 1977 by the Organization of American States
under the title *Desarrollo regional de la Linea Nordoeste*.

In 1984, the United States Geological Survey (USGS)
signed a mapping agreement with the Dominican
Republic. The USGS will collaborate with DGM in
geological mapping, and in mineral and marine
investigations over a 5 year period.

The standard map series of the Dominican Republic are
notoriously difficult to obtain, and we received no replies
to requests for information from the mapping authorities.
However, both the 1:250 000 and 1:50 000 series were
being offered for sale in the mid-1980s by the mapping
jobber GeoCenter. A good general map is provided by the
recent Hildebrands Urlaubskarte of *Hispaniola*. Other
tourist maps have been published by Texaco/Rand
McNally (1976) and by the Santo Domingo publisher
Triunfo.

Addresses

Instituto Geográfico Universitario (IGU)
El Conde No. 4, SANTA DOMINGO, DN.

Dirección General de Mineria (DGM)
Edificio Gubernamental, Av. Mexico 11A, El Huacal, SANTA
DOMINGO.

Triunfo
Arzobispo Portes No. 326, 2do piso, SANTA DOMINGO.

For GUGK, see USSR; for Karto+Grafik, see German Federal
Republic.

Catalogue

General

Gaiti Dominikanskaja Respublika 1:1 500 000
Moskva: GUGK.
In Russian.

Hispaniola 1:816 000
Hildebrands Urlaubskarte.
Frankfurt AM: Karto+Grafik, 1985.

Mapa turístico de la República Dominicana 1:740 000
Santo Domingo: Triunfo.
On reverse: street map of Santa Domingo.

Mapa de la República Dominicana 1:600 000
Santo Domingo: IGU, 1979.

Topographic

Mapa topográfica general 1:250 000
Santa Domingo: IGU, 1970–.
5 sheets, all published.

República Dominicana 1:50 000 Series E733
Santa Domingo: IGU, 1960–.
122 sheets, all published.

Town maps

Planos de ciudades, República Dominicana 1:5000 Series E931
Santa Domingo: IGU.

Grenada

Formerly a British dependency, Grenada acquired full independence in 1974. The national territory includes the more southerly of the Grenadines, in particular Carriacou Island and Petit Martinique (for mapping of the northern Grenadines, see under Saint Vincent).

Topographic mapping was undertaken by the **Directorate of Overseas Surveys (DOS)** (now Overseas Surveys Department, Southampton, OSD) in the 1950s based upon air survey, the first cover being flown in 1951. The 1:25 000 map (DOS 342) in two sheets was first published in 1958. It reached its fourth revised edition in 1979. This map is in five colours with a 50 foot/25 foot contour interval. The projection is Transverse Mercator with BWI grid. The derived 1:50 000 map (DOS 442) was first published in 1966. The current 4th edition (1985) is a tourist map printed in process colours with shaded relief and a 200 foot contour interval. It is on a Transverse Mercator projection with BWI grid. The DOS has also mapped the St George's area at 1:2500, commencing in 1972 and reaching completion in 1978 with the issue of 39 sheets available as diazo prints.

Carriacou Island is covered by sheet 5 of DOS 344/ E803, the only sheet now maintained in this series.

Mapping is now carried out by OSD in cooperation with the **Lands and Surveys Department**, St George's.

Addresses

Lands and Surveys Department
Office of the Prime Minister, Botanical Gardens, ST GEORGE'S.

Overseas Surveys Directorate (DOS/OSD)
Ordnance Survey, Romsey Rd, Maybush, SOUTHAMPTON SO9 4DH, UK.

Catalogue

General

Grenada (with Carriacou) 1:150 000 DOS 995 Edition 3
Tolworth: DOS, 1980.

Topographic

Grenada 1:50 000 DOS 442 (E703) Edition 4
Southampton: OSD for Government of Grenada, 1985.
Inset maps of south-west peninsula (1:25 000) and St George's (1:10 000).

Grenada 1:25 000 DOS 342 (E803) Edition 4
Tolworth: DOS for Government of Grenada, 1979.
2 sheets, both published.

Grenadines 1:25 000 DOS 344 (E803) Edition 3
Tolworth: DOS for Government of Grenada, 1978.
Sheet 5 Carriacou Island.

Guadeloupe

Administratively, Guadeloupe is an overseas Department of France, and topographic mapping was initiated by, and remains the responsibility of, the **Institut Géographique National (IGN)**, Paris. The IGN first obtained an air photo cover in 1946, and this was followed by triangulation, levelling and photogrammetric work in the period 1947 to 1952. A basic 1:20 000 map series was then issued in 36 sheets, and completed in 1957. It was in the style of the French metropolitan mapping of that scale (and period), and was printed in four colours with contours at 10 or 5 m intervals. The series, which covers also the neighbouring islands of Marie Galante, La Desirade, and St Martin and St Barthelemy, was issued in a second, revised edition in 1969. In the mid-1980s it was further revised and converted to a 1:25 000 series in only eight sheets. The 1:50 000 derived map in six sheets was first issued in 1958 and revised in 1969. This is in five colours with shaded relief. The road and tourist map at 1:100 000 is similar in style to that of Martinique, with shaded relief (but no contours or grid) and with insets of the main towns. The fourth edition of this map was published in 1982.

Guadeloupe was mapped geologically in the early 1960s by the **Bureau de Recherches Géologiques et Minières (BRGM)**, Orleans, and the islands are covered in five

sheets. A new geological map of La Desirade at 1:25 000 was published in 1980, and of *Saint Barthelemy et ses islets* (1:20 000) in 1983. These two monographic maps are accompanied by explanatory notes. There is also a geological survey office, **Arrondissement Mineralogique**, on Guadeloupe.

Soil mapping was carried out by F. Colmet Daage and A. Leveque for the **Office de la Recherche Scientifique et Technique Outre Mer (ORSTOM)**, Paris. The principal series, at 1:20 000, is in colour and is issued as a set of 18 sheets covering Basse-Terre. Grande Terre is covered by a 1:10 000 monochrome soils map in 54 sheets and Marie Galante by a single sheet at 1:20 000 issued in 1961, but now out of print.

The atlas of Guadeloupe is one of a series covering the overseas departments of France, prepared by the Centre d'Etudes de Géographie Tropicale of the **Centre National de Recherches Scientifiques (CNRS)**, at Bordeaux-Talance, together with geographers at ORSTOM, the University of Bordeaux III and the Centre Universitaire des Antilles et la Guyane. The atlas has 36 map sheets in colour at a principal scale of 1:150 000, accompanied by 80 pages of text, and covering themes concerned with the physical environment and with economic and social geography.

Addresses

Arrondissement Mineralogique
BP 448, POINTE-A-PITRE.

Institut Géographique National (IGN)
107 rue la Boetie, 75008 PARIS, France.

Bureau de Recherches Géologiques et Minières (BRGM)
Service Géologique National, BP 6009, 45060 ORLEANS-CEDEX, France.

Librarie du CNRS (CNRS)
295 rue St Jacques, 75005 PARIS, France.

Office de la Recherche Scientifique et Technique Outre Mer (ORSTOM)
70–74 Route d'Aulnay, 93140 BONDY, France.

Catalogue

Atlases and gazetteers

Atlas des Départements Français d'Outre Mer. III. La Guadeloupe G. Lasserre
Paris: CNRS/ORSTOM, 1982.
pp. 150.

General

Carte touristique 1 : 100 000 Edition 4 Guadeloupe
Paris: IGN, 1982.
Insets of major towns.

Topographic

Carte générale de la Guadeloupe 1 : 50 000
Paris: IGN, 1969.
6 sheets, all published.

Carte de la Guadeloupe 1 : 25 000
Paris: IGN, 1985.
8 sheets, all published.

Geological and geophysical

Carte géologique de la Guadeloupe 1 : 50 000
Orleans: BRGM, 1961–65.
5 sheets, all published.

Environmental

Resources en eau de surface de la Guadeloupe 1 : 100 000
Bondy: ORSTOM, 1982.
In 4 sheets: I, Géomorphologie; II, Carte des sols; III, Végétation et occupation des sols; IV, Reseaux hydrométrique et pluviométrique—isohyètes interannuelles—periode 1929–78.

Carte des sols de la Guadeloupe 1 : 20 000
Bondy: ORSTOM, 1970.
18 sheets published + legend.

Carte des cultures et d'utilisation du sol de la Guadeloupe
1 : 20 000 F. Colmet Daage
Fort-de-France: ORSTOM, 1980.
28 sheets, published in atlas format.
Diazo maps.

Haiti

The Republic of Haiti occupies the western third of the mountainous island of Hispaniola. It has been independent since 1804. Topographic mapping is the responsibility of the **Service de Géodésie**, Port-au-Prince. With IAGS cooperation, a triangulation network was implemented and tied to the Dominican Republic and to Cuba and Jamaica. Large-scale topographic mapping appears to have become publicly available recently. There is a complete cover of the country at 1 : 100 000 and 1 : 50 000, and a partial cover (of the eastern part) at 1 : 25 000. These maps are in the US Army Map Service series and were prepared photogrammetrically over the period 1960 to 1965 from air photography flown in 1956. The 1 : 50 000 series is on a Transverse Mercator projection with UTM grid and has 20 m contours. A five-colour sheet map of Port-au-Prince and six provincial town maps have also been published. A 1 : 250 000 topographic cover of almost the entire country is available in three sheets published as part of the PAIGH Americas programme (Sheets NE 18-4, NE 18-7 and NE 18-8, published 1983), while an adequate general map is provided by Hildebrands Urlaubskarte of *Hispaniola*.

Geological mapping has been carried out by the **Ministère des Mines**, Port-au-Prince. The most recent geological synthesis was published in 1979 as *Carte géologique en édition provisoire 1 : 200 000*.

A number of resource surveys have been undertaken under the auspices of international or foreign agencies, including the UN Food and Agriculture Organization (1969), the Organization of American States (1972) and the United Nations Development Programme (1980), which all included the production of various ecological, hydrological, soil and geological maps for limited distribution. In 1981–82, an environmental mapping project was undertaken for the **Direction de l'Aménagement du Territoire et de la Protection de l'Environnement (DATPE)**, Port-au-Prince, by the Bureau pour le Developpement de la Production Agricole, Paris through an external aid programme. The mapping made extensive use of interpretation from infrared air photography flown in 1978, and covered seven themes: land use, soil capability, erosion hazard, surface water, population density, communication network and urban centres. Mapping scales of 1 : 100 000 and 1 : 250 000 were used. The maps were prepared as stable-base transparencies for monochrome diazo reproduction, although the 1 : 250 000 scale maps were also issued in small print runs as full-colour editions. The base maps prepared for the thematic overlays also show 1978 administrative areas down to commune level.

A national population census of the 132 communes of Haiti was carried out in 1982, and base maps for this census were prepared by **l'Institut Haitiën de Statistique et d'Informatique (IHSI)**. These include maps at 1 : 2000 for small towns and 1 : 4000 for the larger ones. They are issued in diazo form.

The capital, Port-au-Prince, has been mapped in a series of 51 sheets at 1 : 5000 by the **Bureau Cadastral de Port-au-Prince**, covering the whole metropolitan region, and based on 1980 air photography. There is also an orthophotomap series at 1 : 1000.

A national *Atlas d'Haiti* has been prepared by the **Centre d'Etudes de Géographie Tropicale (CEGET)** at the University of Bordeaux, Talence, and was published in 1985. It includes 32 colour map plates, with maps mostly at 1:1 000 000 scale, and each with an accompanying text. These cover a range of historical, physical, biogeographical, demographic, economic, administrative and cultural themes. There is also a false-colour satellite mosaic of the country, and a short index of toponyms referenced to the map of *Relief*.

An *Atlas critique d'Haiti*, covering mainly social and economic themes, was published in 1982 by the **Groupe d'Etudes et de Recherches Critiques d'Espace (ERCE)** at the Université du Québec at Montreal.

Further information

A full report of the thematic mapping carried out in 1981–82 has been published as *Cartographie thematique d'Haiti, notice explicative* by DATPE.

Addresses

Service de Géodésie
Rue Joseph Janvier, PORT-AU-PRINCE.

Ministère des Mines
Delmas 19, PORT-AU-PRINCE.

Direction de l'Aménagement du Territoire et Protection de l'Environnement (DATPE)
Ministère du Plan, rue du Marron Inconnu No. 11, PORT-AU-PRINCE.

Ministère de l'Agriculture, des Ressources Naturelles et du Developpement Rural
Route Nationale, No. 1 Damine, PORT-AU-PRINCE.

Institut Haïtien de Statistique et d'Informatique (IHSI)
Angle rue Joseph et blvd Harry Truman, PORT-AU-PRINCE.

Bureau Cadastral de Port-au-Prince
Delmas 16 No. 1, PORT-AU-PRINCE.

Centre d'Etudes de Géographie Tropicale (CEGET)
Domaine Universitaire de Bordeaux, 33405 TALENCE-CEDEX, France.

Groupe d'Etudes et de Recherche Critiques d'Espace (ERCE)
Département de Geographie, Université du Québec à Montreal, Case postale 8888, Succursale A, MONTREAL, Qb H3C 3P8, Canada.

For Karto+Grafik, see German Federal Republic; for DMA, see United States.

Catalogue

Atlases and gazetteers

Atlas d'Haiti
Talence: CEGET, 1985.
pp. 146, including 32 plates of maps.

Atlas critique d'Haiti   Georges Anglade
Montreal: ERCE, 1982.

Haiti: official standard names approved by the United States Board on Geographic Names   2nd edn
Washington, DC: DMA, 1973.
pp. 211.

General

Hispaniola: Haiti, Dominikanische Republik   1:816 000
Hildebrand's Urlaubskarte
Frankfurt AM: Karto+Grafik, 1985.
Tourist map with text in German.

Topographic

Haiti   1:100 000
Port-au-Prince: Service de Géodésie
28 sheets, all published.

Haiti   1:50 000   Series E 732
Port-au-Prince: Service de Géodésie, 1960–65.
78 sheets, all published.

Environmental

Occupation de l'Espace   1:250 000
Port-au-Prince: DATPE, 1982.

Resources forestières   1:250 000
Port-au-Prince: DATPE, 1982.

Resources en eau   1:250 000
Port-au-Prince: DATPE, 1982.

Risques d'érosion   1:250 000
Port-au-Prince: DATPE, 1982.

Jamaica

Jamaica, independent from the United Kingdom since 1962, has its own survey department, which has worked in cooperation with the British Directorate of Overseas Surveys (DOS) (now Overseas Surveys Directorate, Southampton, OSD) and other agencies in the mapping of the country.

The original 1:50 000 map of Jamaica was the first major mapping project of the Directorate of Colonial Surveys, as it was initially called when it was formed in 1946, and it was based on work done before and immediately after World War II by the Royal Marine Survey Co. The provisional edition of this map was issued in 1952–53 (DCS 1) and has been followed by several regular editions (DOS 410). Despite its scale, however, this map was a feet-and-inches map in contrast to the current metric edition, in 20 sheets, printed in process colours, the first of which were issued in 1982. This series is being prepared jointly by OSD and the **Jamaica Survey Department (JSD)**, and is based on air photography flown in 1968 and 1980. It has contours at 20 m intervals up to

80 m, and thereafter at 40 m, and it shows the Jamaica metre grid. The map includes much information about vegetation, including differentiation of the different kinds of plantation agriculture.

The modern basic scale mapping is a contoured series at 1:12 500. Pilot sheets at this scale of the Montego Bay area were published by DOS in 1970, and the series has been extended to cover the country in 232 sheets through the aid of a UNDP mapping project. The first edition sheets (Jam 201) are in four colours with a 25 or 50 foot contour interval, but most sheets are currently in a preliminary one- or two-colour edition (Jam 200). There is also a series of planimetric maps of principal towns at 1:1250, while cadastral maps cover the island at the scale of 20 chains to one inch. Kingston is covered by a 1:10 000 contoured map in six sheets with an additional special sheet of the central area.

There is a special map sheet of Montego Bay at 1:25 000 (DOS 301/1), published in 1971.

Geological mapping is in the hands of the **Geological Survey Division (GSD)**, Kingston. A coloured geological map at 1:250 000 of the whole island was published in 1959 by DOS but this is now out of print. A monochrome map at the same scale by N. McFarlane was published in 1977, and a new colour edition was being printed for GSD in 1986. The country has also been mapped in a 1:50 000 geological series initiated by DOS in cooperation with GSD in 1972, the latter supplying the geological information and the former the topographic base and the printing of the sheets. Only five sheets remain unpublished, covering the eastern end of the island, but several published sheets have gone out of print. Most sheets (1–21) are in a provisional edition with geology printed in black on a topographic base, but sheets 22, 23 and 25 are in full colour. The sheets have explanatory texts printed in the margins.

Addresses

Jamaica Survey Department (JSD)
PO Box 493, 23 Charles Street, KINGSTON.

Geological Survey Division (GSD)
PO Box 141, 189, 191, Hope Gardens, KINGSTON 6.

For Karto+Grafik, see German Federal Republic; for Macmillan, see United States.

Catalogue

General

Jamaica. Hildebrands Urlaubskarte   1:400 000
Frankfurt AM: Karto+Grafik, 1983.
Insets of Caymans, Kingston and Montego Bay.

Jamaica   1:250 000   Ja 11
Kingston: JSD, 1972.

Shell road map of Jamaica   1:250 000
New York: Macmillan Publishers, 1985.
Kingston 1:30 000 on reverse.

Topographic

Jamaica   1:50 000 (metric edition)   Edition 1
Tolworth: DOS/Kingston: JSD, 1982–.
20 sheets, 10 published. ■

Geological and geophysical

Geological map of Jamaica   1:250 000   Pub. 162   N. McFarlane
Kingston: GSD, 1977.
Monochrome map.

Jamaica geological sheets   1:50 000   DOS 1177
Tolworth: DOS/Kingston: GSD, 1972–.
30 sheets, 25 published. ■

Town maps

Road map of Kingston and St Andrew   1:20 000
Kingston: JSD.

Kingston   1:10 000   DOS 201 (E922)   Edition 3
Tolworth: DOS, 1972.
7 sheets, including special *Central* sheet, all published.

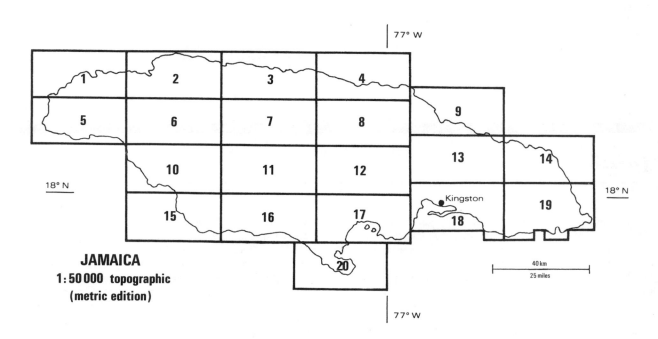

JAMAICA
1:50 000 topographic
(metric edition)

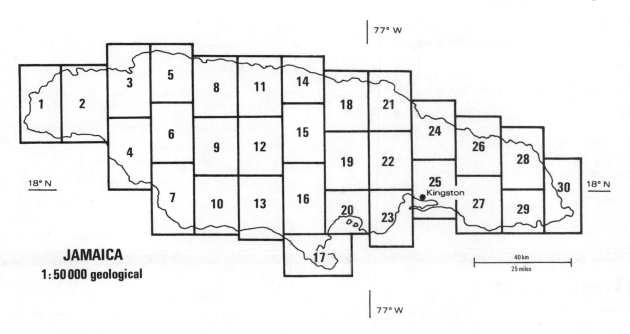

JAMAICA
1 : 50 000 geological

Martinique

Martinique is an overseas Department of France, and topographic mapping is the responsibility of the **Institut Géographique National (IGN)**, Paris. Air photo cover was obtained in 1947 and, with subsequent precise levelling and revision of an earlier triangulation network, modern topographic series at the basic scale of 1:20 000 and the derived scale of 1:50 000 were initiated. The two series were in the standard IGN style of the time (*Type 1922*), with contours respectively at 20 m and 10 m intervals, relief shading, and four- and five-colour printing. The projection and grid are UTM. In 1985, the 1:20 000 map was superseded by a new 1:25 000 scale map in only four sheets. This forms part of the French *série bleue*. The projection remains UTM and contours are at 10 m intervals.

IGN also publishes a general map of the island at 1:100 000, and the 5th edition was issued in 1982. This is a tourist map with no contours (relief is shown by spot heights and shading) or grid. It has a legend in both French and English, and gives information about the *Parc Naturel Regional* and tourist circuits of the island.

In 1976 IGN published the *Atlas de la Martinique*, one of a series of atlases of the overseas Departments of France, prepared by the Centre d'Etudes de Géographie Tropicale of the **Centre National de Recherche Scientifique (CNRS)**. It is in the style of a National Atlas, with 37 map folios at 1:500 000 on themes concerned with the natural environment, population, economic geography and cultural problems, and 100 pages of text and supplementary maps and graphics.

A geological map at 1:50 000 was published by the **Bureau de Recherches Géologiques et Minières (BRGM)** in 1962 together with an explanatory notice, and in 1983 a 1:20 000 scale geological map of Mount Pelée was published.

Soil and land use mapping of Martinique was undertaken by F. Colmet Daage for the **Office de la Recherche Scientifique et Technique Outre Mer**

(ORSTOM) and published in 1970. The land use series is a monochrome diazo production. A multi-coloured *carte écologique* was published in 1978 by the **Laboratoire de Biologie Végétale**, University of Grenoble. The map is described in *Documents de Cartographie Ecologiques* 20 (1978).

Addresses

Institut Géographique National (IGN)
Service Vente et Edition, 107 rue la Boetie, 75008 PARIS, France.

Librarie du CNRS
295 rue St Jacques, 75005 PARIS, France.

Bureau de Recherches Géologiques et Minières (BRGM)
BP 6009, Avenue de Concyr, F-45060 ORLEANS-CEDEX, France.

Office de la Recherche Scientifique et Technique Outre Mer (ORSTOM)
70-74 Route d'Aulnay, 93140 BONDY, France.

Laboratoire de Biologie Végétale, Université I de Grenoble
Bibliothèque, BP 68, 38402 ST-MARTIN-D'HERES, France.

Catalogue

Atlases and gazetteers

Atlas des Départements Français d'Outre-Mer. II. La Martinique
Paris: IGN/CNRS, 1977.

General

Martinique: Parc Naturel Regional, Carte touristique   1 : 100 000
Edition 5
Paris: IGN, 1982.
Insets: Fort de France, St Pierre.

Topographic

Carte du Département de la Martinique 1:50 000
Paris: IGN, 1958.
4 sheets, all published.

[Martinique] Série bleue 1:25 000 Edition 1
Paris: IGN, 1982.
4 sheets (4501–4504M) cover the island, all published.

Geological and geophysical

Carte géologique de la Martinique 1:50 000
Orléans: BRGM, 1962.
2 sheets, both published.
With text.

Environmental

[Ensemble sols—modèle—occupation des terres] 1:100 000
F. Colmet Daage
Bondy: ORSTOM, n.d.

Carte écologique de la Martinique 1:75 000 Jacques Portecop
Grenoble: Laboratoire de Biologie Végétale, 1978.

Carte des sols de la Martinique 1:20 000 F. Colmet Daage
Bondy: ORSTOM, 1970.

Montserrat

Montserrat, a British Crown Colony, has been mapped by the **Directorate of Overseas Surveys (DOS)** (now Overseas Surveys Directorate, Southampton, OSD), the principal map being a single sheet at 1:25 000 (DOS 359). This was first published in 1963 with photogrammetric compilation from air photography of the 1950s. The latest editions of this map have incorporated revisions from subsequent large-scale mapping and from local information. They have also been redesigned in a tourist style and printed in process colours. The projection is Transverse Mercator, and the UTM grid appears on Edition 5 (1978), while Edition 6 is a BWI grid version, with arbitrary alphanumeric labelling of the squares for local referencing. The relief is shown by 50 foot contours supplemented by hill shading and layer tints.

Large-scale contoured mapping was undertaken by DOS in the early 1970s at 1:2500 (Plymouth and the western part of the island) and 1:5000 (eastern part). Since 1978 there has been some new mapping at 1:1250 associated with the Land Registration Project, and in 1983 two new large-scale mapping projects were started to provide full revision of existing sheets and new mapping for the remainder of the island. Scales will be 1:2500 for the more developed coastal areas and 1:5000 for the rest of the island. Aerial triangulation of the island was completed in 1984.

Addresses

Lands and Surveys Department
Parliament Street, PLYMOUTH.

Overseas Surveys Directorate (DOS/OSD)
Ordnance Survey, Romsey Rd, Maybush, SOUTHAMPTON SO9 4DH, UK.

Catalogue

Tourist map of Montserrat 1:25 000 DOS 359 (E803) Edition 6
Tolworth: DOS, 1983.
Inset: Plymouth 1:5000.

Netherlands Antilles

The Netherlands Antilles comprise two widely separated groups of islands: one, situated off the coast of Venezuela, includes the principal islands of Aruba, Curaçao and Bonaire; the other, in the Windward Islands east of Puerto Rico, includes Sint Eustatius, Saba and Sint Maarten (only the south part of this island, the north end is administered by France).

The Netherlands Antilles form an autonomous part of the Kingdom of the Netherlands, in which Aruba acquired separate entity status in 1986. The islands were mapped by the Dutch air survey company KLM–Aerocarto in the period 1960–63 and published by the **Dienst van het Kadaster van de Nederlandse Antillen**, Willemstad (printed at the Topografische Dienst of The Netherlands in 1971). The maps of Aruba, Bonaire, Curaçao and Sint Maarten were at 1:25 000 and those of Sint Eustatius and Saba at 1:10 000. There was also a 1:10 000 series of Curaçao and a 1:10 000 town map of Willemstad. All these maps are currently out of print. However, KLM–Aerocarto prepared new editions in 1975–82 although these have not yet been printed.

Geomorphological and terrain classification maps of the volcanic islands of Sint Eustatius and Saba were published by the **International Institute for Aerial Survey and Earth Sciences (ITC)**, Enschede, in 1977 based on a 1972 survey by H.Th. Verstappen. They were issued as a supplement to the *ITC Journal*, but may be obtained as a separate.

Addresses

Dienst van het Kadaster van de Nederlandse Antillen
Building 'Soerabaja', Araratwijk, CURAÇAO.

Topografische Dienst
Bendienplein 5, Postbus 115, 7800 AC EMMEN, The Netherlands.

For CIA and General Drafting, see United States; for ITC, see Netherlands.

Catalogue

General

Netherlands Antilles 1 : 1 000 000
Washington, DC: CIA, 1982.

Aruba road map (Esso) 1 : 39 000
Convent Station, NJ: General Drafting Co., 1976.
Street map of Oranjestad on the reverse.

Geological and geophysical

St Eustatius 1 : 25 000
Enschede: ITC, 1977.
Two maps on same sheet.
Maps of geomorphology and terrain classification.

Saba 1 : 25 000
Enschede: ITC, 1977.
Two maps on same sheet.
Maps of geomorphology and terrain classification.

Puerto Rico

Puerto Rico is a self-governing commonwealth in union with the US. Both topographic and geological mapping have been undertaken by the **United States Geological Survey (USGS)** in association with the Puerto Rico government. There is a local **Puerto Rico Geological Survey** at San Juan, while topographic mapping is distributed by the **Photogrammetry Office** of the Highway Authority. In 1964, Puerto Rico was described as one of the most thoroughly mapped areas of the world. A corollary of this has been a lack of much entirely new mapping in the succeeding period.

The modern topographic mapping forms part of the standard USGS programme although the basic scale is 1 : 20 000 (instead of the more usual 1 : 24 000 or 1 : 25 000). The scale originally chosen was 1 : 30 000 and the sheets covering the islands of Culebra and Vieques are still at this scale (but that of Mona is at 1 : 20 000). The 64 sheets which cover the main island conform to 7.5' sheet lines. The maps are printed in three colours with contours at 5 or 10 m intervals, and besides the usual cultural details they also show the municipal and barrio boundaries. The series was initiated in 1935, the first sheets being compiled from plane table survey, but the current editions are from photogrammetric surveys in the 1960s, and although there have been no new editions since 1972, the majority of the sheets were photorevised in 1982.

Smaller-scale derived maps are published at 1 : 120 000 and 1 : 240 000. They are in Spanish, and available in two-colour base or five-colour topographic versions. The former has no contours; the latter has contours at 50 m or 100 m according to scale. There is also a shaded relief version of the 1 : 240 000 scale map.

Geological and other resource surveys were undertaken as early as the 1910s and 1920s as part of a Scientific Survey sponsored by the New York Academy of Sciences, the Puerto Rico government and several other agencies. In 1952, USGS began a project in cooperation with the Puerto Rican Industrial Development Corporation to assess the mineral resources of the island. This led to a detailed basic geological mapping programme and the production of a 62-sheet, multi-colour series, now complete, at 1 : 20 000 scale. These maps are published as part of the USGS Miscellaneous Investigations Series. Puerto Rico is thus geologically one of the best mapped islands in the Caribbean. A derived geological map at 1 : 100 000 has also been compiled, and is currently in review. It may be published soon. Meanwhile, the 1 : 240 000 Provisional Geologic Map of 1964 by R.P. Briggs is still available.

A series of *Marine geologic maps of the Puerto Rico Insular Shelf* 1 : 40 000 has been in progress, and since 1981 sheets have been published in the USGS Miscellaneous Investigations Series. The maps show isobaths as well as underwater geology.

The USGS Water Resources Division has undertaken numerous hydrologic investigations and the main cartographic output has been a series of 1 : 20 000 scale flood maps published in the *Hydrologic Investigations Atlases series*.

Soil mapping was carried out by the Soil Survey of the US Department of Agriculture in cooperation with the Agricultural Experiment Station of the University of Puerto Rico in the 1930s and a 1 : 50 000 soil map in four sheets was published in 1942 (*Soil Survey of Puerto Rico* by R.C. Roberts *et al.*). More detailed mapping was undertaken in the 1960s by The Soil Conservation Service of the US Department of Agriculture.

A *Gazetteer of Puerto Rico* by Henry Gannett was published in 1901 (USGS *Bulletin* 183) but is currently out of stock.

Further information

A summary of *Cartography in Puerto Rico* by Rafael Pico was published in 1964 for the International Geographical Congress held in London that year. More recent mapping is notified in the *List of Geological Survey Geologic and Water-Supply Reports and Maps for Puerto Rico and Virgin Islands*, in the *Geologic Map Index of Puerto Rico* and in the *Topographic Map Index of Puerto Rico and Virgin Islands*, all available from USGS.

Addresses

US Geological Survey (USGS)
Map Distribution, Federal Center, Building 41, Box 25286, DENVER, Colorado 80225, USA.

Photogrammetry Office
Highway Authority, SAN JUAN, Puerto Rico.

Puerto Rico Geological Survey
Department of Natural Resources, PO Box 5887, Puerto De
Tierra Sta, SAN JUAN, Puerto Rico 00906.

Institute of Tropical Forestry
US Forest Service, PO Box AQ, RIO PIEDRAS, Puerto Rico 00928.

Catalogue

Atlases and gazetteers

Nuevo Atlas de Puerto Rico   J.A. Toro-Sugranes
Rio Piedras: Editorial Edil Inc., 1982.
pp. 16 maps + pp. 128 text.

General

Puerto Rico e Islas Limitrofes   1:240 000
Reston: USGS, 1951–52.
Available in base, topographic and relief versions.

Mapa de carreteras estatales de Puerto Rico   1:205 000
San Juan: Highway Authority, 1983.

Puerto Rico e Islas Limitrofes   1:120 000
Reston, VA: USGS, 1951.
Available in base and topographic versions.

Topographic

Puerto Rico   1:20 000
Reston, VA: USGS, 1957–.
64 sheets, all published. ■

Geological and geophysical

Provisional geologic map of Puerto Rico and adjacent islands
1:240 000   I-392   R.P. Briggs
Reston, VA: USGS, 1964.

Hydrogeologic map of Puerto Rico and adjacent islands   1:240 000
R.P. Briggs and J.P. Akers
Reston, VA: USGS, 1965.

Metallogenic map of Puerto Rico   1:240 000   I-721   D.P. Cox
and R.P. Briggs
Reston, VA: USGS, 1973.

Natural gamma aeroradioactivity map of Puerto Rico   1:240 000
GP-525   J.A. MacKallor
Reston, VA: USGS, 1965.

Geology. Quadrangle maps   1:20 000
Reston, VA: USGS, 1960–.
62 sheets, all published. ■

Environmental

*Ecological life zones of Puerto Rico and the United States Virgin
Islands*   1:250 000   J.J. Ewel and J.L. Whitmore
Rio Piedras: Institute of Tropical Forestry, 1973.

*Map showing landslides and areas of susceptibility to landsliding in
Puerto Rico*   1:240 000   I-1148   W.H. Monroe
Reston, VA: USGS, 1979.

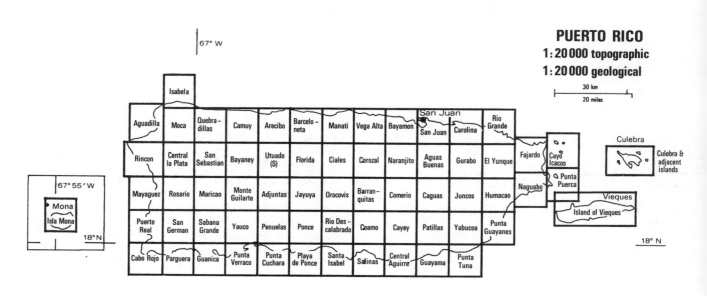

St Christopher–Nevis

Formerly these two islands were associated with Anguilla as a British colony, and were mapped as such by the **Directorate of Overseas Surveys (DOS)** (now Overseas Surveys Directorate, Southampton, OSD). St Christopher (St Kitts) and Nevis achieved independence as a sovereign democratic federal state within the British Commonwealth in 1983. Anguilla, which had long fought for separate status, was formally separated from the group in 1980.

St Christopher (or St Kitts) and Nevis are represented by two 1:25 000 sheets of DOS 343 (E803), originally a three-sheet series which included Anguilla. The first edition of this series was published in 1959–60 based on air photography flown in 1946 by USAAF and in 1956 by Hunting Aerosurveys. The current fifth edition was revised from 1982 air photography with field revision by the **Survey Division** of St Kitts. The sheets are in five colours with 50 foot contours, and the projection is Transverse Mercator with BWI grid. (There is also a version with UTM grid published as Edition 6.)

A 1:50 000 map of the two islands was first published in 1979 (DOS 443). It is in tourist style with 200 foot contours.

A series of 83 1:2500 contoured maps was begun in 1982 (DOS 043). This will cover the coastal lowlands of both islands, and many of the sheets have now been published.

Addresses

Survey Division
Central Housing and Planning Authority, PO Box 190, BASSETERRE, St Christopher.

Overseas Surveys Directorate (DOS/OSD)
Ordnance Survey, Romsey Rd, Maybush, SOUTHAMPTON SO9 4DH, UK.

Catalogue

Topographic

Lesser Antilles   1:50 000   DOS 443 (E703)   Edition 1
St Christopher and Nevis
Tolworth: DOS, 1979.
Insets: Basseterre and Charlestown 1:12 500.

Lesser Antilles   1:25 000   DOS 343 (E803)   Edition 5
St Christopher
Southampton: OSD for St Christopher and Nevis Government, 1984.

Lesser Antilles   1:25 000   DOS 343 (E803)   Edition 5   Nevis
Southampton: OSD for St Christopher and Nevis Government, 1984.

St Lucia

St Lucia, formerly a British colony, became independent in 1979, remaining a member of the Commonwealth. Mapping was undertaken by the **British Directorate of Overseas Surveys (DOS)** (now Overseas Surveys Directorate, Southampton, OSD) and is now maintained by OSD on behalf of the Government of St Lucia. The principal series is at 1:25 000 (DOS 345), first published in 1958 and subsequently revised from more recent air photo sorties and new field information. The latest edition is printed in process colours and has contours at 25 foot intervals (50 foot above 250 feet). The projection is Transverse Mercator and the grid BWI.

The 1:50 000 (DOS 445) map has been derived from the latest edition of the 1:25 000. It is a tourist map with layer tinted and shaded relief as well as 200 foot contours. It has an arbitrary alphanumeric grid, inset map of the capital, Castries, and a descriptive text.

Densification and adjustment of the survey network was carried out in the early 1980s, some new air photography acquired in 1981, and a partly new, partly revised series of 1:2500 scale maps had been published by 1984.

Addresses

Lands and Survey Department
Walcott Building, Upper Jeremie Street, CASTRIES.

Overseas Surveys Directorate (DOS/OSD)
Ordnance Survey, Romsey Rd, Maybush, SOUTHAMPTON SO9 4DH, UK.

Catalogue

General

Saint Lucia   1:125 000   DOS 945   Edition 2
Tolworth: DOS for Government of St Lucia, 1982.

Topographic

Saint Lucia Tourist Map   1:50 000   DOS 445 (E703)   Edition 4
Tolworth: DOS for Government of St Lucia, 1982.

Saint Lucia   1:25 000   DOS 345 (E803)   Edition 5
Tolworth: DOS for Government of St Lucia, 1981.
3 sheets, all published.

St Vincent and the Grenadines

St Vincent and the Grenadines became an independent state within the British Commonwealth in 1979. Mapping has been undertaken by the **Directorate of Overseas Surveys (DOS)** (now Overseas Survey Directorate, Southampton, OSD), with the first edition 1:25 000 appearing in 1959 and the 1:50 000 in 1961.

The latest topographic map is the sixth edition of the 1:50 000 (DOS 417) in a redesigned tourist format. It is a shaded relief map incorporating some text and an inset map of Kingstown. Contours are at 500 foot intervals, the projection is Transverse Mercator, and the BWI grid is used. This map is printed in process colours. Two further sheets have been added in the same style to cover the St Vincent Grenadines (Edition 1). The 1:25 000 map (DOS 317), covering St Vincent only, has contours at 25 foot or 50 foot intervals, and is on the Transverse Mercator projection with BWI grid.

The Grenadines were originally mapped together in a 1:25 000 scale series of six sheets (DOS 344/E803) and a 1:10 000 series (DOS 244/E802). Following independence, sovereignty was split between St Vincent (the northern Grenadines) and Grenada (chiefly Ronde Island and Carriacou). Of the 1:25 000 series, only Sheet 5 is maintained (see under Grenada), while the 1:10 000 mapping of the St Vincent Grenadines has been reorganized into a five-sheet series (DOS 217). These sheets are in several colours, and have a 20 foot contour interval. The projection is Transverse Mercator with the BWI grid.

DOS has also mapped St Vincent and the Grenadines in 1:2500 scale series, used for property valuation purposes, and a revision programme was in progress in the 1980s.

Addresses

Lands and Survey Department
KINGSTOWN.

Overseas Surveys Directorate (OSD/DOS)
Ordnance Survey, Romsey Rd, Maybush, SOUTHAMPTON SO9 4DH, UK.

Catalogue

General

Saint Vincent and the Grenadines 1:200 000
Tolworth: DOS for Government of St Vincent, 1984.

Topographic

St Vincent 1:50 000 DOS 417 (E703) Edition 6
Tolworth: DOS for Government of St Vincent, 1984.
Inset: Kingstown 1:10 000.

St Vincent Grenadines 1:50 000 DOS 417 (E703)
Tolworth: DOS for Government of St Vincent, 1984.
2 sheets, both published.

Saint Vincent and the Grenadines 1:25 000 St Vincent DOS 317 (E803) Edition 5
Tolworth: DOS for Government of St Vincent, 1983.
2 sheets, both published.

St Vincent—Grenadines 1:10 000 DOS 217
Tolworth: DOS for Government of St Vincent, 1983.
5 sheets, all published.

Trinidad and Tobago

Until 1962 Trinidad and Tobago, now a Republic (in which Tobago has semi-autonomous status), was a British Crown Colony and topographic mapping was undertaken by the Directorate of Overseas Surveys (formerly Colonial Surveys) from about 1950, continuing after independence until 1977. The present mapping authority is the **Lands and Survey Division (LSD)**, Port-of-Spain.

DOS produced a rather mixed bag of topographic mapping for the Trinidad and Tobago Government, and until recently no uniform series covered Trinidad. The most recent DOS series is the 1:25 000 series DOS 316/1 which covered the north and south parts of Trinidad in 30 sheets. This programme terminated in 1977, but subsequently the eight sheets covering the central area were also completed by LSD with the assistance of the Canadian Government. DOS 316/1 is on the Transverse Mercator projection with UTM grid and 50 foot contours and is litho printed in seven colours.

Tobago is presently covered in three sheets in an old 1:25 000 series (DOS 307) on the Cassini projection, with 25 foot contours. Tobago and central Trinidad were also mapped at 1:10 000 in this period (1962–66). DOS 307 is now under revision and the projection is being changed to UTM.

LSD has issued general road maps of Trinidad and Tobago and a tourist map of Port-of-Spain (to be periodically revised), and further small-scale general maps at 1:150 000 and 1:300 000 are in preparation. Several small-scale resource maps (land use, land capability and soil) have also been issued. Urban and suburban mapping at 1:1250 and 1:2500 respectively is available in diazo print form from **Mapping and Control** at Richmond Street. The topographic maps in our catalogue can be obtained from LSD, Knox Street.

In the mid-1970s, the Government of Trinidad and Tobago set up a national atlas committee to consider specification, content and production of a national atlas for the territory. This was still under consideration in the mid-1980s, when firm proposals were being formulated.

The **Ministry of Energy and Natural Resources**

(MENR) is responsible for geological investigations, and maps by Kugler (1969) and Maxwell (1965) are available as diazo prints.

Further information

LSD issues a list of published maps of Trinidad and Tobago.

Addresses

Lands and Survey Division (LSD)
Knox Street, Red House, PORT-OF-SPAIN.

Lands and Survey Division
Mapping and Control Section, 2B Richmond Street, PORT-OF-SPAIN.

Ministry of Energy and Natural Resources (MENR)
4th Floor, Salvatori Building, Independence Square, PORT-OF-SPAIN.

Catalogue

General

Map of Trinidad 1:150 000
Port-of-Spain: LSD, 1983.

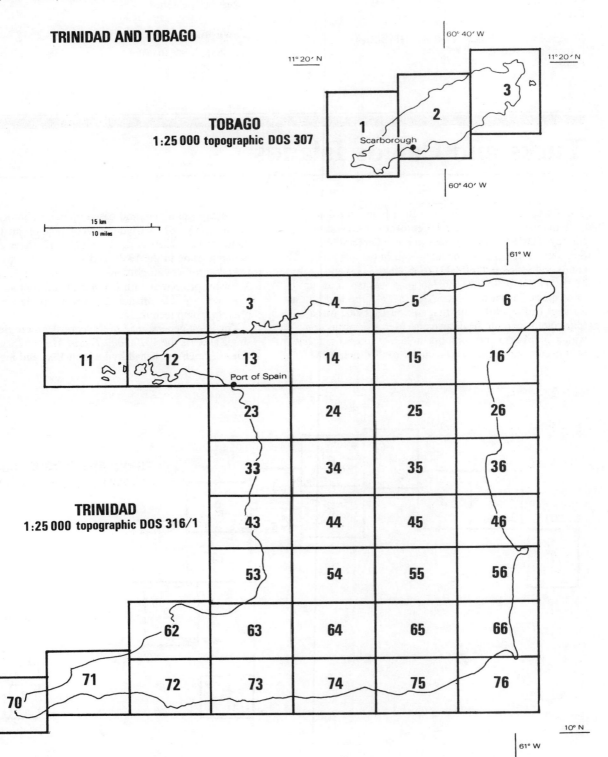

TRINIDAD AND TOBAGO

TOBAGO
1:25 000 topographic DOS 307

TRINIDAD
1:25 000 topographic DOS 316/1

Tobago 1:50 000
Port-of-Spain: LSD, 1977.

Topographic

Trinidad 1:25 000 DOS 316/1 (E804)
Tolworth: DOS/Port-of-Spain: LSD, 1970–.
38 sheets, all published. ■

Tobago 1:25 000 DOS 307 (E8410)
Tolworth: DOS/Port-of-Spain: LSD, 1962–63.
3 sheets, all published. ■

Geological and geophysical

Geological map of Trinidad 1:100 000 H.G. Kugler
Port-of-Spain: MENR, 1959.
Diazo.

Geological map of Trinidad 1:50 000 H.G. Kugler
Port-of-Spain: MENR, 1959.
Diazo.

Generalized geology of Tobago 1:25 000 After Maxwell
Port-of-Spain: MENR, 1965.
Diazo.

Environmental

Trinidad land use 1:150 000
Port-of-Spain: LSD, 1970.

Trinidad land capability 1:150 000
Port-of-Spain: LSD, 1972.

Trinidad soil 1:150 000
Port-of-Spain: LSD, 1971.

Soil map of Central Trinidad 1:50 000 DCS (Misc) 55
Tolworth: DCS, 1954.
4 sheets, all published.

Town maps

Port-of-Spain 1:10 000
Port-of-Spain: LSD, 1986.

Turks and Caicos Islands

The Turks and Caicos Islands, a British crown colony, were mapped by the British **Directorate of Overseas Surveys (DOS)** (now Overseas Surveys Directorate, Southampton, OSD) in the late 1960s in two parallel contoured series at 1:10 000 (DOS 209/E8113) and 1:25 000 (DOS 309/E8112). Mapping was based on air photographs flown by the USAAF in 1943 and by Hunting Surveys Ltd in 1961. This first edition 1:25 000 was a four-colour map on the Transverse Mercator projection with a UTM grid and contours at 25 foot intervals. It has now been superseded by a colourful new edition using

photomap techniques and based upon air photography flown in 1980–81 by Clyde Surveys. The original monochrome sheets of the 1:10 000 series have also been under revision in the 1980s and some sheets have been issued in full-colour editions. These are on a Transverse Mercator projection with UTM grid and contours at 10 foot intervals. The offshore coral reefs are shown by a photomapping technique.

The latest general map of the two island groups on a single sheet is the 1:200 000 *Tourist Map of Turks and Caicos Islands* published by DOS in 1984 and including an

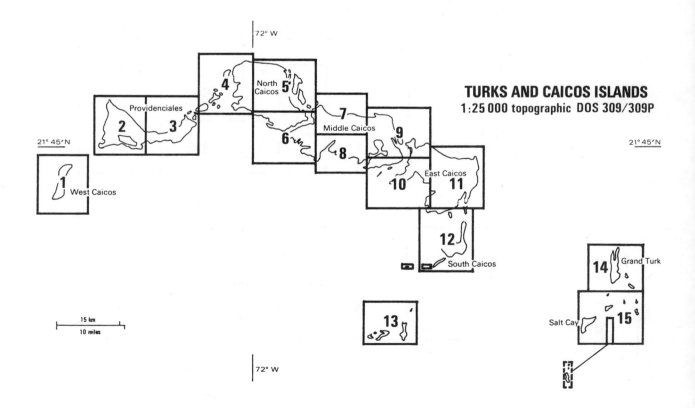

inset map of the capital Cockburn Town (on Grand Turk), and some text giving tourist information.

Large-scale maps of Cockburn Town and Cockburn Harbour have been produced at 1:2500 (1970).

Addresses

Land Registration and Survey Department
PO Box 66, GRAND TURK.

Overseas Surveys Directorate (DOS/OSD)
Ordnance Survey, Romsey Rd, Maybush, SOUTHAMPTON SO9 4DH, UK.

Catalogue

General

Tourist map of Turks and Caicos Islands 1:200 000 DOS 609 Edition 2
Tolworth: DOS, 1984.
Inset: Grand Turk (Cockburn Town) 1:10 000.

Topographic

Turks and Caicos Islands 1:25 000 DOS 309P (E8112) Edition 2
Southampton: OSD, 1984–87.
15 sheets, all published. ■

Virgin Islands (British)

The British Virgin Islands are a British colony with internal self-government. They have been mapped by the **British Directorate of Overseas Surveys (DOS)** (now Overseas Surveys Directorate, Southampton, OSD). Some early mapping was done in the late 1940s using USAAF photography, but the principal series at 1:25 000 (DOS 346) was compiled in 1959 using 1953 air photography, and was published in 1960. The current sheets in this series are in their third or fifth editions, having been revised from 1981 photography, from incomplete 1:2500 scale mapping and from field compilations. Sheet 6 *Anegada*, a low-lying coral island, is a colour photomap published in 1977; the other five sheets are conventional five-colour line maps all published in 1984. They have 25 foot contours and the projection is Transverse Mercator with a UTM grid.

In 1982, a derived tourist map of the whole island group was published at a scale of 1:63 360 (DOS 446), superseding an earlier general map at 1:100 000. This map has 250 foot contours and the relief is enhanced with layer tints and relief shading. There is an alphanumeric grid based on the UTM, and an inset map of Road Town.

An up-to-date cover of the main islands at 1:2500 is currently in progress to replace and extend the rather rudimentary mapping originally prepared for the Caribbean Regional Cadastral Survey.

Addresses

Survey Department
PO Box 142, ROAD TOWN, Tortola.

Overseas Surveys Directorate (OSD/DOS)
Ordnance Survey, Romsey Rd, Maybush, SOUTHAMPTON SO9 4DH, UK.

Catalogue

General

British Virgin Islands 1:63 360 DOS 446 Edition 1
Tolworth: DOS, 1982.
Inset: Road Town 1:10 000.

Topographic

Lesser Antilles 1:25 000 British Virgin Island DOS 346 (E803) Editions 3/5
Tolworth: DOS, 1977–84.
6 sheets, all published.

Virgin Islands (US)

The US Virgin Islands are an 'unincorporated territory' of the US, purchased from Denmark in 1917. The main islands are St Thomas, St John and St Croix. They have been mapped by the **United States Geological Survey (USGS)** at the scale of 1:24 000. The maps are in five colours with contours at 40 foot or 20 foot intervals according to relief gradients. The projection is polyconic. These sheets were published in 1955–58 with photorevisions in 1982.

In 1983, route maps of the islands were published privately by C.B. Mitchell.

Address

Branch of Distribution (USGS)
US Geological Survey, Box 25286, Federal Center, DENVER, Colorado 80225, USA.

Catalogue

General

Map of St Croix route system 1:95 000
C.B. Mitchell, 1983.

Map of St John route system 1:20 000
C.B. Mitchell, 1983.

Map of St Thomas route system 1:50 000
C.B. Mitchell, 1983.

Topographic

Virgin Islands of the United States 1:24 000
Reston, VA: USGS, 1955–.
8 sheets, all published.

SOUTH AMERICA

Detailed mapping of most of South America is of very recent origin, and although many countries established national mapping organizations early in the twentieth (or in some cases in the nineteenth) century, progress was at first slow, unsystematic and hampered by the lack of survey control and the multitude of problems associated with raw survey of undeveloped and often unexplored lands.

In the 1930s there was a concerted attempt, under the auspices of the **American Geographical Society (AGS)** to produce a uniform map of the whole of Latin America at 1:1 000 000 scale. The 107 sheets of this *Map of Hispanic America* took 28 years to complete. It was essentially a compiled map, and, due to lack of information about large areas of Amazonia and elsewhere, the quality was by no means uniform, and indeed the mapping of many areas, even at this small scale was extremely rudimentary. Nevertheless, it did provide a complete small-scale continent-wide series, which continued to be of some value into the 1980s. Most sheets, however, are now out of print, and the AGS is no longer active in promoting this kind of mapping. A few countries have published sheets of their territories to the IMW standard, notably Brazil which issued a complete cover of the country in 49 sheets in 1946, and has subsequently revised them.

Another uniform series of continent-wide mapping is provided by the 1:2 500 000 *Karta Mira* series prepared by Soviet and East European mapping organizations. The first edition of the South America sheets was completed by the German Democratic Republic in 1975. A special set of 30 sheets of this world series covering South and Central America with a toponymic index is available from **Cartographia**.

The development of modern national programmes of topographic mapping has in many countries been closely related to mapping agreements made with the Inter-American Geodetic Survey, created by the US Government in 1946 to help Latin American countries develop an accurate geodetic survey control network and to carry out first-time topographic mapping (see also the introduction to Central America). Only Argentina and Uruguay have not participated with IAGS, but both these countries have their own mature survey programmes.

In the cloudy inter-tropical regions, the use of SLAR imagery, introduced from 1969, has had a major effect not only in facilitating first-time topographic survey, but also in promoting geological prospecting and resource inventory. The outstanding example of the use of this technique in the production of multiple thematic maps has

been the Projeto RADAMBRASIL in Brazil, but SLAR has also been used extensively in Colombia and Venezuela, for example.

More recently, Landsat imagery has begun to prove its value for providing first-time map cover of the remoter jungle and mountain areas, and in 1984 a German contractor provided the Peruvian government with a complete 1:250 000 first-time cover of the country in the form of colour satellite image maps. However, the amount of information given away to this particular satellite's multi-spectral scanning system by a continuous jungle canopy is not great!

At the time of writing, seven of the 13 South American countries were in the PAIGH 1:250 000 Unified Hemispheric Mapping programme (which is described in the introduction to Central America) and small numbers of sheets have been distributed from Ecuador, Argentina, Bolivia, Colombia, Uruguay and Venezuela, while a complete set of the Peruvian satellite image maps mentioned above has also been included in the programme.

Numerous West European countries have also undertaken cooperative mapping projects in the continent, usually as part of their overseas aid programmes. Thus Great Britain has been active in geological investigations in the Andes, as have also the French. The Federal Republic of Germany has worked in both an active and advisory capacity in Argentina, Bolivia, Brazil, Colombia, Ecuador and Peru.

There are several earth science thematic maps of South America, mostly published under the aegis of the **Commission for the Geological Map of the World (CGMW/CCGM)**. They are available from various outlets, but most may be obtained from CCGM in Paris. Further maps are in preparation, including a new geology map at 1:5 000 000, and energy resources and metamorphic maps at the same scale.

The *Vegetation map of South America* was prepared by the **Institut de la Carte Internationale du Tapis Végétal (ICITV)**, Toulouse and published by UNESCO. Compilation was at 1:1 000 000, and classification was carried out with the aid of satellite imagery. There is also a *Mapa de la vegetación de América del Sur* by Kurt Hueck, which was first published in 1972. The map itself is in Spanish and was reissued in 1981, with 90 pages of German text and a German translation of the legend, by **Gustav Fischer Verlag**.

Good general maps of the continent, some of which also include Central America, are published by several commercial houses. Among those listed in the catalogue is the unusual one by **Kevin Healey**, which has many descriptive annotations, a table of statistical data and small ancillary maps of Easter Island, the Galapagos, the Falkland Islands (Malvinas) and some of the continental metropolitan and tourist areas.

Further information

The evolution of topographic mapping in Latin America is described country by country in Larsgaard, M.L. (1984) *Topographic Mapping of the Americas, Australia and New Zealand*, Littleton, Colorado: Libraries Unlimited, pp. 180.
  'A tectonic map of South America 1:5 000 000' is described in *Nature and Resources*, **16** (1980) 33.

Addresses

Kevin Healey
4 Page Street, ALBERT PARK, Victoria 3206, Australia.

For NGS, AGS and GSA, see United States; for Bartholomew, see Great Britain; for Wagner, Gustav Fischer and VWK, see German Federal Republic; for Haack, see German Democratic Republic; for Cartographia, see Hungary; for GUGK and Ministerstvo Geologii, see USSR; for IGM, see Chile; for Instituto de Geociências da UFRGS, see Brazil; for DGMG, see Venezuela; for UNESCO and CCGM/CGMW, see World.

Catalogue

General

South America   1:10 700 000
Washington, DC: NGS, 1983.

South America   1:10 000 000
Edinburgh: Bartholomew, 1982.

Südamerika   1:10 000 000   Wagnerkarte
Berlin: Kartographischer Verlag Wagner, 1982.

Südamerika   1:9 000 000
Gotha: Haack, 1985.
4 language legend.
49 pp. index.

South America   1:8 000 000
Bern: Kümmerly + Frey, 1984.

Südamerika   1:8 000 000
Obertshausen: VWK, 1986.

Mapa de América del Sur   1:8 000 000
Santiago: IGM.

A contemporary reference map of South America   1:5 000 000
Victoria, Australia: Kevin Healey, 1981.
2 sheets, both published.

South America. Karta Mira   1:2 500 000
DDR Berlin: Department of Geodesy and Cartography, 1966–.
Special set of 30 sheets in a slip case with 93 pp. index.

Geological and geophysical

Geological map of South America   1:5 000 000
Boulder, CO: GSA, 1964.
2 sheets, both published.

Geologicheskaja karta lujhnoi Ameriki   1:5 000 000
Moskva: Ministerstvo Geologii, 1979.
6 sheets, all published.
+ 136 pp. text.
Separate legend in English and Spanish.

Tectonic map of South America   1:5 000 000
Brasília: DNPM/Paris: UNESCO/CCGM, 1978.
2 sheets, both published + 23 pp. text.

Mapa metalogénico de America del Sur   1:5 000 000
Caracas: DGMG/Paris: UNESCO/CCGM, 1983.
2 sheets, both published + 112 pp. text.

Gravimeticheskaja Karta lujhnoi Ameriki   1:5 000 000
Moskva: Ministerstvo Geologii, 1981.
14 sheets, all published.

General structural geology map of the north part of South America   1:3 275 000
Caracas: DGMG, 1981.
+ 81 pp. text.

Mapa tectónico norte de America del Sur   1:2 500 000
Caracas: DGMG, 1978.
2 sheets, both published + legend.

Southern South America continental margin
Structural map/basins/South Atlantic opening periods
Porte Alegre: Instituto de Geociências da UFRGS.
3 sheets (monochrome), all published.

Environmental

Climatic atlas of South America, Volume 1
Budapest: Cartographia/Paris: UNESCO/Geneva: WMO, 1975.

Klimaticheskaja karta lujhnoi Ameriki   1:16 000 000
Moskva: GUGK.

Mapa de vegetación de América del Sur   1:8 000 000   2nd edn
Stuttgart: Gustav Fischer Verlag, 1981.
+ 90 pp. text.

Vegetation map of South America   1:5 000 000
Toulouse: ICITV/Paris: UNESCO, 1980.
2 sheets, both published + text.

Administrative

Südamerika: politisch   1:10 000 000
Berlin: Kartographischer Verlag Wagner, 1984.
Provincial boundaries shown in red.

Argentina

The official topographic mapping authority in Argentina is the **Instituto Geográfico Militar (IGM)**, Buenos Aires, which is a technical branch of the Argentine army. It has its origins in the Oficina Topográfica Militar founded in 1879.

The principal tasks of the IGM are set out in a government law (*Ley de la carta—Ley 12.696*) which requires it to carry out the geodetic and topographic survey necessary to produce a uniform map cover of the national territory, and to coordinate all geotopographic work required by the state. IGM has also carried out mapping of the areas claimed by Argentina in the Antarctic (see the section on Antarctica).

Topographic mapping is carried out at scales ranging from 1:500 000 to 1:25 000, and for all these series (but excluding the Antarctic mapping) a Gauss–Krüger conformal cylindrical projection is used, and the country is divided into seven zones, each 3° longitude wide.

The sheet systems of the various series are graticule-based with the sheet lines of the larger scales nesting within the smaller. The 1:500 000 scale sheets each cover an area of 3° longitude by 2° latitude. 1:250 000 sheets cover one-quarter of this area, while the 1:100 000 sheets fit nine to a 1:250 000 sheet. The graphic index shows only the 1:500 000/1:250 000 system. The 1:100 000 series uses the 1:500 000 sheet numbers with an additional number (1 to 36) to identify each sheet. 1:50 000 sheets each cover a quarter of a 1:100 000 sheet.

The principal modern topographic series are at 1:100 000 and 1:50 000. Of all the series, however, only the 1:500 000 is complete, and progress of postwar new specification maps at the larger scales has been rather slow, although about half the country is complete at 1:250 000 and 1:100 000, and the whole of Buenos Aires province (where 39 per cent of the population lives) at 1:50 000. The 1:500 000 series is of varying quality. Relief is usually by layer colouring and a graticule and grid is printed on the sheets. The 1:250 000 series has a 10 km grid and is contoured, the interval varying from mountain to plain. The 1:100 000, with 4 km grid, and 1:50 000 sheets also have a varying contour interval (which is as close as 1.25 m in very flat areas), and the more recent sheets have been surveyed by aerial photogrammetry and are printed in five colours.

Large-scale plans of Buenos Aires (at 1:10 000, 1:20 000 and 1:30 000) are produced by the municipal authority (**Municipalidad de la Ciudad de Buenos Aires**).

Geological mapping is undertaken by the **Servicio Geológico Nacional (SGN)**, Buenos Aires, founded in 1963, and an extensive series of maps, the *Carta geológico-economica*, at 1:200 000 has been published. These are based on an old series of topographic maps at that scale (see the graphic index on page 246).

Soil, land use and land capability maps are prepared by the Centro de Investigaciones de Recursos Naturales originally founded in 1944 but in 1970 becoming part of the **Instituto Nacional de Tecnologia Agropecuaria (INTA)**. Since then, there has been an active programme of 1:100 000 scale soil maps published on a photomap base. In addition the Centre has published a range of monographic soil and resource maps (soil, climate, vegetation, geomorphology, soil capability) of small areas or of individual provinces.

The *Mapa ecológico de la República Argentina* is published by the **Fundación Agro Palermo**, a private organization concerned with investigation and dissemination of knowledge about the agricultural potential of the country. The map divides the country into 100 ecological areas whose soils, climate and land capability are briefly described in an accompanying booklet.

The **Servicio Meteorológico Nacional** has mapped climatic variables and published two atlases: *Atlas Agrometeorológico Argentina* and *Atlas Climatológico Argentina*, the latter containing 1:10 000 000 maps of monthly means dated 1954 and 1962. Both atlases are currently out of print, and it is not certain if or when they will be reprinted.

A methodology for census mapping has been developed since 1970 by the **Instituto Nacional de Estadística y Censos (INDEC)**, but no detailed population maps have been published. Small-scale maps appear however in the *Atlas de la población Argentina*, published in 1982, and forming part of Volume 3 of the *Atlas Total*.

The *Atlas Total de la República Argentina*, published by the **Centro Editor de América Latina (CEAL)** in a series of volumes since 1981, has the breadth of scope of a National Atlas. The first two volumes, *Atlas físico*, comprise text, maps, photos and Landsat images, with the first given to provincial descriptions and the second adopting a thematic approach. Volume 3, *Atlas político*, includes an index of toponyms and demographic maps based on the 1980 census. Volumes 4 and 5 form an *Atlas económico*, while volume 6 is an *Atlas de la actividad económica*.

The **Automovil Club Argentino (ACA)** publishes a range of motorist and tourist maps, including maps of each province (scales of these vary from 1:1 000 000 to 1:125 000) showing primary and secondary roads, road distances and various tourist information, and incorporating street maps of major cities. ACA also publishes some city maps (*cartas especiales*), a *Guía turística* of the whole country, and another of Buenos Aires and surroundings (*Guía turística de Buenos Aires, La Plata y Alrededores Tomo I*). Road maps and city maps are also published by **Automapa SRL**.

Maps of several of the National Parks of Argentina have been prepared by the **Servicio Nacional de Parques Nacionales** and printed by IGM.

Further information

A thorough listing and graphic indexing of map series (both topographic and thematic) up until 1981 is provided in Instituto Panamerico de Geografía e Historia (1983) *Guía de la República Argentina para investigaciones geográficas*, prepared by the Instituto Geográfico Militar Argentino, 299 pp.

The development of census mapping is described by C. Reboratti in Nag, P. (1984) editor, *Census mapping survey*, New Delhi: Concept Publishing Company (Chapter 2 'Argentina').

INTA issues a list of maps and publications, as does also the ACA.

Addresses

Instituto Geográfico Militar (IGM)
Avenida Cabildo 301, 1426 BUENOS AIRES.

Municipalidad de la Ciudad de Buenos Aires
Avenida de Mayo 525, 1084 BUENOS AIRES.

Servicio Geólogico Nacional (SGN)
Avenida Santa Fe 1548, 1060 BUENOS AIRES.

Instituto Nacional de Tecnologia Agropecuaria (INTA)
Centro de Investigaciones de Recursos Naturales, Avenido
Rivadavia 1439, 1033 BUENOS AIRES.

Fundación Agro Palermo
Sarmiento 1183 6° Piso, 1041 BUENOS AIRES.

Servicio Meteorológico Nacional (SMN)
25 de Mayo 658, 1002 BUENOS AIRES.

Instituto Nacional de Estadística y Censos (INDEC)
Hipólito Yrigoyen 250, 1310 BUENOS AIRES.

Centro Editor de América Latina (CEAL)
Junín 981, BUENOS AIRES.

Automovil Club Argentino (ACA)
Avenida del Libertador 1850, 1425 BUENOS AIRES.

Servicio Nacional de Parques Nacionales
Avenida Santa Fe 690, 1059 BUENOS AIRES.

Automapa SRL
Bulnes 2094, 1425 BUENOS AIRES.

Catalogue

Atlases and gazetteers

Atlas total de la República Argentina
Buenos Aires: CEAL, 1981–83.
6 volumes.

Atlas de la República Argentina   2nd edn
Buenos Aires: IGM, 1983.
80 pp. (including 40 maps, 20 overlays).

General

República Argentina   1 : 10 000 000
Buenos Aires: IGM, 1981.

República Argentina   1 : 5 000 000
Buenos Aires: IGM, 1980.

República Argentina. Red caminero principal   1 : 4 000 000
Buenos Aires: ACA, 1982.
Road map with provincial boundaries.
Distance map on the reverse.

República Argentina   1 : 2 500 000
Buenos Aires: IGM, 1986.
2 sheets, both published.

República Argentina   1 : 2 000 000
Buenos Aires: ACA.

Topographic

Carta topográfica de la República Argentina   1 : 500 000
Buenos Aires: 1939–.
70 sheets, all published. ■
Series includes special sheet of the Malvinas.

Carta topográfica de la República Argentina   1 : 250 000
Buenos Aires: IGM, 1951–.
231 sheets, 90 published. ■

Carta topográfica de la República Argentina   1 : 100 000
Buenos Aires: IGM, 1911–.
ca 1900 sheets, 714 published.

Carta topográfica de la República Argentina   1 : 50 000
Buenos Aires: IGM, 1906–.
7197 sheets, 1710 published.

Geological and geophysical

Mapa geológico de la República Argentina   1 : 5 000 000
Buenos Aires: SGN, 1964.

Mapa hidrogeológico de la República Argentina   1 : 5 000 000
Buenos Aires: SGN, 1963.

Mapa geológico de la República Argentina   1 : 2 500 000
Buenos Aires: SGN, 1982.

Mapa geotécnico de la República Argentina   1 : 2 500 000
Buenos Aires: SGN, 1978.
2 sheets, both published.

Mapa metalogenética de la República Argentina   1 : 2 500 000
Buenos Aires: SGN, 1970.
3 sheets, all published.

Carta geológico-económica de la República Argentina   1 : 200 000
Buenos Aires: SGN, 1932–.
ca 700 sheets, *ca* 25% published. ■
Descriptive texts for each sheet.

Environmental

Mapa ecológico de la República Argentina   1 : 5 000 000
Buenos Aires: Agro-Palermo, 1983.

Carta de suelos de la República Argentina   1 : 100 000
Buenos Aires: INTA, 1970–.
27 sheets published.

Town maps

Buenos Aires y Alrededores   1 : 35 000
Buenos Aires: ACA.
Reverse has 6000 entry street index.

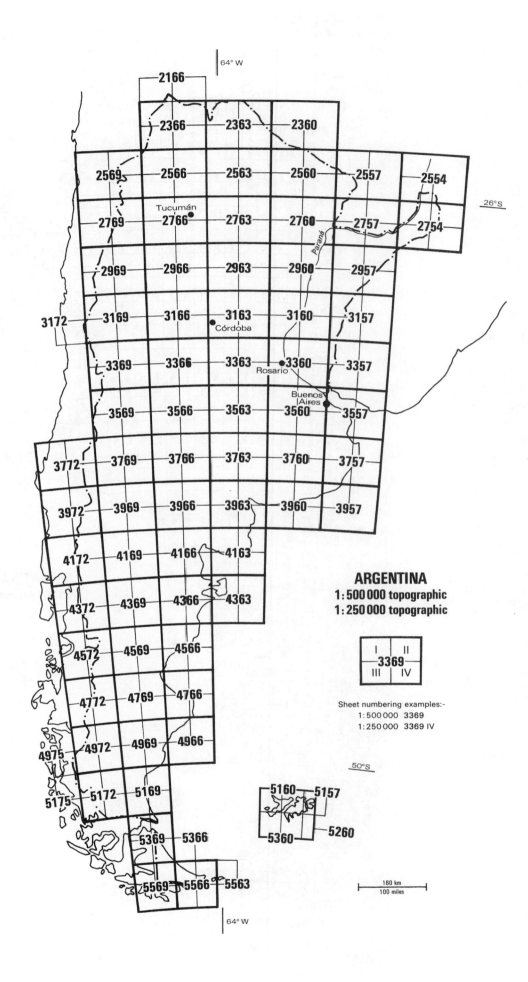

ARGENTINA
1:500000 topographic
1:250000 topographic

I	II
III 3369	IV

Sheet numbering examples:-
1:500000 3369
1:250000 3369 IV

160 km
100 miles

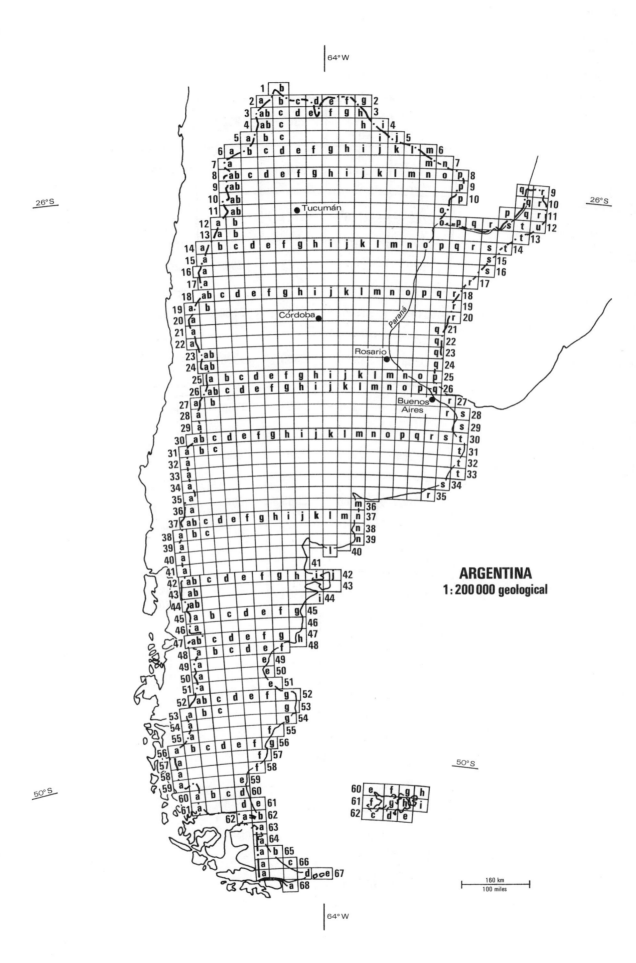

ARGENTINA
1 : 200 000 geological

Bolivia

Topographic mapping in Bolivia is the responsibility of the **Instituto Geográfico Militar (IGM)**, La Paz. Although the IGM was founded in 1936, detailed topographic mapping did not get under way until the postwar period, mainly under the stimulus of the Agrarian Reform Act which resulted in the establishment of a National Planning Secretariat. In the 1960s, IGM was provided with modern air survey and photogrammetric equipment, and the current national series at 1:50 000 and 1:250 000 date from this time. These series are based on subdivisions of the International Map of the World, but are on the Transverse Mercator projection. The basic series is at 1:50 000, and sheets each cover an area of 10' × 15'. The map is in five colours with a 20 m contour interval. So far, only the southern half of the country has been covered, and some of these sheets are in an American (AMS) edition. The 1:250 000 series covers the same southern area. This is a six-colour series with 100 m contours.

In the late 1970s, the Bolivian Government became interested in the use of satellite data for surveying its natural resources and, with the aid of the Inter-American Development Bank and cooperation with Purdue University, US, has set up a Bolivian Geographical Information System using a grid data structure, which can be organized into four levels of resolution to meet different requirements. Initially the system has been implemented for the Oruro Department.

Geological and resource surveys are carried out by the **Servicio Geológico de Bolivia (GEOBOL)**, La Paz. A 1:100 000 scale series was initiated in the 1960s and a number of sheets were published over that decade in the south-west of the country. No new sheets have appeared since 1972. GEOBOL now also functions as a centre for remote sensing investigations (**El Centro de Investigación y Aplicación de Sensores Remotos, CIASER**) and is concerned with the application of satellite technology to resource surveys.

In 1976 a cooperative mapping project, *Proyecto Precámbrico*, was started between GEOBOL and the **British Geological Survey (BGS)**. Initially the aim was a systematic survey of the Precambrian Shield area of eastern Bolivia, and ultimately to delimit areas favourable for the location and exploitation of minerals. The first two phases of the project ran until 1983 when a third phase extended the reconnaissance mapping to north-east Bolivia. The survey has produced much new information about the structural evolution of the Andes. A series of 14 colour maps have been published at 1:250 000 as the DOS Series 1224, and a 1:1 000 000 compilation map of the whole area appeared in 1984. An Overseas Memoir 'The geology and mineral resources of the Bolivian Precambrian Shield' by M. Litherland was published in 1986 to accompany the 1:1 000 000 map. A *Geochemical Atlas of Eastern Bolivia* by J.D. Appleton and A. Llanos was published in 1985.

A National Atlas, *Atlas de Bolivia*, was published in 1985, and contains a wide range of thematic mapping as well as general and departmental maps. Numerous Landsat image maps are also used.

Addresses

Instituto Geográfico Militar (IGM)
Gran Cuartal General, Avenida Saavedra final, Casilla Postal 7641, LA PAZ.

Servicio Geológico de Bolivia (GEOBOL)
Calle Federico Zuazo 1673, Esq. Reyes Ortiz, Casilla Postal 2729, LA PAZ.

British Geological Survey (BGS)
Keyworth, NOTTINGHAM NG12 5GG, UK.

Ministerio de Asuntos Campesinos y Agropecuarios (MACA)
LA PAZ.

Editions de l'ORSTOM
Service Diffusion, 70–74 Route d'Aulney, 93140 BONDY, France.

Instituto Nacional de Estatística
Casilla 6129, LA PAZ.

Editorial 'Los Amigos del Libro'
Casilla Postal 4415, LA PAZ.

Catalogue

Atlases and gazetteers

Atlas nacional de Bolivia
La Paz: IGM, 1985.
pp. 236.

Diccionario geográfico Boliviano
(Enciclopedia Boliviana 64)
René Gonzales Moscoso
La Paz/Cochabamba: Editorial 'Los Amigos del Libro', 1984.
257 pp. + departmental maps.

General

Mapa de la República de Bolivia 1:4 000 000 3rd edn
La Paz: IGM, 1980.

Mapa de comunicaciones de la República de Bolivia 1:3 000 000
La Paz: IGM, 1984.

Mapa de comunicaciones de la República de Bolivia 1:3 000 000
La Paz: IGM, 1984.

Foto-mosaico LANDSAT de Bolivia 1:2 000 000
La Paz: GEOBOL, 1980.

Mapa de la República de Bolivia 1:1 500 000 2nd edn
La Paz: IGM, 1980.
2 sheets, both published.

Mapa de la República de Bolivia 1:1 000 000 2nd edn
La Paz: IGM, 1980.

Mapa de la República de Bolivia. Carta Preliminar 1:500 000
La Paz: IGM, 1974.

Topographic

Carta Nacional de Bolivia 1:250 000 Series H531
La Paz: IGM, 1967–.
84 sheets, *ca* 50% published. ■

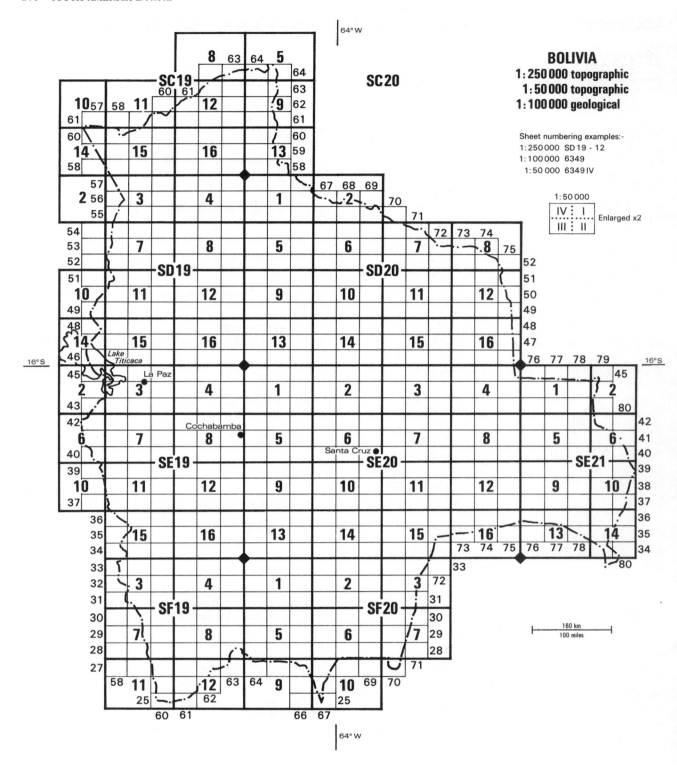

BOLIVIA
1:250 000 topographic
1:50 000 topographic
1:100 000 geological

Sheet numbering examples:-
1:250 000 SD 19 - 12
1:100 000 6349
1:50 000 6349 IV

Carta Nacional de Bolivia   1:50 000   Series H731
La Paz: IGM, 1955–.
2357 sheets, *ca* 50% published. ■

Geological and geophysical

Carte tectonique de Bolivie   1:5 000 000
Bondy: ORSTOM, 1973.
Legend in French and Spanish.
Accompanies *Cahiers ORSTOM Sér. Géologie* IV (2), 1972.

*Carte tectonique des terrains palaeozoiques et precambriens du Perou et
du Bolivie*   1:5 000 000
Bondy: ORSTOM, 1971.
Accompanies *Cahiers ORSTOM Sér. Géologie* III (1).

Mapa gravimétrico de Bolivia   1:2 500 000
La Paz: IGM, 1972.
Free air anomalies.

Carta isoclina de Bolivia. Epoca 1975.0   1:2 500 000
La Paz: IGM, 1976.

Carta intensidad total de Bolivia. Epoca 1975.0   1:2 500 000
La Paz: IGM, 1976.

Mapa geológico de Bolivia   1:1 000 000
La Paz: GEOBOL, 1978.
4 sheets, all published.
+ text.

Carte structurales des Andes septentrionales de Bolivie   1:1 000 000
Bondy: ORSTOM/La Paz: GEOBOL, 1978.

Mineralización de los Andes Bolivianos en relación con la placa de nazca   1:1 000 000
La Paz: GEOBOL, 1979.

Mapa geológico del area del Proyecto Precámbrico   1:1 000 000
Keyworth: BGS for Government of Bolivia, 1984.

Proyecto Precámbrico. Mapas geológicos   1:250 000   DOS 1224
Keyworth: BGS for Government of Bolivia, 1979–87.
14 sheets, all published.

República de Bolivia. Hojas geológicas   1:100 000
La Paz: GEOBOL, 1962–72.
53 sheets published. ■

Environmental

Mapa hidrográfico de Bolivia   1:1 000 000   1st edn
La Paz: IGM, 1985.
9 sheets, all published.
Includes *mapa de Isohyetas* 1951–81.

Mapa de cobertura y uso actual de la tierra   1:1 000 000
La Paz: GEOBOL, 1978.
4 sheets, all published.
+ a general land use map 1:4 000 000 and text.

Mapa de complejos de tierra del Oriente Boliviano   1:1 000 000
La Paz: GEOBOL, 1979.

Mapa ecológico de Bolivia   1:1 000 000
La Paz: MACA, 1975.
9 sheets, all published.
With 312 pp. text.

Human and Economic

Atlas censal de Bolivia   Edn 1
La Paz: Instituto Nacional de Estadística, 1982.

Town maps

Mapa urbano de La Paz   1:10 000
La Paz: IGM.

Brazil

The principal topographic mapping agencies in Brazil are the **Instituto Brasileiro de Geografia e Estatística (IBGE)**, Rio de Janeiro, and the **Diretoria de Serviço Geográfico (DSG)** of the Ministério do Exército, Brasília. IBGE was formerly the Conselho Nacional de Geografia, set up in 1937 to plan and expedite the mapping of the country, beginning with a modern version of the 1:1 000 000 map constructed from new trimetrogon photography and completed in 1946. The present agency is responsible for the collection, analysis and production of census statistics as well as geodetic and topographic survey. Today, in order to avoid overlaps, IBGE and DSG work cooperatively on the systematic mapping of the country.

Administratively, Brazil is divided into four territories (*Territórios*) and 21 states (*Estados*), and consequently much mapping is carried out on a regional basis by state governmental institutions. However, all the mapping activities in Brazil are coordinated by the Brazilian Commission of Cartography, COGAR, which was created in 1967 for that purpose and for regulating cartographic standards over this vast territory between the various mapmakers both national and regional.

At present the largest-scale series to cover the country completely is the 1:1 000 000. This series, drawn to the IMW specification, was published in a complete second edition in 1971–72, but since then many sheets have been up-dated *inter alia* from Landsat and radar imagery. The bound set of these maps which was formerly available is no longer in print, but the sheets may be obtained individually. The series forms a base for many thematic series (see below). Sheets each cover an area 6° by 4° and are on a Lambert conformal projection. From 1946 until 1965, a 1:500 000 series covering the eastern part of the country was in production, but has been discontinued.

Practically the whole country has now been mapped at 1:250 000, partly by conventional line maps published by IBGE and partly by line, radar and satellite image maps by DSG. The radar and satellite maps are controlled, or semi-controlled mosaics combined with planimetric detail.

Sheets each cover 1° 30' longitude by 30' latitude. This scale is the largest at which it is currently intended to cover the entire country. Originally the polyconic projection was used, but today the UTM is favoured. Contours are at 100 or 50 m intervals.

1:100 000 scale mapping, executed partly by IBGE, partly by DSG and partly by regional mapping organizations, is intended to reach areas of average population size and where detailed resource assessment is required. Sheets cover 30' latitude by 30' longitude on the UTM projection and the contour interval is generally 50 m. In the north-east, this mapping has been executed by SUDENE (Superintendência de Desenvolvimento do Nordeste) with 40 m contours. By 1986, very few areas remained untouched by this scale of mapping, the main exception being the rainforest area north of the Amazon.

The 1:50 000 series is intended for areas with relatively great population density and a more complex socioeconomic infrastructure. Sheets, each 15' × 15', cover mainly coastal areas in the south-east, and are based on photogrammetric survey. Contours are at 20 m and the projection is UTM. As well as IBGE and DSG, several state mapping organizations contribute significantly to the 1:50 000 programme, e.g. the **Instituto Geográfico e Geológico de São Paulo (IGGSP)** for the series *Região Sul do Brasil*. The latter series is photogrammetric, compiled from air survey in the early 1960s and with 10 m contours.

A limited amount of mapping at 1:25 000 is undertaken by DSG for military purposes.

Since 1978 there has been an effort to revitalize the systematic mapping of the country through a programme coordinated by COGAR and known as DINCART. During this period new mapping has averaged 5 per cent of the country annually, and by the end of 1984 80 per cent cover had been achieved. This mapping has proceeded in step with similar progress in aerial photographic and aerial geophysical survey and the expansion of the geodetic system, which is tied to the Inter-American network.

BRAZIL
States (estados) and territories*(territórios)

In addition to the systematic mapping programme, small but significant parts of the country have been mapped by private firms contracted by governmental organizations requiring special project mapping. The addresses and maps listed below are necessarily selective. Large-scale metropolitan mapping has been a recent development, but some states, e.g. São Paulo, have well advanced programmes at 1:10 000 and larger. The area of Greater São Paulo has also been covered by a suite of thematic maps at 1:100 000 by Empresa Metropolitana de Planejamento da Grande São Paulo.

Geological mapping and mineral resource survey is carried out by the **Departamento Nacional da Produção Mineral (DNPM)** and its Divisão de Geologia e Mineralogia. DNPM has published a number of small-scale geoscientific maps of the whole country, the most recent being the *Mapa Hydrogeológico do Brasil* 1:5 000 000 and the *Mapa Geológico do Brasil* 1:2 500 000. The latter is a major synthesis of current geological knowledge, and includes mapping of the adjacent ocean floor as well as the land territory of Brazil (incorporating data from the REMAC Project—a Global Reconnaissance of the Brazilian Continental Margin). The map accompanies a 500-page monograph *Geologia do Brasil*.

A series of geological maps of states and territories of Brazil is also in progress. Scales vary between 1:100 000 and 1:1 000 000 according to the size of the particular state. They are generally published by state authorities

with the cooperation of DNPM, and are accompanied by explanatory texts. Recent in-print state geological maps are included in our catalogue, and several more are in preparation. State geological surveys have also published numerous more detailed series. For example, the State of Rio de Janeiro has had a 1:50 000 scale series in progress since 1978; Paraná also published a number of sheets in a 1:50 000/75 000 series during the 1960s. Geological institutes in Brazilian universities have also conducted significant mapping programmes, in particular the **Instituto de Geociências, Universidade Federal do Rio Grande do Sul**, and the **Instituto de Geociências Aplicadas**, Minas Gerais.

A series of metallogenic and mineral resource maps is in preparation at the scale of 1:250 000 (1:1 000 000 in areas of low economic interest). These are compiled from both old and new geological, geophysical and geochemical survey data supplemented by interpretation from radar and satellite imagery. Much of the eastern part of the country has been covered by this mapping.

Considerable interest has arisen in the bathymetry and tectonics of the continental margin off the coast of Brazil. A series of bathymetric maps at varying scales is being published the **Diretoria de Hidrografia e Navegação (DHN)**, Rio de Janeiro, while a series of small-scale thematic maps were published as part of Projeto Remac by **Petrobrás**.

An important contribution to resource surveys in Brazil has come through the use of side looking airborne radar

(SLAR) which provided cloud-penetrating imagery of the country in the 1960s and 1970s. This was initially used in the project known as RADAM for the production of 1:1 000 000 scale thematic maps of the Amazon basin. Generally, for each IMW sheet, five thematic maps and a numbered monograph are published. The map themes are geology, geomorphology, soils, vegetation and land capability. The first 19 volumes, covering the Amazon basin, were issued between 1973 and 1980. Some have subsequently gone out of print. But such was the success of the project that in 1975 it was renamed **RADAMBRASIL** with a remit to cover the whole country. Five regional centres were set up to carry out the intensive fieldwork needed to support the imagery interpretation, with headquarters at Salvador, Bahia. Since 1980, a further 11 volumes have been issued, edited by DNPM, and extending the cover to the south of Brazil. A further eight volumes, some in process of publication, will complete the series. Landsat imagery has also been used in the preparation of these maps.

Soil mapping, besides being a part of the RADAMBRASIL programme, is also undertaken by a number of regional organizations and by the **Serviço Nacional de Levantamento e Conservação de Solos (SNLCS)**.

Vegetation and forest surveys have been undertaken by the Brazilian Institute of Forest Development, **Instituto Brasileiro de Desenvolvimento Florestal (IBDF)** in cooperation with the **Instituto das Pesquisas Espaciais (INPE)**. INPE has cooperated with many other government agencies over the last decade in developing methodologies for survey and monitoring of natural resources, e.g. the evaluation of deforestation in Amazonia, and the spectral discrimination of soils and of mineralized areas.

Regional cartographic authorities include super-intendencies for development in each region, with the acronyms SUDENE (for the Northeast), SUVALE (for the São Francisco Valley), SUDAM (for Amazonia), SUDECO (for the Centre-west) and SUDESUL (for the South). All these authorities participate on a contract basis in regional mapping.

Demographic and economic censuses are the responsibility of IBGE. Demographic censuses are carried out every 10 years, and the information is published mainly in the form of statistical tables. However, a number of population distribution maps have been published based upon the 1950 census (e.g. in the Atlas do Brasil of 1959), and on subsequent censuses (e.g. in state atlases such as the *Atlas de Ceará*), and a general distribution map from 1970 is still available, as are also three 1:1 000 000 sheets (*Rio de Janeiro/São Paulo, Paraná River* and *Rio Grande do Sul*) from 1973.

In 1966, the first, general volume of an *Atlas Nacional do Brasil* was issued. It was in a loose-leaf format and based essentially on 1960 data. Four further regional volumes were anticipated. The first volume, *Parte Geral*, is out of print, but the North East, the first regional volume, was published in 1985.

Further information

The IBGE issues a catalogue of its publications and maps (an English language version is available), and DSG supplies a general index map showing the progress of topographic mapping at the various scales. A *Catálogo de publicações* is available from Radambrasil.

Brazilian cartography has been described by J.K. Fialho (1975) in *Auto-Carto 2, 2nd International Symposium on Computer Assisted Cartography*, pp. 556–569. An article by M.J. de A. Gardini (1979), translated into English as 'Brazilian map sources' appeared in the *Bulletin of the Society of University Cartographers*, **13**, 42–51, and gives a useful overview of the range of organizations involved in Brazilian mapping.

An excellent listing of Brazilian maps published in the postwar period is provided by Carvalho, M.B.P. de (1983–84) *Mapas e outras materiais cartográficos no biblioteca central do IBGE*, a two-volume catalogue of Brazilian mapping held in the central library of IBGE.

Project RADAM and its successor RADAMBRASIL are well documented in a number of languages. An illustrated account in English is Bittencourt, O. (1982) Sidelooking airborne radar as a tool for natural resource mapping—the Brazilian approach, in *Resources for the Twenty-First Century* (USGS Professional Paper 1193), edited by F.C. Whitmore Jr and M.E. Williams, pp. 317–335.

A description of the progress in basic geological documentation up to 1981 is given by Trompette, R. *et al.* (1982) La documentation de base sur la géologie du Brésil, *Cah. ORSTOM, sér. Géol.* **12**, 165–169.

Census mapping is described by Rossini, R.E. and Simielli, M.E.R. (1984) in *Census Mapping Survey*, edited by Prithvish Nag, pp. 46–58.

Addresses

Instituto Brasileiro de Geografia e Estatística (IBGE)
Diretoria de Administração, SERGRAF—Departamento de Distribuição, Av. Brasil 15.671—Lucas, 21.241 RIO DE JANEIRO, RJ.

Diretoria de Serviço Geográfico (DSG)
QGEX Bloco F, 2° Andar SMU, 70630 BRASILIA, DF.

Instituto Geográfico e Geológico de São Paulo (IGGSP)
Cidade Universitaria, Butanta, 05508 SAO PAULO SP.

Departamento Nacional de Produção Mineral (DNPM)
Divisão de Geologia e Mineralogia, Sector de Autaquias Norte, Quadra 1, Bloco B, 70040 BRASILIA DF.

Instituto de Geociências
Universidade Federal do Rio Grande do Sul, Rua General Viterino 255, 90 000 PORTO ALEGRE, Rio Grande do Sul.

Instituto de Geociências Aplicadas
Universidade Federal de Minas Gerais, Cidade Universitária, Pampulha, Caixa Postal 1621–1622, 30 000 BELO HORIZONTE, Minas Gerais.

Diretoria de Hidrografia e Navegação (DHN)
Ministerio da Marinha, Ilha Fiscal, RIO DE JANEIRO, RJ.

Petrobrás
Divisão da Informação Técnica, CP 809, 20 000 RIO DE JANEIRO, RJ.

Projeto Radambrasil (RADAMBRASIL)
Av. Antônio Carlos Magalhães 1.131, 4° andar, Itaigara, CEP 40 000, SALVADOR.

Serviço Nacional de Levantamento e Conservação de Solos (SNLCS)
Rua Jardim Botânico 1024, 22 460 RIO DE JANEIRO, RJ.

Instituto Brasileiro de Desenvolvimento Florestal (IBDF)
Av. L-4 Norte, Setor de Areas Isoladas NT, S/N-Ed. Sede, CEP 70.040, BRASILIA DF.

Instituto de Pesquisas Espaciais (INPE)
Caixa Postal 515, 12 200 SAO JOSE DOS CAMPOS SP.

Instituto de Pesquisas Tecnológicas (IPT)
Cidade Universitária Armando de Salles Oliveira, Caixa Postal 7141, 01 000 SAO PAULO, SP.

Universidade Federal de Sergipe
Rua Lagarto 952, ARACAJU, Sergipe.

Sociedade Paulista de Produçoes Cartograficas
Rua Bitencourt Rodrigues 88, 5° andar, SAO PAULO, SP.

For CIA, see United States; for Centro de Ciencia Tropical, see Costa Rica.

Catalogue

Atlases and gazetteers

Atlas nacional do Brasil: Região Nordeste
Rio de Janeiro: IBGE, 1985.
53 maps at 1:4 000 000.

Índices dos topônimos da carta do Brasil ao Milionésimo—1971
Rio de Janeiro: IBGE.
Index of names on the 1:1 000 000 map of Brazil.
pp. 322.

General

Brazil 1:11 800 000
Washington, DC: CIA, 1977.
With 5 thematic inset maps.

Brasil rodoviário polimapas 1:7 352 000
São Paulo: Sociedade Paulista de Produções Cartográficas, 1985.

Mapa polivisual do Brasil 1:6 000 000
São Paulo: Sociedade Paulista de Produções Cartográficas, 1985.

República Federativa do Brasil 1:5 000 000
Rio de Janeiro: IBGE, 1985.

Brasil 1:2 500 000
Rio de Janeiro: IBGE, 1982.
4 sheets, all published.

Topographic

Carta do Brasil 1:1 000 000
Rio de Janeiro: IBGE, 1947–.
46 sheets, all published. ■

Carta do Brasil 1:250 000
Rio de Janeiro: IBGE/Brasilia: DSG, 1949–.
401 sheets, *ca* 360 published. ■

Carta do Brasil 1:100 000
Rio de Janeiro: IBGE/Brasilia: DSG, 1965–.
1544 sheets, *ca* 1200 published.

Carta do Brasil 1:50 000
Rio de Janeiro: IBGE/Brasilia: DSG, 1963–.
10 495 sheets in series.

Bathymetric

Brasil—Carta batimétrica scales vary
Rio de Janeiro: DHN, 1977–.
22 sheets, 10 published.

Brasil, margem continental: mapa batimétrica 1:5 592 000 Projeto Remac 11
Rio de Janeiro: Petrobrás, 1979.

Brasil, margem continental: mapa de relêvo 1:5 592 000 Projeto Remac 11
Rio de Janeiro: Petrobrás, 1979.

Brasil, mapa de recursos naturais superficiais da plataforma continental brasileira 1:3 500 000 Projeto Remac 10–11
Rio de Janeiro: Petrobrás, 1979.

Brasil, margem continental . . . 1:3 500 000 Projeto Remac
Rio de Janeiro: Petrobrás, 1979.
Series of bathymetric, physiographic and sediment facies maps of Brazilian continental shelf.

Geological and geophysical

Mapa tectônico do Brasil 1:5 000 000
Brasília: DNPM, 1971.

Mapa metalogenético do Brasil 1:5 000 000
Brasília: DNPM, 1973.

Mapa hidrogeológico do Brasil 1:5 000 000
Brasília: DNPM, 1983.

Mapa geologico do Brasil e da área oceânica adjacente, incluindo depositos minerais 1:2 500 000
Brasília: DNPM, 1981.
Legend in Portuguese and English.
Accompanies 501 pp. text *Geologia do Brasil*.

Environmental

Provisional ecological map of the Republic of Brazil 1:10 000 000
J.A. Tosi Jr
San José, Costa Rica: Centro de Ciencia Tropical, 1983.

Brasil. Climas, Precipitação, Temperatura 1:5 000 000
Rio de Janeiro: IBGE, 1978.
Set of three maps.

Projeto RADAMBRASIL 1:1 000 000
Salvador: RADAMBRASIL, 1973–.
38 volumes, 34 published. ■
Each volume generally includes maps of geology, geomorphology, soils, vegetation and land use capability.

Projeto mapas metalogenéticos e de previsão minerais 1:1 000 000/250 000
Rio de Janeiro: DNPM.

Administrative

República Federativa do Brasil 1:5 000 000
Rio de Janeiro: IBGE, 1980.

Divisão municipal do Brasil. Microrregiões homogêneas do Brasil 1:5 000 000
Rio de Janeiro: IBGE, 1968.

República Federativa do Brasil 1:2 500 000
Rio de Janeiro: IBGE, 1973.
4 sheets, all published.

Divisão administrativa do Brasil 1:2 500 000
Rio de Janeiro: IBGE, 1968.
4 sheets, all published.

Human and economic

Brasil. População. Densidade demográfica 1970 1:5 000 000
Rio de Janeiro: IBGE, 1971.
Population density.

Brasil. População urbana e rural
Rio de Janeiro: IBGE, 1972.

A selection of maps by states and territories

Acre

Mapa do Estado do Acre 1:1 000 000
Rio de Janeiro: IBGE, 1982.

Alagoas

Mapa geológico do Estado de Alagoas 1:150 000
Brasília: DNPM, 1984.
With text.

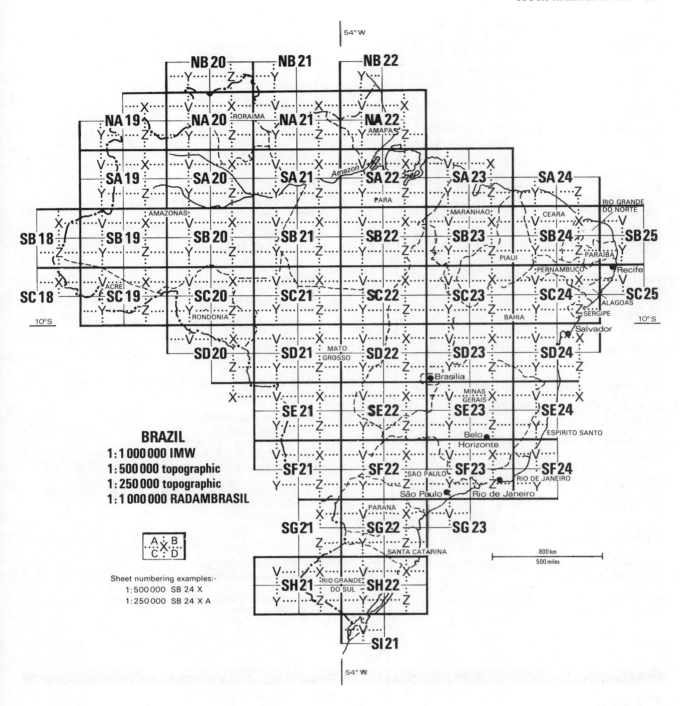

Sheet numbering examples:-
1:500 000 SB 24 X
1:250 000 SB 24 X A

BRAZIL
1:1 000 000 IMW
1:500 000 topographic
1:250 000 topographic
1:1 000 000 RADAMBRASIL

Aptidão agrícola das terras: Estado de Alagoas   1:400 000
Maceió: Ministério da Agricultura, 1978.

Amapá

Mapa do Estado do Amapá   1:1 000 000
Rio de Janeiro: IBGE, 1981.

Bahia

Mapa geológico do Estado da Bahia   1:1 000 000
Salvador: Secretaria de Minas e Energia, 1978.
With text.

Ceará

Mapa geológico do Estado do Ceará   1:500 000
Brasilia: DNPM, 1983.

Goiás

Atlas geográfico do Estados de Goiás
Goiânia: Secretaria de Planejamento e Coordenação, 1979.

Mapa geológico do Estado de Goiás   1:1 000 000
Brasilia: DNPM, 1986.

Maranhão

Atlas do Estado do Maranhão
Rio de Janeiro: IBGE.

Mapa do Estado do Maranhão   1:1 000 000
Rio de Janeiro: IBGE, 1983.

Maranhão   1:100 000
Brasilia: DNPM, 1986.

Mato Grosso

Mapa do Estado de Mato Grosso   1:1 500 000
Rio de Janeiro: IBGE, 1983.

Paraíba

Mapa do Estado da Paraíba   1:500 000
Rio de Janeiro: IBGE, 1970.

Mapa geológico do Estado da Paraíba 1:500 000
Brasília: DNPM, 1982.
With text.

Aptidão agrícola dos solos: Estado da Paraíba 1:1 000 000
Recife: SUDENE, 1970.

Pernambuco

Mapa geológico do Estado de Pernambuco 1:500 000
Brasília: DNPM, 1980.
With text.

Aptidão agrícola dos solos do Estado de Pernambuco 1:600 000
Recife: SUDENE, 1973.

Rio de Janeiro

Atlas do Estado do Rio de Janeiro
Rio de Janeiro: Fundação de Amparo a Pesquisa do Estado do Rio
de Janeiro, 1982.

Rio de Janeiro, básico 1:400 000
Rio de Janeiro: IBGE, 1975.

Rio de Janeiro, político 1:400 000
Rio de Janeiro: IBGE, 1975.

Rio de Janeiro—Região metropolitana 1:200 000
Rio de Janeiro: IBGE, 1975.

Rio Grande do Sul

Mapa geológico do Estado do Rio Grande do Sul 1:1 000 000
Brasília: DNPM, 1986.

Rio Grande do Sul. Geomorfologia 1:750 000
Rio de Janeiro: IBGE, 1972.

*Rio Grande do Sul. Hidrologia—deficiência de umidade nos
solos* 1:750 000
Rio de Janeiro: IBGE, 1972.

Rio Grande do Sul. Capacidade de uso dos solos 1:750 000
Rio de Janeiro: IBGE, 1972.

Rio Grande do Sul. Uso da terra 1:750 000
Rio de Janeiro: IBGE, 1972.

Rio Grande do Sul. Sócio-econômico. Setor primário 1:750 000
Rio de Janeiro: IBGE, 1972.

Rondônia

Rondônia 1:1 000 000
Rio de Janeiro: IBGE, 1982.

Roraima

Atlas de Roraima
Rio de Janeiro: IBGE.

Roraima 1:1 000 000
Rio de Janeiro: IBGE, 1970.

Santa Catarina

Mapa rodoviário do Estado de Santa Catarina 1:750 000
Salvador: RADAMBRASIL.

Mapa geológico do Estado de Santa Catarina 1:500 000
Brasília: DNPM, 1986.

São Paulo

Estado de São Paulo 1:1 000 000
São Paulo: IGC, 1982–84.
Administrative map in 4 sheets.

São Paulo—geológico 1:1 000 000
Rio de Janeiro: IBGE, 1974.

Mapa geomorfológico do Estado de São Paulo 1:1 000 000
São Paulo: IPT, 1981.

Hipsometria do Estado de São Paulo 1:1 000 000
São Paulo: IGCSP, 1982.

Mapa geológico do Estado de São Paulo 1:500 000
São Paulo: IPT, 1981.

Carta de utilização da terra do Estado de São Paulo 1:250 000
São Paulo: IGCSP, 1980–.
19 sheets, 8 published.

Sergipe

Atlas de Sergipe 1:500 000
Aracajú: Universidade Federal de Sergipe, 1979.

Mapa geológico do Estado de Sergipe 1:250 000
Brasília: DNPM, 1983.

Chile

The official topographic mapping organization in Chile is
the **Instituto Geográfico Militar (IGM)**, Santiago,
founded in 1922 and responsible for all matters relating to
the geography, survey and mapping of the national
territory.

Although some large-scale topographic mapping had
been undertaken in the first decades of this century, the
first complete topographic cover of the country was
achieved in 1955 with a 102-sheet series at 1:250 000, the
Carta Preliminar, compiled from trimetrogon photography
flown in 1944. This series was on a Lambert conformal
conic projection with each sheet covering an area of 1°
latitude by 2° longitude. The relief was shown by form
lines. This mapping, together with more detailed mapping
from limited previous surveys, formed the basis of the
Carta Nacional at 1:500 000 issued in the early 1970s.

This series is complete in 41 sheets with an additional
special sheet at 1:10 000 000 showing the Chilean
Antarctic Territory. It is in six colours, also on the
Lambert conformal projection and with relief shown by
form lines.

Since 1976 a new *Carta Regular* at 1:250 000 has been
in progress in 90 sheets each covering an area of 1° latitude
by 1.5° or 2° of longitude. The series is due for completion
in 1988. It is on a UTM projection, has a kilometric grid,
and is printed in seven colours. Contours are at 50 or
100 m. IGM has been considering the use of this map as a
base for thematic mapping, and a pilot project was
initiated in 1978.

Until recently, the above series were the most detailed
generally available to the public, but since 1958
considerable attention has been given to the production of

a modern photogrammetric series of 15′ maps at 1:50 000, from which the smaller-scale public mapping is being derived. Production of this series began in the north of the country, and the first 950 sheets, extending south to latitude 43.30° S, were completed in 1978. Since then, the mapping has been extended into the less accessible southern part of the country, using the latest techniques of satellite geodesy for position fixing and control. This series, and the 1:25 000 series mentioned below, are now available for public sale.

Simultaneously to the 1:50 000 programme, 1:25 000 mapping has also been in progress in areas of prime importance for socioeconomic development, following densification of the existing triangulation network.

IGM also publishes the National Atlas, *Atlas de la República de Chile*, first issued in 1966. In 1983, a revised and enlarged edition was published. It includes a 1:1 000 000 general cover of the country occupying 13 double pages, as well as an extensive range of general and regional thematic maps. The latter are organized into the 13 new administrative regions of the country. There is an index of placenames.

In 1984, IGM issued a *Listado de Nombres Geográficos* in two volumes, comprising 65 000 toponyms contained in the 1:50 000 and 1:250 000 mapping of the area between 17° 35′ S and 42° 55′ S.

IGM has published several small-scale maps of Chile, including physical maps at 1:1 000 000 (in IMW style) and 1:2 000 000 and maps of the new administrative regions at 1:2 000 000 and 1:3 000 000.

In 1976 the government approved a National Plan for Cartography and Photogrammetry which provided resources for modern surveying and photogrammetric equipment. At the same time, agreements for technical cooperation were made with France and Spain. IGM has also been experimenting with geographical databanks and automated cartography, with the use of Landsat images for small-scale mapping and, since 1979, with the production of orthophotomaps (at 1:20 000).

Cadastral mapping has been carried out by the Cadastral Branch of IGM, and large-scale cadastral plans have been made of metropolitan areas.

Geological and magnetic maps are published by the **Servicio Nacional de Geología y Minería (SNGM),** Santiago (formerly Instituto de Investigaciones Geológicas). Geological maps form a numbered series initiated in 1958, with publication scales ranging from 1:50 000 to 1:500 000. Some of these are out of print. They are accompanied by monographs. In addition, there is a standard series at 1:250 000 (*Mapas geológicos preliminares de Chile*) with several new sheets appearing annually, and a new general geological map (1982) at 1:1 000 000 in six sheets. A series of magnetic maps, originally published at 1:1 000 000, is now being continued at 1:250 000 scale.

The **Servicio Nacional de Turismo (SNT)** has issued a road map of Chile at 1:2 000 000, and a number of tourist maps of cities, including a small one of central Santiago. A larger, indexed street map of Santiago is published by **Informaciones Unidas para América Latina (INUPAL).**

Further information

IGM issues a *Cátalogo de Ventas* which includes detailed index maps and listings of all available sheets in the various topographic series. Due to the size and complexity of indexes to the larger-scale topographic series, we have not attempted to reproduce them here.

SNGM has issued a *Cátalogo de Publicaciones 1957–81* with subsequent updates.

Addresses

Instituto Geográfico Militar (IGM)
Nueve Santa Isabel 1640, SANTIAGO.

Servicio Nacional de Geología y Minería (SNGM)
Agustinas 785, 6o piso, Casilla 10465, SANTIAGO.

Servicio Nacional de Turismo (SNT)
Catedral 1165, Casilla 14082, SANTIAGO.

Informaciónes Unidas para América Latina (INUPAL)
Portal Fernandez Concha 960, (Plaza de Armas) Of. 432, SANTIAGO.

For DMA and CIA, see United States.

Catalogue

Atlases and gazetteers

Atlas de la República de Chile 2nd edn
Santiago: IGM, 1983.
pp. 351.

Atlas regionalizado de Chile 2nd edn
Santiago: IGM, 1981.
pp. 64.

Listado de nombres geográficos
Santiago: IGM, 1984.
2 volumes.

Chile. Official standard names approved by the US Board on Geographic Names
Washington, DC: DMA, 1967.
pp. 591.

General

Chile 1:8 500 000
Washington, DC: CIA, 1972.
Inset maps of population and administrative divisions, vegetation and economic activity.

Mapa de Chile 1:2 000 000
Santiago: IGM, 1977.

Gran mapa caminero de Chile geográfico y turístico 1:2 000 000
Santiago: SNT, 1984.
Folded road map with distance table.

Mapa de Chile 1:1 000 000
Santiago: IGM, 1977.
6 sheets, all published.

Topographic

Chile 1:500 000
Santiago: IGM, 1971–.
41 sheets, all published. ■

Carta preliminar 1:250 000
Santiago: IGM, 1945–.
102 sheets, all published.

Carta regular 1:250 000
Santiago: IGM, 1976–.
90 sheets, 46 published. ■

CHILE
1 : 500 000 topographic

Carta topográfica   1 : 100 000
Santiago: IGM, 1968–.

Carta topográfica   1 : 50 000
Santiago: IGM, 1958–.

Geological and geophysical

Volcanes activos de Chile   1 : 6 000 000
Santiago: IGM.

Mapa temático: geomorfológico   1 : 6 000 000
Santiago: IGM.

Mapa tectónico de Chile   1 : 5 000 000
Santiago: SNGM.
3 sheets.
With text.

Mapa geológico de Chile   1 : 1 000 000
Santiago: SNGM, 1982.
6 sheets, all published.

Carta geológica de Chile
1 : 500 000/1 : 250 000/1 : 100 000/1 : 50 000
Santiago: SNGM, 1959–.
Numbered series at various scales, each published with a
monograph.

Carta magnética de Chile   1 : 250 000/1 : 100 000
Santiago: SNGM, 1980–.
14 sheets published.

Environmental

Mapa temático: fitogeográfico   1 : 6 000 000
Santiago: IGM.

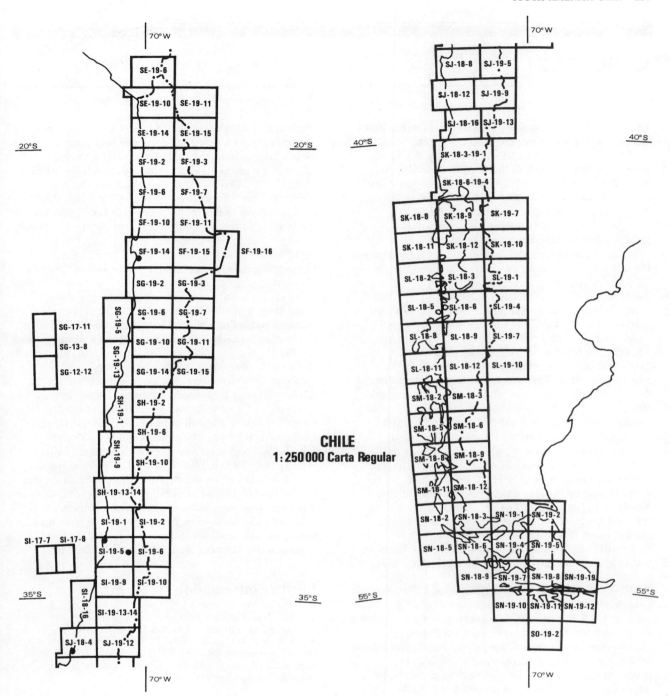

CHILE
1:250 000 Carta Regular

Mapa temático: suelos   1:6 000 000
Santiago: IGM.

Mapa temático: hydrográfico   1:6 000 000
Santiago: IGM.

Administrative

Mapa político de Chile con regionalización   1:6 000 000
Santiago: IGM.

Mapa temático: Chile regional y provincial   1:6 000 000
Santiago: IGM.

Mapa de Chile: regionalizada   1:3 000 000
Santiago: IGM, 1983.

Human and economic

Mapa temático: Chile asentamientos urbanos   1:6 000 000
Santiago: IGM.

Mapa temático: actividad industrial y provincial   1:6 000 000
Santiago: IGM.

Town maps

Plano de Santiago: guía de calles   1:25 000
Santiago: INUPAL.
With street index on reverse.

Santiago centro   1:25 000
Santiago: SNT.

Colombia

The central mapping agency in Colombia is the **Instituto Geográfico 'Agustín Codazzi' (IGAC)**, Bogotá, founded in 1935, and acquiring its present name in 1950. It is under the administration of the Ministerio de Hacienda. IGAC is responsible for basic topographic mapping, the inventory and classification of soils, studies of physical, economic, social and urban geography, and for management of cadastral survey; and there are five *Subdirecciónes* for handling these various tasks.

Systematic topographic mapping of the country began in 1945, in cooperation with the Army Map Service of the US through a mapping agreement with the IAGS. The basic series are at 1 : 25 000 and 1 : 50 000. The 1 : 25 000 *Carta Preliminar* was issued from 1946 and most of the northern third of the country has been covered at this scale. There are two 1 : 50 000 series, one produced by the AMS (in five colours, with 25 m contours, and on a Transverse Mercator projection) the other by the IGAC on a Gauss projection. These series cover the northern half of the country. Many sheets in all these series are available only as monochrome copies, and it is necessary to obtain permission in advance from the Ministry of Defence (Comando General, Fuerzas Militares, Ministerio de Defensa Nacional en Bogotá, DE) before seeking to acquire any mapping at scales of 1 : 200 000 and larger.

Medium-scale mapping at 1 : 100 000 has been produced by IGAC and by the **Departamento Administrativo Nacional de Estadística (DANE)**. The maps produced by the latter are planimetric maps, but they cover a more extensive area than the IGAC series which is a five-colour map with 25 m contours on a Gauss projection. Cover at this scale, however, is confined to the north of the country.

Smaller-scale mapping includes a DMA series at 1 : 250 000 (Series 1501), an IGAC series at 1 : 200 000, and a recently started 1 : 500 000 series.

From 1969, radar imagery was obtained of many parts of the country and in particular of the West Pacific Departments and in the southern *Comisarias/Departamentos* of Amazonas, Caquetá, Vaupes and Quainia where persistent cloud cover made conventional photography impossible. This imagery has provided 1 : 200 000 scale radar maps of the south of the country where detailed survey is otherwise lacking.

IGAC has also issued a series of maps at various scales each covering one of the *Departamentos* of the country. These are coloured maps in recent editions, and include inset maps of major cities. There is also a series of city maps published mainly at scales of 1 : 5000 or 1 : 10 000.

There is a National Atlas of Colombia and a series of four regional atlases which together cover the country.

Geological surveying is carried out by the **Instituto Nacional de Investigaciones Geológico-Mineras (INGEOMINAS)**, which dates from 1916 but got its present name in 1968, when the Servicio Geológico Nacional was amalgamated with several related institutions. Present responsibilities include geological mapping, hydrogeology, geochemical and mineral reconnaissance, remote sensing, geophysics and investigating geothermal resources. Future programmes may include work in marine and environmental geology.

From 1964 until 1972 there was a cooperative programme with the United States Geological Survey to produce a systematic geological survey and an evaluation of mineral resources. These surveys have been reported in the *Boletín Geológico* published by INGEOMINAS. In addition, a series of quadrangle maps has been published. initially at 1 : 200 000 (mostly described as *Edición preliminar*) but subsequently (after 1967) at 1 : 100 000. A number of new sheets in this series have been published in the 1980s, and these use the sheet system of the 1 : 100 000 *Carta general*. Several geological maps of individual departments have also been published, and a number of hydrogeological maps are included in reports in the *Boletín Geológico*.

Surveying of soils and land use began in the 1940s. From 1968, a modified form of the US Department of Agriculture land capability classification came into use. Surveys are carried out by the **Subdirección Agrológico** of the IGAC, using a modified form of the USDA Soil Conservation Service taxonomy. Scales vary with the nature of the specific investigation, and publication is in the form of texts with associated maps. These surveys are arranged by Department—there is no regular quadrangle series—and thus although numerous cannot satisfactorily be summarized in the catalogue. A full list is available from IGAC. A general soils map, with memoir, for the whole country was published in 1983. The Subdirección also undertakes ecological surveys, forest inventories, and land evaluation for cadastral purposes. A major resource study covering the southern Amazonian provinces and based on radar survey was published in 1979, and includes geological, soil and forestry maps at 1 : 500 000 scale (*La Amazonia Colombiana y sus Recursos—PRORADAM*).

Further information

The full range of mapping in Colombia up to the early 1970s is summarized in Martinson, T.L. and Showalter, G.R. (1975) *Research Guide to Colombia*, PAIGH. Map indexes and lists of current mapping are issued by IGAC and INGEOMINAS, and there is a catalogue *Actividades y realizaciones* issued by the Subdirección Agrológica of IGAC.

Addresses

Instituto Geográfico 'Agustín Codazzi' (IGAC)
Carrera 30 No. 48 51, Oficina 212, BOGOTÁ.

Departamento Administrativo Nacional de Estadística (DANE)
Centro Administrativo Nacional, Avenida Eldorado, BOGOTÁ.

Instituto Nacional de Investigaciones Geológico-Mineras (INGEOMINAS)
Diagonal 53 No. 34 53, BOGOTÁ.

For CIA, see United States.

Catalogue

Atlases and gazetteers

Atlas de Colombia 3rd edn
Bogotá: IGAC, 1977.
pp. 286.

Atlas regional Andino
Bogotá: IGAC, 1982.
pp. 168.

Atlas regional del Caribe
Bogotá: IGAC, 1978.
pp. 121.

Atlas regional Orinoquia-Amazonia
Bogotá: IGAC, 1983.
pp. 162.

Atlas regional Pacífico
Bogotá: IGAC, 1983.
pp. 96.

Diccionario Geográfico de Colombia   2nd edn
Bogotá: IGAC, 1980.
2 volumes.

General

Colombia   1:4 000 000
Washington, DC: CIA, 1980.
Inset maps of population, land use and economics.

Mapa Landsat de Colombia   1:2 000 000
Bogotá: INGEOMINAS, 1980.
Uncontrolled Landsat mosaic.

Mapa vial y turístico de Colombia   1:1 500 000
Bogotá: IGAC, 1981.

República de Colombia. Mapa físico-político   1:1 500 000
Bogotá: IGAC, 1983.

Topographic

Colombia. Mapas de Departamentos/Intendencias/Comisarias
1:100 000–1:500 000
Bogotá: IGAC, 1978–.
ca 24 sheets available.

República de Colombia   1:500 000
Bogotá: IGAC, 1979–.
26 sheets, 13 published. ■
Sheets in south of country based on radar imagery.

Carta topográfica   1:250 000   Series 1501
Bogotá: IGAC/Washington, DC: DMATC, 1971–.
96 sheets, 11 published.

Carta topográfica   1:200 000
Bogotá: IGAC, 1977–.
172 sheets, 76 published. ■
Sheets in south of country based on radar imagery.

República de Colombia. Carta general   1:100 000
Bogotá: IGAC, 1948–.
ca 570 sheets, *ca* 220 published. ■

Carta topográfica   1:50 000   Series E772
Bogotá: IGAC, 1958–.
2996 sheets, *ca* 50% published.

Carta topográfica   1:25 000
Bogotá: IGAC.
7937 sheets, *ca* 35% published.

Geological and geophysical

Mapa metalogénico de Colombia   1:5 000 000
Bogotá: INGEOMINAS, 1976.

Mapa de riesgo sísmico   1:5 000 000
Bogotá: Instituto Geofísico, 1977.

Mapa geológico de Colombia   1:1 500 000
Bogotá: INGEOMINAS, 1976.

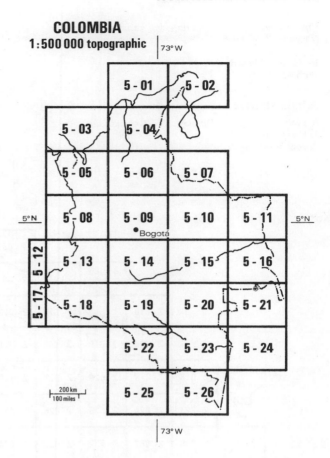

COLOMBIA
1:500 000 topographic   73°W

República de Colombia. Mapa gravimétrico   1:1 500 000   2nd edn
Bogotá: IGAC, 1983.

República de Colombia. Carta isogónica Epoca 1980.0   1:1 500 000
Bogotá: IGAC, 1980.

*Mapa geológico-tectónico de los Andes más septentrionales,
Colombia*   1:500 000
Bogotá: INGEOMINAS, 1971.
Accompanies Volume 19 (2) *Boletín Geológico.*

Mapa geológico de Colombia   1:200 000
Bogotá: INGEOMINAS, 1956–.
12 sheets published. ■
Some sheets out of print.

Mapa geológico de Colombia   1:100 000
Bogotá: INGEOMINAS, 1967–.
ca 520 sheets, 16 published. ■

Environmental

Mapa de suelos de Colombia   1:1 500 000
Bogotá: IGAC, 1983.
With 86 pp. text.

Mapa de estaciones hidrométricas   1:1 500 000
Bogotá: IGAC, 1974.

Mapa de estaciones meteorológicas   1:1 500 000
Bogotá: IGAC, 1974.

Mapa de estaciones pluviométricas   1:1 500 000
Bogotá: IGAC, 1974.

República de Colombia. Carta ecológica   1:500 000
Bogotá: IGAC, 1977.
21 sheets, all published + 238 pp. text.

República de Colombia. Mapa de bosques   1:500 000
Bogotá: IGAC, 1981–.

Programa nacional de inventario y clasificación de tierras
(Proclas) 1:500 000
Bogotá: IGAC, 1973.
19 sheets published.
With 42 pp. text.

Administrative

República de Colombia. División política-administrativa
1:1 500 000
Bogotá: IGAC, 1979.

Town maps

Mapa de la Sabana de Bogotá 1:100 000
Bogotá: IGAC, 1982.

Plano de la ciudad de Bogotá 1:25 000
Bogotá: IGAC, 1979.

Mapa turístico de Bogotá y sus alrededores
Bogotá: IGAC, 1973.
In Spanish and English.

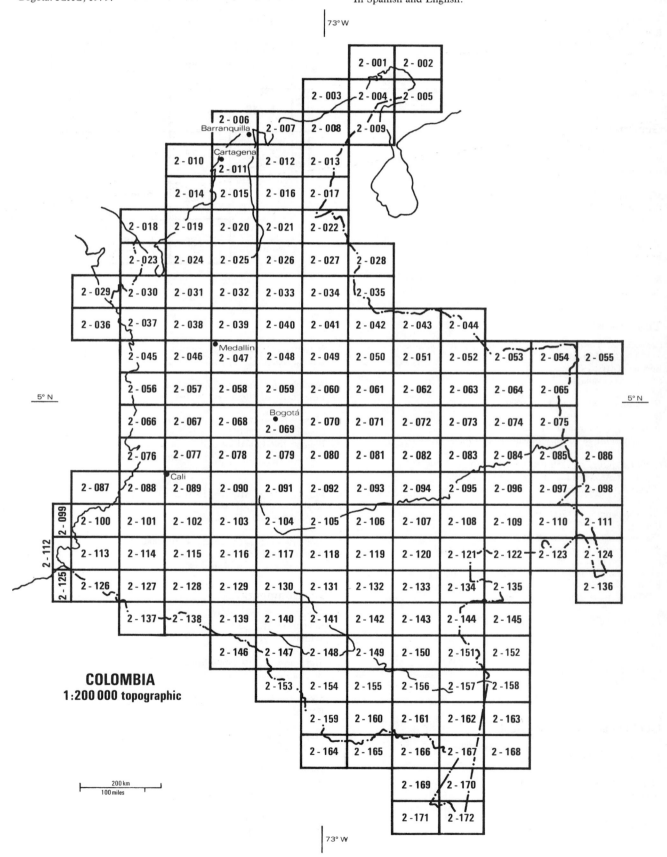

COLOMBIA
1:200 000 topographic

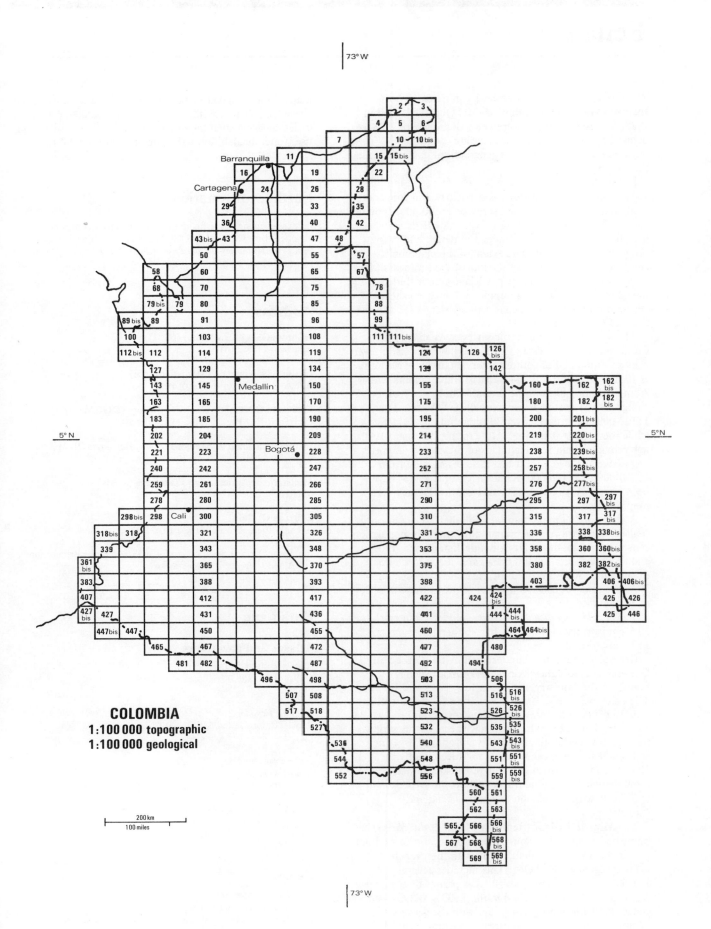

73° W

Barranquilla

Cartagena

Medallin

Bogotá

Cali

5° N

5° N

COLOMBIA
1:100 000 topographic
1:100 000 geological

200 km
100 miles

73° W

Ecuador

The topographic mapping authority in Ecuador is the **Instituto Geográfico Militar (IGM)**, Quito, founded in 1928. Its responsibilities were defined in a law of 1978 as being those of surveying and mapping the national territory and the archiving of geographical data concerning the country.

Little progress was made with basic mapping until the 1940s. Since then, Ecuador has participated in cooperative mapping with the IAGS, and series at 1:100 000, 1:50 000 and 1:25 000 are under way. By the mid-1980s more than half the country had been mapped in detail, and current progress is quite rapid. The extent of the mapping is confined to the area west of the line of the Protocol of Rio de Janeiro 1942. The Galapagos Islands are included in the topographic mapping programme, although no sheets have yet been published (for the Galápagos Islands see also the Pacific Ocean section).

The three main topographic series cover mainly the Costa and western Sierra. The 1:100 000 series is in seven colours with an 80 m contour interval (auxiliary contours at 40 m), and the 1:50 000 series is in five colours with 20 m contours (40 m in the Sierra). The series are on a Transverse Mercator projection with UTM grid. The 1:100 000 sheets each cover 30′ longitude by 20′ latitude, while the 1:50 000 are 15′ × 10′. In the 1970s and 1980s a few provincial and regional maps of previously mapped areas have also been published at scales ranging from 1:300 000 to 1:100 000. Some sheets of Ecuador have been issued as part of the PAIGH Unified Hemispheric Mapping Series.

A series of 260 photomaps at 1:50 000 scale (*Las cartas croquis planimétricas*) was prepared for the 1974 population census by the **Instituto Nacional de Estadística (INE)**, and there is a series of 108 urban area maps (*Los planos croquis*) at 1:5000 prepared for the same purpose. The IGM has also published a number of urban maps at scales of 1:1000 to 1:15 000.

The *Atlas Geográfico del Ecuador*, published to mark the 50th anniversary of the IGM, has 86 thematic maps at 1:2 000 000 scale.

Geological mapping is undertaken by the **Dirección General de Geología y Minas (DGGM)**, which is a branch of the Ministry of Natural Resources and Energy. The principal series, *Mapa geológico del Ecuador*, is mainly at 1:100 000, although a few sheets have been produced at 1:50 000 and 1:25 000. Many new sheets in this series have been issued in the late 1970s and the 1980s. A new 1:1 000 000 geological map was produced jointly by DGGM and the British Institute of Geological Sciences (now **British Geological Survey, BGS**) in 1983. It is in two sheets with correlation diagram, legend and explanatory text.

The **Ministerio de Agricultura y Ganadería (MAG)** has published a number of soils maps and has also carried out resource mapping in collaboration with the French overseas agency ORSTOM. This includes several 1:1 000 000 scale maps published as part of the *Programa Nacional de Regionalización Agraria* (PRONAREG), and concerned with hydrogeology and with the various climatic parameters affecting agriculture. A bioclimatic and an ecological map have also been published at this scale together with a text by Luis Canadas Cruz describing both

maps. More detailed mapping includes an inventory of the Costa (the fertile coastal lowlands) at 1:200 000, a 14-sheet series covering such themes as geomorphology, soils, land use, water availability, irrigation and land capability.

Further information

Indexes of topographic mapping and a list of maps may be obtained from IGM, while DGGM issues a list of geological maps.

A PAIGH guide *Guía para investigadores de Ecuador* (117 pp.) was published in 1982.

Addresses

Instituto Geográfico Militar (IGM)
Apartado 2435, QUITO.

Instituto Nacional de Estadística (INE)
Avenida 10 de Agosto 229, QUITO.

Dirección General de Geología y Minas (DGGM)
Carrion 1016 y Paez, Casilla 23A, QUITO.

Instituto Nacional de Meteorología y Hidrologia (INMH)
Calle Daniel Hidalgo 132 y 10 de Agosto, QUITO.

Ministerio de Agricultura y Ganadería (MAG)
QUITO.

Ediguias C. Ltda
Avenida 10 de Agosto 9614, QUITO.

For Jeune Afrique, see France; for BGS, see Great Britain; for DMA and CIA, see United States.

Catalogue

Atlases and gazetteers

Atlas geográfico del Ecuador
Quito: IGM, 1983.
pp. 89.

Atlas del Ecuador
Paris: Editions Jeune Afrique, 1982.
pp. 80.

Indice toponómico
Quito: IGM.
12 volumes.

Ecuador. Official standard names approved by the US Board on Geographic Names
Washington, DC: DMA, 1957.
pp. 189.

General

Ecuador 1:2 500 000
Washington, DC: CIA, 1973.
Inset maps of population, economic activity, vegetation.

República del Ecuador. Mapa físico 1:2 000 000
Quito: IGM, 1983.

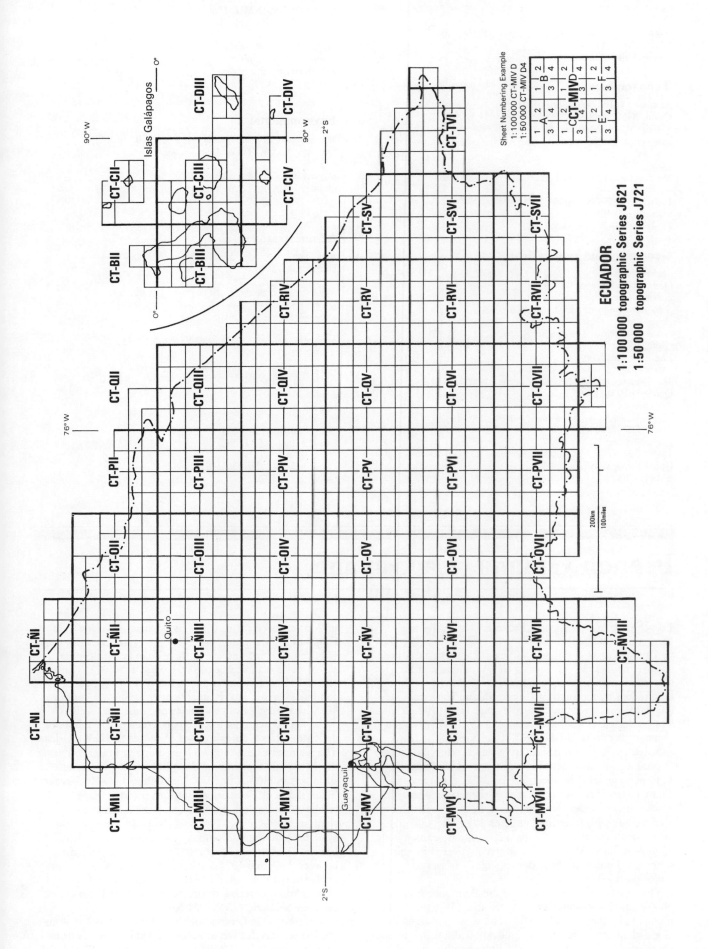

Islas Galápagos

ECUADOR

1:100 000 topographic Series J621
1:50 000 topographic Series J721

Sheet Numbering Example
1:100 000 CT-MIV D
1:50 000 CT-MIV D4

200km
100miles

República del Ecuador 1 : 1 000 000
Quito: IGM, 1985.

República del Ecuador. Mapa geográfico 1 : 500 000
Quito: IGM, 1979.
4 sheets, all published.

Topographic

Ecuador 1 : 100 000 Series J621
Quito: IGM, 1968–.
151 sheets, 40 published. ■

Ecuador 1 : 50 000 Series J721
Quito: IGM, 1962–.
588 sheets, 221 published. ■

Ecuador 1 : 25 000 Series J821
Quito: IGM, 1966–.
2352 sheets, 488 published.

Geological and geophysical

Mapa geológico nacional de la República del Ecuador 1 : 1 000 000
Quito: IGM/London: IGS, 1983.
2 sheets + text.
Also available in English language edition.

Mapa gravimétrico de anomalies Bouguer 1 : 1 000 000
Quito: IGM.
English–Spanish edition.

Mapa mineralógico del Ecuador 1 : 1 000 000
Quito: DGGM, 1969.

Mapa metalogénico del Ecuador 1 : 1 000 000
Quito: DGGM, 1980.

Mapa hidrogeológico nacional del Ecuador 1 : 1 000 000
Quito: DGGM, 1983.
+ 14 pp. text.

Mapa geológico del Ecuador 1 : 100 000
Quito: DGGM/INMH, 1968–.
151 sheets, 55 available.

Mapa de rocas industriales 1 : 100 000
Quito: DGGM, 1980–.
147 sheets, 2 published.

Environmental

Mapa bioclimático del Ecuador 1 : 1 000 000
Quito: MAG/ORSTOM, 1978.

Mapa ecológico del Ecuador 1 : 1 000 000
Quito: MAG/ORSTOM, 1978.

Administrative

República del Ecuador. Mapa político 1 : 2 000 000
Quito: IGM, 1984.

República del Ecuador. Mapa político 1 : 1 000 000
Quito: IGM, 1983.

Town maps

Plano de la ciudad de Quito 1 : 15 000
Quito: IGM, 1983.
2 sheets.
English edition.

Guia informativa de Quito 1 : 15 000
Quito: Ediguias C. Ltda.

Plano de la ciudad de Quito 1 : 10 000
Quito: IGM, 1979.

French Guiana (La Guyane)

French Guiana—La Guyane Française—is an overseas Department of France.

Although a small military survey office was established at Cayenne in 1934 (Le Service Géographique de l'Inini), systematic modern survey dates from 1945, and was undertaken by the **Institut Géographique National (IGN)**, Paris. A reconnaissance map at 1 : 500 000 scale was published in 1950 and revised in 1963. Trimetrogon photography flown by the USAAF in World War II was adjusted by IGN to rudimentary ground control and used to produce the planimetric sketch maps which form the 1 : 100 000 series. The sheets are on a Gauss projection. Those covering the northern part of the territory are described as *esquisses photogrammétriques* and are in three or four colours with form lines at 25 or 50 m intervals; those in the south are in a provisional monochrome edition with 50 m form lines. Most of the country is covered by this series and some sheet lines depart slightly from the uniform grid shown here.

The 1 : 200 000 series is a set of mostly monochrome photogrammetric sketch maps based upon IGN air photography flown in the 1950s. The series was designed as a skeleton map for the manipulation and location of air photographs, and not as a finished topographic map. The projection and grid are UTM, with sheets covering 1° of longitude and latitude. Sheets 10 and 11 in this series were published in 1965 and are in three colours with contours.

The basic map was begun in 1957 at 1 : 50 000. This is on a UTM projection, and is in four colours with a 20 m contour interval. This is an active series, not yet complete, but with a number of new sheets published in the 1980s.

Geological mapping has been undertaken by the **Bureau de Recherches Géologiques et Minières (BRGM)**, Orléans. The main series is a multi-coloured one at 1 : 100 000 using the 1 : 100 000 topographic sheet lines. The most recent sheet was published in 1981. The **Office de la Recherche Scientifique et Technique Outre Mer (ORSTOM)** published geological sketch maps of the north and the south in, respectively, 1949 and 1953 and these are still available with their accompanying monographs. ORSTOM has also published a few 1 : 50 000 soils maps, issued between 1968 and 1974, each accompanied by an explanatory notice, and a 1 : 100 000 *Carte des sols des Terres Basses*, covering the lowlands south and east of Cayenne (*Memoirs ORSTOM* No. 3, 1962).

The *Atlas de la Guyane*, published in 1979, is No. 4 in the series Atlas des Départements d'Outre-Mer, and was prepared jointly by ORSTOM and the **Centre d'Etudes de**

Géographie Tropicale (**CEGET**), Bordeaux-Talence, and is sold through the distribution office of CNRS. Its 36 map folios with principal scales of 1:1 000 000 and 1:350 000 are accompanied by commentaries and supplementary illustrations, and cover environmental, cultural and economic topics.

Further information

A graphic showing topographic mapping progress of all French overseas territories is available from IGN, Paris.

Addresses

Institut Géographique National (IGN)
Service Vente et Edition, 107, rue la Boetie, 75008 PARIS, France.

Bureau de Recherches Géologiques et Minières (BRGM)
BP 6009, Avenue de Concyr, F-45060 ORLEANS-CEDEX, France.

Bureau de Recherches Géologiques et Minières (BRGM)
BP 42, CAYENNE, La Guyane.

Office de la Recherche Scientifique et Technique Outre Mer (ORSTOM)
Service Diffusion, 70–74 route d'Aulnay, 93140 BONDY, France.

Editions du CNRS (CNRS)
15, quai Anatole France, 75700 PARIS, France.

For CEGET, see France; for DMA, see United States.

Catalogue

Atlases and gazetteers

Atlas de la Guyane
Talance: CEGET/Bondy: ORSTOM, 1979.
36 folios.

French Guiana. Official standard names approved by the US Board on Geographic Names
Washington, DC: DMA, 1974.
pp. 62.

General

Carte de la Guyane   1:1 500 000
Paris: IGN, 1964.

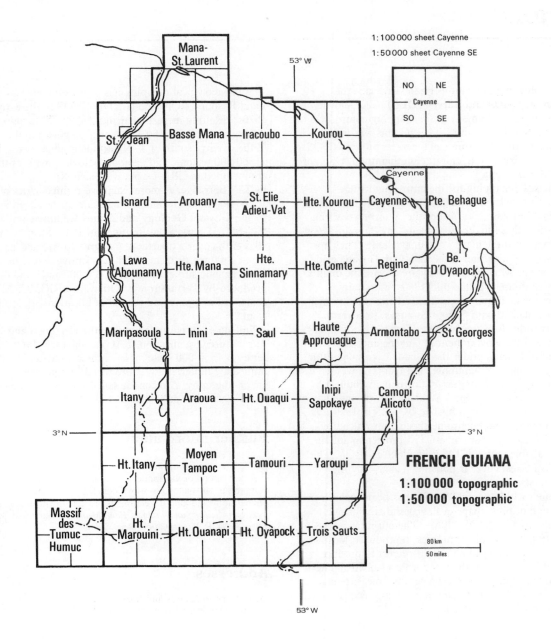

Guyane: carte touristique et routière 1 : 500 000 edition 1
Paris: IGN/Office Departemental du Tourisme de la Guyane,
1980.
Inset of Cayenne area at 1 : 100 000.

Topographic

Carte de la Guyane 1 : 200 000
Paris: IGN, 1954–.
11 sheets, all published.
Most sheets monochrome.

Carte de la Guyane 1 : 100 000
Paris: IGN, 1947–.
37 sheets, 31 published. ■
Esquisses photogrammétriques.
Monochrome sheets in south, coloured in north.

Carte de la Guyane 1 : 50 000
Paris: IGN, 1946–.
125 sheets, 59 published. ■

Geological and geophysical

Carte géologique de Guyane 1 : 500 000
Orléans: BRGM, 1960.
2 sheets, both published.

Carte géologique de Guyane 1 : 100 000
Orléans: BRGM, 1956–.
35 sheets, 20 published. ■

Environmental

Carte pédologique de la Guyane 1 : 50 000
Bondy: ORSTOM, 1969–.
125 sheets, 9 published. ■

Town map

Cayenne, Kourou 1 : 100 000
Paris: IGN, 1985.
Regional map with ancillary street maps of Cayenne and Kourou.

Ile de Cayenne (ville et environs) 1 : 20 000
Paris: IGN, 1957.

Guyana

The topographic mapping authority in Guyana is the
Department of Lands and Survey (DLS), Georgetown.
The original geodetic framework for mapping control was
established by the IAGS, and from 1967 this was improved
and densified by the Directorate of Overseas Surveys (DOS)
(now Overseas Surveys Directorate, Southampton, OSD) of
the British Government.

The basic scale mapping for the entire country is
1 : 50 000, and provisional mapping at this scale had been
issued by DOS for about a third of the country—covering
mainly the densely settled coastal plain—by the mid-1960s.
Stocks of some of these maps are still available. They are
on the Transverse Mercator projection with Guyana grid
and are in one or two colours. They are of variable quality
and are not contoured. In the mid-1960s, a new aid
programme was instituted to produce a better quality map
at the same scale and with 50 foot contours, to cover the
remaining two-thirds of the country. This was a joint
project between the Government of Guyana and the
Government of Canada under the Canada Commonwealth
Caribbean Assistance Programme, and mapping was
plotted from infrared photography flown at 15 000 feet by
Terra Surveys, Ottawa, using a Wild superwide angle
camera. These maps were published by DLS in 1970–77,
mostly as contoured monochrome sheets reproduced by
the diazo process. They are on a UTM projection and an
edition is available with a UTM grid added by DOS. The
dense tropical forest cover is not shown except by the
annotation 'dense forest'.

During the early 1980s, DOS/OSD has been engaged in
the production of photomaps of the coastal area using
infrared photography (DOS 240M). The output is in the
form of 1 : 10 000 scale photomosaics, reproduced locally
by diazo. Work also began on a joint project to produce
new 1 : 50 000 mapping of the northern coastal area. The
original geodetic control, based on aerodist measurements,
was unsatisfactory, and the mapping is being reconstructed
using new control measurements.

Some geological mapping has been conducted by the
Land Resources Division (LRD) of DOS in the early
1970s, resulting in particular in a 1 : 200 000 coloured
photogeological map by J.P. Berrange covering the whole
of the country south of 4° N latitude in nine sheets. The
1 : 500 000 tectonic and geomorphological maps of this area
are included in LRD *Overseas Memoir* No. 4 (1977), also
by J.P. Berrange. A more extensive printed series of
1 : 200 000 monochrome geological maps has been issued
by the **Guyana Geology and Mines Commission
(GGMC)**, Georgetown, founded in 1933. Sheets of this
atlas each cover a quarter of a degree square and have
about 1500 words of marginal descriptive text. The colour
aeromagnetic survey maps at 1 : 200 000 and 1 : 50 000 were
produced by Terra Surveys, Ottawa, for GGMC. All
geological mapping and related publications are available
from GGMC.

Soil survey was carried out by the UN Food and
Agriculture Organization (FAO) in the early 1960s. A
series of 1 : 60 000 scale maps of the coastal zone and
immediate hinterland was produced by interpretation from
air photography. A reconnaissance soil map of the whole
country was also published at 1 : 1 000 000.

Further information

GGMC issues a list of publications and maps.
A description of the application of air photography to resource
mapping in Guyana is given by Daniel, R.K. (1986) Aerial
photography and land development projects in Guyana, Chapter 3
in *Remote Sensing and Tropical Land Management*, edited by M.J.
Eden and J.T. Parry, Chichester: Wiley.

Addresses

Department of Lands and Survey (DLS)
22 Upper Hadfield Street, GEORGETOWN.

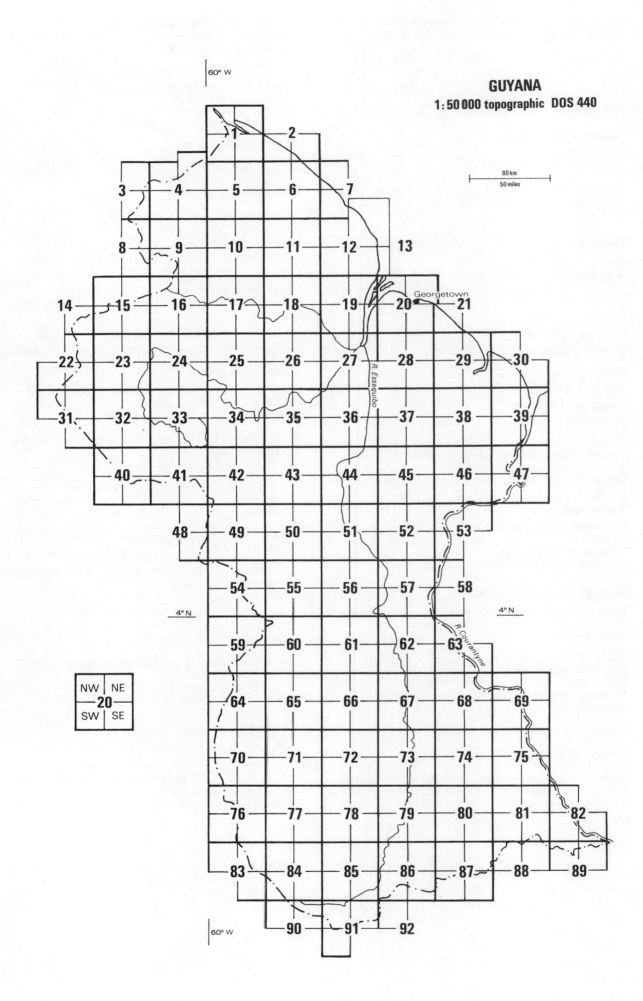

60° W

GUYANA
1:50 000 topographic DOS 440

80 km
50 miles

Georgetown

R. Essequibo

R. Courantyne

4° N

4° N

NW	NE
20	
SW	SE

60° W

Guyana Geology and Mines Commission (GGMC)
PO Box 1028, GEORGETOWN.

For DMA and CIA, see United States; for FAO, see World.

Catalogue

Atlases and gazetteers

Gazetteer of Guyana
Georgetown: DLS, 1974.

Guyana, official standard names approved by the USBGN
Washington, DC: DMA, 1976.
pp. 123.

General

Guyana 1:2 500 000
Washington, DC: CIA, 1973.
Small shaded relief map.
Insets: vegetation, ethnic groups, economic activity, population.

Map of Guyana 1:2 174 000
Georgetown: DLS, 1974.

Map of Guyana 1:1 584 000
Georgetown: DLS, 1971.

Guyana 1:1 000 000
Georgetown: DLS, 1975.
Monochrome map.

Guyana 1:500 000 DOS (Misc)17a(E491) Edition 6
Tolworth: DOS for Guyana Government, 1966–72.
4 sheets, all published.
Only the NE sheet is 1972, others 1966 and uncontoured.

Topographic

Guyana 1:50 000 DOS 440 (E791)
Tolworth: DOS/Georgetown: DLS, 1953–.
325 sheets, all published. ■
Monochrome or two-colour maps.

Geological and geophysical

Provisional mineral map showing production of economic minerals from 1968 to 1981 1:2 000 000
Georgetown: GGMC.

Provisional geological map of British Guiana 1:1 000 000
Georgetown: GGMC, 1962.

Residual magnetic anomaly map of Guyana 1:1 000 000
Georgetown: GGMC, 1973.

Southern Guyana: tectonic–geological map 1:500 000 DOS (Geol)1186
Tolworth: DOS for Guyana Government, 1973.

Southern Guyana: Provisional geomorphological map 1:500 000 (approx.) DOS (Geol)1186A J.P. Berrange
Tolworth: DOS for Guyana Government, 1974.

Geological atlas of British Guyana 1:200 000
Georgetown: GGMC, 1961–62.
92 sheets, 36 published. ■

Southern Guyana 1:200 000 DOS (Geol)1182
Tolworth: DOS for Guyana Government, 1973–74.
9 sheets, all published. ■

Aeromagnetic survey maps 1:200 000
Georgetown: GGMC, 1973.
15 sheets published. ■

Aeromagnetic survey maps 1:50 000
Georgetown: GGMC, 1973.
140 sheets published. ■

UN Aeromagnetic maps 1:50 000
Georgetown: GGMC, 1963.
83 sheets published. ■
Diazo prints.

UN Electromagnetic maps 1:50 000
Georgetown: GGMC, 1964–65.
36 sheets published. ■
Diazo prints.

Environmental

Map of roadmaking materials in Guyana 1:1 000 000 DOS (Misc)442
Tolworth: DOS for Guyana Government, 1967.

Resource map of Cooperative Republic of Guyana 1:1 000 000
Georgetown: DLS, 1976.
Shows forest and mineral resources.

General soil map of British Guiana 1:1 000 000
Rome: FAO, 1964.
Diazo map.

Engineering soils on part of the coastal plain of Guyana 1:250 000 DOS (Misc)441
Tolworth: DOS for Guyana Government, 1967.
Accompanies Road Research Technical Paper 81, *Road Making Materials in the Caribbean*.

Administrative

Administrative map Cooperative Republic of Guyana 1:1 000 000 Edition 5
Georgetown: DLS, 1981.
Shows regional and sub-regional boundaries and names.

Town maps

City of Georgetown 1:12 000
Georgetown: DLS, 1969.

Greater Georgetown 1:4800 DOS (Misc)301 Edition 2
Tolworth: DOS for Guyana Government, 1967.
2 sheets, both published.

Paraguay

The official mapping agency for Paraguay is the **Dirección del Servicio Geográfico Militar (DSGM)** (formerly Instituto Geográfico Militar), Asunción, founded in 1941. All detailed topographic mapping is of recent origin. A 1:100 000 photomap series of the south of the country was prepared by the United States Army Map Service in the 1960s, but the present national series are at 1:50 000 and 1:250 000. These are being produced in association with the IAGS. Neither is complete, but new sheets are being issued rapidly. The 1:50 000 series is in five colours with 10 m contours and the 1:250 000 series is also in five colours, with 50 m contours. The projection is Transverse Mercator. There is also a current series of maps of the 19 departments—*Mapa departamental*—at 1:200 000. These have 50 m contours and are also on the Transverse Mercator Projection with UTM grid. The small-scale general maps are on a Gauss–Krüger projection and have been through several revisions.

A series of city maps has also been in progress since about 1966 (Series H941). These are of varying format. Scale is usually about 1:10 000.

We have included the address of the **Dirección de Recursos Minerales**, but have not been able to ascertain that any geological maps have been published. A 1:1 000 000 geological map was published by the United States Geological Survey in 1959 and a 1:2 000 000 soil map in the same year. In 1975, several resource maps at 1:500 000 were published covering the *Region nororiental del Paraguay*. This survey was sponsored and published by the OAS, and the map themes included geology, soils, vegetation, soil capability and population.

The **Touring y Automovil Club Paraguayo** publishes route maps for motorists.

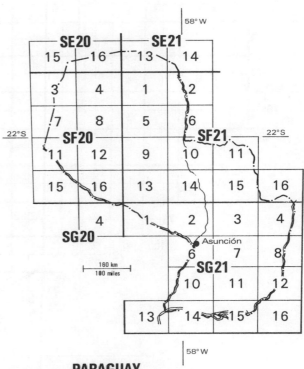

PARAGUAY
1:250 000 topographic Series H501

Addresses

Dirección del Servicio Geográfico Militar (DSGM)
Avenidas Artígas y Via Ferrea, ASUNCION.

Dirección de Recursos Minerales
Calle Alberdi y Oliva, ASUNCION.

Touring y Automovil Club Paraguayo
ASUNCION.

For DMA, see United States.

Catalogue

Atlases and gazetteers

Paraguay. Official standard names approved by the US Board on Geographic Names
Washington, DC: DMA, 1957.
pp. 32.

General

Paraguay 1:2 000 000 Edition 3
Asunción: DSGM, 1984.

Paraguay 1:1 000 000 Edition 4
Asunción: DSGM, 1984.
Available in 1 or 4 sheets.

Topographic

Carta Nacional de Paraguay 1:250 000 Series H501/H541
Asunción: DSGM, 1978–.
38 sheets, 14 published. ■

Carta Nacional de Paraguay 1:50 000 Series H741
Asunción: DSGM, 1968–.
ca 500 sheets, 253 published. ■

Town maps

Gran Asunción 1:25 000 Series H741
Asunción: DSGM, 1980.
4 sheets + street index.

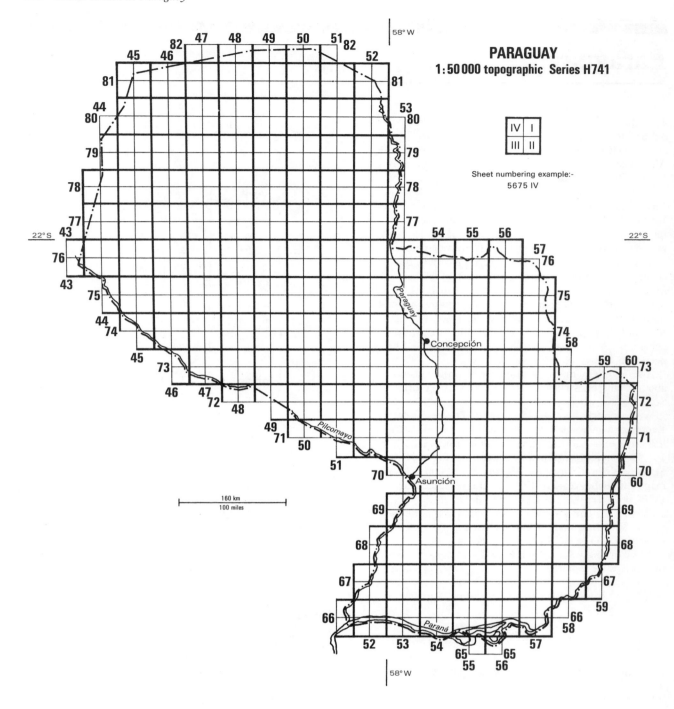

PARAGUAY
1:50 000 topographic Series H741

Sheet numbering example:-
5675 IV

160 km
100 miles

Peru

The topographic mapping authority in Peru is the
Instituto Geográfico Nacional (IGN), Lima. The origins
of this institute go back to the first decade of the century,
and there have been numerous changes of name (most
recently from Instituto Geográfico Militar). There is a
separate National Aerial Photographic Service (SAN)
which supplies IGN with the photographic cover required
for photogrammetric map making.

The first national map to be undertaken was the *Carta
Nacional* 1:200 000. This was a four-colour map issued
from 1922 to 1958, but only about 35 per cent of the

country (mainly the coastal zone or *costa*) was covered. A
1:100 000 map, issued over the same time period, was
merely an enlargement of this map.

In 1948, a mapping agreement was signed with the
IAGS and 1:60 000 scale air photographs of most of the
country were obtained in the late 1950s and early 1960s
from sorties flown by the USAAF, by a private company
HYCON and by SAN. This photography has been used to
compile the current national map series.

Thus in 1958, a new national base map at 1:100 000
could be initiated. This is a photogrammetric map

prepared by IGN in collaboration with the IAGS. It is printed in five colours with a 50 m contour interval. The projection is Transverse Mercator, and sheets each cover an area of 30' by 30'. However, so far this series also covers only the *costa*. A more limited area of the coastal zone is also covered by the 1:50 000 national map. Sheets in this series are also in five colours and the contour interval is 25 m.

For many inland areas lacking conventional mapping, the air photography has been used to provide semi-controlled photomaps at 1:100 000 and 1:50 000. In the early 1970s, SLAR imagery was flown in the eastern forests and in the inter-Andean valleys, and this imagery was made into 1:250 000 or (in the case of the inter-Andean valleys) 1:100 000 scale mosaics.

Recently, a German company was contracted to produce a series of 1:250 000 Landsat maps of Peru. This project was completed in 1985, and 88 of these map sheets, covering virtually the whole country, have subsequently become available as part of the PAIGH Unified Hemispheric Mapping Series (see introductory section to South America). These are planimetric satellite image maps (they have no heights or contours), but they are printed in colour with UTM grid and some toponyms and added linework, and they provide for the first time a uniform map cover of the whole country at an intermediate scale.

A good modern general map cover of the country is also provided by the series of Departmental maps, editions of most sheets having been published in the 1980s.

An extensive series of 1:25 000 maps, again covering mainly the settled areas of the *costa*, have been prepared by the Rural Cadaster Office of the Ministry of Agriculture. These are monochrome maps with 25 m contours.

Geological mapping is carried out by the **Instituto Geológico Minero y Metalúrgico (INGEMMET)**, Lima. INGEMMET was formed in 1979 by the fusion of two geological and mining institutes, but its origins go back to 1902. Its responsibilities include geological field survey, prospecting for minerals and metals, and the inventory and evaluation of mineral resources. A numbered series of geological quadrangle maps at 1:100 000 is in progress, each accompanied by a text. These form the *Boletines Serie 'A' Carta Geológica Nacional*. Although only some 38 bulletins have been published, some of the quadrangles are grouped together and so cover, in published or preliminary form, is fairly complete for the Pacific coastal area. Maps are also included in other bulletin series, including the *Boletines Serie 'D' Estudios especiales*, which cover irregular regions of the country.

INGEMMET is also responsible for some of the radar (SLAR) surveys mentioned above. These surveys are available in a planimetric format, in a geological interpretation or as photographic copies of the imagery. The scale is 1:100 000 or 1:500 000.

Foreign earth science agencies have also aided in the mapping of Peru, particularly in the tectonic and geological mapping of the Andes, and 1:500 000 scale geological maps of the Peruvian Cordillera were published by the French ORSTOM in 1978 (*Travaux et Documents de l'ORSTOM* Nos 93–95, *Mémoires ORSTOM* No. 86) and by the British Government (*Geological map of the Western Cordillera of Northern Peru*, DOS 1187, two sheets, 1973). Recently, the British Geological Survey has been involved in the Puno Geological Project which entails mapping 35 000 km² of southern Peru as a basis for assessing mineral potential.

An organization with primary responsibility for the inventory and evaluation of natural resources is the **Oficina Nacional de Evaluación de Recursos Naturales (ONERN)**, which provides an information and advisory service for social and economic development, and to that end has carried out a number of integrated surveys of natural resources, and has issued a variety of geological, ecological, soils and land use maps.

Census, urban and development mapping are carried out by the Instituto Nacional de Estadística and the Ministério de Vivienda y Construcción.

Commercial map publishers include **Cartográfica National SA**, Lima, which publishes a road map of the country, maps of individual Departments, and a street map of Lima, and the Touring y Automovil Club del Perú, which publishes tourist maps, route guides and highway diagrams.

High-quality mapping of the Cordillera Blanca and the Cordillera Huayhuash was undertaken during scientific expeditions by the German Alpine Club (**Deutsche Alpenverien**) in the 1930s and the three sheets (two at 1:100 000 and one at 1:50 000) are still available.

Further information

IGN issues indexes of topographic series and INGEMMET issues a catalogue of publications.

There is a PAIGH *Guide to cartographic and natural resources information of Peru*, available in English or Spanish and published in 1979 by the National Office for the Evaluation of Natural Resources (ONERN).

Addresses

Instituto Geográfico Nacional (IGN)
Av. Andrés Aramburú No. 1198, LIMA 34; or Apartado Postal 2038, LIMA 100.

Instituto Geológico y Metalurgico (INGEMMET)
Pablo Bermúdez No. 211, LIMA 11; or Apartado Postal 889, LIMA 100.

Oficina Nacional de Evaluación de Recursos Naturales (ONERN)
Apartado Postal 4992, LIMA 100.

Cartográfica Nacional SA
Jr. Carabaya 719, Of. 206, LIMA.

Universidad Nacional Agraria
Apartado 456, La Molina, LIMA 2000.

Office de la Recherche Scientifique et Technique Outre Mer (ORSTOM)
Service Diffusion, 70–74 Route d'Aulney, 93140 BONDY, France.

Deutsche Alpenverien
Praterinsel 5, D-8000 MÜNCHEN 22, German Federal Republic.

For DMA and CIA, see United States.

Catalogue

Atlases and gazetteers

Peru. Official standard names approved by the US Board on Geographic Names
Washington, DC: DMA, 1955.
pp. 609.

Gazetteer 1983
Lima: IGN, 1983.
pp. 123.
Relates to toponyms on the 1:100 000 map.

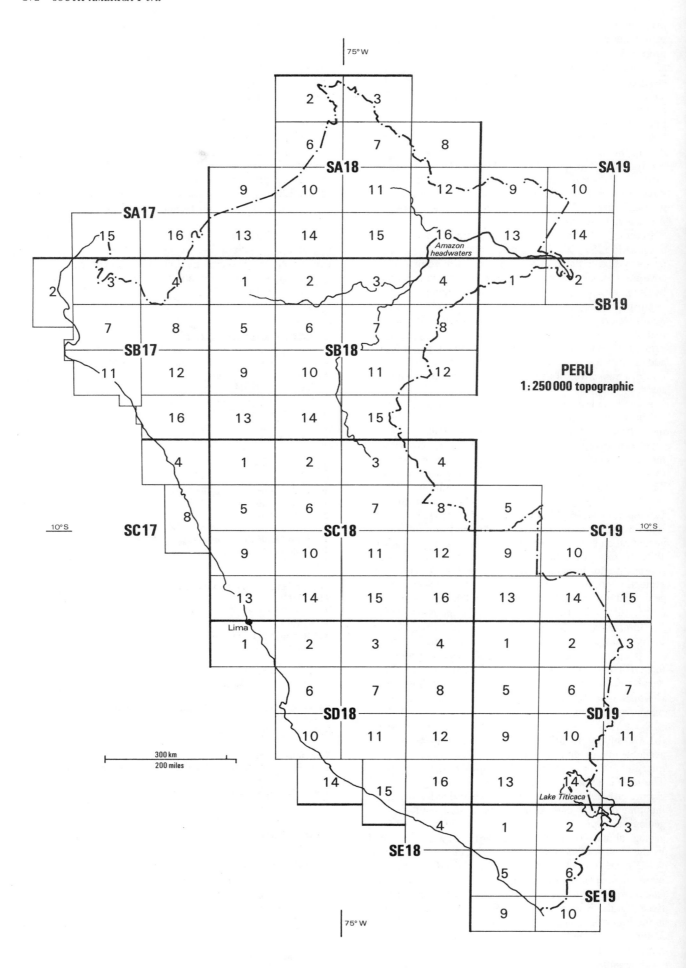

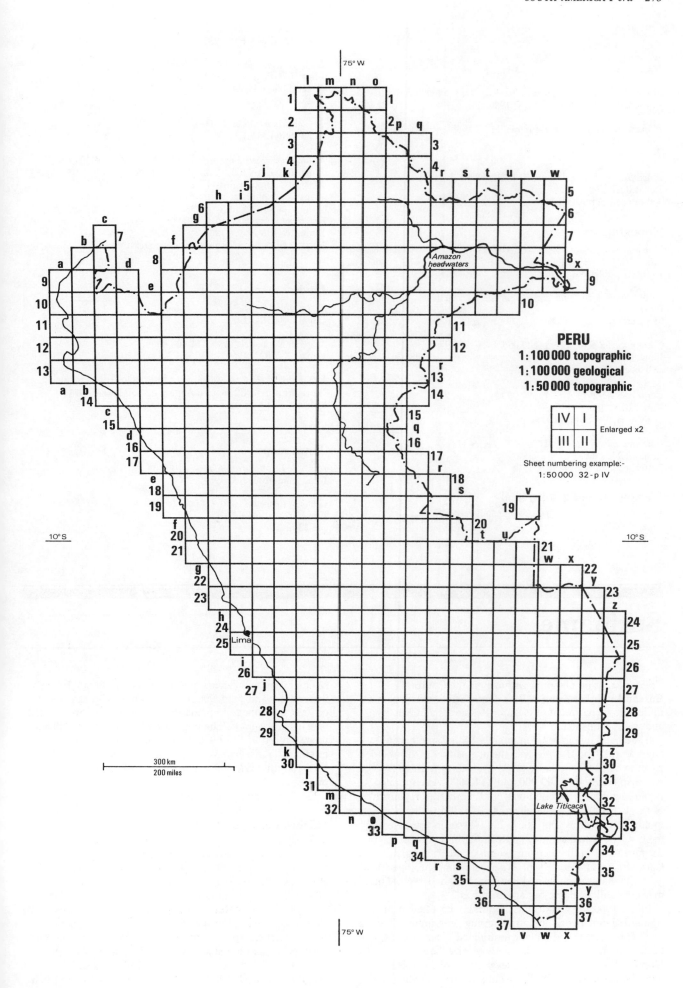

75° W

PERU
1 : 100 000 topographic
1 : 100 000 geological
1 : 50 000 topographic

IV	I
III	II

Enlarged x2

Sheet numbering example:-
1 : 50 000 32 - p IV

Amazon headwaters

10° S

10° S

Lima

Lake Titicaca

300 km
200 miles

75° W

General

Peru 1 : 4 000 000
Washington, DC: CIA, 1983.
Inset maps of vegetation, population, economy.

Mapa político del Perú: caminero turístico 1 : 2 500 000
Lima: Cartográfica Nacional SA, 1986.

Mapa físico político y vial del Perú 1 : 2 200 000
Lima: IGN, 1982.
Town maps on the reverse.

Mapa físico-político del Perú 1 : 1 000 000
Lima: IGN, 1982.
4 sheets, all published.

Topographic

Mapas departamentales . . . scales vary
Lima: IGN, 1980–.
18 sheets published.

Mapa planimétrico de imagenes de satelite 1 : 250 000
Lima: IGN, 1984.
95 sheets, 88 available. ■

Carta nacional del Perú 1 : 200 000
Lima: IGN, 1922–58.
281 sheets, 98 published.

Carta nacional del Perú 1 : 100 000
Lima: IGN, 1960–.
504 sheets, 194 published. ■

Carta nacional del Perú 1 : 50 000 Series J731
Lima: IGN, 1968–.
2016 sheets, *ca* 148 published. ■

Geological and geophysical

Yacimientos no metálicos Perú 1 : 2 500 000
Lima: INGEMMET.

Mapa tectónico del Perú 1 : 2 500 000
Lima: INGEMMET, 1982.

Mapa geológico del Perú 1 : 1 000 000
Lima: INGEMMET, 1977.
4 sheets, all published.
With 44 pp. text.

Yacimientos metálicos del Perú 1 : 1 000 000
Lima: INGEMMET.

Yacimientos no metálicos del Perú 1 : 1 000 000
Lima: INGEMMET, 1982.
14 sheets, all published.
+ text in 3 volumes.

Cuadrángulos geológicos 1 : 100 000
Lima: INGEMMET, 1960–.
Numbered series covering the coastal areas.

Environmental

Mapa ecológico del Perú 1 : 1 000 000
Lima: ONERN, 1976.

Mapa de Suelos del Perú 1 : 1 000 000
Lima: ONERN, 1969.

República del Perú. Mapa forestal 1 : 1 000 000
Lima: Universidad Nacional Agraria de La Molina, 1975.
8 sheets + 161 pp. text.

Town maps

Plano de Lima Metropolitana *ca* 1 : 40 000 Edition 5a
Lima: Lima 2000 SA, 1984.

Guide to Lima streets 1 : 25 000
Lima: IGN, 1983.
pp. 136.

Suriname

The topographic mapping authority is the **Centraal Bureau Luchtkartering (CBL)**, Paramaribo, which comes under the Department of Natural Resources and Energy and was established in 1948.

Systematic topographic mapping of Suriname began after World War II, following the acquisition of air photo cover at 1 : 40 000 scale by KLM–Aerocarto from 1947, and map series at 1 : 100 000 and 1 : 40 000 were initiated. A stereographic projection was used. Only the area north of the 4° parallel was completed at 1 : 40 000 (as a mainly monochrome series) but the 1 : 100 000 series was extended to cover the whole country and printed in monochrome in the north and in three colours south of the 4° parallel. Contours are at 25 or 50 m. A derived monochrome series at 1 : 200 000 was published between 1960 and 1966. These maps are still available but are unrevised.

In the 1960s, it was decided to undertake a new primary triangulation network and a remapping programme. The 1 : 40 000 series was therefore terminated. The triangulation (using Aerodist methods and Satellite Doppler Position Fixing) and levelling were completed over the period 1968–78. Following aerotriangulation, air

photography was carried out at scales of 1 : 30 000 in the flatter areas and 1 : 55 000 in the mountains. A new series of 1 : 50 000 topographic mapping is now in progress based on this new survey. The maps are plotted at 1 : 25 000 and then reduced to 1 : 50 000 for scribing. The projection is a modified Transverse Mercator and there are nearly 400 sheets in this series, of which about half are so far published.

Geological mapping is undertaken by the Suriname Government Geological and Mining Service, **Geologisch Mijnbouwkundige Dienst (GMD)**, Paramaribo, founded in 1943. The basic scale for geological mapping was 1 : 100 000, with maps produced primarily for internal use, but a number of monographic maps at scales of 1 : 100 000 and 1 : 200 000 have been published as part of the *Mededelingen* series of GMD. A series of five 1 : 100 000 geological sheets covering the north-east of the country were also published in the 1950s and are still available. Aeromagnetic maps were prepared in 1960–65 by the Aero Service Corporation at scales ranging from 1 : 40 000 to 1 : 500 000. The best modern geological synthesis is provided by the 1 : 500 000 scale geological map published

in 1977 and described by Kroonenberg *et al.* (1984) in
Contributions to the Geology of Suriname, **8**. Previously (in
1966) a general photogeological map and a
geomorphological map at 1:500 000 had been published.

Soils mapping was pioneered in the 1950s by J. J. van
der Eyck, and in 1958 the **Dienst Bodemkartering (DBK)**
was set up. A *First approximation of the reconnaissance soil
map of Suriname* 1:4 000 000 (a very small map), based on
photo interpretation was published in 1963 by DBK. This
was followed by reconnaissance survey at 1:100 000, and
the whole area north of the 5° parallel has been mapped at
this scale. 1:200 000 scale mapping has extended the
survey southwards to 4°, although due to the limited
number of field observations in this area, the soil
information is tentative. More detailed surveys at
1:40 000, 1:20 000 and 1:10 000 are taking place as new
roads open up access to new land. A 1:500 000 coloured
soil map of the northern half of the country has also been
published.

An ecological map of the northern lowlands was
published in 1978 by P. A. Teunissen and is available from
the **Stichting Natuurbescherming Suriname (STINASU)**.

Cadastral surveying is the responsibility of the Dienst
der Domeinen and 1:5000 scale air photography has been
obtained for the production of cadastral maps of urban
areas at 1:1000 and rural areas at 1:2000 or 1:2500.

The capital city, Paramaribo, was mapped by CBL in
the late 1950s at 1:1000 and in 1963 a six-sheet map at
1:5000 and a single sheet at 1:12 500 were issued. The
1:5000 map has not been maintained, but the 1:12 500
has become the basis for commercial city street maps, of
which the latest was published by **VACO NV** in 1978.

Plans for a National Atlas were revived in 1978 and
funds were granted for its publication. It is to be a four-
colour production of about 35 sheets with maps ranging in
scale from 1:500 000 to 1:5 000 000. Publication was
expected in 1986.

Gazetteer work, with the aim of standardizing
geographical names, has been carried out by the
Surinamese Cartographic Committee, established by the
government in 1970. At present there are no plans to
publish a gazetteer of official names.

Further information

The new triangulation and mapping is described (in Dutch) by
Wekker, J.B.C. (1980) in Kaarteringen in Suriname, *Geodesia*, **22**,
14–20; while Koeman, C. (1979) gives a good review (also in
Dutch) of both topographic and thematic mapping up to the late
1970s in De kartografie van Suriname, *Kartografisch Tijdschrift*,
5(4), 17–23.

A *Bibliography of printed maps of Suriname 1671–1971* by C.
Koeman has been published by Theatrum Orbis Terrarum,
Amsterdam, 1973.

Some information on recent activities can be found in the
*Technical Papers of the 2nd UN Cartographic Conference for the
Americas* held in Mexico City in 1979 (E/CONF.71/3/Add.1).

Geological and geophysical maps of Suriname (up to 1971) are
described and listed by Bosma, W. (1971) in *Contributions to the
Geology of Suriname*, **2**, 137–143.

DBK publishes an index map of soil mapping.

Addresses

Centraal Bureau Luchtkaartering (CBL)
PO Box 971, PARAMARIBO.

Geologisch Mijnbouwkundige Dienst van Suriname (GMD)
Kleine Waterstraat 2–6, PARAMARIBO.

Dienst Bodemkartering (DBK)
Hoek Coppenamestraat/Commissaris Weytinghweg, PARAMARIBO.

Stichting Natuurbescherming Suriname (STINASU)
Corn. Jongbawstraat 14, Postbus 436, PARAMARIBO.

VACO NV
Domineestraat 26, PARAMARIBO.

For DMA, see United States.

Catalogue

Atlases and gazetteers

*Suriname. Official standard names approved by the US Board on
Geographic Names*
Washington, DC: DMA, 1974.
pp. 65.

General

Kaart van de Republiek Suriname 1:1 000 000 H.N. Dahlberg
Paramaribo: C. Kersten and Co., n.d.
Legend in Dutch, English and Spanish.

Kaart van Suriname 1:1 000 000
Paramaribo: CBL, 1974.

Suriname 1:500 000
Paramaribo: CBL, 1977.
In 2 or 4 sheets, all published.

Topographic

Suriname 1:200 000
Paramaribo: CBL, 1960–66.
29 sheets, all published. ■
Monochrome maps.

Suriname 1:100 000
Paramaribo: CBL, 1960–.
101 sheets, all published. ■
Monochrome sheets in north, three-colour in south.

Suriname 1:50 000
Paramaribo: CBL.
385 sheets, *ca* 50% published. ■

Geological and geophysical

Metallogenic map of Suriname 1:1 250 000
Paramaribo: GMD, 1975.
Explanatory note by E.H. Dahlberg published as *Contributions to
the Geology of Suriname*, **5**, 1976.

Geological map of Suriname 1:500 000
Paramaribo: GMD, 1977.
2 sheets, both published.
Explanation by S.B. Kroonenberg *et al.* published in *Contributions
to the Geology of Suriname*, **8**, 1984.

Environmental

Reconnaissance soil map of northern Suriname 1:500 000
Paramaribo: DBK.
Colour map covering country between 4° and 6°.

Reconnaissance soil map of northern Suriname 1:200 000
Paramaribo: DBK, 1977–.
29 sheets, 18 published. ■
Monochrome maps.

Overzichtskaart Surinaamse Laagland Ecosystemen 1:200 000
P.A. Teunissen
Paramaribo: STINASU, 1978.
8 sheets including legend sheet.
In Dutch and English.

Reconnaissance soil map of Northern Suriname   1 : 100 000
Paramaribo: DBK, 1977–.
101 sheets, 31 published. ■
Monochrome maps.

Town maps

Kaart Paramaribo   1 : 12 500
Paramaribo: VACO, 1978.

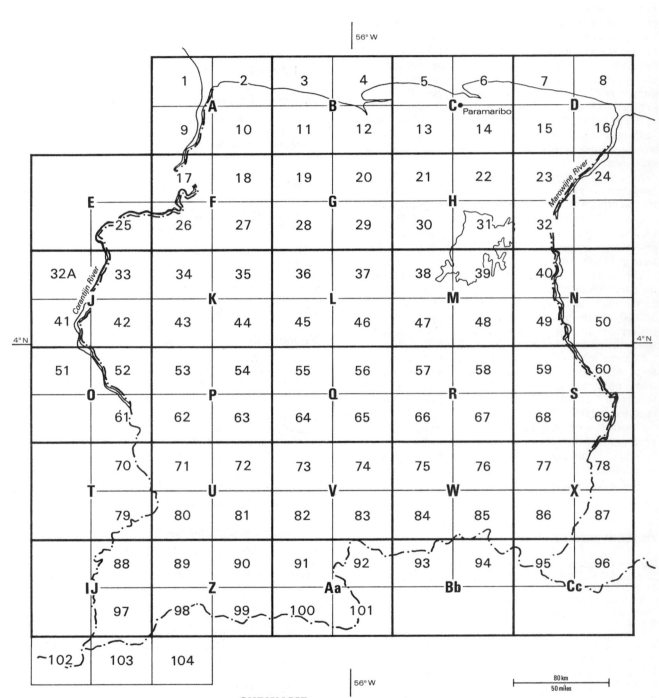

SURINAME
1 : 200 000 topographic
1 : 100 000 topographic
1 : 200 000 soils
1 : 100 000 soils

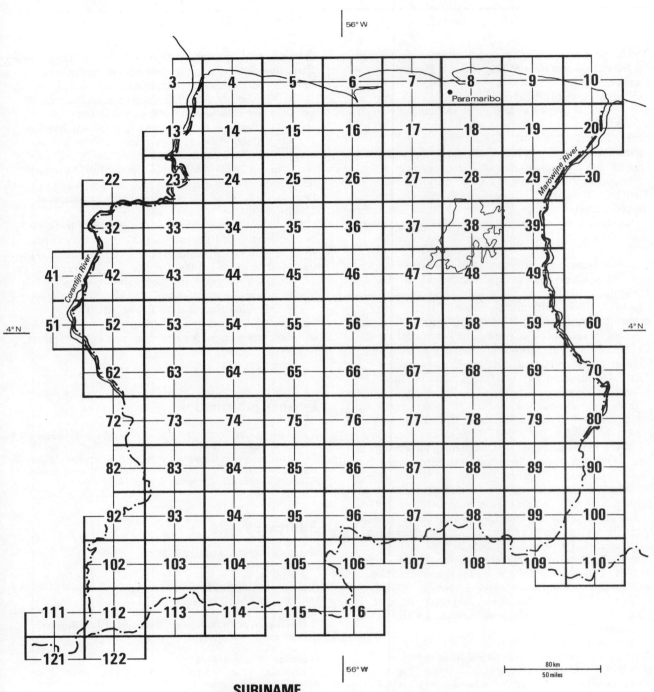

SURINAME
1 : 50 000 topographic

Sheet numbering example:
18A

Uruguay

Topographic mapping in Uruguay is the responsibility of the **Servicio Geográfico Militar (SGM)**, Montevideo, founded in 1913.

Early work by SGM concentrated on the establishment of a geodetic network for the country and the initiation of a 1:50 000 series (first sheet issued 1927) and a 1:200 000 series (from 1933). Until 1943, only 10 per cent of the country had been mapped in detail, but in that year the USAAF flew a complete cover of Trimetrogon air photographs, from which a 1:500 000 map was compiled. Then from 1966 to 1967 air photography was acquired for the production of a series of 1:50 000 photomaps covering the whole country and sharing the 1:50 000 topographic sheet system. The sheets were issued between 1967 and 1973.

Today the national cartographic plan allows for the production of topographic maps at scales of 1:20 000, 1:50 000, 1:100 000 and 1:200 000. These maps are based on a Gauss conformal projection. The mapping system is planned to form a base for thematic information collected by the Cadastral Survey, the Direccion Forestal and the Dirección de Suelos. For this purpose, CONEAT (Comisión Nacional de Estudio Agro-Economico de la Tierra) has prepared a 1:20 000 base map from the 1:50 000 photomap.

In the early 1980s, the 1:100 000 planimetric series was completed. It is in three colours with the road network correct to 1971–74. The 1:200 000 series, published in 1978, is a reduction of the 1:100 000 map, but only about half the country (the northern half) is so far covered by this series. From 1980, a new five-colour 1:50 000 series has been in rapid progress, with 10 m contours. This will progressively replace the earlier series (editions of 1930 to 1962), the 79 sheets of which are still available, as coloured originals or monochrome reprints. 1:50 000 sheets each cover an area of 0.40 grad longitude by 0.20 grad latitude. A new edition of the 1:100 000 series also began to appear in 1984. This is in five colours, on the Gauss projection, and has 20 m contours. One sheet of a new 1:250 000 series with UTM grid (the Montevideo sheet) was also issued in 1984 and is available through the PAIGH programme.

Geological mapping is the responsibility of the **Dirección Nacional de Minería y Geología (DNMG)**—previously the Instituto Geológico 'Ing. E. Terra Arocena'—Montevideo, founded in 1912. The principal series is at 1:100 000, based upon the topographic sheet lines. These maps have been produced cooperatively with the **Universidad de la República**, from which they may be obtained. Only four sheets, with explanatory texts, have so far been published, although further sheets are in manuscript. A new edition of the 1:500 000 geological map of Uruguay (first published 1957) was issued in 1985, and an *Atlas del inventario de materias primas no metálicas en el Uruguay* was prepared in 1982 in a collaborative programme with the Bundesanstalt für Geowissenschaften und Rohstoffe, Hannover.

A first approximation of an inventory of soils of Uruguay was produced in 1962 as part of a National Agricultural Development Plan, and in 1965 a full soil mapping programme got under way through cooperation between the Ministry of Agriculture, the University and SGM.

Currently the organization responsible for soil survey is the **Dirección de Suelos** of the Ministerio de Agricultura y Pesca **(MAPDS)**, and a reconnaissance soils map at 1:100 000 is in progress. This shows soil associations, as well as geological and geomorphological features, plotted on a photomap base reduced from the 1:50 000 fotoplano. The maps are grouped by Department, and the first set, covering the Departments of Canelones and Montevideo, was issued in 1982. MAPDS has also published a map of soil erosion (*Erosion actual*) at 1:200 000 for these Departments (1985), and is undertaking assessments of land capability.

The 1:20 000 map of Montevideo published by SGM is in six colours with 10 m contours. There is also an urban series at 1:10 000 in 38 sheets and with 2 m contours. The small-scale maps at 1:1 000 000 and 1:500 000 are revised periodically. Both are in six colours with contours respectively at 100 m and 50 m.

The **Dirección Nacional de Topografía (DNT)**, Montevideo, issues large-scale urban maps and also a series of departmental maps, some of them very old, but others revised in the 1980s.

Further information

A 24-page booklet describing the work of SGM and including indexes of map cover is available, while Volume 7 of the *Boletin del Servicio Geográfico Militar* (1984) includes information on the organizational structure of SGM and papers on the development of topographic, cadastral and urban surveys in Uruguay.

Addresses

Servicio Geográfico Militar (SGM)
Av. 8 de Octubre 3255, MONTEVIDEO.

Dirección Nacional de Minería y Geologia (DNMG)
Hervidero 2861, MONTEVIDEO.

Departamento de Publicaciones, Universidad de la República
Avda 18 de Julio 1824, MONTEVIDEO.

Dirección de Suelos (MAPDS)
Ministerio de Agricultura y Pescos, Avda Eugenio Garzon 456, MONTEVIDEO.

Dirección Nacional de Topografía (DNT)
Ministerio de Transporte y Obras Publicas, MONTEVIDEO.

For DMA and CIA, see United States.

Catalogue

Atlases and gazetteers

Atlas geográfico de la República del Uruguay   U. Rubens Grub
Montevideo: 1980.
pp. 56.

Uruguay. Names approved by the US Board on Geographic Names
Washington, DC: DMA, 1955.
pp. 126.

General

Uruguay 1:1 490 000
Washington, DC: CIA, 1974.
Inset maps of population and administrative divisions, vegetation and economic activity.

Carta regional. República Oriental del Uruguay 1:1 000 000
Montevideo: SGM, 1983.

Uruguay: mapa turístico y plano de Montevideo 1:800 000
Montevideo: ANCAP SA, 1981.
Reverse has street map of Montevideo 1:23 500.

Carta geográfica República Oriental del Uruguay 1:500 000
Montevideo: SGM, 1985.
2 sheets, both published.

Topographic

Cartas departamentales 1:200 000/1:100 000
Montevideo: DNT, 1937–.
17 sheets of individual departments published.

Carta planimétrica 1:200 000
Montevideo: SGM, 1978.
25 sheets, 13 published. ■

Carta planimétrica 1:100 000
Montevideo: SGM, 1978.
87 sheets, all published. ■

Carta topográfica 1:50 000
Montevideo: SGM, 1980–.
300 sheets, 78 published. ■
Sheets from earlier series also available.

Geological and geophysical

Carta geo-estructural del Uruguay 1:2 000 000
Montevideo: DNMG, 1979.
Monochrome map + 62 pp. text.

Carta geológica del Uruguay 1:1 000 000
Montevideo: MAPDS, 1975.

URUGUAY
1:200 000 topographic
1:100 000 topographic
1:100 000 geological

100 km
60 miles

URUGUAY
1:50 000 topographic

56° W

100 km
60 miles

32° S

32° S

56° W

Montevideo

Carta gravimétrica de la República Oriental del Uruguay 1:1 000 000
Montevideo: SGM, 1973.

República Oriental de Uruguay. Carta geológica 1:500 000 2nd edn
Montevideo: DNMG, 1985.
2 sheets, both published.

Carta geológica del Uruguay 1:100 000
Montevideo: DNMG/Universidad de la República, 1969–.
4 sheets + texts published. ∎

Environmental

Carta de reconocimiento de suelos del Uruguay 1:1 000 000
Montevideo: MAPDS, 1976.

Carta da reconocimiento de suelos de la República Oriental del Uruguay 1:100 000
Montevideo: MAPDS, 1982–.

Town maps

Plano de Montevideo 1:20 000
Montevideo: SGM, 1977.
4 sheets, all published.

Venezuela

The topographic mapping agency for Venezuela is the **Dirección de Cartografía Nacional (DCN)**, Caracas, founded in 1935 as the Servicio Aerocartográfico Permanante. DCN now falls within the administration of the Ministerio del Ambiente y de los Recursos Renovables. It is responsible for cartographic, geodetic, cadastral, geophysical and marine surveys.

Photogrammetric survey of the country began in 1938, and a basic mapping scale of 1:25 000 was selected with contours surveyed at 20 m intervals. In 1946, technical assistance from the IAGS was negotiated, and map series at 1:100 000 and 1:250 000 were also started at about this time.

The current topographic series are at 1:25/50 000 (the basic series), 1:100 000, 1:250 000 and 1:500 000. The 1:25 000 series, based partly on old mapping, has been reconstituted and extended since 1962 on a Transverse Mercator projection. It is a one-colour map with 20 m contours, and covers much of the country except for the far south. Since 1979, an alternative scale of 1:50 000 has been introduced for some areas. Sheets in the 1:25 000 series cover one-twelfth of a 1:100 000 scale sheet, and those at 1:50 000 cover one-quarter.

The 100 000 series (which is a new version started in 1962 to replace the earlier preliminary edition) covers approximately the same area as the basic scale mapping but is slightly less extensive. This is a good-quality five-colour map on Transverse Mercator projection with UTM grid and 40 m contours.

The current (post-1970) 1:250 000 map is almost complete. This is in seven colours with 50 or 80 m contours, the projection is Transverse Mercator, and sheets are divisions of the International Map of the World system. Sheets in the south of the country are SLAR image maps, described as 'picto-radar'. There is also a complete cover of radar imagery at 1:250 000. The 1:500 000 map, issued since 1977, is a multi-colour map on a Lambert conformal projection. This too is almost complete.

The sheet indexes of Venezuela include the disputed territory on the eastern side of the country—the *Zona en reclamación*—although there are no published Venezuelan maps of this area.

DCN has published a number of large-scale maps of cities and a complete set of general maps of each state or federal territory.

The first edition of the National Atlas was prepared in 1969, and compilation of the current second edition took place between 1973 and 1976. It was edited by DCN and comprises 260 maps giving a 'retrospective and dynamic vision of the historical, geographical, economic and social–cultural evolution of the country'. It incorporates statistical data from the 1971 population census. The main maps are at a scale of 1:4 000 000, and there is also a set of larger-scale provincial maps.

DCN is responsible for the national register of geographical names, and since 1967 there has been a programme to compile toponymic gazetteers. A series of regional gazetteers began to be issued in 1969, but is not yet complete.

Geological mapping is carried out by the **Dirección General de Minas y Geología (DGMG)** of the Ministério de Energía y Minas. A wide range of mapping is available (though mostly not as full-colour editions, since much is out of print and available only as monochrome copies). A high proportion of the geological mapping is monographic in character, i.e. special-purpose mapping covering specific areas rather than forming part of a standard series. Available mapping of this kind is listed under the individual states in the DGMG catalogue mentioned below. For many states there is also a general geological map for the whole state, at a scale varying with size of state from 1:100 000 to 1:1 000 000. There is a structural map of the whole country at 1:500 000 printed in colour and available as a complete set of 30 sheets, and a very few quadrangle maps at 1:100 000, some of which conform to the sheet system of the old and superseded topographic map of that scale.

Thematic mapping by several other government institutions has been undertaken in the past. A climatic atlas was published by the Venezuelan Airforce in 1957, and an *Atlas Agrícola de Venezuela* in 1960 by the Ministério de Agricultura y Cría. In 1961, an *Atlas Forestal de Venezuela* was published by the Dirección de Recursos Naturales with the Universidad de los Andes. These atlases are all out of print and no new thematic mapping has been publicized in recent years.

Further information

An index of map, air photo and radar image cover was published by DCN in 1985. DGMG also issues a catalogue of maps and geological publications.

Addresses

Dirección de Cartografía Nacional (DCN)
Dirección General de Información e Investigación Renovables, Esquina de Camejo, Edificio Camejo, CARACAS 1010.

Dirección General de Minas y Geología (DGMG)
Torre Oeste de Parque Central, Piso 8 Caigeomin, CARACAS.

Ministério de Transporte y Comunicaciones (MTC)
Dirección General de Vialidad Terrestre, CARACAS.

Dirección de Suelos, Vegetación y Fauna
Edificio y Esquina de Camejo, Mezzanina, CARACAS 1010.

For DMA and CIA, see United States; for ORSTOM, see France.

Catalogue

Atlases and gazetteers

Atlas de Venezuela 2nd edn
Caracas: DCN, 1979.
pp. 331.

Venezuela. Official standard names approved by the US Board on Geographic Names
Washington, DC: DMA, 1961.
pp. 245.

General

Venezuela 1:3 200 000
Washington, DC: CIA, 1972.
Inset maps of population, vegetation, economic activity and petroleum.

Mosaicos de imágenes de satélites Landsat 1 y 2 de Venezuela 1:3 000 000
Caracas: DGMG, 1977.

Mapa físico y político de la República de Venezuela 1:2 000 000
Caracas: DCN, 1980.

Mapa físico de la República de Venezuela 1:2 000 000
Caracas: DCN, 1975.

Mapa vial de Venezuela 1:1 000 000
Caracas: MTC, 1981.
Inset: *Mapa político de la República de Venezuela* 1:4 000 000.

Topographic

República de Venezuela. Mapas de las Entidades Federales scales vary
Caracas: DCN, 1969–.
22 sheets published.

República de Venezuela 1:500 000
Caracas: DCN, 1977–.
29 sheets, 22 published. ■
Some sheets available only as heliotypes.

República de Venezuela 1:250 000
Caracas: DCN, 1970–.
ca 90 sheets, 42 published. ■

República de Venezuela. Imágenes de radar 1:250 000
Caracas: DCN.
ca 90 sheets, 75 published.

República de Venezuela 1:100 000
Caracas: DCN, 1962–.
635 sheets, 314 published. ■

República de Venezuela 1:50 000/1:25 000
Caracas: DCN, 1962–.
7260 sheets, *ca* 5000 published.

Geological and geophysical

Mapa metalogénico de Venezuela 1:3 700 000
Caracas: DGMG, 1976.

Mapa de recursos minerales no metálicos y facilidades industriales Venezuela 1:3 700 000/*Mapa metalogénico de Venezuela* 1:2 500 000
Caracas: DGMG, 1981.
2 sheets + text.

Mapa geológico estructural de Venezuela 1:2 500 000
Caracas: DGMG.

Carta isogónica 1:2 000 000
Caracas: DCN.

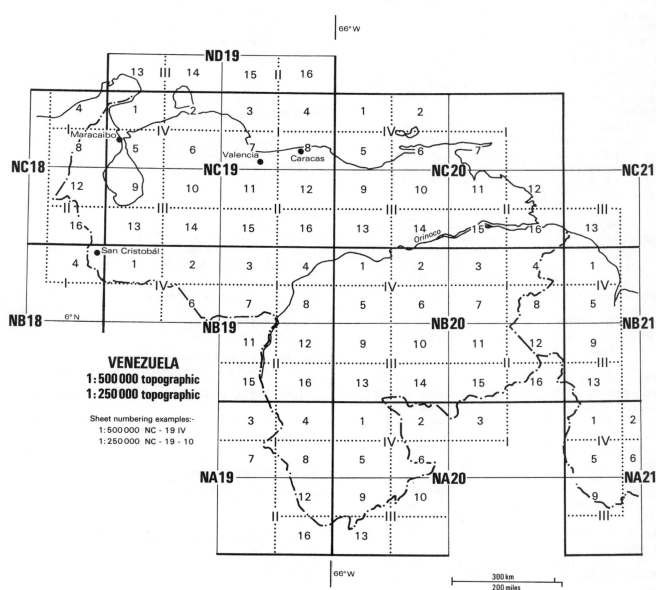

Mapa sísmico 1930–80   1 : 2 000 000
Caracas: DCN, 1981.

Mapas geológicas y de recursos minerales de los estados
Caracas: DGMG, 1950–.
Mostly monochrome maps of individual *estados*.

Mapa geológico estructural de Venezuela   1 : 500 000
Caracas: DGMG, 1976.
30 sheets, all published.

Mapa de anomalías gravimétricas de Bouguer   1 : 500 000
Caracas: DGMG.

Environmental

Mapa de suelos de Venezuela   1 : 2 000 000
Caracas: DGMG, 1947.

Mapa ecológico de Venezuela   1 : 2 000 000
Caracas: DCN, 1976.

Mapa de vegetación de Venezuela   1 : 2 000 000
Caracas: DCN, 1960.

Mapa fitogeográfico preliminar   1 : 2 000 000
Caracas: DCN.

Administrative

Mapa de regionalización administrativa   1 : 2 000 000
Caracas: DCN.

Human and economic

[Venezuela] Densidad de la población por distritos   1 : 3 000 000
Bondy: ORSTOM, 1979.
2 sheets: I, Evolucion comparativa . . .; II, Establiceda en base al
censo de la población y vivienda de 1971.
+ 734 pp. text *Dinamica de la población—casa de Venezuela.*

Town maps

Caracas   1 : 20 000   3rd edn
Caracas: DCN, 1978.

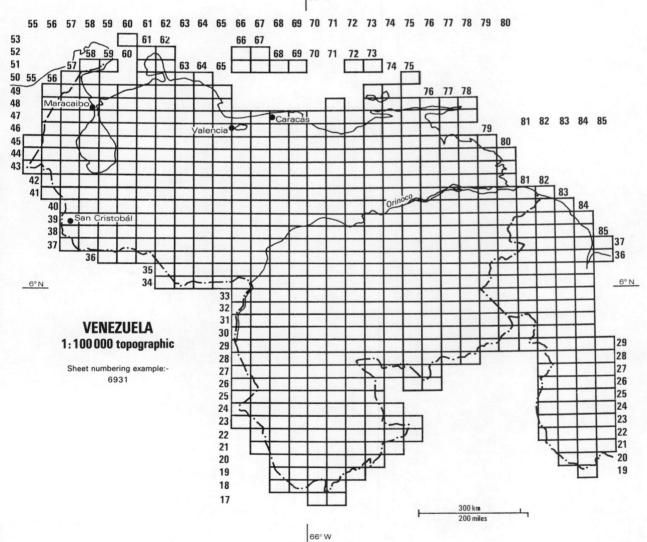

ASIA

This discussion of the mapping of Asia is divided into two distinct sections: a brief consideration of general continental mapping, followed by a more detailed evaluation of the standards, systems and organizations that have contributed to the map bases of different subcontinental blocks within Asia. This is reflected in the choice of catalogue headings we have used: the mapping of the Middle East forms one discrete section, followed by maps of South East Asia and finally by maps of the Indian Subcontinent. Other areas of Asia are covered in the relevant country sections, i.e. under USSR or China.

There are few continent-wide mapping standards or organizations. No Asia-wide topographic series is now published (we have listed international series covering the area under World). Different regions have their own families of similar mapping and these are covered in the relevant sections. The **Economic and Social Commission for Asia and the Pacific (ESCAP)** in Bangkok has compiled a number of Asia-wide economic maps, and earth science mapping of the ESCAP countries at small scales has also been published. The **Commission for the Geological Map of the World (CGMW)** in Paris has compiled a variety of smaller-scale geological and other earth science maps in conjunction with several other international or UN agencies which cover the whole continent. These include geological mapping (a new third edition is forthcoming) and tectonic and metamorphic maps of the southern parts of the continent. Other continent-wide earth science mapping has been compiled by Russian and Chinese agencies, including 1:5 000 000 scale complete Chinese language geological mapping, and complete 1:5 000 000 scale fault tectonics and metamorphic mapping compiled by the **Ministerstvo Geologii SSSR (MGSSSR)**. The southern areas of Asia have been mapped at small scales by the **International Rice Research Institute (IRRI)** to show agroclimatic distributions and rice cultivation, and other resources and economic mapping of large parts of the continent has appeared in the **Royal Thai Survey Department** regional economic atlas.

The Middle East

The Middle East is defined here as including all the countries from Turkey to Iran, between the Indian and the USSR border. As such it includes all the Asian Arab states, Iran and Israel. Much of this area was covered in British or American military mapping established in the early years of the twentieth century; the Survey of India under the British covered large parts of the Eastern Middle East. However, little accurate mapping of the whole area was available until after World War II. The development of mapping after the war has been much more fragmented and has reflected the independence of States, their political affiliations and socioeconomic conditions, and available technologies. Thus western surveying and mapping expertise has been used by pro-western countries. The various Gulf States, Iran before the Islamic revolution and Saudi Arabia have all used oil money in the 1970s to establish modern map series. This has been accomplished either through aid agreements with official mapping agencies, e.g. Saudi Arabia, or through contracts let to the private sector, e.g. Bahrain and Qatar. In contrast Syria and Iraq have used Eastern bloc assistance to help in the creation of their newer mapping programmes. Other States have continued to develop their own standards, both Turkish and Israeli mapping reflects a continuing indigenous mapping tradition, and the French influence on the Lebanese map base is still clear. There has been little attempt to develop common mapping systems. An exception is the Arabic language *Inter Arab Map Series*, introduced in the early 1970s by Lebanon, Syria and Egypt. Maps in this series were compiled at 1:100 000 (104 sheets), 1:250 000 (36 sheets) and 1:500 000 (15 sheets) to cover the central part of the Middle East area. Like much of Middle Eastern topographic mapping this series was not publicly available. Apart from this series there has been little standardization in terms of scales, styles or specifications. There are also great disparities between the amount of resources allocated to mapping by different states (compare the total lack of modern mapping of South Yemen with the 1 year revision cycle on the very detailed mapping of Bahrain or with the vast Saudi surveying and mapping programmes). Different countries have established their basic scale mapping at very different dates, e.g. the Lebanese 1:20 000 survey dating substantially from the early 1960s in comparison with the OSD-produced 1:50 000 map base for the Yemen Arab Republic, which is currently being published. One common factor for much of the Middle East in terms of topographic mapping is a general restriction placed upon the availability of basic scale mapping: only Lebanese, Israeli, Saudi and some Gulf State mapping is available and some of these countries do not release all scales to the public. In view of the continuing political turmoil and conflict in this strategic area this restriction is perhaps not too surprising, yet paradoxically in Lebanon, the most unstable state in the region, mapping is still advertised as being available.

The geological coverage of the Middle East has developed in the postwar period along similar lines to topographic mapping. Mineral rich areas are well mapped, indeed Saudi compiles topographic editions of the 1:250 000 scale quads, rather than the other way around, thus emphasizing the crucial importance of minerals and especially oil to the Middle Eastern economies. In contrast there is very little geological mapping of those areas with few mineral reserves. One exception is the Israeli 1:50 000 geological mapping programme. No general geological mapping of this whole region is published (see Asian section for small-scale coverage).

There is little other general thematic mapping of the area, apart from some oilfield maps. The significant exception to this generalization is the *Tübinger Atlas des Vorderen Orients*. This major thematic atlas was begun in 1969 and will when complete comprise a total of 379 maps on 285 folio sheets. Areas covered in the project extend from Turkey to Afghanistan and from the Persian Gulf to Egypt, and maps are published in instalments in one of two series: A, Geography; and B, History. A range of scales and themes is covered. The programme is the result of cooperation between 14 departments in five faculties of the three Universities of Tübingen, Mainz and Köln, and monographs are also published in two series to supplement the atlas sheets.

Numerous commercial houses compile general maps of the region. These include the Israeli **Carta**, the Iranian **Sahab** and several European cartographic houses: **Kümmerly + Frey, Ravenstein, Bartholomew, VWK Ryborsch, Hallwag. Geoprojects**, with offices in both Lebanon and Great Britain, is perhaps the most important commercial publisher for the region, with its useful *Arab Map World* series of expensive country maps. The Russian **Glavnoe Upravlenie Geodezii i Kartografii (GUGK)** compiles maps of both the general area and individual Middle Eastern countries, as does the American **Central Intelligence Agency (CIA)**.

Indian Subcontinent

The Indian Subcontinent comprises India, Pakistan, Bangladesh, Sri Lanka, Nepal, Bhutan and Afghanistan. The whole area was covered by mapping compiled by the **Survey of India**, and following independence the mapping standards established in the early twentieth century have continued to ensure a degree of standardization for maps of the Subcontinent that is not found in the mapping of the Middle East. Some new series have been established after independence but most have closely followed SI conventions in terms of style. Exceptions have been the new 1:50 000 Sri Lankan mapping, using new sheet lines and a land use base, and the new Nepali series. Most of the Indian subcontinent is now covered in modern 1:50 000 scale mapping but much of the basic scale mapping is officially unavailable for export.

In some cases very little current mapping indeed is available, because of military or political restrictions (e.g. Afghanistan) or because very little is compiled (e.g. Bhutan).

Smaller-scale official mapping of most of the countries in the area is published and available.

The geological surveying of South Asia is very patchy.

Small-scale coverage has been compiled and the earth science surveying agencies are involved in the **Commission for the Geological Map of the World (CGMW)** mapping of the area, but apart from India there are no indigenous systematic geological large-scale programmes. Aid agencies have compiled geological cover of specific areas, e.g. DOS in the Salt Ranges in Pakistan and the Canadians in Nepal.

Only India is really actively involved in other thematic mapping of the region. Both the **National Atlas and Thematic Mapping Organization** and the **Census of India** compile very useful environmental and socioeconomic mapping of the subcontinent, whilst the **Institut Français de Pondichéry (IFP)** is responsible for the compilation of a variety of vegetation mapping. This includes both larger-scale mapping of South India, and 1:1 000 000 coverage of most of the subcontinent. IFP has also produced small-scale bioclimatic mapping of the whole subcontinental area.

Amongst Western agencies producing maps of the area are the **Central Intelligence Agency (CIA)**, **Nelles Verlag** and **Bartholomew**, whilst the Russian **Glavnoe Upravlenie Geodezii i Kartografii (GUGK)** issues general maps of several states. Western agencies produce mapping of the Himalayas, including the Schneider East Nepal series, for the climbing and trekking market.

South East Asia

South East Asia comprises the mainland Asian countries of Vietnam, Laos, Kampuchea, Thailand and Burma, together with Malaysia and Indonesia, the Philippines and Papua New Guinea. There are no official series covering the whole region and little standardization in mapping systems or specifications. There is little information about any topographic or earth science mapping of most of mainland South East Asia, apart from Thailand. American series covered the area and date from the American involvement in the Vietnamese conflict but are no longer available. Only Thai official mapping is currently available to the west. In contrast, Indonesia and Malaysia both have very well developed official mapping systems, but restrict the availability of their products. The more recent colonial past of Papua New Guinea led to the establishment with Australian aid of a modern official mapping infrastructure and series that may be readily acquired. Geological and resources mapping of the region also reflects this division into countries with available mapping and active programmes and those where almost no mapping is available. Thus Thailand, Malaysia, Papua New Guinea and Indonesia have all produced large amounts of earth science and resources mapping in the last 20 years, whereas Burma, Kampuchea, Laos and Vietnam have almost no available maps.

Small-scale mapping of the region as a whole has been carried out by American agencies, notably the **United States Geological Survey (USGS)** (listed under Indonesia), and by the **American Association of Petroleum Geologists (AAPG)**. The **Institut Français de Pondichéry (IFP)** has compiled vegetation and bioclimatic mapping of the area. Other thematic mapping of the region is compiled by The Economic and Social Commission for Asia and the Far East and is listed under Asia.

Among the commercial agencies that have compiled maps of South East Asia are **Bartholomew, Karto+Grafik**, and **Nelles Verlag/Apa**, while the American **Central Intelligence Agency (CIA)** and the

Russian **Glavnoe Upravlenie Geodezii i Kartografii (GUGK)** both publish a variety of country maps and some thematic coverage of the region.

Further information

The best sources of information about Asian mapping are the papers presented at the *United Nations Regional Cartographic Conferences for Asia and the Pacific*, organized by the United Nations Department of Technical Cooperation and Development.

An early synthetic work was Hale, G.A. (1969) Maps and atlases of the Middle East, *Middle East Association Bulletin*, **3**(3), 17–39. There are no other adequate current overviews of mapping systems of the whole continent.

For further information about the *Tübinger Atlas des Vorderen Orients* see Denk, W. and Rollig, W. (1979) Der Tübinger Atlas des Vorderen Orients, *International Yearbook for Cartography*, 54–77.

Addresses

Economic and Social Commission for Asia and the Pacific
United Nations Building, Rajdamnern Ave., Bangkok 2,
THAILAND.

Tübinger Atlas des Vorderen Orients (TAVO)
Dr Dietrich Reichart Verlag, WIESBADEN, German Federal Republic.

For Carta see Israel; for Sahab see Iran; for Kümmerly + Frey and Hallwag see Switzerland; for Ravenstein, VWK Ryborsch and Nelles Verlag see German Federal Republic; for CIA, NGS, DMA, AAPG, PennWell and USGS see United States; for GUGK and MGSSSR see USSR; for IFP, Census of India, NATMO, SI and GSI see India; for CGMW, WMO and UNESCO see World; for KIER see Korea; for Academy of Geological Sciences and CPH see China; for Bartholomew, Geoprojects and OPL see Great Britain; for IRRI see Philippines; for Royal Thai Survey Department see Thailand.

Catalogue

General

Eurasia 1 : 15 000 000
Edinburgh: Bartholomew, 1976.

Azija 1 : 12 500 000
Moskva: GUGK.

Asien 1 : 12 000 000
Bern: Kümmerly + Frey, 1983.

Evrazija: fizicheskaja karta 1 : 8 000 000
Moskva: GUGK.
4 sheets, all published.
In Russian.
Physical map of Eurasia.

Geographic base map of South and East Asia 1 : 5 000 000
Paris and Calcutta: CGMW and GSI, 1979.
5 sheets, all published.
Includes 1 : 5 000 000 scale map (4 sheets) and 1 : 10 000 000 scale map (1 sheet).

Geological and geophysical

Metamorphic map of South and East Asia 1 : 10 000 000
Paris and Seoul: CGMW and KIER, 1986.
With accompanying explanatory notes.

Yazhou dadi gouzao tu/Tectonic map of Asia 1 : 8 000 000
Beijing: Academy of Geological Sciences.
6 sheets, all published.
In Chinese and English.
With short accompanying explanatory note.

Yazhou dizhen gouzao tu/Seismotectonic map of Asia 1 : 8 000 000
Beijing: Academy of Geological Sciences.
4 sheets, all published.
Chinese language with English and Chinese legend.
With Chinese language explanatory text (English summary).

Yazhou dizhi tu/Geological map of Asia/Geologische Karte Asien 1 : 5 000 000
Beijing: Academy of Geological Sciences and CPH, 1975.
20 sheets, all published.
Chinese language, with English language legend sheet.

Geological map of Asia and the Far East 1 : 5 000 000 2nd edn
Paris: UNESCO and ESCAFE, 1971.
4 sheets, all published.
With accompanying explanatory notes.

Tectonic map of South and East Asia 1 : 5 000 000
Paris and Calcutta: CGMW and GSI, 1982.
7 sheets, all published.
Includes 1 : 5 000 000 scale map (4 sheets), 1 : 10 000 000 scale map (2 sheets), legend sheet and explanatory notes.

Karta razlomoj tektoniki juga Azii/Fault tectonics map of the south of Asia 1 : 5 000 000
Moskva: MGSSSR, 1983.
8 sheets, all published.
In Russian.
Includes Russian and English legend sheet.
With accompanying explanatory booklet.

Metamorphic map of Asia 1 : 5 000 000
Moskva: MGSSSR, 1978.
9 sheets, all published.
Prepared for the CGMW Subcommission on Mapping Metamorphic belts of the world.

Gravity anomaly map of Western ESCAP Region 1 : 5 000 000
Bangkok: ESCAP, 1976.

Gravity anomaly map of the Eastern ESCAP Region 1 : 5 000 000
Bangkok: ESCAP, 1982.
2 sheets, both published.
With accompanying explanatory text.

Karta chetvertichych otlozhenij Aziatskoj chasti regiona ESKATO/Map of Quaternary deposits of the Asian part of the ESCAP region/Cartes des formations quaternaires de la region CESAP 1 : 5 000 000
Moskva: MGSSSR, 1983.
6 sheets, all published.
With accompanying Russian language explanatory booklet.
3 language legend.

Mineral distribution map of Asia and the Far East 1 : 5 000 000 2nd edn
New York and Bangkok: ESCAP, 1979.
4 sheets, all published.
With accompanying explanatory text.

Environmental

Climatic atlas of Asia/Atlas climatique de l'Asie/Klimaticheskij Atlas Azii
Geneva: World Meteorological Organization, 1981.
pp. 28.

Klimaticheskaja karta Azii 1 : 8 000 000
Moskva: GUGK, 1969.
4 sheets, all published.
In Russian.
Includes inset maps.
Educational map.

Pochnennaja karta Azii/Soil map of Asia   1:6 000 000
Moskva: Akademija Nauk, 1971.
6 sheets, all published.

*Agroclimatic and dry season maps of South, South East and East
Asia*   1:6 336 000
Manila: IRRI, 1982.
5 sheets, all published.
With climatic diagrams on sheet and accompanying explanatory
booklet.

*Rice area by type of culture: South, Southeast and East
Asia*   1:4 500 000
Manila: IRRI, 1982.
3 sheets, all published.
With accompanying explanatory booklet.

Human and economic

Regional economic atlas of Asia and the Far East
Bangkok: Royal Thai Survey Department, 1970–.

Oil and natural gas map of Asia   1:5 000 000
Bangkok: ESCAP, 1975.
Excludes USSR and Middle East.

INDIAN SUBCONTINENT

Atlases and gazetteers

Historical atlas of South Asia   J.E. Schwartzberg
Chicago: University of Chicago Press, 1978.
pp. 352.

General

India and adjacent countries   1:12 000 000
Dehra Dun: SI, 1972.

Indian subcontinent   1:4 000 000
Edinburgh: Bartholomew, 1984.
Includes 5 1:200 000 scale city area maps.

Indija, Nepal, Butan, Bangladesh, Shri-lanka   1:3 700 000
Moskva: GUGK.
In Russian.

Southwest Asia   1:3 000 000
Washington, DC: CIA, 1980.
Covers Afghanistan, Pakistan and Iran.

Geological and geophysical

Seismotectonic map of Iran, Afghanistan and Pakistan   1:5 000 000
Paris and Calcutta: CGMW and GSI, 1984.
With accompanying explanatory notes.

Environmental

*Carte des bioclimats du sous-continent indien/Map of bioclimate of the
Indian Sub-Continent*   1:2 534 400
Pondichéry: IFP, 1965.
4 sheets, all published.
With accompanying explanatory text.

Human and economic

Peoples of South Asia   1:7 000 000
Washington, DC: NGS, 1984.

MIDDLE EAST

Atlases and gazetteers

Tübinger Atlas des Vorderen Orients
Wiesbaden: Ludwig Reichart Verlag für Sonderforschungsbereich
19 TAVO der Universität Tübingen, 1977–.
2 series.
A: Geography 118 sheets.
B: History 172 sheets.
In progress.

The Daily Telegraph Atlas of the Arab World   M.W. Dempsey
London: Daily Telegraph, 1983.
pp. 124.

General

Middle East   1:6 700 000
Washington, DC: NGS, 1984.

Mittlere Osten   1:5 000 000
Frankfurt AM: Ravenstein.

Naher Osten   1:5 000 000
Frankfurt AM: Ravenstein.

The Middle East   1:4 500 000
Washington, DC: CIA, 1980.

The Middle East   1:4 000 000
Edinburgh: Bartholomew, 1985.
With insets of Socotra and Southern Arabia.

The Middle East: road map   1:4 000 000
Obersthausen: VWK Ryborsch, 1980.
Legend and title in English, French, German and Arabic.

Yugo-zapadnaja Azija: fizicheskaja karta   1:4 000 000
Moskva: GUGK.
2 sheets, both published.
In Russian.
Relief map of the Middle East.

Road map of Turkey and the Near East   1:3 000 000
Bern: Hallwag, 1980.
English, French and German legend.

Rand McNally international map of the Middle East   1:3 000 000
Chicago: Rand McNally.

Near East   1:2 000 000
Wien: Freytag-Berndt.
Road map.
Legend in German, English, French, Arabic and Turkish.

The Persian Gulf   1:1 600 000
Washington, DC: CIA, 1980.

Middle East briefing map   1:1 500 000
Washington, DC: DMATC.

Middle East road map   1:1 000 000   Series 1310
Washington, DC: DMA, 1983.
6 sheets, all published.

Arabian Gulf   1:1 000 000
Edinburgh: Bartholomew, 1977.

Political

Carta's map of the Middle East   1:6 000 000
Jerusalem: Carta, 1980.
Layer coloured by State.
Includes thematic insets on sheets.

Yugo-zapadnaja Azija   1:5 000 000
Moskva: GUGK.
In Russian.
Political map of the Middle East.

Naher Osten/Middle East/Proche Orient/Vicino Oriente 1:5 000 000
Bern: Kümmerly + Frey, 1984.
Political base.
4 language legend.

Geological and geophysical

Tectonic map of the Middle East 1:5 000 000
Geneva: Petroconsultants, 1977.

Human and economic

Middle East oil and gas 1:4 500 000
Tulsa: PennWell, 1983.
With 1:1 600 000 scale inset of Persian Gulf area.

Middle East area: oilfields and facilities 1:4 500 000
Washington, DC: CIA, 1980.

Yugo-zapadnaja Azija: ekonomicheskaja Karta 1:4 000 000
Moskva: GUGK.
2 sheets, both published.
In Russian.

The Middle East: oil and gas activity and concession map
1:3 000 000
Ledbury: OPL, 1985.
2 sheets, both published.

SOUTH EAST ASIA

Atlases and gazetteers

Atlas of South East Asia
Canberra: Allen and Unwin, 1983.

Atlas for marine policy in South East Asian seas J.R. Morgan and
M.J. Valencia
Berkeley: University of California Press, 1983.
pp. 144.

General

Asia South-East 1:5 000 000
Edinburgh: Bartholomew, 1985.

Yugo-zapadnaja Azija: fizicheskaja karta 1:5 000 000
Moskva: GUGK.
2 sheets, both published.
In Russian.
Relief map of South East Asia.

*Thailand Burma Malaysia mit Singapur: Hildebrands
urlaubskarte* 1:2 800 000
Frankfurt AM: Karto+Grafik, 1985.
Includes tourist information.
Available in German and English language versions.

South-East Asia briefing map 1:2 000 000
Washington, DC: DMATC, 1981.

Geological and geophysical

Geothermal gradient map of South East Asia 1:5 000 000
Singapore and Jakarta: SEAPEX and IPA, 1981.
With accompanying explanatory text.

Tectonic map of South and East Asia 1:5 000 000 and
1:10 000 000
Tulsa, OK: AAPG, 1982.
7 sheets, all published.
Set includes 4-sheet tectonic map at 1:5 000 000; gravity anomaly,
epicentre and heat flow map at 1:10 000 000; tectonic units,
dislocations and fault plane solutions of earthquakes at
1:10 000 000; legend sheet and explanatory text.

Environmental

Vegetation map of Malesia 1:5 000 000
T.C. Whitmore
Oxford: Commonwealth Forestry Institute, 1984
Published with *Journal of Biogeography*.

*Bioclimats du Sud Est Asiatique/Bioclimates of South-East
Asia* 1:2 534 400
Pondichéry: IFP, 1967.
With accompanying explanatory text.
Several climatic diagrams and 2 small-scale inset maps also on
sheet.
Covers Indochina, Malaya and Thailand.

Political

Indokitai 1:3 000 000
Moskva: GUGK.
In Russian.
Political map of Indochina.

Human and economic

The oilman's SE Asia oil and gas map ca 1:7 000 000
London: 1983.

Yugo-vostochnaja Azii: ekonomicheskaja karta 1:5 000 000
Moskva: GUGK.
2 sheets, both published.
In Russian.
Economic map of South East Asia.

The South-East Asia oil and gas activity map 1:3 000 000
Ledbury: OPL.

Afghanistan

The **Afghan Geodesy and Cartography Office (AGCO)**
carries out topographic and cadastral survey activities in
Afghanistan. The entire area of the country is covered in
1:50 000, 1:100 000 and 1:250 000 mapping dating from
1957 to 1959. These series replaced the earlier Survey of
India quarter-inch coverage and were compiled by the US
Army Map Service (now Defense Mapping Agency). A
UNDP-funded project was in operation from 1976 until
1982. This had the aim of strengthening AGCO by the

introduction of aerial triangulation methods, by training
and installation of modern geodetic and computer
processing hardware, and by renewing cadastral surveying
activity upon completion of land reform proposals. The old
1:50 000 map is being up-dated through the production of
a new 1:25 000 series being compiled by photogrammetric
methods followed by ground verification. The 1:25 000
map has 5, 10 or 20 m contours and uses UTM projection:
it will only cover areas under development or likely to be

developed. Larger-scale series are published for development projects. None of these topographic maps is at present available for sale.

AGCO also produces small-scale maps, including a recent political map of the country and two earlier physical maps. The National Atlas was compiled in conjunction with AGCO by the Polish overseas survey agency **Geokart**, whilst **Sahab** produced a general atlas and map of the country with English, French and Persian text in the early 1970s. Other small-scale mapping of Afghanistan has been compiled by American **Central Intelligence Agency (CIA)**, and Russian **Glavnoe Upravlenie Geodezii i Kartografii (GUGK)**.

The **Afghan Ministry for Mines and Industry** has been responsible for the compilation of mining and geological mapping, but none of their maps is still available. The best available geological coverage is a single-colour small-scale Sahab-produced map and 4-sheet 1:500 000 full-colour coverage of central and south Afghanistan, published with an explanatory monograph by the German **Bundesanstalt für Geowissenschaften und Rohstoffe (BfGR)**.

The French **Office de la Recherche Scientifique et Technique Outre Mer (ORSTOM)** published soil maps of parts of the country in 1977.

Other thematic mapping of the country has appeared in the **Tübinger Atlas des Vorderen Orient (TAVO)** series.

Further information

Afghanistan (1977) Cartographic activities in Afghanistan, paper presented by Afghanistan to the *7th UN Regional Cartographic Conference on Asia and the Pacific*, New York: UN.

Addresses

Afghan Geodesy and Cartography Office (AGCO)
Pashtunistan Watt, KABUL.

Afghan Ministry for Mines and Industry
Darulaman, KABUL.

Office de la Recherche Scientifique et Technique Outre Mer (ORSTOM)
70–74 route d'Aulnay, 93140 BONDY, France.

For Sahab see Iran; for BfGR and TAVO see German Federal Republic; for DMA and CIA see United States; for GUGK see USSR.

Catalogue

Atlases and gazetteers

Atlas of the Democratic Republic of Afghanistan
Warszawa: Geokart, 1985.
73 sheets.

General atlas of Afghanistan
Tehran: Sahab, 1974.
pp. 200.
In English, French and Persian.

Historical and political gazetteer of Afghanistan   L.W. Ademec
Graz: Akademische Druck und Verlagsanstalt.
6 Vols.
Arranged by regions. Vols also contain complete quarter-inch coverage.

General

Afghanistan   1:2 500 000
Washington, DC: CIA, 1980.

Afganistan   1:2 500 000
Moskva: GUGK, 1985.
In Russian.
With 3 small ancillary thematic insets and placename index.

Physical map of Afghanistan   1:2 000 000
Kabul: AGCO, 1973.

Physical map of Afghanistan   1:1 500 000
Kabul: AGCO, 1967.

Geological and geophysical

Afghanistan: geological   1:2 500 000
Tehran: Sahab.
In English and Persian.

Geologische Karte von Zentral und Sud-Afghanistan   1:500 000
Hannover: BfGR, 1973.
4 sheets, all published.

Administrative

Political map of Afghanistan   1:2 000 000
Kabul: AGCO, 1984.

Afghanistan general   1:2 000 000
Tehran: Sahab, 1975.
In Persian and English.
With insets: town plan of Kabul and economic map.

Human and economic

Afghanistan: Städte und Marktorte   1:4 000 000
Wiesbaden: TAVO, 1981.

Afghanistan: ethnische Gruppen/Afghanistan ethnic groups   1:2 500 000
Wiesbaden: TAVO, 1983.

Afghanistan economical   1:2 500 000
Tehran: Sahab, 1974.
In Persian, English and French.

Bahrain

Modern mapping of Bahrain has been carried out by the Survey Directorate of the Bahrain **Ministry of Housing (MH)**. The Directorate was established in the old Public Works Department, and joined with part of the Land Registration Section of the Ministry of Justice to form the present administrative set-up in the Ministry of Housing in the late 1970s. Fairey (now Clyde) Surveys flew aerial coverage of Bahrain at four different scales in both colour and black and white in the late 1970s and much of the current map base of Bahrain has been derived from this. The basic mapping of the whole state is at 1:10 000 scale: a 28-sheet series on UTM projection is available in either orthophoto editions dated 1977, or two-colour line map editions with 2 and 5 m contours, dated 1985–86 and derived from recent photogrammetric plots. In the north of the island in the areas of main urban development there are two larger-scale series of line maps drawn at 1:1000 (449 sheets) and 1:2000 (220 sheets) scales. This planimetric data is held digitally using the Intergraph mapping system and is also used for cadastral mapping purposes. There is a revision programme in progress; most sheets are dated 1985–86. It is intended to introduce a 1 year revision cycle for this large-scale coverage, which is available as diazo prints. The four-colour, seven-sheet 1:25 000 series is the largest scale publicly available series and covers the main island only. Most sheets date from 1977; one has been revised to 1985. A three-sheet, six-colour 1:50 000 series was issued in new editions in 1986, and single-sheet 1:100 000 and 1:200 000 scale maps are also available. All of these maps include both English and Arabic information and use UTM projection. Work is in hand to up-date all the 1970s mapping. MH also carries out hydrographic charting activities.

Thematic mapping of Bahrain has been carried out as part of the 1980 Surface Materials Resources Study by a team of British geomorphologists. This 1:50 000 programme maps as single sheets the following themes: geology, geomorphology and superficial materials, drainage, soils and land capability for agriculture. A provisional sheet to cover the Hawar Islands is also available and a detailed explanatory text is issued by Geo Books. The maps are also available separately from MH. Other commercial mapping of Bahrain has been carried out including a map issued in the *Arab World Map Series* and distributed by **Geoprojects SARL**.

Further information

Fairey Surveys (1976) Bahrain contract: survey and mapping of the State of Bahrain, *Fairey Surveys Newsletter*, **16**, November 1976.

MH issues a short catalogue.

For background information on the Surface Materials Study see Brunsden, D., Doornkamp, J.C. and Jones, D.K.C. (1979) The Bahrain Surface Materials Resources Survey and its application to regional planning, *Geographical Journal*, **145**(1), 1–35.

Addresses

National Survey Directorate, Ministry of Housing (MH)
PO Box 802, AL MANAMA.

For Geoprojects see Lebanon; for DMA see United States.

Catalogue

Atlases and gazetteers

Bahrain, Kuwait, Qatar, and the United Arab Emirates: official standard names approved by the United States Board on Geographic Names
Washington, DC: DMA, 1976.
pp. 157.

General

The Oxford map of Bahrain   1:310 000   2nd edn
Beirut: Geoprojects, 1985.
Double-sided.
Includes maps of Northern Bahrain 1:57 750 and Central Manama 1:10 000.

The State of Bahrain   1:200 000
Manama: MH, 1977.
In English and Arabic.

The State of Bahrain   1:100 000
Manama: MH, 1978.
In English and Arabic.

Topographic

The State of Bahrain   1:50 000
Manama: MH, 1986.
3 sheets, all published.
In English and Arabic.

The State of Bahrain   1:25 000
Manama: MH, 1977–.
7 sheets, all published.
Does not cover Hawar Islands.
In English and Arabic.

Environmental

The geology, geomorphology and pedology of Bahrain
1:50 000   J.C. Doornkamp, D. Brunsden and D.K.C. Jones
Norwich: Geo Books, 1980.
6 sheets, all published.
With accompanying text.
Single-sheet thematic set covering: geology, geomorphology, drainage, soils, land capability for agriculture and Hawar Islands geomorphology and superficial deposits.
Only geology and geomorphology maps are full-colour editions.
Map sheets also available individually.

Bangladesh

The **Survey of Bangladesh (SB)** is responsible for geodetic and topographic surveys of the country and produces various scales of topographic maps. Prior to independence Bangladesh was mapped as part of the Survey of India programme and, between the partition of the subcontinent in 1947 and the establishment of an independent Bangladesh in 1973, by the Survey of Pakistan. The basic scale series is still the 1:50 000 map, which covers the country in 266 sheets. This series is complete, and is being revised from 1:30 000 aerial photographs flown by the Canadians in 1975. Sheets are printed in six colours, use the Transverse Mercator projection and show relief by 50 foot contours. From this map are derived smaller-scale series, including the 27-sheet 1:250 000 map, six-sheet 1:500 000 scale, and four-sheet 1:1 000 000 scale maps. It is hoped to complete the full revision of the 1:50 000 and 1:250 000 series by 1991. There are plans to produce a 1:25 000 map and to convert toponymic information from the current English forms to Bengali. Project mapping at large scales is also published. None of the above-mentioned maps is available outside the country. Small-scale English language general maps and town plans are available for export and listed in our catalogue section, as are the 1:250 000 scale Bengali language district maps.

A very well developed programme of cadastral mapping is carried out by the **Directorate General of Land Resources and Survey**.

Geological mapping activities are the responsibility of the **Bangladesh Geological Survey**, but there is no information about any current mapping programmes.

Small-scale thematic mapping of the country has been carried out by the **World Bank**. It published a set of 1:500 000 scale single-sheet thematic maps in the early 1970s, covering administrative divisions, hydrology, land use associations, population, transportation and land development units. The World Bank also issued Landsat image maps of the country overprinted with land use and land cover information. These are distributed by **International Mapping Unlimited**.

Within the country, the **Bangladesh Space Research and Remote Sensing Organisation (SPARRSO)** coordinates the use of remotely sensed imagery.

The **International Rice Research Institute (IRRI)** has recently also published a thematic map set.

Other foreign mapping of the country has been carried out by **Glavnoe Upravlenie Geodezii i Kartografii (GUGK)** and the American **Central Intelligence Agency (CIA)**. For other mapping of Bangladesh see South Asia and India.

Further information

Background to the development of the mapping of Bangladesh is contained within Bangladesh (1977 and 1983) Cartographic activities in Bangladesh: papers presented to the *8th and 10th United Nations Regional Cartographic Conference for Asia and the Pacific*, New York: UN.

SB issues a short catalogue.

Addresses

Survey of Bangladesh (SB)
Office of the Surveyor General, Tejgaon Industrial Area, DHAKA 8.

Directorate General of Land Resources and Survey
Tejgaon Industrial Area, DHAKA 8.

Geological Survey of Bangladesh
Pioneer Road, Segun Bagicha, DHAKA.

Bangladesh Space Research and Remote Sensing Organisation (SPARRSO)
Cabinet Secretariat, DHAKA.

For GUGK see USSR; for IRRI see Philippines; for CIA, World Bank, DMA and International Mapping Unlimited see United States.

Catalogue

Atlases and gazetteers

Bangladesh: official standard names approved by the United States Board on Geographic Names
Washington, DC: DMA, 1976.
pp. 526.

General

General map of Bangladesh   1:1 000 000
Dhaka: SB.

Map of Bangladesh showing police stations   1:1 000 000
Dhaka: SB.

Bangladesh   1:1 000 000
Moskva: GUGK, 1979.
In Russian.
With climatic and economic inset maps and placename index.

Environmental

Bangladesh   1:500 000
Washington, DC: World Bank, 1981.
3 sheets, all published.
Landsat-based land cover and land use set, showing reflections, classification of land use and land cover indications.

Thematic maps of Bangladesh   1:500 000
Washington, DC: World Bank, 1971.
6 sheets, all published.
Thematic set.
Sheets cover: political divisions, population, hydrology, transportation, land use associations, land development units.

Administrative

District maps   1:250 000
Dhaka: SB.
16 sheets, all published.
In Bengali.

Human and economic

Map series of Bangladesh   1:750 000
Manila: International Rice Research Institute, 1983.
4 sheets, all published.
Thematic set.
Sheet titles: Base map; Rice area planted by season; Rice area planted by culture type; Farm size distribution.

Town maps

Dhaka guide map   1:20 000
Dhaka: SB, 1979.

Bhutan

The Himalayan mountain kingdom of Bhutan is not well mapped. Official survey activities are the responsibility of the **Survey Department** in the **Department of Trade and Industries**, but no published series are available. Overseas agencies have been involved in the publication of maps of the country. These include a two-sheet 1:250 000 scale map published by the **Survey of India**, which is uncontoured and shows relief by hypsometric tints, and the Landsat image map published by the World Bank and distributed by **International Mapping Unlimited**. Other general coverage of the country has been compiled by the **Swiss Foundation for Alpine Research**. Their 1:500 000 scale topographic sketch map is based upon field sketches and Landsat Imagery. The map supplement issued in 1965 by the **Association of American Geographers (AAG)** is still available.

Addresses

Survey Department
PO Tashichhodzong, THIMPU.

Survey of India (SI)
Surveyor General's Office, PO Box 37, Hathibarkala Estate, DEHRA DUN 248001, India.

International Mapping Unlimited
4343 39th Street, NW, WASHINGTON DC 20016, USA.

Association of American Geographers
1710 16th Street, NW, WASHINGTON DC 20009, USA.

Catalogue

General

Bhutan–Himalaya 1:500 000
Zurich: Swiss Foundation for Alpine Research.

Kingdom of Bhutan 1:253 440
Washington, DC: AAG, 1965.
Issued as map supplement 5 with Annals of the AAG, 55 (4).

Bhutan 1:250 000
Dehra Dun: SI, 1972.
2 sheets, both published.

Bhutan: landcover, soil and water reflections 1:250 000
Washington, DC: World Bank, 1982.

Brunei

Brunei became a fully independent territory in 1984. The national mapping agency is the **Survey Department Brunei** which has built upon the map base compiled by the British Directorate of Overseas Surveys. The new national basic scale map is a 14-sheet national series compiled on the Brunei National Grid. This series is still in progress, and sheets in the older DOS-based 1:50 000 series T735 are still current for some areas of the country. This older series was compiled upon 15' graticule quads and showed relief with 50 or 100 foot contour intervals. Smaller-scale maps compiled by the Survey Department include a 1:100 000 map arranged on district sheet lines, which was redesigned in 1982, and a two-sheet 1:250 000 scale map. A 1:12 500 scale map with 20 foot contour interval covers the main town areas and there is also a four-chain series of maps covering Tutong and Muara. None of these topographic or photomap products is available outside of Brunei government organizations. A recently published tourist map of the country is the only available map of Brunei, with the exception of smaller-scale maps of the whole of Sarawak and Northern Borneo (see Malaysia).

Geological mapping of the country is the responsibility of the **Government Geologist**, but the 1:50 000 and 1:125 000 geological sheets published in the 1960s are no longer available and there is no new work in progress.

Further information

The Survey Department publishes a map catalogue.

Addresses

Survey Department Brunei (Juru Ukor Agong Negen)
BANDAR SERI BEGAWAN.

The Government Geologist
BANDAR SERI BEGAWAN.

Catalogue

General

Tourist map of Brunei 1:250 000
Bandar Seri Begawan: Survey Department Brunei, 1983.
Includes town plans on back of sheet.

Burma

The official mapping agency is the **Survey Department** in Rangoon but there is no information about the current mapping of Burma. The British-produced 1:63 360 scale Hind and Survey of India mapping, compiled towards the end of World War II, is still the best mapping of the country known to the west.

Geological mapping of the country is the responsibility of the **Department of Geological Survey and Exploration (DGSE)** in the Ministry of Mines. A general geological map of the country was compiled in the late 1970s and distributed through **Myanma Oil Corporation**. The **British Geological Survey (BGS)** compiled some mapping of the northern Shan States in the late 1970s, at scales of 1:100 000 and 1:125 000. These were published with accompanying explanatory monographs. There is no information about current programmes.

Small-scale general coverage of Burma has been published by the American **Central Intelligence Agency (CIA)**, by the Russian **Glavnoe Upravlenie Geodezii i Kartografii (GUGK)** and by the commercial agency **Nelles Verlag**. The **World Bank** published a false colour image map of parts of the south-west of the country, showing land cover associations, which is still available and is distributed by **International Mapping Unlimited**.

Addresses

Survey Department
Ministry of Agriculture and Forestry, Thiramingala Lana, Kaba Aye Pagoda Rd, RANGOON.

Department of Geological Survey and Exploration (DGSE)
Ministry of Mines, Kanbe Rd, Yankin PO, RANGOON.

Myanma Oil Corporation
604 Merchant St, PO Box 1049, RANGOON.

For World Bank see World; for DMA, CIA and International Mapping Unlimited see United States; for BGS see Great Britain; for GUGK see USSR; for Nelles Verlag see German Federal Republic.

Catalogue

Atlases and gazetteers

Burma: official standard names approved by the United States Board on Geographic Names
Washington, DC: DMA, 1966.
pp. 726.

General

Burma 1:3 999 000
Washington, DC: CIA, 1972.
With land use, population and economic insets.

Birma 1:2 000 000
Moskva: GUGK, 1983.
In Russian.
With placename index, climatic, economic and population inset maps.

APA map of Burma 1:1 500 000
München: Nelles Verlag, 1986.

Geological and geophysical

Geological map of the Socialist Republic of Burma 1:1 000 000
Rangoon: DGSE, 1977.
3 sheets, all published.

Environmental

Burma: land cover–land use associations 1:1 000 000
Washington, DC: World Bank, 1976.
Satellite image map.

China

Since 1956 the national survey authority in the People's Republic of China has been the **National Bureau of Surveying and Mapping (NBSM)**. It is responsible for geodetic, gravimetric, aerophotographic, cartographic and map publishing operations in the country and coordinates the work of numerous mapping agencies at national or state level. Very few of NBSM products are available outside of China.

There is now complete modern photogrammetrically derived topographic coverage of the country in several scales. The national mapping programme aimed to give basic scale coverage at 1:25 000 for major industrial and urban centres, 1:50 000 coverage for remaining densely populated and developed regions and 1:100 000 scale coverage for deserts, high-altitude and mountain areas by the end of 1967. Almost 25 000 1:50 000 sheets were

required for complete coverage. These maps use 5, 10 or 20 m contour intervals for plains, hills and mountains. These first generation topographic survey targets were probably attained during the early 1970s; complete 1:100 000 coverage has been maintained since 1969. The 1:100 000 series is derived from 1:60 000 aerial coverage, whilst the 1:50 000 map is compiled from 1:40 000 air photos. These maps use UTM projection and character script, and a unified sheet numbering and division system.

Since the completion of the national programme, efforts have been concentrated on revision. A 1:1 000 000 series has been revised and issued as photomaps derived from corrected Landsat images: this revision programme was completed in the early 1980s. 1:500 000 and 1:200 000 scale derived series have been issued. There is a continuous programme for the revision of the three basic

CHINA

Provinces

scales using a combination of photogrammetry and field checking. From the early 1970s NBSM has devoted an increasing amount of resources to larger-scale series: a major 1:10 000 mapping programme is under way in the eastern areas of the country. This series uses 1, 2 and 5 m contour intervals depending upon the nature of the terrain. There is coverage of the larger cities at scales of 1:5000, 1:2000, 1:1000 and 1:500. None of these topographic or planimetric maps is available outside China.

Smaller-scale mapping of the country is also carried out by NBSM. A five-volume National Atlas project is currently under way. It is intended to publish this work in several different versions: popular, Chinese phonetic alphabetic and foreign language.

The **Cartographic Publishing House (Ditu Chubanshe) (CPH)** in Peking is administratively part of NBSM and acts as the publishing agent for the publicly available NBSM mapping. CPH was established in 1954 and issues a range of maps very similar to the publicly available material distributed by GUGK in Russia. A great variety of smaller-scale maps is available, in addition to tourist and town mapping. Available maps include Chinese language political maps of the provinces and autonomous regions, and a wide range of smaller-scale wall maps of the whole country. These are published as either political or topographic bases in several different language editions. A series of 43 country maps similar in design to American CIA and Russian GUGK mapping is also issued: we have not listed these in country sections because of the language barrier of character script. Several important atlases are published, including a Pinyin script administrative atlas, an English language tourist atlas of China, and a major six-volume historical atlas. A major new national atlas project is under way, to produce five thematic volumes of national map coverage. The general and agricultural volumes in this programme are both completed. Thematic mapping covering a variety of topics is distributed: most maps are at

scales of 1:4 000 000 or smaller, and very few are published with full English language or Pinyin transcriptions. Tourist maps and town plans are published for all the major open cities in China in both Chinese and English language editions. In addition to this reference and tourist mapping a wide variety of educational maps are published by CPH, including topographic and political maps of each province, a wide variety of educational wall maps and wall maps in minority languages, e.g. Mongolian, Uygur, Korean and Kazak. A large number of historical wall maps are also published. The distribution of CPH mapping outside the country is carried out by the **China International Book Trading Corporation**.

At a provincial level there are also surveying and mapping organizations, whose work is coordinated by NBSM. Few of their maps are publicly available. An exception is the **Lands and Survey Department Guangdong**, which issues a variety of mapping of south China, including town plans of major cities, general provincial maps and several atlases.

Thematic mapping activities in China are mainly carried out by other agencies, such as the **Ministry of Geology** and the various branches of the **Chinese Academy of Sciences**, especially the **Earth Sciences Branch**. For instance, 1:200 000 scale geological mapping has been compiled on a regional level and 1:1 000 000 scale geological coverage has been published to give nationwide coverage. This has allowed the publication of a variety of geological and hydrogeological atlases and smaller-scale geological maps. Many of the smaller-scale geological surveys are distributed by CPH but other regional earth science mapping is also publicly available through the **Geological Publishing House (GPH)**. GPH is distributing the 1:500 000 scale regional geology studies and some other small-scale mapping, including a new 1:4 000 000 scale metamorphic map of the country published in 1986.

Another recent important publication from the Academy

of Sciences is distributed by **AGS Management Consultants**. This is a nine-sheet 1:2 500 000 scale tectonic systems map of China that uses the topographic map sheet lines and is published with English and Chinese explanatory notes. A 1:5 000 000 scale version of the same map is distributed by **Science Press** in Peking. Regional geomorphological maps have also been compiled at medium scales and used for the publication of a variety of smaller-scale geomorphological sheets. Soil and vegetation maps have also been prepared. The **Institute of Remote Sensing Applications** has compiled a variety of Landsat and other satellite-based maps, including a Landsat Image Atlas (see below).

In Hong Kong **Asian Research Service** has published a series of single-sheet English language thematic maps in the *China current data maps series*. It also acts as a distributor for maps and atlases from the Republic, including Science Press publications like the recent atlas of geoscience analysis of Landsat imagery, and complete 1:500 000 false colour coverage of the country in 533 sheets.

English language sketch maps of the Chinese Provinces were published by Aarhus Universitet Kulturgeografisk Institut, but are no longer available. Many other western commercial cartographic houses have published general maps of China. These include a recent collaboration between **Esselte** and CPH, and useful maps from **Bartholomew, Karto + Grafik, Kümmerly + Frey** and **Falk**.

Further information

For the background to the development of current mapping in China see Williams, J.F. (1976) The development of modern mapping in China, *Surveying and Mapping*, **36**(3), 231–239.

For a current critical overview see Brandenberger, A.J. (1983) Surveying and mapping (S&M) in China, *Surveying and Mapping*, **43**(2), 153–160.

For the Chinese viewpoint, see China (1980, 1983, 1985 and 1987) Surveying and mapping activities of the People's Republic of China: papers submitted by China to the *United Nations Regional Cartographic Conferences for Asia and the Pacific* (1980, 1983 and 1987) *and the Americas* (1985), New York: UN; and National Bureau of Surveying and Mapping (1983) *A Brief Introduction*, Beijing: NBSM.

Catalogues are available from CPH, GPH and Lands and Survey Department Guangdong Province.

Addresses

National Bureau of Surveying and Mapping (NBSM)
Baiwanzhuang, BEIJING.

Cartographic Publishing House (Ditu Chubanshe) (CPH)
3 Baizhifang Xijie, BEIJING.

China International Book Trading Corporation (Guoji Shudian)
PO Box 399, BEIJING.

Lands and Surveys Department Guangdong Province
468 Huanshi Rd East, GUANGZHOU.

Ministry of Geology
BEIJING.

Institute of Earth Sciences
Academia Sineca, BEIJING.

Geological Publishing House (GPH)
Xi-Si, BEIJING.

Science Press
137 Chaoyangmennei St, BEIJING.

Institute of Remote Sensing Applications
Academia Sineca, BEIJING.

For Asian Research Service, AGS Management Consultants and Petroleum News South East Asia see Hong Kong; for Esselte see Sweden; for Karto+Grafik and Falk see German Federal Republic; for Quail, Bartholomew and Pergamon Press see Great Britain; for Kümmerly + Frey see Switzerland; for CIA and DMA see United States; for GUGK see USSR.

Catalogue

Atlases and gazetteers

China (mainland): official standard names approved by the United States Board on Geographic Names   2nd edn
Washington, DC: DMA, 1968.
2 vols.

Zhongguo diminglu: zhongguo renmin gongheguo dituji diming suoyin/ Gazetteer index to the atlas of the People's Republic of China
Beijing: CPH, 1983.
pp. 316.
Character and Pinyin script.

Atlas of Cancer mortality in the People's Republic of China   compiled by the National Cancer Control Office and the Nanjing Institute of Geography
Shanghai: CPH, 1979.
pp. 109.
Also available as Chinese language edition.

Zhonghua renmin gongheguo dituji   2nd edn
Beijing: CPH, 1984.
pp. 75.
Chinese language National Atlas.

Tourist atlas of China
Beijing: CPH, 1985.
Also available as Chinese language edition.

Historical atlas of China/Zhongguo lishi dituji   Q.X. Tan
Beijing: CPH, 1982–.
8 vols, 3 published.
Introduction, legend and contents in English and Chinese.

Zhongguo zirandili tuji
Beijing: CPH, 1984.
pp. 210.
Chinese language physical atlas of China.

Atlas of false colour Landsat images of China/Zhongguo ludi weijing jia caise yingxiang tuji   1:500 000
Beijing: Institute of Geography Academia Sineca, 1983.
3 vols.
553 images.
Chinese description, introduction index and title also in English.

General

Map of China   1:10 000 000
Washington, DC: CIA, 1979.

Map of the People's Republic of China   1:6 000 000
Beijing: CPH, 1981.
Topographic map.
Also available as Chinese and Pinyin editions.

China   1:5 000 000
Bern: Kümmerly + Frey, 1986.
Physical base.

China: Hildebrands Urlaubskarte   1:4 500 000
Frankfurt AM: Karto+Grafik, 1985.
Available as English or German edition.

Map of the People's Republic of China   1:4 000 000
Stockholm and Beijing: Esselte and CPH, 1984.
English and Chinese description.
Physical base.

Geological and geophysical

Zhonghua renmin gongheguo shuiwen dizhi tuji
Beijing: CPH, 1979.
pp. 68.
Chinese language hydogeological atlas.

*Zhongguo ludi xianxing gouzao tu/Tectonic map of the linear structures
on the Territory of China*   1:6 000 000
Beijing: Chinese Academy of Sciences, 1981.
Chinese text, with English language legend.
With accompanying Chinese explanatory text.

*The marine and continental tectonic map of China and its
environs*   1:5 000 000
Beijing: Science Press, 1983.

Zhongguo renmin Gongheguo dizhi tu   1:4 000 000
Beijing: CPH, 1976.
Chinese language geological map.
English language legend available from Department of Geography
and Geology, University of Hong Kong.

Zhongguo dadi gouzao tu/Tectonic map of China
1:4 000 000   Huang Jiqing
Beijing: CPH, 1979.
Chinese language with English legend sheet.

Zhongguo dadi gouzao tu   1:4 000 000
Beijing: CPH, 1977.
Chinese language tectonic systems map.

Zhonghua renmin gongheguo dizhen gouzao tu   1:4 000 000
Beijing: CPH, 1979.
Chinese language seismotectonic map.
With accompanying Chinese text and English language legend and
summary.

Zhongguo quiang dizhen zhenzhong fenbu tu   1:4 000 000
Beijing: CPH, 1976.
Chinese language epicentral distribution map of strong
earthquakes.
Also available at 1:6 000 000 and 1:8 000 000 scale.

*The marine and continental tectonic map of China and its
environs*   1:5 000 000
Beijing: Chinese Academy of Sciences, 1983.
6 sheets, all published.

*Tectonic systems map of the People's Republic of China and adjacent
sea areas*   1:2 500 000
Beijing: Institute of Geomechanics, Chinese Academy of
Geological Sciences, 1984.
9 sheets, all published.
Chinese language with Pinyin transcriptions.
With accompanying explanatory notes in English and Chinese.

Environmental

Zhongua renmin gongheguo qihou tuji
Beijing: CPH, 1979.
pp. 226.
Chinese language climatic atlas.

Zhongguo nian jiangshuiliang   1:6 000 000
Beijing: CPH, 1976.
Chinese language map of annual precipitation.

Zhongguo tudi liyong xianzhuang gaitu   1:6 000 000
Beijing: CPH, 1982.
Chinese language land use map.

Zhonghua renmin gongheguo zhibeitu   1:6 000 000
Beijing: CPH, 1979.
Chinese language vegetation map.
With accompanying Chinese text and Pinyin script index.

Zhonghua renmin gongheguo turang tu   1:4 000 000
Beijing: CPH, 1978.
Chinese language soil map, with English summary and legend.
With accompanying Chinese text.

Administrative

Zhonghua renmin gongheguo fen sheng dituji (hanyu pinyinban)
Beijing: CPH, 1977.
pp. 170.
Pinyin edition of Provincial atlas of China.

Map of the People's Republic of China   1:9 000 000
Beijing: CPH, 1981.
Political map with placename index on back.
Also available as Chinese character and Esperanto editions.

Map of the People's Republic of China   1:6 000 000
Beijing: CPH, 1981.
Political map.
Also available as Chinese, Japanese and Pinyin editions.

China and Mongolia   1:6 000 000
Edinburgh: Bartholomew, 1985.
Political shading of provinces, with placename index on verso.

China   1:5 000 000
Washington, DC: CIA, 1981.
Administrative and population map.
With accompanying Pinyin and Character gazetteer.

Kitai   1:5 000 000
Moskva: GUGK.
In Russian.

Sheng ditu   various scales
Beijing: CPH, 1975–.
31 sheets, all published.
Chinese language provincial maps.

Human and economic

China current data maps   1:9 000 000
Hong Kong: Asian Research Services, 1984–.
11 sheets, all published.
Thematic series: covers human resources, hydrocarbon potential,
water power development, coal mining industry, railway network,
growth of industry, agricultural economy, urban development,
coastal cities (2 sheets), special economic zones.
With accompanying current data report on human resources.

China: railway map   1:8 000 000   2nd edn
Exeter: Quail, 1985.

Peoples of China   1:7 150 000
Washington, DC: NGS, 1980.
With 1:6 000 000 scale map on back.

China oil and gas map   1:6 000 000
Kowloon: Petroleum News South East Asia, 1985.

Zhonghua renmin gongheguo minzu fenbu luetu   1:4 000 000
Beijing: CPH, 1981.
Chinese language ethnographic map.

Town maps

Tourist town plans of China   various scales
Beijing: CPH, 1978–.
20 sheets, all published.
English and Chinese language editions available for 20 major
cities.

Map of China with city maps of Peking, Canton, Shanghai various scales
Hamburg: Falk.
Folded plan includes 1:4 500 000 scale physical map of China, indexes, and tourist information and town maps.

Beijing 1:15 000
Washington, DC: CIA, 1982.
6 sheets, all published.
Includes land use and population information.
Also available as single-sheet indexed 1:35 000 scale plan.

Shanghai 1:12 500
Washington, DC: CIA, 1982.
Includes land use and population information.
Index on back.

Canton 1:12 500
Washington, DC: CIA, 1984.
Includes land use and population information.
Index on back.

Cyprus

The **Department of Lands and Surveys (DLS)** is responsible for the topographic and cadastral surveying of the island of Cyprus. From 1969 DLS has had the capability to produce its own mapping using photogrammetric methods. Current series use UTM projection and a kilometre grid. The basic scale series is the *National topographic map series* 1:5000 DLS 17. This was begun in 1969 and is being produced with the assistance of the British **Directorate of Overseas Surveys (DOS)** (now Overseas Surveys Directorate, Southampton, OSD). Complete coverage would require 1293 sheets but it is unlikely that the map will ever cover the northern areas of the island controlled by Turkey. The four-colour sheets show relief by 2 m contour intervals. This map is under continuous revision. Other partial coverage of the southern areas is available in two DOS series dating from the 1960s: DOS 255 (K818) a 1:10 000 contoured series, and in the south 1:25 000 DOS 355 (K8110). For complete coverage of the island it is necessary now to use the four-sheet 1:100 000 scale DLS 18. This map is photographically derived from a 1:50 000 series no longer available. It shows relief with 100 m interval contours: English and Greek editions are available. The nine-colour 1:250 000 administration and road map shows relief by layer tints and depicts the ethnic composition of population distribution. This is published in English, Greek and Turkish language editions. A tourist edition of this map is available with town plans, and tourist information on the verso. DLS also issues a 1:500 000 scale map of the island, and a wide variety of large-scale street names maps and cadastral maps are available as sunprints.

Thematic mapping activities are carried out by other government bodies, though DLS issues some of their mapping. The **Geological Survey Department** was founded in 1950 and has produced a variety of earth science mapping, mostly with bilateral aid from British or German agencies. 1:250 000 scale maps are published to show geology, mineral resources and hydrogeology, and larger-scale coverage exists for regions of geological or mineralogical interest. The British DOS produced eight sheets of 2 inch to the mile scale geological coverage in 1959–60, which is still available. Current programmes include 1:10 000 scale engineering geology maps, a 1:25 000 scale groundwater map of Nicosia, and 1:100 000 scale groundwater quality maps of the Nicosia area produced with the technical cooperation of the German **Bundesanstalt für Geowissenschaften und Rohstoffe (BfGR)**.

The **Department of Agriculture** carries out soil survey and land use mapping activities. Small-scale single-sheet coverage of the island is available to show state forests, soils, land use and precipitation. Some larger-scale series have been published prior to the Turkish invasion, such as the soil map and the land capability series both published at 1:25 000 on the DOS 355 sheet lines. Neither of these maps gives more than partial coverage; some new sheets were published in the 1980s but there is little on programme at present.

Several commercial concerns produce tourist mapping of the islands, including **Clyde Surveys**, **Geoprojects**, **Hallwag**, **Karto+Grafik** and **Freytag Berndt Artaria**.

Further information

Cyprus (1977, 1980, 1983) Surveying and mapping of Cyprus: papers presented to the *United Nations Regional Cartographic Conferences on Asia and the Pacific*, New York: UN.

For historical background see Stylianou A. and Stylianou, J.A. (1980) *The History of the Cartography of Cyprus*, Nicosia: Cyprus Research Centre.

DLS and the Geological Survey Department both issue catalogues and annual reports.

Addresses

Department of Lands and Surveys (DLS)
Ministry of the Interior, PO Box 5598, Archbishop Makarios III Avenue, NICOSIA.

Geological Survey Department
Ministry of Agriculture and Natural Resources, NICOSIA.

Department of Agriculture
Ministry of Agriculture and Natural Resources, PO Box 809, NICOSIA.

For Clyde Surveys, Geoprojects and DOS see Great Britain; for Freytag Berndt Artaria see Austria; for Karto+Grafik see German Federal Republic; for Hallwag see Switzerland.

Catalogue

Atlases and gazetteers

A concise gazetteer of Cyprus
Nicosia: Cyprus Permanent Committee for the Standardization of Geographical Names, 1982.
pp. 63.
Greek and English placename information.

General

General use map of Cyprus 1:500 000 DLS 20
Nicosia: DLS, 1976.

Cyprus 1:316 800
Maidenhead: Clyde Surveys, 1983.
With town maps of Nicosia, Limassol, Larnaca and Paphos and a
map of the Troodos mountains.
Text in English, French and German.

Cyprus touring map 1:250 000
Nicosia: DLS, 1981.
Double-sided.
Town plans and tourist information on verso.

Geoprojects Cyprus 1:250 000
Beirut: Geoprojects, 1982.
Double-sided, with city maps of Nicosia 1:7500, Limassol and
Larnaca 1:15 000 and Paphos 1:20 000, and indexes.

Topographic

Topographical map 1:100 000 DLS 18
Nicosia: DLS, 1984.
4 sheets, all published.
Also available as Greek edition (DLS 6).

Topographical map 1:25 000 DOS 355 (K8110)
Tolworth: DOS, 1960–66.
59 sheets, 21 published. ■
Only available for south-west and south-east areas.

Geological and geophysical

Geological map 1:250 000
Nicosia: Geological Survey Department, 1979.

Hydrogeological map of Cyprus 1:250 000
Nicosia: Geological Survey Department, 1970.

Mineral resources map of Cyprus 1:250 000
Nicosia: Geological Survey Department, 1982.

Environmental

Cyprus: state forests map 1:250 000 DLS 4
Nicosia: DLS, 1984.

Land use map of Cyprus 1:250 000
Nicosia: Department of Agriculture, 1975.

Cyprus: average annual precipitation map (1951–80) 1:250 000
Nicosia: Department of Agriculture, 1983.

Cyprus: general soil map 1:200 000
Nicosia: Department of Agriculture, 1970.

Cyprus soil series 1:25 000
Nicosia: Department of Agriculture, 1965–.
59 sheets, 8 published. ■

*Cyprus soil series: land suitability classification for irrigation
purposes* 1:25 000
Nicosia: Department of Agriculture, 1968–.
59 sheets, 8 published. ■

Administrative

Cyprus: administration and road map 1:250 000 DLS 14–6
Nicosia: DLS, 1984.
Available flat or folded, as English, Greek or Turkish editions.

Human and economic

*Cyprus distribution of population by ethnic group 1960 and positions of
the invading Turkish Forces* 1:250 000
Nicosia: DLS, 1976.

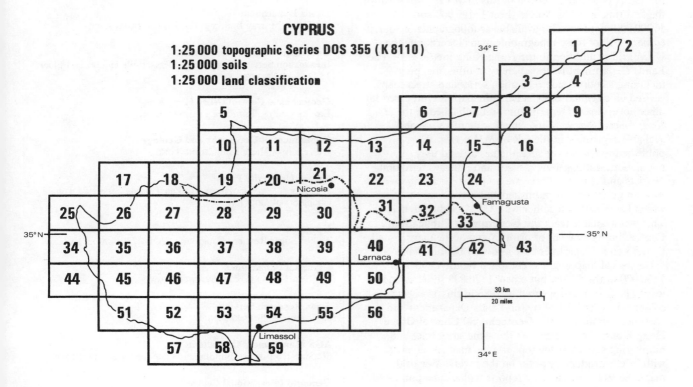

Hong Kong

Hong Kong is probably the best mapped territory in Asia. Responsibility for cartographic activities falls to the **Lands Department**, which has since 1971 been publishing and compiling its own topographic coverage of Hong Kong Island and the New Territories. The first modern basic surveys were British military maps dating from the 1920s and 1930s (one of the earliest examples of a photogrammetrically derived series). GSGS series provided medium-scale mapping until the 1960s. Early large-scale basic plans, compiled by plane tabling, were in progress until the early 1960s at 1:600 or 1:1200 scale. Aerial photography was used to produce new series at these scales in the 1960s and in 1965 the British **Directorate of Overseas Surveys (DOS)** (now Overseas Surveys Directorate, Southampton, OSD) agreed to produce 1:10 000 and 1:25 000 series, also using imperial measures. All these series are now either obsolete or being replaced. The decision to metricate mapping was taken in 1973 and an overall grid system and Gauss conformal projection for all Hong Kong's mapping were introduced. The basic scale is now the 1:1000 plan, which will cover the whole territory in 3000 sheets: only about 500 sheets, mainly in country park areas, remain to be converted to the new metric specifications. From this map are derived medium-scale plans at 1:2500, 1:5000, 1:10 000 and 1:15 000. Topographic map series are issued at 1:20 000 (HM20C); a multi-colour, 16-sheet series, with 20 m contours completed in 1977 and under continuous revision. There is also a two-sheet multi-colour bilingual 1:50 000 series with relief by hill shading and elevation tinting. A two-colour version of this map was published for the first time in 1986. Single-sheet 1:100 000 and 1:200 000 maps are also available as topographic or tourist editions. All of these topographic maps incorporate the UTM grid in addition to the Hong Kong metric grid. The Lands Department issues a variety of other mapping, including a countryside recreation series and street maps, revised on a regular basis. Thematic maps are prepared for other government Departments and published in the *Hong Kong Annual Report*. These small-scale maps are also available separately in the AR or MM series: sheets published to date cover such themes as land use, population, trade, geology, roads and railways, electoral districts and climate. There are no plans for a National Atlas: a gazetteer was published and revised up to 1978, which listed names together with UTM grid reference. This was available from the **Government Publications Centre**, but is now out of print. Many of the maps issued by the Lands Department are also sold by this Centre.

Geological mapping was compiled by the DOS at 1:50 000 in the 1970s, but a new 1:20 000 series is under way. The first map appeared in 1986 in this 15-sheet full-colour series, published by the Lands Department and compiled jointly with the **Geotechnical Control Office Hong Kong**. This series uses the same sheet lines and numbering system as the topographic map and is printed with brief introductory notes on the reverse side of the maps. Separate explanatory reports will also be published. Completion of this series is expected to take several years.

Other maps are published by the commercial sector, including school atlases by **OUP** and **Longman**, and general tourist maps and town plans by a number of firms including **Universal Publications**. Universal also produces town and tourist mapping of some neighbouring areas of the People's Republic. Hong Kong is also the location for several commercial agencies specializing in the compilation of mapping of mainland China. These include **Asian Research Services**. Other Hong Kong agencies distribute People's Republic mapping, including **AGS Management Consultants** and **Geocarto International Centre**. **Petroleum News SE Asia** in Kowloon produces hydrocarbon mapping of the region.

Further information

A useful review of the development of Hong Kong map series is provided by leaflets issued by the Lands Department and revised on a regular basis, e.g. Lands Department (1986) *Hong Kong: the Facts Mapping*, Hong Kong: Government Information Services.

Other published accounts are Lands Department (1979) Mapping in Hong Kong, *New Zealand Cartographic Journal*, **8**(2), 27–32; and Lands Department (1983) Report on survey and mapping activities in Hong Kong 1980–3, paper presented to the *10th UN Regional Cartographic Conference on Asia and the Pacific*, New York: UN.

Addresses

Lands Department
5th Floor, Murray Building, Garden Rd, HONG KONG.

Government Publications Centre
Information Services Department, Baskerville House, 2nd Floor, 13 Duddell St, HONG KONG.

Geotechnical Control Office Hong Kong
Empire Centre, Tsim Sha Tsui, KOWLOON.

Department of Geography and Geology
University of Hong Kong, HONG KONG.

Oxford University Press
18th Floor, Warwick House, Taikoo Trading Estate, 28 Tong Chong St, QUARRY BAY.

Longman
Taikoo Sugar Refinery Compound, POB 223, QUARRY BAY.

Universal Publications
709 New St, PO Box 78669, SHEUNNGWEN.

Asian Research Services
GPO Box 2232, HONG KONG.

AGS Management Consultants
5D Sing-Ho Finance Building, 168 Gloucester Rd, HONG KONG.

Geocarto International Centre
GPO Box 4122, HONG KONG.

Petroleum News SE Asia Ltd
10th Floor, 146 Prince Edward Rd, KOWLOON.

Catalogue

Atlases and gazetteers

Hong Kong and Macao: official standard names approved by the United States Board on Geographic Names
Washington, DC: DMA, 1972.
pp. 52.

General

Hong Kong, Kowloon and the New Territories 1:200 000 HM 200 CL 10th edn
Hong Kong: Lands Department, 1986.
In English and Chinese.

Hong Kong, Kowloon and the New Territories 1:100 000 HM 100 CL 6th edn
Hong Kong: Lands Department, 1985.
In English and Chinese.

Hong Kong: official guide map 1:30 000 7th edn
Hong Kong: Lands Department, 1984.
In English and Chinese.

Topographic

Hong Kong, Kowloon and the New Territories 1:50 000 HM 50 CL 6th edn
Hong Kong: Lands Department, 1984.
2 sheets, both published.
In English and Chinese.

Hong Kong 1:20 000 HM 20C
Hong Kong: Lands Department, 1974–.
16 sheets, all published. ■
In English and Chinese.

Geological and geophysical

Geological map of Hong Kong, Kowloon and the New Territories 1:50 000 DOS 1184 A
Hong Kong: Lands Department, 1972.
2 sheets, both published.
In English and Chinese.
With accompanying explanatory text.

Geological map of Hong Kong 1:20 000
Hong Kong: Lands Department, 1986–.
15 sheets, 1 published. ■
In English and Chinese.
With accompanying explanatory texts.

HONG KONG
1:20 000 topographic Series HP20C
1:20 000 geological

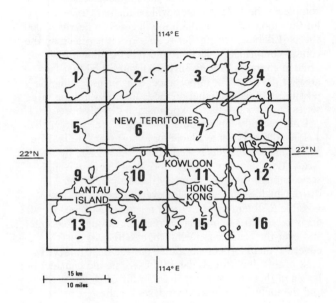

Town maps

Hong Kong: streets and places
Hong Kong: Lands Department, 1978–.
2 vols.
Maps of non-urban areas at 1:20 000 and 1:40 000.

Hong Kong guide maps various scales
Kowloon: Universal, 1982.
Atlas, gazetteer and timetables.

Hong Kong: street maps various scales
Kowloon: Universal.
5 published sheets.

India

The **Survey of India (SI)** is the oldest scientific Department in the Government of India; its history can be traced back to the appointment of James Rennell as Surveyor General of Bengal in 1767. Survey departments were set up in Bengal, Bombay and Madras by the first decade of the nineteenth century and the Survey of India grew from the Great Trigonometrical Survey, established in 1818. Throughout the nineteenth century complete quarter-inch coverage of the subcontinent was attained in the monochromatic Indian Atlas series, and revenue and cadastral mapping was carried out. No one subcontinental standard system of topographic survey was established until 1905, when the 1-inch base scale and quarter-degree sheet line system, with full-colour printing and contoured

relief, was standardized throughout the Indian Empire. A quarter-inch series was to be the basic scale for remote areas.

After independence SI became responsible for the mapping of India. By then 3355 1-inch sheets had appeared and 266 maps at quarter-inch scale had been published as a result of post-1905 surveys. The metric system was introduced from 1956, and the 1 inch map became the 1:50 000 series, on the same sheet lines. This series is still current and remains the basic scale for much of the country. Complete coverage of the country with 5084 metric sheets was attained in 1979. Relief is shown by 5, 10 or 20 m contours, depending on the nature of the terrain. In cartographic style these sheets still resemble the remarkable early Survey of India topographic sheets, though they are now produced photogrammetrically and are regularly revised. The 1:250 000 series covers India in 394 sheets and uses a 100 or 200 m contour interval. The 1:25 000 series has been in progress since 1956, but substantial numbers of new sheets have only begun to appear since the completion of the 1:50 000 series. When completed the 1:25 000 map will be the basic scale series for the whole country, which it will cover in nearly 20 000 sheets. Like the two smaller-scale series it is drawn on the conical conformal projection, and relief is shown by 5, 10 or 20 m contours. Sheets have been published first for areas of development importance. SI is investigating the possibility of using digital methods to produce these three series, and has produced experimental digital sheets.

Until recently topographic coverage at scales larger than 1:1 000 000 was not available for export from India. The 1:250 000 scale sheets issued in series AMS 502 by the US Army Map Service were therefore then the best available scale, though these were simply based upon earlier quarter-inch SI mapping. However it recently became possible to obtain maps in the three major SI series for all areas except coastal or frontier regions, through Geocenter and over the counter at Delhi, though SI does not advertise them as being available outside the country. In view of this doubtful availability, users should consider our listing of these topographic maps in the catalogue section, and the inclusion of graphic indexes for these series, with caution.

Amongst the other maps produced by SI are small-scale road and tourist maps of the country, 1:1 000 000 scale state maps, and the Indian sheets in the International Map of the World series and the World Aeronautical Chart Series. The eight-colour 1:2 500 000 scale road map is revised every 2 years. SI also produces large-scale urban mapping, including town guide maps of all the major urban centres, and large-scale mapping for special projects. 1:15 000 scale forest maps are produced for the State Forest Departments and cartographic and geodetic backup is provided to other government and state agencies.

Geological mapping of India is carried out by the **Geological Survey of India (GSI)**. A 1:250 000/1:253 440 scale geological quadrangle series will cover India in 394 sheets and use five colours. This series uses the same sheet lines as SI's 1:250 000 map and generalizes the base topographic information from this source. A geological atlas of India at 1:1 000 000 scale has been in progress since 1976. Mineralogical and geological maps of the states have been published at 1:1 000 000 and 1:2 500 000 scales.

The **National Geophysical Research Institute (NGRI)** has produced small-scale gravity surveys of the country.

The **Indian National Atlas and Thematic Mapping Organization (NATMO)**, founded in 1956, is the main central agency responsible for the compilation of small scale thematic mapping of the country. A preliminary Hindi edition of the *National Atlas* comprising 26 sheets was issued in 1957, and is now out of print. Since then, work has been progressing on the vast full edition. This eight-volume work was completed in 1982, and comprises 274 maps on 220 sheets, covering a complete thematic range in scales between 1:1 000 000 and 1:6 000 000. Some maps cover states, others are on regular sheet lines. Individual sheets were published when they were completed and several interim bound cumulations of the work were issued. Single sheets may still be purchased and are the best available thematic mapping of the country, though some sheets, especially those covering population and economic data, are now somewhat dated. Much of the emphasis of the National Atlas is upon agriculture;

industrial coverage is more patchy. Other single-theme atlases have been published, covering such topics as irrigation, tourism, water resources, agriculture and forest resources.

The **Office of the Registrar General and Census Commissioner** carries out census mapping activities. Since 1961 mapping of the results of the decennial censuses of India has been undertaken, but not until the 1971 census did a well developed atlas and map publication system evolve. Prior to each census over 1.2 million maps of enumeration areas are prepared, on a standard size, with English/Hindi legends. These national maps are not published but are used to prepare jurisdictional, urban schematic and SUA maps to assist in the huge data collection phase of the census. These maps are essentially administrative and cover districts, taluks and urban data collection units in about 4800 maps for the 1981 census, at scales of 1:250 000 and larger.

Many of the results of the census are presented in cartographic form, as smaller-scale, single-colour, demographic and socioeconomic maps, published in census atlases. These atlases are being issued for the 1981 census for each State and Union Territory, with data presented at levels down to the taluk and tahsil level in maps at 1:1 000 000 and 1:2 500 000, and as a national volume with 1:5 000 000 mapping down to district level. For the first time the 1981 national volume is also to contain larger-scale regional summary mapping.

Other useful publications of the Census include very detailed gazetteers of each of the States, and a national volume.

Other organizations at national level in the public domain producing mapping include the **Town and Country Planning Organization**, the **India Meteorological Department** (climatological and rainfall atlases), the **Department of Agriculture** (agricultural atlases), the **Central Board on Irrigation and Power** (irrigation maps) and the **All India Soil and Land Use Survey**. Hydrographic and bathymetric charting is carried out by the **Naval Hydrographic Office** (including major contributions to GEBCO), whilst remote sensing and the production of satellite image maps is the responsibility of the **National Remote Sensing Agency**. This organization has also been involved in providing technical backup to SI and in the production of small-scale land use maps for some of the states. A series of vegetation maps is published on the sheet lines of the International Map of the World by the **Institut Français de Pondichéry (IFP)** to give coverage of much of the Indian peninsula. IFP has also begun the 1:250 000 scale forest mapping of South India: this series will give 15-sheet coverage of areas to the south of 16° N and is compiled in collaboration with the different state governments. 1:250 000 scale forest mapping derived from Landsat imagery has also been published for two areas of south India. Other areas of south-east and south Asia have also been mapped by IFP, in conjunction with its associated vegetation survey organizations in France and Indonesia. All these vegetation maps may be obtained from the French **Institut de la Carte Internationale du Tapis Végétal (ICITV)**.

Since 1905 the different states have had separate survey organizations responsible for revenue and cadastral mapping activities, but unlike other large federal countries topographic and thematic mapping activities in India are largely centralized. Significant exceptions to this are the excellent regional agricultural, resources and planning atlases issued by many states in collaboration with University Geography Departments in the 1970s and early 1980s, e.g. the excellent *Resources atlas of Kerala*, published in 1984 by the **Trivundrum Centre for Earth Science Studies**.

Commercial map producers publish a variety of cheap tourist maps, mostly either administrative layer coloured maps of the different states or town plans. These include **TT Maps & Publications Private Ltd (TT)**, formerly Tamilnad Printers and Traders, whose range of products closely mirror the state and town maps produced by SI. TT maps are readily available in the west, and most are published as folded guide books and include indexes, but the quality of these maps is rather poor.

Amongst other commercial mapping of India are sheets published by **Bartholomew**, **Karto+Grafik** and **Nelles Verlag**. The American **Central Intelligence Agency (CIA)** and the Russian **Glavnoe Upravlenie Geodezii i Kartografii (GUGK)** both also publish small-scale coverage of the country.

The other distinguishing feature of Indian map production is the very long history of detailed 'gazetteer' production. These gazetteers are descriptions of places in an area or district, rather than simply placename finding tools. Many are distributed by **DK Agencies**.

Further information

Catalogues are issued by SI, GSI, National Atlas Organization and the Census. These are the best source of information about the four major map-producing agencies in India.

Information about the general current state of the art in India is contained in Agarwal, G.C. (1984) *National report on Cartography in India 1980–84*, presented at the 12th international Conference of the ICA, Perth, August 1984, Perth: ICA; and in various papers in the *7th United Nations Regional Cartographic Conference on Asia and the Pacific* (1977) New York: UN.

Historical background information is contained in Cook, A. (1979) Maps, chapter in *South Asian Bibliography: A Handbook and Guide*, edited by J.W. Pearson, London: South Asia Library Group.

Other background articles may be found in the journal *Indian Cartographer*, Hyderabad: Indian National Cartographic Association, 1979–.

Two other sources on the National Atlas and Census are: Dutt, A.K. (1984) The National Atlas of India, *Geographical Review*, **74**(1), 94–100; and Roy, B.K. (1982) Indian Census cartography, *Geojournal*, **6**(3), 213–224.

Addresses

Survey of India (SI)
Surveyor General's Office, Hathibarkala Estate, PO Box 37, DEHRA DUN 248001.

Geological Survey of India (GSI)
27 Jawaharlal Nehru Rd, CALCUTTA 700016.

National Geophysical Research Institute (NGRI)
Uppal Rd, HYDERABAD 500007.

National Atlas and Thematic Mapping Organization (NATMO)
49 Gariahat Rd, CALCUTTA 700019.

Office of the Registrar General and Census Commissioner of India
NEW DELHI 110022.

Town and Country Planning Organization
c/o Ministry of Health, Family Planning, Works, Housing and Urban Development, NEW DELHI.

India Meteorological Department
Lodi Rd, NEW DELHI 110011.

Department of Agriculture
NEW DELHI.

All India Soil and Land Use Survey
NEW DELHI.

Central Board on Irrigation and Power
Kasturba Gandhi Marg, NEW DELHI 110001.

Naval Hydrographic Office
Post Box No. 75, DEHRA DUN 1, Uttar Pradesh.

National Remote Sensing Agency
Department of Science and Technology, No. 4 Sardar Patel Rd, PO Box 1519, SECUNDERABAD 500003, Andhra Pradesh.

Institut Français de Pondichéry (IFP)
PB Box No 33, PONDICHERY 605001.

Trivundrum Centre for Earth Science Studies
PB 2235, TRIVUNDRUM 695010.

TT Maps & Publications Private Ltd (TT)
328 GST Rd, Chromepet, MADRAS 600044.

DK Agencies
H-12 Bali Nagar, NEW DELHI 110015.

For Bartholomew see Great Britain; for Karto+Grafik and Nelles Verlag see German Federal Republic; for GUGK see USSR; for CIA and DMA see United States; for ICITV see France.

Catalogue

Atlases and gazetteers

National Atlas of India
Calcutta: NATMO, 1982.
8 vols.
Individual sheets also available separately, or as thematic collections.

India: official standard names approved by the United States Board on Geographic Names
Washington, DC: DMA, 1952.
2 vols.

Alphabetical list of towns in Indian Census 1981
New Delhi: Office of the Registrar General and Census Commissioner of India, 1983.
pp. 139.

General

India's roads 1:5 300 000
Madras: TT, 1985.
Also available as railway edition.

Physical map of India 1:4 500 000
Dehra Dun: SI, 1974.
Available as either English or Hindi language edition.

India: Hildebrands Urlaubskarte 1:4 255 000
Frankfurt AM: Karto+Grafik, 1984.

Indian subcontinent 1:4 000 000
Edinburgh: Bartholomew, 1984.
Includes five city area inset maps at 1:200 000.

Railway map of India 1:3 500 000
Dehra Dun: SI, 1982.
Available as either English or Hindi language edition.

Road map of India 1:3 500 000
Dehra Dun: SI, 1986.
2 sheets, both published.

India and adjacent countries 1:2 500 000
Dehra Dun: SI, 1979.
4 sheets, all published.
Also available as Hindi language version, and as uncoloured version.

Road map of India 1:2 500 000
Dehra Dun: SI, 1977.
2 sheets, both published.

General maps of India 1:2 000 000
Calcutta: NATMO, 1965–74.
5 sheets, all published.
Extracts from the National Atlas.

India 1:2 000 000
Madras: TT, 1983.
4 sheets, all published.

APA maps of India 1:1 500 000
München: Nelles Verlag, 1985.
5 sheets, all published.

India 1:1 000 000
Calcutta: NATMO, 1960–81.
16 sheets, all published.
Physical maps.

International Map of the World 1:1 000 000
Dehra Dun: SI.
21 sheets, all published.

Topographic

India 1:250 000
Dehra Dun: SI, 1970–.
394 sheets, all published. ■
Not available for coastal and border areas.

India 1:50 000
Dehra Dun: SI, 1956–.
5084 sheets, all published. ■
Not available for border and coastal areas.

Geological and geophysical

Coastal landforms of India 1:5 000 000
Calcutta: GSI, 1972.

Gravity map series of India 1:5 000 000
Hyderabad: NGRI, 1974–78.
5 sheets, all published.
Thematic set with 1 accompanying text per theme.

Mineralogical maps of India 1:5 000 000
Calcutta: GSI, 1972–.
4 sheets, all published.
Topics covered in set include bauxite deposits, raw materials for
fertiliser industry, base metal deposits, coal and lignite fields.

Geological and mineral maps of the States 1:2 250 000
Calcutta: GSI, 1969–.
15 sheets, all published.
Some sheets with accompanying explanatory texts.

Geological map of India 1:2 000 000
Calcutta: GSI, 1962.
4 sheets, all published.

Metallogenic-minerogenetic map of India 1:2 000 000
Calcutta: GSI, 1963.
4 sheets, all published.

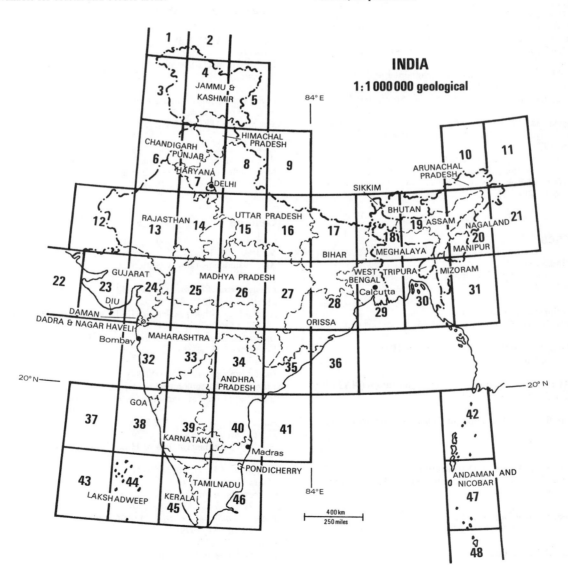

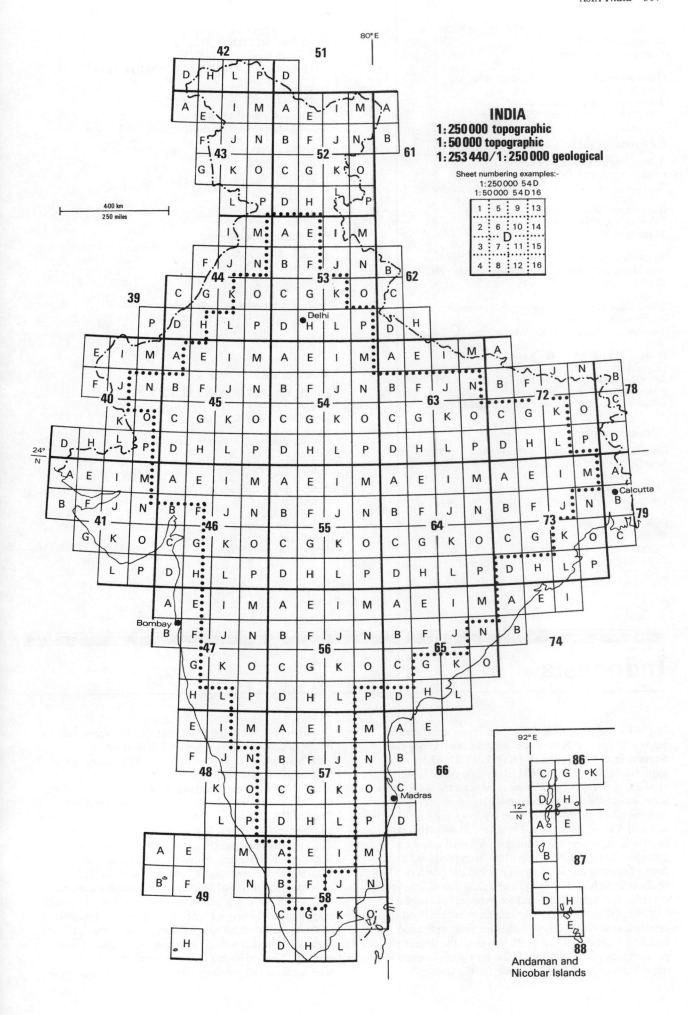

INDIA
1:250 000 topographic
1:50 000 topographic
1:253 440/1:250 000 geological

Sheet numbering examples:-
1:250 000 54 D
1:50 000 54 D 16

Andaman and
Nicobar Islands

Geological and mineral atlas of India   1 : 1 000 000
Calcutta: GSI, 1976–.
48 sheets, 20 published. ■

Geological quadrangle map of India   1 : 250 000
Calcutta: GSI, 1977–.
394 sheets, *ca* 30 published. ■
Some sheets published at 1 : 253 440 scale.

Environmental

Atlas of Forest resources of India
Calcutta: NATMO, 1976.
36 plates.

Atlas of water resources of India
Calcutta: NATMO, 1986.
25 plates.

Agroclimatic atlas of India
New Delhi: India Meteorological Department, 1978.
91 plates.

*International map of vegetation and environmental
conditions*   1 : 1 000 000
Toulouse: Institut de la Carte Internationale du Tapis Végétal,
1961–.
21 sheets, 13 published. ■
Legend and title in English and French.
Accompanying thematic maps on each sheet at 1 : 5 000 000 cover
administrative divisions and hypsometry, geology and lithology,
soils, bioclimates, vegetation types and agriculture.

Administrative

Census of India 1981: administrative atlases
New Delhi: Office of the Registrar General and Census
Commissioner of India, 1982–.
21 vols, all published.
1 national volume, the rest to cover states and union territories.

India   1 : 4 100 000
Madras: TT, 1986.
Political map.

Political map of India   1 : 4 000 000
Dehra Dun: SI, 1983.
Title also in Hindi.
Also available as 1 : 4 500 000 Hindi language edition.

State maps of India   1 : 1 000 000
Dehra Dun: SI, 1971–.
14 sheets, all published. ■
Some states covered in English, some in Hindi language editions.

State map guidebooks   various scales
Madras: TT, 1986.
16 published sheets.

Human and economic

Agricultural atlas of India
Calcutta: NATMO, 1981.
pp. 36.

India: administration, population and communication   1 : 3 500 000
Madras: TT, 1984.

Town maps

The city atlas of India   S. Muthia and R.P. Arya
Madras: TT, 1981.
60 city maps.

City map guidebooks   various scales
Madras: TT, 1986.
13 published sheets.

Guide maps   scales vary
Dehra Dun: SI, 1969–.
15 published sheets.
Series of town plans covering major cities mostly at 1 : 10 000 or
1 : 20 000 scales.
Street indexes on sheets.

Indonesia

In 1969 at the beginning of the first 5 year development
plan or 'Pelita' the **National Coordination Agency for
Surveying and Mapping (BAKOSURTANAL)** was
founded. BAKOSURTANAL is now responsible for the
production of topographic series, for geodetic work and for
aeronautical charting of Indonesia. It also acts as the
national advisory and coordinating agency for all surveying
and mapping activities in the country. Prior to that date
there was no systematic Indonesian national mapping
programme: topographic maps had been produced by the
Army Topographic Service, on a polyconic projection with
series at 1 : 50 000, 1 : 100 000 and 1 : 250 000 scales, but
coverage was very patchy and for many areas colonial
Dutch maps or wartime American surveys remained the
best available scales. There had been little systematic
thematic mapping. Since 1969, however, the development
programmes of the different Pelita have each stressed the
importance of a modern map base for the country's

development. There has been considerable investment in
mapping projects and numerous bilateral aid
arrangements. All this now makes Indonesia one of the
best mapped third world countries.

The National Base Mapping Programme is carried out
by BAKOSURTANAL and the **Army Topographic
Service (JATOP)**. It is seen as essential to provide a base
for resource evaluation as part of the development process.
The Programme comprises the 1 : 50 000 base mapping of
Sumatera, Kalimantan, Sulawesi and Maluku; the
1 : 100 000 base mapping of Irian Jaya and the 1 : 25 000
base mapping of Jawa, Bali and Nusa Tenggara islands. A
new specification was drawn up in 1969 for the 1 : 50 000
map, which now has a UTM grid, shows relief with 25 m
contours, is issued on quarter-degree sheet lines and
requires 3198 sheets to give complete coverage of the
country. The 1 : 100 000 series conforms to similar
specifications and requires 800 sheets to cover the whole

country. The first sheets in the new programme were issued in 1976 and following rescheduling it is expected that photogrammetric compilation will be complete by 1993. To meet this target, over 355 new sheets will need to be published every year. The maps are derived from 1:100 000 high-altitude aerial cover flown under a Canadian aid agreement and more recently by Indonesian state and private sector contractors. Aerial coverage for some areas has been flown at larger scales. Maps are published in two stages: first as provisional editions, with no field completion, as either enhanced orthophotomaps, or as three-colour line maps; the second stage is the publication of full-colour field checked maps.

In addition to the basic scales BAKOSURTANAL also compiles smaller-scale derived series. The 30 1:1 000 000 sheets covering Indonesia in the International Map of the World series have been revised to 1983, and a 1:500 000 series is in progress which will be completed in 85 sheets. 1:250 000 scale mapping on a UTM grid will cover the country in 140 sheets and information from this series is being digitized to create a digital base for resources information. A geo-referenced Resource Information System is now installed, though there has been little direct application of digital techniques to map production itself, apart from research work and tests using Landsat imagery for revision purposes for the smaller-scale series. BAKOSURTANAL has been involved in the production of small-scale resource mapping. An *Atlas of Indonesian Resources* has been produced which presents national data at 1:7 500 000 scale for 16 themes. Individual islands are mapped in this atlas at larger scales, depending on the size and importance of the island, ranging from Bali at 1:250 000 to Sumatra at 1:2 500 000. Resource mapping at larger scales has been compiled by BAKOSURTANAL, using satellite data as the prime source material. An atlas of indexes is also published.

Larger-scale maps for development projects are also compiled. For instance the vast transmigration schemes involving the resettlement of 500 000 families from Java to other islands has meant that 1:5000, 1:10 000 and 1:20 000 scale plans of potential settlement sites have been produced. 1:1000 line or orthophotomaps of the 100 major cities are to be produced for urban planning purposes, with assistance from the World Bank. The capital Jakarta is mapped at 1:5000 and 1:1000 by a separate survey authority, the **Jakarta Mapping Center**. Agricultural projects have required 1:5000 or 1:1000 scale mapping, especially of irrigation schemes.

None of these maps produced by BAKOSURTANAL is at present available outside of Indonesia, but BAKOSURTANAL has recently advertised topographic products in a brief priced listing, so availability may be changing.

Geological mapping of Indonesia is carried out by the **Direktorat Geologi (DGI)** in the Ministry of Mines and Energy. Prior to 1969 very little geological mapping was carried out, but the first 5 years of the first Pelita saw **United States Geological Survey (USGS)** involvement in the setting up of an active geological mapping programme. Small-scale mapping of the whole country at 1:5 000 000 scale was compiled by USGS for several earth science themes, and larger-scale systematic programmes begun. Java and Madura are to be covered at 1:100 000 scale in 57 full-colour half-degree sheets. The rest of the islands are to be covered in about 100 1:250 000 scale sheets, each covering 1° 30′ by 1°. A 16-sheet 1:1 000 000 scale series was also begun on an IMW base. Other geological mapping activities carried out by DGI include the compilation of single-colour gravity maps at 1:100 000, 1:250 000 and 1:1 000 000 scales, and a single-sheet 1:2 000 000 scale gravity sheet. In the last decade DGI's Directorate of Environmental Geology has been receiving assistance from the German **Bundesanstalt für Geowissenschaften und Rohstoffe (BfGR)** in the compilation and publication of hydrogeological mapping. This 1:250 000 scale series now covers Java, and a small-scale hydrogeological map has recently been published. Other DGI mapping has included the publication of some engineering geology maps. DGI maps are distributed through the **Pusat Penelitian dan Pengembangan Geologi (PPPG)**, and appear in map jobbers' listings, but no replies were received from the PPPG or DGI so the availability of mapping listed in the catalogue section is in question.

Soil maps of Indonesia are prepared by the **Soil Research Institute (Lembaga Penelitan Tanah)**. These are mostly three-colour maps derived from aerial coverage. Most of the effort has so far been devoted to the production of 1:100 000 and 1:250 000 reconnaissance maps of Java and Southern Sumatera.

The **Directorate of Land Use** has compiled land use mapping at a variety of large scales to cover much of Indonesia. Most maps are single-colour ungridded editions compiled since the early 1970s and published at 1:50 000, 1:100 000 and 1:250 000 scales. Bali and Java are covered in full-colour land use mapping at 1:250 000 scale. These series are compiled from 1:25 000 and 1:12 500 scale survey information. The current availability of this vast and very important resources mapping project is very doubtful and we have neither listed maps nor provided graphics for these series.

Vegetation and bioclimatic mapping of Indonesia has been compiled by **Institut de la Carte Internationale du Tapis Végétal (ICITV)** in association with the Indonesian **Regional Center for Tropical Biology (BIOTROP)**. Sumatra is covered in 1:1 000 000 scale full-colour vegetation mapping on the model of other ICITV maps and a four-sheet vegetation map of the whole region with accompanying explanation was prepared in the 1970s.

The **Directorate General of Forestry** has carried out reconnaissance type forest mapping, derived from aerial photography, the new national base maps, and satellite imagery. Sulawesi is also mapped by a land cover series at 1:250 000 scale compiled from satellite imagery.

Various foreign agencies have also produced mapping of Indonesia. These include the British **Land Resources Development Centre (LRDC)** who published mapping compiled by the British DOS on several resources themes in the 1970s. Sheets issued included 1:50 000 scale soil mapping of Bali, and 1:250 000 scale soils and land suitability mapping of parts of Sumatra for the transmigration settlement studies. Most of this mapping is no longer available.

The Dutch **International Institute for Aerospace Survey and the Earth Sciences (ITC)** has also produced maps and provided technical assistance. Maps include 1:250 000 scale geomorphological mapping of a basin in Central Java. Other general mapping of the country is published by **Nelles Verlag**, the Russian **Glavnoe Upravlenie Geodezii i Kartografii (GUGK)**, **Karto+Grafik** and **Falk Verlag**.

There are numerous private sector surveying and mapping agencies in Indonesia. The most important is **PT Pembina**, who compile small-scale tourist maps at a variety of scales to cover the different islands and provinces of Indonesia. Pembina also issues town mapping.

Further information

Much Indonesian mapping is restricted and not available but the mapping activities are very well documented. A useful overview of both historical background and current position is Asmoro, P. (1982) Developments in mapping in Indonesia, *ITC Journal*, **1982**(2), 200–206.

The different Indonesian mapping agencies have regularly contributed reports to the *UN Regional Cartographic Conferences for Asia and the Pacific*. The 1977 (11 papers), 1980 (six papers), 1983 (six papers) and 1987 (eight papers) conferences all include useful summaries of the current position.

For a detailed consideration of the new 1:50 000 mapping programme see Mastra, R. and Brown, M.H. (1981) Proposed specification for the new 1:50 000 topographic map series of Indonesia, *Cartographica*, **18**(3), 66–73.

The best recent summary of geological mapping activities in Indonesia is Voskuil, R.P.G.A. (1982) Nieuwe geologische kaarten voor Indonesie, *Kartografisch Tijdschrift*, **VIII**(1), 35–39.

DGI also publishes a regular *Geosurvey newsletter, Berita Geologi*, which contains current information about the state of geological cartography in Indonesia.

Addresses

National Coordinating Body for Survey and Mapping (BAKOSURTANAL) (Balan Koordinasi Survai dan Pemetaan Nasional)
Jalan Dr. Wahidan 1/11, JAKARTA PUSAT; and Jalan Raya Jakarta km 46, Cibinong, BOGOR.

Army Topographic Service (JATOP) (Jawatan Topografi Angkatan Darat)
Jalan Gunung Sahari 90, JAKARTA.

Jakarta Mapping Center
JAKARTA.

Geological Survey of Indonesia (DGI) (Direktorat Geologi) Seksi Publikasi (Pusat Penelitian dan Pengenbangan Geologi) (PPPG)
Jalan Diponegoro 57, BANDUNG.

Soil Research Institute (Lembaga Penelitian Tanah)
Ministry of Agriculture, Jalan Ir. H Juanda 98, BOGOR.

Directorate of Land Use (Direktorat Tata Guna Tanah)
Ministry of Internal Affairs, Jalan Sisingamangaraja 2, Jakarta Selatan, Kebayoran, JAKARTA.

Regional Center for Tropical Biology (BIOTROP)
PB 17, Jalan Raya, Tajur, BOGOR.

Agency for Inventorization and Forest Utilization (Badan Inventarisari dan Tata Guna Hutan)
Jalan Ir. H. Juanda 100, BOGOR.

PT Pembina
Jalan DI Panjaitan No 45, JAKARTA.

For LRDC and DOS see Great Britain; for ITC see Netherlands; for Falk, Karto+Grafik and Nelles Verlag see German Federal Republic; for GUGK see USSR; for IFP see India; for DMA and USGS see United States.

Catalogue

Atlases and gazetteers

Indonesia: official standard names approved by the United States Board on Geographic Names
Washington, DC: DMA, 1982.
2 vols.

General

Indonezija 1:7 000 000
Moskva: GUGK, 1974.
In Russian.
With placename index and economic and population inset maps.

Indonesia wawasan nusantara 1:4 500 000
Jakarta: Pembina, 1986.
Physical map with administrative divisions.
Includes list of provinces and kabupaten and their administrative centres.

APA maps of Indonesia 1:1 500 000
München: Nelles Verlag, 1985.
5 sheets, all published.
Sheets cover Sumatra, Java, Kalimantan, West Indonesia and Bali (1:180 000).

Indonesia: province and island maps various scales
Jakarta: Pembina.
13 published sheets.
Bali 1:160 000 1981.
Bali 1:256 000 1979.
Java 1:500 000 1985 (3 sheets).
Java Madura 1:1 100 000 1985.
Kalimantan 1:1 500 000 1979.
Kalimantan selatan 1:400 000 1980.
Maluku dan Irian Jaya 1:2 250 000 1978.
Nusa Tengara Barat and Timor 1:1 330 000 1976.
Sulawesi 1:1 000 000 1985.
Sulawesi Selatan 1:625 000 1979.
Sumatera 1:1 750 000 1984.
Timor 1:400 000 1980.

Geological and geophysical

Bouguer anomaly map of Indonesia 1:5 000 000
Bandung: DGI, 1979.
Includes explanatory text on sheet.

Map of the sedimentary basins of the Indonesian region 1:5 000 000
Reston, VA: USGS, 1974.
Also available as Earthquake map, Tectonic map and Bathymetric map.

Hydrogeological map of Indonesia/Peta hidrologi Indonesia 1:2 500 000
Bandung: DGI, 1984.
2 sheets, both published.

Geologic map of Indonesia/Peta geologi Indonesia 1:2 000 000
Bandung: DGI and USGS, 1965.

Geologic map of Indonesia/Peta geologi Indonesia 1:1 000 000
Bandung: DGI, 1975–.
16 sheets, 3 published.

Geologic map of Indonesia/Peta geologi Indonesia 1:250 000
Bandung: DGI, 1973–.
ca 100 sheets.

Bouguer anomaly map/Peta anomali Bouguer 1:250 000
Bandung: DGI, 1979.
ca 100 sheets.

Hydrogeological map of Indonesia/Peta hidrogeologi Indonesia 1:250 000
Bandung: DGI, 1983–.
ca 150 sheets, 11 published.

Peta geologi/Geological map 1:100 000
Bandung: DGI.
57 sheets.

Administrative

Indonesia 1:5 300 000
Jakarta: Pembina, 1984.
Political colouring by province.
Includes Indonesian description of administrative divisions in margin.

Environmental

Bioclimats du monde indonesien/Bioclimates of the Indonesian Archipelago 1:2 534 400
Pondichéry: IFP, 1978.
With accompanying explanatory text.

Town maps

Pembina: town maps various scales
Jakarta and Surabaya: Pembina.
5 sheets published.
Maps published for Jakarta, Semerang, Surabaya, Surakarta and Yogyakarta.

Jakarta 1:17 500
Hamburg: Falk, 1986.

Iran

Until World War II Iran was mapped by the British Survey of India, in quarter-inch series, and for some areas in half-inch series. The **National Cartographic Centre** and National Geographic Organization were established in 1953 and assumed responsibility for the provision of official geodetic and topographic survey information and the compilation of topographic series mapping. These two bodies were merged in 1973 and continued to compile new photogrammetrically derived 1:1 000 000, 1:250 000 and 1:50 000 scale series. Only the 1:1 000 000 scale series was completed.

Following the Islamic Revolution in 1979 the **National Cartographic Centre (NCC)** altered the emphasis of its mapping programmes. The priority project became the preparation of a 1:25 000 basic scale map, which required new and more precise levelling and the creation of a full geodetic network. Some progress has been made in the south of the country but there is little information about either the scope of the programme or its rate of progress. Other current projects include preparation of large-scale plans of Iranian cities at 1:2000 and 1:5000, large-scale surveys of areas with development potential, and forest maps at 1:25 000 of parts of the north of the country. The **Army Geographic Department (AGD)** has prepared coverage of the country in maps at 1:50 000 and 1:250 000.

None of these topographic maps produced by NCC or AGD is currently available outside the country. Perhaps the best currently available Iranian detailed topographic mapping is contained in the four-volume *Historical gazetteer of Iran*. This major work includes a detailed gazetteer of placenames derived from best available sources, together with 1:300 000 scale mapping. These maps are reduced from a variety of often inconsistent source material.

Geological mapping of Iran has been carried out by oil companies and by the **Geological Survey of Iran (GSI)**. GSI was founded in 1958 with assistance from the United Nations. A very wide range of geological coverage has been compiled since the 1970s. National full-colour series at 1:250 000 and 1:100 000 scale were compiled and the former will give complete coverage of the country when the 41 sheets currently in press are issued. Some sheets are accompanied by explanatory texts. The 1:100 000 coverage on topographic sheet lines is very incomplete: a large block was compiled for the Kerman province area by the Jugoslav Geological Survey but much of the rest of the country is not yet covered. About 200 sheets in this series are now either published, in press or in preparation. 1:250 000 scale aeromagnetic coverage is available for almost all of Iran, using green and red overprints on a grey topographic background. National coverage at 1:1 000 000 scale was completed by the National Iranian Oil Company in the 1970s and GSI is currently compiling a total magnetic intensity map at this scale. Single-sheet 1:2 500 000 scale geological, mineral, seismotectonic and tectonic sheets were published in the 1970s and several are still available from GSI. Two recent additions to maps at this scale are a new edition of the geological map and a metamorphic map, while a palaeomagnetic basement map is under preparation. Many other monographic sheets of important areas were also published.

There has been little other thematic mapping of the country. Some small-scale national coverage has appeared as part of the **Tübinger Atlas des Vorderen Orients (TAVO)** and for Azerbayejan province the **Geographische Institut Universitäts Bern** has compiled regional suitability mapping.

In view of the difficulties in obtaining officially produced mapping the commercial sector in Iran is perhaps more important than the quality of its mapping would suggest.

Sahab Geographic and Drafting Institute is a major commercial producer of maps of both Iran and other Middle Eastern countries. The company was established in 1888 and has published a wide variety of mostly fairly poor-quality maps, for the tourist and educational markets. These include over 100 educational maps of Iran, the continents and the World: physical and political bases are used and some pictorial thematic maps are also published. Tourist maps of cities and provinces of Iran are published, many in English language editions. A recent project is the publication at a standard scale (1:500 000) of relief-based maps of Iran which will be published to give nine-sheet complete coverage as either English or Persian language editions. A wide range of historical maps and facsimile publications has also been issued and Sahab has compiled and published several atlases, including an important atlas of the White Revolution, issued in the early 1970s and now out of print, which described the Shah's economic liberalization and westernization programmes. The catalogue concentrates on the English language versions of Sahab mapping.

Gita Shenassi is the other major commercial map producer in Iran. Founded 12 years ago its output closely mirrors the Sahab maps. It too concentrates on the tourist and educational sector of the market, but there are fewer English language versions of its maps available. Where a Sahab published English language map exists, the Gita Shenassi Persian equivalent is omitted. For instance a Persian language series of provincial maps is issued by Gita Shenassi, which closely parallels the administrative series compiled in English and Persian by Sahab. Other mapping of Iran is compiled by **Bartholomew** and by the Russian **Glavnoe Upravlenie Geodezii i Kartografii (GUGK)**.

Further information

For the best published summary of the historical background to Iranian mapping, see Cook, A.S. (1983) Geography and maps, in *Bibliographical Guide to Iran and the Middle East*, pp. 114–159, Brighton: Harvester Press.

Catalogues and background information about GSI, Sahab and Gita Shenassi are available.

Addresses

National Cartographic Centre (NCC)
PO Box 1844, TEHRAN.

Army Geographic Department
Azadi Square—Mevage ave., PO Box 11365-5167, Cab Senca, TEHRAN.

Geological Survey of Iran (GSI)
PO Box 13185-1494, TEHRAN.

Gita Shenassi
PO Box 14155-3441, TEHRAN.

Sahab Geographic and Drafting Institute
PO Box 11365-617, TEHRAN.

For Bartholomew see Great Britain; for TAVO see German Federal Republic; for GUGK see USSR; for DMA see United States; for Akademische Drück und Verlagsanstalt see Austria; For Geographisches Institut Universitäts Bern see Switzerland.

Catalogue

Atlases and gazetteers

Historical gazetteer of Iran   L.W. Ademec
Graz: Akademische Drück und Verlagsanstalt, 1976–.
4 vols, 2 published.
Published volumes cover northern half of Iran.
Includes 1 : 300 000 scale mapping.

Iran: official standard names approved by the United States Board on Geographic Names
Washington, DC: DMA, 1956.
pp. 578.

General

Iran   1 : 2 500 000
Edinburgh: Bartholomew, 1977.

Road map of Iran   1 : 2 500 000
Tehran: Sahab, 1977.
In English and Persian.
With English index.
Available with hypsometric shading or physical background.

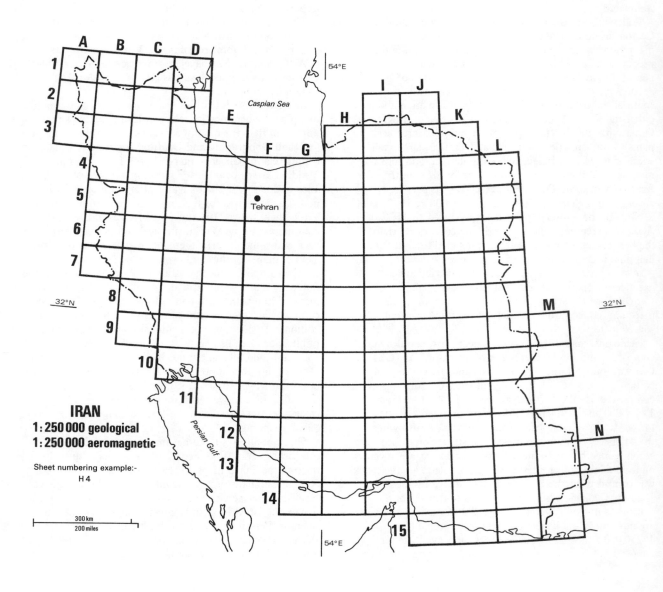

IRAN
1 : 250 000 geological
1 : 250 000 aeromagnetic

Sheet numbering example:-
H 4

300 km
200 miles

Iraq 1:2 500 000
Moskva: GUGK.
In Russian.
With economic and climatic insets and placename index.

Iran road map 1:2 200 000
Tehran: Gita Shenassi, 1985.

Regional maps of the provinces of Iran various scales
Tehran: Sahab, 1976–.
10 sheets, all published.

Topographic

Province maps: relief 1:500 000
Tehran: Sahab, 1985–.
9 sheets, 1 published.
English or Persian editions.

Geological and geophysical

Geological map of Iran 1:4 000 000
Tehran: Sahab, 1975.
In English and Persian.
Single colour.

Geological map of Iran 1:2 500 000
Tehran: GSI, 1984.

Metamorphic map of Iran 1:2 500 000
Tehran: GSI, 1986.

Mineral distribution map of Iran 1:2 500 000
Tehran: GSI, 1976.

Tectonic map of Iran 1:2 500 000
Tehran: GSI, 1973.
English and Persian text.

Seismotectonic map of Iran 1:2 500 000
Tehran: GSI, 1976.
English and Persian text.

Geological map of Iran 1:1 000 000
Tehran: NIOC, 1975.
6 sheets, all published.

Geological quadrangle map of Iran 1:250 000
Tehran: GSI, 1969–.
124 sheets, 83 published. ■

Aeromagnetic map of Iran 1:250 000
Tehran: GSI, 1977.
124 sheets, 104 published. ■

Geological map of Iran 1:100 000
Tehran: GSI, 1971–.
134 published sheets.

Administrative

Iran: administrative divisions 1:2 500 000
Tehran: Sahab, 1978.
With two population inset maps.

Map of the Islamic Republic of Iran 1:1 600 000
Tehran: Gita Shenassi, 1985.

Human and economic

Iran: Binnenwanderung/Iran internal migration
1:4 000 000 TAVO A VIII 5.3
Wiesbaden: TAVO, 1983.

Iran: railways 1:2 500 000
Tehran: Sahab, 1975.

Town maps

City maps various scales
Tehran: Sahab.
23 sheets, all published.
14 cities covered in English and Persian bilingual maps.
5 other cities available only as Persian versions.
4 additional Persian language maps of Tehran available.

City maps various scales
Tehran: Gita Shenassi.
5 sheets, all published.
English and Persian bilingual editions cover Tehran, Esfahan, Tabriz, Rasht and Kashan.

Iraq

The national mapping organization in Iraq is the **State Establishment for Surveying**, now part of the Ministry of Irrigation. A 1:250 000 scale series giving complete coverage of the country in 45' by 1° 30' quads was compiled in the 1960s. The current map base of the country is being established with technical assistance from Geokart, the Polish overseas survey agency. A complete geodetic and cartographic network for all the country has been set up and a modern topographic map at 1:25 000 covering much of the country has been compiled. Compilation took place between 1974 and 1978 from 1:50 000 scale aerial coverage. 1107 sheets were taken to preliminary publication stage—they show relief with 10 m contours and sheet lines follow the international system. Larger-scale basic coverage of the capital Baghdad has also been established by the Poles. No current officially produced Iraqi mapping is available to the West.

Geological mapping of the country is the responsibility of the **Directorate General of Geological Surveys**.

Recent small-scale publications include single-sheet tectonic and geological maps.

The Iraqi Meteorological Organization compiled a small-scale climatic atlas of the country in the late 1970s that is no longer available.

Foreign general mapping of Iraq is compiled by **Sahab**, **Gita Shenassi**, the American **Central Intelligence Agency (CIA)** and the Russian **Glavnoe Upravlenie Geodezii i Kartografii (GUGK)**. A tourist map of Baghdad is published and compiled by **Engineering Surveys Reproduction Ltd (ESR)**.

Further information

Kwiatkowski, H. (1985) Polish geodesy in Iraq, *Przeglad geodezyjny*, **2/85**, 1–4 gives Polish language information about the recent 1:25 000 mapping programme.

Addresses

State Establishment for Surveying
Ministry of Irrigation, PO Box 5813, Gailani Sq., BAGHDAD.

Directorate General of Geological Surveys (DGGS)
State Organization for Minerals, PO Box 986, Alwiyah, BAGHDAD.

For Sahab and Gita Shenassi see Iran; for CIA and DMA see
United States; for GUGK see USSR; for ESR see Great Britain.

Catalogue

Atlases and gazetteers

Iraq: official standard names approved by the United States Board on
Geographic Names
Washington, DC: DMA, 1957.
pp. 175.

General

Irak   1:1 500 000
Moskva: GUGK, 1980.
In Russian.
With 3 ancillary small thematic insets and index.

Geological and geophysical

Tectonic map of Iraq   1:1 000 000
Baghdad: DGGS, 1984.
In Arabic and English.
With 3 insets showing sedimentary cover, structure and
earthquake epicentres.

Administrative

Guide map of Iraq   1:2 000 000
Tehran: Sahab, 1971.
In English and Arabic.
With inset town plans of Baghdad, Erbil, Mosul and Karbala.

Map of Iraq   1:1 200 000
Tehran: Gita Shenassi, 1986.
In English and Arabic, layer coloured by province.

Town maps

Baghdad   1:36 000
Tehran: Sahab, 1980.
In English and Persian.

Baghdad   1:30 000
West Byfleet: ESR, 1983.
Includes Baghdad 1:10 000, Baghdad environs 1:750 000, index
and tourist information on back.

Israel

Geodetic and topographic surveys were begun by the
mandatory Survey of Palestine after World War I. The
Survey of Israel (SI) was established in 1949 and took
over as the national mapping and surveying agency. It
carries out topographic, cadastral and engineering surveys
and maintains geodetic networks. Official topographical,
topocadastral and thematic maps are printed by SI which
also carries out contract work for the private sector. All
topographic maps published by SI use the Cassini–Soldner
Transverse Cylindrical projection and are printed with the
kilometric Israeli grid, with an indication of UTM grid on
the margins. The basic scale for all developed areas of the
country is a 340-sheet 1:10 000 scale topocadastral map.
This series was started in the 1950s as a six-colour edition,
with 10 m contours, with a cadastral information
overprint. Sheets cover 5 km quads. It now uses three
colours for developed areas and appears as a two-colour
edition for the northern Negev. An experimental 1:10 000
scale orthophotomap series might in the future replace the
conventional line map format used for this series. An 88-
sheet 1:50 000 map is the basic scale series for the whole
country. This map was started in the 1950s, is
photogrammetrically derived and uses 10 m contour
intervals; sheets cover 20 km quads. The current
specification uses four-colour printing and Hebrew names.
There is a regular revision programme for this map, but
like the topocadastral map it is no longer publicly
available. The largest scale national series that is still freely
available is the 1:100 000 map which covers Israel
including the West Bank in 22 sheets. The history of this
series can be traced back to a 16-sheet map compiled by
the Survey of Palestine, which was republished in 1958 to
give 26-sheet coverage of the country. Recast in the 1970s
it now includes tourist information, is derived from the

1:50 000 map and shows relief with 25 m contours. This
series is also available with a partial overprint of names in
Latin script. The revision cycle for this series is between 2
and 5 years depending on the area and the demand. SI also
publishes a variety of smaller-scale maps, mostly available
as either English or Hebrew language editions. These
include a two-sheet 1:250 000 tourist map available with
gazetteer, and a two-sheet layer tinted and hill shaded
relief map, with 100 m contours at the same scale. Both are
available as either English or Hebrew editions. These
1:250 000 scale maps are revised on an annual basis. A
1:400 000 scale single-sheet tourist map was compiled in
the late 1970s, including photographically reduced urban
detail derived from the 1:250 000 series. A Hebrew
language administrative map is published and smaller-scale
earth science mapping is issued: satellite photomap
editions of the 1:500 000 geological maps of the country
and of Sinai are also available from SI and a duotone
brown 1:750 000 satellite image map of the whole country
is also published. Small-scale soil maps are also issued.
Large-scale town mapping is issued to cover the major
cities; the current scale for Hebrew language versions is
1:12 500. There have been several changes in specification
for these maps which have seen them change from
essentially topographic special sheets to more specialized
town plans. The latest change has been the incorporation
of a street and institution index on the back of the sheet.
Nine sheets are available at 1:12 500, five at the older
1:10 000 scale. These maps are also available with
computer-generated registration block overprints. English
language editions are issued for Jerusalem, Tel-Aviv–Yafo
and Nazerat. SI has also recently published the third
edition of the *Atlas of Israel*, which is available in Hebrew
version or as an English language publication, distributed

by Macmillan. This major work provides excellent small-scale thematic coverage of the country. A computerized bi-scriptual gazetteer is held by SI and it is planned to publish this in the near future.

Geological mapping of Israel is the responsibility of the **Geological Survey of Israel (IGS)**, established in 1949. A 1 : 100 000 series was started, which is still available for some areas. This used the sheet lines of the old 1 : 100 000 topographic series. 1 : 250 000 mapping was completed for the whole country in the 1960s, and has since been revised. The current programme, begun in 1970, involves the remapping of Israel at 1 : 50 000 scale, with publication of geological and structural maps and explanatory notes. In July 1984 nine full-colour sheets had been published, six more were available as provisional editions and a further 15 were nearing completion. It is intended to provide complete national coverage in this series by 1996. Smaller-scale IGS maps are issued by SI, and include geological mapping superimposed on satellite imagery to cover Israel and the Sinai. IGS has also compiled bathymetric mapping of the coastal areas. The other organization responsible for the compilation of earth science mapping is the **Institute for Petroleum Resources and Geophysics (IPRG)**, which recently compiled a full-colour 1 : 500 000 scale earthquake map.

Other useful thematic mapping of the area has appeared as sheets in the **Tübinger Atlas des Vorderen Orients (TAVO)**. The American **Central Intelligence Agency (CIA)** has published several current maps of the area, especially relating to boundary and settlement issues on the West Bank. Commercial mapping of Israel is carried out by **Carta**. Their publications concentrate upon the tourist market but include a very useful gazetteer and series of historical atlases. Many western commercial houses publish mapping of Israel including **Kümmerly + Frey**, **Bartholomew**, **Hallwag**, **Ravenstein** and **Karto+Grafik**.

Further information

The best English language source of information about surveying and mapping of Israel are sections 6, 7 and 8 of the new edition of the National Atlas Survey of Israel (1985) *Atlas of Israel*, London: Macmillan.

SI issues Hebrew language Cartographic, Geodetic and Photogrammetric papers with English abstracts which give further information on the state of cartography in Israel.

A recent book gives a very comprehensive Hebrew language overview: Kadmon, N. and Shmueli, A. (1982) *Maps and Mapping*, Jerusalem: Keter Publishing.

A useful guide to the publication of geological mapping of the country is Ginzburg, D. and Pekarevich, I. (1986) *An inventory of geological maps of Israel and Sinai*, Jerusalem: IGS, 2 vols.

SI and IGS both publish Annual Reports and map catalogues; Carta issues a regularly updated catalogue.

Addresses

Survey of Israel (SI)
1 Lincoln St., TEL AVIV 61141, POB 14171.

Geological Survey of Israel (IGS)
30 Malkhe Yisrael St., JERUSALEM 95501.

Institute for Petroleum Research and Geophysics (IPRG)
POB 1717, HOLON.

Carta
4/6 Yad Harutzim St., POB 2500, JERUSALEM.

For TAVO and Karto + Grafik see German Federal Republic; for Kümmerly + Frey and Hallwag see Switzerland; for CIA see United States; for Bartholomew see Great Britain.

Catalogue

Atlases and gazetteers

The Atlas of Israel 3rd edn
Tel-Aviv: SI and Macmillan, 1985.
pp. 232 (400 maps).
Also available as Hebrew language version distributed by Carta.

Carta's Official Guide to Israel: complete gazetteer to all the sites in the Holy Land
Jerusalem: Carta, 1985.
pp. 472.

General

Satellite photomap of Israel 1 : 750 000
Tel Avia: SI, 1979.

Israel: road map 1 : 750 000
Bern: Kümmerly + Frey, 1986.

Israel 1 : 400 000
Tel-Aviv: SI, 1982.

Israel: touring map 1 : 250 000
Tel-Aviv: SI, 1980.
2 sheets, both published.

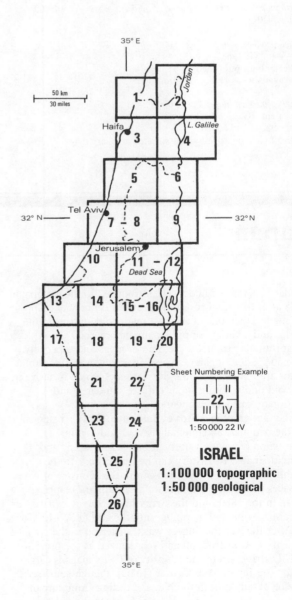

ISRAEL
1:100 000 topographic
1:50 000 geological

Topographic

Israel topographic 1 : 250 000
Tel-Aviv: SI, 1980.
2 sheets, both published.
Also available as Hebrew edition.

Israel 1 : 100 000
Tel-Aviv: SI, 1975–.
22 sheets, all published. ■
In Hebrew with English placenames overprinted.

Bathymetric

Bathymetric charts of the Mediterranean Coast of Israel 1 : 250 000
Jerusalem: GSI, 1983.
With accompanying explanatory text.

Geological and geophysical

Israel: geological map 1 : 500 000
Tel-Aviv: SI, 1979.

Israel: geological photomap 1 : 500 000
Tel-Aviv: SI, 1979.

Israel: geomorphological map 1 : 500 000
Tel-Aviv: SI, 1978.

Earthquake epicentres in Israel and adjacent areas 1 : 500 000
Holon: IPRG, 1983.

Israel: geological map 1 : 250 000
Tel-Aviv: SI, 1965.
2 sheets, both published.
Also available as Hebrew edition.

Israel: geological map 1 : 100 000
Jerusalem: IGS, 1949–.
24 sheets, 16 published.
With English overprint and associated ancillary maps.

Israel: geological map 1 : 50 000
Jerusalem: IGS, 1972–.
88 sheets, 30 published. ■
With English and Hebrew legends, structural map and
accompanying explanatory booklet.

Environmental

Soil map of Israel 1 : 500 000
Tel Aviv: SI, 1975.

Distribution of soils affected by salinity 1 : 500 000
Tel-Aviv: SI, 1969.

Südliche Levant: Landnutzung 1 : 500 000
Wiesbaden: TAVO, 1981.

Südliche Levant: Vegetation 1 : 500 000
Wiesbaden: TAVO, 1981.

Südliche Levant: Landnutzung zu 1880 1 : 500 000
Wiesbaden: TAVO, 1981.

Administrative

Administrative map of Israel 1 : 250 000
Tel-Aviv: SI, 1972.
2 sheets, both published.
Hebrew language edition only.

Human and economic

Südliche Levant: ethnische Gruppen 1 : 500 000
Wiesbaden: TAVO, 1984.

Town maps

Town map series 1 : 14 000
Tel-Aviv: SI, 1973–.
3 maps, all published.
English editions cover Jerusalem, Tel-Aviv–Yafo and Nazerat
(1 : 7500).

Japan

Modern survey methods were introduced to the country in
1868 after the Meiji restoration: over a century of mapping
has made Japan one of the best mapped countries in the
world, and a variety of public and commercial
organizations produce a plethora of maps and especially of
thematic series. Japanese mapping activities in the public
sector are regulated by the Survey Law, which is designed
to avoid undue duplication, enforce standards and
coordinate cartographic work.

The **Geographical Survey Institute (GSI)** is responsible
for the national fundamental surveys and for the
production of a wide variety of both topographic and
thematic maps. Its history can be traced back to the
establishment of a survey division in the Ministry of Civil
Services in 1869. Systematic modern military mapping
began in the 1880s and the Army Land Survey carried out
basic surveying and mapping until the end of World War
II. After the war the Survey was transferred to the civilian
Ministry of Construction and reorganized as the present
GSI. Distribution of the majority of GSI's maps is carried
out by the **Japan Map Center (JMC)**, though some series
may be obtained from the **Nagai Trading Company** or
other commercial agents. Most of GSI's topographic maps

are published as very small sheets, which accounts for the
very large numbers needed to complete coverage of Japan.
The maps are published with a standard hierarchical sheet
designation system and use the UTM projection. Unless
otherwise stated all of the written information on the
sheets, with the exception of the sheet designation, is in
Japanese characters.

The large-scale topographic mapping programme begun
by GSI in 1960 consists of cyclical aerial photography at 3
year intervals for urban areas and 5 year intervals for the
remainder of the lowland: on the basis of this aerial
coverage large-scale plans are produced at 1 : 2500 and
1 : 5000 scale, some published as photomaps.

The 1 : 25 000 scale topographic map is now the basic
map for the whole of the country, which it covers in 4455
sheets. It was begun in 1964 and complete coverage was
attained in 1979. Over 90 per cent of these maps have been
completed by private companies under contract to GSI: the
maps they compile are derived from 1 : 40 000 colour aerial
coverage. This series shows relief with 10 m contours and
appears as one-, two- or three-colour editions, which are
revised at intervals of 3 years for urban areas, 5 years for
intermediate areas and 10 years for mountainous regions,

on the basis of the amount of unit change in the map. Complete recompilation is carried out after three or four revisions.

On completion of the 1:25 000 series it was decided to introduce a new basic scale map for urban areas and suburbs. The scale chosen was 1:10 000, and four 1:10 000 sheets cover the area of a 1:25 000 quad. The new series is compiled from up-dated locally produced 1:2500 scale mapping where this is available and is printed in four colours. Relief is shown by 2 m contours, building heights are distinguished, and a large number of functional symbols used. It is intended to cover the Tokyo, Osaka and Nagoya areas by 1994.

The 1:50 000 scale topographic map was completed in 1925 by the Army Map Survey and has been constantly revised since then. It is now compiled from 1:25 000 mapping, and about 100 sheets are revised every year, simultaneously with the revision of the corresponding 1:25 000 sheets. The 1:50 000 series of 1249 sheets is now mostly published as a four-colour map, with 20 m contours. The six-colour 1:200 000 scale regional map covers Japan in 129 sheets, whilst an eight-sheet 1:500 000 scale district map series is available as either a four-colour edition or a nine-colour version with layer tinted relief. In 1983 a new plastic relief 1:250 000 scale map was published by GSI to cover Japan in 43 sheets. Special sheets in accordance with the International Map of the World specifications are issued for the Japanese archipelago at 1:1 000 000 scale: this three-sheet series is also available as an English language edition published in 12 colours, and is derived from the district and regional maps. GSJ also issues a 12-colour English language version of the 1:3 000 000 scale map of Japan and her surroundings.

GSI also published the National Atlas in 1977, and coordinated much of the input of data to this large volume, which is available in English language and Japanese versions. A new edition is currently being prepared and is expected to be available some time in 1986 or 1987: it is hoped to revise this work at 10 year intervals in the future. Another recent atlas compiled by GSI and distributed by the Japan Map Centre is the *Regional Planning Atlas of Japan*. This six-part work was issued in 1984 in loose-leaf format and includes 1:2 500 000 scale maps of the whole of Japan, together with larger-scale coverage of the big cities.

Unlike many other national survey organizations GSI is very involved in the production of thematic map series. A 1:25 000 *Land condition map series* is published by GSI to map landforms, elevation and the location of public facilities. Though mainly a geomorphological series this map is also intended to aid in disaster planning. It is as yet very incomplete in coverage. The 1:25 000 *Land condition map of coastal areas* is issued by GSI to provide detailed data on land and sea bottom conditions in the nearshore zone; this series is issued parallel to the 1:25 000 *Topographic map of coastal areas*, which presents both topographic and bathymetric information. Other natural hazard mapping is carried out by GSI to map all main natural disasters at large scales. A series of *Lake charts* and a 1:50 000 scale *Water use map* of the major rivers are also available to cover the principal river systems.

The **National Land Agency (NLA)** was established in 1974 and is responsible for coordinating with other public mapping agencies in the compilation of a variety of thematic map series, as well as carrying out cadastral survey functions. Land use mapping of Japan began in 1947 with the compilation of a 1:800 000 scale map to aid postwar reconstruction. A 1:50 000 series was compiled between 1953 and 1972 to cover nearly half of the country.

GSI in collaboration with NLA began the current land use base map in 1973. This six-colour 1:25 000 *Land use series* maps 35 land use units, and uses the same sheet lines and topographic base as the 1:25 000 scale map series. It is intended to cover the whole country. Lowland areas were given priority and are now all covered, and about 1500 sheets have been issued since 1975, at a rate of about 150 new sheets a year. The *Land Classification map series* comprises three thematic maps—land classification, subsurface geology and soil—at 1:50 000, and will be issued for each of the first-order administrative units of the country. A 1:200 000 scale *Land classification map* of each of the 46 prefectures has also been issued by NLA and was completed in 1979, to give national coverage. This series comprises a wide variety of thematic bases and overlays including geomorphology, soil, geology, soil production, drainage systems, slopes, potential land use, etc. The NLA is also involved in the creation of a National Land Information System, which provides access to digitally stored spatial data, derived from digitized 1:25 000 scale mapping and a variety of other data sources. Much of the data was captured using an automatic colour pattern scanning system. It is possible to produce computer generated mapping according to a wide variety of specifications from this system.

A new national land use series at 1:200 000 scale has been compiled by GSI and NLA to give complete coverage of the country on the topographic map sheet lines. This map is derived from the 1:25 000 topographic and land use series, from other thematic mapping from aerial coverage and from the National Land Information System.

The **Geological Survey of Japan (GSJ)** is responsible for geological mapping of Japan. Over half the country is mapped in three medium-scale base series; about 800 sheets out of a planned total of 1219 in the scale ranges 1:50 000 to 1:100 000 have now been published. The *1:50 000 geological series* was begun in 1951 and is published with explanatory texts in Japanese with English summaries. The maps themselves have bilingual legends and are printed on a topographic base. Coverage is best for Hokkaido. Other smaller-scale geological series are published at 1:200 000 and 1:500 000, and derived from the basic scale coverage. Also at 1:500 000, using the same sheet lines, is a neotectonic map series. There is complete geological coverage of the country at 1:1 000 000 scale, and a single-sheet 1:5 000 000 map. Marine geological mapping is published by GSJ at 1:200 000 and 1:1 000 000 scales, and there are series of hydrogeological maps of important areas. Other series include total intensity aeromagnetic maps at 1:200 000, detailed geological coverage of coalfield and oil and gas field areas, a 1:25 000 volcanic areas series, a small-scale tectonic series and isogonal contour maps of geothermal areas. GSJ also publishes the important *1:2 000 000 Geological map series*. This small-scale series contains a variety of bilingual geoscience maps of the country, including mineralogical coverage. GSJ publications may also be purchased from **Maruzen Co. Ltd Export Department**.

The **Maritime Safety Agency (MSA)** of the Hydrographic Department carries out oceanographic surveys in Japanese waters. Several 80-sheet chart series are published at 1:200 000 scale on the same sheet lines; these are the *Submarine structural chart*, the *Gravity anomaly chart*, the *Total magnetic intensity chart*, and the *Bathymetric chart*. The more important information on these charts is also presented in English. Other smaller-scale bathymetric mapping is also carried out: there are plans for a 1:500 000 scale chart. Larger-scale surveys at 1:10 000 and 1:50 000 are published as basic scale bathymetric coastal charting.

The **Bureau of Statistics (BS)** has carried out thematic mapping of the results of the different population censuses in Japan, which are taken every 5 years. These 1:1 000 000 and 1:1 500 000 scale sheets plot a variety of social and demographic statistics and BS has also used digital methods to produce larger-scale regional choropleth mapping of census results. 1980 census mapping is currently available; 1985 mapping is expected soon. It has also been responsible for the compilation of an English and Japanese language placename listing derived from the Census results.

The **Remote Sensing Technology Center** was established in 1978 and acts as the Japanese focal point for remote sensing research and applications.

Soil maps in Japan have been produced for cultivated areas by the **Agricultural Administration Bureau** at 1:50 000 scale, and for forest areas by the **Forestry Agency** at 1:20 000 scale. Both series are now complete. Large-scale surveying of upland forest areas is also carried out by the Forestry Agency, who are producing a 1:5000 scale basic forest series.

The nature conservation section of the **Environment Agency** has been responsible for the compilation of three kinds of mapping. These present the results of the second national survey of the natural environment. The *Natural Environment of Japan* (an environmental atlas) is distributed by the Government Publications Service Centre and contains small-scale computer-generated mapping of the distribution of plants and animals and other ecological factors. 1:200 000 scale plant and animal distribution maps compiled on a prefecture base giving complete cover, and a 1:50 000 vegetation map giving about 50 per cent national coverage in 600 sheets are distributed by the **Japan Wildlife Research Center**.

Other public agencies which have carried out mapping projects include the **Road Bureau**, which prepares traffic volume maps, **Japanese National Railways**, and a variety of local government bodies.

There are over 20 commercial map producers in Japan, but most of their products are only issued in Japanese language versions. Exceptions to this include the **International Society for Educational Information** which published a useful atlas in 1974, and some of the tourist and road mapping companies. These include the **Japan Guide Map Company**, who publish a useful seven-sheet English language *Handy map of Japan*, with tourist information and town plans printed on the back of each sheet. They also issue English language town plans. The **Japan Topography Association (Nippon Kokuseisha)** produces the regularly revised *Great Map of Japan* series and **Teikoku Shoin** publishes English language atlases. Other commercial map publishers include **Shobunsa**, **Buyodo** and **The Japan Travel Bureau**.

Many foreign publishers produce mapping of Japan. We have included the **Mairs**, **Hallwag**, **Falk**, **NGS** and **Bartholomew** sheets in our listings and also referred to the physical and economic maps prepared by the Russian **Glavnoe Upravlenie Geodezii i Kartografii (GUGK)**.

Further information

Catalogues are issued by most organizations listed above: the Japan Map Center is particularly useful as a source of information about Japanese mapping. English language items are cited first.

For the latest information about cartography in Japan, see nine papers presented to the *11th United Nations Regional Cartographic Conference for Asia and the Pacific* held in Bangkok in 1987, New York: UN.

Bulletin of the Geographical Survey Institute, Tokyo: GSI 1948–. English language journal reporting on GSI activities, contains reports on National mapping presented to UN Regional Cartographic Conferences, and summeries of the state of topographic mapping in Japan, together with many other useful articles on the state of the art.

Kanakubo, T., Saito, S. and Eqawa, Y. (1984) 1:10 000 map series in Japan, pp. 167–182, in *Technical papers of the 12th Conference of the International Cartographic Association, Perth Australia, August 6–13, 1984*, Perth: ICA Conference Committee.

Takasaki, M. (1984) The regional planning atlas of Japan, pp. 439–446, in *Technical papers of the 12th Conference of the International Cartographic Association, Perth Australia, August 6–13, 1984*, Perth: ICA Conference Committee.

Geojournal (1980) **4**(4). Special issue of journal with articles on methods and mapping in Japan, good summaries on land use mapping, natural hazard mapping, National Land Information, thematic maps, etc.

Tanaka, K. and Inamura, Y. (1980) Geological maps published by the Geological Survey of Japan, *International Yearbook of Cartography*, **xx**, 170–179.

Geographical Survey Institute (1977) Thematic mapping in Japan, *Bulletin of the Geographical Survey Institute*, **22**(1), 42–68. An excellent summary of the variety of thematic series available, similar to article in *World Cartography* (1977) with the same title and authorship.

Chizu senta nyusa (Map Center News), Tokyo: Japan Map Center, 1972–. Japanese language report on mapping activities in the country.

Chizu (Map), Tokyo: Chizu Kyokai, 1958–. Japanese language cartography journal, English abstracts.

Addresses

Geographical Survey Institute (GSI)
Kitasato—1, YATABE-MATI, Tsukuba-Gun, Ibaraki-Ken, 305.

Japan Map Center (JMC)
4–9–6 Aobadai, Meguro-ku, TOKYO 153.

Nagai Trading Company
7–3 Hiroo 1 chome, Shibuya-ku, TOKYO 150.

National Land Agency (NLA)
Kasumigaseki, TOKYO 100.

Geological Survey of Japan (GSJ)
1–1–3 Higashi, YATABE MACHI, Tsukuba-gun, Ibaraki-ken 305.

Geoscience Information Center
No. 1–18–16, Sakae-cho, Fuchi-shi, TOKYO 183.

Maruzen Co. Ltd Export Department
PO Box 5050, TOKYO International, 100-31.

Maritime Safety Agency (MSA)
Hydrographic Department, 5–3–1 Tsukiji, Chuo-ku, TOKYO 104.

Bureau of Statistics
92 Wakamatsucho, Shinjuko, TOKYO 162.

Remote Sensing Technology Center
7F Yuni-Roppongi Building, 7–5–17 Roppongi, Minato-ku, TOKYO 106.

Agriculture Administration Bureau
Ministry of Agriculture and Forestry, 1–2–1 Kasumigaseki, Chiyoda-ku, TOKYO 100.

Forestry Agency
Ministry of Agriculture and Forestry, 1–2–1 Kasumigaseki, Chiyoda-ku, TOKYO 100.

Environment Agency
3–1–1 Kasumigasaki, Chiyoda-ku, TOKYO 100.

Japan Wildlife Research Center
39–12, Hongo 3-chome, Bunkyo-ku, TOKYO.

Road Bureau
Ministry of Construction, 2–1–3 Kasumigasaki, Chiyoda-ku,
TOKYO 100.

Japanese National Railways
1–6–5 Marunouchi, Chiyoda-ku, TOKYO 100.

Japan Topography Association (Nippon Kokuseisha)
18–20 Kohinata, 1-chome, Bunkyo-ku, TOKYO 112.

International Society for Educational Information
Kikuei Building No. 7–8, Shintomi 2-chome, Chuo-ku, TOKYO.

Japan Guide Map Co. Ltd
3F Dainichi Bldg, 10–8 2-chome, Ginza, Chuo-ku, TOKYO.

Teikoku Shoin Co. Ltd
3–2–9 Jimbo-cho, Kanda, Chiyoda-ku, TOKYO 101.

Shobunsa
4–2–11 Kudan-kita, Chiyoda-ku, TOKYO 102.

Buyodo
3–8–16 Nihonbashi, Chuo-ku, TOKYO 103.

Japan Travel Bureau (Publication Department)
7F Oki Building, 3–3 Kanda-kaji-cho, Chiyoda-ku, TOKYO 101.

For Mairs and Falk see German Federal Republic; for NGS and
DMA see United States; for Bartholomew see Great Britain; for
GUGK see USSR; for Hallwag see Switzerland.

Catalogue

Atlases and gazetteers

The National Atlas of Japan
Tokyo: JMC, 1977.
pp. 367.
English or Japanese language versions available.

Atlas of Japan: physical, economic and social
Tokyo: International Society for Educational Information, 1974.
pp. 128.
Supplement issued in 1979.

Teikoku's complete atlas of Japan   8th edn
Tokyo: Teikoku, 1984.
pp. 58.

1980 population census of Japan. Prompt report of basic findings
Tokyo: BS, 1981.
2 vols.
Japanese and English language gazetteer.

*Japan: official standard names approved by the United States Board
on Geographic Names*
Washington, DC: DMA, 1955.
pp. 731.

General

Japan and her surroundings   1:3 000 000
Tokyo: GSI, 1972.
Also includes Manchuria, Korea and Ryukyu Group. English and
Japanese.

Shell road map of Japan   1:3 000 000
Ostfildern: Mairs, 1981.

Japan   1:2 982 000
Washington, DC: NGS, 1984.

Japan   1:2 500 000
Edinburgh: Bartholomew, 1980.

Japonija: fizicheskaja karta   1:2 000 000
Moskva: GUGK.
In Russian.

Great map of Japan   1:1 200 000
Tokyo: Nippon Kokuseisha, 1983.

International map of the world: special sheets   1:1 000 000
Tokyo: GSI, 1980.
3 sheets, all published.
Available as either English or Japanese version.

Japan: handy map series   scales vary
Tokyo: Japan Guide Map Co., 1985.
7 sheets, all published.
English language tourist maps, town plans and tourist information
on back of sheets.
Okinawa 1:150 000; Kanto 1:500 000; Kyushu 1:500 000; Kansai
1:600 000; Chugoku-Shikoku 1:600 000; Tohoku 1:700 000;
Hokkaido 1:700 000.

Topographic

All topographic maps listed here contain Japanese character script
information, only sheet designations are in English language.

District map   1:500 000
Tokyo: GSI, 1974–75.
8 sheets, all published.

Raised relief map   1:250 000
Tokyo: GSI, 1983–.
43 sheets, all published.

Regional map   1:200 000
Tokyo: GSI, 1953–.
129 sheets, all published. ■

Topographic map   1:50 000
Tokyo: GSI, 1925–.
1249 sheets, all published. ■

Topographic map   1:25 000
Tokyo: GSI, 1964–.
4455 sheets, all published. ■
Also available on the same sheet lines as 1:50 000 series, and
covering some coastal areas, is a larger sheet 1:25 000 *Topographic
map of coastal areas.*

Bathymetric

Depth curve chart of the adjacent seas of Japan   1:8 000 000
Tokyo: MSA, 1971.
In English and Japanese.

Bathymetric chart of the adjacent seas of Nippon   1:3 000 000
Tokyo: MSA, 1968.
4 sheets, all published.
In English and Japanese.

Bathymetric chart   1:200 000
Tokyo: MSA, 1969–.
80 sheets, all published.
Japanese language.

Geological and geophysical

Unless otherwise mentioned, main places, legends and titles are in
English and Japanese for geological maps.

Geological map of Japan   1:5 000 000   4th edn
Yabate-machi: GSJ, 1982.

Marine geological map around the Japanese Islands   1:3 000 000
Yatabe-machi: GSJ, 1983.

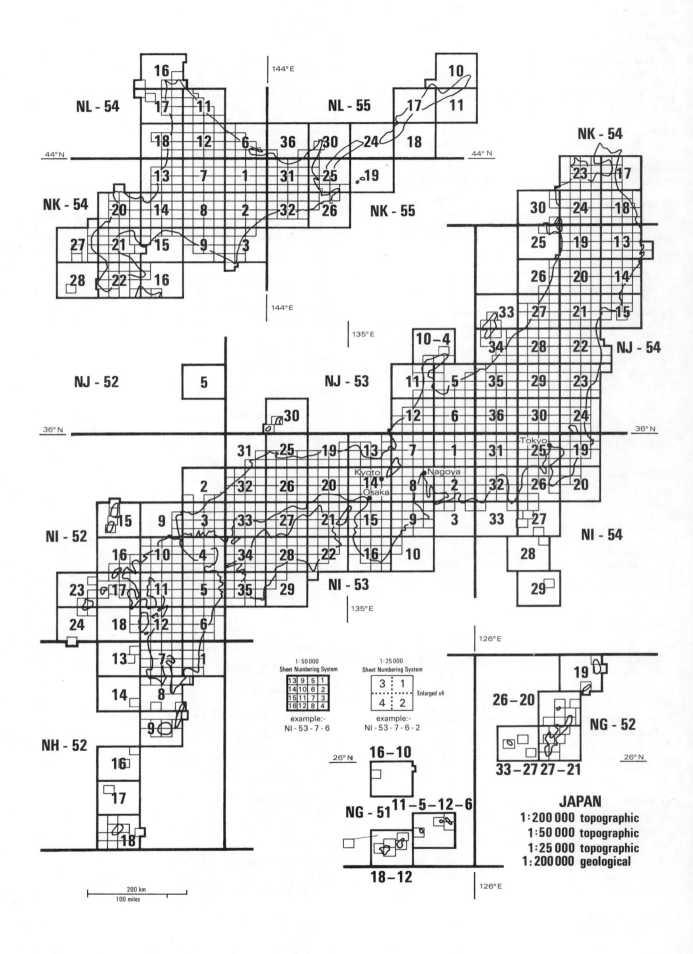

NL - 54

NL - 55

NK - 54

NK - 54

NK - 55

NJ - 54

NJ - 52

NJ - 53

10-4

NI - 52

NI - 54

NI - 53

NH - 52

1:50 000
Sheet Numbering System

13	9	5	1
14	10	6	2
15	11	7	3
16	12	8	4

example:-
NI - 53 - 7 - 6

1:25 000
Sheet Numbering System

3	1
4	2

Enlarged x4

example:-
NI - 53 - 7 - 6 - 2

16-10

NG - 51 11-5-12-6

18-12

19

26-20

NG - 52

33-27 27-21

JAPAN
1:200 000 topographic
1:50 000 topographic
1:25 000 topographic
1:200 000 geological

200 km
100 miles

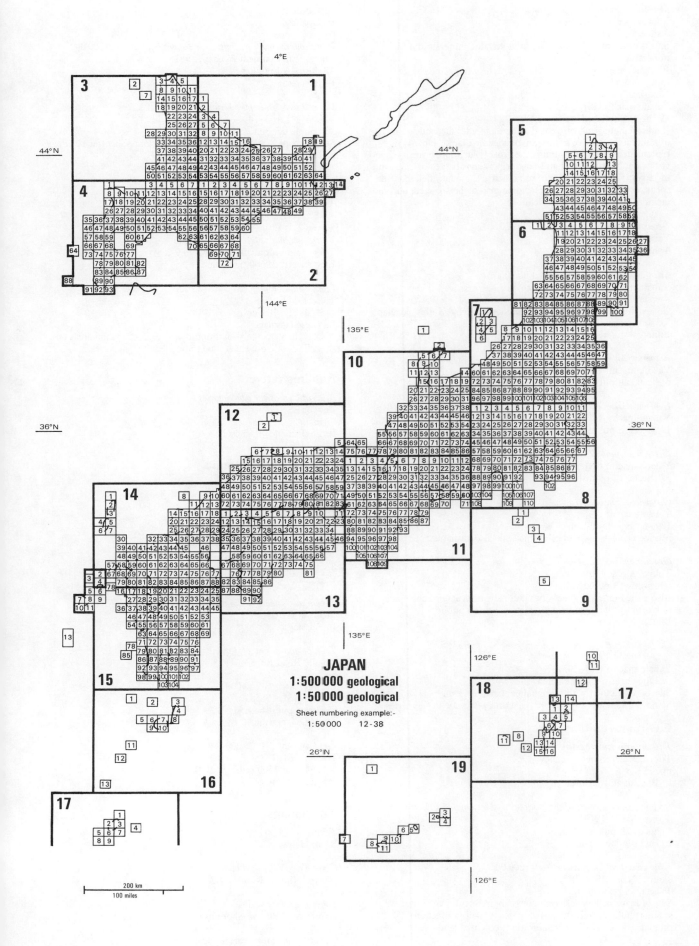

JAPAN
1:500 000 geological
1:50 000 geological

Sheet numbering example:-
1:50 000   12-38

Mineral distribution map of Japan
Yabate-machi: GSJ, 1960.
3 sheets, all published.
Sheets cover metallic minerals, non-metallic minerals and fuel minerals.
Bilingual map.

Geological map series 1:2 000 000
Yabate-machi: GSJ, 1953–.
29 sheets published.
English language sheets issued to date include: distribution maps of mineral deposits (6 sheets); metallogenic map of Japan; plutonism and mineralization map; coal fields; hot springs; oil and gas fields; hydrothermal and hot spring deposits; geological map of Japan; volcanoes of Japan; radiometric age map, metamorphic rocks; active faults in Japan; tectonic map of Japan; metamorphic facies of Japan; geothermal fields.

Geological map of Japan 1:1 000 000
Yabate-machi: GSJ, 1978.
4 sheets, all published.

Marine geology map series 1:1 000 000
Yabate-machi: GSJ, 1977–.
8 sheets, all published.
Geomagnetic, free gravity and Bouguer anomaly overlays available for some sheets.
Bilingual descriptions.

Aeromagnetic map off the coasts of Japan 1:1 000 000
Yatabe-machi: GSJ, 1982.
3 sheets, all published.

Geological Survey of Japan 1:500 000
Yabate-machi: GSJ, 1950–.
17 sheets, 6 published. ■

Neotectonic map series 1:500 000
Yatabe-machi: GSJ, 1982–.
17 sheets, 12 published. ■

Geological Survey of Japan 1:200 000
Yabate-machi: GSJ, 1956–.
129 sheets, 40 published. ■
Marine geology map series at this scale also available.

Gravity anomaly chart 1:200 000
Tokyo: MSA, 1969–.
80 sheets, 53 published.
Also issued on same sheet lines: submarine structural chart (all published); total magnetic intensity chart (69 published).
Bilingual scripts.

Geological map of Japan (regional geology quadrangle series) 1:50 000
Yabate-machi: GSJ, 1951–.
ca 1400 sheets, 450 published. ■
Sheets now issued with bilingual explanatory texts.
Some sheets issued at 1:75 000 or 1:100 000 scale in this series.

Environmental

The Natural Environment in Japan
Tokyo: Environment Agency, 1981–.
pp. 123.
In English and Japanese.

Bottom sediment chart of the adjacent seas of Japan 1:1 200 000
Tokyo: MSA, 1949.
4 sheets, all published.
In English and Japanese.

Land classification map series 1:200 000
Tokyo: NLA, 1954–1979.
46 sets, all published.
Sets of maps in Japanese only for each prefecture, covering: geomorphology, soil; subsurface geology; soil production; drainage systems; slope classification; potential land use.

National land use map 1:200 000
Tokyo: NLA, 1985–.

129 sheets.
In Japanese.

Plant and animal distribution maps by prefecture 1:200 000
Tokyo: Environment Agency, 1981.
47 sheets, all published.
In Japanese.

Land classification map series 1:50 000
Tokyo: NLA, 1974–.
300 published sets.
Sets of maps available for land classification, subsurface geology and soil.
Japanese language.

Soil map of Japan 1:50 000
Tokyo: Agricultural Administration Bureau, 1959–.
1249 sheets, *ca* 500 published.

Vegetation maps of Japan 1:50 000
Tokyo: Environment Agency, 1980–.
1250 sheets, *ca* 600 published.

Land condition map series 1:25 000
Tokyo: GSI, 1963–.
4455 sheets, 70 published. ■
Japanese language.
Also available is land condition maps series of coastal areas on sheet lines of large-sheet topographic equivalent series.

Administrative

Atlas of Census tract boundaries
Tokyo: BS, 1980.
87 maps.
In English and Japanese.

Human and economic

Regional planning atlas of Japan
Tokyo: JMC, 1984.
pp. 126 (66 maps).

Population maps of Japan: 1980 population census of Japan 1:1 000 000–1:5 000 000
Tokyo: BS, 1984.
8 series (96 sheets, all published).
Bilingual.

Japonija: ekonomicheskaja karta 1:2 000 000
Moskva: GUGK.
In Russian.

Traffic volume map of Japan 1:1 200 000
Tokyo: Road Bureau, 1978.

Land use map 1:50 000
Tokyo: GSI, 1965–72.
53 sheets published. ■
Japanese language.

Land use map 1:25 000
Tokyo: GSI, 1975–.
4455 sheets, 1500 published. ■
Japanese language.

Town maps

Tokyo 1:25 000
Bern: Hallwag, 1985.

Great Tokyo: detailed map 1:15 000
Tokyo: Nippon Kokuseisha.

Tokyo various scales
Hamburg: Falk and Shobunsa.

Japan: town maps various scales
Tokyo: Japan Guide Map Co.
7 sheets, all published.

Jordan

The **Jordan National Geographic Centre (JNGC)** was established in 1975 as the national mapping agency, with French and later American aid. It is responsible for the geodetic and topographic survey activities, for the production of national basic mapping and for the provision of maps to other government departments. JNGC remains under military control but maps are produced with the aim of meeting development requirements. None of the current mapping is, however, available for export.

Modern topographic mapping of Jordan was established by the allied forces during World War II. The geodetic network and mapping remained unrevised until the mid-1970s when, following establishment of JNGC, primary and secondary networks were established. Work on a tertiary net was begun in 1984. The basic scale map for the whole country is a multi-colour 1:50 000 series which covers the country in 184 sheets. This map uses 25 m contours and the Transverse Mercator projection and is compiled to full photogrammetric standards. A revision programme is in progress, producing about six new sheets a year, using 1981 flown aerial coverage. This map is used to derive an 84-sheet 1:100 000 series with 20 m contours and a 17-sheet 1:250 000 series with 100 m contours. A hypsometric tinted 1:750 000 scale map of the country is also compiled by JNGC. A National Atlas Project is under way: this work will be published in parts according to theme and will appear sequentially. Earlier agricultural and climatic atlases are no longer available. An atlas of the Palestinian case was published in a new edition in 1982. Other plans include a 1:500 000 scale tourist map and a new gazetteer of areas and placenames.

None of these current official mapping products is at present available outside the country.

JNGC began in 1985 to produce a 1:50 000 scale geological series for the **Natural Resources Authority**. The country was covered in five full-colour 1:250 000 scale geological sheets with English, German and Arabic text in the late 1960s by the German Bundesanstalt für Bodenforschung (now **Bundesanstalt für Geowissenschaften und Rohstoffe**) (**BfGR**). 1:100 000 coverage in 14 sheets produced under the same aid agreement was compiled as full-colour editions for only three areas, the remainder of the country being taken only as far as single-colour compilations. Geomorphological maps are also compiled by JNGC.

Cadastral mapping of the country is the responsibility of the **Department of Lands and Surveys**, which uses JNGC base mapping.

Small-scale general maps have been produced by a variety of overseas agencies including the **Central Intelligence Agency (CIA)**, **Bartholomew** and **Geoprojects**. The latter is perhaps most useful because of the city mapping it also contains. Maps of Israel now cover the West Bank and other western parts of the country (see also Israel and Middle East).

Further information

Jordan National Geographic Centre (1985) National report: paper submitted by Jordan to the *3rd United Nations Regional Cartographic Conference for the Americas*, New York: UN.

Jordan National Geographic Centre (1977) Jordan country report in technical papers submitted to the *9th United Nations Regional Cartographic Conference for Asia and the Pacific*, New York: UN.

Addresses

Jordan National Geographic Centre (JNGC)
PO Box 20214, AMMAN.

Department of Lands and Surveys
PO Box 70, AMMAN.

Department of Geological Resources and Mining
National Resources Authority, PO Box 39, AMMAN.

Natural Resources Authority
PO Box 7, 39 or 2220, AMMAN.

For Geoprojects and Bartholomew see Great Britain; for BfGR see German Federal Republic; for CIA and DMA see United States.

Catalogue

Atlases and gazetteers

Jordan: official standard names approved by the United States Board on Geographic Names
Washington, DC: DMA, 1971.
pp. 419.

General

The Hashemite Kingdom of Jordan　1:750 000
Amman: JNGC, 1979.

Oxford map of Jordan　1:730 000　2nd edn
Beirut: Geoprojects, 1984.
Double-sided.
Includes indexed town maps of Amman 1:25 000 and Central Amman 1:10 000. Also small inset maps of Queen Alia International Airport, Jeresh, Aqaba and Petra, tourist information and business guide.

Israel with Jordan　1:350 000
Edinburgh: Bartholomew.

Geological and geophysical

Geologische Karte von Jordanien/Geological map of Jordan　1:250 000
Hannover: Bundesanstalt für Bodenforschung, 1969.
5 sheets, all published.

Kampuchea

Kampuchea has been mapped by various colonial authorities, notably by the Americans during the years of the Vietnam war. The American mapping (no longer available) is still the best published coverage of the country. Series were compiled at scales of 1:100 000 (L 605) and 1:250 000 (L 509) from aerial photographs to give complete coverage. The current position in the country is very uncertain—the **Kampuchean National Geographic Service** is responsible for surveying and mapping of the country but its equipment and premises have been destroyed in the war between Vietnam and the various Democratic Kampuchean forces. Thematic mapping of the country has been carried out by the **United States Geological Survey (USGS)** (hydrogeological maps) and by the **Institut Français de Pondichéry (IFP)** (vegetation maps). Small-scale maps of the country have been produced by both the American **Central Intelligence Agency (CIA)** and the Russian **Glavnoe Upravlenie Geodezii i Kartografii (GUGK)**.

Addresses

Service Géographique de Kampuchea
BP 515, Bd Monivong, PHNOM PENH. NB: no mail is currently getting into Kampuchea.

For CIA, DMA and USGS see United States; for GUGK see USSR; for Institut Français de Pondichéry see India.

Catalogue

Atlases and gazetteers

Cambodia: official standard names approved by the United States Board on Geographic Names 2nd edn
Washington, DC: DMA, 1971.
pp. 392.

General

Kampuchea 1:1 400 000
Washington, DC: CIA, 1983.

Kampuchija 1:1 000 000
Moskva: GUGK, 1984.
In Russian.
With ancillary climatic inset map and placename index.

Geological and geophysical

Ground water resources of Cambodia 1:1 000 000
Washington, DC: USGS, 1977.
2 sheets, both published.
With accompanying text *The ground water resources of Cambodia.*
Maps cover geology and location of wells.

Environmental

Carte internationale du tapis végétale et des conditions écologique: Cambodge 1:1 000 000
Pondichéry: IFP, 1971.
With accompanying explanatory text.

Korea (North)

There is very little information about the current official mapping of the People's Republic of North Korea. Some mapping is carried out by the **Geography and Geology Research Institute of the Academy of Sciences**, in Pyongyang, but no published mapping is available and there is no information about current specifications or series. Many South Korean series were originally designed to include coverage of the areas to the north of the 51st parallel, but no recent larger-scale maps cover the north.

Several small-scale maps provide general coverage. These include the 1:1 100 000 scale administrative and physical base map of the whole peninsula, revised on a regular basis by **Chungang Atlas Co.** and the Russian reference map by **Glavnoe Upravlenie Geodezii i Kartografii (GUGK)**. The American **Central Intelligence Agency (CIA)** published in 1972 a general map covering just North Korea.

Addresses

Geology and Geography Research Institute
Academy of Sciences, Mammoon-Dong, Central District, PYONGYANG.

For Chungang see South Korea; for CIA and DMA see United States; for GUGK see USSR.

Catalogue

Atlases and gazetteers

North Korea: official standard names approved by the United States Board on Geographic Names
Washington, DC: DMA, 1982.
pp. 598.

General

Koreja 1:1 500 000
Moskva: GUGK.
In Russian.
With ancillary thematic insets and placename index.
Also covers South Korea.

North Korea 1:1 220 000
Washington, DC: CIA, 1972.
With small-scale land use, population and economic insets.

Korean Peninsula 1:1 200 000
Washington, DC: CIA, 1980.
Also covers South Korea.

Administrative

The map of Korea 1:1 100 000
Seoul: Chungang, 1983.
In English and Korean.
Also covers South Korea.

Korea (South)

The **National Geography Institute (NGI)** of the Ministry of Construction is responsible for all geodetic surveys in the Republic of Korea and produces topographic map series of the country. By 1974 NGI had completed its topographic map programme which comprises four scales of mapping. The 1:25 000 scale covering South Korea in 763 sheets is still the national base map; complete coverage is also available in the 293-sheet 1:50 000 series and the 13-sheet 1:250 000 series, and NGI compiles a 1:1 000 000 map. All these full-colour series are compiled on the Transverse Mercator projection using metric contour intervals and are photogrammetrically derived. A systematic revision programme is being carried out at 2, 5 or 7 year intervals depending on the number of units of change in the mapped area. A large-scale programme of 1:5000 mapping is in progress and will eventually cover the entire country in about 15 000 sheets. This map is available in either diazo, mono- or multi-colour prints, depending on demand. It has 5 m contours, is derived from 1:20 000 aerial coverage, and so far about half of South Korea has been covered. Like the Japanese GSI, NGI produces a coastal topographic map series at 1:25 000 scale: this programme will be completed by the end of 1986.

NGI is also involved in thematic mapping projects, especially in the field of land use series. A 1:25 000 map on the same sheet lines as the topographic map, depicting 12 land use groups and 45 subgroups, is issued by NGI. This series has been under revision since 1982. At 1:250 000 NGI publishes a land use series derived from computer analysis of satellite image data. Other mapping activities have involved the production of a National Atlas, and large-scale orthophotomaps of metropolitan new development areas, as well as large-scale mapping backup for other government agencies.

The public availability of NGI mapping outside Korea depends on authorization obtainable from the Director General of NGI. We have listed the topographic series in the catalogue section but have not provided a graphic index.

The Republic of Korea Army Map Service (ROKAMS) produces military mapping of the country, and bathymetric and marine chart production is carried out by the Hydrographic Office. Other thematic series mapping uses NGI series as a base. The Office of Forestry produces 1:25 000 series showing forest type, land use, planning information and soils. The **Korean Institute of Energy and Resources (KIER)** prepares 1:50 000 and 1:250 000 geological mapping, and the Plant Environment Research Institute prepares a 1:50 000 soil series. All of these series are published in Korean language editions.

The commercial sector in Korean map publishing is like in Taiwan perhaps more important than the quality of its mapping would suggest, in view of the difficulty of obtaining official map series. **Chungang Atlas Co.** has published a small-scale map of the whole peninsula, including North Korea, which shows administrative divisions. The same company also issues a regularly revised relief based tourist map of South Korea at 1:500 000 scale. **Asia Aero Survey Co.** publishes an English language tourist map at 1:750 000 scale which contains useful town maps of the 11 cities on the verso. Town maps of Seoul are also published by these organizations, by the Seo Jean Industrial Company (on a contoured relief base) and by the **Korean National Tourist Office**. Amongst foreign agencies publishing mapping of South Korea are the American **Central Intelligence Agency (CIA)**, the Russian **Glavnoe Upravlenie Geodezii i Kartografii (GUGK)** and the German commercial house **Karto+Grafik**.

Further information

Kim, W. (1985) Mapping in Korea, *ICA Newsletter* May 1985, **5**, 6–7.

Dege, E. (1981) Ämtliche und halbämtliche Kartenwerke der Republic Korea: eine Bestandsaufnahme, *International Yearbook of Cartography*, **21**, 53–74.

Mapping in Korea: paper presented to the *11th United Nations Regional Cartographic Conference for Asia and the Pacific*, New York: UN.

Addresses

National Geography Institute (NGI)
Ministry of Construction, 43-1 Hwikyeong-dong, Dongdaemun-ku, SEOUL.

Korean Institute of Energy and Resources (KIER)
219-5 Garibong-dong, Youngdeungpo-gu, SEOUL 150-06.

Asia Aero Survey and Consulting Engineers
429 Map'O-Gu, Sinsu-Dong, SEOUL.

Chungang Atlas Co.
199 34 2-ga, Eulijiro, Jung gu, SEOUL.

Korean National Tourist Corporation
3rd Floor, Kukdong Building, 60-1 3ka, Chungmuro, Chung-ku, SEOUL.

For CIA and DMA see United States; for GUGK see USSR; for Karto+Grafik see German Federal Republic.

Catalogue

Atlases and gazetteers

South Korea: official standard names approved by the United States Board on Geographic Names
Washington, DC: DMA, 1966.
pp. 370.

General

Koreja 1:1 500 000
Moskva: GUGK.
In Russian.
Covers both North and South Korea.
Includes placename index and 3 ancillary thematic inset maps.

South Korea 1:1 220 000
Washington, DC: CIA, 1973.
Includes vegetation, economic and population insets.

Korean peninsula 1:1 200 000
Washington, DC: CIA, 1980.
Covers both North and South Korea.

Korea Sud: Hildebrands Urlaubskarte 1:800 000
Frankfurt AM: Karto+Grafik, 1985.
Available as English or German edition.

Tourist map of Korea 1:750 000
Seoul: Asia Aero Survey, 1983.
English language.
Includes town plans of Seoul, Pusan, Taegu Suwon, Ch'unh'on,
Taejon, Kwangju, Inch'on, Chonju, Chongju and Cheju on back.

Tourist map of Korea 1:500 000
Seoul: Chungang, 1985.
In English and Korean.

Topographic

Republic of Korea 1:250 000
Seoul: NGI.
13 sheets, all published.
In Korean.

Republic of Korea 1:50 000
Seoul: NGI.
239 sheets, all published.
In Korean.

Republic of Korea 1:25 000
Seoul: NGI.
762 sheets, all published.
In Korean.

Geological and geophysical

Geological map of Korea 1:250 000
Seoul: KIER, 1974–.
13 sheets, all published.
In Korean.

Geological map of Korea 1:50 000
Seoul: KIER, 1960–.
239 sheets, *ca* 100 published.
In Korean.

Administrative

The map of Korea 1:1 100 000
Seoul: Chungang, 1983.
In English and Korean.
Also covers North Korea.

Town maps

Seoul 1:60 000
Seoul: Asia Aero Survey Co., 1986.
In English and Korean.
Contoured town plan.
Includes 1:9000 scale downtown Seoul map on verso.

Complete map of Seoul 1:40 000
Seoul: Seo Jean Industrial Company, 1981.
In English and Korean.
Contoured town plan.
Includes 1:12 000 scale Central Seoul inset.

Tourist map of Seoul 1:36 000
Seoul: Korea National Tourist Commission, 1981.
Double-sided.
1:6000 scale map of downtown Seoul on verso.

Kuwait

Topographic, cadastral and utility mapping and surveying
activities are carried out by the **Survey Department**, one
of the technical departments of the Municipality of
Kuwait. Until the 1950s there was very limited mapping.
The Survey Department grew from a surveying unit
established in the Public Works Department in 1951 and
Kuwait is now very well mapped indeed, but the official
series are not available, except for approved projects,
outside the country. The whole state is covered in line
maps at 1:100 000, 1:50 000 and 1:25 000, and there is a
photomap series at 1:10 000 scale which also gives
complete coverage in about 650 sheets. These maps are
regularly revised. One kilometre square 1:2000
topographic line maps sheets cover settled areas of the
state. A 1:500 scale series is compiled for major built-up
areas. All these series use UTM grid, are derived from
aerial photography and are revised on a regular basis.
There are at present plans to introduce a digital mapping
system to integrate cadastral, utility and topographic maps
using a Kuwait Transverse Mercator grid. This KUDAMS
project began in 1983 and includes digital data capture at
1:500 scale. A useful general map of the country was
compiled in the late 1970s by Geoprojects in the Arab
World Map series and also included tourist notes and town
plans of Kuwait and the Kuwait Urban area. The second
edition of this map is now distributed and published by
Gulf Union Company. **Sahab** has also published a tourist
map of Kuwait.

Geological coverage of Kuwait at 1:250 000 scale was
compiled for the Ministry of Commerce and Industry in
the late 1960s but is no longer available.

Addresses

Survey Department
Kuwait Municipality, PO Box 10, Safat, KUWAIT.

Gulf Union Company
PO Box 2911, Safat, KUWAIT.

Further information

Background to current developments is contained in Kuwait
(1980) Cartographic activities in Kuwait: paper presented to the
*9th United Nations Regional Cartographic Conference for Asia and
the Pacific*, New York: UN; and Kuwait (1983) Computer assisted
mapping in Kuwait: paper presented to the *10th United Nations
Regional Cartographic Conference for Asia and the Pacific*, New
York: UN.

Catalogue

General

The map of Kuwait 1:500 000 2nd ed.
Kuwait: Gulf Union Company, 1983.
Includes tourist information and town plans of Kuwait 1:10 000
and Kuwait Urban area 1:70 000.

Kuwait general map 1:500 000
Tehran: Sahab, 1976.

Laos

The national mapping agency of Laos is the **Service Géographique National (SGN)** but current mapping of Laos is not available and there is very little information about what has been compiled. American maps published during US involvement in South East Asia in the 1960s are still the best published topographic series. These were compiled at scales of 1:250 000, 1:100 000 and 1:50 000 and derived from aerial coverage, but they are no longer available. SGN compiled town plans, general maps and administrative maps of the country in the 1960s, but none is still available. No earth science mapping is published, but the Russian **Glavnoe Upravlenie Geodezii i Kartografii (GUGK)** has produced a small-scale map of the country, which is available. For small-scale thematic coverage of the country see Asia, South Eastern and Asia.

Addresses

Service Géographique National (SGN)
Office of the Council of Ministers, BP 167, VIENTIANE.

Direction de la Géologie et des Mines
Ministère de l'Industrie l'Artisanat et Forets, VIENTIANE.

For GUGK see USSR.

Catalogue

General

Laos 1:1 250 000
Moskva: GUGK, 1978.
In Russian.
With accompanying ancillary climate, population and economic maps and placename index.

Town maps

Atlas de la ville de Louang Phrabang
Vientiane: UN, 1973.
pp. 257.

Lebanon

The **Directorate of Geographic Affairs (DAG)**, part of the Lebanese Ministry of Defense, was established in 1962 to bring together a variety of separate mapping organizations into a single official mapping agency. DAG is now responsible for geodetic, cadastral, topographic and aerial surveying of the country and for the publication of a variety of official maps. The basic scale map is still a five-colour 1:20 000 scale series introduced in the early 1960s to replace an older French colonial 1:50 000 series published by the Bureau Topographique du Levant. This 1:20 000 map uses 10 m contour intervals and the oblique stereographic Lambert pseudogrid and is published in Arabic and French. It was completed in 1974 and covers the country in 121 sheets. A limited revision programme has taken place. The 1:50 000 old series is still available and some sheets were revised by DAG in the early 1960s but this map has been discontinued. DAG has also published seven sheets in a new French language large-sheet 1:50 000 map, derived from the 1:20 000 survey: this 12-sheet series has not been completed. There are plans to introduce a new 1:25 000 scale series on the same specifications as the French *Série bleue* and to introduce digital mapping systems. Other French aid has been given to revise the 1:250 000 and 1:500 000 scale mapping of the country. These, however, are both contingent upon the political situation in the Lebanon. Large-scale monochrome coverage of towns at 1:1000 and 1:2000 scales has also been published by DAG.

Other mapping carried out by DAG includes soil, land cover, water resources and erosion maps and a variety of thematic coverage at 1:200 000 scale. Some DAG mapping is still available but no new sheets have been compiled since the publication of the 1980 map catalogue.

The 1:20 000 soil maps using the topographic base and sheet lines were prepared in the 1960s by Institut de Recherches Agronomique. Current plans include a new 27-sheet 1:50 000 soil series to be compiled by **Centre de Recherche Agronomique** in the Ministry of Agriculture. These maps are not currently available.

Geological mapping of Lebanon was carried out in the 1950s and 1960s by L. Dubertret: some sheets were issued on old series 1:50 000 scale sheet lines. A new version of the 1:200 000 geological map was published in 1978 but is already unavailable. The **Ministry of Hydraulics** is now responsible for geological surveying and the **National Council for Scientific Research** has requested the Geology Department of the American University of Beirut to compile new 1:20 000 scale geological coverage of central parts of the country.

Other mapping of the country is compiled by the American **Central Intelligence Agency (CIA)**, by **Bartholomew** and as sheets issued in the *Tübinger Atlas des Vorderen Orients* (TAVO).

Commercial mapping in the Lebanon is regulated by the Army HQ and publishers must seek approval from DAG before the publication and distribution of maps. **Geoprojects SARL**, based in Beirut, is the major commercial house and issues maps in the useful *Arab World Map Library* series to cover most Arab countries. **Guide Stephen** and **Boulos** in Beirut compile tourist mapping of the Lebanon.

Further information

DAG publishes a 1980 catalogue.

Addresses

Directorate of Geographic Affairs (DAG)
Ministry of Defense, Yarze, BEYROUTH.

Centre de Recherche Agronomique
Ministry of Agriculture, BEYROUTH.

National Council for Scientific Research
c/o Prime Minister's Office, BEYROUTH.

Ministry of Hydraulics
BEYROUTH.

American University of Beirut
BP 1786, BEYROUTH.

Geoprojects SARL
PO Box 113-5294, BEYROUTH.

Bureau G Stephen
Imm Matta, Hazmieh, BP 116-5239, BEYROUTH.

Boulos
BEYROUTH.

For Bartholomew see Great Britain; for CIA and DMA see United States; for TAVO see German Federal Republic.

Catalogue

Atlases and gazetteers

Lebanon: official standard names approved by the United States Board on Geographic Names
Washington, DC: DMA, 1970.
pp. 676.

General

Lebanon 1:400 000
Washington, DC: CIA, 1979.
With economic, land use and population inset maps.

Lebanon 1:250 000
Washington, DC: CIA, 1983.
Includes refugee and other settlements on topographic base.

Map of Lebanon 1:200 000
Beyrouth: All Prints Distributors and Publishers/Geoprojects, 1985.
Double-sided.
Includes indexed city map of Beirut 1:10 000, business guide, tourist guide and descriptive text.
Also available as French language edition.

Map of Lebanon and Beirut 1:200 000
Beyrouth: Boulos, 1984.
Includes 1:10 000 scale Beirut townplan.

Lebanon 1:200 000
Edinburgh: Bartholomew, 1980.

Liban 1:200 000
Beyrouth: Guide Stephen, 1983.

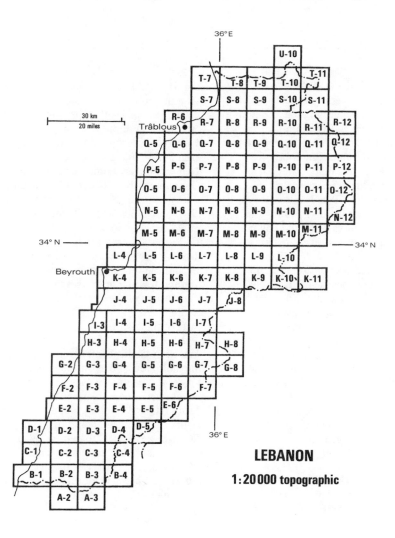

LEBANON

1:20 000 topographic

Topographic

Carte du Liban   1 : 100 000
Beyrouth: DAG, 1964.
6 sheets, all published.

Carte du Liban   1 : 50 000
Beyrouth: DAG, 1938–66.
27 sheets, all published.
Partly replaced by new 12-sheet series.

Carte du Liban   1 : 20 000
Beyrouth: DAG, 1963–74.
121 sheets, all published. ∎

Human and economic

Lebanon population and religious affiliation   1 : 250 000
Washington, DC: CIA, 1985.

Libanon Religionen/Lebanon religions   1 : 350 000
Wiesbaden: TAVO, 1979.
In English and German.

Libanon Christentum/Lebanon christianity   1 : 350 000
Wiesbaden: TAVO, 1979.
In English and German.

Town maps

Beyrouth et banlieu   1 : 10 000
Beyrouth: Guide Stephen, 1982.

Macao

Macao is a small Portuguese territory near to Hong Kong. Mapping of the peninsula and islands of Taipa and Coloane is the responsibility of the **Serviço de Cartografia e Cadastro (SCC)**, formerly Missão de Estudos Cartográficos de Macao (MECM), which was established as an independent agency in 1975. It carries out topographic, geodetic and cadastral activities. The territory is covered in large-scale plans—1 : 500, 1 : 1000 and 1 : 2000 scales are compiled and are used to derive smaller-scale mapping. Current single-sheet 1 : 25 000 coverage is published with 10 m contours, and three-sheet series are published and regularly revised by SCC at 1 : 5000 and 1 : 10 000 scale.

Two small full-colour land use maps were compiled by Richard Edmonds at the University of Hong Kong, and are available from the author at the **School of Oriental and African Studies**, London.

An indexed tourist town plan is published by **Universal Publications**, Hong Kong. The territory is also covered on many of the smaller-scale maps of Hong Kong, see Hong Kong.

Further information

DECRETO-LEI NO.102/84/M (1984) Governo de Macau *Boletim oficial de Macau*, No. 36. Portuguese and comprehensive administrative background to SCC.

Macau (1977, 1980) Cartographic activities in Macao: papers presented to the *8th and 9th United Nations Regional Cartographic Conferences for Asia and the Pacific*.

Addresses

Divisão de Topografia e Cartografia, Serviço de Cartografia e Cadastro (SCC)
CP 3018, MACAU.

Universal Publications
PO Box 78669, HONG KONG.

School of Oriental and African Studies
University of London, Malet St, LONDON WC1 7HP, UK.

For DMA see United States.

Catalogue

Atlases and gazetteers

Hong Kong and Macao: official standard names approved by the United States Board on Geographic Names
Washington, DC: DMA, 1972.
pp. 52.

Topographic

Territorio de Macau   1 : 25 000
Macau: MECM, 1983.

Macau   1 : 10 000
Macau: SCC, 1985.
3 sheets, all published.

Macau   1 : 5000
Macau: SCC, 1985.
3 sheets, all published.

Environmental

Mapa de Macau: utilizacão de terrenos   1 : 8000   Richard Edmonds
Hong Kong: University of Hong Kong, 1983.
2 sheets, both published.

Town maps

Map of Macau and Zhuhai   1 : 6000
Hong Kong: Universal Publications, 1986.
With Chinese and English indexes and tourist notes on verso.
Includes 4 small-scale inset location maps.

Malaysia

Malaysia comprises two areas with different cartographic histories and organizations. Sabah and Sarawak on the island of Borneo, which were British colonies until independence in 1963, were mapped by British agencies, whilst peninsular Malaysia has a much longer history of independent mapping.

The **Directorate of National Mapping Malaysia (PPNM)** is responsible for topographic and geodetic surveys of peninsular Malaysia, and coordinates survey and mapping of Sabah and Sarawak. In 1984 it also assumed cadastral surveying responsibilities and was reconstituted within the Ministry of Land and Regional Development. The first survey office in the country was established in 1909 as the Federated Malay State Survey Department and the early specialist map publication was carried out in conjunction with Britain and the Survey of India. By the beginning of World War II 60 per cent of Malaya was covered in 1-inch scale topographical maps and the series was completed by 1953 from aerial photographs. The New Standard Mapping 1:63 360 scale series L7010 was introduced in 1950. Sheets cover quarter-degree squares and complete coverage of peninsular Malaysia was attained in 1974 with revisions of sheets continuing until 1977. Series L8010 at 1:25 000 scale was also started in the 1950s and six sheets covered a single 1-inch map. 377 sheets in this basic scale series were published but this map has also been discontinued. Following independence in 1963 PPNM was established in its current form. The current series are the 1:50 000 metric series L7030 and its four 1:25 000 component sheets. These photogrammetrically derived multi-colour maps use 25 m contour intervals and are plotted on a Rectified Skew orthomorphic projection around a central meridian of 324°. It is intended to publish the 211 sheets covering peninsular Malaysia over a 10 year period and the project was launched in 1979. Larger-scale urban mapping projects include series L905 which will map 110 towns at 1:5000 and 1:10 000 scale, and a new 1:10 000 scale series L808. In the 1986–1990 period government approval has been given for the use of digital techniques in the production of topographic series.

All of these topographic and large-scale maps are restricted and may not be sold without security clearance.

Smaller-scale coverage is available without restriction. General maps of the whole of Malaysia are published at 1:760 000, with either an administrative or a physical base. Both are revised on a regular basis. At larger scales PPNM maps only the peninsula. State-based coloured land use maps, published at different scales depending on the size of the state, are issued with accompanying explanations, and in 1979 the first unrestricted town map of a city was published, which covers Kuala Lumpur at 1:15 000 scale.

Other activities of PPNM include the publication of cartographic works for the Geological Survey, Soil Survey and Forestry Departments.

Geological mapping of peninsular Malaysia is carried out by the **Geological Survey of Malaysia (Penyiasatan Kajibumi Malaysia) (PKM)**. 1:63 360 scale maps cover about 60 per cent of peninsular Malaysia, and have been published with accompanying explanatory bulletins. A further 10 per cent of the country is currently being surveyed at this scale. Since 1973 this has been issued on the new 1:63 360 topographic map sheet lines. Smaller-scale coverage is also published, including 1:500 000 scale hydrogeological, geological and mineral maps of West Malaysia.

Amongst the commercial map publishers issuing coverage of Malaysia are the **Automobile Association of Malaysia**, **Bartholomew**, **Nelles Verlag** and **Karto+Grafik**.

Sarawak and Sabah

Sabah and Sarawak are covered by mapping compiled by the **Directorate of Lands Sarawak (Jabatan Tanah dan Ukur) (JTU)** and the **Directorate of Lands and Survey Sabah**. The basic scale is a 1:50 000 map series T735, which is published by PPNM. This six-colour series uses British DOS style with contours at 50 foot intervals: it was first established as a monochrome uncontoured map in the early 1950s. Only since 1960 have more ground control points been established and photogrammetric methods used to produce the full-colour editions. It is still produced with the assistance of the British **Overseas Surveys Directorate (OSD)**, and gives complete coverage of Sarawak (210 sheets) and of much of Sabah. There are plans to introduce a new 1:50 000 series with a metric contour interval. Some areas are mapped at larger scales, including a 1:25 000 series T 834 and a 1:10 000 scale town map series. 1:250 000 series T 1501 coverage is also compiled to JOG specifications and is used as a base for 1:250 000 scale population and land use maps, which each cover Sarawak in 10 sheets.

All these larger-scale official maps of Sabah and Sarawak are restricted, but thematic mapping is available for the two states at small scales, as are general maps of Sarawak and Sabah.

Geological mapping of Sabah and Sarawak is carried out by PKM through its offices in Kota Kinabalu (Sabah) and Kuching (Sarawak). Large-scale programmes have resulted in the publication of 1:50 000, 1:125 000 and 1:250 000 scale geological mapping with accompanying monographs for large areas of North Borneo. Smaller-scale coverage is also available, including igneous rock distribution maps at 1:500 000, 1:2 000 000 scale geological cover and 1:500 000 scale full-colour geological mapping of Sarawak.

Resources mapping of Sabah has been carried out by the DOS for the British **Land Resources Development Centre (LRDC)**. Using Joint Operations Graphic sheet lines, 10 sheets were published for both soils and land capability classification as part of the *Land Resource Studies* series in the mid-1970s. These series give complete coverage of Sabah and are still available with accompanying textual volumes.

Further information

Brief lists of available mapping are issued by PPNM, PKM and the Departments of Lands and Surveys Sabah and Sarawak.

Useful background to the development of Malaysia's mapping is in Malaysia (1980) Cartographic activities in Malaysia 1977–1979, paper presented to the *9th Regional Cartographic Conference for Asia and the Pacific*, New York: UN; Malaysia (1983) An outline to the development of cartography in Malaysia, paper presented to the *10th United Nations Regional Cartographic Conference for Asia and the Pacific*, New York: UN; Malaysia (1987) Cartographic activities in Malaysia 1984–86: paper presented to the *11th United Nations Regional Cartographic Conference for Asia and the Pacific*, New York: UN.

Addresses

Directorate of National Mapping (PPNM) (Direktorat Pemataan Negara Malaysia)
Bangunan Ukor, Jalan Gurney, KUALA LUMPUR 15-02.

Geological Survey Department (PKM) (Penyiasatan Kajibumi)
PO Box 1015, IPOH, Perak.

Automobile Association of Malaysia
3 Jalan 8/ID, PO Box 34, PETALING JAYA.

Lands and Survey Department (JTU) (Jabatan Tanah dan Ukor)
KUCHING, Sarawak.

Lands and Survey Department
KOTA KINABALU, Sabah.

Geological Survey Department Sabah (PKM)
Beg Berkunci, Jalan Penampang, KOTA KINABALU, Sabah.

Geological Survey Department Sarawak (PKM)
Jalan Wan Abdul Rahman, Peti Surat 560, KUCHING, Sarawak.

For GUGK see USSR; for Nelles Verlag and Karto+Grafik see German Federal Republic; for DOS/OSD, LRDC and Bartholomew see Great Britain; for DMA see United States.

Catalogue

Atlases and gazetteers

Malaysia, Singapore and Brunei: official standard names approved by the United States Board on Geographic Names
Washington, DC: DMA, 1970.
pp. 1014.

General

Malajzija 1:2 500 000
Moskva: GUGK, 1982.
In Russian.
With climatic and economic inset maps and placename index.

Malaysia 1:2 000 000 Series 1208 5th edn
Kuala Lumpur: PPNM, 1980.
2 sheets, both published.
Relief base with bilingual legend.

APA map of Malaysia 1:1 500 000
München: Nelles Verlag, 1984.
Double-sided road map.
With inset town plans of Bandar Seri Begawan, Kuala Lumpur and Singapore.

Malaysia–Singapore 1:1 000 000
Edinburgh: Bartholomew, 1983.
Includes 1:150 000 scale inset of Singapore.

Semenanjung Malaysia 1:760 000 Series 1307
Kuala Lumpur: PPNM, 1980.
3 sheets, all published.
Relief base.

Administrative

Semanjung Malaysia 1:760 000 Rampaian 45
Kuala Lumpur: PPNM, 1978.
3 sheets, all published.
Political map.

Town maps

Kuala Lumpur 1:15 000
Kuala Lumpur: PPNM, 1979.

PENINSULAR MALAYSIA

General

Semenanjung Malaysia 1:500 000 Series L 4010
Kuala Lumpur: PPNM, 1977.
Relief map of peninsular Malaysia.

Geological and geophysical

Peta kajibumi semenanjung Malaysia/Geological map of peninsular Malaysia 1:2 000 000
Ipoh: PKM, 1973.

Peta kajibumi semenanjung Malaysia/Geological map of peninsular Malaysia 1:1 000 000
Ipoh: PKM, 1984.

Peta kajibumi Malaysia Barat/Geological map of West Malaysia 1:500 000
Ipoh: PKM, 1973.
2 sheets, both published.

Peta hidrogeologi Semanjung Malaysia/Hydrogeological map of peninsular Malaysia 1:500 000
Ipoh: PKM, 1975.
2 sheets, both published.

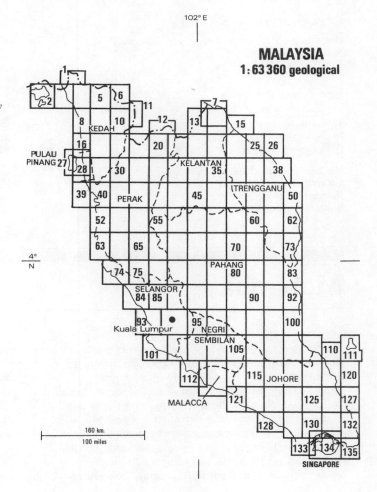

Peta taburan galian semenanjung Malaysia/Mineral distribution map of peninsular Malaysia 1:500 000
Ipoh: PKM, 1976.
2 sheets, both published.

Peninsular Malaysia: geological map 1:63 360
Ipoh: PKM, 1950–.
135 sheets, *ca* 30 published. ■
With accompanying explanatory bulletins and memoirs.

Environmental

Malaya: land utilisation map 1:760 320
Kuala Lumpur: PPNM, 1970.

Malaysia Barat Semenanjung Malaysia 1:253 440–1:63 360
Kuala Lumpur: PPM, 1961–.
16 sheets, all published.
Land use maps of West Malaysian States.
With bilingual text.

Administrative

Semenanjung Malaysia 1:500 000 Rampain 81
Kuala Lumpur: PPNM, 1977.
Administrative map of peninsular Malaysia.

SABAH

See also Sarawak for mapping covering both states.

General

Sabah 1:1 425 000
Tolworth: DOS, 1964.

Sabah 1:500 000
Kuala Lumpur: PPPM, 1976.
Includes administrative divisions.
Also available with forest use overprint.

Geological and geophysical

Peta taburen galien Sabah/Mineral resources map of Sabah 1:500 000
Kota Kinabalu: PKM, 1976.

Environmental

The soils of Sabah 1:250 000
Tolworth: DOS for LRDC, 1975.
10 sheets, all published.
With accompanying land resources study (5 vols).

The land capability classification of Sabah 1:250 000
Tolworth: DOS for LRDC, 1976.
10 sheets, all published.
With accompanying land resources study (4 vols).

SARAWAK

General

Sarawak and Brunei 1:1 670 000 DOS 973
Tolworth: DOS, 1964.

Sarawak 1:1 000 000 SS7 6th edn
Kuching: JTU, 1983.
Relief base.

Sarawak 1:500 000 SS8 5th edn
Kuching: JTU, 1983.
2 sheets, both published.
Relief base.

Geological and geophysical

Geology of Sarawak and Sabah 1:3 300 000
Kuching: PKM, 1975.

Geological map of Sarawak, Brunei and Sabah 1:2 000 000
Kuching: PKM, 1964.

Mineral resources of Sarawak and Sabah Malaysia 1:2 000 000
Kuching: PKM, 1965.

Peta hasil galien Sarawak/Mineral resources map of Sarawak Malaysia 1:1 000 000
Kuching: PKM, 1976.

Geological map of the British territories in Borneo 1:1 000 000
Kuching: PKM, 1955.

Peta Kajibumi Sarawak/Geological map of Sarawak 1:500 000
Kuching: PKM, 1982.
2 sheets, both published.

The igneous rocks of Sarawak and Sabah 1:500 000
Kuching: PKM, 1968.
2 sheets, both published.
With accompanying text.

Geological Survey of North Borneo/Sarawak/Sabah 1:250 000,
1:125 000 and 1:50 000
Kuching: PKM, 1954–.
ca 40 published sheets.
Sheets issued with accompanying reports, memoirs and bulletins.

Environmental

Sarawak Malaysia Timor: land use 1:1 500 000
Kuching: JTU, 1970.

Administrative

Sarawak Malaysia Timor 1:2 000 000 SS13
Kuching: JTU, 1970.

Mongolia

The national mapping agency in Mongolia is the **State Geodesy and Cartography Board (SGCB)**. Before its establishment in 1970 various state agencies were responsible for geodetic and topographical surveying and for the production of official mapping. From 1935 until 1955 work progressed on the creation of a geodetic network for the country using Soviet expertise and resources. A 1:100 000 scale topographic series, modelled on Russian specifications, was compiled by stereo-topographic methods from aerial coverage between 1946

and 1955. From this basic scale were derived 1:200 000, 1:500 000 and 1:1 000 000 scale mapping. Only after the rationalization of surveying and mapping in the 1970s has there been any attempt to produce larger-scale coverage. SGCB now has photogrammetric programmes at 1:10 000, 1:25 000 and 1:50 000 scales for the developed areas of the country. Important thematic series at 1:1 000 000 scale have been compiled from remotely sensed data for use by the various planning agencies in the State. There are also revision programmes in operation to up-date the 1:100 000 scale series, and atlas compilation projects. Recent publications include an ethnographic atlas and an economic atlas. All of this mapping activity continues to be carried out in association with the Soviet mapping agencies and sometimes in association with other Mongolian scientific agencies such as the **Institute of Geography and Permafrost Studies**. All is published in Mongolian or Russian language editions and none is available to the western world.

Small-scale mapping of the country is published by several foreign organizations, often in conjunction with mapping of China, e.g. by **Bartholomew**. The Russian **Glavnoe Upravlenie Geodezii i Kartografii (GUGK)** has published Russian language 1:3 000 000 scale sheets with thematic insets of the country.

For other available mapping of the country see World, Asia, USSR and China.

Further information

The only source on the mapping of Mongolia was recently issued: Sanzhaazhamts, Zh. (1985) Geodesy and cartography in Mongolia, *Geodeziya i Kartografiya* 2/85, Russian language.

Addresses

State Geodesy and Cartography Board (SGCB)
ULAANBAATAAR.

Institute of Geography and Permafrost Studies
Academy of Sciences, ULAANBAATAAR.

For DMA see United States.

Catalogue

Atlases and gazetteers

Mongolia: official standard names approved by the United States Board on Geographic Names
Washington, DC: DMA, 1970.
pp. 256.

General

Mongolskaja narodnaja Respublika 1:3 000 000
Moskva: GUGK.
In Russian.
With economic and climatic inset maps and placename index.

Administrative

China and Mongolia 1:6 000 000
Edinburgh: Bartholomew, 1985.
Political shading of provinces, placename index on verso.

Nepal

The central organization for surveying and mapping activities in Nepal is the **Survey Department (NSD)**. There are four branches responsible for topographical survey, geodetic survey, cadastral survey and training but it has only been in the last decade that this organization has begun to produce its own mapping. Until then Nepal was mapped by **Survey of India** series at 1 inch and quarter-inch scales. A 1-inch series begun under the Colombo Plan in the 1950s was based upon field surveys that were completed by NSD in 1962. It was to cover the country in 266 15' quads and used 100 m contours on a polyconic projection. This series was completed around 1980 but was never publicly available and was restricted like other large-scale Survey of India series at the time. It was possible to obtain poor black-and-white copies of sheets from this series from Nepal Graphic Art in Kathmandu. The 1:250 000 AMS series U502, based upon Survey of India mapping, also gave complete coverage of the country, in 23 sheets. In 1976, United Nations Development Programme funding enabled the full establishment of the topographical survey branch of the NSD. Some revision of the 1-inch series was carried out and interim photographic enlargement of the sheets to 1:50 000 scale was undertaken. New 1:50 000 scale aerial coverage was flown in 1977–78 and it was agreed to produce a new map at 1:50 000 scale on the UTM projection using the new Nepal National Grid, to compile sheets photogrammetrically, and to complete the series to cover the country in 266 sheets by 1987. There are at present no plans for other national topographic series, though larger-scale topographical mapping of development

sites has been carried out. No graphic indexes of any NSD maps are issued, but the maps are advertised as being available so we have listed them in the catalogue section.

Smaller-scale mapping is also carried out by NSD, including a four-sheet bilingual 1:2 000 000 scale multi-colour map of the country. A glossary of geographical names of Nepal is under compilation and NSD has recently published an atlas of economic development of the country. Administrative mapping of regions (1:500 000, five sheets), zones (1:250 000, 15 sheets) and districts (1:125 000, 75 sheets) was begun in 1976. Only the regional maps are available in English and Nepali: the other series are published in Devanagari script.

Other mapping projects have been carried out on a joint basis with various foreign mapping agencies including the British **Directorate of Overseas Surveys (DOS)** (now Overseas Surveys Directorate, Southampton, OSD), the US Agency for International Development and especially the Canadian **Surveys and Mapping Branch of the Department of Energy, Mines and Resources (DEMR)**. This Canadian aid project was begun in 1977 and aims to produce a variety of monochrome diazo land resources series mapping: three series are being published at 1:50 000 scale to cover land utilization, land systems (soils), and land capability (land potential): all give complete coverage in 266 sheets. A 1:125 000 scale geological map is being prepared, and a seven-sheet climatological map was completed in 1982. Geological mapping of Nepal is the responsibility of the **Geological Survey of Nepal**, but has been carried out mainly by various foreign agencies. No information about the sheet lines of these series is available. Not until the Canadian aid project was there any large-scale coverage available of the country, but 1:1 000 000 scale single-colour geological and tectonic maps were compiled in the 1970s.

An important high-quality mapping project has been carried out since 1955 by the Arbeitsgemeinschaft für Vergleichende Hochgebirgsforschung. These 'Schneider' maps in the Nepal Kartenwerk series are issued by **Freytag Berndt Artaria** and **Nelles Verlag** at 1:50 000 scale and have German and English information with 40 m contour intervals. It is planned to issue 20 sheets, to cover the major Himalayan mountain areas in the east of the country; so far 11 sheets have been published, including the Kathmandu valley and town maps of Kathmandu and Patan. These maps are now distributed by **Nelles Verlag**.

The World Bank has issued satellite image mapping of the country, which is distributed by **International Mapping Unlimited**: a new 1984 edition recently became available.

Vegetation mapping at 1:250 000 and 1:50 000 has been carried out by the French **Centre National de la Recherche Scientifique (CNRS)**. This multi-colour series began in 1970 and completed in 1985, and covers most of the country. Explanations are also available with the maps.

Census mapping in Nepal is carried out by the **Nepal Central Bureau of Statistics** which prepared enumeration district mapping of the country's 3800 village Panchayatts, mostly at 1:5000 or 1:10 000 scale for the 1981 census. None of these maps was published or publicly available, and the exact relationship of the mapping to the demographic statistics that were collected is rather doubtful.

Another aid-funded mapping project was begun in 1977. The Mountain Hazards Mapping Project was funded by the United Nations University and has produced prototype land use, hazards, geomorphic damage and base maps of three test areas in different zones of the country.

Commercial mapping has been published in recent years with the increasing popularity of the Himalayas as a tourist destination. Good small-scale general maps of the country have been published by **Nelles Verlag** and **Astrolabe**, and several Nepali companies (such as **Lakoul Press**) issue low-quality trekking maps aimed at the tourist market.

Further information

Useful summaries are provided by Gurung, H. (1983) *Maps of Nepal: Inventory and Evaluation*, Bangkok: White Orchid; Nepal (1977, 1980, 1983, 1987) Country reports presented to the *8th, 9th, 10th and 11th United Nations Regional Cartographic Conferences on Asia and the Pacific*, New York: UN; and Fiese, A. and Welsh, H. (1982) Vermessungs- und Kartenwesen von Nepal, *Kartographische Nachrichten*, **32**(1).

Information on specific publishers is available from: Ives, J.D. and Messerli, B. (1981) Mountain hazards mapping in Nepal: introduction to an applied mountain research project, *Mountain Research and Development*, **1**(3–4), 223–230; and Shrestha, C.F. (1984) Nepal, in *Census Mapping Survey*, edited by Nag, P., New Delhi: Concept Publishing Co., pp. 137–145.

Very basic catalogues are issued by NSD.

Addresses

Survey Department HMG Nepal (NSD)
Dilli Bazar, KATHMANDU, Nepal.

Survey of India (SI)
PO Box 37, DEHRA DUN 248001, India.

Overseas Surveys Directorate (DOS/OSD)
Ordnance Survey, Romsey Rd, SOUTHAMPTON SO9 4DH, UK.

Surveying and Mapping Branch, Department of Energy, Mines and Resources (DEMR)
615 Booth Street, OTTAWA, Ontario K1A 0E9, Canada.

Geological Survey of Nepal
Department of Mines and Geology, Lainchaur, KATHMANDU.

Freytag-Berndt Artaria
Schottenfeldgasse 62, A1071, WIEN, Austria.

International Mapping Unlimited
4343 39th Street NW, WASHINGTON, DC 20016, USA.

Centre National de la Recherche Scientifique (CNRS)
15 Quai Anatole, 75700 PARIS, France.

Central Bureau of Statistics
Ramshah Path, Thapathali, KATHMANDU.

Lakoul Press
Palpa—Tanben.

For Nelles Verlag see German Federal Republic; for Astrolabe see France; for Kümmerly + Frey see Switzerland.

Catalogue

Atlases and gazetteers

Nepal: atlas of economic development
Kathmandu: NSD, 1984.

Glossary of (geographical names of) Nepal
Kathmandu: NSD, 1986.

General

Nepal: relief map   1:1 400 000
Bern: Kümmerly + Frey, 1986.

Nepal   1:1 000 000
Paris: Astrolabe, 1986.

Nepal   1:1 000 000
Kathmandu: NSD, 1984.

APA map of Nepal   1:1 500 000
München: Nelles Verlag, 1983.
Double sided.
East Nepal 1:500 000 on back.

Nepal   1:500 000   NEP 500
Tolworth: DOS, 1982.
2 sheets, both published.
Landsat-based image map.

Nepal   1:500 000
New York: World Bank, 1984.
2 sheets, both published.
Satellite image map.

Topographic

Nepal   1:50 000
Kathmandu: NSD, 1981–.
266 sheets, all published.

Nepal Kartenwerk   1:50 000
München: Nelles Verlag, for Arbeitsgemeinschaft für Vergleichende Hochgebirgsforschung, 1974–.
20 sheets, 11 published.
Covers Himalayan areas of East Nepal.

Geological and geophysical

Geological map of Nepal   1:125 000
Kathmandu: NSD, 1981–.
82 sheets, all published.

Environmental

Carte écologique du Nepal 1:250 000 and 1:50 000
Paris: CNRS, 1971–85.
11 sheets, all published.
With explanatory texts.

Climatological maps of Nepal 1:250 000
Kathmandu: NSD, 1981.
7 sheets, all published.

Administrative

Latest political map of Nepal 1:1 250 000
Delhi: SI, 1981.

Regional maps of Nepal 1:500 000
Kathmandu: NSD.
5 sheets, all published.

Town maps

Kathmandu city 1:10 000
München: Nelles Verlag, 1979.

Patan city 1:7500
München: Nelles Verlag, 1980.

Oman

The Sultanate of Oman is one of the oldest established
states in the world. The national mapping agency is the
National Survey Authority in the Ministry of Defence:
there is no information about the current official
topographic surveying and mapping of the country and no
mapping is available. A general map of the whole country,
with useful indexed maps of the major towns, is published
in the Arab World Map Series by **Malt International**. The
Musandam Peninsula is covered in a 1:100 000 scale four-
colour map published by the **Royal Geographical Society
(RGS)**, which shows relief by 100 m contours.

Geological mapping of Oman is carried out by the
Ministry of Petroleum and Minerals. Some mapping has
been published in conjunction with western agencies,
including 1:250 000 scale geological and gravity mapping
of the northern areas of the country and 1:100 000
geological mapping of ophiolite areas by the British **Open
University**.

Addresses

National Survey Authority
Ministry of Defence, PO Box 113, MUSCAT.

Ministry of Petroleum and Minerals
PO Box 551, MUSCAT.

Malt International
PO Box 8357, BEYROUTH.

For Open University and RGS see Great Britain; for DMA see
United States.

Catalogue

Atlases and gazetteers

*Oman: official standard names approved by the United States Board
on Geographic Names*
Washington, DC: DMA, 1976.
pp. 97.

General

The Sultanate of Oman 1:1 320 000 2nd edn
Beirut: Malt International, 1983.
Double-sided, with city maps of Muscat, Mutrah, Qurm, Ruwi,
Salalah and the capital area, indexes, introductory notes and
businessman's guide.

Geological and geophysical

Northern Oman: geological map 1:250 000
Milton Keynes and Muscat: Open University and Ministry of
Petroleum and Minerals, 1983.
Also available on same sheets lines as Bouguer gravity anomaly
map.

Pakistan

The surveying and mapping of Pakistan are carried out by
the **Survey of Pakistan (SP)**, which was created from the
Survey of India following independence in 1947. Like
other countries of the subcontinent the mapping of
Pakistan derives from the substantial base provided by the
Survey of India. The basic scale map is now the 1:50 000
series which covers the country in about 1350 sheets, on
the same sheet lines as the earlier SI 1-inch series. This
map uses metric contour intervals and the conical

conformal projection. In 1983 a block of about 80 sheets
had not yet been completed. There is a revision
programme updating coverage, but many sheets still date
from 1950 surveys. The 1:250 000 map covers Pakistan in
121 sheets. Both of these larger-scale topographic series
have only recently become available and may certainly not
be obtained for large areas of the country near to borders
that are still restricted. A 26-sheet 1:500 000 scale series
with layer coloured relief was begun in 1968 and is now

completed. We have listed these series in the catalogue section but doubt remains about their availability.

SP also carries out general mapping activities: political maps of the country in English and Urdu are published at a variety of scales, and province maps at 1:1 000 000 scale are available and are regularly revised. A city and town guide series is issued to provide large-scale town maps of the major urban centres.

The **Geological Survey of Pakistan (GSP)** is responsible for the geological mapping and surveying of the country. Small-scale geological and tectonic coverage has recently been compiled and the British **Directorate of Overseas Surveys (DOS)** (now Overseas Surveys Directorate, Southampton, OSD) compiled full-colour 1:50 000 geological mapping of the Salt Ranges in the early 1980s. The DOS maps are no longer available. GSP has compiled some full-colour 1:50 000 mapping on topographic sheet lines for a few areas.

Soils and land use mapping of the country at 1:253 440 was compiled in the 1950s under a Colombo Scheme aid project by the **Central Soil Conservation Organization**; this series, too, is no longer available.

Foreign agencies have produced general mapping of the country. These include the German firms **Nelles Verlag** and **Karto+Grafik** (under preparation) and the American **Central Intelligence Agency (CIA)**.

Further information

Survey of Pakistan *General map catalogue*, Rawalpindi: SP, 1983.

Addresses

Survey of Pakistan (SP)
PO Box 10, Murree Rd, RAWALPINDI.

Geological Survey of Pakistan (GSP)
PO Box 15, QUETTA.

For DOS/OSD see Great Britain; for CIA and DMA see United States; for Nelles Verlag and Karto+Grafik see German Federal Republic.

Catalogue

Atlases and gazetteers

Pakistan: official standard names approved by the United States Board on Geographic Names 2nd edn
Washington, DC: DMA, 1978.
pp. 522.

General

Map of Pakistan 1:6 000 000
Rawalpindi: SP, 1985.

Map of Pakistan: general information 1:2 000 000
Rawalpindi: SP, 1982.

Road map of Pakistan 1:2 000 000 3rd edn
Rawalpindi: SP, 1984.

Map of Pakistan 1:1 500 000
Rawalpindi: SP, 1984.
Relief map available as English or Urdu edition.

APA map of Pakistan 1:1 500 000
München: Nelles Verlag, 1985.
With small inset map of Karachi.

Topographic

Pakistan 1:500 000
Rawalpindi: SP, 1978–.
24 sheets, all published.

Pakistan 1:250 000
Rawalpindi: SP, 1947–.

Pakistan 1:50 000
Rawalpindi: SP, 1947–.

Geological and geophysical

Geological map of Pakistan 1:2 000 000
Quetta: GSP, 1979.

Tectonic map of Pakistan 1:2 000 000
Quetta: GSP, 1982.

Administrative

Map of Pakistan showing political divisions 1:3 000 000
Rawalpindi: SP, 1981.
Also available in Urdu edition.

Map of Pakistan showing administrative divisions 1:2 500 000
Rawalpindi: SP, 1983.

Province maps 1:1 000 000
Rawalpindi: SP, 1979–.
5 maps, all published.

Human and economic

Pakistan: main industries, minerals and fuels 1:2 000 000
Rawalpindi: SP, 1983.

Town maps

Islamabad and Rawalpindi guide map 1:30 000
Rawalpindi: SP, 1984.

Karachi guide map 1:40 000
Rawalpindi: SP, 1984.

Lahore guide map 1:25 000
Rawalpindi: SP, 1981.

Quetta guide map 1:20 000
Rawalpindi: SP, 1982.

Peshawr guide map 1:15 000
Rawalpindi: SP, 1984.

Papua New Guinea

Before Papua New Guinea became independent in 1975, Australia had been responsible for establishing the modern mapping infrastructure and the **Royal Australian Survey Corps (RASC)** continues to be involved in bilateral aid projects between the two countries. The **National Mapping Bureau (NMB)** has been responsible for surveying and mapping activities in Papua New Guinea since independence. The basic scale series is the 280-sheet *Papua New Guinea 1:100 000 topographic survey* which was begun in 1977 and completed in 1981. This five-colour map is very similar in specification to Australian series. It is drawn on the Transverse Mercator projection and shows relief by 40 m contour intervals. It was compiled from 1:80 000–1:100 000 aerial coverage but the series is based on two different datums, the mainland on the Australian Geodetic Datum and the offshore islands on the World Geodetic System 1972. Amongst current plans are selected revision of sheets in this series by NMB. A 1:250 000 scale series is also published by NMB, according to Joint Operations Graphic specifications. This gives complete coverage in 73 sheets, and has UTM grid and 200 m contours (50 m in lowland areas). Some larger-scale maps have also been compiled. 1:50 000 and 1:25 000 scale coverage is available for the Port Moresby area and parts of Bougainville and New Ireland, and 1:10 000 scale aerial cover has been flown of 150 major towns and villages. Smaller-scale series are at present maintained by the Australian RASC: these include 1:1 000 000 scale coverage. NMB also produces a four sheet road map and provincial boundary mapping. The availability of a complete and high-quality medium-scale basic series has, however, allowed the NMB to concentrate on improving larger-scale mapping resources. The current programme aims to provide 1:2000 mapping of provincial centres, including the National Capital, and to extend the geodetic network to determine a PNG National Spheroid. There are also plans to produce a revised National Gazetteer based on names appearing on the 1:100 000 map.

Geological mapping of Papua New Guinea is the responsibility of the **Geological Survey (GSPNG)**. As with the topographic mapping of the country, Australian agencies have been responsible for the establishment of the series. The **Bureau of Mineral Resources (BMR)** in Canberra began the 1:250 000 geological map in the 1970s on JOG sheet lines. Most of the country is now covered in this full-colour map but some sheets are available only as preliminary editions. Explanatory booklets are published with sheets which are now issued by GSPNG. Larger-scale coverage of important mineral areas such as Ok Tedi is also published: 1:100 000 scale mapping of New Ireland and Sepik is currently on programme, and single-sheet 1:2 500 000 scale and four-sheet 1:1 000 000 scale maps are issued. A 1:1 000 000 scale metallogenic map of the country is at present being compiled.

The **Department of Primary Industry** is currently preparing soils mapping of the country.

Other thematic mapping of Papua New Guinea has been carried out by the Australian **Commonwealth Scientific and Industrial Research Organization (CSIRO)**. As part of its *Land research series* of maps, CSIRO published many medium- to small-scale land resources surveys of about 40 per cent of Papua New Guinea in the 1960s and 1970s.

These culminated in the publication of the three 1:1 000 000 scale sets of four maps covering geomorphology, vegetation and land limitation and agricultural land use in 1974 and 1975. These maps were accompanied by explanatory notes and provided a useful comprehensive overview of the resource potential of the country on independence.

CSIRO continues to be involved in resource mapping. A current aid programme involves the establishment of a Papua New Guinea Resources Information System for the Department of Primary Industry in order to up-date the resource information base and to improve access to this information. This has required the preparation of a resource inventory for the whole country at 1:500 000 scale and the establishment of a geographic information system with the country divided into 4300 resource mapping units.

The National Atlas of Papua New Guinea was compiled in the mid-1970s but is no longer available. A useful recent atlas of full-colour maps of Papua New Guinea is published by the Australian firm of **Robert Brown and Associates**. The same company also issues a town and tourist map and a general map of the country showing the provincial boundaries. All are distributed by **Gordan and Gotch Pty**.

Further information

Background to the mapping of Papua New Guinea is contained in the following articles: Australia (1983) Technical assistance to the National Mapping Bureau of Papua New Guinea: paper presented to the *10th United Nations Regional Cartographic Conference for Asia and the Pacific*, New York: UN; Done, P. (1983) Surveying and development potential in Papua New Guinea: technical paper L3 presented to the *2nd South East Asian Survey Congress*, Hong Kong: Hong Kong Institute of Land Surveyors, 1984; and Done, P. (1985) Surveying, charting and mapping problems in Papua New Guinea, *Geographical Journal*, **151**(3), 371–378.

Background to CSIRO mapping of Papua New Guinea is in CSIRO (1986) *Division of Water and Land Resources Research report, 1983–85*, Melbourne: CSIRO.

Addresses

National Mapping Bureau (NMB)
PO Box 5665, BOROKO.

Geological Survey Papua New Guinea (GSPNG)
PO Box 778, Port Moresby, PORT MORESBY, National Capital District.

Department of Primary Industry
PO Box 417, KONEDOBU.

Gordan and Gotch Pty Ltd
PO Box 107, BOROKO.

For RASC, BMR, CSIRO and Robert Brown and Associates see Australia.

Catalogue

Atlases and gazetteers

Papua New Guinea atlas
Bathurst: Robert Brown and Associates, 1985.
pp. 113.

General

Papua New Guinea   1 : 5 000 000
Boroko: NMB, 1979.

Papua New Guinea   1 : 2 500 000
Boroko: NMB, 1984.
Physical base with placename index on verso.

Papua New Guinea: road system   1 : 1 000 000
Boroko: NMB, 1983.
4 sheets, all published.
Physical base.

Topographic

Papua New Guinea Joint Operations Graphic   1 : 250 000   Series 1501
Boroko: NMB, 1975–.
73 sheets, all published. ■

Papua New Guinea   1 : 100 000
Boroko: NMB, 1977–.
280 sheets, all published. ■

Geological and geophysical

Geology of Papua New Guinea   1 : 2 500 000
Port Moresby: GSPNG, 1976.

Geological map of Papua New Guinea   1 : 1 000 000
Port Moresby: GSPNG, 1972.
4 sheets, all published.

Geomorphology of Papua New Guinea   1 : 1 000 000   E. Loeffler
Melbourne: CSIRO, 1974.
4 sheets, all published.
With explanatory booklet.

Papua New Guinea geological map   1 : 250 000
Port Moresby: GSPNG, 1980–.
73 sheets, *ca* 60 published. ■

Administrative

Papua New Guinea   *ca* 1 : 1 450 000
Bathurst: Robert Brown and Associates, 1985.
Administrative broadsheet map.

Earth resources

Mineral resources of Papua New Guinea   1 : 2 500 000
Canberra: BMR, 1974.
With accompanying explanatory bulletin.

Vegetation of Papua New Guinea   1 : 1 000 000   K. Paijmans
Melbourne: CSIRO, 1975.
4 sheets, all published.
With explanatory booklet.

Land limitation and agricultural land use potential of Papua New Guinea   1 : 1 000 000   P. Bleeker
Melbourne: CSIRO, 1975.
4 sheets, all published.
With explanatory booklet.

Land systems of Papua New Guinea   1 : 250 000
Melbourne: CSIRO, 1964–76.
11 areas, all published.
For each area 1 : 250 000 scale land systems maps published with accompanying texts.
For some areas forest types, vegetation, land use intensity or forest resources mapping at a variety of scales accompanies land systems maps.

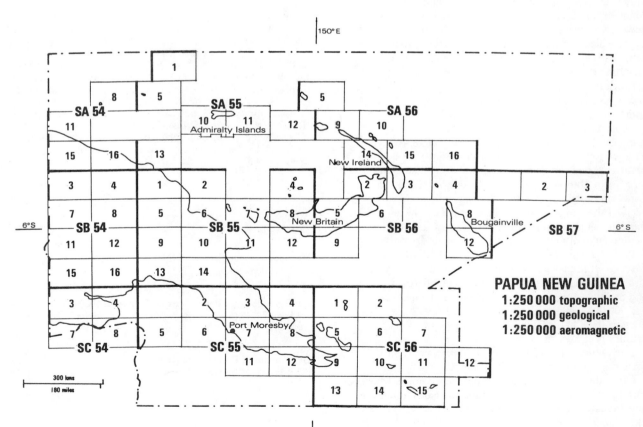

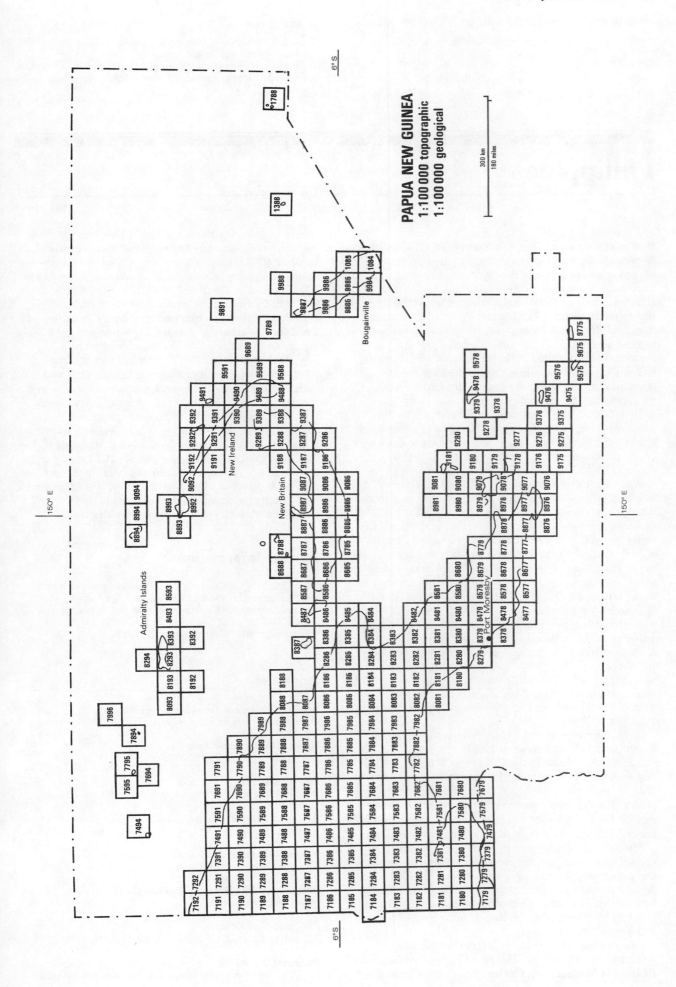

PAPUA NEW GUINEA
1:100 000 topographic
1:100 000 geological

300 km
180 miles

Human and economic

*Papua New Guinea: population distribution by
province* 1 : 2 350 000
Boroko: NMB, 1981.

Town maps

Town and tourist map of Papua New Guinea various scales
Bathurst: Robert Brown and Associates, 1986.
Includes town plans of Port Moresby, Arawa, Goroka, Kieta, Lae,
Madang, Mt Hagen, Rabaul and Wewak, with tourist map of
PNG.

Philippines

The national mapping agency in the Philippines is the
Bureau of Coast and Geodetic Survey (BCGS), which is
responsible for both topographic and hydrographic
surveying and mapping of the country and for the
maintenance of geodetic networks. Its functions are to be
taken over by a new **National Cartographic Authority**,
but currently BCGS is still the major civilian Philippine
map publisher. Modern mapping of the Philippines was
begun under American colonial rule and a geodetic
network was established in the first quarter of this century.
Current series are compiled to full photogrammetric
standards and the islands are covered by modern aerial
photographs. The national basic scale series is the 976-
sheet 1 : 50 000 map. Completed in the early 1970s this
five-colour series uses Transverse Mercator projection and
shows relief by 20 m contours. Apart from some areas of
Luzon there has been little revision of the 1 : 50 000
coverage. A 1 : 250 000 series covers the islands in 55
sheets, with 100 m contours and hill shading. Current
mapping projects have centred upon providing new more
detailed coverage of specific areas. An aid project with the
Japanese has provided 1 : 25 000 coverage of the Cagayan
valley area, 76 1 : 25 000 topographic maps have been
published and 19 1 : 10 000 orthophotomaps of urban areas
in the project have also been issued. Other larger-scale
programmes associated with development projects are also
under way, most using newly flown aerial cover. UNDP
has funded a reequipment exercise for the survey
authority. A major urban mapping programme in Manila is
under way, in association with Japanese agencies.
Compiled from 1 : 32 000 scale aerial coverage it involves
the publication of 1 : 10 000 scale coverage of the capital
area, issued as topographic, planimetric, land use and land
condition editions. Line maps at scales of 1 : 25 000,
1 : 50 000, 1 : 100 000 and 1 : 200 000 scales are to be derived
from this mapping.
Cadastral mapping of the Philippines is the
responsibility of the **Bureau of Lands**.
Geological mapping in the Philippines is the
responsibility of the **Bureau of Mines and Geosciences
(BMG)**. Full-colour geological mapping of the country was
prepared in the 1970s: an eight-sheet 1 : 1 000 000 scale
map on IMW sheet lines was published and single-colour
mineral and geological mapping of many provinces at
1 : 250 000 scale were issued as supplements to reports of
investigations monographs. A small-scale mineral map was
also published. More recently a full-colour 1 : 50 000 scale
geological map using the topographic series sheet lines has
been started. The first sheets were published in 1981 and
about 20 have been issued to date to cover parts of Luzon
and Marinduque islands.
Other official mapping agencies include the military
Armed Forces Philippines Mapping Center, the **National
Resources Management Centre**, the National Irrigation

Administration, the Bureau of Forest Development, and
the Bureau of Soils. The **International Rice Research
Institute (IRRI)** in Manila was established in 1960 and has
compiled a variety of small-scale agricultural mapping of
different parts of South and South East Asia. This includes
a 1 : 2 500 000 scale agroclimatic map of the Philippines.
National Atlases were compiled in the 1970s by the
Fund for Assistance to Private Education and the BCGS,
but are no longer available. **Yale University Press**
published Harold Concklin's important thematic atlas
relating to the province of Ifuguo in Luzon. This work
assessed the role of environment, culture and society in the
province and includes many full-colour resource maps.
General maps of the Philippines are published by
various commercial agencies including **Nelles Verlag,
Lansdowne Press, Karto+Grafik** and **Bartholomew**. The
American **Central Intelligence Agency (CIA)** compiled a
useful small-scale map with thematic insets in the early
1970s. City maps of Manila have been published by several
small Philippine commercial houses including the
Philippine Map Co. and **Philippine Guides Inc**.

Further information

Philippines (1977, 1980, 1983 and 1987) Philippines country
reports submitted to the *8th, 9th, 10th and 11th United Nations
Regional Cartographic Conferences for Asia and the Pacific*, New
York: UN.
Catalogues are available from BCGS.

Addresses

Bureau of Coast and Geodetic Survey (BCGS)
PO Box 1620, Barraca St, St Nicolas, MANILA 2807.

National Cartographic Authority
Fort Bonifacio, METRO MANILA.

Bureau of Lands
PO Box 617, Plaza Cervantes, Binondo, MANILA.

Bureau of Mines and Geosciences (BMG)
Pedro Gill St, Ermita, PO Box 1595, MANILA 2801.

National Resources Management Center
Triumph Condominium, Quezon Avenue, QUEZON CITY 3008.

International Rice Research Institute (IRRI)
PO Box 583, MANILA.

Philippine Map Company
738 Aurora Blvd Cor., Balete Dr., Cubao, QUEZON CITY.

Philippine Guides Inc.
511 Federation Center Building, Muelle de Binondo, MANILA.

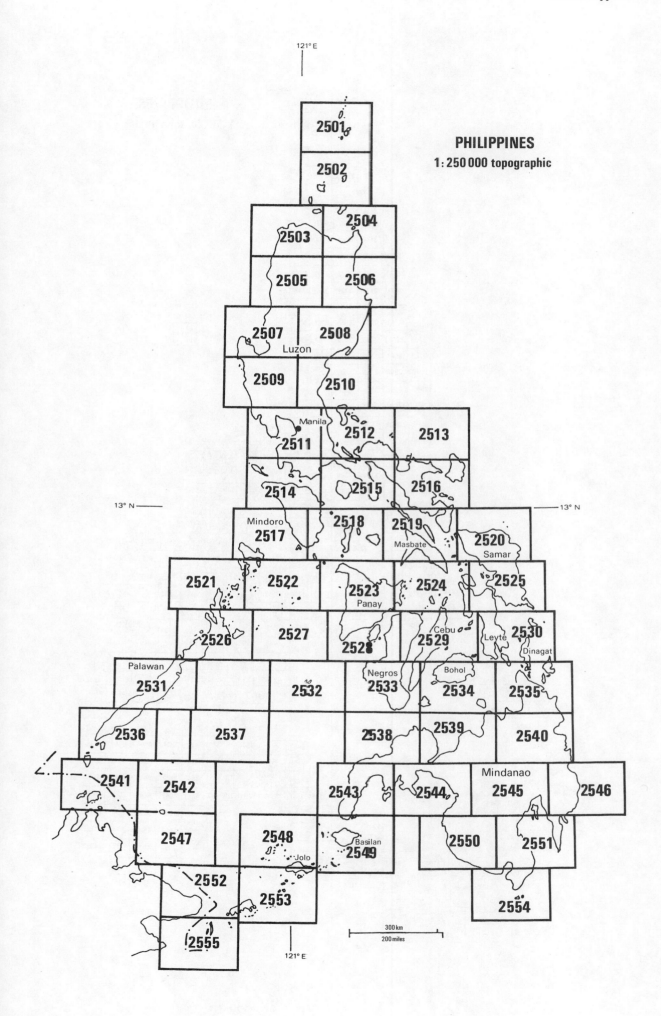

121° E

2501

2502

2503

2504

2505

2506

2507

2508

Luzon

2509

2510

Manila

2511

2512

2513

2514

2515

2516

13° N

Mindoro

2517

2518

2519

Masbate

2520

Samar

2521

2522

2523

Panay

2524

2525

2526

2527

2528

Cebu

2529

Leyte

2530

Dinagat

Palawan

Negros

2531

2532

2533

Bohol

2534

2535

2536

2537

2538

2539

2540

Mindanao

2541

2542

2543

2544

2545

2546

2547

2548

Basilan

2549

2550

2551

Jolo

2552

2553

2554

2555

PHILIPPINES
1:250 000 topographic

13° N

300 km
200 miles

121° E

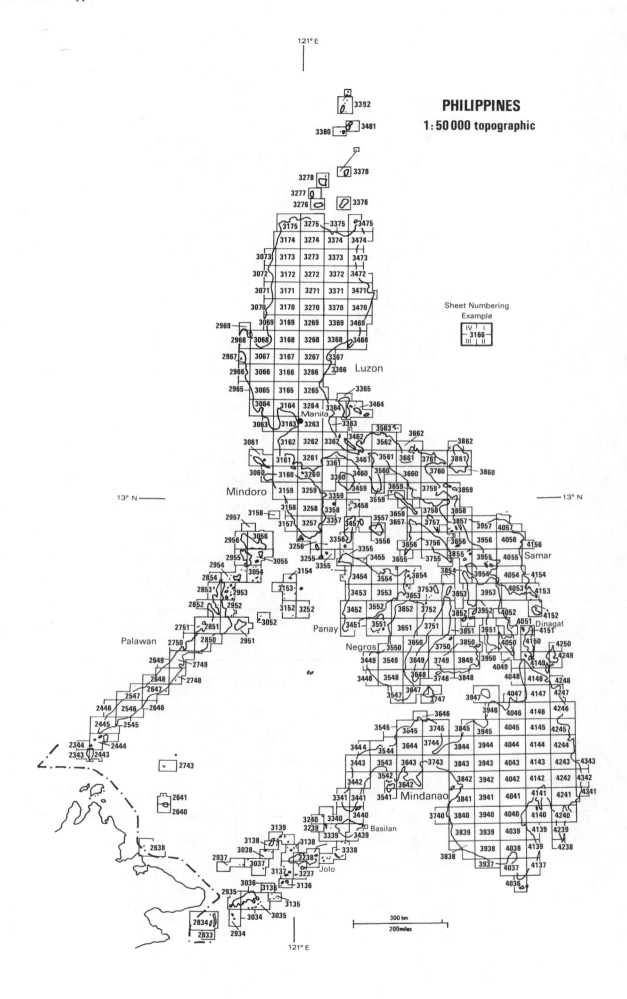

PHILIPPINES
1:50 000 topographic

For Lansdowne Press see Australia; for CIA and DMA see United States; for Bartholomew see Great Britain; for Karto+Grafik and Nelles Verlag see German Federal Republic.

Catalogue

Atlases and gazetteers

Philippine Islands: official standard names approved by the United States Board on Geographic Names
Washington, DC: DMA, 1953.
2 vols.

General

Pocket map of the Philippines and Celebes   1:5 000 000
Edinburgh: Bartholomew, 1985.

Filippiny   1:3 000 000
Moskva: GUGK, 1986.
In Russian.
With economic, climatic and ethnographic insets and placename index.

Philippines: Hildebrands Urlaubskarte   1:2 860 000
Frankfurt AM: Karto+Grafik, 1985.
Published as German or English language versions.
Tourist information on sheet.

Republic of the Philippines   1:2 000 000
Manila: BCGS, 1978.

Robinson's Philippines   1:2 000 000
Dee Why West: Lansdowne Press, 1985.

APA map of the Philippines   1:1 500 000
München: Nelles Verlag, 1984.
Includes town sketch of the Philippines.

Philippines   1:1 000 000
Manila: BCGS, 1976.
2 sheets, both published.

Topographic

Philippines   1:250 000
Manila: BCGS, 1954–.
55 sheets, all published. ■

Philippines   1:50 000
Manila: BCGS, 1964–.
976 sheets, all published. ■

Geological and geophysical

Mineral distribution map of the Philippines   1:2 500 000
Manila: BMG, 1978.
With accompanying explanatory text.

Geological map of the Philippines   1:1 000 000
Manila: BMG, 1963.
8 sheets, all published.

Geology and mineral resources of the Philippines   1:250 000
Manila: BMG, 1974–.

Administrative

Republic of the Philippines   1:3 651 400
Manila: BCGS, 1981.
Shows administrative divisions.

Environmental

Agroclimatic map of the Philippines   1:2 500 000
Manila: IRRI, 1980.

Town maps

Metro Manila: street directory   1:10 000
Quezon City: Philippine Map Co., 1985.
pp. 60.
Indexed town atlas.

Qatar

The national mapping agency in Qatar is the **Engineering Services Department** in the Ministry of Public Works. The first basic scale modern mapping of the country was produced by **Hunting Surveys** in the 1970s. Huntings have been involved in the aerial surveying and mapping of Qatar since 1947. In the 1970s they carried out new aerial coverage and established a new geodetic network for the State: the Qatar National Grid was set up (derived from the Transverse Mercator projection). Mapping produced since then has been based on this standard and all sheet lines of all series are bounded by grid lines and number from the grid coordinates of the south-west corner of each sheet. Placenames on all maps are transcribed in both Arabic and English. The basic scale mapping for urban areas is at scales of 1:5000, 1:2000, 1:1000 and 1:500 depending on the area. 1:5000 and 1:2000 plans show relief with 1 m contours, the larger scales are uncontoured and use spot heights. Some of the large-scale plans are also available in digital form as magnetic tapes and the contract for the digital mapping of Doha the capital at 1:1000 scale has recently been let. Maps were published at 1:10 000 scale to UTM specifications in the 1970s, and developed areas have been revised using 1980s aerial coverage. 176

sheets in this series have been published to cover most of Qatar; contour information is available for the whole Qatar Peninsula at 2 m intervals. A 15-sheet 1:50 000 map printed in four colours with 2 and 4 m contour intervals is derived from these basic scale series and gives complete coverage in 15 sheets. A four-sheet 1:100 000 scale map was published in 1982, with 4 m contours and English and Arabic legends. The most useful smaller-scale publicly available map of the State is the 1:200 000 scale sheet produced by Huntings for the Ministry of Information, which gives complete coverage and has a town map of Doha and tourist notes on the back. **Geoprojects** also compiled a general map of Qatar in its Arab World Map Series. A Landsat image atlas of Qatar was published in 1983 by the **Ministry of Information** in association with the **University of Qatar**. This atlas contains topographic, geological, water depth, and sea bathymetry and sediment mapping to accompany 1:250 000 scale false colour Landsat imagery. It is published with an index of geographical names and is the second volume of a projected National Atlas series.

There is no information about geological survey activities in Qatar.

Further information

Background to the establishment of modern mapping of Qatar is contained in Leatherdale, J. and Kennedy, R. (1975) Mapping Arabia, *Geographical Journal*, **141**(2), 240–251.

The Engineering Services Department publishes the *Map Users Handbook* (1982), which gives information about scales, extent, format and sheet numbering systems used in the mapping of Qatar. It is the best modern guide.

Addresses

Engineering Services Department (ESD)
Ministry of Public Works, PO Box 38, DOHA.

Ministry of Information Press and Public Relations Department
PO Box 5147, DOHA.

University of Qatar (Jamiyat Qatar)
PO Box 2713, DOHA.

For Hunting Surveys see Great Britain; for Geoprojects see Lebanon; for DMA see United States.

Catalogue

Atlases and gazetteers

Atlas of Qatar from Landsat Images   M.A.A. Yehia
Doha: Qatar University and Ministry of Information, 1983.
pp. 166.
In English and Arabic.

Bahrain, Kuwait, Qatar and the United Arab Emirates: official standard names approved by the United States Board on Geographic Names
Washington, DC: DMA, 1976.
pp. 145.

General

The Oxford map of Qatar   1 : 270 000
Beirut: Geoprojects, 1980.
Double-sided.
With city map of Doha and tourist information.

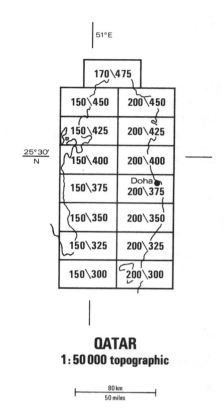

QATAR
1 : 50 000 topographic

80 km
50 miles

Map of Qatar   1 : 200 000
Doha: Ministry of Information Press and Publicity Department.
Printed on both sides, with city map of Doha *ca* 1 : 20 000 scale, index and tourist notes on back.

Topographic

Qatar   1 : 100 000
Doha: ESD, 1982.
4 sheets, all published.

Qatar   1 : 50 000
Doha: ESD, 1982
15 sheets, all published.

Saudi Arabia

The national mapping agency of Saudi Arabia is the **Aerial Survey Department** of the **Ministry of Petroleum and Mineral Resources (MPRM)**. It is responsible for the compilation of both thematic and topographic map series and carries out a range of geodetic, topographic, photogrammetric and earth science surveys. The **Deputy Ministry for Mineral Resources (DMMR)** is responsible for the compilation of earth science mapping and the publication of both geographic and geologic editions of the Ministry's maps. Military mapping activities are carried out by the **Military Survey Department**.

We have listed under Saudi Arabia maps that cover the Arabian peninsula, as well as mapping of Saudi Arabia alone.

There were no modern maps of Saudi Arabia until the discovery of oil and the rapid development of the 1960s that followed. The country was therefore able to take advantage of a late start and establish an accurate and rapid geodetic control, fly modern aerial coverage and

prepare a systematic topographic mapping programme over a short time period. The first mapping to be produced was compiled by the **United States Geological Survey (USGS)**. 1 : 500 000 regional scale mapping was carried out between 1958 and 1963 by USGS, with the cooperation of the **Arabian American Oil Company (ARAMCO)** to give complete coverage of the country in 21 sheets. These were published as either geographic or geologic editions, with Arabic and English text. Both were generalized to form 1 : 2 000 000 scale geographic and 1 : 2 000 000 scale geologic maps. The 1 : 500 000 scale maps are currently under revision and are being republished by DMMR.

1 : 50 000 and 1 : 100 000 topographic programmes were established in the 1960s and maps published as either six-colour photomaps or full-colour line maps. Coverage at these scales was first published for areas of economic development in conjunction with a variety of western contracting agencies, the aim being eventually to complete coverage of the country. Larger-scale series have been

published for some areas, including a 1:25 000 survey of the Jeddah region, and urban mapping at large scales has also been produced. None of these larger-scale topographic maps is available outside Saudi Arabia.

A 1:100 000 scale geological series was started by DMMR in 1963, with the aim of giving complete coverage of the Precambrian mineral rich Arabian shield area. The 253 sheets required are now all available in at least manuscript form. They are being published as 30' quads, or issued as internal DMMR Open File reports or Technical Records and use an image background.

USGS has been using satellite data for the last decade in the DMMR mapping programmes. Geodetically controlled two-colour Landsat image mosaics at scales of 1:250 000 and 1:500 000 have been used as bases for the compilation of both geographic and geologic maps. 1:250 000 scale image maps cover 1° latitude by 1.5° longitude quadrangles, and are published in three different editions: a full-colour basic geology edition overprinted on a Landsat image base, a map comprising only the geologic information without the base, and a geographic edition with settlement communication and drainage information on the image base. These maps are derived from the six 1:100 000 scale quads, supplemented by field checking. The 1:250 000 geographic series will ultimately be available as full four-colour image maps to cover the whole country. Coverage is limited at present to the more developed central and northern areas. For the geologic series effort is being concentrated upon the publication of 54 full-colour 1:250 000 quads for the whole of the Arabian Shield, but on completion of coverage of the Shield mapping will be extended to other areas. These maps are accompanied by explanatory notes.

Other 1:250 000 scale earth science mapping is published by DMMR using the same sheet lines. Series compiled and in progress include hydrogeological and aeromagnetic mapping, issued as Open File reports.

The Remote Sensing Centre of DMMR has been involved with ARAMCO and DMA in the publication of a new 1:2 000 000 scale geographic map of the Arabian Peninsula. This map uses a Landsat mosaic base overprinted with cultural and physical features.

A national population atlas was compiled in 1981 by the National Atlas Committee: this was published only as an Arabic language edition. A historical gazetteer of Arabia was started in the late 1970s under the editorship of A.A. Scoville, and following the pattern of the Iranian and Afghanistani multi-volume works also published by **Akademische Druck und Verlagsanstalt**. Only a single volume has appeared to date.

Some thematic sheets covering the peninsula have been published, mostly at 1:2 000 000 scale in the *Tübinger Atlas des Vorderen Orients* series, and a variety of small-scale mapping of Saudi Arabia is also currently available for purchase. General maps are published in the Arab Map World series by **Geoprojects**, by **Bartholomew**, by **Clyde Surveys** for the Ministry of Communications, and by the **Middle East Information Company**. A number of commercially published town maps of Riyadh, Jeddah and Mecca may be purchased. Some of these western commercial maps are compiled for official Saudi bodies and are likely to be difficult to acquire.

Further information

The best overall source about both current official mapping and its development is Deputy Ministry for Mineral Resources (1985)

Saudi Arabia Mineral Resources Annual Report AH 1404–05 (AD 1984–85), Jiddah: The Deputy Ministry.

Other information is contained in Leatherdale, J. and Kennedy, R. (1975) Mapping Arabia, *Geographical Journal*, **141**(2), 240–251; and in United States Geological Survey (1981, 1984) *USGS 1980 and 1983 Annual Reports: international activities*, pp. 58–59 and 89–90.

For a more bibliographic and historical treatment see the thesis compiled by Al-Harbi, A. (1983) *Maps and mapping activities in Saudi Arabia: annotation and cartobibliography*, Washington: University of Seattle. For information about the Jeddah municipal mapping project see Hannigan F.L. (1984) Municipal mapping in Saudi Arabia, *Computer Graphics World*, **7**(7), 37–41.

Addresses

Aerial Survey Department, Ministry of Petroleum and Mineral Resources (MPRM)
PO Box 247, RIYADH.

Deputy Ministry for Mineral Resources (DMMR)
Directorate of Technical Affairs, PO Box 345, JIDDAH.

Military Survey Department
PO Box 8652, RIYADH 11492.

Arabian American Oil Company (ARAMCO)
POB 1389, DHARRAN.

Ministry of Communications
Airport Rd, RIYADH 11148.

For CIA, DMA, USGS and Middle East Information Company see United States; for BRGM see France; for Bartholomew, Geoprojects and Clyde Surveys see Great Britain; for TAVO see German Federal Republic; for Akademische Druck see Austria.

Catalogue

Atlases and gazetteers

Population atlas of the Kingdom of Saudi Arabia
Riyadh: National Atlas Committee, 1981.
pp. 34.
Arabic language.

Saudi Arabia: official standard names approved by the United States Board on Geographic Names
Washington, DC: DMA, 1978.
pp. 374.

Gazetteer of Arabia edited by A.A. Scoville
Graz: Akademische Druck und Verlagsanstalt, 1979–.
4 vols, 1 published.

General

Saudi Arabia 1:6 000 000
Washington, DC: CIA, 1979.
With economic, population and vegetation insets.

Topographic map of the Arabian peninsula 1:4 000 000
Jiddah: DMMR and USGS, 1972.
Also available as 1974 published Arabic edition.

The Kingdom of Saudi Arabia 1:4 000 000
Trumbull: Middle East Information Co., 1979.
In English and Arabic.

Arabian Peninsula 1:3 000 000
Edinburgh: Bartholomew, 1983.

Oxford map of Arabia 1:3 000 000 2nd edn
Beirut: Geoprojects, 1980.
Double-sided.
With index, descriptive text and country statistics on back.

Road map of the Kingdom of Saudi Arabia 1:3 000 000
Jiddah: Ministry of Communications, 1981.
Double-sided.
In Arabic and English.
Includes town maps of 10 cities on verso.

Oxford map of Saudi Arabia 1:2 600 000
Beirut: Geoprojects, 1981.
Double-sided.
With businessman's guide, index and tourist information.

Arabian peninsula 1:2 250 000 Series 5211
Washington, DC: DMA, 1985.

Geographic map of the Arabian peninsula 1:2 000 000 compiled
by USGS and ARAMCO
Jiddah: DMMR, 1963.
Available in English or Arabic edition.

Geographic map of Saudi Arabia 1:500 000
Jiddah: DMMR, 1973–.
21 sheets, all published. ■
Some sheets only available as 1963 USGS editions.

Geological and geophysical

Tectonic map of the Arabian Peninsula 1:4 000 000
Jiddah: DMMR, 1972.

Geologic map of the Arabian peninsula 1:2 000 000 compiled by
USGS and ARAMCO
Reston, VA: USGS, 1963.

Geologic map of the Kingdom of Saudi Arabia 1:500 000
Jiddah: DMMR, 1979–.
21 sheets, all published. ■
Some sheets only available as 1963 USGS editions.

Town maps

Jeddah map and guide 1:25 000
Trumbull: Middle East Information Co., 1983.

Riyadh map and guide 1:25 000
Trumbull: Middle East Information Co., 1983.

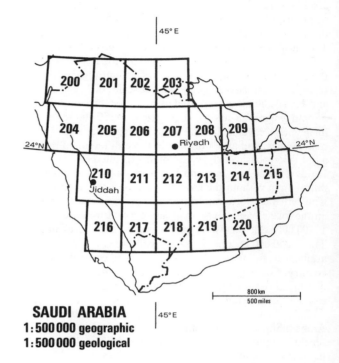

SAUDI ARABIA
1:500 000 geographic
1:500 000 geological

Singapore

Prior to independence in 1965 the mapping of Singapore was the responsibility of British and Malay agencies. After that date responsibility fell to authorities on the island itself. Mapping of Singapore is compiled by the **Defense Mapping Unit (DMU)** which acts as the national military mapping agency and compiles all topographic mapping of the island. The **Singapore Survey Department (SSD)** is the civilian mapping and surveying authority and distributes DMU maps for public sale, as well as acting as the cadastral and boundary surveying agency. The basic scale for Singapore is 1:5000; of the 114 line maps needed for coverage at this scale 55 had been issued late in 1986. A 30-sheet four-colour 1:10 000 series with a 5 m contour interval is published by DMU derived from 1:2500 plots and 1974 aerial photography, but the series is not publicly available, with the exception of three sheets of the city area. From this map are derived an eight-sheet, six-colour 1:25 000 series revised to 1986–7, and the various 1:50 000 scale maps. These are compiled on the Cassini Rectangular Soldner Spherical projection: the topographic edition is seven-colour; recreation and locality editions are also available. Outline maps of Singapore are also produced. A road map at 1:25 000 covers the island in four sheets. All these series are under continuous revision and are issued to metric standards. Electoral division mapping and population census maps are also produced by

SSD and a major part of its resources are devoted to title surveying activities.

The **Geological Unit** of the Public Works Department has compiled 1:25 000 full-colour geological coverage of Singapore, using the 1:10 000 topographic series as a base. This eight-sheet map was compiled with assistance from the UN and New Zealand, and was published with the accompanying geological report.

Hydrogeological mapping of the island at 1:100 000 scale was compiled in the 1970s by the German **Bundesanstalt für Geowissenschaften und Rohstoffe (BfGR)** and published with explanation in *Geologisches Jahrbuch*.

A **New Zealand Soil Survey (NZSB)** 1-inch soil map published in 1977 remains the best available soil mapping of the island. A 1:50 000 scale vegetation map of Singapore was prepared by R. D. Hill and the Department of Geography, University of Singapore and issued in the *Journal of Tropical Geography*, but is no longer available.

Street maps of Singapore city are published by **Roger Lascelles** and **Nelles Verlag**.

The important commercial map publisher **APA Productions** (whose maps are distributed in the west by Nelles Verlag) is based in Singapore.

Smaller-scale coverage and other thematic mapping of Singapore is listed under Malaysia or under South East Asia.

Further information

Singapore (1977) Cartographic activities in Singapore 1973–76, paper presented by Singapore to the *1977 UN Regional Cartographic Conference on Asia and the Pacific*, New York: UN, 1980. Singapore (1987) Report on survey and mapping activities in the Republic of Singapore 1983–86, paper presented to the *11th United Nations Regional Cartographic Conference for Asia and the Pacific*, New York: UN.

Addresses

Survey Department (SSD)
National Development Building, 5 Maxwell Rd, SINGAPORE 0106.

Defence Mapping Unit (DMU)
Ministry of Defence, Dover Road Camp, Dover Cres., SINGAPORE 0513.

Geological Unit
Public Works Department, National Development Building, 5 Maxwell Rd, SINGAPORE 0106.

APA Productions
5 Lengkong Satu, SINGAPORE 1441.

For Roger Lascelles and Clyde Surveys see Great Britain; for BfGR and Nelles Verlag see German Federal Republic; for NZSB see New Zealand.

Catalogue

General

Outline map of Singapore with main roads   1 : 100 000
Singapore: DMU, 1981.

Clyde leisure map of Singapore   1 : 50 000
Maidenhead: Clyde Surveys, 1983.
Includes tourist information and plan of Singapore city.

Topographic

Topographic map of Singapore   1 : 50 000   3rd edn
Singapore: DMU, 1983.
Also available as recreation map (1981 edn), locality map (1985 edn) and outline map (1981 edn).

Road map of Singapore   1 : 25 000   4th edn
Singapore: DMU, 1984.
4 sheets, all published.

Geological and geophysical

Hydrogeologic map of Singapore   1 : 100 000
Hannover: BfGR, 1975.
With accompanying explanatory text.

Geological maps of Singapore   1 : 25 000
Singapore: Geological Unit, 1976.
8 sheets, all published.
With accompanying explanatory text.

Environmental

Soil map of the Kingdom of Singapore   1 : 63 360
Lower Hutt: NZSB, 1977.
Also available with explanation.

Town maps

Singapore   1 : 22 500
München: Nelles Verlag.

Topographic map of Singapore city   1 : 10 000
Singapore: DMU, 1980.
3 sheets, all published.

Singapore town plan   ca 1 : 7500
Brentford: Lascelles, 1985.

South Yemen

Prior to independence in 1967 the British colony of Aden had been mapped by the British **Directorate of Overseas Surveys (DOS)** (now Overseas Surveys Directorate, Southampton, OSD). The basic scale map for southern and western areas of the country was a 1 : 100 000 scale photomap series, compiled to UTM specifications using standard DOS sheet numbering, which was published in 31 sheets. This three-colour series was never extended to the rest of the country, and was not issued as publicly available revised editions by the host survey after independence. Some sheets are no longer available from OSD either. In 1978 the **Royal Geographical Society (RGS)** compiled a 1 : 125 000 scale topographic map of the island of Socotra, showing relief with 100 m contours.

The official mapping of South Yemen is now the responsibility of the **Ministry of Public Works**. There is no information about current mapping activities in the country.

Geological mapping of South Yemen was also carried out by the British DOS in the 1960s, but only the 1 : 500 000 scale map of the South Mahra area is still available from

Southampton. 1 : 1 000 000 and 1 : 250 000 full-colour maps are no longer available. There is no information about any current indigenous programmes.

Small-scale mapping of South Yemen is compiled by the American **Central Intelligence Agency (CIA)**. The country is covered by many of the maps compiled to cover the whole Arabian Peninsula, including sheets published in the **Tübinger Atlas des Vorderen Orients (TAVO)**. We have listed this material under Saudi Arabia.

Addresses

Ministry of Public Works
Khormaksar, ADEN.

Overseas Surveys Directorate (DOS/OSD)
Ordnance Survey, Romsey Rd, SOUTHAMPTON SO9 4DH, UK.

For RGS see Great Britain; for CIA and DMA see United States; for TAVO see German Federal Republic.

Catalogue

Atlases and gazetteers

Yemen: People's Democratic Republic (Aden): official standard names approved by the United States Board on Geographic Names
Washington, DC: DMA, 1976.
pp. 204.

General

Yemen (Aden) 1:3 000 000
Washington, DC: CIA, 1979.
With economic, population and Aden area inset maps.

Topographic

Aden protectorate 1:100 000
Tolworth: DOS, 1958–64.
85 sheets, 31 published.

Geological

Aden: geology of the South Mahra area 1:500 000 DOS (Geol) 1147A
Tolworth: DOS, 1963.

Sri Lanka

The **Sri Lanka Survey Department (SLSD)** is responsible for topographic and cadastral surveying activities in the country. It has a very long history which can be traced back to the establishment of a survey office in Colombo in 1800. The nineteenth century mapping of the island included quarter-inch topographic and larger-scale cadastral series, and following reorganization of the Department in 1897 a 1-inch topographic series was started. This was completed in 1924, and until recently revisions of these maps were still the basic scale topographic map for the island. The revisions of this 72-sheet seven-colour series have used aerial photographic information, and relief was plotted using 100 foot contour intervals. The 1-inch map was also available as single-colour orographic editions. A new photogrammetrically derived 1:50 000 base map in 92 sheets is replacing the 1-inch series. This uses Transverse Mercator projection and is being prepared as part of the Agricultural Base Mapping Project to serve as both a topographic series, with relief shown by 10 m contours in flat areas and 20 m contours in hilly regions, and also as a land use series, with a detailed breakdown of land use and cover categories. At present priority is being given to the production of maps in this 92-sheet multi-colour series. Also in the Agricultural Base Map Project are plans to produce a 1:10 000 scale map to cover the island in 1835 sheets. Sheets show relief by 5 m contours in hilly regions and 2 m intervals on the plains, and like the 1:50 000 map give detailed multi-colour land use breakdown. At present 1:10 000 maps are being issued for areas where development projects are being carried out.

SLSD also issues larger-scale project mapping and is responsible for some thematic maps. A National Atlas is underway and will contain 45 1:1 000 000 scale loose-leaf sheets. Soil surveys have been carried out by SLSD and 1:63 360 scale single-colour forest and soil maps are published, as well as smaller-scale full-colour single-sheet thematic maps of the island. Administrative maps are also published by SLSD which also compiles town maps and tourist guides to major settlements.

Larger-scale series mapping published by SLSD is no longer available for public distribution and we do not list either 1-inch or 1:50 000 scale series in the catalogue.

The **Sri Lanka Centre for Remote Sensing** is compiling with Swiss aid 1:100 000 scale full-colour land use maps of districts in conjunction with SLSD, and a 1:500 000 scale forest cover map has been prepared from satellite imagery.

The **Geological Survey Department (SLGSD)** is responsible for geophysical and geological mapping activities in Sri Lanka. Systematic geological mapping of the country began in 1955, using the 1-inch topographic series as a base. By 1975 the whole of Sri Lanka was covered in this unpublished reconnaissance mapping programme and the manuscript maps were reduced to a single 1:250 000 scale provisional sheet. This was further reduced to 1:500 000 scale and published in 1982 with Australian technical and financial assistance. In 1983 mineral resources, tectonic and metamorphic maps were also published at 1:500 000. Some aeromagnetic and radiometric contour maps have also been prepared, and a 1:1 000 000 scale gravity map was issued as a *Professional Paper* in 1975. It is envisaged that a detailed geological mapping programme at 1:32 000 and a mineral resources related series may be launched.

Numerous commercial western agencies have produced tourist maps of Sri Lanka, including **Nelles Verlag**, **VWK Ryborsch** and **Astrolabe**.

Further information

Catalogues are issued by SLSD and a good summary of historical development and current position is provided by Sri Lanka (1983) National Report of Sri Lanka: presented to the *10th United Nations Regional Cartographic Conference*, New York: UN.

Addresses

Survey Department (SLSD)
PO Box 506, Park Rd, COLOMBO 5.

Geological Survey Department (SLGSD)
48 Sri Jinaratana Rd, COLOMBO 2.

Sri Lanka Centre for Remote Sensing
COLOMBO

For Astrolabe see France; for Nelles Verlag, Karto+Grafik and VWK see German Federal Republic; for IFP see India; for GUGK see USSR; for DMA see United States.

Catalogue

Atlases and gazetteers

National atlas of Sri Lanka
Colombo: SLSD, 1983–.
45 sheets, 4 published.
Atlas will contain 1 : 1 000 000 scale mapping.

Ceylon: official standard names approved by the United States Board on Geographic Names
Washington, DC: DMA, 1960.
pp. 359.

General

Sri Lanka: physical map   1 : 1 000 000
Colombo: SLSD, 1978.

Sri Lanka: Hildebrands Urlaubskarte   1 : 750 000
Frankfurt: Karto+Grafik, 1985.

Sri Lanka   1 : 750 000
Moskva: GUGK, 1971.
In Russian.
With climatic, economic and ethnographic insets and placename index.

Sri Lanka road map   1 : 500 000
Colombo: SLSD, 1982.

APA map of Sri Lanka   1 : 450 000
München: Nelles Verlag, 1983.
Road map with tourist information.

Strassen und Touristenkarte Sri Lanka und Maldiven   1 : 400 000
Obertshausen: VWK Ryborsch, 1986.
Double sided with town plans and 4 language tourist information on back.

Sri Lanka   1 : 253 440
Colombo: SLSD, 1977.
4 sheets, all published.

Geological and geophysical

Geological map of Sri Lanka   1 : 500 000
Colombo: SLGSD, 1982.

Mineral resources map of Sri Lanka   1 : 500 000
Colombo: SLGSD, 1983.

Tectonic map of Sri Lanka   1 : 500 000
Colombo: SLGSD, 1983.

Metamorphic map of Sri Lanka   1 : 500 000
Colombo: SLGSD, 1983.

Environmental

Climate map of Sri Lanka   1 : 1 000 000
Colombo: SLSD, 1972.

Sri Lanka: land utilization   1 : 1 000 000
Colombo: SLSD, 1972.

International map of vegetation and environmental conditions: Sri Lanka   1 : 1 000 000
Pondichéry: IFP, 1964.
With accompanying explanatory booklet.

Sri Lanka: general soil map   1 : 506 880
Colombo: SLSD, 1975.

Sri Lanka: forest cover   1 : 500 000
Colombo: SLSD, 1981.

Land use map series: districts   1 : 100 000
Colombo: SLSD, 1979–.
5 sheets published.

Administrative

Sri Lanka: province and district   1 : 1 000 000
Colombo: SLSD, 1978.

Town maps

Colombo and suburbs   1 : 12 500
Colombo: SLSD, 1976.

Syria

The Syrian national mapping agency is the **Service Géologique de l'Armée (SGA)** which was founded in 1955. Since that date new geodetic and levelling networks have been established and a modern basic scale map at 1 : 25 000 has been compiled for all the non-desert areas of the country. This map is compiled by photogrammetric methods from newly flown aerial cover and sheets cover $7\frac{1}{2}°$ quads. For the remaining desert areas a 1 : 50 000 basic scale series has been completed: this map is compiled from 1 : 70 000 scale aerial coverage. Smaller-scale maps are derived from the two basic scales by generalization and give complete coverage of the country at 1 : 100 000, 1 : 200 000 (27 sheets) and 1 : 500 000 (four sheets). Neither of the basic scale series is available outside Syria, but some of the smaller-scale mapping is listed in the catalogue, having become available in the last few years via GeoCenter.

Geological mapping of Syria is the responsibility of the **General Establishment of Geology and Mineral Resources (GEGMR)**. From 1958 to 1962 Soviet geologists were active in Syria and compiled a number of smaller-scale sheets and series that are still available (at a price) from GEGMR. These maps are all English language full-colour editions and include single-sheet 1 : 1 000 000 scale geological, mineral, tectonic, Quaternary sediment and hydrogeological maps, a four sheet 1 : 500 000 scale series and complete coverage in a 19-sheet 1 : 200 000 scale map.

A 1 : 50 000 scale series was also begun by the Soviets: GEGMR restarted work on this scale in the late 1970s and a further 12 sheets were issued between 1978 and 1984. Much of the developed area of the country is on programme in this series, which uses a topographic base with 20 m contours and is printed in Arabic with some English annotation.

The national focal point for research in remote sensing and for image mapping is the **National Remote Sensing Centre**.

Town maps of the major cities in Syria are published by the **Ministry of Tourism**. The best available general map of the country is published in the Arab Map World series by **Geoprojects** SARL and includes town maps of Damascus and Halab: other commercially published maps covering Syria are listed in the Middle Eastern section of this book.

Further information

For background information to the official mapping of Syria see Syrian Arab Republic (1980) Rapport sur les activités cartographiques en République Arabe Syrènne, *United Nations Regional Cartographic Conference for Asia and the Pacific, Vol. 2, Technical Papers*, p. 82.

GEGMR issues a useful catalogue and index for its maps.

Addresses

Service Géographique de l'Armée (SGA)
Ministère de la Défense, BP 3094, DAMASCUS.

General Establishment for Geology and Mineral Resources (GEGMR)
PO Box 7645, DAMASCUS.

National Remote Sensing Centre
Ministry of Electricity, Sultan Salim St, DAMASCUS.

Ministry of Tourism
Abou Firas El Hamadani St, DAMASCUS.

For Geoprojects SARL see Lebanon; for DMA see United States.

Catalogue

Atlases and gazetteers

Syria: official standard names approved by the United States Board on Geographic Names
Washington, DC: DMA, 1983.
pp. 734.

General

Syrie et territoires adjacents   1:2 000 000
Damascus: SGA, 1984.
In Arabic: major towns in English.

Syrie: carte routière et touristique   1:1 000 000
Damascus: SGA, 1981.

Suriya—Lubnan   1:1 000 000
Damascus: SGA, 1982.
In Arabic.

The Oxford Map of Syria   1:1 000 000
Beirut: Geoprojects, 1980.
Double-sided map, includes businessman's guide, indexes and city maps for Damascus and Halab, also map of Palmyra and descriptive text.

Syria and Lebanon   1:750 000
Damascus: SGA, 1980.
Arabic language with title also in English.

Geological and geophysical

Geological map of Syria   1:1 000 000
Damascus: GEGMR, 1964.

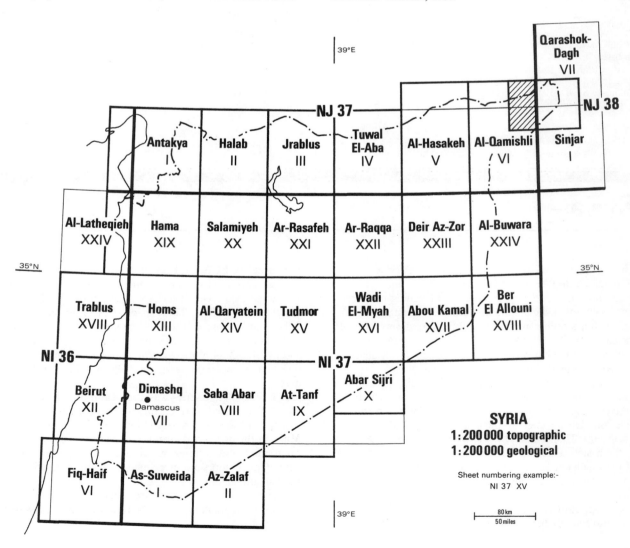

Tectonic map of Syria 1:1 000 000
Damascus: GEGMR, 1964.

Schematic map of Quaternary sediments of Syria 1:1 000 000
Damascus: GEGMR, 1964.
With accompanying explanatory text.

Map of the mineral resources of Syria 1:1 000 000
Damascus: GEGMR, 1964.
On geological base.

Geological map of Syria 1:500 000
Damascus: GEGMR, 1964.
4 sheets, all published.

Geological map of Syria 1:200 000
Damascus: GEGMR, 1964.
19 sheets, all published. ■
Each with accompanying explanatory note.

Geological map of Syria 1:50 000
Damascus: GEGMR, 1978–.
53 sheets, 15 published.
Arabic language: each with accompanying explanatory notes.

Taiwan

The National Mapping Agency of Taiwan, the Republic of China, is the **China Map Service Topographic Section** in the Ministry of Interior. There is no information about current mapping programmes or specifications. The Ministry of Interior has published two atlases which are available outside Taiwan. These are the Taiwan atlas, a national atlas with all names printed in Chinese and English which was published in 1981, and the Administrative atlas of Taiwan showing 16 county maps and five city maps, published in 1982. Both these atlases are available through **Youth Cultural Enterprise**. A recently available contoured map of the island was prepared by the **Chinese Society of Photogrammetry and Remote Sensing** for the **Council of Agriculture**. This map is bilingual and includes numerous smaller-scale maps printed on the verso to show the distribution of various research organizations associated with different aspects of the agricultural sciences.

Geological mapping of Taiwan is carried out by the **Central Geological Survey**, which had compiled two-sheet 1:250 000 scale geological and hydrogeological maps and single-sheet 1:500 000 maps to show geology, tectonics and metamorphic facies, all with accompanying English and Chinese language explanations.

Nun Hua Publishing produces a variety of commercial maps. These include maps of the counties of Taiwan, published on a topographic base in Chinese language at scales that vary according to the size of the unit; maps of 16 different cities in the country, a variety of general tourist, road and other small-scale maps of the island; as well as maps of mainland China and of the world. The more popular titles are published as English language editions, but most are available only as Chinese versions. We have listed the English language edition and the Province series, in view of the lack of any other available topographic mapping.

Overseas agencies producing maps of Taiwan include the People's Republic of China **Cartographic Publishing House (CPH)** and the German commercial agencies **Karto+Grafik** and **Nelles Verlag**.

Addresses

China Map Service Topographic Section
Ministry of the Interior, Roosevelt Road, TAIPEI.

Youth Cultural Enterprise
71 Yen Ping, South Rd, TAIPEI 100.

Chinese Society of Photogrammetry and Remote Sensing
TAIPEI.

Council of Agriculture
TAIPEI.

Central Geological Survey
PO Box 968, TAIPEI.

Nun Hua Publishing Co.
No. 3 Lane 278, Kang Ting Rd, Lung Shing District 108, TAIPEI.

For CPH see China; for Karto+Grafik and Nelles Verlag see German Federal Republic; for DMA see United States.

Catalogue

Atlases and gazetteers

Taiwan atlas, Republic of China
Taipei: Ministry of the Interior, 1981.
pp. 81 + 41 maps.

Republic of China: official standard names approved by the United States Board on Geographic Names
Washington, DC: DMA, 1974.
pp. 789.

General

Relief map of Taiwan 1:750 000
Taipei: Nun Hua, 1975.
In Chinese.

Taiwan: sheng ditu 1:750 000
Beijing: CPH, 1980.
In Chinese.

Hildebrands Urlaubskarte Taiwan 1:700 000
Frankfurt AM: Karto+Grafik, 1984.
Available in English and German editions.

Map of Taiwan 1:500 000
Taipei: Nun Hua, 1986.

Map of Republic of China on Taiwan 1:400 000
Taipei: Council of Agriculture, 1986.
Double sided.
With agricultural, forestry, fishery and animal organization, and national park and forest distribution maps on back.

Taiwan 1:400 000
München: Nelles Verlag, 1985.
With small inset plans of Taipei, Taichung, Tainan and Kaohsing.

Topographic

Map of the counties in Taiwan various scales
(1:30 000–1:150 000)
Taipei: Nun Hua, 1975–.
21 sheets, all published.
In Chinese.

Geological and geophysical

Geological map of Taiwan 1:500 000
Taipei: Central Geological Survey, 1974.
In English and Chinese.
With accompanying explanatory text.

Tectonic map of Taiwan 1:500 000
Taipei: Central Geological Survey, 1978.
In English and Chinese.

Metamorphic facies map of Taiwan 1:500 000
Taipei: Central Geological Survey, 1983.
In English and Chinese.
With accompanying explanatory text.

Geological map of Taiwan 1:250 000
Taipei: Central Geological Survey, 1974.
2 sheets, both published.
In English and Chinese.
With accompanying explanatory text.

Hydrogeological map of Taiwan 1:250 000
Taipei: Central Geological Survey, 1969.
2 sheets, both published.
In English and Chinese.

Administrative

Minute traffic map of Taiwan 1:360 000
Taipei: Nun Hua, 1986.
Road map on political background.
With highway map of Taiwan 1:360 000 on reverse.

Town maps

Map of Taipei city 1:33 000 and 1:18 000
Taipei: Nun Hua, 1982.

Map of Taichung city 1:19 800
Taipei: Nun Hua, 1977.

Thailand

Cartographic activities in Thailand have been carried out by the **Royal Thai Survey Department (RTSD)** since its creation in 1885. The Department is responsible for ground and aerial surveys for the production of topographic and thematic maps; it also engages in geodetic and geophysical work. The first modern base mapping of the country was a 1161-sheet 1:50 000 series L 708, produced in the 1950s under a joint mapping agreement with the US, which was compiled using photogrammetric methods and covered all the country north of latitude 7° N. This series was recast as series L 7017 from 1964 to give 772 quarter-degree sheets by 1973; 58 border sheets in other series have been added to L 7017 to give 830 sheets covering the whole country. This multi-colour map is bilingual, uses UTM grid and shows relief by 20 m contours. A continuous revision programme is in operation and priority is being given to the recompilation to higher standards of accuracy of a block of 262 sheets in the north-eastern and border areas of the country, which will be completed by 1988. There is also a programme for the production of a bilingual multi-colour 1:25 000 series L 8019, which has been undertaken since 1970. This map covers areas of development projects and other particular needs and is not intended to give complete coverage. It shows relief by 10 m contours and over 300 sheets have now been produced. A full-colour bilingual city map series at 1:12 500 (L 9013) is being carried on concurrently with the 1:50 000 series. It will cover the vicinities of primary and secondary administrative centres and a continuous revision programme is in operation. The 1:50 000, 1:25 000 and 1:12 500 series are all restricted.

The largest scale available coverage of Thailand is a 1:250 000 scale 52-sheet multi-colour bilingual map series 1501 S. Other publicly available mapping of the country includes a 1:20 000 scale, 20-sheet series of Bangkok, and

a variety of general-purpose and thematic small-scale maps, including the *National Resources Atlas of Thailand*, a 51-sheet bilingual loose-leaf collection of thematic maps of the country. RTSD has also compiled toponymic information from series 1501S for inclusion in a new US Board Gazetteer of the country.

The **Department of Mineral Resources of Thailand (DMRT)** is responsible for the production of geological mapping and for mapping activities in support of mining industries in the country. Since 1966 DMRT has been carrying out 1:250 000 scale geological mapping, with aid from the West German **Bundesanstalt für Geowissenschaften und Rohstoffe (BfGR)**. This systematic programme is intended to cover the whole country. From 1977 limited 1:50 000 and 1:25 000 programmes of areas of mineralogical importance have been surveyed. Other geological maps of Thailand are compiled, including 1:1 000 000 and 1:500 000 series. Hydrogeological and some geochemical maps are produced. A 1:2 500 000 scale thematic series of 14 mineral maps was published by DMRT in 1975. BfGR is to publish new 1:2 000 000 scale tectonic, mineral resources and Cenozoic sediments and basalt maps in 1987.

The **Soil Survey Division** of the Department of Land Development has been responsible for soil survey activities in Thailand since 1963. A 1:100 000 scale series of reconnaissance diazo soil maps using RTSD base data has been compiled and issued on provincial sheet lines. These maps have also been used to derive a soil suitability for rubber growing series for the south and south-east areas of the country. Multi-colour maps have been published at 1:1 000 000 scale to show general soil distribution and potential land use for crop production. The Division has also produced some geomorphological maps of Thailand,

and detailed project based soil maps. The **Land Classification Division** of the same Department carries out land capability mapping, including a black-and-white 1:250 000 scale land capability map of the whole country, and a provincial 1:100 000 map showing capability for upland crops or paddy crops. 1:500 000 maps derived from Landsat imagery show regional land use patterns and a programme for a 1:100 000 provincial land use series is currently under way. A multi-colour 1:250 000 scale land type classification map has also been published and the Division also produces larger-scale project mapping. Few of these larger-scale maps are available outside Thailand.

Other agencies involved in map production include the **Royal Irrigation Department**, which compiles large-scale line maps and orthophotomaps of irrigation projects from specially flown aerial cover. The **Royal Thai Forest Department** compiles land use and forest type maps. These have included 1:50 000 four-colour maps of much of Southern Thailand derived from 1:15 000 aerial coverage, and smaller-scale forest maps of the northern and eastern regions derived from Landsat imagery. Cadastral surveying activities are carried out by the **Cadastral Surveying and Mapping Division**, and hydrographic charting is the responsibility of the Hydrographic Department. The Agricultural Land Reform Office uses RTSD maps and aerial coverage to produce large-scale line maps for the planning and implementation of land reform, and the **National Energy Administration** produces large-scale plans for energy resource development areas.

The **Thailand Remote Sensing Center** has encouraged applications of remote sensing technologies to a variety of fields and has issued some small-scale thematic maps derived from Landsat data.

Several foreign commercial mapping agencies have published general maps of the country: these include the useful **Nelles Verlag** APA map. Amongst the Thai commercial sector publishers is T and R United who publish a double-sided and indexed English language town plan of Bangkok with a route map of Thailand on the back. **Recta Foldex** also publishes a Bangkok town map.

Further information

Catalogues are issued by the major organizations and a regular newsletter by the Thailand Remote Sensing Centre.

Background information on Thai mapping is contained in Thailand (1977, 1980, 1983) Cartographic activities in Thailand, papers presented to the *8th*, *9th* and *10th United Nations Regional Cartographic Conferences for Asia and the Pacific*, New York: UN.

Addresses

Royal Thai Survey Department (RTSD)
Kalayanamaitree Rd, BANGKOK 2.

Department of Mineral Resources (DMRT)
Rama VI Rd, BANGKOK 10600.

Soil Survey and Classification Division
Department of Land Development, Phaholyothin Rd, Bangkhen, BANGKOK 10900.

Royal Irrigation Department
811 Samser Rd, BANGKOK 10300.

Royal Forest Department
Phaholyothin Rd, BANGKOK 10900.

Cadastral Surveying and Mapping Division
Phrapipit Rd, BANGKOK.

National Energy Administration
Kasatsuyk Bridge, BANGKOK 10500.

Thailand Remote Sensing Center
National Research Council, 196 Phaholyothin Road, BANGKOK 10900.

For BfGR and Nelles Verlag see German Federal Republic; for Recta Foldex see France; for DMA see United States.

Catalogue

Atlases and gazetteers

National resources atlas of Thailand
Bangkok: RTSD, 1974–.
51 sheets, all published.

Thailand: official standard names approved by the United States Board on Geographic Names
Washington, DC: DMA, 1966.
pp. 675.

General

Road map of Thailand   1:2 000 000
Bangkok: RTSD, 1985.
In English and Thai.

Highway map   1:1 000 000
Bangkok: RTSD, 1986.
4 sheets, all published.
In English and Thai.

APA map of Thailand   1:1 500 000
München: Nelles Verlag, 1983.
Relief based road map with townplan insets and tourist information.

Map of Thailand   1:1 000 000
Bangkok: RTSD, 1979.
2 sheets, both published.
In English and Thai.

Topographic

Thailand   1:250 000   Series 1501 S
Bangkok: RTSD, 1973–.
52 sheets, all published. ■
In English and Thai.

Geological and geophysical

Mineral resources of Thailand   1:2 500 000
Bangkok: DMRT, 1975.
14 sheets, all published.
Single-sheet thematic series, 1 sheet per mineral.

Geological map of Thailand   1:1 000 000
Bangkok: DMRT, 1984.
2 sheets, both published.

Hydrogeological map of Thailand   1:1 000 000
Bangkok: DMRT, 1983.
2 sheets, both published.

Hydrogeological map of Thailand   1:500 000
Bangkok: DMRT, 1977–83.
8 sheets, all published.

Map of Thailand showing locations of wells drilled   1:500 000
Bangkok: DMRT, 1980.

Geological map of Thailand   1:250 000
Bangkok: DMRT, 1971–.
52 sheets, *ca* 40 published. ■

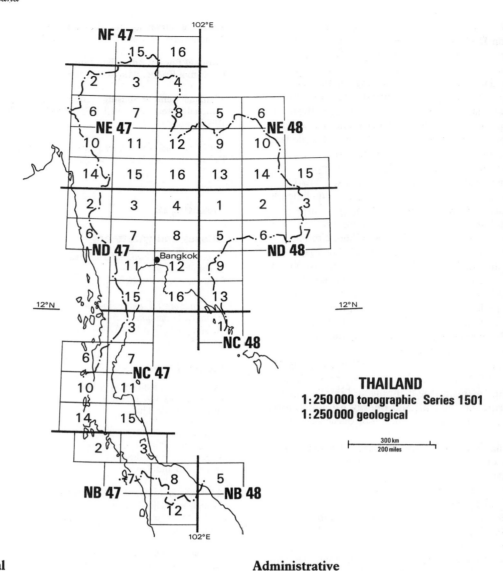

THAILAND
1:250 000 topographic Series 1501
1:250 000 geological

300 km
200 miles

Environmental

Main landforms of Thailand as related to soil formation   1:2 500 000
Bangkok: Soil Survey Division, 1972.

*Kingdom of Thailand: soil grouping based upon a combination of
particle size, mineralogy, reaction and moisture regime
classes*   1:2 500 000
Bangkok: Soil Survey Division, 1983.

General soil map of Thailand   1:1 000 000
Bangkok: Soil Survey Division, 1979.
3 sheets, all published.
With accompanying explanatory text.

Administrative

Administrative division map of Thailand   1:2 500 000
Bangkok: RTSD, 1971.
In Thai.

Town maps

City map of Krung Thep (Bangkok)   1:20 000   Series L9013 S
Bangkok: RTSD, 1978–82.
20 sheets, all published.
In Thai.

The TRU Bangkok map with bus routes   1:25 000
Bangkok: T and R United, 1983.
With Thailand route map on back.

Turkey

The modern mapping of Turkey began in 1895 with the
establishment of a mapping section in the general staff for
production of maps of the whole country. Prior to that
date mapping activities had been concentrated upon
European Turkey. In 1909 the 1:25 000 basic scale series
was begun and the General Directorate of Mapping
continued work on this series. The **General Command of**

Mapping (GCM) (Harita Genel Komutanligi) has since
1983 been responsible for provision of mapping, geodetic
data and aerial coverage of the country. The 1:25 000 map
is now complete and in the process of revision to full
photogrammetric standards. A 1:50 000 series is in
progress and is derived from the basic scale, as is a new
revised second edition of the 1:100 000 scale series.

1:250 000 scale coverage is prepared to JOG specifications and a revised edition is in progress. Smaller-scale series include an 18-sheet 1:500 000 scale compiled on Series 1404 sheet lines and specifications, and the Turkish sheets in the International Map of the World. A variety of other smaller-scale mapping of the country is also produced, including administrative and physical wall maps at 1:1 000 000 scale. 1:5000 scale plans are prepared for development purposes.

Topographic mapping at scales larger than 1:500 000 is not available for export from Turkey.

Cadastral mapping activities are carried out by **Tapu ve Kadastro**.

Geological mapping of Turkey is carried out by the Mineral Research and Exploration Institute, **Maden Tetkik ve Arama Enstitutu (MTAE)**. A 1:500 000 scale full-colour series was completed in the 1960s and published with accompanying monographs. The texts are no longer available. More recent mapping has concentrated upon smaller-scale coverage of a variety of economic geology themes, including a number of mineral distribution maps.

Other thematic mapping of the country has been published by German organizations. Some maps covering Turkey are available in the *Tübinger Atlas des Vorderen Orients* (TAVO). Two-sheet 1:2 000 000 scale coverage has been published to show geomorphology and natural regions, the former accompanied by one of the useful TAVO German language *Beihefte*. Geological coverage at the same scale in this atlas has also recently been published. Palaeogeographic mapping at 1:1 500 000 scale of Turkey was compiled by the German **Bundesanstalt für Bodenforschung und Rohstoffe (BfGR)** and published as a palaeogeographic atlas of the country in the mid-1970s.

Overseas commercial publishers of maps of Turkey include **Kümmerly + Frey, Ravenstein, Karto+Grafik, Hallwag**, and **Reise und Verkehrsverlag**, whilst town maps are also published by **Falk** and **Hallwag**.

Further information

Background information is available from GCM about its mapping activities.

Addresses

General Command of Mapping (Harita Genel Komutanligi) (GCM)
Ministry of National Defence, TC, MSB, Cebeci, ANKARA.

Tapu ve Kadastro
ANKARA

Mineral Research and Exploration Institute of Turkey (Maden Tetkik ve Arama Enstitutu) (MTAE)
Eskisehir Yolu Ustu, ANKARA.

For TAVO, BfGR, Falk, Reise und Verkehrsverlag, Karto+Grafik and Ravenstein see German Federal Republic; for Kümmerly + Frey and Hallwag see Switzerland; for GUGK see USSR; for CIA and DMA see United States.

Catalogue

Atlases and gazetteers

Turkey: official standard names approved by the United States Board on Geographic Names
Washington, DC: DMA, 1960.
pp. 651.

Yeni Turkiye atlasi
Ankara: GCM, 1977.
pp. 172.

General

Turzija 1:2 000 000
Moskva: GUGK, 1984.
In Russian.
With index and economic inset map.

Turkiye fiziki karayollari haritasi 1:1 000 000
Ankara: GCM, 1981.
3 sheets, all published.
Hypsometric tinted road map.

Topographic

Dunya 1:500 000 Series 1404
Ankara: GCM, 1981.
18 sheets, all published.

Geological and geophysical

Tectonic map of Turkey 1:2 500 000
Ankara: MTAE.

Metallogenic map of Turkey 1:2 500 000
Ankara: MTAE, 1970.
Also available as Turkish language edition.

Turkei Geologie/Turkey geology 1:2 000 000
Wiesbaden: TAVO, 1985.
2 sheets, both published.

Turkei: geomorphologie/Turkey geomorphology 1:2 000 000
Wiesbaden: TAVO, 1981.
2 sheets, both published.
Also available with *Geomorphologie der Turkei*, pp. 265.

Turkiye'nin bilinen maden ve mineral kaynaklari/Known ore and mineral resources of Turkey 1:2 000 000
Ankara: MTAE, 1981.
With accompanying explanatory text.

Turkiye ekonomik maden yataklari haritasi 1:2 000 000
Ankara: MTAE, 1981.

Palaeogeographic atlas of Turkey 1:500 000
Hannover: BfGR, 1976.
7 sheets, legend + 64 pp. text.

Turkiye jeoloji haratasi/Geological map of Turkey 1:500 000
Ankara: MTAE, 1961–.
18 sheets, all published.

Environmental

Turkei: Naturräumliche Gliederung/Turkey: natural regions 1:2 000 000
Wiesbaden: TAVO, 1982.
2 sheets, both published.

Administrative

Turkiye mulki idare bolumleri 1:1 000 000
Ankara: GCM, 1981.
3 sheets, all published.
Layer coloured by provinces.

Human and economic

Turkiye ekonomik yataklari haritasi 1:2 000 000
Ankara: MTAE, 1981.
Map of economically useful minerals.

Town maps

Ankara 1:19 500
Ankara: GCM, 1976.

Istanbul 1:10 560
Bern: Hallwag, 1986.

Istanbul 1:12 500
Hamburg: Falk, 1986.

United Arab Emirates

The United Arab Emirates is a federation of seven internally self-governing emirates that has been independent since 1971. The official mapping agency is the Survey Section of the **Ministry of Public Works and Housing**. There is no information about the official mapping of the Emirates. The best available map of the country is published in the Arab World Map Series by Geoprojects SARL and is distributed by **All Prints Distributors and Publishers** in Abu Dhabi. It contains both a general map of the Emirates and nine town plans of the most important cities, with tourist notes.

Geoprojects also compiles town maps of three other cities in the UAE in the same series, to cover Al Ain, Dubai and Abu Dhabi.

A full-colour image map at 1:100 000 scale using a Landsat 5 Thematic Mapper image base, overprinted with cultural and hypsographic detail covering the Dubai area, has been published by **Environmental Remote Sensing Applications Centre (ERSAC)**.

Geological and geochemical mapping of Trucial Oman was compiled by the British **Directorate of Overseas Surveys (DOS)** in the late 1960s and is still available from Overseas Surveys Directorate, Southampton (OSD).

Addresses

Survey Section, Ministry of Public Works and Housing
Abu Khabi, ABU DHABI.

All Prints Distributors and Publishers
Hamdan St, PO Box 857, ABU DHABI.

For DOS/OSD and ERSAC see Great Britain; for Geoprojects see Lebanon; for DMA see United States.

Catalogue

Atlases and gazetteers

Bahrain, Kuwait, Qatar and the United Arab Emirates: official standard names approved by the United States Board on Geographic Names
Washington, DC: DMA, 1976.
pp. 145.

General

Geoprojects United Arab Emirates 1:750 000
Abu Dhabi: All Prints Distributors and Publishers, 1986.
Double-sided, also includes city maps of Abu Dhabi, Dubai, Sharjah, Ajman, Umm al Qaiwain, Ras al Khaimah and Fujairah; also Dubai Island 1:138 000 and Road connections Dubai, Sharjah and Ajman 1:108 000.

Dubai: image map 1:100 000
Livingston: ERSAC, 1984.

Geological and geophysical

Photogeological map of the Trucial Oman 1:100 000 DOS Geol 1168AB
Tolworth: DOS, 1969.
2 sheets, both published.

Geochemical and mineral location map of Trucial Oman 1:100 000 DOS Geol 1169A
Tolworth: DOS, 1969.
2 sheets, both published.

Town maps

Geoprojects Al Ain city map and guide 1:50 000
Beirut: Geoprojects, 1980.
Double-sided, also includes Central Al Ain 1:10 000, United Arab Emirates 1:750 000, index and business information.

Geoprojects Abu Dhabi city map and guide 1:25 000
Beirut: Geoprojects, 1980.
Double-sided, also includes Abu Dhabi environs 1:100 000, Central Abu Dhabi 1:7300, indexes and business information.

Geoprojects Dubai city maps 1:25 000
Beirut: Geoprojects, 1980.
Double-sided, also includes Road connections Dubai Sharjah and Ajman 1:108 000, Central Dubai 1:10 000, Mina Jebel Ali 1:37 000 and indexes.

USSR

The national mapping authority in the USSR is the Chief Administration of Geodesy and Cartography, **Glavnoe Upravlenie Geodezii i Kartografii (GUGK)**. This is now an autonomous body, having formerly been part of the Ministry of the Interior. GUGK is responsible for geodetic, photogrammetric and topographic surveying activities throughout the Union and plays a coordinating role where other agencies have been involved in the production of mapping. Topographic series all use the Gauss conformal projection, a standard sheet division and numbering system and a unified system of levelling and geodetic control. The basic scale for most of the country is 1:100 000. This series was completed in the 1960s. 26 000 six-colour maps give complete cover, and the enormous task of revision is in progress to maintain the currency of this map base: about 1000 revised editions are published each year. This series is produced by photogrammetric methods, from medium-scale aerial coverage, and shows relief with 20 m contours for plains or 40 m contours for mountainous areas. From it is derived a 1:200 000 map, which covers the USSR in about 6000 sheets. The specifications for 1:1 000 000 scale maps were set after World War II and this series was the first modern map to give complete coverage of the country.

Following the completion of the 1:100 000 series effort has been concentrated upon the production of a national 1:50 000 map. This project is now nearly complete and will cover the country in about 90 000 six-colour maps. A 1:25 000 programme has also been started, and large-scale programmes to cover urban and development areas using scales of 1:500, 1:1000, 1:2500, 1:5000 and 1:10 000.

None of the topographic or large-scale maps is available outside the USSR.

There are plans to introduce an automated cartographic system, based on digital terrain data representation, but little information is available about the state of these plans. Some use is now made of remotely sensed imagery for the compilation and revision of medium-scale topographic series of the more remote areas of the country, e.g. the Pamirs, Tien Shan, the polar areas and the extreme North East.

GUGK also publishes a range of smaller-scale and tourist products which are available to the west through authorized agents and are listed in the annually produced catalogue *Karty i atlasy* which is issued by **Mezhdunarodnaja Kniga** publishing house. These include political and physical maps of the whole Union published at 1:8 000 000 in Russian, English, French, German and Spanish language editions, revised on an annual basis and at 1:4 000 000 in four sheets. Atlases of the USSR and world are published: these include the regularly revised Atlas SSSR, a road atlas and a rail atlas. In the 1960s and 1970s GUGK published political maps of the Oblasts, Krais and Republics, at different scales according to the size of the region covered: many are still available. Much of GUGK's production is aimed at the home educational market and a variety of useful simplified thematic maps are available: these include the *uchebnaja karta* series, which comprises 20 different 1:10 000 000 or 1:5 000 000 maps of the USSR, revised on a regular basis, each concerned with a specific theme. A specialist series of historical educational maps is also produced. Town plans and

transport maps of the major cities are produced for the tourist market: English, French and German language editions are published for seven cities, and another 20 cities are published as Russian language editions. GUGK also has an extensive overseas publication programme producing general maps of other countries to a standard formula of a general map with hypsometric tinted relief, accompanying small-scale thematic inset maps and placename index, in the *spravochnaja karta* series. Most of these small-scale products are published in Russian language editions only, but most are at least available in the west. The first country-based English language maps in the spravochnaja karta series, for Angola and Libya, were published in 1985 and there are plans to extend this translation programme.

Information about GUGK mapping is available from the **Information Centre for Scientific Cartography** in Moscow.

Geological mapping of the USSR is the responsibility of GUGK in collaboration with the Soviet Ministry of Geology (**Ministerstvo Geologii SSSR, MGSSSR**), and its subsidiary bodies, especially the All Union Geological Research Institute (**Vsesojuznnoi Nauchno-Issledovatel'skii Geologicheskii Institut, VSEGEI**). VSEGEI now acts as the publisher of MGSSSR mapping. More specialized earth science mapping is produced by, or in collaboration with, the relevant branch of the Academy of Science. An example of such an agency is the Research Institute in Arctic Geology (**Nauchno-Issledovatel'skii Institut Geologii Arktiki, NIIGA**) in Leningrad. Some regional earth science mapping is carried out at the region rather than at the central/federal level—organizations here mirror the national scientific establishment, with regional academies of science collaborating with national level agencies in the production of specialist maps of their own Republic for their own subject area. MGSSSR is also involved in the geological and geophysical surveying of other areas of the world through its All Union Research Institute for Foreign Countries Geology and the **Subcommission for the Tectonic Map of the World** is based in the Geology Research Institute of MGSSSR in Moscow (see World section for details of its publications).

The history of geological mapping in the USSR can be traced back to the Geological Committee established in 1882 and charged with the publication of a geological map of European Russia. This was published in 1894. Not until 1922 was a geological map of the whole Union (at 1:5 000 000 scale) published.

Since then a vast range of earth science mapping has been compiled, both in terms of a wide variety of scales used and series published and also in terms of the variety of data mapped: over 30 different kinds of earth science map are published. As with topographic mapping, very little of this output is available to the west, with the exception of smaller-scale products. There is now complete geological coverage of the USSR in series using the same sheet lines as the topographic maps at scales of 1:1 000 000 and 1:200 000. All the important mineral districts are to be mapped at 1:50 000 scale and this programme is also well advanced; by 1983 30 per cent of the country was covered in these medium-scale geological maps. In the 1970s and 1980s a new series of State earth science maps at

1 : 1 000 000 scale was compiled to include coverage of pre-Quaternary formations, Quaternary sediments and magnetic anomalies, gravimetry and mineral resources. For some regions hydrogeological and geomorphological maps have been added to these sets. Much of the information currently used for the compilation of small-scale thematic mapping by MGSSSR and by GUGK is derived from the extensive Russian earth resources satellite data programme.

All Union small-scale and general earth science maps are issued at 1 : 10 000 000, 1 : 7 500 000, 1 : 5 000 000 and 1 : 2 500 000. These are usually the official editions of a ministry or department and are revised on a regular 10 or 15 year cycle. 1 : 2 500 000 scale maps are published in 16-sheet editions, with varying numbers of additional sheets to accommodate explanatory text where needed. MGSSSR issued a 1 : 2 500 000 geological map of the whole Union in the 1960s. This has been followed by sets at the same scale covering such themes as planation surfaces and weathering crusts, magmatic formations, tectonics, hydrogeology, sedimentary and volcanogenic formations and metallogenic regions. The geological base to these editions is revised regularly, the latest (fourth) edition appeared in 1981, containing for the first time the structure of water area floors. Smaller-sized 16-sheet editions at 1 : 7 500 000 have been compiled to cover geology, geomorphology, tectonics, hydrogeology, geochemistry and mineral deposits. An important 1 : 10 000 000 scale geological atlas was compiled by VSEGEI between 1975 and 1981 to include 17 different earth science thematic maps. All the maps in this atlas are accompanied by explanatory texts and an English abstract.

For information about the various Soviet geological mapping programmes contact the **All Union Geological Library** in Leningrad.

Bathymetric and thematic mapping of large areas of the world's oceans is carried out by **Glavnoe Upravlenie Navigatsii i Okeanografii (GUNO)** and by various divisions of the Soviet Ministry of Defence, especially the Voenno Morskoi Flot. For details of its worldwide work see the various ocean sections in this book.

Other state publishing houses who issue mapping in the USSR include **Mysl** and the Science Publishing House **Nauka**.

Many foreign commercial houses publish general mapping of the USSR. These include **Haack**, **Hallwag**, **Bartholomew** and **Karto+Grafik**. The American **Central Intelligence Agency (CIA)** publishes many useful maps of the Soviet Union, including town plans of Moscow and Leningrad and a 1982 produced series of small thematic maps of the country. The 1974 published CIA *USSR Agriculture Atlas* is still available.

Further information

For information about the organization of surveying and mapping in the USSR see Brandenberger, A.J. (1975) Surveying and mapping in the USSR, *Surveying and Mapping*, 35(2), 137–145.

For progress reports on the state of the art of cartography in the USSR see the following papers presented at UN Regional Cartographic Conferences for Asia and the Pacific and at ICA conferences: Committee of Soviet Cartographers (1977) Topography, geodesy and cartography in the Union of Soviet Socialist Republics, 1973–1976, paper presented at the *8th United Nations Regional Cartographic Conference on Asia and the Pacific*, New York: UN; Committee of Soviet Cartographers (1977) Some characteristic features of the compilation of geological maps in the Union of Soviet Socialist Republics, paper presented at the *8th United Nations Regional Cartographic Conference on Asia and the Pacific*, New York: UN; USSR (1980) Topography, geodesy and cartography in the Union of Soviet Socialist Republics, paper presented at the *9th United Nations Regional Cartographic*

Conference on Asia and the Pacific, New York: UN; USSR (1983) Thematic mapping: the base of the space natural studies, paper presented at the *10th United Nations Regional Cartographic Conference on Asia and the Pacific*, New York: UN; and Committee of Soviet Cartographers (1984) Cartographic works in the USSR 1981–1984, paper submitted to the *7th ICA General Assembly*, Perth 1984.

GUGK issues a regular Russian language catalogue listing publicly available mapping distributed through its authorized agents in the west. This is *Katalog Karty Atlasy*, published as part of Sovetskie Knigi, on an annual basis, by Mezhdunarodnaja Kniga.

For detailed information about geological mapping of the USSR see Nikiforova, G.Ya. (1984) *Maps of the geological contents of the Union of Soviet Socialist Republics at Geomap-84*, prepared for the XXVII session of the International Geological Congress, Moscow, 1984, Moskva: Ministry of Geology of the USSR.

Addresses

Main Administration for Geodesy and Cartography (CNIIGIK Glavnoe Upravlenie Geodezii i Kartografii pri Sovete Ministrov SSSR) (GUGK)
Ul. Krzhizhanovskogo 14, Korp. 2, V-218 MOSKVA.

V/O Mezhdunarodnaja Kniga (International Publishing House)
MOSKVA G-200.

Information Centre for Scientific Cartography
Novoshchukinskaya 11, 123098 MOSKVA D-98.

Ministry of Geology USSR (MGSSSR)
Bolshaya Gruzinskaya 4, 123242, MOSKVA D-242.

All Union Geology Scientific Research Institute (VSEGEI)
Sredniy Prospekt 71, 199026 LENINGRAD.

Research Institute of Arctic Geology, Nauchno-issledovatel'skii Institut Geologii Arktiki (NIIGA)
Sredniy Prospekt 71, 199026 LENINGRAD.

All Union Geological Library
Sredniy Prospekt 72B, 199026 LENINGRAD.

Chief Administration for Navigation and Oceanography, Glavnoe Upravlenie Navigatsii i Okeanografii (GUNO)
Ministetstva Oborony, 8111 Iliniya B34, LENINGRAD.

Izdatel'stvo Mysl'
Leninskij prosp 15, 117071 MOSKVA.

Izdatel'stvo Nauka
Profsajuznaja ul. 90, 117485 MOSKVA.

For Bartholomew and Quail see Great Britain; for Haack see German Democratic Republic; for Karto+Grafik see German Federal Republic; for Hallwag see Switzerland; for CIA and DMA see United States; for Sub-Commission for the Tectonic Map of the World see World.

Catalogue

Atlases and gazetteers

USSR: Official standard names approved by the United States Board on Geographic Names
Washington, DC: DMA, 1970.
7 vols.

Atlas SSSR 3rd edn
Moskva: GUGK, 1983.
pp. 260.
Russian language.

General

Sojuz Sovetskich Sotsialisticheskich Respublik 1 : 15 000 000
Moskva: GUGK, 1985.

Sowjet Union: Landkarte 1 : 9 000 000
Gotha: Haack, 1982.
With accompanying explanatory text.

Union of Soviet Socialist Republics 1 : 8 000 000
Moskva: GUGK, 1986.
Available as Russian, English, French, German or Spanish
edition.

USSR and adjacent areas 1 : 8 000 000 Series 5104
Washington, DC: DMA, 1974.

Sojuz Sovetskich Sotsialisticheskich Respublik 1 : 4 000 000
Moskva: GUGK, 1986.
4 sheets, all published.

Geological and geophysical

Atlas of geological and geophysical maps of the USSR 1 : 10 000 000
Leningrad: VSEGEI, 1975–81.
17 maps, all published.
Single-sheet thematic coverage for 17 geological and geophysical
themes.

Geologicheskoje strojenie SSSR 1 : 7 500 000
Leningrad: VSEGEI, 1967.
16 sheets, all published.
Sheets published in 7 different thematic sets: geological,
geomorphological, tectonic, metallogenic, hydrogeological and
geochemical.

Karta novejshej tektoniki SSSR 1 : 4 000 000
Leningrad: VSEGEI, 1984.
4 sheets, all published.

Tektonicheskaja karta neftegazonosnych oblastej SSSR 1 : 2 500 000
Leningrad: VSEGEI, 1970.
16 sheets, all published.

Karta magmaticheskich formaciij SSSR 1 : 2 500 000
Leningrad: VSEGEI, 1971.
17 sheets, all published.

*Karte poverchnostej vyravnivanija i kor vyvetrivanija
SSSR* 1 : 2 500 000
Leningrad: VSEGEI, 1972.
17 sheets, all published.

Metallogenicheskaja karta SSSR 1 : 2 500 000
Leningrad: VSEGEI, 1971.
18 sheets, all published.

Environmental

USSR Agriculture atlas
Washington, DC: CIA, 1974.
pp. 59.

*Karta pochevenno-geografícheskogo rajonirovanija
SSSR* 1 : 8 000 000
Moskva: GUGK, 1983.
2 sheets, both published.

Administrative

Soviet Union 1 : 8 750 000
Washington, DC: CIA, 1985.
Administrative regions map.

Union of Soviet Socialist Republics 1 : 8 000 000
Moskva: GUGK, 1986.
Political map.
Available as Russian, English, French, German or Spanish
editions.

Sojuz Sovetskich Sotsialisticheskich Respublik 1 : 4 000 000
Moskva: GUGK, 1986.
4 sheets, all published.
Political map.

Politiko-administrativnoie karti various scales
Moskva: GUGK, 1967–70.
20 maps, all published.

Human and economic

Atlas zheleznych dorog SSSR Passazhirskoe soobychehenie
Moskva: GUGK.
pp. 188.
Russian language railway atlas.

Soviet Union: thematic maps
Washington, DC: CIA, 1982.
11 sheets.
Maps published to cover: general relief, administrative divisions
1981, coal, electric power, land use, machine building and metal
working, metallurgy, nationalities, petroleum refining and
chemical industry, population.

Uchebnaja karta 1 : 5 000 000
Moskva: GUGK, 1985.
20 sheets, all published.
Thematic set.

Town maps

Skhematicheskie plany gorodov various scales
Moskva: GUGK, 1975–.
43 sheets, all published.
16 sheets published in foreign language editions.

Skhemy gorodskogo passadzirskogo transporta various scales
Moskva: GUGK, 1975–.
16 sheets, all published.

Moscow 1 : 35 000
Washington, DC: CIA, 1980.
Index on verso.
Also available in same format as Moscow Central 1 : 15 000.

Vietnam

The national mapping agency of the Socialist Republic of
Vietnam is the **National Geodetic and Cartographic
Service (NGCS)** in Hanoi. It was established in 1959 and
is responsible for geodetic, topographic and
photogrammetric surveying of the country and for the
compilation of official and educational mapping. Until
1975 it compiled mapping of the North of the country but
following the end of the war and the withdrawal of
American forces it became responsible for the compilation
of the map base of the whole Republic. The map base was
established by French and American occupying powers
during the 1950s and 1960s—the French compiled various

scales of mapping and American Indochina series at
1 : 100 000, 1 : 50 000 and 1 : 25 000 scales covered the whole
of South Vietnam and were issued from the National
Geographic Directorate in Dalat. This organization was
also responsible for the compilation of smaller-scale
coverage of South Vietnam and for some thematic
mapping. None of these maps is still available. Since 1974,
NGCS has produced modern basic scale
photogrammetrically derived mapping of about one-third
of the area of North Vietnam, at scales of 1 : 10 000 and
1 : 25 000.

Current mapping of the country is not available to the
west, and only Russian-produced maps compiled by
Glavnoe Upravlenie Geodezii i Kartografii (GUGK),
which give a general overview of the country, may be
acquired.

Further information

Thai, N.V. (1980) Rozwoj geodezii i kartografi w Wietnamie,
Przeglad Geogezyjny, **52**(4/5), 169–173. le The, T. (1984) A
terkepezes helyzete Vietnamban, *Geodezia es Kartografia Budapest*,
35(1), 53–54. Only Hungarian and Polish language articles about
the mapping of Vietnam are available!

Addresses

National Geodetic and Cartographic Service (NGCS)
Cuc Do Dag Ban Do Nha Nuoc, Dong-Da, HANOI.

For DMA see United States; for GUGK see USSR.

Catalogue

Atlases and gazetteers

*South Vietnam: official standard names approved by the United States
Board on Geographic Names*
Washington, DC: DMA, 1971.
pp. 337.

*North Vietnam: official standard names approved by the United States
Board on Geographic Names*
Washington, DC: DMA, 1964.
pp. 311.

General

V'etnam: fizicheskaja karta   1 : 2 000 000
Moskva: GUGK, 1979.
In Russian.

V'etnam   1 : 2 000 000
Moskva: GUGK.
In Russian.
With economic and climate insets and placename index.

Administrative

V'etnam   1 : 2 000 000
Moskva: GUGK, 1979.
In Russian.
Political coloured map.

Human and economic

V'etnam: ekonomicheskaja uchebnaja karta   1 : 2 000 000
Moskva: GUGK, 1982.
In Russian.
Economic educational wall map.

Yemen Arab Republic

The Yemen Arab Republic has been mapped recently in
aid projects involving cooperation with British, American,
German and Swiss agencies. The national surveying and
mapping organization is the **Survey Authority**, which has
been cooperating in the last 7 years with the British
Overseas Surveys Directorate (OSD) in the production of
the first modern basic scale series of the country. This
series *Yemen Arab Republic* 1 : 50 000 DOS/YAR 50
requires about 230 sheets to give complete coverage. It
uses UTM projection, is compiled from aerial coverage
with ground checking and shows relief by 20 m contours.
Toponymic information and legend is given in both Arabic
and English forms. The first sheets appeared in 1979 and a
block of territory in the more developed south-western
areas has now been covered by this series. This map is no
longer available from OSD but may be obtained by direct
application to Sana'a. Other British participation has
resulted in the production of an eight-sheet 1 : 250 000 scale
map, which showed relief with hypsometric tints and
100 m contours (no longer available) and a single-sheet
1 : 500 000 map. The **United States Geological Survey
(USGS)** has also produced maps of the YAR, including a
1 : 500 000 scale image map based on false colour composite
imagery from Landsat 1.

USGS has also been involved in geological mapping and
their 1 : 500 000 scale geological map has recently been
reprinted. The German **Bundesanstalt für
Geowissenschaften und Rohstoffe (BfGR)** has compiled
1 : 250 000 scale geological coverage of the YAR which
gives complete coverage in four sheets and uses Landsat
imagery for topographic background. This map was
completed in 1984.

Other thematic mapping has been produced through
Swiss aid projects, funded by the **Swiss Technical
Cooperation Service**, including 1 : 2 000 000 scale thematic
coverage, a 1 : 500 000 scale population, land use and
administrative map, and large-scale photomaps of various
sites. Most of these maps were compiled by the
Geographisches Institut of the **Universität Zürich**, in
conjunction with the Survey Department. The *Tübinger
Atlas des Vorderen Orients* (TAVO) also includes several
sheets covering different themes for the YAR or for
Southern Arabia. General TAVO coverage of Arabia is
listed under Saudi Arabia.

Other small-scale mapping of the country has been
compiled by the **Deutsch-Jemenitsche Gesellschaft**, by
Macmillan for the Survey Authority, and by the American
Central Intelligence Agency (CIA).

YEMEN ARAB REPUBLIC
1:50 000 topographic
Series YAR 50

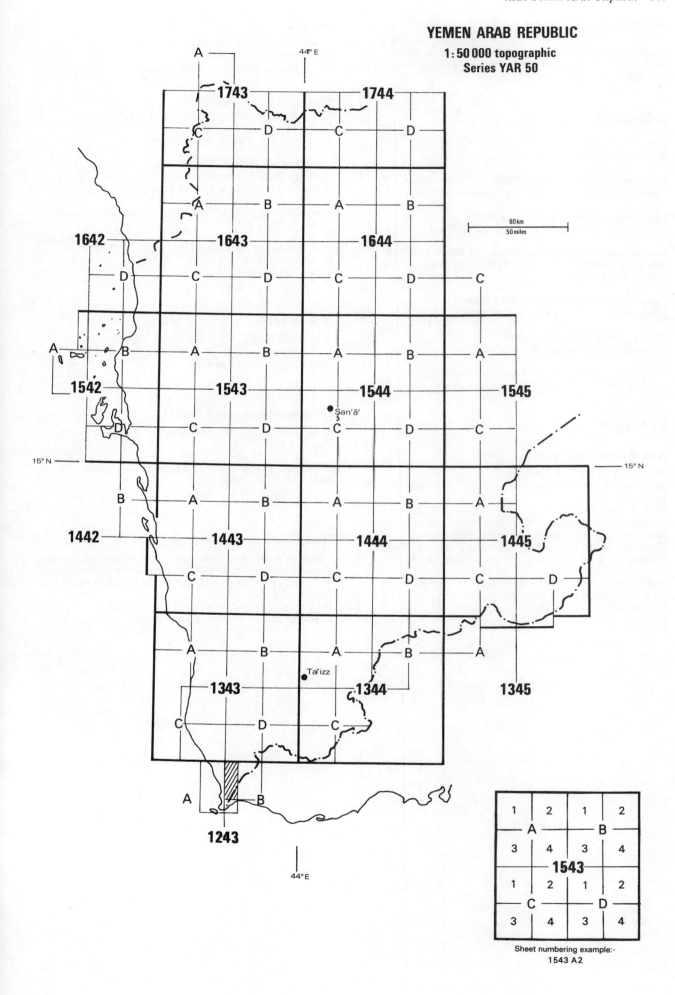

80 km
50 miles

Sheet numbering example:-
1543 A2

Further information

See DOS (from 1984 to 1985 OS) Annual Reports for information on the official map series of the YAR produced through joint OSD/Survey Authority projects.

Addresses

Survey Authority
Zubeiri St, PO Box 11137, SANA'A.

Overseas Surveys Directorate (DOS)
Ordnance Survey, Romsey Rd, SOUTHAMPTON SO9 4DH, UK.

United States Geological Survey (USGS)
516 National Center, RESTON, Virginia 22092, USA.

Bundesanstalt für Geowissenschaften und Rohstoffe (BfGR)
Alfred Bentz Haus, Postfach 510153, 3000 HANNOVER 51, German Federal Republic.

Geographisches Institut, Universitat Zürich
Blumlisalpstrasse 10, 8006 ZURICH, Switzerland.

Deutsch-Jemenitsche Gesellschaft
FREIBURG, German Federal Republic.

For TAVO see German Federal Republic; for Macmillan see Great Britain; for CIA and DMA see United States.

Catalogue

Atlases and gazetteers

Yemen Arab Republic 1 : 2 000 000
Zürich and Sana'a: Department of Geography, University of Zurich, 1978.
10 sheets published.
Thematic set: sheets published cover: physical setting, population distribution and relief, population density, emigration distribution, emigration rate, administrative division, geographic division, education standard, settlement size, and base map.

Yemen Arab Republic (Sana): official standard names approved by the United States Board on Geographic Names
Washington, DC: DMA, 1976.
pp. 124.

General

Yemen (Sana'a) 1 : 1 280 000
Washington, DC: CIA, 1973.
With 2 small-scale insets covering population and land use on sheet.

Yemen Arab Republic 1 : 1 000 000
Sana'a: Survey Authority and Macmillan, 1982.
Double-sided.
Town plans and tourist map on verso.

Arabische Republik Jemen 1 : 1 000 000
Freiburg: Deutsch-Jemenische Gesellschaft, 1983.
German text, with tourist notes on back.

Yemen Arab Republic 1 : 500 000 DOS YAR 500
Tolworth: DOS, 1978.

Geographic map of the Yemen Arab Republic 1 : 500 000
Reston, VA: USGS, 1978.
Satellite image map.

Topographic

Yemen Arab Republic 1 : 50 000 DOS YAR 50
Tolworth and Sana'a: DOS and Survey Authority, 1979–.
230 sheets, *ca* 50 published. ∎

Geological and geophysical

Geologic map of the Yemen Arab Republic 1 : 500 000
Reston, VA: USGS, 1978.

Geological map of the Yemen Arab Republic 1 : 250 000
Hannover: BfGR, 1984.
4 sheets, all published.

Human and economic

Population distribution, administrative divisions and land use in the Yemen Arab Republic 1 : 500 000 compiled by the Department of Geography University of Zürich for the Central Planning Organization
Bern: Swiss Technical Cooperation Service, 1977.
English and arabic description and legends.

AUSTRALASIA

Australia

Throughout Australia over 50 government departments, both state and federal, and 30 private organizations are involved in major mapping programmes. Responsibility for the topographic mapping of the 3 000 000 square mile continent of Australia is divided between state and federal government agencies. Coordination of mapping activities is achieved through the operation of the National Mapping Council, which is composed of representatives of each mapping agency under the Chairmanship of the Director of National Mapping.

Until 1950 there was no national coordination between the different states' survey organizations: only after that date have federal and state maps been produced using a standardized projection (UTM) and with a single Australian Map Grid (AMG). Since 1973 all major cartographic establishments have been operating using metric standards. Australia was the last continent to be adequately surveyed: there has, however, in the last 25 years been a vast amount of effort devoted to the production of a wide variety of very high-quality topographic mapping.

The **Division of National Mapping (NATMAP)** is responsible, with the **Royal Australian Survey Corps (RASC)** for the publication of small- and medium-scale topographic mapping of the country. They cooperate with individual state mapping agencies in the publication of sheets covering their areas, though RASC is involved in the production of larger-scale mapping in some states, notably Queensland. At present two series comprise the basic scale mapping of the majority of the continent. The *1:100 000 topographic mapping programme*, begun in 1968, was originally intended to cover the whole country in 3065 seven-colour maps by 1975. However, this proved beyond NATMAP's capabilities and the programme was amended to include publication of 1640 sheets around Australia's seaboard, to cover inhabited areas and areas of potential resource development, leaving the remaining 1425 sheets in the less settled interior to be taken only as far as compilation plots. The full-colour maps in this series include vegetation type and density information and show relief by 20 m contours, spot heights and on NATMAP published sheets additionally by hill shading: they show 1 km Australian Map Grid squares. Since 1972 a 1:100 000 orthophoto mapping programme has been in operation to cover those areas not mapped in the full series and contour and line conversion overlays are available to aid

interpretation of these photomaps. The publication of the full-colour 1:100 000 programme has now been completed: the remainder of the 1:100 000 sheets covering the interior ought to have been compiled by the end of 1987. A systematic revision programme has begun using new aerial photography flown in 1984/85, and from 1987 this programme is expected to occupy most of the Division's resources. NATMAP commenced digital compilation of topographic mapping in 1983, with about 40 1:100 000 digital sheets on programme.

The whole of Australia is covered by 1:250 000 mapping. Together with the state mapping agencies, NATMAP and RASC jointly produce the 544 map sheets needed for coverage at this scale. These have been published in two series. The first, R 502, completed in 1968, was a preliminary series with 250 foot contours or hill shading using the now superseded yard grid. This is being replaced by the *National Topographic Map*, each sheet of which is derived from six 1:100 000 maps or orthophotomaps, which is available as either a Joint Operations Graphic RASC edition or a conventional NATMAP version. All sheets in this series have 50 m contours supplemented by hill shading, and use the Australian Map Grid. Sheets cover an area of 1.5° × 1°. It is expected that this new series will be completed by 1988, when a systematic revision programme will begin.

Also published at this scale is the *1:250 000 Bathymetric Map Series*. NATMAP began a programme of mapping the 2.3 million km² of the Australian Continental shelf in 1971. In July 1984 this programme was 60 per cent complete. The maps in this series show sea depth to a maximum of 300 m and now use a 10 m isobath, with different blue layer tints at 100 m depth intervals. Reconnaissance dyelines only are available for the Great Barrier Reef area.

Smaller-scale topographic mapping of Australia is also carried out by NATMAP. The *1:1 000 000 series* complying with the specifications of the International Map of the World covers Australia in 49 sheets. NATMAP has begun to establish a digital topographic database at this scale with the aims of revising the World Aeronautical Chart series, and assisting in rationalizing between the IMW series and the WAC series; to act as a base for thematic data handling; and to be available for sale. Until 1982 revision of the IMW sheets relied heavily upon the compilation of component 1:250 000 and 1:100 000 scale mapping. Since then, however, 1:1 000 000 sheets have been revised using satellite imagery.

Also published by NATMAP, at this scale and using the same sheet lines, are several thematic series. Three series are available only as diazo prints: the forestry reserves, forest types and nature conservation reserves series are all regularly up-dated. The other three series are issued by NATMAP as printed maps and cover only the developed areas of the country. Land tenure maps show types and ownership of land, whilst a land cover series derived from Landsat imagery identifies nine categories of cover. These two series are combined to produce a land use series using a classification system with nine major categories subdivided into over 30 uses, similar to that used by the US Geological Survey.

Two other major thematic series are produced by NATMAP. The *Australia 1:5 000 000 Map Series* covers a variety of themes; these large multi-coloured single topic maps of the whole continent are available either flat or folded. The *Australia small-scale thematic map series* comprises sheets published at a scale of 1:10 000 000 covering a variety of topics, so far mainly concerned with agriculture, land use and population. Both of these maps are derived from material compiled for the major National Atlas project, the Third Series *Atlas of Australian Resources*. This new edition is being published by NATMAP in 10 bound volumes, three of which have so far been issued. The Second Series of this project, completed in 1977, is at present still available as are some single sheets from the work. NATMAP has also been involved in thematic mapping projects at larger scales for some states, for instance the 1:1 000 000 scale Queensland resource atlases.

Amongst the other publications of NATMAP are a guide to map reading which contains details of all the topographic maps of Australia, and the computer listing Master Names File, a gazetteer regularly updated of all the named features on current 1:100 000 and 1:250 000 scale Australian maps. NATMAP also publishes a series of monographic technical reports relating to its mapping and survey activities.

NATMAP is increasingly involved in digital mapping activities. In addition to the 1:1 000 000 digital programme NATMAP has also digitized the coastline and state boundaries from the 1:250 000 map series. Boundaries used in the 1976 and 1981 Censuses of Population and Housing have been digitized from the maps prepared by the **Australian Bureau of Statistics (ABS)**: many of these files have been used to generate maps in joint publications of the two Departments. For instance the *Atlases of Population and Housing* for the 1976 censuses contain many digitally derived maps, as do the important *Social Atlases for Australian cities* covering the state capitals and Canberra, which present data from the 1981 census. ABS itself publishes statistical area mapping at small scales relating to the 5 yearly censuses. It also issues large-scale 1981 census field maps, which are available for purchase as fiche or hard copy. These 5489 sheets are published at a variety of scales, depending on the size of the census collection unit, and show census information overprinted on a black topographic base.

Earth science mapping in Australia is carried out by the **Bureau of Mineral Resources, Geology and Geophysics (BMR)**, which was established in 1946 and has since 1965 been based at a single site in Canberra. There are currently plans to move to a new purpose-built complex. Until the 1980s BMR had a geological surveying and mining overview role: following the near completion of basic data collection activities the focus has shifted towards more applied geoscientific research.

BMR has been involved in cooperative projects with the various State Geological Surveys and Mining Authorities to produce the *1:250 000 Geological Map*, which is published on the same sheet lines as the topographic map of that scale, and uses it as a base. Each sheet in this series is now published with an explanatory note; some early sheets were issued at a quarter-inch scale. There is now nearly complete coverage, and most sheets may be obtained from BMR, with the exception of those covering New South Wales and South Australia which must be obtained from the state geological surveys. BMR will not be reprinting maps in this series, as responsibility for this is now vested with the State Authorities who have taken over the plates. Amongst the recent interesting developments taking place in the production of this series is the use by both the **Western Australian Department of Mines** and the **Queensland Department of Mines** of laser scanning as a means of reprinting out-of-print maps.

In areas of potential economic importance and where there is complicated geology, BMR and some of the state Geological Surveys have produced a variety of larger-scale geological mapping, especially at 1:100 000 scale. There is also a series of stream sediment geochemical maps published at this scale. Regional mapping to accompany bulletins and reports is also available, especially at 1:500 000 scale, covering parts of Northern Territory and Queensland.

Geophysical mapping in Australia has been centralized and has been carried out by BMR for the different State agencies. The 1:250 000 *Radiometric Maps* and *Aeromagnetic Maps* are two single-colour series jointly published by BMR and the **South Australian Department of Mines and Energy (SADME)**. The radiometric maps often display aerial photographs showing anomaly locations, whilst the aeromagnetic series shows magnetic intensity contours. Many sheets are only available from the **Government Printer Copy Service** as diazo copies; printed sheets are available from BMR and SADME.

The 1:500 000 and 1:250 000 scale gravity maps compiled by BMR were one of the earliest applications of digital methods in thematic cartography. Data from reconnaissance and systematic regional gravity surveys are presented as Bouguer anomaly contour maps over a controlled planimetric base; these are reduced by the computer from 1:250 000 survey information to 1:500 000 mapping. Diazo prints and transparencies of the 250 000 scale originals and the computer-generated maps are available from the Copy Service. With the introduction of the Australian National Gravity Data Base the published 1:500 000 scale maps are being withdrawn, and only 1:250 000 and 1:1 000 000 output from the Copy Service may be acquired.

Other applications of digital technology are being used by BMR in both map production and the establishment of geological and geophysical spatial databases.

Offshore areas are published in the marine geophysical series at a scale of 1:1 000 000 and include free air, water depth and magnetic anomaly profiles. A continental shelf sediments series is published at 1:1 000 000 scale.

Small-scale geological and geophysical mapping is available as single sheets for each of the states at a variety of scales. 1:2 500 000 scale geological coverage in four sheets is regularly revised. The 1:5 000 000 scale is used for single-sheet high-quality geological, gravity, metallogenic, metamorphic and petroleum exploration and development maps of the whole country. BMR also issues at 1:10 000 000 scale the *Earth Science Atlas of Australia*, comprising 15 sheets, and explanations covering a variety of earth science themes.

Various other organizations in the public domain issue maps at a country-wide level. The **Australian Electoral Commission** issues in fiche format every proclaimed Commonwealth electoral map since federation. Consolidated maps showing the federal electoral boundaries are also available in 14 sheets to cover the whole continent.

The **Australian Survey Office (ASO)** in Canberra is responsible for a variety of mostly larger-scale mapping, in addition to its role as the Commonwealth's surveying service authority. ASO is increasingly using autocarto methods and has compiled numerous one-off and series maps in digital format (see below for information on Australian Capital Territory). A digital elevation model of the whole continent is at present being produced.

Small-scale soil maps have been published by the Division of Soils of the **Commonwealth Scientific Industrial Research Organization (CSIRO)**, including the 1 : 2 000 000 scale 10-sheet continental coverage compiled by Northcote in the 1960s. Other divisions of this body are also involved in mapping activities, notably the **Division of Water and Land Resources**. As the Division of Land Research it has been responsible for the publication of a very wide variety of land system surveys, which were issued until 1977 in the Land Research Series. These maps depict at scales between 1 : 250 000 and 1 : 1 000 000 the totality of physical and biological attributes of the areas under consideration, and use aerial photography as the essential basis for their interpretations. This branch of CSIRO is currently using the digitized base of federal electoral divisions for a variety of automated thematic mapping activities, and is applying digital technologies to resource mapping projects at both a continental and a local scale.

The **Bureau of Meteorology** is currently producing a *Climatic Atlas of Australia* at a scale of 1 : 12 500 000. Most small-scale climatic maps of the continent are based on median and percentile values, rather than on means and standard deviations, because of the variability of much of the Australian climatic environment.

Hydrographic charting of Australian waters is the responsibility of the **Royal Australian Navy, Hydrographic Service**.

In the commercial field mapping is carried out by a wide variety of organizations. These include **Jacaranda** (small-scale atlas producers) and several road and town map publishers that are now grouped together under the conglomerate **Universal Press**. Universal maintains offices in Sydney, Melbourne, Adelaide, Brisbane and Perth, and distributes the maps formerly published by **Gregory's Publishing Co.** and by UBD. These include street directories and maps of major cities, state maps, suburb maps and district maps. The Lansdowne Press was until recently the distributor for Robinsons and Broadbent town and road mapping but they have also now been subsumed under the Universal umbrella. There are therefore at least four alternative yet similar road maps for each State, and at least four alternative town maps for each of the major urban centres in the country. **Reader's Digest Australia** is another private sector mapping body, responsible for the compilation of a useful atlas in the 1970s.

State organizations

The different state mapping agencies are responsible for the production of larger-scale topographic mapping of their areas and for the compilation of cadastral and urban mapping, as well as providing a service to other state government agencies and publishing small-scale general state mapping. Earth science mapping is carried out by State authorities as well, in collaboration with the federal organizations. The role of these state mapping agencies where they cooperate with the federal bodies in the production of national series will not be repeated here.

Australian Capital Territory

Mapping of the Australian Capital Territory is the responsibility of the **Australian Survey Office (ASO)**, which publishes a variety of mapping between 1 : 2500 and 1 : 25 000 in scale, as topographic, cadastral, detail and orthophotomaps. Much of this large-scale output is compiled using computer-assisted cartography.

New South Wales

The **Central Mapping Agency (CMA)** of New South Wales issues a variety of large-scale series. A 1 : 50 000 scale series is the standard map for central areas of the state, whilst coastal and southern areas are covered in a 1 : 25 000 series. The topographic sheets in these series are printed in four to six colours, whilst property maps showing rural ownership and boundaries, and cadastral maps showing survey and administrative boundaries, are available only in diazo format. With the 1 : 100 000 scale mapping of the outback areas these mapping programmes have now been completed and give complete state coverage for these scale ranges. A variety of tourist and forestry maps are also issued by CMA, which is also responsible for the compilation of metropolitan and non-metropolitan large-scale mapping. Sheets in these large-scale series are issued as orthophotoplots, and cadastral information is also available as dyelines at the larger scales. Programmes of map revision are well advanced for all these series.

Geological mapping of New South Wales is carried out by the **Department of Mineral Resources (DMR)**. The 1 : 500 000 and 1 : 250 000 series are compiled jointly with BMR, but sold by DMR. A metallogenic series is published at 1 : 250 000, which is still in progress. Blocks of 1 : 100 000, 1 : 50 000 and 1 : 25 000 geological mapping are also available, though coverage is very patchy.

Northern Territory

The **Northern Territory Department of Lands (NTDL)** is the state mapping agency, but no major series at a large scale are published for the Territory. NTDL issues small-scale and tourist maps of the state and street directories of Alice Springs and Darwin. The Northern Territory Geological Survey is carried out by the **Department of Mines and Energy (NTDME)**, which is undertaking a regional mapping programme to provide complete geological coverage of the state at 1 : 100 000 scale. The first sheet of this series appeared in August 1983, and by May 1986 18 sheets were available as compilations and three full-colour sheets had been published. Other mapping is available as paper prints, in exploration and metallogenic series. NTDME also issues small-scale petroleum and mining maps of the State.

We have listed both topographic and geological series under the Australia section.

Queensland

In Queensland the state mapping agency is the
Department of Mapping and Surveying (SUNMAP). It
was established as a separate office in 1975, having
formerly been part of the Department of Lands. A new
headquarters is currently being built in Wooloongabba,
which should be opened in 1987–88. SUNMAP cooperates
with NATMAP in the production of 1 : 100 000 and
1 : 250 000 topographic series which we have listed and
described under the national section. It issues cadastral
coverage to complement the 1 : 250 000 series for the
outback regions of the state, and at 1 : 100 000 where
published 1 : 100 000 scale topographic coverage is
available. A variety of larger-scale mapping is also
published by this body. A 1 : 50 000 topographic series is
available for several blocks within the state. Maps in this
series are compiled from aerial photographs coordinated to
ground surveys, they have 10 m contour intervals, and
each sheet covers 15′ × 15′ area. They are produced on
the Australian Map Grid. Black-and-white cadastral
mapping on the same sheet lines is also published by
SUNMAP. Neither series is intended to give complete
state coverage; both are very incomplete.

1 : 25 000 topographic, cadastral and orthophotomap
series are also available for parts of the state. These series
have replaced an earlier 1 : 31 680 scale map. All are still in
progress and are similar in style and specification to the
1 : 50 000 series: 5 or 10 m contours are used. The
orthophotomaps are derived from black-and-white aerial
photography and are colour enhanced to achieve maximum
definition of detail. SUNMAP is also responsible for the
publication of larger-scale series covering urban areas.
1 : 2500 and 1 : 10 000 have been adopted as the scales for
urban mapping and are replacing an earlier four chain
series. The cadastral data on these 4000 large-scale urban
plans is being captured to form the basis of the digital
Queensland Land Information System, which is to be used
both to produce computer-generated mapping and to
provide spatial referencing to other information systems.

Since 1983 SUNMAP has given a very high priority to
the production of tourist maps and has cut back its
topographic programmes in an effort to both generate sales
and increase public awareness of mapping in the State.
The wide variety of recreational and tourist maps includes
the Amazing Queensland series produced in association
with the Queensland Tourist and Travel Corporation,
satellite maps and photomaps of urban and holiday areas,
street and road maps. SUNMAP publishes a variety of
smaller-scale general state maps, including a newly revised
1 : 2 000 000 scale map, and various administrative and
electoral mapping. It acts as the sales centre for other
Queensland mapping agencies and compiles the
comprehensive *Catalogue of Queensland Maps* for the
Queensland Surveying and Mapping Council.

Larger-scale maps of areas of geological or mineralogical
importance are available from the Queensland **Department
of Mines (QDM)**, which cooperates with BMR to produce
1 : 250 000 geological mapping of the state. There is now
complete coverage in this series. QDM also publishes
mining lease maps and an authority to prospect series, as
well as small-scale state earth science mapping.

The **Department of Forestry (QDF)** is the other major
Queensland map-producing agency. It publishes a variety
of forest mapping, including a five-sheet 1 : 500 000 series
showing state forests, timber reserves, Forestry District
boundaries and topographic information. Larger-scale
forest map series are also available, which use the

Australian Map Grid and are on the same sheet lines as the
topographic and cadastral maps. Important forest areas
have been mapped in these series at either 1 : 25 000 or
1 : 50 000 scale. Amongst the other state organizations
producing maps are the Department of Primary Industries
(which produces land resource maps), the Valuer General,
and the National Parks and Wildlife Service.

South Australia

The South Australia **Department of Lands (SADL)**
provides most of the state's cartographic needs. Its
standard mapping programme produces a variety of the
larger-scale series: topographic, cadastral and scale-
corrected orthophotomaps are published for urban and
closely settled areas, whilst rural and remote areas are
mapped using uncorrected photomaps. The northern
unsettled areas of the state have been mapped at 1 : 100 000
in a joint project with NATMAP, and SADL has added
data to the machine compilations prepared from the
1 : 250 000 scale national series. These maps are available as
transparencies. A 1 : 50 000 series is produced in
accordance with the national six-colour topographic map
specification, covers all the settled areas of the state and
the North Flinders Range in 410 sheets, and has a 10 m
contour interval. Of these 1 : 50 000 maps, 350 are issued as
single-colour cadastral maps. 1 : 10 000 (with 5 m contours)
and 1 : 2500 (with 2 m contours) series are available for
Metropolitan Adelaide, rural cities, principal towns and, in
the case of the 1 : 10 000 map, irrigation areas. Sheet
numbering is derived from the 1 : 100 000 scale
quadrangle—50 1 : 10 000 sheets to each 1 : 100 000, and 16
1 : 2500 sheets to each 1 : 10 000. Both series are
continuously revised on a 10 year cycle and are prepared as
topographic, cadastral and orthophoto components which
are available as separate transparencies; diazo copies of any
combination may be purchased. Computer-assisted
methods of compilation from aerial coverage are now being
used and digital topographic data is available for some
sheets in both series. SADL also produces base maps for
other Departments, and publishes administrative mapping
of the State and of Adelaide.

The **South Australian Department of Mines and
Energy (SADME)** publishes and sells the regional
geological map series at 1 : 250 000 scale (the *South
Australian Geological Atlas Series*). There is complete
geological coverage at this scale, but 24 sheets are so far
only available as preliminary compilations. Some sheets are
available with accompanying explanatory notes. We have
listed these maps under Australia since they correspond to
the national geological series in specification. On the same
sheet lines and scale are available diazo copies of
aeromagnetic, Bouguer gravity and metallic and non-
metallic exploration index maps. At larger scales the old
1 : 63 360 series is being superseded by a 1 : 50 000
geological map, but this series will give only incomplete
coverage, of areas of special geological interest. SADME
also publishes a range of smaller-scale State earth science
mapping (at 1 : 7 300 000, 1 : 2 000 000 and 1 : 1 000 000
scales) and some soils and resources mapping at large
scales of the Adelaide area.

Other mapping agencies in South Australia include the
Department of Environment and Planning (planning maps
and national park maps) and the Department of Recreation
and Sport. The first major atlas of the state was published
to celebrate the 150th anniversary of South Australia in
1986 by the **South Australian Government Printing
Division** in association with Wakefield Press.

Tasmania

In Tasmania the **Department of Lands (TASMAP)** is responsible for the publication of a variety of base map series. Unlike other States there has been almost no commonwealth mapping activity. At 1:100 000 scale complete coverage is available in two different maps. A topographic series (part of the National map) covers the island in 49 sheets. It had been planned not to revise this map (editions date from the late 1970s), but an experimental new edition of two sheets was begun in 1985, to be issued as double-sided maps with corresponding sheets from the other 1:100 000 series. This land Tenure Index Series (LTIS) is to be revised on a 5 year cycle, with 13 revised sheets appearing a year. The Land Tenure map uses a topographic base (without the hill shading) overprinted with property, forest and reserve information, and covers Tasmania in 40 sheets (rationalization of marginal sheets accounts for the reduced numbers in this series).

A new 1:25 000 series was begun by TASMAP in 1980. There is now a 5 year revision cycle, and it is intended that the whole island will be mapped at this scale by 1990. This full-colour series shows both cadastral and topographic detail, and is used as a base for both forest and geological mapping.

Other smaller-scale maps produced by TASMAP include a four-sheet 1:250 000 series with 100 m contours, hill shading and layer tints, and a single-sheet, double-sided 1:500 000 land map showing land tenure information and, on the reverse, administrative, estates and electoral boundaries. Both of these smaller-scale maps are available flat, folded or mounted. Larger-scale mapping of Hobart and the bigger towns is also available, as 1:5000 scale orthophotomaps overlaid with property boundaries and 5 m contours. TASMAP publishes jointly with the Department of Agriculture a dyeline 1:100 000 Land Systems series. Land district and rural valuation charts and town plans at a variety of large scales are published by TASMAP as dyeline copies, and provide a cadastral record. These cadastral maps are slowly being replaced by a dyeline series using the new 1:25 000 topographic base and sheet lines, and by the 1:5000 urban map series. Touring and recreation maps of the national parks and other tourist areas are also available and a useful Holiday Atlas is now revised annually and includes town plans, general mapping and State gazetteer.

The other major state map publisher is the Tasmania **Department of Mines (TDM)**. Two major full-colour geological series are published and are available either with or without explanatory texts. The eight-sheet 1:250 000 series is complete, whilst the 1:50 000 series is in progress and slowly replacing an earlier 1-inch 96-sheet series on the same sheet lines: 39 are now available, work is in hand on a further 12 sheets, and completion is envisaged around the turn of the century. A variety of other smaller-scale earth science maps are also available from TDM.

The **Forestry Commission Tasmania (TFC)** produces 1:25 000 scale native forest type maps and 1:10 000 scale plantation series in diazo and film transparency format. Small-scale full-colour forest maps are published and TFC has invested heavily in computer hardware with a view to data capture at 1:500 000 scale for a variety of themes, and to 1:25 000 scale data capture to set up a forest management information system. Amongst the other map producing agencies in Tasmania is the Hydro Electric Commission which surveys the water resources, especially in the south-west of the island.

Victoria

Topographic mapping in Victoria is the responsibility of the Division of Surveying and Mapping of the **Department of Crown Lands and Survey (VICMAP)**. Its current production is concentrating on the compilation of a 1:25 000 scale five-colour series, which is also available in dyeline cadastral editions. Certain areas of the interior are published as photomosaics. A 1:50 000 series and a superseded 1-inch map are also still available but not in current production. All of these maps concentrate upon the southern and eastern areas of the state and coverage is patchy.

Larger-scale series are also available from VICMAP, as topographic, orthophoto or cadastral editions. All maps now use AMG sheet lines and specifications. Like other state mapping agencies VICMAP also publishes a variety of smaller-scale state maps.

The **Victoria Department of Minerals and Energy** produces earth science mapping of Victoria. There is no series available at a scale larger than the 1:250 000 geological map; single-sheet larger-scale geological maps are available for areas of interest and the current scale of production is 1:50 000. Earlier 1-inch mapping is still available for some areas, as are the now superseded geological parish plans. Other series cover goldfields and deep lead areas.

The **Forest Commission Victoria** produces a 1-inch black-and-white forest series of the east of the state, and a variety of colour maps of forests and tourist areas.

Western Australia

In Western Australia the **Central Map Agency (WACMA)** of the Department of Lands and Surveys carries out cadastral and topographic surveys in the state. All current production uses the national standard AMG specifications, but some older series published on Bonne or UTM projections are still available. A 1:50 000 four-colour topocadastral series will eventually cover all of the farming areas in the south-west of the state, whilst six-colour 1:25 000 coverage is in progress for the coastal plain, for areas to the south of Geraldton. State large-scale series are available for the metropolitan area as dyeline prints in several cadastral and topographical series. WACMA also publishes smaller-scale thematic state maps, road maps and special project mapping.

The **Department of Mines Western Australia (WADM)** produces a range of maps to assist the mining industry in the state. Within the organization, geological mapping is carried out by the Geological Survey Division; all other mineral mapping is the responsibility of the Survey and Mapping Division. The 1:250 000 regional geology series is now complete for the state. WADM is reprinting out-of-print sheets from this series using a laser scan method. Other geological mapping includes an environmental/urban geology series at 1:50 000 and small-scale state geology maps.

The Survey and Mapping Division compiles Mineral and Petroleum Tenement maps, together with a variety of thematic mineral mapping and detailed plans of 800 mines in the state.

The **Forests Department** of Western Australia publishes a six-colour 1:50 000 series, now complete, which covers all the forest areas of the state and shows forest information on a topographic base. A 1:500 000 Forest Atlas map of the South West of the state is also available, as is a small-scale vegetation map of the whole state showing floristic units of natural vegetation prior to settlement, with an explanatory leaflet. The Department also compiles 1:25 000 scale diazo based mapping and is planning to use 1:100 000 topographic bases for the production of a new forest tenure series. Other vegetation mapping has been carried out in the vegetation survey of Western Australia. Complete 1:1 000 000 scale coverage of the State in seven sheets was published by the **University of Western Australia Press** in the 1970s, and **Vegmap Publications** distributes the 1:250 000 scale vegetation survey mapping of the South West that was compiled by J. S. Beard over a similar period.

Further information

Each of the mapping agencies mentioned above produces a catalogue; many also issue an annual report. Particularly useful are the following. *NATMAP information leaflets 1–14* are updated on a regular basis and available free from Division of National Mapping.*Thematic Mapping Bulletin*, Canberra: Division of National Mapping (1970–), is issued every year, and lists all new thematic map issues by NATMAP. *Australian Maps*, Canberra: National Library (1968–) lists all new atlases and maps published in Australia. *Cartography (Journal of the Australian Institute of Cartographers)*, Perth: AIC (1954–) is an essential source of information about the state of the art of Australian map production. *National Mapping Report of Australia (1981–4)*, presented to the 12th International Conference and 7th General Assembly of the International Cartographic Association, Perth: ICA, 1984.

Useful for its background information about the complexity of Australian environmental mapping is: Christian, C.S. (1983) *The Australian Approach to Environmental Mapping*, Canberra: CSIRO (CSIRO Institute of Biological Resources; Technical Memorandum 83/5).

Amongst the state organizations, *Catalogue of Queensland Maps*, prepared by the Department of Mapping and Surveying (SUNMAP), Brisbane: Queensland Mapping and Surveying Advisory Council, which is updated annually, brings together all maps issued by government bodies throughout the state. A vast work!

For a very useful (though already dated) guide to Australian map publishers, see Australian Map Circle (1985) *Checklist of Australian map catalogues and indexes*, 2nd edn, Melbourne: ACT.

Addresses

Division of National Mapping (NATMAP)
Sales Office, PO Box 31, BELCONNEN, ACT 2616.

Royal Australian Survey Corps (RASC)
Campbell Park Offices, Campbell Park, CANBERRA, ACT 2600.

Australian Bureau of Statistics (ABS)
PO Box 10, BELCONNEN, ACT 2616.

Bureau of Mineral Resources, Geology and Geophysics (BMR)
GPO Box 378, CANBERRA, ACT 2601.

Government Printer, Copy Service
GPO Box 84, CANBERRA, ACT 2601.

Australian Electoral Commission
PO Box E201, Queen Victoria Terrace, CANBERRA, ACT 2600.

Australian Survey Office (ASO)
Department of Administrative Services, PO Box 2, BELCONNEN, ACT 2616.

Commonwealth Scientific Industrial Research Organization Water and Land Resources Division (CSIRO)
GPO Box 1666, CANBERRA, ACT 2601.

Bureau of Meteorology
CANBERRA ACT.

Royal Australian Navy, Hydrographic Office
161 Walker Street, NORTH SYDNEY, NSW 2060.

Australian Landsat Station
14–16 Oatley Court, PO Box 28, BELCONNEN, ACT 2616.

Jacaranda-Wiley Ltd
65 Park Rd, MILTON, Queensland 4064.

Gregory's Publishing Co. Pty
1 Unwin Bridge Rd, ST PETER'S, New South Wales 2044.

Universal Press
64 Talavera Rd, NORTH RYDE, New South Wales 2113.

Reader's Digest Australia
Surrey Hills, SYDNEY, New South Wales 2010.

State Organizations

New South Wales

Central Mapping Authority, New South Wales (CMA)
PO Box 143, BATHURST, New South Wales 2795.

NSW Department of Mineral Resources (DMR)
GPO Box 5228, 8–18 Bent Street, SYDNEY, New South Wales 2000.

Northern Territory

Department of Lands Northern Territory (NTDL)
PO Box 1680, DARWIN, Northern Territory 5790.

Department of Mines and Energy (NTDME)
Northern Territory Geological Survey, Minerals House, Esplanade, DARWIN, Northern Territory 5794.

Queensland

Department of Mapping and Surveying, Queensland (SUNMAP)
PO Box 234, BRISBANE NORTH QUAY, Queensland 4000.

Department of Mines, Queensland (QDM)
GPO Box 194, 41 George St, BRISBANE, Queensland 4001.

Department of Forestry (QDF)
GPO Box 944, 41 George St, BRISBANE, Queensland 4001.

South Australia

Department of Lands South Australia (SADL)
GPO Box 1047, ADELAIDE, South Australia 5001.

Department of Minerals and Energy (SADME)
Geological Survey of South Australia, PO Box 151, EASTWOOD, South Australia 5063.

South Australia Government Printing Division
282 Richmond Rd, NETLEY, South Australia 5037.

Tasmania

Department of Lands Tasmania (TASMAP)
GPO Box 44A, HOBART, Tasmania 7001.

Tasmanian Government Publication Centre
GPO Box 44A, HOBART, Tasmania 7001.

Department of Mines (TDM)
GPO Box 56, ROSNY PARK, Tasmania 7018.

Forestry Commission Tasmania (TFC)
GPO Box 207B, 199 Macquarie St, HOBART, Tasmania 7001.

Victoria

Department of Crown Lands and Survey (VICMAP)
2 Treasury Place, MELBOURNE, Victoria 3002.

Victoria Department of Minerals and Energy
Publications Centre, 140 Bourke St, MELBOURNE, Victoria 3000.

Forest Commission Victoria
Map Sales and Publications, Ground Floor Forestry House, 601
Bourke St, MELBOURNE, Victoria 3000.

Western Australia

Central Map Agency (WACMA)
Department of Lands and Survey Western Australia, Cathedral
Ave, PERTH, Western Australia 6000.

Department of Mines Western Australia (WADM)
Mineral House, 66 Adelaide Terrace, PERTH, Western Australia
6000.

Forests Department Western Australia
50 Hayman Rd, COMO, Western Australia 6152.

University of Western Australia Press
NEDLANDS, Western Australia 6009.

Vegmap Publications
6 Fraser Rd, APPLECROSS, Western Australia 6153.

Catalogue

Atlases and gazetteers

Atlas of Australian resources, Third series
Belconnen: NATMAP, 1980–.
10 vols, 4 published.
Available in separate thematic parts, 4 so far published covering
soils and land use, population, agriculture and climate.

Atlas of Australian resources, Second series
Belconnen: NATMAP, 1963–77.
Available complete or as separate sheets.

Reader's Digest atlas of Australia
Sydney: Reader's Digest, 1977.
pp. 288.
Includes complete scale coverage at 1 : 1 000 000.

Master names file
Belconnen: NATMAP, 1986.
Microfiche gazetteer of names from 1 : 250 000 and 1 : 100 000
series up-dated on annual basis.

General

Australia 1 : 10 000 000: general reference map
Belconnen: NATMAP, 1984.
Small-scale thematic map series; 1.

Australia   1 : 8 000 000
Bern: Kümmerly + Frey, 1985.

Australia   1 : 5 000 000
Edinburgh: Bartholomew, 1982.

Australia 1 : 5 000 000: general reference map
Belconnen: NATMAP, 1979.
Australia 1 : 5 000 000 map series; 1.
Includes gazetteer printed on back.

Australia 1 : 5 000 000: relief map   2nd edn
Belconnen: NATMAP, 1979.
Australia 1 : 5 000 000 map series; 2.

Australia 1 : 5 000 000: relief, land and seabed
Belconnen: NATMAP, 1985.
Australia 1 : 5 000 000 map series; 17.

Robinson's Australia road map   1 : 5 000 000
North Ryde: Universal.

Australia 1 : 2 500 000: geographic map   4th edn
Belconnen: NATMAP.
4 sheets, all published.

Robinson's State road maps   various scales
North Ryde: Universal.
7 sheets, all published.
New South Wales 1 : 1 400 000; Northern Territory 1 : 2 000 000;
Queensland 1 : 2 500 000; S. Australia 1 : 1 700 000; Tasmania
1 : 650 000; Victoria 1 : 850 000; Western Australia 1 : 2 800 000.

Topographic

International Map of the World   1 : 1 000 000
Belconnen: NATMAP, 1978–.
49 sheets, all published. ■

Australia   1 : 250 000   national topographic map series
Belconnen: NATMAP, 1969–.
544 sheets, all published. ■
Also available as Joint Operations Graphic edition issued by
RASC.

Australia 1 : 100 000: topographic map
Belconnen: NATMAP, 1968–.
1640 sheets, all published. ■
Covers settled areas only: remainder, i.e. interior, available as
1425 orthophotomaps. Title varies according to which state
organization jointly issues sheets.

Australia   1 : 50 000 topographic map
Canberra: RASC, 1958–.
ca 700 sheets published. ■
Issued by the various state survey authorities and for some areas
by RASC.
Very incomplete coverage.

Bathymetric

Australia national bathymetric series   1 : 250 000
Belconnen: NATMAP, 1973–.
275 sheets, 100 published. ■
Covers continental shelf areas.

Geological and geophysical

Earth science atlas of Australia   1 : 10 000 000
Canberra: BMR 1980–.
19 sheets + commentaries so far published.

Published topics: general geology, solid geology, plate tectonics
(1 : 20 000 000), earthquakes, sedimentary sequences, major
structural elements, Cainozoic cover and weathering, surface
drainage and continental margin, phanerozoic palaeogeography (3
sheets at 1 : 30 000 000), Bouguer gravity anomalies, free-air
anomalies, petroleum and oil shale, coal, Cainozoic geology and
mineral deposits, palaeomagnetism, metamorphism, and main
rock types.

Australia: geology   1 : 5 000 000
Belconnen: NATMAP, 1982.
Australia 1 : 5 000 000 map series; 11.

Gravity map of Australia   1 : 5 000 000
Canberra: BMR, 1976.

Australia and Papua New Guinea: metallogenic map   1 : 5 000 000
Canberra: BMR, 1972.
With accompanying explanatory text.

Metamorphic map of Australia   1 : 5 000 000
Canberra: BMR, 1984.

Geology of Australia   1 : 2 500 000
Canberra: BMR, 1976.
4 sheets + texts.

Magnetic map of Australia 1 : 2 500 000: residuals of total intensity
Canberra: BMR, 1976.
4 sheets, all published.
Explanations on sheets.

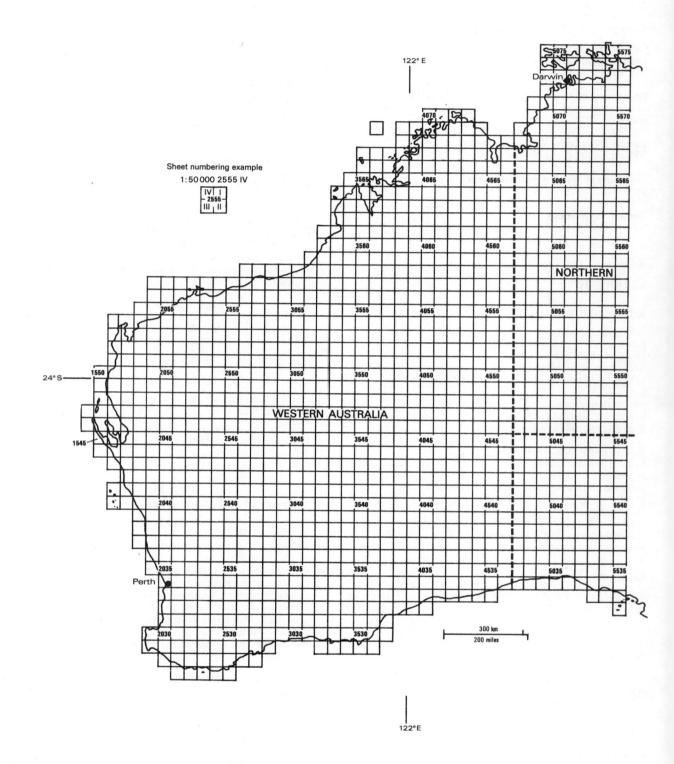

Sheet numbering example
1:50 000 2555 IV

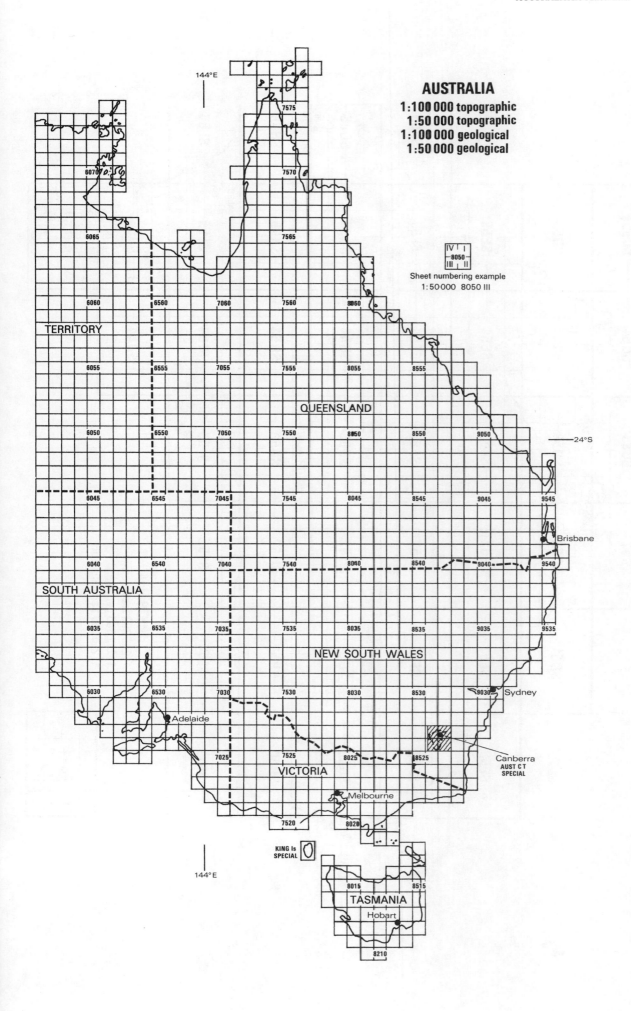

144°E

AUSTRALIA
1:100 000 topographic
1:50 000 topographic
1:100 000 geological
1:50 000 geological

6070

7575

7570

6065

7565

Sheet numbering example
1:50 000 8050 III

TERRITORY

6060   6560   7060   7560   8060

6055   6555   7055   7555   8055   8555

QUEENSLAND

6050   6550   7050   7550   8050   8550   9050

24°S

6045   6545   7045   7545   8045   8545   9045   9545

Brisbane

6040   6540   7040   7540   8040   8540   9040   9540

SOUTH AUSTRALIA

6035   6535   7035   7535   8035   8535   9035   9535

NEW SOUTH WALES

6030   6530   7030   7530   8030   8530   9030   Sydney

Adelaide

Canberra
AUST C T
SPECIAL

7025   7525   8025   8525

VICTORIA

7520   8020   Melbourne

KING Is
SPECIAL

144°E

8015   8515

TASMANIA

Hobart

8210

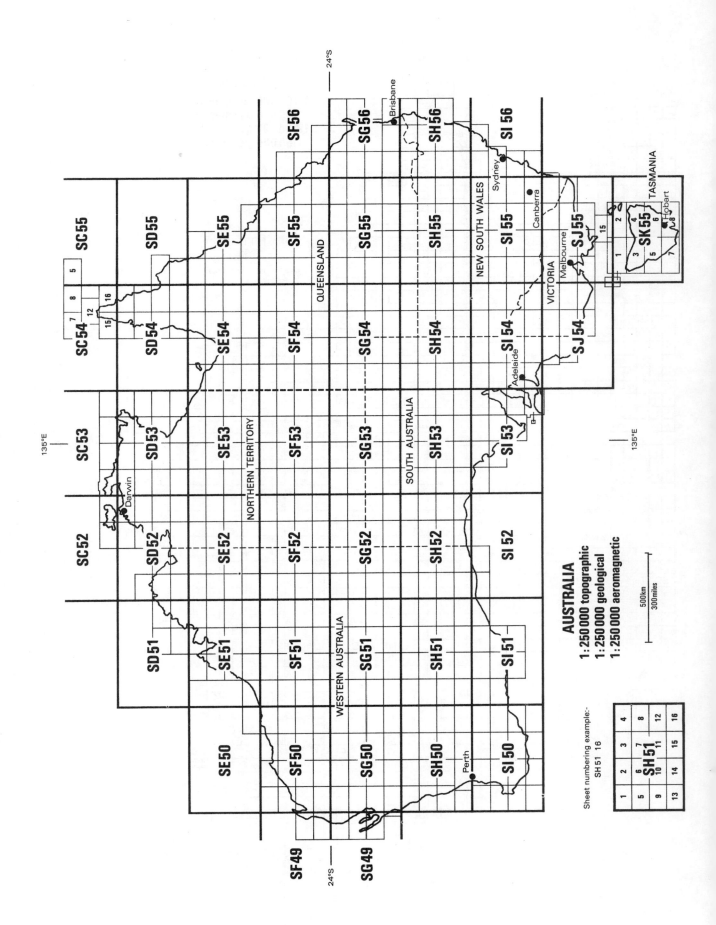

AUSTRALIA
1:250 000 topographic
1:250 000 geological
1:250 000 aeromagnetic

500km
300miles

Sheet numbering example:-
SH 51 16

1	2	3	4
5	6	7	8
9	SH51 10	11	12
13	14	15	16

Geotechnical landscape map of Australia 1:2 500 000
Canberra: CSIRO, 1984.
4 sheets, all published.
With accompanying report by K. Grant, R. Davis and C. de
Visser.

Australia: gravity map 1:1 000 000
Canberra: BMR.
49 sheets, 12 published. ∎

Australia: gravity map 1:500 000
Canberra: BMR.
600 sheets, 550 published. ∎

Australia geological series 1:250 000
Canberra: BMR, 1959–.
544 sheets, 475 published. ∎
Many sheets available with explanatory texts.
Published in association with state geological agencies.
Also available as dyeline and computer-generated gravity editions
and computer-generated aeromagnetic maps.

Australia geological series 1:100 000
Canberra: BMR, 1974–.
3085 sheets, *ca* 75 published. ∎

Environmental

Climatic atlas of Australia 1:12 500 000
Canberra: Bureau of Meteorology, 1975–79.
8 map sets.

Australia 1:5 000 000 map series
Belconnen: NATMAP, 1978–.
17 sheets so far published.
Single topic maps: earth resources sheets issued to date include
dams and storages 2nd edn, soil resources, principal groundwater
resources, nature conservation reserves, land use and forestry
reserves.

Provisional environmental regions of Australia 1:5 000 000
Canberra: CSIRO, 1980.
With accompanying report by P. Laut, D. Faith and T.A. Paine.

Atlas of Australian soils 1:2 000 000
Melbourne: CSIRO and University of Melbourne Press, 1960–68.
10 sheets, all published.

Australia: thematic map series 1:1 000 000
Belconnen: BMR, 1976.
45 sheets, all published. ∎
6 different thematic series issued on the same sheet lines.
Nature conservation, 45 available (dyeline).
Forestry reserves, 25 available (dyeline).
Forest types, 24 available (dyeline).
Land tenure, 9 available.
Land cover, 6 available.
Land use, 6 available.

Continental shelf sediments 1:1 000 000
Canberra: BMR, 1973–80.
15 sheets, all published.
With accompanying bulletins.

Land systems of Australia 1:250 000
Canberra: CSIRO, 1962–77.
11 published studies.
Issued with monographs in the land research series.

Administrative

*Australia 1:10 000 000: Commonwealth electoral divisions and
election results 1980*
Belconnen: NATMAP, 1981.
Small-scale thematic map series; 18.

*Australia 1:10 000 000: Commonwealth electoral divisions and
election results 1983*
Belconnen: NATMAP, 1984.
Small-scale thematic map series; 21.

*Australia 1:5 000 000: statistical divisions and local government areas
1981 census*
Belconnen: NATMAP, 1983.
Australia 1:5 000 000 map series; 7.
Also available for the 1976 census.

Australia 1:5 000 000: public lands
Belconnen: NATMAP, 1983.
Australia 1:5 000 000 map series; 14.

Australia: statistical area maps various scales
Belconnen: ABS, 1982–.
19 sheets.
Shows various levels of 1981 Census divisions.

Australia: electoral areas consolidated maps various scales
Canberra: Australian Electoral Commission, 1985.
14 sheets.

Human and economic

Australia small scale map series 1:10 000 000
Belconnen: NATMAP, 1980–.
21 sheets, in progress.
Single topic maps, human and economic sheets issued to date
include: population change 1971–76, sown pastures and fodder
crops, grazing density, sheep, cattle, wheat, croplands, farm
types, value of agricultural production, overseas-born population
urban areas, population change 1971–76, population distribution
1981.

Australia 1:5 000 000 map series
Belconnen: NATMAP, 1978–.
17 sheets.
Single-topic maps, human and economic sheets issued to date
include: railway systems 1981, aboriginal land and population
distribution, population distribution 1981.

Town maps

Social atlas of Australian cities
Canberra: ABS and NATMAP, 1983–84.
7 vols.
One volume per state capital, with 1981 census data.

Town maps of over 100 Australian towns and cities are distributed
by Universal Publications. Several alternative maps for each State
capital are available.

AUSTRALIAN CAPITAL TERRITORY
Topographic

Australian Capital Territory 1:100 000
Belconnen: ASO, 1976.

Geological and geophysical

Geology of Canberra, Queanbeyan and environs 1:50 000
Canberra: BMR, 1980.
With accompanying explanatory notes.

NEW SOUTH WALES
General

New South Wales: state map 1:1 500 000
Bathurst: CMA, 1982.
With 4 small ancillary thematic inset maps.

New South Wales 1:1 000 000
Bathurst: CMA, 1983.
2 sheets, both published.

Topographic

New South Wales 1:25 000
Bathurst: CMA.
ca 700 published sheets.
Series covers coastal plain of State.

Geological and geophysical

Geology of New South Wales   1:1 000 000
Sydney: DMR, 1972.
4 sheets, all published.

Tectonic map of New South Wales   1:1 000 000
Sydney: DMR, 1974.
4 sheets, all published.
With explanatory text.

New South Wales: geological series   1:500 000
Sydney: DMR, 1967–77.
10 sheets, all published.

New South Wales: metallogenic map series   1:250 000
Sydney: DMR, 1972–.
50 sheets, 14 published.
Issued with map data sheets and notes.

Environmental

Natural vegetation of New South Wales   1:1 520 640
Bathurst: CMA, 1971.

NORTHERN TERRITORY
General

Northern Territory of Australia   1:2 500 000
Belconnen: NATMAP, 1974.

Geological and geophysical

Geology of Northern Territory   1:2 500 000
Canberra: BMR, 1976.
Accompanying explanatory text published separately.

Northern territory offshore–onshore   1:2 500 000
Darwin: NTDME.
Also available with computer printout of petroleum tenure.

Northern Territory: mineral exploration licence map   1:1 000 000
Darwin: NTDME.
3 sheets, all published.
Darwin–Katherine area available at 1:1 500 000.
Also available with computer printout of exploration licences.

QUEENSLAND
General

Queensland   1:5 000 000
Brisbane: SUNMAP.

Queensland   1:2 500 000
Brisbane: SUNMAP.

Queensland   1:2 000 000
Brisbane: SUNMAP, 1986.
2 sheets, both published.
Also available with stock routes and trucking facility overprint.

Queensland   1:1 000 000
Brisbane: SUNMAP.
4 sheets, all published.

Topographic

Queensland   1:25 000
Brisbane: SUNMAP.
ca 200 published sheets.
Orthophoto, topographic and cadastral editions available for some areas.

Geological and geophysical

Queensland: mineral resources   1:5 000 000
Brisbane: QDM, 1976.

Queensland geology   1:2 500 000
Brisbane: QDM, 1976.
Also available as hydrogeological, major administrative areas, gold localities and mineral resources sheets and with a transparent structural elements overlay.

Environmental

Queensland: state forests map   1:5 000 000
Brisbane: QDF.
Includes South East Queensland 1:2 000 000 on back.

Queensland: land use   1:2 534 000
Brisbane: SUNMAP.

Queensland: state forests map   1:2 000 000
Brisbane: QDF, 1977.

Administrative

Queensland: tenure map   1:2 000 000
Brisbane: SUNMAP.
Also available with 5 different boundary overprints.

Queensland: tenure map   1:1 000 000
Brisbane: SUNMAP.
4 sheets, all published.

SOUTH AUSTRALIA
Atlases and gazetteers

Atlas of South Australia   edited by T. Griffin and M. McCaskell
Adelaide: South Australian Government Printing Division, 1986.
pp. 148, 190 maps.

General

South Australia state map   1:2 500 000
Adelaide: SADL, 1977.
Available with or without relief shading.

Geological and geophysical

Geological map of South Australia   1:2 000 000
Adelaide: SADME, 1982.

Tectonic map of South Australia   1:2 000 000
Adelaide: SADME, 1982.

Interpreted depths to magnetic basement map of South Australia   1:2 000 000
Adelaide: SADME, 1982.

Bouguer gravity anomaly map of South Australia   1:2 000 000
Adelaide: SADME.

Total magnetic intensity map of South Australia   1:2 000 000
Adelaide: SADME.

South Australia: geological map   1:1 000 000   2nd edn
Adelaide: SADME, 1983.
4 sheets, all published.
Includes tectonic inset map.

Bouguer gravity anomaly map of South Australia   1:1 000 000
Adelaide: SADME, 1973.
4 sheets, all published.

Total magnetic intensity map of South Australia   1:1 000 000
Adelaide: SADME, 1980.
4 sheets, all published.

Interpreted depths to magnetic basement map of South Australia   1:1 000 000
Adelaide: SADME, 1981.
4 sheets, all published.

Environmental

South Australia: state map showing land utilisation and pastoral areas 1 : 1 000 000.
Adelaide: SADL.

Groundwater resources map of South Australia 1 : 2 000 000
Adelaide: SADME, 1982.

Environmental provinces and environmental regions of South Australia 1 : 2 000 000
Canberra: CSIRO, 1977.
With accompanying handbook by P. Laut *et al.*

Groundwater resources map of South Australia 1 : 1 000 000
Adelaide: SADME, 1980.
4 sheets, all published.

Administrative

South Australia 1 : 1 000 000
Adelaide: SADL.
4 sheets, all published.
Shows hundreds and county boundaries.
Also available with overprint of 1 : 250 000 and 1 : 100 000 scale map indexes.

TASMANIA

Atlases and gazetteers

Holiday atlas of Tasmania
Hobart: TASMAP, 1986.

General

Tasmania land map 1 : 500 000
Hobart: TASMAP, 1986.
Double-sided.

Tasmania 1 : 250 000
Hobart: TASMAP, 1980.
4 sheets.

Topographic

Tasmania land tenure index series 1 : 100 000
Hobart: TASMAP, 1981–85.
40 sheets, all published.

Tasmania topographic/cadastral 1 : 25 000
Hobart: TASMAP, 1980–.
419 sheets, *ca* 200 published.

Geological and geophysical

Tasmania mineral resources map 1 : 1 000 000
Hobart: TDM, 1967.
2 sheets, both published.
Sheets cover lode and alluvial deposits; and stratified and residual deposits, granite and serpentinite.

Geological map of Tasmania 1 : 500 000
Hobart: TDM, 1976.

Structural map of the pre-Carboniferous rocks of Tasmania 1 : 500 000
Hobart: TDM, 1976.

Tasmania geological atlas series 1 : 50 000
Hobart: TDM, 1973–.
96 sheets, 39 published.
Earlier 1-inch sheets still available for some areas.
Issued with explanatory notes.

Environmental

Forestry in Tasmania 1 : 600 000
Hobart: TFC, 1981.
Includes text on back.

Vegetation of Tasmania 1 : 500 000
Hobart: TFC, 1984.

VICTORIA

General

Victoria 1 : 1 000 000
Melbourne: VICMAP, 1976.

Victoria 1 : 500 000
Melbourne: VICMAP, 1966.
4 sheets.

Topographic

Victoria 1 : 25 000
Melbourne: VICMAP.
ca 350 published sheets.
Topographic and cadastral editions.

Geological and geophysical

Geology of Victoria 1 : 2 000 000
Melbourne: VITR, 1982.

Victoria geological map 1 : 1 000 000
Melbourne: VITR, 1977.
4 sheets + text.

Environmental

Mineral location map of Victoria 1 : 2 500 000
Melbourne: VITR.

Mineral location map of Victoria 1 : 1 000 000 2nd edn
Melbourne: VITR, 1981.

Groundwater resources map of Victoria 1 : 1 000 000
Melbourne: VITR, 1982.

Forests of Victoria 1 : 1 000 000
Melbourne: Forest Commission, 1984.

Victoria: vegetation map 1 : 2 000 000
Melbourne: Forest Commission, 1981.

WESTERN AUSTRALIA

General

Western Australia 1 : 3 000 000
Perth: WACMA.
Also available as a local authority edition.

Map of Western Australia with gazetteer 1 : 2 500 000
Perth: WADM.

Western Australia 1 : 2 000 000
Perth: WACMA, 1979.
2 sheets.
Also available as a pastoral edition showing administrative information.

Topographic

Western Australia 1 : 25 000
Perth: WACMA.
ca 120 sheets published.
Series to provide coverage of coastal plain south of Geraldton.

Geological and geophysical

Geological map of Western Australia 1 : 2 500 000
Perth: WADM, 1979.

Environmental

Vegetation map of Western Australia   1 : 3 000 000
Como: Forests Department, 1981.

Mineral deposits of Western Australia   1 : 2 500 000
Perth: WADM, 1982.
Published with mineral deposits overlay.

Vegetation survey of Western Australia   1 : 1 000 000
Nedlands: University of Western Australia Press, 1974–80.
7 sheets, all published.

Vegetation survey of Western Australia   1 : 250 000
Applecross: Vegmap, 1972–80.
27 sheets, all published.
Covers South Western areas only.
Single colour maps.

New Zealand

The **Department of Lands and Survey (NZMS)** issues topographical and cadastral mapping and is responsible for geodetic, topographical and photogrammetric survey activities. As the national survey and mapping organization it provides the mapping data base for all other officially produced New Zealand maps. NZMS also publishes town maps, recreational maps and smaller-scale general coverage, as well as aeronautical plotting charts and certain thematic series. In 1987 as part of a major restructuring of government departments the Department of Lands and Survey's responsibilities are being transferred to a new agency, the **Department of Survey and Land Information**, to be housed in a new location.

In 1970 a decision was made to metricate all the NZMS map output. At present this programme is still under way, so there are both metric and non-metric map series available at various scales. NZMS maps use the New Zealand Map Grid Projection—a minimum error conformal projection.

The six-colour *1 : 63 360 scale NZMS 1* series is still the basic scale map for much of the country. Completed in 1975 this series has been continually revised since the first sheet appeared in 1937 and now sheets are available with 100 foot contour interval and use either North or South Island grid. In 1977 the first sheet in the *1 : 50 000 NZMS 260* series was issued. This six-colour metric topographical series has 20 m contour intervals and is printed on revised 40 × 30 km sheet lines using a single New Zealand map grid. It is designed to replace the 1 : 63 360 series which is being withdrawn as new areas are published at 1 : 50 000. Completion of this series is scheduled for the end of the century. Individual sheets are fully revised on a 10 year cycle. This series is derived from 1 : 25 000 survey information and is produced using conventional photogrammetric methods. A 1 : 10 000 scale series using 5 m contours is also under way.

At 1 : 250 000 scale there were until recently also two topographical series: the non-metric *NZMS 18* with 500 foot contours and relief shading, and the metric *NZMS 262* with revised specifications, larger sheet size and relief shading. The non-metric series has now been withdrawn upon completion of the more modern map but its sheet lines are still used by other thematic series.

There are currently plans for the establishment of a 1 : 250 000 scale digital topographic database in association with the DSIR Science Mapping Unit to act as a base for the display of various spatial data sets.

Amongst the other maps issued by NZMS are recreational sheets published at a variety of scales to cover the state forest parks and national parks. Town plans in both metric and non-metric forms are issued to cover the

major settlements and show streets, bus routes and places of interest; they also contain an index to placenames. Following metrication the specifications of these maps are also being changed to conform to the New Zealand Map Grid.

Cadastral mapping of the whole country is available at 1 : 50 000 (NZMS 261), a series with black base cadastral information and other administrative detail overprinted in blue. Like the topographical series this map was begun in 1977 but unlike the topographic 1 : 50 000 series has now completely replaced an earlier 1-inch cadastral series. Other more detailed cadastral series are also available, at 1 : 1000 and 1 : 2000 for urban areas and 1 : 10 000 for rural regions. There is currently a digitizing programme capturing data from earlier imperial cadastral mapping and establishing a cadastral database. This national land information system should be fully established by 1988. The 1 : 10 000 series has been produced by digital methods since 1981.

Since the late 1970s the NZMS has also been responsible for the compilation of various series of *land inventory maps* at a scale of *1 : 100 000 (NZMS 290)*. This resource mapping has been published for seven themes using combined NZMS 260 sheet lines. Maps are to date published for parts of Northland and the Waitaki basin in South Island. A further four themes are covered with maps published as transparent overlays, including a satellite image relief base registered to the New Zealand Map Grid.

Administrative mapping of New Zealand is published by NZMS at a variety of scales. A new 1 : 250 000 scale administrative map using NZMS 262 sheet lines and topographic base has been compiled, and shows regional, local authority and administrative boundaries in addition to the various categories of parks, reserves and crown lands. Other smaller-scale metric administrative editions of topographic maps have also been published. Numerous other smaller-scale mapping is also issued by NZMS, and the Department was responsible for the compilation of the official gazetteer of New Zealand, issued in 1968. NZMS has also been involved in the preparation of map bases for the NZ territories and other areas in the South Pacific Ocean.

Various divisions of the New Zealand **Department of Scientific and Industrial Research (DSIR)** also carry out mapping activities in the fields of their responsibilities. Some of their products may be obtained from the DSIR publishing centre, but most must be obtained from the four divisions of the organization that publish mapping. The Science Mapping Unit of DSIR has been responsible for most of New Zealand's applications of digital technology to the presentation of map data, including maps issued by the various divisions of the main organization.

The **New Zealand Geological Survey (NZGS)** was established in 1865. Early mapping was carried out in response to mineral prospecting activities, and a 1:63 360 scale series was initiated at the beginning of the twentieth century. Progress was slow, one-third of the country was covered by this series by 1956, and some sheets are still being published at this scale though no new maps are on programme. The 1:50 000 map has since the late 1970s replaced the 1-inch programme; sheets cover half the sheet outline adopted for the metric topographic series NZMS 260, and are published with accompanying explanatory text. Areas mapped are in accord with NZGS policy rather than with the aim of providing complete national cover. There are various derivative series related to the 1:50 000 programme, within urban, industrial, late Quaternary tectonic and miscellaneous series. In the 1950s a decision was taken to produce a 28-sheet 1:250 000 scale series. This full-colour map shows the geology in time units, with explanations on sheets and is the largest scale to give complete coverage. Most sheets date from the early 1960s. There are plans to introduce a new 1:250 000 series to be based on different sheet lines and using lithostratigraphic rather than time stratigraphic divisions. Other smaller-scale geological mapping of New Zealand covers the country at 1:1 000 000 (new edition underway), 1:2 000 000 and 1:5 000 000 scales. NZGS is experimenting with the Intergraph system and digitally produced maps are to be produced for them by the DSIR Science Mapping Unit.

The **Geophysics Division (NZGD)** of DSIR is responsible for publishing magnetic and gravity maps. The 1:250 000 gravity series compiled between 1969 and 1980 was an early example of the use of digital methods to produce contour information. It gives complete cover of New Zealand in 27 sheets, as does the 1:250 000 magnetic map. A full-colour 1:250 000 coastal gravity anomaly series in 30 sheets was begun in 1985. Other mapping produced by NZGD includes electrical resistivity maps at 1:50 000 scale for a few areas, and 1:1 000 000 and 1:4 000 000 scale gravity maps of the country.

The **New Zealand Soil Bureau (NZSB)** publishes both soil maps and interpretative maps based on soil surveys showing areas for potential forestry use, pasture farming, urban development, etc. Single-factor maps of both islands at 1:3 700 000 show the broad variations of various soil properties, and various other small-scale maps are issued, often with explanatory texts, as *NZ Soil Survey Reports*. The more detailed soil surveys carried out by the Bureau include a quarter-inch scale series for each island and many other larger-scale sheets for specific areas, usually issued with *NZ Soil Survey Bulletins*.

The **New Zealand Oceanographic Institute (NZOI)** is responsible for the publication of charts showing bathymetry and sediments. These are published at scales of 1:200 000 and 1:1 000 000 in coastal, oceanic and island series. NZOI also publishes a miscellaneous series which includes a variety of oceanographic mapping of much of the South Pacific. Large-scale hydrographic charts are also published to cover New Zealand lakes.

Various other government agencies also produce important map series. The **National Water and Soil Conservation Authority (NWSCA)** issues two major series. The *Erosion map of New Zealand* at 1:250 000 shows potential erosion severity on a 6-point scale, together with an indication of types and characteristics of erosion and is published on NZMS18 sheet lines. The *Land Resource Inventory Worksheets* compiled by the NWSCA are issued at a scale of 1:63 360 and are the only comprehensive national physical resource base for New Zealand. They cover rock type, soil type, slope, erosion and vegetation, and map some 90 000 units, with an assessment of land use capability. This series is produced using automated cartographic techniques and the data is available in digital form allowing output in hard copy as plotted maps or printouts of various variables at a variety of scales.

Amongst the maps issued by the **New Zealand Forest Service (NZFS)** is a 27-sheet *Forest Class Map* published at 1:250 000 scale by province rather than conventional sheet lines. NZFS uses 1:10 000 scale NZMS topographic photoplots in order to prepare basic scale mapping of forests. These maps are reproduced in dyeline format, are updated in local offices, and show forest information on a topographic base. Slope maps at 1:10 000 scale are produced showing categories of slope on a topographic base. The Service also publishes smaller-scale administrative and vegetation maps, including maps for species distribution, fire control, communications and roading. A 1:250 000 regional conservancy series is also published. A tourist series covering forest parks is under way in conjunction with NZMS. From 1987, following the reorganization of government agencies, the mapping functions of NZFS are absorbed into the new national mapping agency, the **Department of Survey and Land Information**.

The **Civil Aviation Division** of the Ministry of Transport publishes aeronautical charts of New Zealand and adjacent areas, many of which are compiled by NZMS using digital methods.

In 1981 the **Commission for the Environment** issued the *New Zealand Atlas of Coastal Resources*. Originally conceived as a coastal resource inventory in the event of an oil spill, this major thematic work was extended to more general needs and covers the New Zealand coast with 1:500 000 scale mapping.

There are very few commercial map publishers in New Zealand, perhaps because of the cheapness and reliability of the NZMS maps. Tourist, street plans and road mapping is produced by amongst others the **Automobile Association** (including the useful *AA Road Atlas of New Zealand*), **Lansdowne Press** (recently taken over by Universal Publications) and **Wises Publications**, which is also part of the UBD group. Another large publisher of mapping in New Zealand is the **Tourism and Publicity Department**; like the AA many of their maps are issued free.

Further information

Various articles in the *New Zealand Cartographic Journal*, Wellington: New Zealand Cartographic Society (1965–) are of interest, especially Barton, P.L. (1982) Map collections, NZ maps: selection and acquisition, *New Zealand Cartographic Journal*, **11**(2) 7–9; and Marshell, B. (1977) Map production and map collecting in New Zealand, *Bulletin, SLA Geography and Map Division*, **110**, 25–29.

For more recent developments see papers presented at the UN Regional Cartographic Conferences for Asia and the Pacific, e.g. New Zealand (1983) Report of cartographic activities in New Zealand 1980–1982, paper presented by New Zealand to the *10th United Nations Regional Cartographic Conference for Asia and the Pacific*, New York: UN.

Catalogues and annual reports are issued by all of the map producing organizations. Most are free. The exception which is also the most detailed is *NZMS Catalogue of Maps*, priced at NZ$13 including quarterly amendment service.

The *New Zealand National Bibliography*, issued monthly by the Government Printer, lists all published New Zealand maps.

A useful overview of the application of new technologies to different aspects of map production in New Zealand is contained in a special issue (1986) of the *New Zealand Cartographic Journal*, **16**(2). This includes papers on (among other subjects) the Science Mapping Unit, the NZSB and other sections of the DSIR.

Another useful article on New Zealand mapping is Aitkin, G.D. *et al.* (1985) The map market in New Zealand, *New Zealand Cartographic Journal*, **15**(2), 2–9.

Addresses

New Zealand Government Printing Office
Mulgreve St, Private Bag, WELLINGTON.

Department of Lands and Survey (NZMS)
Private Bag, Charles Ferguson Building, Bowen St, WELLINGTON.

Department of Scientific and Industrial Research (DSIR)
Science Information Publishing Centre, PO Box 9741, WELLINGTON.

Maps are available from the four following divisions of DSIR:

1. New Zealand Geological Survey (NZGS)
PO Box 30–368, LOWER HUTT.

2. Geophysics Division (NZGD)
PO Box 1320, WELLINGTON.

3. Soil Bureau (NZSB)
Private Bag, LOWER HUTT.

4. NZ Oceanographic Institute (NZOI)
PO Box 12–346, WELLINGTON NORTH.

National Water and Soil Conservation Authority (NWSCA)
Ministry of Works and Development, PO Box 12–041, WELLINGTON NORTH.

New Zealand Forest Service (NZFS)
Bowen State Building, Bowen St, Private Bag, WELLINGTON.

Civil Aviation Division
Ministry of Transport, Private Bag, WELLINGTON.

New Zealand Commission for the Environment
CPD House, 108 The Terrace, POB 10241, WELLINGTON.

University of Auckland
Private Bag, AUCKLAND.

Automobile Association (Auckland) Inc.
PO Box 5, AUCKLAND 1.

Automobile Association (Wellington) Inc.
PO Box 2472, WELLINGTON.

Wises Publications
Mapping Division, 368 Dominion Rd, AUCKLAND 3.

Lansdowne Press
59 View Rd, AUCKLAND.

New Zealand Tourist and Publicity Department
Private Bag, WELLINGTON.

For Bartholomew and Quail see Great Britain.

Catalogue

Atlases and gazetteers

New Zealand atlas I. Wards
Wellington: Shearer, 1976.
pp. 292.

New Zealand atlas of coastal resources
Wellington: Commission for the Environment, 1981.
pp. 60.

New Zealand railway atlas 3rd edn
Exeter: Quail, 1985.

AA road atlas of New Zealand revised edn
Auckland: Lansdowne Press, 1985.
pp. 72.
Includes 1 : 500 000 scale mapping, streetplans and gazetteer.

Wises NZ guide: a gazetteer of New Zealand edited by J.A. Cullen
Auckland: Wises, 1979.
pp. 518.

Gazetteer of New Zealand Placenames
Wellington: NZMS, 1967.
pp. 576.

General

Map of New Zealand 1 : 4 000 000 NZMS 268
Wellington: NZMS, 1975.
Includes Antarctic dependencies on inset.

New Zealand 1 : 2 000 000
Edinburgh: Bartholomew, 1982.

New Zealand 1 : 2 000 000 NZMS 84
Wellington: NZMS, 1975.
Non-metric version.

New Zealand from Landsat 1 : 2 000 000 NZMS 296
Wellington: NZMS, 1978.

Touring map of New Zealand 1 : 1 700 000 NZMS 238
Wellington: NZMS, 1979.
Includes streetplans of Wellington, Christchurch, Auckland and Dunedin.

New Zealand: wall map 1 : 1 250 000 NZMS 293
Wellington: NZMS, 1985.

New Zealand 1 : 1 000 000 NZMS 265
Wellington: NZMS, 1984.
2 sheets, both published.
Special sheets from the International map of the world series.
Also available with national park overprint as NZMS 265A.

New Zealand 1 : 1 000 000 NZMS 266
Wellington: NZMS, 1984.

New Zealand: topographical map 1 : 500 000 NZMS 242 2nd edn
Wellington: NZMS, 1981.
4 sheets, all published.

Topographic

New Zealand topographical map 1 : 250 000 NZMS 262
Wellington: NZMS, 1977–.
18 sheets, all published. ■

NZ topographical map 1 : 63 360 NZMS 1
Wellington: NZMS, 1937–75.
313 sheets, all published. ■

New Zealand topographical map 1 : 50 000 NZMS 260
Wellington: NZMS, 1977–.
324 sheets, 110 published. ■

Bathymetric

New Zealand region: bathymetry 1 : 6 000 000 2nd edn
Wellington: NZOI, 1980.
9 other oceanographic sheets also available on same sheet lines: winter sea surface temperatures 1956 and 1957; oceanic charts index; sea surface temperatures at the ocean floor; zooplankton biomass 0–200 m, primary productivity surface; primary productivity integrated; surface chlorophyll *a*; reactive phosphorus (October–April) surface.

New Zealand region: physiography   1 : 4 000 000
Wellington: NZOI, 1979.

Bathymetry: New Zealand   1 : 2 191 400
Wellington: NZOI, 1960.
2 sheets, both published.

New Zealand: coastal series bathymetry   1 : 200 000
Wellington: NZOI, 1964–.
37 sheets, 33 published.

New Zealand: coastal series sediments   1 : 200 000
Wellington: NZOI, 1966–.
37 sheets, 14 published.

Geological and geophysical

Gravity map of New Zealand   1 : 4 000 000
Wellington: NZGD, 1965.
Available as either Bouguer anomalies or Isostatic anomalies
edition.

Free air gravity field in the New Zealand region   1 : 3 000 000
Boulder: NGDC, 1983.
With accompanying explanatory text.

Late Quaternary tectonic map of New Zealand   1 : 2 000 000
Lower Hutt: NZGS, 1984.

Geological map of New Zealand   1 : 1 000 000
Lower Hutt: NZGS, 1972.
2 sheets, both published.

Quaternary geology of New Zealand   1 : 1 000 000
Lower Hutt: NZGS, 1973.
2 sheets, both published.

Gravity map of New Zealand   1 : 1 000 000
Wellington: NZGD, 1977–79.
2 sheets available in 4 different single editions showing free air
anomalies, Bouguer anomalies, isostatic anomalies, and isostatic
vertical gravity anomalies.

Geological map of New Zealand   1 : 250 000
Lower Hutt: NZGS, 1960–73.
28 sheets, all published. ■
With explanatory text on each sheet.

Gravity map of New Zealand   1 : 250 000
Wellington: NZGD, 1965–80.
27 sheets, all published.■
Each sheet is available in three types: Bouguer anomalies, isostatic
anomalies, and isostatic vertical gradient anomalies.

Magnetic map of New Zealand: total force anomalies   1 : 250 000
Wellington: NZGD, 1969–84.
27 sheets, all published. ■

Geological map of New Zealand   1 : 63 360
Lower Hutt: NZGS, 1953–.
313 sheets, 55 published. ■
Most sheets published with texts.

Geological map of New Zealand   1 : 50 000
Lower Hutt: NZGS, 1979–.
648 sheets, 16 published. ■
With accompanying explanatory text.

Environmental

New Zealand: single factor soil maps   1 : 3 700 000
Lower Hutt: NZSB, 1962–.
92 sheets, all published.

NEW ZEALAND
1 : 250 000 topographic Series NZMS 262

NEW ZEALAND
1 : 250 000 forest
Series FSMS 6

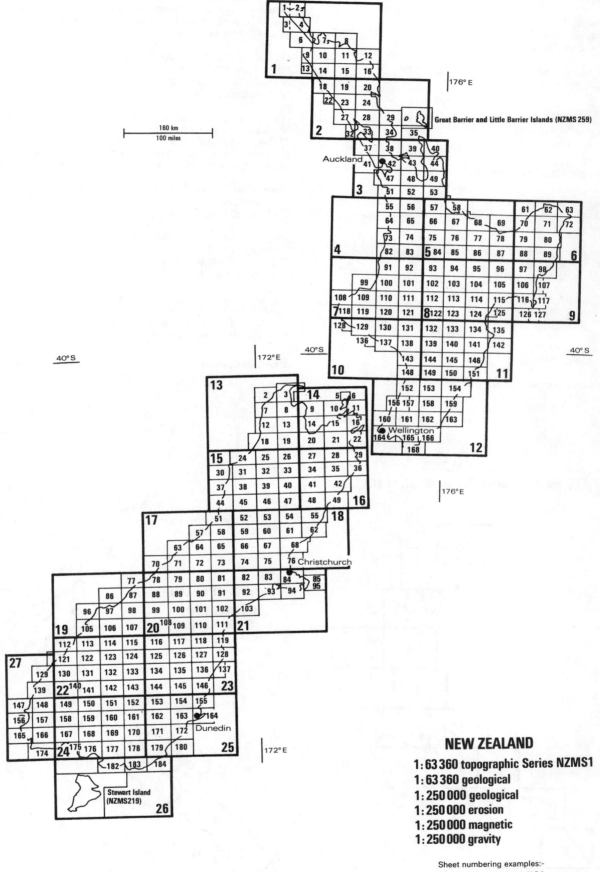

160 km
100 miles

176° E

Great Barrier and Little Barrier Islands (NZMS 259)

Auckland

40°S

172°E

40°S

40° S

176°E

Wellington

Christchurch

Dunedin

Stewart Island
(NZMS219)

172°E

NEW ZEALAND

1 : 63 360 topographic Series NZMS1
1 : 63 360 geological
1 : 250 000 geological
1 : 250 000 erosion
1 : 250 000 magnetic
1 : 250 000 gravity

Sheet numbering examples:-
1 : 63 360 North Island N 84
1 : 63 360 South Island S S127

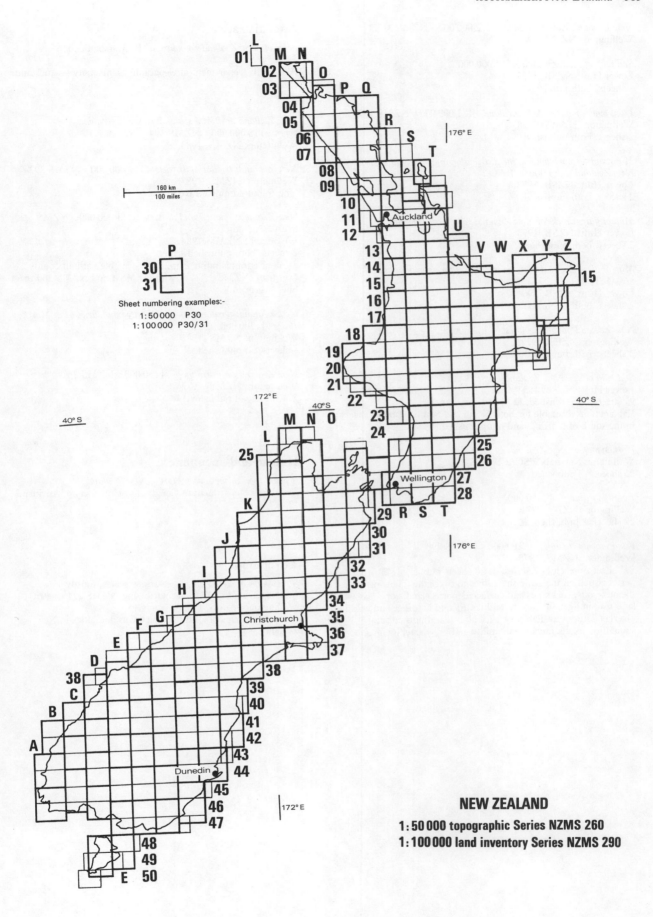

160 km
100 miles

P

30
31

Sheet numbering examples:-
1:50 000   P30
1:100 000 P30/31

176° E

172° E

40° S

40° S

176° E

172° E

NEW ZEALAND

1:50 000 topographic Series NZMS 260
1:100 000 land inventory Series NZMS 290

New Zealand: climatic regions 1 : 2 000 000 NZMS 315/2
Wellington: NZMS, 1983.

Soil map of New Zealand 1 : 1 000 000
Lower Hutt: NZSB, 1973.
2 sheets, both published.

Land tenure maps of New Zealand 1 : 1 000 000 NZMS 187
Wellington: NZMS, 1978.
2 sheets, both published.

Maps and sections showing the distribution and stratigraphy of loess in New Zealand 1 : 1 000 000
Lower Hutt: NZSB, 1973.
2 sheets, both published.

Map of parent rocks of New Zealand soils 1 : 1 000 000
Lower Hutt: NZSB, 1973.
2 sheets, both published.

Map of New Zealand showing soil classes for potential pastoral use 1 : 1 000 000
Lower Hutt: NZSB, 1964–65.
2 sheets, both published.

New Zealand indigenous forests 1 : 1 000 000
Wellington: NZFS, 1974.
2 sheets, both published.

Soil map of . . . New Zealand 1 : 253 440
Lower Hutt: NZSB, 1954–65.
21 sheets, 17 published. ■
Explanations available for both North and South Island. 4 sheets in the south of North Island not published.

Erosion map of New Zealand 1 : 250 000
Wellington North: NWSCA, 1974–84.
25 sheets, all published. ■

Forest class map 1 : 250 000
Wellington: NZFS, 1971–.
27 sheets, 8 published. ■

Land inventory maps 1 : 100 000 NZMS 290
Wellington: NZMS, 1978–.
143 sheets in each of 7 series, *ca* 50 sheets published. ■
Series issued on the same sheet lines to cover the following themes: soils, rocktypes and surface deposits, land slope, existing land use, indigenous forest, land tenure and holdings and wildlife. Overlay editions available for agricultural and horticultural suitability, exotic forestry suitability and relief satellite image.

Administrative

Atlas of New Zealand boundaries B. Marshall and J. Kelly
Auckland: Department of Geography, University of Auckland, 1986.
pp. 145.

New Zealand territorial sea and exclusive economic zones 1 : 3 000 000 NZMS 304 3rd edn
Wellington: NZMS, 1983.

New Zealand: territorial districts 1 : 2 000 000 NZMS 292
Wellington: NZMS, 1983.
One-colour administrative map.

New Zealand: counties and districts 1 : 1 000 000 NZMS 310A 2nd edn
Wellington: NZMS, 1984.
2 sheets, both published.
2-colour administrative map. Also available without county overprint (NZMS 310), or with NZMS 1 and NZMS 260 sheet lines overprint.

New Zealand: territorial local authority boundaries as at March 1983 1 : 500 000 NZMS 242B
Wellington: NZMS, 1983.
4 sheets, all published.

New Zealand cadastral map 1 : 50 000 NZMS 261
Wellington: NZMS, 1977–.
324 sheets, all published. ■

Human and economic

A population cartogram of Atlas of New Zealand
Auckland: Department of Geography, University of Auckland, 1986.

Town maps

Street maps of New Zealand various scales, mostly
1 : 20 000 NZMS 17 (non-metric) and NZMS 271 (metric)
Wellington: NZMS, 1965–.
Full-colour maps of all the major towns and cities in New Zealand.

EUROPE

Europe has a greater quantity and variety of mapping than any other continent, and this is reflected in the texts and catalogues of many of the individual countries which follow. Although this richness extends to both Eastern and Western European countries, the larger-scale topographic and thematic maps of states in the Warsaw Pact are generally very restricted in availability and therefore do not usually appear in the catalogue sections. We have, however, tried to do this material some justice in the texts, since several East European countries have ambitious programmes not only of topographic but also of both large- and small-scale environmental mapping.

Most topographic mapping is carried out by the survey organizations of the individual nation states, and retains many of the idiosyncrasies of style and specification which have developed in those states since the initiation of systematic topographic survey, usually in the nineteenth century. But there has also been a detectable trend in the postwar period towards more conformity between mapping systems. In NATO countries this has been partly the result of the adoption of NATO mapping standards, at least at the 1:50 000 scale, while in the Eastern Bloc countries, postwar topographic mapping standards have been laid down by Moscow. A European Committee of National Mapping Agencies (CERCO) has a membership composed of the directors of some 17 national mapping agencies, and this serves as a forum for the exchange of information of common interest. The growing similarity of mapping styles is also without doubt a reflection of the almost universal change from ground-based survey techniques to methods of photogrammetric compilation from air photographs and to common methods of scribing, phototypesetting and lithographic printing.

Most European countries have embarked on fundamental resurvey of their territories during the postwar period, usually preceded by retriangulation and new levelling programmes. At the same time the metric system of measurement has been universally adopted, and scale families of 1:10 000, 1:25 000, 1:50 000, 1:100 000 and 1:250 000 have become characteristic (with the 1:200 000 scale also retaining some popularity).

Most European government mapping establishments were originally of military origin and function. In most cases the link remains, but there has been a growing trend for them to merge with civilian sectors of government. There is increasing cooperation between government departments with diverse mapping needs, new goals have been identified, and several surveys—of The Netherlands, France, Germany and Austria for example—have recently adopted new larger basic mapping scales, at 1:10 000 or 1:5000, and often in cooperation with the country's cadastral agency. Most West European mapping authorities are now also engaged in major digital mapping programmes, with the aim of producing computer databases which offer great flexibility in meeting the demand for large-scale mapping and terrain modelling for census, cadastral, planning and military purposes. At the smaller end of the topographic scale range, an increasing number of Western European mapping authorities have begun actively to promote the general public use of their maps, and have introduced new specifications to make them more suitable for outdoor recreation and travel.

Although there are few general or topographic series which cover more than one country—even a modern 1:1 000 000 cover of the continent is not available—there are two groups of prewar maps which transgress present-day political boundaries, and which remain partially in print and of some interest in providing cover of areas where modern indigenous mapping is restricted. Firstly, the **Institut für Angewandte Geodäsie (IfAG)**, Frankfurt, continues to make available an extensive 1:100 000 *Karte des Deutschen Reiches*, covering present-day East Germany and a large part of Poland, and a *Topographische Übersichtskarte des Deutschen Reiches* 1:200 000 covering a similar area. Sheets of the old 1:25 000 plane table maps of the German Reich are also available. Secondly, the **Bundesamt für Eich- und Vermessungswesen (BEV)**, Vienna, continues to sell sheets of the *Generalkarte von Mitteleuropa* 1:200 000, although many are now out of print. In view of their age, we have not included these series in our catalogue, but indexes and ordering information may be obtained from the respective survey organizations.

There is a large number of multi-sheet thematic map series covering all—or substantial parts—of Europe. Most notable are the various earth science series sponsored by **UNESCO** and the **Commission for the Geological Map of the World (CGMW/CCGM),** many of which were started in the 1960s. The completion of these series has, in many cases, proceeded very slowly, although progress is still being made, and most of the early sheets remain in print. Several of the geological series are copublished with the **Bundesanstalt für Geowissenschaften und Rohstoffe (BfGR)**, Hannover, or with the **Bureau de Recherches Géologiques et Minières (BRGM)**, Orléans or the Academy of Sciences of the USSR, and are available from these sources, as well as from UNESCO and its distributors. The longest established of all these series, and one of the earliest examples of international scientific cooperation, is the *International Geological Map of Europe 1 : 1 500 000*, which has been compiled under the authority of a special commission of the International Geological Congress set up in 1881. This map is now being issued in its third edition as a joint publication of UNESCO and BfGR. More recently, an *International Hydrogeological Map*, at the same scale but covering a more restricted area, has been in progress. Ten of 30 sheets have been published so far. The sheets are in three languages, usually English, French and German, though others are substituted to reflect the region covered by individual sheets.

The European Geotraverse is a 7-year long international project begun in 1982 to map a 100 km wide swathe of land from North Cape in Norway to Tunisia. The aim is to improve understanding of the tectonic evolution of the European continent. The project is coordinated by the **European Science Foundation** and includes a wide range of geophysical profiling and mapping techniques. At the time of writing no hard copy products have become available.

A geomorphological map of Europe has been in preparation for many years, organized by the Commission for Geomorphological Survey and Mapping of the International Geographical Union. The map will cover the continent as far east as the Urals and will be published in 15 sheets at 1 : 2 500 000. A trial sheet was published in 1976, and further sheets are in an advanced state of preparation, but none is yet available.

The European Community has published a number of thematic maps of the member countries, and has sponsored the production of several atlases. The main three maps are an administrative map, a farming map and a map of forests. These are published in seven to nine different language editions. By the end of 1986, new editions of all three maps had been published covering the enlarged Community of 12 nations.

In 1985, a *Soil map of the European Communities* at 1 : 1 000 000 was published by the **Office for Official Publications of the European Communities (OOPEC)**, Luxembourg. This map draws on the national soil surveys of the various member countries, and attempts to unify them in a common classification (that used for the FAO/UNESCO *Soil Map of the World*). Fifty-one dominant soil units are distinguished in colour on the map, with a further breakdown into 312 numbered soil associations. The map is available as a boxed set or as flat sheets, accompanied in each case by an explanatory text. It had been hoped to extend this map, with the help of UNEP funding, to the whole of Europe.

In 1986, soil specialists in Eastern Europe began plans for a new 1 : 2 500 000 soil map of the territory of CMEA countries. A five-year schedule has been allowed for its preparation.

Groundwater resources of the European Community is published as a series of ten reports with 152 1 : 500 000 scale maps providing an inventory of aquifers and their hydrogeological characteristics. The maps are available separately, either as a complete European Community Atlas or for individual member countries.

Since 1974, the European Community has also been interested in the development of an environmental mapping system which could be used to identify environmentally sensitive areas in the Community. Following experiments in the late 1970s, the need then emerged for a comprehensive digital information system. The Community agreed the first phase of the implementation of such a system during the period 1984–87, and initial feasibility studies, testing the availability and compatibility of data required by the system, have been completed, focusing principally on an evaluation of biomass potential.

Many cooperative thematic mapping projects have also been carried out—or are in progress—in the Eastern European countries. An atlas of the COMECON countries has recently been compiled by specialists from several socialist countries, although at the time of writing we have no bibliographic information about this.

The *Types of Agriculture Map of Europe* 1 : 2 500 000 was prepared by the Polish Academy of Sciences and is distributed through the **Trade Centre of Polish Science (DHN)**. The map uses a method of classifying types of agriculture developed by the Commission on Agricultural Typology of the International Geographic Union.

The *Land Use Map of Europe* 1 : 2 500 000, published by **Cartographia**, Budapest, in 14 colours is the product of cooperation between 22 European countries and makes use of the UN Food and Agriculture Organization's land use categories.

The preparation of a unified vegetation map of Europe is under way, with editorial direction by the Botanical Institute of the Czechoslovakian Academy of Sciences (**Československá Akademie Věd**). The legend and a trial sheet had been prepared by 1985. The map is being compiled at 1 : 2 500 000 and will be in 15 sheets. A *Vegetation map of the Council of Europe* member states at 1 : 3 000 000 was prepared by P. Ozenda and others for the European Committee for the Conservation of Nature and Natural Resources, and published in 1979 by the **Council of Europe**. It accompanies monograph No. 16 in the *Nature and Environment Series*.

There are a number of important thematic atlases covering all or substantial parts of the continent. These include the massive *Atlas sozialökonomischer Regionen Europas*, based on demographic censuses of 1950, 1960 and 1970. This was prepared at the Sociographisches Institut of the Johann Wolfgang Goethe-Universität, Frankfurt AM and was issued over two decades to be finally completed in 1980. The work is in German, French and English, and is published by **Nomos**, Baden-Baden. The most recent small-format economic atlas is the *Atlas Economique de l'Europe*, published in 1987 by the **Société Royale Belge de Géographie** and covering the 18 countries of *Western* Europe. The statistical maps are in two colours and analysis is at the level of provinces, departments or counties (516 territorial units in all). In the earth science and environmental thematic areas, major atlases are the *Geological Atlas of Western and Central Europe* (Peter Ziegler), which has a text volume supplemented with a volume of small-scale tectonic/geological, palaeographic and isopach maps and stratigraphic correlation charts, and the UNESCO *Climatic atlas of Europe*, Volume 1 which contains 1 : 10 000 000 scale precipitation and temperature maps.

A particularly important European regional atlas is the *Atlas der Donauländer*, published by the **Österreichisches Ost- und Südosteuropa-Institut (ÖOSI)**, Vienna, under the editorship of Josef Breu. This atlas, which has been in progress since 1970, will have about 50 map sheets and is being issued in 10 instalments. The maps are at a principal scale of 1 : 2 000 000 and cover a range of physical, economic and social themes. Text sheets are in German, English, French and Russian. The atlas is distributed by **Franz Deuticke**.

A very large number of commercial map publishers are based in Europe, and many of them produce general and regional maps of the continent. We have included only a very small sample of such mapping in the catalogue section.

Further information

Latest information about UNESCO maps is found in the *Unesco publications catalogue*, issued annually, and further background information about many series is given in issues of the UNESCO periodical *Nature and Resources*.

The development of an EEC environmental database is described in Rhind, D. and Briggs, D. (1984) Environmental mapping and the information system on the state of the European environment, in *Further Examples of Environmental Maps*, edited by D. Bickmore, International Cartographic Association/ International Geographic Union, pp. 51–58.

Addresses

United Nations Educational, Scientific and Cultural Organization (UNESCO)
7 place de Fontenoy, F–75700 PARIS, France.

Commission for the Geological Map of the World (CGMW/CCGM)
Maison de la Géologie, 77 rue Claude-Bernard, F–75005 PARIS, France.

European Science Foundation
1 quai Lezay-Marnésia, F–67000 STRASBOURG, France.

Office for Official Publication of the European Communities (OOPEC)
2 rue de Commerce, L–2985 LUXEMBOURG, Luxembourg.

Verlag Th. Schäfer
Tivolistrasse 4, POB 5469, D–3000 HANNOVER 1, German Federal Republic.

Verlag TUV Rheinland GmbH
Konstantin-Wille-Strasse 1, D–5000 KOLN 91, German Federal Republic.

Československá Akademie Věd
Botanický Ústav, 25243 PRŮHONICE U PRAHY, Czechoslovakia.

Council of Europe
Publication Section, BP 431 R6, F–67006 STRASBOURG Cedex, France.

Trade Centre of Polish Science (DHN)
PO Box 410, ul. Miodowa 2, 00 950 WARSZAWA, Poland.

Nomos Verlagsgesellschaft mbH & Co. KG
Waldseestrasse 3–5, Postfach 610, D–7570 BADEN-BADEN, German Federal Republic.

Österreichisches Ost- und Südosteuropa-Institut (ÖOSI)
Josefsplatz 6, A–1010 WIEN, Austria.

Thomas Cook Ltd
PO Box 36, PETERBOROUGH PE3 6SB, UK.

For Wagner, BfGR, IfAG, DBB and Karto+Grafik, see German Federal Republic; for Haack, see German Democratic Republic; for GUGK and VSEGEI, see USSR; for Kümmerly + Frey, see Switzerland; for Bartholomew and NRSC, see Great Britain; for Société Royale Belge de Géographie, see Belgium; for OBB, Deuticke and BEV, see Austria; for Cartographia, see Hungary; for SGU and LMV, see Sweden; for GT, see Finland; for BRGM and IPGP, see France; for GUDS, see Czechoslovakia.

Catalogue

Atlases and gazetteers

Atlas der Donauländer
Wien: ÖOSI, 1970–.
50 sheets.

General

Europe by satellite
Farnborough, UK: NRSC.
Mosaic of images recorded from NOAA weather satellites, 1979–85.

Europa 1 : 10 000 000
Bern: Kümmerly + Frey, 1986.
Available in physical or political edition.

Europa Haack Handkarte 1 : 6 000 000
Gotha: Haack, 1981.
With separate index of place names.

Hildebrands Europakarta 1 : 6 000 000
Frankfurt AM: Karto+Grafik, 1986.

Europa 1 : 5 000 000
Bern: Kümmerly + Frey.
Available in physical or political edition.

Europa. Wagnerkarte 1 : 5 000 000
Berlin: Wagner, 1982.
Available in 3 editions, land cover, road map and political.

Europa fizicheskaja karta 1 : 4 000 000
Moskva: GUGK.

QTH-Radar map of Europe 1 : 3 000 000
Budapest: Cartographia, 1986.

Western Europe 1 : 3 000 000
Edinburgh: Bartholomew, 1983.

Europa. Haack Handkarte 1 : 3 000 000
Gotha: Haack.
4 sheets, all published.

Eastern Europe 1 : 2 500 000
Edinburgh: Bartholomew, 1979.

Scandinavia 1 : 2 500 000
Edinburgh: Bartholomew, 1983.

Nordeuropa 1 : 2 000 000
Gävle: LMV, 1981.

Norden från satellit 1 : 2 000 000
Gävle: LMV, 1981.
Satellite image map of Nordic countries.

Europa: Mittlerer Teil, Haack Handkarte 1 : 1 500 000
Gotha: Haack, 1974.
+ 136 pp. index.

Central Europe 1 : 1 250 000
Edinburgh: Bartholomew, 1982.

Vengrija, Rumynija, Bol'garija, Albanija, Jogoslavija i Grezija Fizicheskaja karta 1 : 1 250 000
Moskva: GUGK.

Satellitenbild Mosaik von Mitteleuropa 1 : 1 000 000
Hannover: BfGR, 1975.

Geological and geophysical

Geological atlas of Western and Central Europe P. A. Ziegler
Amsterdam: Shell Internationale Petroleum Maatschappij, 1982, distributed by Elsevier.
2 volumes (130 pp. text + 40 maps).

Tectonic map of Europe and adjacent regions 1 : 10 000 000
Moskva: GUGK/Paris: CCGM, 1979.

International geological map of Europe and the Mediterranean region 1 : 5 000 000
Hannover: BfGR/Paris: UNESCO, 1971.
2 sheets, both published.

Map of Quaternary deposits of Eurasia 1 : 5 000 000
Moskva: VSEGEI, 1982.
12 sheets, all published.

Carte sismotectonique. Europe et Bassin Mediterranéen 1 : 5 000 000
Paris: IPGP, 1986.

International tectonic map of Europe and adjacent areas
1 : 2 500 000   2nd edn.
Moskva: Academy of Sciences of the USSR/Paris: UNESCO,
1981.
20 sheets, all published.
In Russian and French.
Two volume text in Russian or English–French.

International Quaternary map of Europe 1 : 2 500 000
Hannover: BfGR/Paris: UNESCO, 1967–.
15 sheets + legend sheet, 13 published.

Metamorphic map of Europe 1 : 2 500 000
Leiden: Sub Committee for the Cartography of the Metamorphic
Belts of the World/Paris: UNESCO, 1973.
17 sheets, all published.
In English and French.
Sheet 17 Metamorphic map of the Alps 1 : 1 000 000 + text.

Metallogenic map of Europe 1 : 2 500 000
Orléans: BRGM/Paris: UNESCO, 1968–83.
9 sheets, all published.
+ 9 separate lists of deposits.
+ 560 pp. text.

International map of natural gas fields in Europe
1 : 2 500 000   2nd edn.
Hannover: BfGR/ECE, 1984.
9 sheets, all published.
+ 200 pp. text.

General geological map of the Baltic Shield 1 : 2 500 000
Espoo: GT, 1985.

Aeromagnetic anomaly map of Scandinavia 1 : 2 500 000
Uppsala: SGU, 1983.

Scandinavian Caledonides, Tectonostratigraphic map 1 : 2 000 000
Uppsala: SGU, 1985.
Available also in Swedish edition.

Scandinavian Caledonides. Gravity anomaly map 1 : 2 000 000
Uppsala: SGU, 1985.

Scandinavian Caledonides. Magnetic anomaly map 1 : 2 000 000
Uppsala: SGU, 1985.

International geological map of Europe 1 : 1 500 000   3rd edn
Hannover: BfGR/Paris: UNESCO, 1964–.
49 sheets, 35 published. ■

International hydrogeological map of Europe 1 : 1 500 000
Hannover: BfGR/Paris: UNESCO, 1970–.
30 sheets, 10 published. ■
+ text for each sheet.

Tectonic map of the Carpathian-Balkan system 1 : 1 000 000
Bratislava: GUDS/Paris: UNESCO, 1973.
9 sheets, all published.
+ legend and text.

Environmental

Groundwater resources of the European Community
Hannover: Verlag Th. Schäfer for Commission of the EC.
10 reports and 152 maps.
Maps available separately.

European solar radiation atlas EUR 6577
Köln: Verlag TUV Rheinland for Commission of the EC, 1984.
2 volumes, pp. 297 and 327.

Climatic atlas of Europe Vol. 1
Budapest: Cartographia/Geneva: WMO/Paris: UNESCO, 1970.

EUROPE
1 : 1 500 000 International Geological Map of Europe
1 : 1 500 000 International Hydrogeological Map of Europe

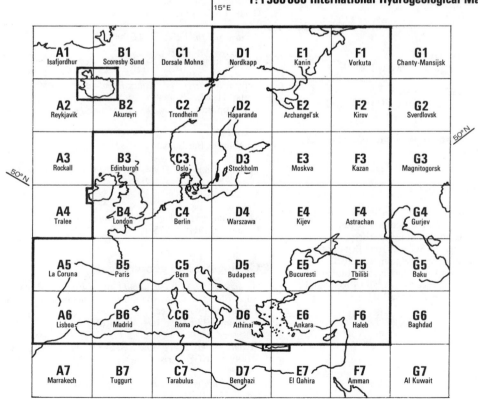

The European Community: forests 1:4 000 000
Luxembourg: OOPEC, 1983.

Vegetation map of the Council of Europe member states 1:3 000 000
Strasbourg: Council of Europe, 1979.
3 sheets, all published.
+ 99 pp. text.

Land use map of Europe 1:2 500 000
Budapest: Cartographia, 1980.
4 sheets, all published.

Soil map of the European Communities 1:1 000 000
Luxembourg: OOPEC, 1986.
7 sheets + 2 legend sheets, all published.

*Klimicheskaja promiyshlenost' sotsialisticheskie stran
Europiy* 1:1 000 000
Moskva: GUGK.
4 sheets, all published.

Administrative

The European Community: member states 1:8 000 000
Luxembourg: OOPEC, 1986.

Europe. Political 1:5 000 000
Edinburgh: Bartholomew, 1984.

Europa politicheskaja karta 1:4 000 000
Moskva: GUGK.

The European Community: member states 1:3 000 000
Luxembourg: OOPEC, 1986.

Human and economic

Atlas économique de l'Europe
Bruxelles: Société Royale Belge de Géographie, 1987.
pp. 164.

An atlas of EEC affairs R. Hudson, D. Rhind and H. Mounsey
London and New York: Methuen, 1984.
pp. 158.

Atlas sozialökonomischer Regionen Europas 1:4 000 000
Baden-Baden: Nomos, 1962–80.
21 sections, all published.
In German, French and English.

(European Community) Regional development Atlas 1981
Luxembourg: OOPEC
6 language edition.

Europa Eisenbahn-Atlas
Bern: Kümmerly + Frey, 1983.
pp. 143.

Thomas Cook rail map of Europe 1985–6 1:5 000 000
Peterborough: Thomas Cook, 1985.

The European Community. Farming 1:4 000 000
Luxembourg: OOPEC, 1986.

Internationale Güterverkehrskarte 1:3 500 000
Frankfurt AM: DBB, 1983.
International rail freight map.

*Eisenbahn Übersichtskarte für den Internationalen
Güterverkehr* 1:3 500 000
Wien: ÖBB, 1985.

*Vengrija, Rumynija, Bol'garija, Albanija, Jugoslavija i Grezija.
Ekonomicheskaja karta* 1:1 250 000
Moskva: GUGK.

Sotsialisticheskie stran Europiy. Ekonomicheskaja karta 1:1 000 000
Moskva: GUGK.
4 sheets, all published.

Albania

Very little information about mapping comes out of
Albania. The country has not responded to the United
Nations questionnaires on topographic mapping, and is not
represented in the International Cartographic Association.

Our catalogue entries are therefore restricted to a very
few general maps which are obtainable through such
sources as GeoCenter (see Chapter 3 on procurement).

The state organization for the publication and
distribution of maps for educational and public use is
**Mjetve Mësimore Kulturore e Sportive 'Hamid Shijaku'
(MMKS 'Hamid Shijaku')**, Tirana. A general map of
Albania at 1:500 000 accompanied by a booklet
introducing the country has been published by this
organization and is distributed overseas by the Book
Distribution Enterprise (**Ndërmarrja e Përhapjes së
Librit**). It is available in English or German.

Albania was covered by sheets of the Austrian 1:200 000
Generalkarte von Mitteleuropa published before 1946, and
these sheets are still available from the Austrian Survey. A
small-scale geological cover is provided by Sheet D6 of the
UNESCO *International Geological Map of Europe
1:1 500 000* (see the Europe catalogue section).

Addresses

**Mjetve Mësimore Kulturore e Sportive 'Hamid Shijaku'
(MMKS 'Hamid Shijaku')**
Byroja Hartografike, Qyteti nxenesve, TIRANE.

Ndërmarrja e Përhapjes së Librit
(Book Distribution Enterprise), TIRANE.

For DMA, see United States; for GUGK, see USSR.

Catalogue

Atlases and gazetteers

Albania. Names approved by the US Board on Geographic Names
Washington, DC: DMA, 1955.
pp. 156.

General

Albanija 1:600 000
Moskva: GUGK, 1982.

PSR of Albania 1:500 000
Tiranë: MMKS 'Hamid Shijaku', 1984.
Physical map with booklet describing the country.
Legend and booklet in English.

Environmental

Harta e Bimësisë te RPS Te Shqipërisë 1:275 000
Tiranë: MMKS 'Hamid Shijaku', 1980.
Distribution of forests and pastureland.

Administrative

Shqipëria: Hartë politiko-administrative 1:500 000
Tiranë: MMKS 'Hamid Shijaku', 1969.

Human and economic

RPSE Shqipërisë (Popullsia) 1:275 000
Tiranë: MMKS 'Hamid Shijaku', 1978.
Population map.

Andorra

The Principality of Andorra, with an area of only 468 km^2
and a population of 42 000, lies between Spain and France
and is included in the national map series of both these
countries. However, there is also a fine selection of maps
published by the Andorran Government (**MI Consell
General**). The principal series is *Valls d'Andorra* at
1:10 000, printed in four colours, with 10 m contours, and
covering the Principality in 19 sheets. This was made from
a photogrammetric survey of 1972 and published in 1976.
A single sheet covering Andorra at 1:50 000 was published
in 1977. This is in seven colours and relief is shown by 20
m contours and shading. Although out of print, it was due
to be replaced by a new edition in 1986. These maps are
on a Lambert conic conformal projection.

MI Consell General also publishes plans of the built-up
areas at 1:1000 and 1:500, and there are seven coloured
town maps at scales of 1:2000 to 1:5000, published in
1986 and covering the main settlements of the seven
Parróquias.

The book *Valls d'Andorra. Geografia i Diccionari
Geografic*, besides offering a brief geography of the
country, has a substantial gazetteer of placenames and
includes a copy of the 1:50 000 map.

The Spanish commercial publisher **Editorial Alpina**,
Granollers, also has a map of the country in its series of
tourist maps. This map is in colour with relief shown by
hypsometric tints and a 20 m contour interval, and is
accompanied by a guide for mountaineers and tourists (in
Catalan). A more up-to-date topographic map with tourist
overprint is provided by Sheet 7 in the 1:50 000 series
Carte de randonnées published by the French **Institut
Géographique National (IGN)**, 1986 edition.

Andorra was mapped geologically by N. Llopis Lladó
for the Instituto de Geologia Economica, Madrid, and an
eight-sheet cover at 1:25 000 was published in 1969 (by
the Real Academia de Ciencias y Artes de Barcelona) but
appears now to be out of print. This map showed solid
geology in colour with Quaternary deposits in black and
white. A 1:50 000 geological map by Sabaris and Lladó
has become available recently.

Environmental and resource mapping of Andorra is
found in a set of three 1:50 000 maps accompanying *El
patrimoni natural d'Andorra*, a book about the natural
systems of Andorra and their utilization. The maps are of
vegetation, geomorphology and land evaluation. This work
is published by **Ketres Editora**, Barcelona, and was
produced with the support of the Department of
Agriculture of the Andorran Government.

Addresses

MI Consell General
Departament d'Obras Publiques, Carrer les Canals 7, 4.rt,
ANDORRA LA VELLA.

Editorial Alpina
Apartat Correus 3, GRANOLLERS (Barcelona), Spain.

Ketres Editora
Diputació 113–115, BARCELONA 15, Spain.

For IGN, see France.

Catalogue

Atlases and gazetteers

Valls d'Andorra. Geografia i diccionari geografic
Andorra la Vella: MI Consell General, 1977.
pp. 286.

Topographic

Valls d'Andorra 1:50 000
Andorra la Vella: MI Consell General, 1986.

Andorra i sectors meridionals fronterers, Engorgs—Puig Pedros—
La Pera, Mapa topographic-excursionista 1:40 000
Granollers: Editorial Alpina, 1980.
Map accompanied by text in Catalan.

Geological and geophysical

Mapa geologico de Andorra 1:50 000 L. Solé Sabrarís and N.
Llopis Lladó
Instituto de Estudios Lerdenses, n.d.

Environmental

El patrimoni natural d'Andorra R. Folch i Guillèn et al.
Barcelona: Ketres Editoria, 1984.
Text volume + 3 1:50 000 maps: Litologia i geomorfologia de les
Valls d'Andorra, Vegetació de les Valls d'Andorra, and Valoració
ecològica i usos territorials postulables.

Austria

Topographic mapping is the responsibility of the **Bundesamt für Eich- und Vermessungswesen (BEV)**, Vienna, founded under that name in 1945, although systematic survey goes back to 1764 with the first *Landesaufnahme*.

The basic scale topographic series is at 1:50 000, and the country is covered in 213 sheets each representing an area of 15' longitude by 15' latitude. Although the series originated in the 1920s, it has been entirely reworked since 1950 on the basis of photogrammetric survey, a programme which has recently been completed. The projection is Gauss–Krüger, and relief is shown by 20 m contours and hill shading. Most sheets are published in four versions: a seven-colour map without footpaths or coloured road infills, an edition with roads classified in colour, an edition with marked walking routes shown in red, and a three-colour base map. Two systems of numbering the sheets in this series are in use: a simple numerical sequence from 1 to 213, and a more recently adopted grid system (*Bundesmeldenetz*) requiring a four-figure reference. Both systems appear on the most recent sheets and both are shown on the accompanying index. The revision interval for the series is on average about 7 years.

The current 1:200 000 series was started in 1961 and replaces the old *Generalkarte von Mitteleuropa*. It will cover the country in 23 sheets each with an area of 1° longitude by 1° latitude. Completion of this series is expected in 1987. Sheets are currently numbered by the system shown in the first of the two index maps. The 1:25 000 and 1:100 000 series are photomechanical enlargements of the 1:50 000 and 1:200 000 respectively. The former has the same sheet system as the 1:50 000, but with the area of each sheet divided into two and printed back-to-back. This map is published as a walker's edition (with the marked walking routes in red) and as a three-colour base map.

In 1973 it was decided to introduce a new larger-scale product, the *Österreichische Luftbildkarte*. This is an orthophotomap fitted to the Gauss–Krüger net. Sheets are printed as black-and-white, cartographically enhanced photo images, each covering an area of 5 × 5 km. A special programme of air photography for the series began in 1976. There are 3630 sheets in the series, but only about 25 per cent of these have so far been published. A by-product of the use of the orthophoto technique has been the production of a digital terrain databank, and by 1984 digital elevation data has been captured for about 76 per cent of the country.

For areas of high priority, a new 1:5000 *Basiskarte* (basic map) was started in 1983, beginning in the St Pölten area. Sheets in this series cover an area of 2.5 × 2.5 km.

Other maps produced by BEV include the 1:500 000 general map on a Lambert conformal conic projection (also published in an administrative and a two-colour outline edition), and a number of *Gebietskarten* (regional maps) of popular tourist areas.

Administrative area maps are published by the **Österreichisches Statistisches Zentralamt (ÖSTZ)**, Vienna. They include a map of urban regions (Stadtregionen) at 1:1 000 000 and maps of the *Gemeinde* at 1:500 000. There is also a series of 1:50 000 scale maps showing the boundaries of enumeration districts for the 1981 national census.

Systematic geological survey in Austria can be traced back to the mid-nineteenth century, with the Imperial Geological Institute (Geologische Reichsanstalt) being founded in 1849. The current organization is the **Geologische Bundesanstalt (GB)**, Vienna. Mapping on the 1:25 000 topographic base maps led to publication (from 1891) of the Special Geological Map (*Geologische Spezialkarte*) at 1:75 000. A few sheets of this series are still available. The use of this scale continued until the early 1950s, when the current 1:50 000 series was launched. This main series is an active one with many sheets published in the 1980s and others in preparation. A few sheets in this series are at 1:25 000 (on the same sheet lines). There are also numerous district maps at various scales. Since 1978, when GB was reorganized, there has been a shift of emphasis towards applied geology, with the production of hydrogeological and geological hazard maps at 1:50 000. As yet, these maps have not been issued in printed form. The aerial magnetic series was flown by Hunting Geology and Geophysics Ltd and covers only the western part of the country. In 1986, a Geochemical Atlas of Austria was in preparation.

Soil mapping is carried out by two institutions: first, at the large scales of 1:2880, 1:5000 and 1:10 000, by the Treasury Board for the tax evaluation of farmland; second, by the **Bundesanstalt für Bodenwirtschaft**, Vienna, a department of the Board of Agriculture and Forestry, which was founded in the late 1950s and has a basic soil mapping programme of agricultural areas at 1:25 000 scale. About 90 per cent of the agricultural area had been mapped by 1986, with fieldwork due for completion in 1990–92. The mapping is organized into about 180 mapping districts which are based on the internal judicial districts (*Gerichtsbezirke*) of the country. Maps are issued together with monographs.

The national *Atlas der Republik Österreich* is a substantial work published over a 20 year period by the Kommission für Raumforschung of the **Österreichische Akademie der Wissenschaften (ÖAdW)**, under the general editorship of Hans Bobek, and completed in 1980. The maps were prepared and printed by Freytag-Berndt und Artaria. The atlas is divided into 12 sections covering the classical range of themes, from physiography via the resource base to settlement, population, economy, trade, culture and administration. The principal map scale is 1:1 000 000.

The preparation of regional atlases of the individual Bundesländer has also been a significant element of Austrian cartography, particularly in the 1950s, when atlases of Niederösterreich, Oberösterreich, Kärnten, Salzburg and Steiermark were all published or started. More recently, the ambitious *Tirol-Atlas* has been in progress, with seven of an expected 12 instalments so far issued. This atlas is being prepared by the Institut für Landeskunde of the University of Innsbruck with the support of the Tiroler Landesregierung. Another recent substantial atlas project is the *Planungsatlas für Wien* published by the Magistrat der Stadt Wien in 12 thematic sections with maps at 1:100 000 (1983–).

The *Geographisches Namenbuch Österreichs* is one of the few gazetteers so far published which are prepared in accordance with the recommendations of the United Nations. It was compiled by members of the Department of Cartographic Toponymy of the Austrian Cartographic Society, and includes names of the more important geographical features of the country as well as settlement names.

The Institut für Kartographie of ÖAdW has supported a number of mapping projects, including experiments with digital satellite data; and two digitally processed Landsat maps printed in false colour and fitted to the BEV 1 : 200 000 topographic series sheet lines were issued in 1983. Subsequently, a new series of satellite image maps at 1 : 200 000 on BEV sheet lines has been launched. They are being published and distributed by **Geospace Verlag**, Bad Ischl. It is planned to publish three or four sheets a year and eventually to cover the whole country. The first sheet (48/14 *Linz*) used Landsat Thematic Mapper imagery. Future sheets will make use of SPOT imagery. Landsat imagery was also used to provide the base maps for thematic overlays in the *Landeskündlicher Luftbildatlas Salzburg* by Lothar Beckel and Franz Zwittkovits (Salzburg: Otto Muller Verlag, 1981), and has been used in the *Linzer Atlas* published by the Kulturamt der Stadt Linz.

There is a history of interest in vegetation mapping in Austria. A recent project is the *Karte der aktuellen Vegetation der Hohen Tauern 1 : 25 000* published by ÖAdW in association with the Hohe Tauern National Park Commission as part of the Austrian contribution to the UNESCO Man and Biosphere Programme. Sheets of a complete vegetation map of the country at 1 : 200 000 have also been in preparation since 1978, although it is not known when these will be published.

The **Österreichisches Ost- und Südosteuropa-Institut**, founded in 1958, publishes documentation about the countries of Southeast Europe, and is responsible for the *Atlas der Donauländer* (described further in the Europe section).

There are several important commercial mapping organizations in Austria. The Austrian Alpine Club (**Österreichischer Alpenverein**), in association with the German Alpine Club, has a 120 year tradition of alpine mapping, and publishes very high-quality topographic maps of the Eastern Alps, mostly at 1 : 25 000 and with a 20 m contour interval. There are about 45 sheets available, and as well as paying close attention to accurate relief representation, they show waymarked walking routes and on some sheets ski routes as well. **Freytag-Berndt und Artaria (FB)** publish an extensive series of hiking maps at 1 : 100 000 and 1 : 50 000, as well as canoing maps, road maps and street maps of Austrian towns. **Fleischmann und Mair** also publish an extensive series of 1 : 50 000 *Kompass Wanderkarten*. **Eduard Hölzel** is mainly a publisher of educational maps and atlases.

Further information

A very thorough description of present day mapping in Austria is given in Arnberger, E. (1984) Editor, *Kartographie der Gegenwart in Österreich*, Vienna: Verlag der Österreichischen Geographischen Gesellschaft; Institute für Kartographie der Österreichischen Akademie der Wissenschaften, pp. 351.

An annual catalogue with sample map extracts and cover indexes is issued by the BEV, Vienna. Geologische Bundesanstalt (GB) issues a catalogue of publications and a separate list of available geological maps. ÖSTZ issue a *Publikationsangebot*, listing and describing their publications.

Addresses

Bundesamt für Eich- und Vermessungswesen (BEV)
Landesaufnahme, Krotenhallergasse 3, A-1080 WIEN.

Geologische Bundesanstalt (GB)
Rasumofskygasse 23-25, Postfach 154, A-1031 WIEN.

Bundesanstalt für Bodenwirtschaft
Denisgasse 31–33, A-1200 WEIN.

Österreichisches Statistisches Zentralamt (ÖSTZ)
Hintere Zollamtstrasse 2b, Postfach 9000, A-1033 WIEN.

Institut für Kartographie der Österreichischen Akademie der Wissenschaften (ÖAdW)
Bäckerstrasse 20, A-1010 WIEN.

Geospace Verlag
Beckel Satellitenbilddaten, Marie-Louisen-Strasse 1A, A-4820 BAD-ISCHL.

Österreichisches Ost- und Südosteuropa-Institut (ÖOSI)
Josefsplatz 6, A-1010 WIEN.

Österreichischer Alpenverein
Wilhelm-Greil Strasse 15, A-6020 INNSBRUCK.

Freytag-Berndt und Artaria KG (FB)
Schottenfeldgasse 62, Postfach 169, A-1071 WIEN.

Fleischmann und Mair
Geographischer Verlag, Kaplanstrasse 2, A-6040 INNSBRUCK.

Franz Deuticke Verlagsgesellschaft mbH
Helferstorferstrasse 4, Postfach 761, A-1011 WIEN.

Eduard Hölzel Verlag
Rudengasse 11, A-1030 WIEN.

Akademische Druck und Verlagsanstalt
Auersperggasse 12, A-8011 GRAZ.

Catalogue

Atlases and gazetteers

Atlas der Republik Österreich
Wien: Kommission für Raumforschung der ÖAdW, 1961–80.
In 12 parts comprising 120 sheets.

Grosser Auto Atlas Österreich   1 : 300 000
Wien: FB, 1979.
pp. 263.
Includes town maps, cultural gazetteer, and placename index.

Geographisches Namenbuch Österreichs   Josef Breu
Wien: Verlag der ÖAdW, 1975.
pp. 323.

Ortsverzeichnis 1981
Wien: ÖSTZ.
Issued in volumes for each of the Bundesländer.
8 volumes + combined index.

General

Übersichtskarte von Österreich   1 : 500 000   OK 500
Wien: BEV, 1978.
Available with or without separate index.

Topographic

Österreichische Karte   1 : 200 000   OK 200
Wien: BEV, 1969–.
23 sheets, 20 published. ■

Österreichische Satellitenbildkarte   1 : 200 000
Bad Ischl: Geospace Verlag, 1984–.
23 sheets, 1 published. ■

Österreichische Karte   1 : 100 000   OK 100 V
Wien: BEV, 1976–.
23 sheets, 20 published. ■
Enlargement of the 1 : 200 000 series.

AUSTRIA
1 : 200 000 topographic (ÖK 200)
1 : 200 000 satellite image maps

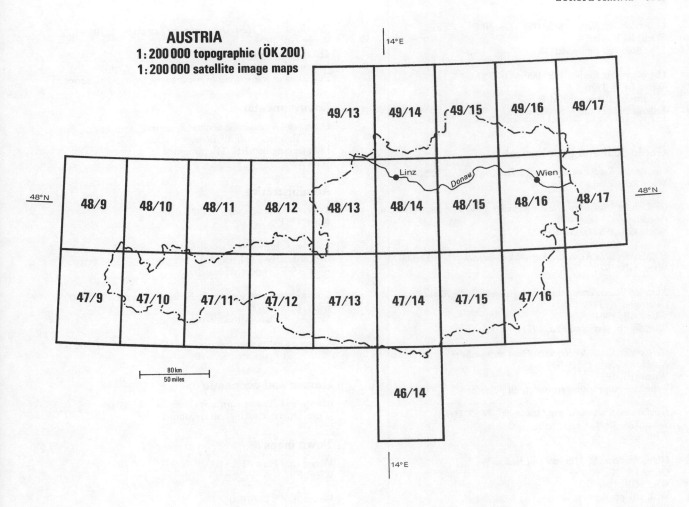

80 km
50 miles

AUSTRIA
1 : 200 000 topographic (ÖK 200)
1 : 50 000 topographic (ÖK 50)
1 : 50 000 geological

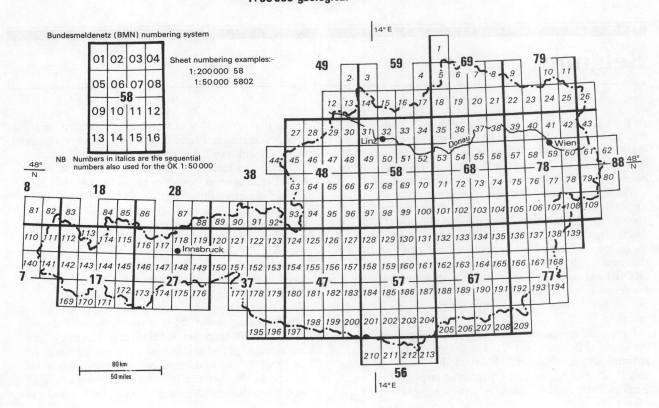

Bundesmeldenetz (BMN) numbering system

01	02	03	04
05	06	07	08
09	10	11	12
13	14	15	16

58

Sheet numbering examples:-
1:200 000 58
1:50 000 5802

NB Numbers in italics are the sequential
numbers also used for the ÖK 1:50 000

80 km
50 miles

Österreichische Karte 1:50 000 OK 50
Wien: BEV, 1969–.
213 sheets, all published. ■

Österreichische Karte 1:25 000 OK 25 V
Wien: BEV, 1976–.
213 sheets, 194 published. ■
Enlargement of the 1:50 000 series.

Geological and geophysical

Geologische Karte von Österreich (ohne Quartär) 1:1 500 000
Wien: GB, 1980.

Geologische Übersichtskarte der Republik Österreich mit tektonischer Gliederung 1:1 000 000
Wien: GB, 1964 (revised 1986).

Hydrogeologische Karte der Republik Österreich 1:1 000 000
Wien: GB, 1970.

Karte der Lagerstätten mineralischer Rohstoffe der Republik Österreich 1:1 000 000
Wien: GB, 1964.
With 94 pp. text (2nd edn, 1977).

Geologische Karte der Republik Österreich und der Nachbargebiete 1:500 000
Wien: GB, 1933.
2 sheets + supplementary sheet, all published.

Landsat-Bildlineamente von Österreich 1:500 000
Wien: GB, 1984.
With text.

Übersichtskarte der Mineral- und Heilquellen in Österreich 1:500 000
Wien: GB, 1966.
With 101 pp. text.

Geologische Karte der Republik Österreich 1:50 000
Wien: GB, 1955–.
213 sheets, 52 published. ■
Some sheets published at 1:25 000.

Aeromagnetische Karte der Republik Österreich 1:50 000
Wien: GB, 1979–.
213 sheets, 56 published. ■
Isoanomalies of total intensity.
Published as transparent overlays to the topographic series.

Environmental

Österreichische Bodenkartierung 1:25 000
Wien: Bundesanstalt für Bodenwirtschaft, 1971–.
180 mapping districts, 110 published.
With texts.

Administrative

Karte der Stadtregionen 1:1 000 000
Wien: ÖSTZ.

Einteilung Österreichs in politische Bezirke und Gerichtsbezirke 1:1 000 000
Wien: ÖSTZ, updated annually.

Übersichtskarte von Österreich: Politische Ausgabe 1:500 000
Wien: BEV, 1978.
Available with or without separate index.

Karte der Gemeindegrenzen der Republik Österreich 1:500 000
Wien: ÖSTZ, updated annually.

Human and economic

Eisenbahn-Übersichtskarte von Österreich 1:500 000
Wien: Österreichische Bundesbahn, 1985.

Town maps

Vienna City Plan 1:25 000
Wien: FB.

Vienna A–Z 1:20 000
Wien: FB.

Wien/Vienna 1:12 500
Bern: Hallwag, 1985.
Includes mass transit map.

Belgium

As the birthplace of Mercator, who produced a map of Flanders in 1540, Belgium has a long and eminent tradition of topographic mapping and large-scale cover had been achieved by Fricx and by Ferraris by 1744 and 1778 respectively. Their maps are still available in facsimile.

The national survey authority owes its origin to the Depôt de la Guerre, founded in 1831, which became successively the Institut Cartographique Militaire (1878), and the Institut Géographique Militaire (1947). Finally, the military authority became a civilian one in 1976 and the present title of **Institut Géographique National (IGNB)** was adopted. The Institut is responsible for geodetic survey, official mapping and air photography. It is not, however, responsible for large-scale urban or cadastral mapping.

Basic scale mapping at 1:20 000 was accomplished for the whole of Belgium by 1871, and derived maps were subsequently published at 1:40 000, 1:100 000, 1:200 000 and 1:320 000. But current topographic mapping of Belgium bears no direct relation to these early series, for it

is entirely post-World War II in origin, compiled using aerial photogrammetry and drawn on a Lambert conformal projection with two standard parallels.

The modern basic scale mapping is at 1:25 000. Work on the preparation of this series began in 1947, and the sheets of the first edition were issued from 1955 to 1970. There is a revision cycle of 12 years, but this period is shorter for areas of rapid change, and some sheets are now in second or third editions. The complete set now comprises 238 sheets (formerly 237). The map is in seven colours, with a contour interval of 1, 2.5 or 5 m (varying according to region), and a kilometric Lambert grid is printed on the face of the sheets. A photographic enlargement of the *carte de base* is used to produce a 1:10 000 scale series in 455 four-colour sheets (2nd edn).

The 1:50 000 map (Series M736) is produced by generalization of the *carte de base* and has contours at 2.5, 5 or 10 m. As well as the nine-colour version, this series may also be obtained in a brown monochrome and in a two-colour relief and drainage edition. The standard full-colour version shows the Lambert grid by marginal ticks, although since 1976 sheets have also been available with the grid printed on the surface of the map.

A 1:100 000 scale series in 24 sheets (Type R) based on prewar mapping was discontinued in 1965, but in 1986 a new specification 1:100 000 series in 19 sheets was launched. This map, generalized from the 1:50 000, has a four-language legend and 20 m contours. The sheets are double-sided with the topographic map on one side and a tourist version on the reverse.

A new digitally produced 1:250 000 general map of Belgium was to be published in 1986. At the time of writing, IGNB was considering a full digital mapping programme.

Geological mapping is undertaken by the **Service Géologique de Belgique (SGB)**, which is a division of the Ministère des Affaires Economiques. Basic mapping at 1:40 000 was completed by 1911, and all sheets of this series are still available. There is also an early geological map in 12 sheets (*Carte de Dumont*) dating from 1900 which may be obtained from IGNB. The modern geological series is at 1:25 000 but sheets do not conform with the modern topographic series at that scale. Only nine sheets have been published so far, all in a French and some also in a Flemish edition. There is at present no general geological map of Belgium published by the SGB, although Plate 8 of the *Atlas de Belgique* (first edition), available as a separate, fulfils this requirement.

The geomorphological mapping programme has been the work of the **Centre National de Recherches Géomorphologiques (CNRG)**, beginning in 1965 under the direction of Professor P. Macar. Nine sheets have been published at 1:25 000, and these conform with the modern topographic series. CNRG is now based at the University of Liège, and the secretary is Professor A. Ozer. The publication scale for future sheets will be 1:50 000.

The *carte des sols* and the *cartes phytosociologiques* are both the work of the **Comité pour l'Etablissement de la Carte des Sols et de la Végétation de la Belgique (CECSVB)**, founded in 1947. Publication of the soil maps is well advanced, but work on the vegetation map has been discontinued.

The first edition of the Belgian national atlas, *Atlas de Belgique*, was issued over the period 1950–75. It is included in our catalogue because, although out of print as a complete volume, many useful individual plates are still available. However, since 1983 a Second National Atlas has been in progress. The new edition is in the same format as the old except that text (in four languages) is now incorporated on the reverse of the map folios (the first edition had accompanying one-language texts in booklet form), and sheets are being issued either flat or guarded and mounted to fit a loose-leaf binder. The new atlas is produced under the direction of a **Commission de l'Atlas National**, with the collaboration of numerous academies, universities and scientific institutes. Provisionally, there are to be 94 map sheets, mostly at a scale of 1:500 000.

The French-speaking provinces are covered by the *Atlas de la Wallonie* which has been issued as a series of thematic sheets since 1979, with some ten published so far. The publisher is the **Societé de Developpement Régional pour la Wallonie (SDRW)**.

Government mapping of various kinds is carried out by the Ministère des Travaux Publics, particularly by its **Service de Topographie et de Photogrammetrie** which publishes colour 1:5000 contour maps of many towns. Other departments within this ministry are responsible for census mapping (using automated cartography) and maps of the road network, and of traffic data. Cadastral mapping is carried out by the **Ministère des Finances.**

In 1977 the Ministry of Public Health and Environment (**Ministère de la Santé Publique et de l'Environnement**) began preparation of a *Carte écologique de la Belgique* which was intended to identify areas of prime ecological importance and to serve as a tool in environmental impact studies. These synthetic maps produced from existing resource data were to be published at 1:25 000. Although a handbook was published in 1979, funding was subsequently frozen, and no sheets have so far appeared.

The **Institut National de Statistique** publishes outline maps of census enumeration districts, and has published a statistical atlas based on the 1981 population census.

The *Atlas économique de la Belgique* was prepared jointly by the **Société Royale Belge de Géographie** and the Laboratory of Human Geography of the Université Libre de Bruxelles. It has a series of 1:1 000 000 computer maps in black and white. The text is in French.

Topographic and especially geological mapping of the African states of Zaire, Rwanda and Burundi has been carried out by the **Musée Royale de l'Afrique Centrale (MRAC)** in Tervuren. MRAC distributes the colonial series compiled by official Belgian agencies, and is currently engaged in geological programmes in association with the earth science survey agencies of these states.

There are several commercial map publishers in Belgium, and among them **Geocart** and **Girault Gilbert** are particularly well known for the production of road maps and street maps of towns. Geocart also publishes a series of provincial maps to a common scale of 1:100 000 which include indexes, and tourist information in textual form, and a series of cycling maps on the same base.

Further information

A descriptive catalogue of topographic maps is issued annually by IGNB, and most other mapping authorities produce lists of their in-print publications.

The Ministère des Travaux Publics publishes a catalogue of maps published by government departments within the Ministry.

The 1:25 000 geological series is described in a brochure by P. Fourmarier and A. Grosjean, *La nouvelle carte géologique de Belgique au 1/25.000*.

The Comité National de Géographie compiles a bibliography of maps published in Belgium biennially.

Addresses

Institut Géographique National (IGNB) Administration
Abbaye de la Cambre 13, B-1050 BRUXELLES.

Institut Géographique National (IGNB) Map sales
Service de Vente, Avenue Louise 306–310, B-1050 BRUXELLES.

Service Géologique de Belgique (SGB)
Rue Jenner 13 (Parc Leopold), B-1040 BRUXELLES.

Centre National de Recherches Géomorphologiques (CNRG)
Géomorphologie et Géologie du Quaternaire, Université de Liège, Place du 20-Aôut 7, B-4000 LIÈGE.

Comité pour l'Etablissement de la Carte des Sols et de la Végétation de la Belgique (CECSVB)
Krijgslaan 281, B-9000 GENT.

Comité National de Géographie
Section de Cartographie, St Denijslaan 76, B-9000 GENT.

Société Royale Belge de Géographie
Campus de la Plaine—CP 246, Boulevard du Triomphe, B-1050 BRUXELLES.

Commission de l'Atlas National
87 av. Ad. Buyl (194/1), B-1050 BRUXELLES.

Société de Developpement Régional pour la Wallonie (SDRW)
Rue Graf 5,
B-5000 NAMUR.

Service de Topographie et de Photogrammetrie
Ministère des Travaux Publics, WTC—Tour 4, Bld. Em
Jacqmain 158, B-1000 BRUXELLES.

Ministère des Finances
Administration du Cadastre, Blvd Pacheco 34, B-1000 BRUXELLES.

Ministère de l'Agriculture
Service Information, Manhatten Center—Office Tower, Avenue
du Boulevard 21, 13eme etage, B-1000 BRUXELLES.

Ministère de la Santé Publique et de l'Environnement
Cite Administrative de l'Etat, Quartier Esplanade 7, B-1010
BRUXELLES.

Institut National de Statistique
Rue de Louvain 44, B-1000 BRUXELLES.

Musée Royale de l'Afrique Centrale (MRAC)
13 Steenweg op Leuwen, B-1980 TERVUREN.

Belfotop-Eurosense NV/SA
Rue J. Vander Vekenstraat, 158, B-1810 WEMMEL.

Geocart
Prins Boudewijnlaan 35, B-2700 SINT-NIKLAAS.

Girault Gilbert
Rue de l'Association 50, B-1000 BRUXELLES.

Uitgeverij J. Van In
Grote Markt 39, B-2500 LIER.

Editions CARTO
Rue Gaucheret 139, B-1210 BRUXELLES.

Catalogue

Atlases and gazetteers

Tweede Atlas van België/Deuxième Atlas de Belgique
Bruxelles: Commission de l'Atlas National, 1983–.

Atlas de Belgique/Atlas van België
Gent: Comité National de Geographie, Section de Cartographie,
1950–77.

Atlas du Survey National, Tome II
Bruxelles: Ministère des Travaux Publics, Administration de
l'Urbanisme et de l'Amenagement du Territoire, 1964–.

Atlas de la Wallonie
Namur: SDRW, 1979–.

General

Le Belgique vue de l'Espace   1 : 350 000
Wemmel: Belfotop Eurosense, 1980.

België: verkeerswegen/Belgique: voies de communication   1 : 300 000
Edition 7
Bruxelles: IGNB, 1973.
3 versions: with provinces tinted, with hypsometric tints, or
without tints.
Road numbers and road distance table.

België/Belgique   1 : 250 000   Series M 534   Edition 3
Bruxelles: IGNB, 1974.
2 sheets, both published.
Legend in 4 languages.
Also available in one-sheet recto-verso.

Topographic

*Topografische kaart van België/Carte topographique de
Belgique*   1 : 100 000
Bruxelles: IGNB, 1986–.
19 sheets, 6 published. ■
Tourist map on the reverse.

België/Belgique   1 : 50 000   Series M 736.
Bruxelles: IGNB, 1969–.
74 sheets, all published. ■

België/Belgique/Belgium   1 : 25 000   Series M 834
Bruxelles: IGMB, 1955–.
238 sheets, all published. ■
Legend in 4 languages.

Geological and geophysical

*Carte géologique et hypsometrique du socle Paleozique de la Belgique
completée par les courbes caracteristiques du Cretace*   1 : 100 000
Bruxelles: SGB, 1952.
10 sheets, all published.

Carte géologique detaillée de la Belgique   1 : 40 000
Bruxelles: SGB, 1893–1911.
226 sheets, 11 published. ■

Carte géologique detaillée de la Belgique   1 : 25 000
Bruxelles: SGB, 1958–.
237 sheets, 9 published. ■
Sheet memoirs in French.

*Geomorfologische kaart van België/Carte géomorphologique de
Belgique*   1 : 25 000
Liège: CNRG, 1965–.
238 sheets, 9 published. ■

Environmental

Carte des sols de la Belgique/Bodemkaart van België   1 : 20 000
Gent: CECSVB, 1958–.
458 sheets, 309 published. ■
Texts in French.

Cartes phytosociologiques/Vegetatiëkaart van België   1 : 20 000
Gent: CECSVB, 1957–.
458 sheets, 27 published. ■
Texts in French or Flemish, according to area.

Administrative

Administratieve kaart/Carte administrative   1 : 500 000
Bruxelles: IGNB, 1983.
Shows boundaries of provinces, arrondissements and communes.

Belgique. Division administrative (1.5.1985)   1 : 350 000
Bruxelles: Editions CARTO, 1985.
Shows boundaries to commune level and lists postcodes.

*België: fusions van gemeenten/Belgique: fusions de
communes*   1 : 320 000
Lier: J. Van In, 1977.
Shows new commune boundaries and language areas.

Administratieve kaart/Carte administrative   1 : 300 000
Bruxelles: IGNB, 1983.
Shows boundaries of provinces, arrondissements and communes.

Belgique administrative   1 : 250 000
Sint-Niklaas: Geocart, 1984.

Belgique: cartes routières régionales par province   1 : 100 000
Sint-Niklaas: Geocart, 1984.
9 sheets, all published.
Series of provincial maps.

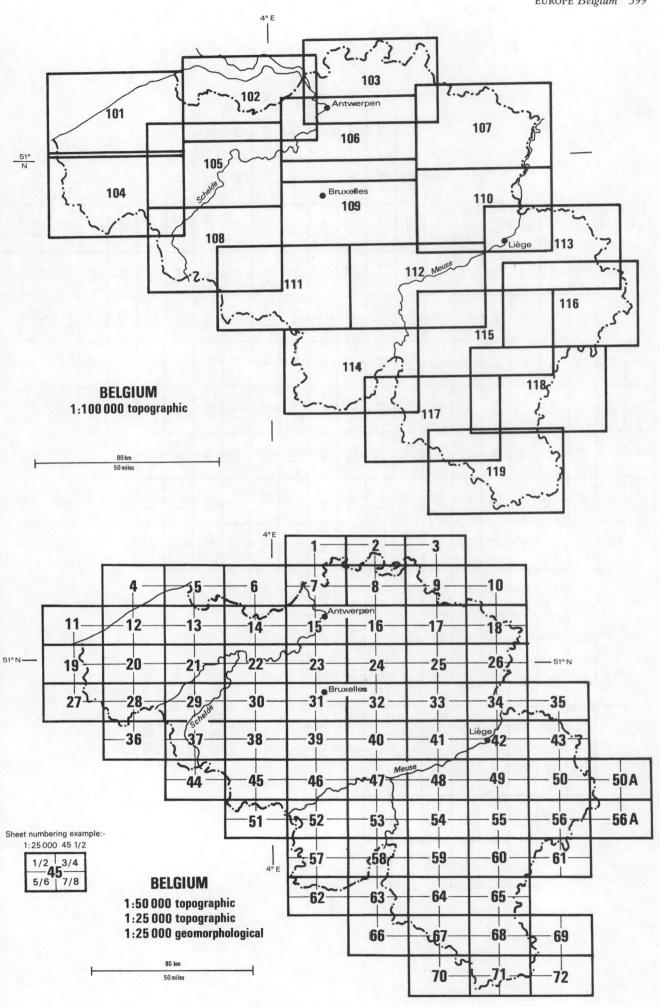

BELGIUM
1:100 000 topographic

80 km
50 miles

Sheet numbering example:-
1:25 000 45 1/2

1/2	3/4
45	
5/6	7/8

BELGIUM
1:50 000 topographic
1:25 000 topographic
1:25 000 geomorphological

80 km
50 miles

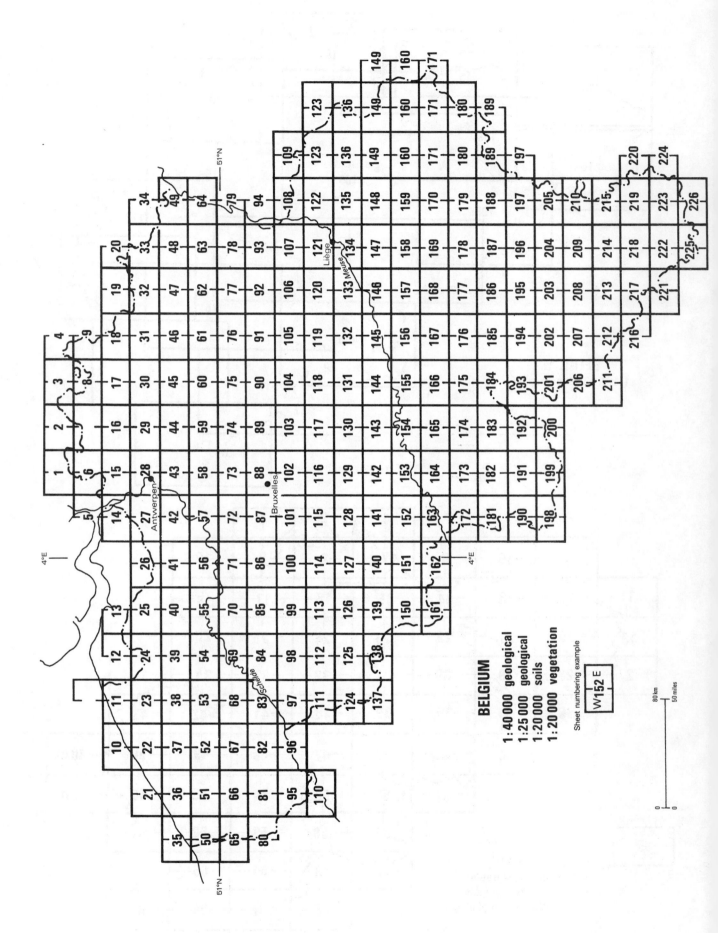

BELGIUM

1:40 000 geological
1:25 000 geological
1:20 000 soils
1:20 000 vegetation

Sheet numbering example

W 152 E

Human and economic

Atlas économique de la Belgique   Christian Vandermotten
Bruxelles: Société Royale Belge de Géographie, 1983.
pp. 112.

*Atlas statistique du rencensement de la population et des logements
1981. Parti 1. Données demographiques.*
Bruxelles: Institut National de Statistique, 1983.

*Belgïe. Kaart van de Verkeerswegen/Belgique. Carte des voies de
communications*   1:420 000   1st edn.
Bruxelles: IGNB, 1986.
For route planning to avoid traffic congestion.

Carte des régions agricoles de la Belgique   1:300 000
Bruxelles: Ministère de l'Agriculture, 1977.

Voies navigables de Belgique   1:125 000
Sint-Niklaas: Geocart, 1984.
Set of 14 sheets.

Town maps

Bruxelles et grande banlieue   1:17 500
Sint-Niklaas: Geocart, 1984.
196 pp. bilingual index.
Postal codes, mass transit network.

Atlas géant du Grand Bruxelles   1:15 000
Bruxelles: Girault Gilbert.

Plan-guide de Bruxelles   1:10 000
Bruxelles: Girault Gilbert.
Separate editions in English, German and Dutch also available.

Bruxelles centre   1:7750
Sint-Niklaas: Geocart, 1984.
Bilingual, with index.

Bulgaria

The national mapping authority in Bulgaria is the General
Board of Geodesy and Cartography (**Glavno Upravlenie
po Geodezia i Kartografia y Kadaster—GUGKK**), Sofia.

Indigenous modern maps of Bulgaria, and information
about them, have proved difficult to obtain, and the slim
text and even slimmer catalogue entries presented here
probably do not reflect the magnitude of mapping
activities, although they do reflect the public availability of
such data. Bulgaria did not respond to the United Nations
questionnaires of 1976 or 1980.

A modern topographic map series at 1:25 000 on the
Gauss–Krüger projection was initiated after 1930, and
continued after the socialist revolution in 1944, although
we do not know whether or when this series was
completed. Derived maps have been produced at 1:50 000
and 1:100 000. The existence of large-scale mapping at
1:5000 and 1:10 000 has also been reported.

A number of thematic maps are also reported to have
been published, including soils, geology, geomorphology
and forest maps at 1:200 000, and soils maps at 1:25 000.
Much effort has also gone into the compilation of complex
regional atlases and monothematic atlases of climate,
hydrology, etc. However, virtually none of this material is
listed in publishers' or retailers' catalogues in the West.

Maps for general, educational, and tourist use are
published by the Institute of Cartography (**Kartproekt**).
About 120 such maps have been published, at a variety of
scales and some of them in several languages. However, no
information has been supplied by this institute, and it
appears to be difficult to obtain a regular supply of these
maps in the West.

Soil mapping is carried out by the **Institut N.
Poushkarov**, but we have no further information.

The **Geological Institute 'Strashimir Dimitrov'** of the
Bulgarian Academy of Sciences has published geological,
tectonic and engineering geological maps of Bulgaria at
scales of 1:1 000 000 or 1:500 000, but we are informed
that all these maps are currently out of print.

The richest source of publicly accessible mapping is
contained in the national atlas, *Atlas Narodna Republika
Balgarija*, published in 1973 following 10 years of
preparation. The atlas is the result of cooperative work

between GUGKK, the Bulgarian Academy of Sciences, the
Institute of Cartography and various academic and
scientific institutions. The work has 168 pages of maps,
including 253 maps in all, with a wide range of themes
covering the resource base through economic to social and
political issues. Most maps are at 1:1 000 000 or
1:1 500 000 scale. The foreword and contents are given in
English and Russian as well as Bulgarian. This atlas is still
readily available.

Small-scale maps of the country are issued by several
foreign publishing houses, including **Karto+Grafik**,
Cartographia and **GUGK**.

Addresses

**Glavno Upravlenie po Geodezia, Kartografia y Kadaster
(GUGKK)**
U1. Musala 1, SOFIA 18.

Kartproekt
Bul. 9 Septemvri 291, SOFIA.

Geologicheski Institut 'Strashimir Dimitrov'
Bulgarska Akademija na Naukite, Str. Akademik G. Bonchev,
Blok II, SOFIA 1113.

Committee on Geology
Ministry of Mineral Resources, 22 Georgi Dimitrov Blvd, SOFIA.

Institut N. Poushkarov
Chaussee Bankya 5, 3 Iskar Str. SOFIA.

For Karto+Grafik, see German Federal Republic; for CIA, see
United States; for Cartographia, see Hungary; for PPWK, see
Poland; for GUGK, see USSR.

Catalogue

Atlases and gazetteers

Atlas Narodna Republika Bâlgarija
Sofia : GUGKK/Bâlgarska Akademija na Nautike, 1973.
pp. 186

Pâtno-turisticeski atlas na Bâlgarija
Sofia: Kartproekt, 1981.
pp. 182.
Road atlas of Bulgaria   1 : 500 000.

General

Bulgarien, Rumänien   1 : 2 000 000
Frankfurt AM: Karto+Grafik, 1983.

Bulgaria   1 : 1 000 000
Washington, DC: CIA, 1972.
Small inset maps of population/administration divisions, economic activity and land utilization.

Bulgaria   1 : 1 000 000
Warszawa: PPWK.

Bolgarija   1 : 750 000
Moskva: GUGK.

Bâlgarija—patna karta   1 : 600 000
Sofia: Kartproekt, 1985.
Road map with distance charts and town maps on reverse.

Bâlgarija—turisticeska karta   1 : 500 000
Sofia: Kartproekt, 1984.

Bulgarien Touristische Karte   1 : 500 000
Sofia: Komitée für Architektur und Städteplanung, 1977.

Administrative

Teritorialno-administrativno ustrojstvo NR Bâlgarija   1 : 500 000
Sofia: Kartproekt, 1979.

Town maps

Sofia
Budapest: Cartographia/Hamburg: Falk, 1983.
Street index on the reverse.

Czechoslovakia

Czechoslovakia has a long history of topographic mapping beginning with the so-called 'Josephinian mapping' of the eighteenth century at 1 : 28 000 scale. Modern topographic mapping is the responsibility of the Central Office of Geodesy and Cartography (**Utstredni Sprava Geodézie a Kartografie, USGK**), Prague, but is shared also with the **Slovak Office of Geodesy and Cartography (SUGK)** at Bratislava. Maps are on a transverse conformal cylindrical Gauss projection, with sheets based on subdivisions of the IMW system. The country is completely covered by modern contoured maps at 1 : 25 000, 1 : 50 000, 1 : 100 000 and 1 : 200 000. These are coloured maps with the contour interval ranging from 10 m on the larger scales to 40 m on the smaller. A 1 : 10 000 national map is also in progress with a 2 m contour interval, and there are also 'technical-economic' maps at scales of 1 : 1000, 1 : 2000 and 1 : 5000. These topographic maps and plans are not, however, available to the general public or for export.

Czech cartography has moved into the automated era, and for large-scale mapping the Digicart interactive mapping system (used also in other East European countries) is in operation.

Two state publishing houses are dominant in the production of maps for public and educational use. These are **Geodeticky a Kartograficky Podník v Praze (GKP)** in Prague and **Slovenská Kartografia (SK)** in Bratislava. The former is currently publishing a new series at 1 : 200 000 entitled *Getting to know Czechoslovakia*, with legends and tourist information in Czech, Russian, German, French and English, and also publishes town maps, and recreation maps at scales mainly of 1 : 100 000 or 1 : 50 000. SK also publishes a wide range of maps for public and educational use including road maps and atlases, town maps of Slovakia, and local recreation maps. Various kinds of thematic maps—on geological, transport, geobotanical and ecological topics—are also published by GKP and SK, usually on behalf of other organizations. There are two export agencies for maps: **Artia** in Prague and **Slovart** in Bratislava.

The National Geological Institute (**Utstrední Ustav Geologicky, UUG**), Prague, was founded in 1918. New geological mapping at 1 : 200 000, with separate editions for Quaternary and bedrock geology, was undertaken in the 1960s and completed in 1969, and currently progress is being made with a 1 : 25 000 scale series. In Slovakia, a state geological institute was founded in 1940 (now **Geologicky ústav Dionyz Stúr, GUDS**), taking over the responsibility of geological mapping in the Slovak Socialist Republic from UUG. Geological and mineral resource maps at 1 : 200 000 were published in the 1960s. There are a number of smaller-scale maps (1 : 500 000 and 1 : 1 000 000), some of which are published in separate English editions.

Vegetation mapping in Czechoslovakia dates from 1947, but the present programme was initiated by the Botanical Institute of the Czech Academy of Sciences in 1954. Mapping of Bohemia took place during 1955–60 while that of Slovakia began in 1962, carried out by the Botanical Institute of the Slovak Academy of Sciences. The mapping, compiled at 1 : 75 000 or 1 : 50 000, was generalized to 1 : 200 000 for publication. Small-scale general maps have also been published such as the 1 : 1 000 000 map of reconstructed natural vegetation in the CSR by Moravec and Neuhäusl.

The first National Atlas was published in 1935, and the second, and current, *Atlas Ceskoslovenske Socialisticke Republiky* is a bound volume formulated to represent the 'contemporary, scientifically generalized geographical knowledge of the nature, population and national economy of Czechoslovakia'. Work on the atlas was initiated in 1952 by the **Czechoslovak Academy of Sciences (CSAV)** and it was published jointly with USGK in 1966. Included in 58 folios are 433 maps with texts in Czech and summaries in English and Russian. Most maps are at the scale of 1 : 1 000 000.

More recently, a regional atlas of Slovenia has been published, *Atlas Slovenskej Socialistickej Republiky*. This was produced jointly by the Slovak Academy of Sciences and the Slovak Office of Geodesy and Cartography, Bratislava. The atlas was published in 1980 and includes more than 875 maps at scales of 1 : 500 000 or smaller, and covering environmental, socioeconomic and industrial subjects. It has been issued in both bound and loose formats.

Mapping of 1980 census data of the CSR at a principal scale of 1:500 000 has been carried out by the Czech Academy of Sciences (CSAV) using digital techniques, and published in an atlas of 30 map sheets in the Czech language.

Further information

A catalogue of Czechoslovak publications and maps is issued by the Geological Survey, Prague, and the two state publishing houses, GKP and SK, issue catalogues.

Some aspects of mapping progress in the country are discussed in *Cartography in the Czechoslovak Socialist Republic*, published by the National Cartographic Committee of the Czechoslovak Scientific and Technical Society, 1982.

Progress in vegetation mapping is described by Neuhausl, R. (1982) Die Vegetationskarte der CSSR 1:200 000 und ihre Interpretation, *Arch. Naturschutz und Landschaftsforschung, Berlin*, **22**, 145–150.

Addresses

Utstredni Sprava Geodézie a Kartografie (USGK) (Czech Office of Geodesy and Cartography)
CS-110 00 PRAHA 1, Nove Mesto, Hybernská 2.

Geodetický a Kartograficky Podnik v Praze (GKP)
CS-170 30 PRAHA 7, Kostelní 42.

Ustredni Ustav Geologicky (UUG) (Czech Geological Survey)
CS-118 21 PRAHA 1, Malostranské nám. 19.

Geograficky Ustav CSAV (CSAV) (Czech Academy of Sciences)
Pracoviste Praha, CS-120 00 PRAHA, Wenzigova 7.

Artia Czech map exports
Foreign Trade Corporation, CS-111 27 PRAHA 1, Ve Smečkach 30.

Slovak Office of Geodesy and Cartography (SUGK)
CS-BRATISLAVA, Besrucová 7.

Slovenská Kartografia NP (SK)
nositel' Radu Práce, CS-834 07 BRATISLAVA, Pekná cesta 17.

Geologicky ústav Dionyz Stúr (GUDS) (Slovak Geological Survey)
CS-814 73 BRATISLAVA, Dúbravská cesta 9.

Slovart AG Slovak exports
CS-814 73 BRATISLAVA, Gottwaldova nám. 6.

Catalogue

Atlases and gazetteers

Atlas Ceskoslovenska Socialisticke Republiky
Praha: Ceskoslovenska Akademie Ved/USGK, 1966.
58 pp. map folios + 9 pp. names index.

Atlas Slovenskej Socialistickej Republiky
Bratislava: Slovenska Academia Ved/SUGK, 1980.
pp. 322 in 15 sections.
An English text is available (1983).

Atlas CSSR 8th edn
Bratislava: SK, 1984.
pp. 56.

Vzitá ceská vlastní jména geografická I. Caslavka *et al.*
Praha: USGK/Bratislava: SUGK, 1981.
pp. 210.
UN Gazetteers of Geographical Names: CSSR.

General

Ceskoslovenská Socialistická Republika. Všeobecnozemepisná map 1:500 000
Bratislava: SK, 1986.

Ceskoslovenská Socialistická Republika 1:1 000 000
Praha: GKP, 1983.

Automapa CSSR 1:800 000 7th edn
Praha: GKP, 1983.

Soubor map 'Poznáváme Svet' Ceskoslovensko 1:750 000 6th edn
Praha: GKP, 1972.
100 pp. text with separate physical and administrative maps and several smaller-scale thematic maps.

Topographic

Poznáváme Ceskoslovensko 1:200 000
Praha: GKP, 1983–.
17 sheets, 4 published. ■
Tourist map with 5-language legend.
Also available as *Automapa* without administrative boundaries or tourist overprint.

Geological and geophysical

Geomorfologické členenie SSR a CSSR 1st edn
Bratislava: SK, 1985.
Geomorphological divisions of the SSR and CSSR.

Preledná geologická mapa CSSR 1:1 500 000 6th edn
Praha: UUG, 1982.

Geologická mapa CSSR, odkrytá 1:1 000 000 2nd edn
Praha: UUG, 1982.
Bedrock geological map.
Available also in English edition.

Geologická a paleogeografická mapa karbonu a permu CSSR 1:1 000 000
Praha: UUG, 1981.
Geological and palaeogeographical map of the Carboniferous and Permian of Czechoslovakia.
Available also in English edition.

Lázne, zridla a minerálni premeny CSSE 1:1 000 000 2nd edn
Praha: UUG, 1982.
Maps of spas and mineral springs.

Mapa minerálnych vod CSSR 1:500 000
Bratislava: GUDS.

Tektonická mapa CSSR 1:500 000
Bratislava: GUDS.

Geologická mapa kvartéru Slovenska 1:500 000
Bratislava: GUDS.

Metalogenetická mapa CSSR 1:500 000.
Praha: UUG, 1981.
2 sheets, both published.
48 pp. text also available.

Map of mineral waters in Czechoslovakia 1:500 000
Bratislava: GUDS, 1983.

Geologická mapa CSSR 1:200 000
Praha: UUG/Bratislava: GUDS, 1961–.
35 sheets, *ca* 11 published. ■

Environmental

Geobotanická mapa Ceské Socialistické Republiky. Mapa rekonstruované prirozené vegetace 1:1 000 000
Praha: GKP, 1976.
In Czech and English.

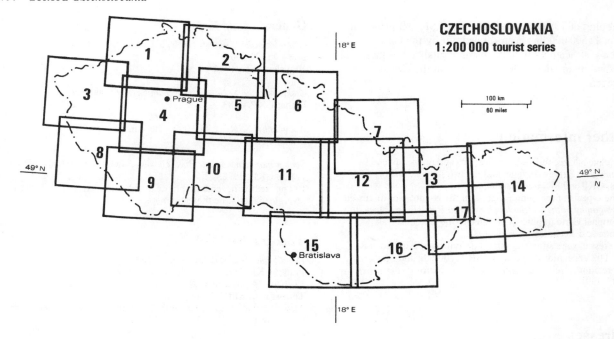

CZECHOSLOVAKIA
1:200 000 tourist series

100 km
60 miles

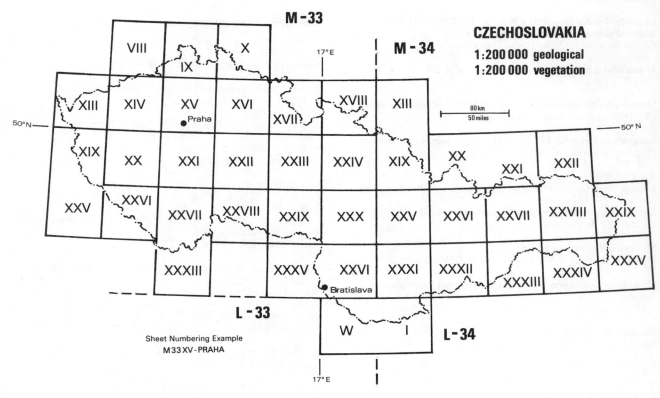

M-33

M-34

CZECHOSLOVAKIA
1:200 000 geological
1:200 000 vegetation

80 km
50 miles

L-33

Sheet Numbering Example
M 33 XV - PRAHA

L-34

Chránená územi prirody CSSR 1:750 000
Praha: GKP, 1982.
Map of nature reserves.

Soubor map fyzicko geografické regionalizace CSR 1:500 000
Brno: Geograficky ústav CSAV, 1971–.
Series of maps of physico-geographical regionalization of the CSR.
In Czech and English.

Geobotanická mapa CSSR 1:200 000
Praha: GKP, 1968.
37 sheets, 21 published. ■

Administrative

Ceskoslovenská Socialistická Republika: Politická mapa 1:1 500 000 3rd edn.
Bratislava: SK, 1986.

Slovenská Socialistická Republika spravne rozdelenie 1:500 000
Bratislava: SK, 1986.
Administrative divisions and postcodes.
With separate index of places.

Human and economic

Atlas ze Scitaní Lidu, Domu a Bytu 1980, CSR
Praha: Geograficky ústav CSAV, 1985.
Atlas of population and housing 1980 of the CSR.
In Czech.

Town maps

Praha 1:20 000 3rd edn
Praha: GKP, 1983.
81 pp. maps + 57 pp. tourist information.
In 5 languages.

Praha, plan stredu mesta 1:15 000 4th edn
Praha: GKP, 1983.
Map of city centre.

Bratislava, orientacná mapa 3rd edn
Bratislava: SK, 1985.

Bratislava, stred mesta 1st edn
Bratislava: SK, 1983.
Map of city centre.

Denmark

Topographic mapping in Denmark is the responsibility of the **Geodætisk Institut (GID)**, Copenhagen, which was established in 1928 by bringing together the Topographic Branch of the army with *Den Danske Gradmaaling*. GID undertakes all official geodetic and topographic work, and has also carried out the mapping of Greenland and the Faeroe Islands. Cadastral mapping is administered separately by the Land Registry (**Matrikeldirektoratet**).

Early detailed topographic survey, begun in the nineteenth century, was carried out by plane table methods, and a basic scale series was established at 1:20 000 with a derived series at 1:40 000. These so-called *Maalebordsblade* remained the standard map series until the 1970s, but in 1966 a new photogrammetric survey was initiated, and this led to the production of modern, high-quality series at 1:25 000 and 1:50 000, which have replaced the plane table maps. The projection was also changed from the Lambert conformal conic to UTM. Some of the sheets in these series, although on new sheet lines, are still based provisionally on the old mapping pending completion of the photogrammetry. The 1:25 000 basic series is in four colours with UTM grid and a 2.5 m contour interval, while the 1:50 000 map has a similar specification but with 5 m contours. The planned revision cycle for both series is about 10 years. The new survey will be fully completed in the early 1990s. There are also modern series at 1:100 000 (with a 5 year revision cycle) and a touring map in four sheets at 1:200 000. There are several regional maps compiled from the 1:100 000 series, and a special map of Bornholm. GID is in the process of creating a digital elevation model of Denmark.

GID also publishes the *Topografisk Atlas*, a bound set of the standard 1:100 000 topographic map with a 22 000 entry place name index and additional maps of the new road numbering system and the *amt* and *politikredse* boundaries. The 1:200 000 *Færdselskort* is also available as a map book with a 12 000 entry index. GID has developed a computer file of 120 000 topographic names which is used for automatic typesetting and for production of the index of the *Topografisk Atlas*.

The Land Registry comes under the Ministry of Agriculture. Cadastral maps of the whole of Denmark are maintained at a variety of large scales. Since 1960, photogrammetric methods have been in use, and currently the registry is experimenting with orthophoto mapping and with a digital cadastral mapping programme.

The *Atlas over Danmark* is a long-term National Atlas project undertaken by the Royal Danish Geographical Association (**Det Kongelige Danske Geografiske Selskab**, **DKDGS**. The first series (Serie I), originally intended to be in five volumes, was not completed: two volumes were published, the first (on landscape types) in 1949 and the second (on population) in 1960. Only Volume 2 remains in print. A second series commenced in 1976, and four volumes have so far been published. The three in-print volumes are listed in our catalogue. Volume 2 *Topografisk atlas Danmark* is out of print. A fifth volume is planned. The *Atlas over Danmark* is distributed by **CA Reitzels Forlag**.

Geological mapping is the responsibility of **Danmarks Geologiske Undersøgelse (DGU)**, Copenhagen. DGU was founded in 1888 and since then has compiled a series of published maps at 1:100 000 showing surface geology at 1 m depth. These maps are therefore records mainly of Quaternary deposits rather than bedrock geology. Fifty per cent (35 sheets) of the country has been published in this series with completion of the field survey for a further 10 sheets, but most sheets are now out of print. Following proposals made in 1977 by a working group on the future of geological mapping in Denmark, a new series of Quaternary and pre-Quaternary maps is to be published at 1:50 000 on the sheet lines of the topographic series. The first sheet was due to be published in 1986. Preliminary maps have been prepared at 1:25 000 from field survey and archival material and are available as black-and-white copies. In fulfilment of the requirements of the 1975 Water Act, and of other land use planning requirements, DGU has also compiled hydrochemical, hydrogeological and mineral resource maps, and cyclogram maps summarizing borehole information. Most of these maps are issued in short print runs and are for use by people working in water planning. They are not on sale to the public. A general geomorphological/peat deposits map at 1:100 000 has also been initiated. The first sheet is in colour, but the remainder will probably be in black and white. Publications of the DGU are also distributed by CA Reitzel.

A major soil survey of Denmark was begun in 1975 by a newly created branch of the Ministry of Agriculture called the Bureau of Land Data (**Arealdatakontoret, ADK**), based at Vejle. Soils were sampled at 35 000 sites. By 1980, the *Basisdatakort* 1:50 000 was complete. This is a colour map which classifies agricultural land into eight mapping units based on soil texture at 0–20 cm depth. Steep slopes which present difficulties for mechanized tillage are shown by hatching, and dominant geology at 1 m depth is indicated by symbols placed in each 25 ha grid cell. The maps are litho-printed on a base map reduced from the GID 1:25 000 topographic series. All soil data has also been placed in a digital database, which permits both the statistical manipulation of the data, and the production of 'customized' digital maps. Further survey data placed on file since 1980 has made it possible to produce soil maps combining many different attributes. In addition, monothematic mapping of soil drainage characteristics and of the distribution of acid sulphate soils is under way. A further product of ADK has been a nine-category landscape map of Denmark at 1 : 100 000 based on geomorphological criteria.

Large-scale urban mapping is undertaken by the municipal authorities of Copenhagen and Frederiksberg.

Among commercial publishers of maps are **Politikens Forlag**, whose cultural and travel guides include city maps, and who publish a road map of Denmark, and **Krak**, who have produced four books of town maps (one of Jutland, one of the islands, one of Århus, and one of Copenhagen) and a recent new map of Copenhagen in folded format. The educational publishers **Geografforlaget**, Brenderup, have some rather unusual maps on their list, including a four-sheet geomorphological map of Denmark, outline maps showing the boundaries and names of communes, and a tourist map of North and South Schleswig which has history and geology as its dual themes.

Further information

GID issues a detailed map catalogue, available also in an English language version, which includes indexes and extracts of map series. In 1986, DGU was in process of publishing a new catalogue of its maps and monographs.

Other sources of information include Gertsen, W.M. (1970) Danish topographic mapping, *Cartographic Journal* **7**, 113–121; Madsen, H.B. (1984) Soil mapping in Denmark, *Soil Survey and Land Evaluation*, **4**, 57–62; and Sørensen, H. and Nielsen, A.V. (1978) Den geologiske kortlægning af Danmark—Den hidtidige kortlægning og den fremtidige (The geological mapping of Denmark—present and future), *DGU Serie A*, 80 pp.

Addresses

Geodætisk Institut (GID)
Rigsdagsgarden 7,
DR-KØBENHAVN K.

Danmarks Geologiske Undersøgelse (DGU)
Thoravej 31, DK-2400 KØBENHAVN NV.

Det Kongelige Danske Geografiske Selskab (DKDGS)
Øster Voldgade 10, DK-1350 KØBENHAVN K.

Arealdatakontoret (ADK)
Landbrugsministeriet, Enghavevej 2, DK-7100 VEJLE.

Skovstyrelsen
Strandvejen 863, DK-2930 KLAMPENBORG.

Matrikeldirektoratet
13 Titangade, DK-2200 KØBENHAVN N.

CA Reitzels Forlag A/S
Nørregade 20, DK-1165 KØBENHAVN K.

Geografforlaget
Fruerhøjvej 43, DK-5464 BRENDERUP.

Politikens Forlag A/S
Vestergade 26, DK-1456 KØBENHAVN K.

Krak
Nytorv 17, DK-1450 KØBENHAVN K.

Catalogue

Atlases and gazetteers

Danmark 1 : 100 000 Topografisk Atlas
København: GID, 1982.
pp. 161.

Atlas over Danmark Serie I, Bind 2, Befolkningen (The Population) A. Aagensen
København: DKDGS, 1961.
Separate atlas and text volumes.

Atlas over Danmark Serie II, Bind 1 Opgivne og tilplantede landbrugsarealer i Jyland K.M. Jensen
København: DKDGS, 1976.
pp. 48 + 5 plates.

Atlas over Danmark Serie II, Bind 3 Danske byers vækst C.W. Matthiessen
København: DKDGS, 1985.
pp. 163 including 18 plates.

Atlas over Danmark Serie II, Bind 4 Landbrugsatlas Danmark K. M. Jensen and A. Reenberg
København: DKDGS, 1986.
pp. 120.

General

Danmark 1 : 1 000 000
København: GID, 1974.
2-colour base map.

Danmark 1 : 750 000
København: GID, 1982.
Coloured road map.
Available with/without UTM grid.

Danmark 1 : 500 000
København: GID, 1981.
Coloured road map.
Insets of Faeroes and Greenland.

Danmark 1 : 300 000
København: GID, 1986.
4 sheets, all published.

Danmark Færdselskort 1 : 200 000
København: GID, 1985.
4 sheets, all published.
Coloured touring map.
Also available as 36 pp. book with index.

Topographic

Danmark 1 : 100 000
København: GID, 1977–.
33 sheets, all published. ■
Folded edition has UTM grid.

Danmark 1 : 50 000
København: GID, 1966–.
110 sheets, all published. ■
UTM grid.

Danmark 1 : 25 000
København: GID, 1952–.
405 sheets, all published. ■
UTM Grid.

Geological and geophysical

Landskabskort over Danmark 1 : 350 000 Per Smed
Brenderup: Geografforlaget, 1979–82.
4 sheets, all published.
Geomorphological map.
Legend in Danish and English.

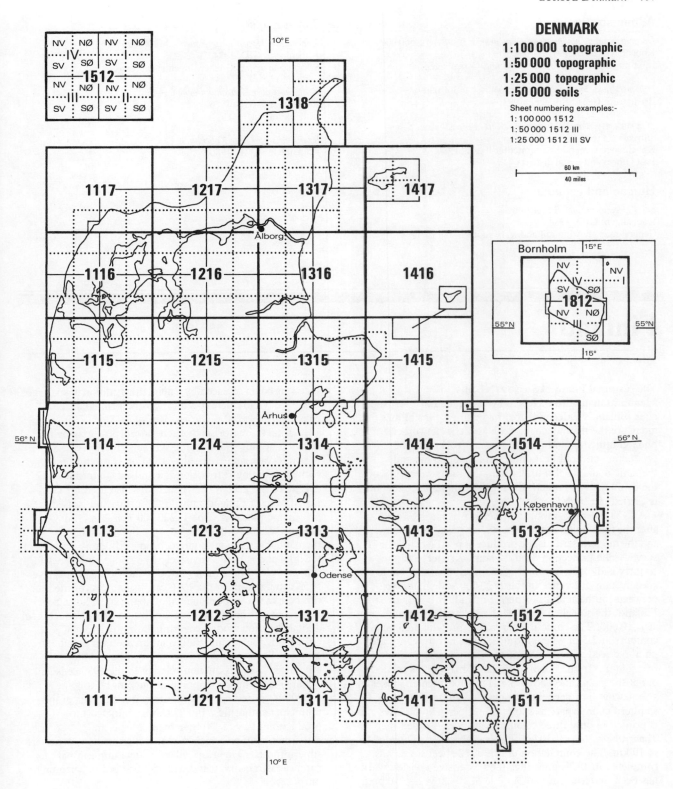

DENMARK

1:100 000 topographic
1:50 000 topographic
1:25 000 topographic
1:50 000 soils

Sheet numbering examples:-
1:100 000 1512
1:50 000 1512 III
1:25 000 1512 III SV

60 km
40 miles

Geologisk kort over Sønderjylland 1:125 000 A. Jessen
København: DGU, 1935.
With 8 pp. text.

Geologisk kort over Danmark 1:100 000
København: DGU, 1893–.
Most sheets out of print.

Geomorfologisk/blødbundskort 1:100 000
København: DGU, 1984–.
33 sheets, 1 published.

Geologisk kort over Danmark 1:50 000
København: DGU, 1986–.
110 sheets, 1 published. ▪

Environmental

Jordbundskort over Danmark 1:500 000 C.H. Bornebusch and
K. Milthers
København: DGU, 1935.
Soil map of Denmark.
Text out of print. Map reprinted 1970.

Skovene i Danmark—Skovregistreringen 1:500 000
Vejle: ADK/Klampenborg: Skovstyrelsen, 1985.
Forested areas.

Jordklassificering Danmark: Basisdatakort 1:50 000
Vejle: ADK, 1975–.
400 sheets, all published.
Classification of soils by surface texture.

Administrative

Danmark amter, kommune og sogne m.v. 1.1.1981   1 : 500 000
København: GID, 1981.
Inset maps of Faeroes, Greenland and Copenhagen.

Danmark med amts- og kommunegrænser   1 : 300 000
Brenderup: Geografforlaget, 1983.

*Danmark, Grønland og Farøerne—Kort over posthuse og post
distrikter*   1 : 610 000
København: Generaldirektoratet for P & T, 1986.
Post Offices and postal districts.

Human and economic

DSB Danmarkskort   1 : 750 000
København: GID, 1984.
General map showing rail and ferry routes.

Town maps

København. Hovedstadsregionen   1 : 100 000
København: GID, 1984.
Small-scale map of Greater Copenhagen region.

Kraks kort over København og omegn '86
København: Krak, 1986.
In book format.

Kraks turistkort over København
København: Krak, 1986.

Finland

The **National Board of Survey (NBS),
Maanmittaushallitus**, Helsinki was founded in 1919, soon
after Finland's independence. Its main responsibilities
today are the maintenance of the geodetic network,
photogrammetric and remote sensing work, and the
production and publication of both topographic and
thematic maps. It also undertakes the cadastral survey of
the country. Thus practically all Finnish map production
is carried out by NBS.

A 1 : 20 000 topographic map series was initiated soon
after the foundation of the Survey, but progress was slow,
and the modern series is the product of post-World War II
survey. The projection used is Gauss–Krüger, and the
country is divided into four zones, each 3° wide. There is a
specification difference in the 1 : 20 000 scale mapping
between southern and northern Finland. In southern
Finland, the so-called *Basic Map* is prepared in manuscript
at 1 : 10 000. The contour interval is 5 m, with auxiliary
contours at 2.5 m where required. The fair drawings are
photoreduced to 1 : 20 000 for publication and printed in
five colours. In the north of Finland, mapping is carried
out at the publication scale (i.e. 1 : 20 000), which makes it
a somewhat more generalized map. In contrast to its
southern counterpart, it shows no property boundaries and
is printed in only three colours. This map is called the
Topographic Map. 1 : 20 000 sheets each cover an area of 10
× 10 km. The compilation of this 3712-sheet series was
completed in 1977, and since then extensive revision work
has been undertaken, with a cycle of 10–20 years. Mapping
is based largely on photogrammetric methods, and there is
a regular programme of new aerial photography flown for
revision purposes. However, the mean age of the Basic
Map is 9 years and of the Topographic Map 14, and so
there are plans to accelerate this revision programme.

The 1 : 50 000 and 1 : 100 000 scale series are produced
respectively by photoreduction and generalization of the
1 : 20 000 map. Complete published cover of the country at
1 : 50 000 was achieved in 1986. Sheets in this series are in
five colours and each covers an area of 20 × 30 km. The
1 : 100 000 series, completed in 1982, is printed in seven to
nine colours with each sheet covering an area of 30 × 40
km.

The General Map at 1 : 400 000, currently available in 31

sheets, is being replaced by a new, third edition of six large
sheets drawn to a new specification and printed in six
colours. The 1 : 200 000 *Road Map of Finland GT* is an
adaptation for motorists of a general-purpose map by the
addition of tourist and traffic symbols. It is revised at 3–5
year intervals, and particular attention is given to this
regular updating. Projection is Lambert conformal conic
and there is a grid system which was also used for the
Postal Map of Finland (now out of print) and is used for
placename indexing. The 1 : 200 000 map has been
designed also as a base map for the overprinting of
thematic information (a monochrome edition is available
for this) and is used extensively by various authorities for
this purpose. Population grid, dot and isoline maps have
been generated from the 1970 census and overprinted on
this base map using automated cartography. The map's
20 m contours have also been digitized for the production of
digital elevation models covering the entire country.

In 1981, the administrative subdivisions of about 350
municipalities were digitized, and will be used for
automated map production relating to the 1980 census.

Large-scale urban mapping in the range 1 : 500 to
1 : 2000 is carried out by the survey departments of the
municipal authorities. In 1979, NBS initiated the
production of a 1 : 5000 base map. This is an
orthophotomap on which digital real estate boundaries are
superimposed. This map will provide a uniform base for
rural cadastral survey and will also be used for planning
purposes.

Recreational mapping is also carried out by NBS, and
there is an extensive range of local touring maps (scale
range 1 : 100 000 to 1 : 400 000) and hiking maps (scales
1 : 20 000 to 1 : 100 000).

The NBS has been steadily automating its mapmaking
processes and is also working towards greater cooperation
with other government bodies which hold files of spatially
related data, with the aim of establishing a unified land
information system. At the time of writing, a long-term
plan for cartographic activities up to the year 2000 is in
preparation.

The **Geological Survey of Finland (Geologian
Tutkimuskeskus, GT)** is a department of the Ministry of
Trade and Industry and is responsible for geological,

geophysical and geochemical investigations. Mapping has been carried out since the 1870s, and current series include detailed maps of bedrock geology, of Quaternary deposits and of geophysical data. The latter are particularly important since only 3–5 per cent of bedrock is exposed in Finland. Regional geochemical maps are also published covering the whole country. In 1979, an agreement was signed between GT, NBS and the Agricultural Research Centre to collaborate in the production of an important new series of 1:20 000 scale Quaternary deposit maps. This mapping is being carried out concurrently with the revision of the Basic Map. By 2006, a six-colour (A-type) map will cover an extensive area of southern Finland, while a five-colour (B-type) map will cover the rest of southern Finland as well as the county of Oulu and part of Lapland. The rest of Lapland will be covered with 1:50 000 scale mapping (designated C- and D-type). The mapping is being done cooperatively by NBS and GT, and the sheets have an explanatory text printed on the reverse. Geological maps and accompanying notes are distributed by NBS, but the geophysical and geochemical maps should be ordered from GT.

Finland prides itself on having produced the first National Atlas (in 1899), and since then four further editions have appeared, culminating in the fifth edition, the first folio of which was issued in 1977 and which is planned for completion in 1990. The complete work will include over 3000 thematic maps and diagrams organized into 26 folios with closely integrated texts, available also in English and Swedish translation. The *Atlas of Finland (Suomen Kartasto)* is published jointly by the Geographical Society of Finland and NBS, and is supervised by an editorial committee under the chair of Professor Stig Jaatinen of the University of Helsinki. The aim of the project is to provide a comprehensive view of the physical and biological features of the land, the population, economic life, society and culture.

There are also many excellent single-sheet thematic maps of Finland, mostly at the 1:1 000 000 scale. These are prepared by various government departments, but for the most part are published and distributed by NBS.

The atlas *Fennia Suuri-kartasto* is a joint publication of Weilin+Goos and NBS. It utilizes the 1:200 000 GT Road Map reduced to 1:250 000 and supplemented with maps of town centres and an index of 9000 placenames.

Further information

NBS issues a folder of index maps showing the status of topographic mapping which is updated annually. NBS also supplies indexes of geological map cover, and a number of brochures (in English) about its mapping activities, including one on *Maps and information supply* and a generously illustrated one on the *Atlas of Finland*.

The Atlas of Finland is also described in Jaatinen, S. (1982) The National Atlases of Finland. The fifth edition, its background and structure, *GeoJournal*, **6**, 201–208.

The Geological Survey publishes a publicity brochure in English and a series of map indexes (also available from NBS).

Addresses

National Board of Survey (Maanmittaushallitus) (NBS)
Map Centre Pasila, Opastinsilta 12, B, Box 85, SF-00521 HELSINKI 52.

Geological Survey of Finland (Geologian Tutkimuskeskus) (GT)
Kimiehentie 1, SF-02150 ESPOO.

Catalogue

Atlases and gazetteers

Suomen kartasto/Atlas över Finland
Helsinki: NBS, 1976–.
5th edition of National Atlas in 26 folios.

Fennia: suuri Suomi-kartasto/Finland in maps   1:250 000
Helsinki: NBS/Weilin+Goos, 1979.
pp. 224.

Suomi Avarudesta
Helsinki: Tähtitieteellinen yhdistys Ursa, 1984.
pp. 176.
Satellite image atlas of Finland.
In Finnish only.

General

Maanteiden yleiskartta/General road map/Översiktskarta över landsvägarna 1986–87   1:1 500 000
Helsinki: NBS, 1986.

Suomi korkeusvyöykekartta/Finland höjdskiktkarta   1:1 000 000
Helsinki: NBS, 1985.
Relief map.

Suomi/Finland   1:1 000 000
Helsinki: NBS, 1975.
IMW-style with hypsometric relief.
General maps with hypsometric relief also available at 1:4 500 000, 1:3 000 000 and 1:2 000 000.

Suomi satelliittikuva/Finland från satellit   1:1 000 000
Helsinki: NBS, 1980.
Monochrome satellite image map.

Suomi ja Pohjoiskalotti: autoiljan tiekartta/Motoring road map of Southern Finland and the North Calotte   1:800 000
Helsinki: NBS, annual.

Suunnittelukartta/Planeringskartan   1:500 000
Helsinki: NBS, 1979.
4 sheets, all published.

Yleiskartta/Generalkarta   1:400 000
Helsinki: NBS, 1984–.
6 sheets, 4 published.

Topographic

Suomen tiekartta/Vägkarta över Finland/Road map of Finland GT   1:200 000
Helsinki: NBS, 1983–.■
19 sheets, all published.■
Also available in boxed set.
Also available as index map of uniform coordinate system.

Topografinen kartta/Topografisk karta   1:100 000
Helsinki: NBS, 1956–.
349 sheets, all published.■

Topografinen kartta/Topografisk karta   1:50 000
Helsinki: NBS, 1956–.
615 sheets, all published.■

Topografinen kartta/Topografisk karta   1:20 000
Peruskartta/Grundkarta   1:20 000
Helsinki: NBS, 1955–.
3712 sheets, all published.■

Geological and geophysical

Suomi Finland. Eranto/missvisning/declination 1972   1:3 000 000
Helsinki: NBS/Ilmatieteenlaitos.

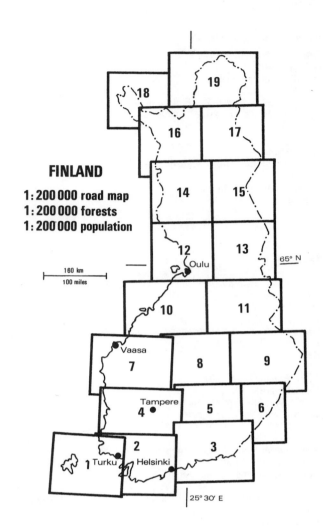

FINLAND

1 : 200 000 road map
1 : 200 000 forests
1 : 200 000 population

Suomen kallioperä 1 : 2 000 000
Espoo: GT, 1971.
Pre-Quaternary rocks of Finland.

Suomen maaperä 1 : 2 000 000
Espoo: GT, 1968.
Quaternary deposits of Finland.

Suomen aeromagneettinen kartta 1 : 2 000 000
Espoo: GT, 1980.
Aeromagnetic map. Total intensity.

Suomen kallioperä 1 : 1 000 000
Espoo: GT, 1982.
Pre-Quaternary rocks of Finland.

Suomen maaperä 1 : 1 000 000
Espoo: GT, 1984.
Quaternary deposits of Finland.

Geomorfologinen yleiskartta Suomi 1 : 1 000 000
Helsinki: NBS, 1982.
General geomorphological map.

Maanpinnan korkokuva 1 : 1 000 000
Helsinki: NBS, 1983.
Terrestrial relief of Finland.

Suomen geologinen yleiskartta: kivilaikartta 1 : 400 000
Espoo: GT, 1900–.
22 sheets, all published.
Most with texts.
General geological map. Pre-Quaternary rocks.

Suomen geologinen yleiskartta: maaperäkartta 1 : 400 000
Espoo: GT, 1903–.
26 sheets, 21 published.
Some with texts.
Different sheet system from north of country.
General geological map. Quaternary deposits.

Suomen geologinen kartta: kallioperakartta 1 : 100 000
Espoo: GT, 1949–.
ca 349 sheets, 134 published. ■
Some with texts.
Geological map of Finland. Pre-Quaternary rocks.

Suomen geologinen kartta: maaperäkartta 1 : 100 000
Espoo: GT, 1950–.
ca 349 sheets, 60 published. ■
Some with texts.
Geological map of Finland. Quaternary deposits.

Suomen geologinen kartta: maaperäkartta 1 : 20 000/1 : 50 000
Espoo: GT, 1973–.
ca 3712 sheets, 150 published. ■
Most with texts.
Geological map of Finland. Quaternary deposits.

Environmental

Suomen vesistöalueet 1 : 1 000 000
Helsinki: NBS/Vesihallitus, 1972.
River basins of Finland.

Vesistökartta Suomi/Hydrografisk karta Finland 1 : 1 000 000
Helsinki: NBS, 1973.
Hydrological map of Finland.

Peltojen levinneisyys Suomi 1 : 1 000 000
Helsinki: NBS, 1983.
Distribution of arable land in Finland.

Suomen valtionmetsäin kartta 1 : 1 000 000
Helsinki: NBS, 1981.
Finnish state forests.

Soran ja hiekan määrä peruskarttalehteä kohti 1 : 1 000 000
Helsinki: NBS, 1979.
Gravel and sand deposits in Finland.

Suomen malmiesiintymät 1 : 1 000 000
Helsinki: NBS, 1977.
Ore deposits in Finland.

Suomen malmilohkarelöydöt 1 : 1 000 000
Helsinki: NBS, 1984.
Ore boulder discoveries in Finland.

Vesivaret Suomi 1 : 1 000 000
Helsinki: NBS, 1984.
Water resources of Finland.

Suomen Suot 1 : 1 000 000
Helsinki: NBS, 1976.
Peat deposits in Finland.

Administrative

Aluepoliittiset-Aluejaot 1982 1 : 1 000 000
Helsinki: NBS, 1982.
Development areas of Finland.
Also available at 1 : 3 000 000 and 1 : 4 000 000.

Suomen maantieverkko ja kunnanrajat 1.1.1985 1 : 1 000 000
Helsinki: NBS, 1985.
Road network and commune boundaries.
Also available at 1 : 600 000.

Suomen maanomistustilanne 1979 1 : 1 000 000
Helsinki: NBS, 1980.
Land ownership in Finland.

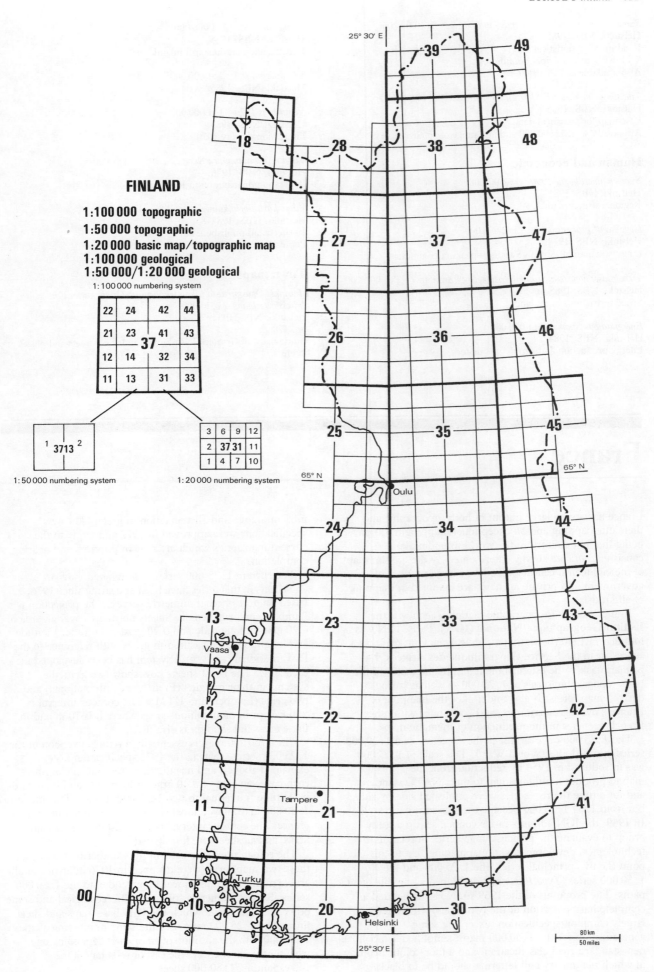

FINLAND

1:100 000 topographic
1:50 000 topographic
1:20 000 basic map/topographic map
1:100 000 geological
1:50 000/1:20 000 geological

1:100 000 numbering system

22	24	42	44
21	23	41	43
12	14	32	34
11	13	31	33

37

1:50 000 numbering system

1 3713 2

1:20 000 numbering system

3	6	9	12
2	37	31	11
1	4	7	10

25° 30' E

65° N

80 km
50 miles

Suomen maantieverkko/Finlands landsvägnät   1 : 1 000 000
Helsinki: NBS, 1985.
Road map for statistical data.
With/without municipal boundaries.
Also available at 1 : 600 000.

Tilaston pohjakartta/Baskartan för statistik   1 : 1 000 000
Helsinki: NBS, 1985.
Base map for statistical data.
Also available at 1 : 600 000 and at a range of smaller scales.

Human and economic

Suomen aluerakenne 1990   1 : 2 000 000
Helsinki: NBS, 1981.
Regional structure for 1990.

Kielalueet/Språkområdena   1 : 2 000 000
Helsinki: NBS, 1978.
Linguistic map of Finnish/Swedish-speaking communes.

Tiekohtaiset nopeusrajoitukset/Hasighetsbegränsningar   1 : 2 000 000
Helsinki: NBS, 1985.
Speed restrictions in Finland.

Energiatalous Suomi   1 : 1 000 000
Helsinki: NBS, 1984.
Energy production and supply in Finland.

Taajamat 1980   1 : 1 000 000
Helsinki: NBS, 1985.
Urban agglomerations in Finland 1980.

Väentiheys 1980   1 : 1 000 000
Helsinki: NBS, 1985.

Väentiheys 1970   1 : 1 000 000
Helsinki: NBS, 1983.
Population density 1970.

Väestön Jakautuminen Suomessa 1970   1 : 1 000 000
Helsinki: NBS, 1976.
Population distribution in Finland 1970.

Väestökartta/befolkningskarta 1980   1 : 200 000
Helsinki: NBS, 1984–85.
19 sheets, all published. ∎
Population grid maps 1980.

Town maps

Helsinki. Opaskartasto ja osoitehakemisto/Guide kartor och Adressförteckning   1 : 15 000
Helsinki: NBS, 1982.
pp. 178.
Guide maps of Greater Helsinki with index of streets and places of interest.

France

France has a very large output of both topographic and thematic mapping and an exceptional mapping tradition, being the first country in the world to undertake a consistent *national* survey (represented by the Cassini maps of the eighteenth century). Today, a large number of government and private agencies are involved in mapping of all kinds.

France's official topographic mapping agency is the **Institut Géographique National (IGN)**, Paris. IGN was founded in 1940 (becoming a public agency in 1967), as a successor to the Service Géographique des Armées. Its primary task is the establishment and maintenance of the basic topographic series at 1 : 25 000 as well as various derived map series. It also undertakes the mapping of France's overseas territories, and cooperates with other French agencies in the production of thematic maps.

The 1 : 25 000 *carte de base* is the successor to a 1 : 20 000 series initiated after World War I. The scale of 1 : 25 000 was introduced in 1956 by reduction from the 1 : 20 000 map to produce a military map to meet NATO standards, and for a time the two series were published side by side, but from 1964 1 : 25 000 became the standard basic scale. In 1969, the IGN administrative council set up a study group to evaluate the principal map series in relation to technological developments in mapmaking. Up to this point the two principal series, the 1 : 25 000 and the 1 : 50 000 series (*Type 1922*) had developed along divergent paths. The proposals of the 1969 study recommended a complete harmonization of the two series, with a new, largely common specification (excepting some generalization for the 1 : 50 000 map), simplified symbols, a new standard road classification, and a four-colour printing in which planimetry and lettering should be in black, drainage and other water detail in blue, contours and road

infill in orange and all vegetation in green. The new specifications were approved in 1972 and apply to the current map series which are now in progress and nearing completion.

The current 1 : 25 000 series is designated the *Série bleue* and sheets in this series have been appearing since 1976, based on new photogrammetric survey. The projection is a Lambert conformal conic. Sheets normally cover an area of 0.20 grad in longitude and 0.20 grad in latitude. Latitude and longitude are measured in grads with reference to the Paris meridian, and a grid system has been developed for each zone. The latest sheets now show two graticule systems in their margins (French and International) and two grids (Lambert and UTM). The contour interval is 5 m except in mountainous areas where it is 10 m and in these areas hill shading is also used.

The 1 : 50 000 map is essentially a reduced version of the 1 : 25 000 series, sharing the 1972 specification (as explained above), and has the same revision cycle. Sheets each cover an area of 0.40 grad in longitude and 0.20 grad in latitude (i.e. equal to two 1 : 25 000 sheets). This modern 1 : 50 000 map, now almost complete, is dubbed the *Série orange*. The contour interval is 10 m (20 m in mountain areas) and all sheets are hill shaded.

The roman numerals used in the numbering system for these two map series were dropped in 1976 in favour of all-arabic numeration, as shown in our index to the 1 : 50 000 series. Roman numerals still appear on geological and some other thematic series which use IGN base maps and sheet lines. The 1 : 25 000 map uses the same numbering as the 1 : 50 000 with the addition of *est* or *ouest* depending on whether the sheet covers the east or west half of the corresponding 1 : 50 000 sheet.

IGN also publishes map series at 1 : 100 000 and

1 : 250 000. The 1 : 100 000 was initially published 1954–58 in 293 sheets, has been kept under revision since then, and is also available in a military version with UTM grid. However, it is now mainly used as a base map for thematic series, since a parallel series of tourist maps covering France and Corsica in 74 much larger sheets was introduced in 1970. This is the *Série verte*, which shows the complete road network and road distances, has a trilingual legend and is updated on a 4–5 year cycle. The 1 : 250 000 series also originated as a series of small sheets (45), but this series was discontinued in favour of the *Série rouge* in just 16 sheets, with an overprint of tourist information and a 2 year revision cycle.

The IGN has given much attention to public interest in its maps for recreational use, and currently a number of specially formatted mountain maps are available, while the standard *Série bleue* sheets of popular areas now have marked walking routes, mountain huts, etc., overprinted on them. There are also several 1 : 1 000 000 maps designed for tourists, including *Routes–autoroutes*, *Richesses artistiques* and *Sentiers de grande randonnée*. Some private publishers, such as Didier and Richard and Club Vosgien, also use the IGN maps as a base for their own recreational overprint.

Large-scale planimetric mapping has been undertaken by IGN in a few urban areas of France, for example in Paris for the Service de l'Urbanisme, but in most areas large-scale mapping has been carried out by local cadastral surveyors (*Geomètres Experts*) at scales ranging from 1 : 500 to 1 : 5000. Cadastral survey is coordinated by a central authority, the **Sous-Direction du Cadastre**, in Paris. In 1986 a new national map series at 1 : 5000 was launched in a cooperative programme between IGN and the Cadastre, and a cadastral map of uniform scale (1 : 2000) and specification is also being introduced for urban areas.

The IGN has been in the forefront of developments in automated cartography, particularly in the application of raster data capture. Computer-assisted cartography was introduced in the 1970s, initially for production of large-scale (1 : 1000 to 1 : 5000) digital base maps. More recently the IGN has been considering developing a topographic database for editing and updating the 1 : 25 000 and 1 : 50 000 series. Over the coming decade a digital database will also be set up for the series at 1 : 100 000 and smaller. This will be designed for a resolution compatible with the SPOT remote sensing system which began operation in 1986, and which is described further in Chapter 5. Early in 1985, IGN completed a project to digitize the relief of France using raster technology and embracing 200 million pixels.

In 1976, attention turned to automated methods in thematic mapping, and in 1979 an operational system called SEMIO was set up by the Service des Applications Nouvelles at Saint-Mandé. Among the projects serviced by this system have been the post-census mapping represented by two publications, *Population Française* (1980) and *Activité et Habitat* (1981), presented as folios of statistical maps in colour with commentaries in both French and English. These map folios are the result of cooperation between IGN and the **Institut National de la Statistique et des Etudes Economiques (INSEE)**. Another special mapping project has been the production of inventory mapping of the French coastal areas (**Inventaire Permanent du Littoral**). This too uses SEMIO, and although output is in the form of four-colour process printed maps, the data (concerning land use and the legal status of land) is held in a relational database in which spatial data can be matched to census statistics, and various derived products may be generated. The standard

printed maps consist of 147 large-format 1 : 25 000 scale maps of land use and 23 sheets at 1 : 100 000 showing legal status. Land use was interpreted from 1977 air photography, but it is planned to repeat the inventory on a 5 year cycle, and so a second generation of maps is being prepared using 1982 photography for up-dating the land use archive. IGN has also been involved in automation of the new agricultural land evaluation maps described below.

IGN cooperates in the development of thematic mapping from digital satellite data with the **Centre National d'Etudes Spatiales (CNES)**.

Geological mapping also has a long history in France: Napoleon III established a geological mapping agency (Service de la Carte Géologique) in 1868, and a complete mapping of the country was undertaken at 1 : 80 000 and almost finished in the following 50 years. A 1 : 50 000 geological series was started in 1925. The national organization now responsible for geological mapping is the **Bureau de Recherches Géologiques et Minières (BRGM)**, Orleans, founded in 1959. The basic series remains the 1 : 50 000, the sheets being issued with separate explanatory notes, and cover is currently about 80 per cent complete. Sheets of the old 1 : 80 000 series, based on the nineteenth century topographic *Carte de l'Etat-Major*, are still available, and there are also a number of gravimetric and magnetic maps at this scale.

Many smaller-scale geological maps are also published by BRGM, including a new 1 : 250 000 series which will replace the old 1 : 320 000 map, and a number of small-scale synthesizing maps, of which the most recent are published at a scale of 1 : 1 500 000 and include a geological map which takes in the continental shelf as well as the land area of France. In addition to the series mapping listed in our catalogue there are also numerous hydrogeological and geothermal maps and atlases of particular regions of France.

A series of maps of slope failure hazard, the *Carte Zermos (zones exposées à des risques liés aux mouvements du sous-sol)* cover areas in the French Alps at scales which vary from 1 : 10 000 to 1 : 25 000, and were published by BRGM jointly with the Direction de la Securité Civile.

The **Centre d'Etudes et de Réalisations Cartographiques Géographiques (CERCG)**, Paris, is one of a number of cartographic institutions and research units which come under the authority of the **Centre National de la Recherche Scientifique (CNRS)**. It is concerned with the graphical realization of thematic cartography and has a growing interest in computer-assisted methods. Among its more traditional output has been a series of 1 : 50 000 geomorphological maps which has been in progress since 1971. The sheets, on IGN sheet lines, are issued in folded format, each with an explanatory text. Nineteen have been published so far, and several more are in preparation.

A wide range of soils mapping at various scales is carried out by the **Service d'Etude des Sols et de la Carte Pédologique de France (SESCPF)** whose centre of research is at Orleans. SESCPF is part of the **Institut National de Recherche Agronomique (INRA)**, and the maps are sold by INRA Publications at Versailles. The principal published series is at a scale of 1 : 100 000, and has been in progress since 1969, but although about a third of the country has been surveyed, only some 15 sheets have been issued in the series. Recently a soils map of the Paris area has been published at 1 : 250 000. A new edition of the soil map of France at 1 : 000 000 is also in preparation. The soils maps are accompanied by descriptive monographs.

A series of land capability maps is in progress departmentally, in fulfilment of the French Agricultural

Development Act of 1980 which requires each Department to prepare maps showing the available agricultural land. These maps classify land into six main yield-capacity classes on the basis of both physical and socioeconomic criteria. Publication is in colour at 1:50 000 scale on IGN base maps. Map information is being computerized, using the IGN SEMIO system, so that various derived maps can be produced by automated cartography and the database itself rapidly revised as the controlling criteria for the land classes change. The series is being coordinated by the **Ministère de l'Agriculture**, Paris.

An extensive programme of vegetation mapping was initiated in 1947 by the Service de la carte de la vegetation, Toulouse, a laboratory of CNRS and now called the **Centre d'Ecologie des Ressources Renouvelables (CERR)**. This series shows actual land cover, whether natural or man induced, together with potential vegetation and ecological regions. The series is practically complete, with all unpublished sheets in press or in active preparation. Some sheets may be out of print. The laboratory is now employing techniques of ecological mapping, remote sensing and process modelling to investigate the dynamics of ecosystems, and problems relating to renewable resources.

The **Service de l'Inventaire Forestier National (IFN)**, founded in 1960, carries out a permanent inventory of forest resources by Department. The first inventory was completed in 1984, and a new cycle is now in progress. The main types of map produced are simple monochrome 1:500 000 scale maps defining the forested areas, and more complex four-colour departmental stock maps at 1:200 000 or 1:250 000.

Another specialist group within the CNRS, the Equipe de recherche or ER 30, was formed in 1967, initially to prepare a *Carte climatique detaillée de la France*. Publication of this map has been in progress since 1972. The IGN 1:250 000 series is used for the base maps (the sheet lines have been slightly modified), and the sheets are printed in eight or nine colours and issued in a folder which incorporates explanatory text. Future sheets covering areas of low relief and characterized by less rapid variations in climate are to be published at 1:500 000 (the first sheet at this scale will be Dijon). Recent work by ER 30 has focussed on solar energy, with a greater emphasis on modelling techniques, rather than on direct cartographic expression. The climatic maps are published by **Editions Ophrys**.

The **Direction de la Météorologie** of the Ministry of Transport has its own printing unit at Trappes and in addition to the preparation of weather charts has published a climatic atlas of France. The original edition was issued in 1969, but the current atlas, published in 1981, is a shortened edition with the maps at a reduced scale.

There is currently no National Atlas of France, the *Atlas de France Metropole* of the 1950s having been long out of print, but discussions began in 1984 on plans for a new atlas, which will use the latest techniques in information technology. There is an impressive range of regional atlases, begun in the 1960s, when a number were funded by DATAR (Délégation a l'Aménagement du Territoire et à l'Aménagement Regional) prepared by various university institutes and published by **Berger-Levrault**. Further atlases have appeared through the 1970s and 1980s and only four regions now lack one.

Two regional atlases (of Aquitaine and the Loire) are published by **Editions Technip**. This publishing house was founded in 1956 primarily to handle publication of the scientific and technical work of the **Institut Français du Pétrole (IFP)**, and a number of maps also appear in its

catalogue, including the recent *Carte bathymétrique de la marge continentale au large du delta du Rhone. Golfe du Lion* (1984). Mapping is carried out by the cartographic unit of BEICIP (**Bureau d'Etudes Industrielles et de Coopération de l'Institut Français du Pétrole**), which has published many bathymetric, soil and geological maps of areas overseas. The publication list of **Elf Aquitaine** also has some maps including a 1:250 000 scale *Carte géologique des Pyrénées Occidentales*.

Marine charting is the responsibility of the **Service Hydrographique et Océanographique de la Marine (SHOM)**. SHOM has its charting establishment (EPSHOM) at Brest, which is also the location of the **Institut Français de Recherche pour l'Exploitation de la Mer (IFREMER)**. IFREMER has absorbed the Centre National pour l'Exploitation des Oceans (CNEXO), an organization which has produced many bathymetric and geophysical maps of sea areas. CNEXO has also cooperated with BRGM on the production of geological maps of the English Channel and the Gulf of Gascogne. Also at the Centre de Brest de l'IFREMER, a Bureau National des Données Océaniques (BNDO) has recently been established.

Several French semi-government organizations are involved in resource mapping overseas. Most important is the **Institut Français de Recherche Scientifique pour le Développement en Coopération (ORSTOM)**, which has published a vast range of soils and other resource inventory mapping in thirty different countries. The **Institut de la Carte Internationale du Tapis Végétal (ICITV)** is concerned with the study and mapping of vegetation in tropical regions. The **Institut d'Elevage et de Médecine Vétérinaire des Pays Tropicaux (IEMVT)** aims to promote animal production in tropical countries and has a cartographic section and a remote sensing laboratory. It has published pasture and range monitoring maps, and maps of tsetse fly distribution. In addition both IGN and BRGM have considerable overseas mapping programmes. The **Groupement pour le Développement de la Télédetection Aérospatiale (GDTA)** is concerned with the applications of remotely sensed data and produces, *inter alia*, hard-copy thematic image maps of vegetation indices and of diurnal or nocturnal temperature variations derived from NOAA satellites.

There are many private cartographic publishers in France of which the best known internationally are **Michelin**. Michelin have been producing maps since 1910, and their speciality is road maps, which for France includes the long established *cartes detaillées* 1:200 000 (40 sheets) and the more recent and larger-format *cartes régionales* 1:200 000 (17 sheets). In 1987, the complete 1:200 000 scale Michelin series was included in a *Michelin Road Atlas of France*, published in association with the Hamlyn Publishing Group and available in both hard and softcover format. There is also a range of maps of Paris (1:10 000 for the city centre, 1:50 000 for the suburbs, 1:100 000 for the *environs* and a new set of 1:15 000 suburban maps), and at 1:1 000 000 of France as a whole. Michelin also publish some very useful road maps of other countries. All Michelin maps are revised annually.

Recta-Foldex are publishing an increasingly interesting range of maps. The core series is the *Cart'index regional* 1:250 000, a set of 15 road maps covering the country and including an extensive index on the reverse of each sheet. Other offerings include an administrative map of France, a 1:550 000 map in four sheets (designed for assembly as a wall map), and several satellite image maps of regions of France in simulated natural or false colour and using imagery from Landsat 4 and 5.

Plans-Guides Blay, Paris, specialize in the production of an excellent list of street maps and indexes of more than 120 towns and cities in France. Some sheets include an English text for visitors, and all are frequently updated.

Editions Géographiques et Touristiques Gabelli, Paris, produce road and leisure maps of France, and plans of Paris including a six-sheet cover of the suburbs at 1:15 000.

Oberthur provides a comprehensive *Index atlas de France*. The gazetteer section is arranged by Departments, and the volume also contains maps of each Department and street maps (with street indexes) of a large number of towns.

Further information

French mapping is exceptionally well documented. The IGN produces general brochures on its services, and regularly up-dated indexes to map cover of the main map series. BRGM issues an annual catalogue and map index, and most significant governmental map publishers also have lists or catalogues of maps and publications.

IGN publishes a technical *Bulletin d'Information*, and No. 36 (1978/2) in this series was devoted to the 1972 specifications developed for the current 1:25 000 and 1:50 000 series. Useful documentation is also provided in the CNRS *Intergéo bulletin* (e.g. 72/1983 is a bibliography of atlases and thematic maps published during 1976–83). Also, the *Bulletin* of the Comité Français de Cartographie is a useful source of new developments, and Fascicule 99–100 (1984) consists of a report for the ICA on national cartography 1980–83, which describes recent work by the main mapping organizations.

The work of the group ER 30 in climatic modelling and mapping has been described in Peguy, Ch.P. *et al.* (1984) Cartographie et modelisation des éléments du climat en France, *Annales de Géographie*, **93**, 204–217.

Geomorphological mapping is described in Joly, F. (1986) Les cartes géomorphologiques, *Géochronique*, **19**, 23–26.

Information on geological mapping also appears from time to time in the periodical *Géochronique*.

Addresses

Institut Géographique National (IGN)
Direction générale, 136 bis, rue de Grenelle, F-75700 PARIS.

Institut Géographique National (IGN) Orders
Service vente et édition, 107, rue la Boëtie, F-75008 PARIS.

Institut Géographique National (IGN) Services, information
2, avenue Pasteur, F-94160 SAINT-MANDE.

Sous-Direction du Cadastre et de la Publicité Foncière
1, rue des Mathurins, F-75436 PARIS Cedex 09.

Institut National de la Statistique et des Etudes Economiques (INSEE)
18, boulevard Adolphe Pinard, F-75675 PARIS Cedex 14.

Inventaire Permanent du Littoral
64 rue de la Fédération, F-75015 PARIS.

Centre National d'Etudes Spatiales (CNES)
129, rue de l'Université, F-75007 PARIS.

Bureau de Recherches Géologiques et Minières (BRGM)
BP 6009, F-45060 ORLEANS.

Centre d'Etudes et de Réalisations Cartographiques Géographiques (CERCG)
191, rue St Jacques, F-75005 PARIS.

Centre National de la Recherche Scientifique (CNRS)
15, quai Anatole, F-75007 PARIS.

Service d'Etude des Sols et de la Carte Pédologique de France (SESCPF)
Centre de Recherches d'Orléans, Ardon, F-45160 OLIVET.

Service des Publications de l'INRA (INRA)
Route de Saint-Cyr, F-78000 VERSAILLES.

Carte des Terres Agricoles, Ministère de l'Agriculture
30, rue Las Cases, F-75007 PARIS.

Centre d'Ecologie des Ressources Renouvelables (CERR)
29, rue Jeanne Marvig, F-31055 TOULOUSE.

Inventaire Forestier National (IFN)
1ter, avenue de Lowendal, F-75007 PARIS.

Editions Ophrys
10, rue de Nesle, F-75006 PARIS.

Berger-Levrault
5, rue Auguste Comte, F-75006 PARIS.

Direction de la Météorologie, Ministère des Transports
77, rue de Sèvres, F-91106 BOULOGNE-BILLANCOURT Cedex.

Editions Technip
27, rue Ginoux, F-75737 PARIS Cedex 15.

Elf Aquitaine Edition
Centre Micoulau, F-64018 PAU Cedex.

Institut Français du Pétrole (IFP)
1–4, av. de Bois-Preau, BP 311, F-92506 RUEIL-MALMAISON Cedex.

Bureau d'Etudes Industrielles et de Coopération de l'Institut Français du Pétrole (BEICIP)
232, avenue Napoleon Bonaparte, F-92500 RUEIL-MALMAISON.

Service Hydrographique et Océanographique de la Marine (SHOM)
3, avenue Octave Gréard, F-75200 PARIS.

Service Hydrographique et Océanographique de la Marine (EPSHOM) Orders
13, rue de Chatellier, BP 426, F-29275 BREST Cedex.

Institut Français de Recherche pour l'Exploitation de la Mer (IFREMER/CNEXO)
Centre de Brest, BP 337, F-29273 BREST.

Institut Français de Recherche Scientifique pour le Développement en Coopération (ORSTOM)
Service de vente,
70–74, route d'Aulnay, F-93140 BONDY.

Institut de la Carte Internationale du Tapis Végétal (ICITV)
Université de Toulouse III, 39, allées Jules Guesde, F-31400 TOULOUSE.

Institut d'Elevage et de Médecine Vétérinaire des Pays Tropicaux (IEMVT)
10, rue Pierre Curie, F-94700 MAISONS-ALFORT.

Bureau pour le Développement de la Production Agricole (BDPA)
27, rue Louis-Vicat, F-75738 PARIS Cedex 15.

Centre d'Etudes de Géographie Tropicale (CEGET)
Domaine Universitaire, F-33415 TALENCE.

Institut de Recherches en Agronomie Tropicale (IRAT)
Direction générale et services scientifiques, 45 bis, avenue de la Belle-Gabrielle, 94130 NOGENT-SUR-MARNE and BP 5035, 34032 MONTPELLIER Cedex.

Groupement pour le Développement de la Télédetection Aérospatiale (GDTA)
18, avenue Edouard-Belin, F-31055 TOULOUSE Cedex.

Institut de Physique du Globe de Paris (IPGP)
4, Place Jussieu, F-75252 PARIS Cedex 05.

Michelin
Service de Tourisme, 46 avenue de Breteuil, F-75341 PARIS Cedex 07.

Recta-Foldex
27, rue Trebois, BP 94, F-92303 LEVALLOIS-PERRET.

Plan-Guides Blay (Blay)
14, rue Favart, F-75002 PARIS.

Editions Géographiques et Touristiques Gabelli (Gabelli)
141, rue de Picpus, F-75012 PARIS.

Oberthur
78, rue de Paris, F-25000 RENNES.

Librairie Larousse
17, rue du Montparnasse, F-75006 PARIS.

Laboratoire de Géographie Physique
Université de Besançon, 30, rue Mégavaud, F-25030 BESANÇON Cedex.

Societé Nationale des Chemins de Fer (SNCF)
Service de l'Information et des Relations Publiques, 86–88, rue Saint-Lazere, F-75001 PARIS.

Gaz de France
23, rue Philibert Delorme, F-75840 PARIS.

Taride
2bis, place du Puits de l'Ermite, F-75005 PARIS.

Astrolabe
46, rue de Provence, F-75009 PARIS.

Editions Delroisses
113, rue de Paris, F-92100 BOULOGNE-BILLANCOURT.

Editions Jaguar
3, rue Requepine, F-75008 PARIS.

Editions Albin-Michel
22, rue Huyghens, F-75014 PARIS.

Catalogue

Atlases and gazetteers

L'index atlas de France
Rennes: Oberthur, 1984.
pp. 996.

Dictionnaire géographique de la France   René Oizon
Paris: Librarie Larousse, 1979.
pp. 915.

Dictionnaire national des communes de France bases 1982 census
Paris: Albin Michel/Berger Levrault, 1984.

General

La France vue de satellite   1:2 000 000
Orléans: BRGM, 1978.

France physique   1:1 750 000
Paris: IGN, 1979.

La France vue de satellite   1:1 300 000
Paris: IGN, 1978.
Colour-infrared satellite montage.

France   1:1 250 000
Edinburgh: Bartholomew, 1982.

Carte de la France   1:1 000 000   5th edn
Paris: IGN, 1985.
In style of IMW.

La France vue de satellite   1:1 000 000
Orléans: BRGM/Rueil-Malmaison: BEICIP, 1974.

Scène Landsat sur la France entière   1:500 000
Orléans: BRGM, 1985.

Topographic

Carte topographique. Type World   1:500 000
Paris: IGN, 1961–.
10 sheets, all published.
Series covers France except north of 50° and east of 7°.

Carte touristique   1:250 000   Série rouge
Paris: IGN, 1973–.
16 sheets, all published. ■
Revised every 2–3 years.

Carte topographique de la France   1:100 000   Série verte
Paris: IGN, 1970–.
74 sheets, all published.
Revised every 2–3 years.

Carte topographique de la France   1:100 000
Paris: IGN, 1952–.
293 sheets, all published. ■

Carte topographique de la France   1:50 000   Série orange
Paris: IGN, 1976–.
1094 sheets, *ca* 950 published. ■
Remaining sheets available in older series.

Carte topographique de la France   1:25 000   Série bleue
Paris: IGN, 1976–.
2024 sheets, *ca* 1650 published.■
Remaining sheets available in older series.

Geological and geophysical

Cartes des tremblements de terre en France. Intensités maximales et épicentres connus   1:2 500 000
Orléans: BRGM, 1981.

Carte géologique de la France et de la marge continentale
1:1 500 000
Orléans: BRGM, 1980.
With/without 102 pp. text.

Carte hydrogéologique de la France: systèmes aquifères   1:1 500 000
Orléans: BRGM, 1980.
With 36 pp. text.

Carte minière de la France   1:1 500 000
Orléans: BRGM, 1980.
With 20 pp. text.

Carte géologique de la France   1:1 000 000
Orléans: BRGM, 1969.
2 sheets, both published.
Available only with overprint of 1:50 000 sheet lines.

Carte géologique du Quaternaire et des formations superficielles de la France   1:1 000 000
Besançon: Université de Besançon, 1974.
4 sheets, all published.

France. Curiosités géologiques   1:1 000 000
Paris: IGN/Orléans: BRGM, 1985.

Carte tectonique de la France   1:1 000 000
Orléans: BRGM, 1980.
2 sheets, both published.
With 52 pp. text. (*BRGM Mémoire* 110).

FRANCE
1:100 000 topographic
1:100 000 soils

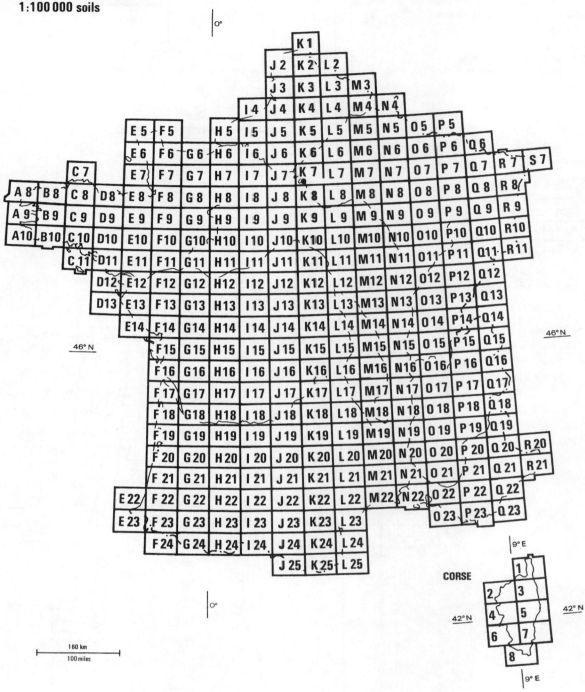

CORSE

160 km
100 miles

Carte sismotectonique de la France 1:1 000 000
Orléans: BRGM, 1981.
With 20 pp. text (*BRGM Mémoire* 111).

Carte des linéaments de la France d'après les images des satellites Landsat (1972 à 1976) 1:1 000 000
Orléans: BRGM, 1980.
With text.

Cartes des gisements de fer de la France 1:1 000 000
Orléans: BRGM, 1962.
2 sheets, both published.

Carte de la qualité chimique des eaux souterraines de la France 1:1 000 000
Orléans: BRGM, 1978.
With 12 pp. text.

Carte du débit moyen des nappes d'eau souterraine de la France 1:1 000 000
Orléans: BRGM, 1970.

Carte des eaux minérales et thermales de la France 1:1 000 000
Orléans: BRGM, 1975.

Carte gravimétrique de la France 1:1 000 000
Orléans: BRGM, 1974–75.
2 sheets, both published.

Carte magnétique de la France 1:1 000 000
Orléans: BRGM, 1980.
2 sheets, both published.

Carte des gîtes minéraux de la France 1:500 000
Orléans: BRGM, 1979–86.
8 sheets, all published.
Each sheet has accompanying text with list of mineral deposits.

FRANCE

1:50 000 topographic Série Orange
1:25 000 topographic Série Bleue
1:50 000 geological
1:50 000 geomorphological
1:50 000 carte départementale
   des terres agricoles

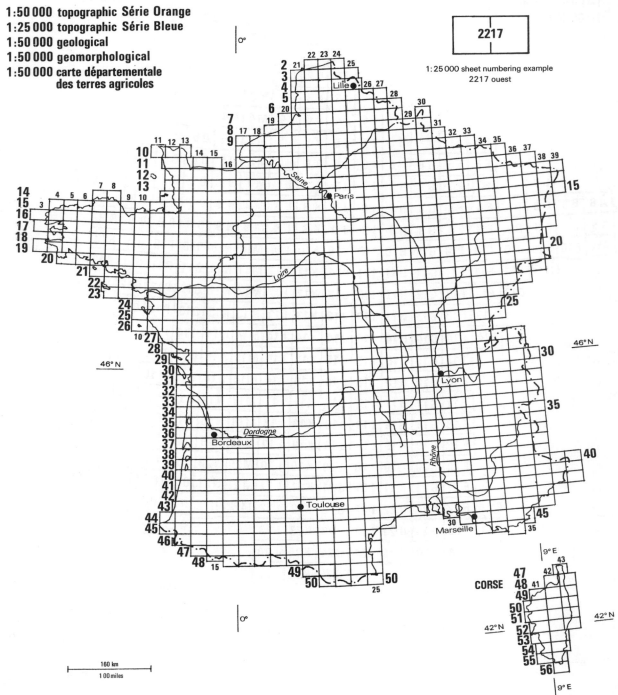

1:25 000 sheet numbering example
2217 ouest

160 km
100 miles

Carte géologique de la France 1:320 000
Orléans: BRGM, 1892–.
21 sheets, 12 published.

Carte gravimétrique de la France 1:320 000
Orléans: BRGM.
21 sheets, 6 published.

Carte géologique de la France 1:250 000
Orléans: BRGM, 1979–.
44 sheets, 12 published. ■

Carte magnétique de la France 1:250 000
Orléans: BRGM, 1974–.
44 sheets, 7 published. ■
Some sheets on irregular sheet lines.

Carte 'géologie et structure' de la marge continentale 1:250 000
Orléans: BRGM, 1970–.
19 sheets, 6 published.
Continental shelf maps.

Carte gravimétrique de la France 1:200 000
Orléans: BRGM.
66 sheets, 38 published. ■

Carte de la nature des depôts meubles sous-marins 1:100 000
Orléans: BRGM, 1970–.
48 sheets, all published.
Continental shelf maps.

Carte géologique de la France 1:80 000
Orléans: BRGM, 1868–.
256 sheets, 214 published.

FRANCE
1:250 000 geological
1:250 000 climatic

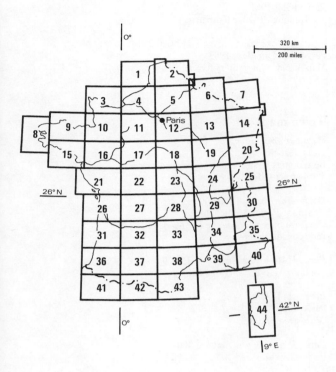

FRANCE
1:200 000 gravimetric
1:200 000 vegetation

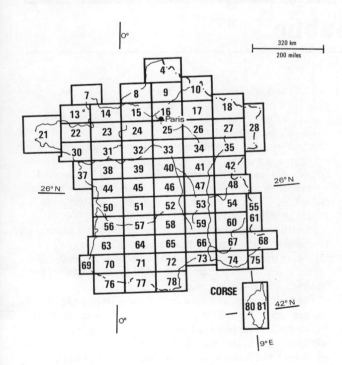

Carte gravimétrique de la France 1:80 000
Orléans: BRGM.
256 sheets, 75% published.

Carte magnétique détaillée de la France 1:80 000
Orléans: BRGM, 1969–.
256 sheets, 12 published.

Carte géologique de la France 1:50 000
Orléans: BRGM, 1928–.
ca 1094 sheets, *ca* 780 published. ■
Text with each sheet.

Carte géomorphologique detaillée de la France 1:50 000
Paris: CERCG, 1971–.
ca 1094 sheets, 19 published.■
Text with each sheet and separate legend booklet available.

Carte hydrogéologique de la France 1:50 000
Orléans: BRGM, 1966–.
ca 1094 sheets, 18 published. ■

Carte magnétique detaillée du champ total 1:50 000
Orléans: BRGM.
ca 1094 sheets, 60 published. ■
Cover Central Massif.

Carte de vulnerabilité des eaux souterraines à la pollution 1:50 000
Orléans: BRGM, 1980–.
ca 1094 sheets, 29 published. ■

Carte des substances utiles 1:50 000
Orléans: BRGM, 1969–.
ca 1094 sheets, 10 published. ■

Environmental

Atlas climatique de la France
Paris: Direction de la Météorologie, 1981.
29 plates.

Atlas agroclimatique de la France
Paris: Direction de la Météorologie, 1980.
30 maps + text.

La France solaire 1:1 000 000
Comité National du Jour du Soleil, 1979.

Carte des regions forestières de la France 1:1 000 000
Paris: IFN, 1983.
Diazo map.

Carte pédologique de la France 1:1 000 000 J. Dupuis
Versailles: INRA, 1967.
2 sheets, both published.
With 56 pp. text.

Carte des zones exposées à des glissements, écroulements, effondrements et affaissements de terrain en France 1:1 000 000
Orléans: BRGM, 1983.
+ text.

Carte climatique detaillée de la France 1:250 000
Paris: Ed. Ophrys, 1971–.
42 sheets, 11 published.■

Inventaire forestier national 1:250 000/1:200 000
Paris: IFN.
Series of departmental maps.

Carte de la végétation de la France 1:200 000
Toulouse: Service de la Carte de la Végétation CERR, 1947–.
64 sheets, 53 published.■

Cartes pédologiques de la France 1:100 000
Versailles: INRA, 1969–.
293 sheets, 15 published.■
Each sheet + text.

Carte de l'environnement et de sa dynamique 1:50 000
Paris: CNRS, 1975–.
1102 sheets, 5 published.■
Each sheet + text.

Administrative

Carte départementale de la France 1:1 400 000
Paris: IGN, 1978.

Carte administrative de la France 1 : 1 400 000
Paris: IGN, 1980.

Régions agricoles 1 : 1 400 000
Paris: INSEE, 1983.

Parcs naturels et réserves de France 1 : 1 400 000
Paris: IGN, 1980.
Plus 8 pp. text, *Les parcs naturels régionaux*.

Carte de France administrative 1 : 1 000 000
Paris: Recta-Foldex, 1985.
Includes index of places with population and postcodes.

Human and economic

Atlas de la France rurale
Paris: Comité National de Géographie, 1984.
pp. 157.

Population française/Population of France
Paris: IGN/INSEE, 1980.
22 map sheets + texts.
In French and English.
Includes maps of overseas departments.

Activité et habitat/Economic activity and habitat
Paris: IGN/INSEE, 1981.
22 map sheets + texts.
In French and English.

Cartes des infrastructures de transport 1 : 1 000 000
Paris: IGN/Ministère des Transports, 1981.

Le réseau SNCF 1 : 1 000 000
Paris: IGN/SNCF, 1982.
Map of rail network.

Réseau de transport et exploitations gazières 1 : 1 000 000
Paris: Gaz de France, 1984.

Sentiers de grande randonnée 1 : 1 000 000 Edition 3
Paris: IGN, 1981.
Map of long-distance footpaths.

France des rivières et canaux 1 : 1 000 000
Paris: Recta-Foldex, 1983.
Map of navigable waters for canoeists.

Town maps

Paris par arrondissement
Paris: Gabelli.
Bound atlas of street maps by *arrondissement*.
With street index.

Banlieue de Paris 1 : 50 000
Clermont-Ferrand: Michelin.
Covers suburbs outside the *Périphérique*.
Index of communes.

Paris. RAC Visitor's map 1 : 15 000
London: Map Productions Ltd.

Paris 1 : 14 000
Paris: Blay.
With street index.

Paris. Petit plan géant 1 : 12 500
Paris: Gabelli.

Plan de Paris 1 : 10 000
Clermont-Ferrand: Michelin.

German Democratic Republic

Detailed topographic mapping of the German Democratic Republic is the responsibility of the **Verwaltung für Vermessungs- und Kartenwesen (VVK)** of the Ministry of Internal Affairs (Ministerium des Innern), Berlin, and there has been an extensive programme of postwar mapping. This conforms to the standards laid down by the Soviet mapping authorities for all Warsaw Pact countries, and complete cover of the country in 6150 sheets has been published at the basic mapping scale of 1 : 10 000 (the *Grundkarte*). This map is printed in five colours with contours at 1 m or 2.5 m intervals. Derived series have also been completed at 1 : 25 000, 1 : 50 000, 1 : 100 000 and 1 : 200 000. The sheet systems are all subdivisions of the International Map of the World sheet lines. However, all these map series are produced primarily for military use and as classified documents are not available to the public.

VVK has developed an automated cartographic system called Digicart which processes both photogrammetric and field data for the production of large-scale maps, cadastral maps and digital terrain models.

Prewar topographic mapping of the former German Reich covering what is now the DDR is still available, unrevised, from the **Institut für Angewandte Geodäsie (IfAG)**, at scales of 1 : 200 000, 1 : 100 000 and 1 : 25 000 (the so-called *Messtischblätter* or plane table maps). This mapping is not included in our catalogue, but indexes and ordering information are available from IfAG, Aussenstelle Berlin, German Federal Republic.

Maps and atlases for use by the general public are published principally by two organizations in the DDR, namely **VEB Hermann Haack (Haack)**, Gotha, and **VEB Tourist Verlag (Tourist Verlag)**, formerly Landkartenverlag, Berlin and Dresden. The former concentrates mainly, though not exclusively, on educational maps and atlases, while the latter produces (1) a series of street maps and guides of some 30 major towns and cities, (2) a series of recreational maps and walking guides at scales varying from 1 : 150 000 to 1 : 30 000, (3) road maps and atlases, including the 1 : 200 000 series in nine sheets, and (4) maps of administrative areas, including the 14-sheet *Verwaltungskarte* at 1 : 200 000. Most of the Tourist Verlag maps have multi-lingual legends, and in most cases English is one of the languages.

The Haack maps include a new series of large-scale regional maps at 1 : 100 000, which will cover selected areas showing the complete road and settlement network, administrative boundaries down to the *Kreis* level, detailed hydrology and forest cover, and shaded relief. The first sheets were issued in 1986, and cover Thüringer Wald, Thüringer Becken and Thüringer Schiefergebirge.

Haack also collaborated in the production of the national *Atlas Deutsche Demokratische Republik* published by the **Akademie der Wissenschaften der DDR**, and completed in 1981. This prestigious atlas has 58 map plates with 106 maps ranging in scale from 1 : 750 000 to 1 : 2 000 000 and covering a wide range of themes in physical, economic and social geography. The map legends are translated into Russian, English, French and Spanish. The atlas is distributed by Haack and many sheets are also available as separates. The atlas is to be followed up by supplementary sheets in a series called *Karten zur Volks-, Berufs-, Wohnungs- und Gebäudezahlung* (KVBWGZ).

Geological mapping is the responsibility of the **Zentrales Geologisches Institut (ZGI)**, Berlin. ZGI issued new general geological maps at 1:500 000 in 1962 and 1973, but these may not be available at present. We have no information on new detailed mapping. However, many sheets of the *Geologische Karte von Preussen und benachbarten deutschen Ländern 1:25 000* dating mainly from the 1920s and 1930s are still available from the **Bundesanstalt für Geowissenschaften und Rohstoffe (BfGR)**, Hannover, German Federal Republic.

Much thematic mapping, especially of soils, geomorphology and land evaluation, has been undertaken by the cartographic sections of universities, by the Academy of Sciences of the DDR, the Academy of Agricultural Sciences and the **Sächsische Akademie der Wissenschaften**, Leipzig. Much of this mapping has been large scale and of an experimental nature. One important general series is the nationwide cover in 63 sheets of the *Mittelmassstabigen Landwirtschaftlichen Standortkartierung* (MMK) at 1:100 000. This is a land classification map which combines information about soil type, soil moisture and relief. The map was produced during 1976–80 by the Forschungszentrum für Bodenfruchtbarkeit Müncheberg of the Academy of Agricultural Sciences for the Ministry of Land, Forestry and Food. It is designed as a management tool for medium- and long-range planning and optimization of land use. The map series is linked to a computerized soil information system. A map of physiographic regions at the 1:100 000 scale is also in preparation, *Naturraumtypenkarte der DDR* (NK 100). It is unlikely that these thematic map series will be readily obtained in the West.

Further information

Both Haack and Tourist Verlag produce comprehensive annual catalogues.

Descriptions and evaluations of thematic mapping in the DDR are frequently found in the periodicals *Petermanns Geographische Mitteilungen* and *Geographische Berichte*, and technical information about topographic mapping in the periodical *Vermessungstechnik*.

A useful summary of maps and atlases is given in Sperling, W. (1978) *Landeskunde DDR: eine annotierte Auswahlbibliographie Band 1*, München: K.G. Saur, and a discussion of recent developments in cartography in the supplementary volume by the same author subtitled *Eine kommentierte Auswahlbibliographie. Ergänzungsband 1978–1983*, published by Saur in 1984.

Addresses

Verwaltung für Vermessungs- und Kartenwesen (VVK)
Mauerstrasse 29–32, DDR-1086 BERLIN.

VEB Hermann Haack (Haack)
Geographisch-Kartographische Anstalt Gotha, Justus-Perthes-Strasse 3–9, DDR-5800 GOTHA.

VEB Tourist Verlag (Tourist Verlag)
Neue Grünstrasse 17, DDR-1020 BERLIN.

Buchexport Export agency for public maps
Volkseigener Aussenhandelsbetrieb der DDR, Postfach 160, DDR-7010 LEIPZIG.

Zentrales Geologisches Institut (ZGI)
Invalidenstrasse 44, DDR-104 BERLIN.

Geographisches Institut der Deutschen Akademie der Wissenschaften zu Berlin
Georgi-Dimitroff-Platz 1, DDR-701 LEIPZIG.

Sächsische Akademie der Wissenschaften zu Leipzig
Goethestrasse 3–5, DDR-701 LEIPZIG.

Institut für Angewandte Geodäsie (IfAG)
Aussenstelle Berlin, Stauffenbergstrasse 11–13, D-1000 BERLIN 30.

Bundesanstalt für Geowissenschaften und Rohstoffe (BfGR)
Postfach 51 01 53, D-3000 HANNOVER 51.

For Westermann, see German Federal Republic.

Catalogue

Atlases and gazetteers

Atlas Deutsche Demokratische Republik
Gotha: Haack/Akademie der Wissenschaften der DDR, 1981.

Autoatlas DDR
Berlin: Tourist Verlag, 1986.
Includes 15 000 placename index.

Verzeichnis der Gemeinden und Ortsteile der DDR
Berlin: Staatsverlag, 1966.
pp. 642.
+ supplement published 1983.

Ortslexikon der DDR
Berlin: Staatsverlag der DDR, 1986.
pp. 352.

Diercke Lexikon Deutschland. DDR and Berlin (Ost)
Braunschweig: Westermann Verlag, 1986.
A descriptive gazetteer of cities in the DDR.

General

Reiseland DDR 1:600 000
Berlin: Tourist Verlag, 1986.
Tourist map.
Legend in German, Russian and English.

Deutsche Demokratische Republik 1:750 000 Edition 1
Gotha: Haack, 1984.
Administrative map on one side, physical on the reverse.
28 pp. text includes placename index and statistics.

Autokarte DDR 1:500 000
Berlin: Tourist Verlag, 1986.
Motoring map with legend in 4 languages.

Reise- und Verkehrskarte DDR 1:200 000
Berlin: Tourist Verlag, 1985–86.
9 sheets, all published.
Legend in 4 languages.

Geological and geophysical

Geologische Karte der DDR. Karte ohne känozoische Bildungen 1:500 000
Berlin: ZGI, 1962.
2 sheets, both published.
Pre-Caenozoic geology.

Geologische Karte der DDR. Karte der quartären Bildungen 1:500 000
Berlin: ZGI, 1973.

Administrative

Bezirke und Kreise der DDR   *ca* 1:1 000 000
Berlin: Tourist Verlag, 1986.
2-colour general administrative map.
Also available reduced to A3 and A4 format.

Verwaltungskarte der DDR   1:600 000
Berlin: Tourist Verlag.
Shows Bezirke and Kreise.
Legend in 3 languages.

DDR. Verwaltungskarte   1:200 000
Berlin: Tourist Verlag, 1981–.
14 sheets, all published.
Maps of individual Bezirke, Kreis and Gemeinde boundaries.

Human and economic

Gestaltung des sozialistischen Landeskultur in der DDR   1:750 000
Gotha: Haack, 1984.
Main map shows land use; supplementary maps show
physiography, economic regions and areas of landscape
conservation.

Sehenswürdigkeiten der DDR   1:600 000
Berlin: Tourist Verlag.
Legend in 8 languages.

Town maps

Stadtplan Berlin
Berlin: Tourist Verlag, 1986.
Covers entire urban area; inner city at 1:10 000
40 pp. text + street index.
4 languages.

German Federal Republic

The cartographic production of the German Federal
Republic is prolific, and it was estimated in the late 1970s
that more than 1100 institutions were engaged in official
mapping of some kind, while there were additionally about
350 commercial cartographic enterprises. Official mapping
is strongly regionalized, with each of the states, or *Länder*,
having its own *Landesvermessungsamt* (land survey
department), entirely responsible for the topographic series
at 1:100 000 and larger, while also contributing to smaller-
scale federally produced series. There are also survey
offices in Berlin and in the *Freie und Hansestädte* of
Hamburg and Bremen. In spite of this decentralization,
the various map series are drawn to common specfications
and the sheet numbers form a nationwide sequence. We
have therefore treated all the topographic series as national
series in both catalogue and indexes. Maps at scales of
1:100 000 and larger should, however, be ordered from
the individual Länder authorities, whose addresses are all
given below. Coordination of activities between the Länder
is achieved through the **Arbeitsgemeinschaft der
Vermessungsverwaltungen der Länder der
Bundersrepublik Deutschland (AdV)**, a non-government
committee of all the Länder administrations.

Thematic mapping is also decentralized, and most
Länder have separate geological and soil map producing
agencies as well as offices producing a variety of other
thematic cartography. As this thematic material is more
idiosyncratic in character, we have listed it under the
individual Länder.

The primary federal mapping agency is the **Institut für
Angewandte Geodäsie (IfAG)** under the Federal Ministry
of the Interior, and is based at Frankfurt AM, with offices
also in Berlin and a satellite observation station at
Kotzting. As well as geodetic, cartographic and
photogrammetric research, IfAG also publishes the
smaller-scale official maps of the Federal Republic. These
include the *Topographische Übersichtskarte* 1:200 000, and
the general maps at 1:500 000 and 1:1 000 000. IfAG also
holds stocks of, and continues to supply, map series of the
former German Reich covering areas outside the Federal
Republic. These maps, at scales of 1:25 000, 1:100 000,
1:200 000 and 1:300 000, have not been revised, but are

for the most part still available in their prewar editions.

The *Topographische Übersichtskarte 1:200 000* (TÜK
200) in 44 sheets (12 of them prepared by the Bavarian
land survey department) was started in 1973 and is now in
its second revision. The sheets are available in versions
with or without relief shading and colour infilling for
roads, and there is also a version showing only hydrology,
relief and the toponyms of these physical features. This
map is on the Gauss–Krüger projection, with graticule and
Gauss–Krüger net shown in the margins. The contours are
at 25 or 12.5 m intervals. A digitally processed satellite
image version of the *Frankfurt AM-West* sheet was
published in 1981. There are also four large format
Umgebungskarten based on this series which include
through-route maps of towns. TÜK 200 has been the
subject of research into digital methods of updating in
raster mode using the Scitex System. The use of digital
raster methods for colour separation has also been tried on
this series.

The 1:500 000 *Übersichtskarte* (ÜK 500) is available as
four standard sheets of the military Series 1404, issued in
1977, but these do not cover the whole country, and so a
special-format four-sheet series (the *Grossblätter*) has been
published. The projection of this map is Lambert
conformal conic, and it has a UTM grid.

IfAG has also published a *Gazetteer of the Federal
Republic of Germany* prepared in accordance with United
Nations recommendations. This gazetteer is in two parts
contained in a single volume. The first part is an
introduction in German and English, the second (in 738
pages) is the gazetteer itself.

A Scitex System has been installed at IfAG for digital
cartography, and since 1982 all the graphical data
processing activities have been placed in a new centralized
computer graphics unit.

The principal topographic series published by the
Länder authorities are at the scales of 1:25 000, 1:50 000
and 1:100 000. A complete cover is available at all three
scales, although in Bavaria the 1:100 000 series is available
only in a provisional edition (*Behelfsausgabe*) photoreduced
from the 1:50 000. These series are all drawn on a
Gauss–Krüger projection, and sheet lines are along lines of

latitude and longitude. Colour codes are used for the covers of the series: green for 1:25 000, blue for 1:50 000 and yellow-orange for the 1:100 000 and 1:200 000.

Until recently, the 1:25 000 topographic map (TK 25) was the basic scale series, and in most Länder the smaller scales continue to be derived from it. The map has contours at 10 m. The series is issued in several different forms, the main variations being a *Normal* three-colour edition (TK 25N), an edition with forestry overprint (TK 25Nw) available for most areas except Schleswig-Holstein, and a monochrome outline edition (*Arbeitskarte*, TK 25A). A water and contours edition (TK 250H) is published for Baden-Württemberg and Saarland, an edition with administrative boundary overprint (*Ausgabe mit Verwaltungsgrenzen*) is available for Hessen, and a relief shaded edition with walking routes printed in red for the southern Alpine areas of Bayern. 1:25 000 sheets each cover an area of about 11 × 11 km, and are identified by a four-digit coordinate reference (note that the *first* two digits in this coordinate are the northings, and the last two the eastings).

The 1:50 000 series (TK 50) covers the whole of the Federal Republic in 558 sheets and also has a 10 m contour interval. As with the 1:25 000 series, there are several variant editions. The *Normal* edition is in four colours and includes woodland shown in green. There are also editions with coloured road infilling (TK 50Str), with shaded relief (TK 50Sch), and with walking and cycling routes and other tourist information (TK 50W + RW). Water and contour (TK 500H) and outline (TK 50A) editions are also available. Finally, the 1:50 000 is also published in a military edition (Series M745) which has UTM grid and trilingual legend. As with the 1:25 000 scale series, not all these variations are available for all areas. The 1:50 000 sheets each cover an area of about 22 × 22 km (equal to four sheets of the 1:25 000 series) and are identified by the prefix L and a four-digit coordinate.

The current 1:100 000 series was launched in the early 1960s as the preparation of the new 1:50 000 neared completion. This map has 20 m contours. Again, there is a choice of editions, with a four-colour *Normal* edition, a relief shaded edition, and a water and contour edition being the main types. A test sheet in the 1:100 000 series was issued in 1984 exploring new ways of showing settlement density and the location of industrial areas. The 1:100 000 sheets each cover an area of about 45 × 45 km and represent the area covered by four 1:50 000 sheets. They are identified by the prefix C and a four-digit coordinate.

The topographic series are now revised by almost all Länder on a 5 year cycle.

In addition to the standard series just described, the Länder surveys also publish a wide range of other maps, including general maps (*Übersichtskarten*) and specially formatted maps of recreation areas, nature parks and the environs of towns. Numerous administrative maps and maps of road networks and distances are also published. Some Länder have a series of *Kreiskarten* formatted to the individual administrative *Kreise*. Thus, for example, Nordrhein-Westfalen is covered completely by such a series of 1:50 000 sheets (SK 50K), each of which has full-colour topographic detail plus an overprint of the *Kreis* and *Gemeinde* boundaries. Descriptive texts are also sometimes included on the maps in this series.

Since 1967, a larger-scale basic map for the whole of the Federal Republic has been in progress. This *Deutsche Grundkarte* (DGK 5) is a contoured map at a scale of 1:5000. It is currently about 60 per cent complete for the country as a whole. In Niedersachsen it is complete and is

now used as the true base map from which all the smaller scales are derived. There is also a series of orthophotomaps (*Luftbildkarte*) at the same scale which was started in 1973.

The Länder survey authorities are also responsible for the property cadaster. Cadastral mapping scales vary, and are in some cases as large as 1:1000 and 1:500. Since 1977, a policy of holding all property boundaries and associated planimetric data in digital form has been adopted. The aim is to integrate these digital files with those pertaining to the tax roll and property register. Large cities also have their own cadastral authorities producing large scale so-called *Flurkarten*.

Many mapping authorities of German cities have a prolific output of maps, ranging from large-scale planimetric maps, via thematic maps of building and land use, pollution and traffic noise to smaller-scale recreation, tourist and public facility maps.

The **Deutsches Hydrographisches Institut (DHI)** is responsible for marine charting, and its publications include some bathymetric charts including the 1:1 000 000 plotting sheets for two sheets of the international GEBCO series.

Geological mapping of the country at the federal level is carried out by the **Bundesanstalt für Geowissenschaften und Rohstoffe (BfGR)**, Hannover. The main product, apart from some small-scale general maps, is the *Geologische Übersichtskarte* 1:200 000 which is being produced in collaboration with the provincial geological survey organizations. It is about 50 per cent complete. A completely revised edition of the 1:1 000 000 geological map is also in preparation. The 1:1 000 000 map of near-surface mineral resources (*Karte der oberflächennahen mineralischen Rohstoffe*), published in 1982, has its thematic content stored in digital form.

BfGR has also published a geochemical atlas of the Federal Republic (*Geochemischer Atlas der Bundesrepublik Deutschland*) which summarizes the findings of a systematic multi-element geochemical survey of the GFR begun in 1977. The main sequence of maps plots heavy metal concentrations in colour at 1:2 000 000 with a three grid square resolution, and the atlas serves as a tool both for environmental protection and for mineral prospecting. The data are stored in the GEOMULDAT database, from which large-scale distribution maps may also be plotted, and some examples of these are also included in the atlas. The atlas is distributed by **E. Schweizerbart'sche Verlagsbuchhandlung**.

BfGR also collaborates with UNESCO in the production of several earth science map series of Europe, and carries out many cooperative resource mapping programmes overseas.

Larger-scale geological mapping is carried out by the geological surveys of the Länder and there are series at 1:25 000, 1:50 000 and 1:100 000 as well as single-sheet general maps of each state. The series cover is very patchy, however, and many sheets are out of print. Although we have included entries for the main products in the catalogue section, it is necessary to consult the catalogues issued by the Länder surveys to appreciate the full range of mapping available.

Soil surveying is also carried out by the geological surveys of the individual Länder, but is coordinated through a pedological working group, which has published a guide to the preparation of soil maps which sets uniform standards for the description and mapping of soil units. Systematic soil survey was originally carried out at 1:25 000 or 1:50 000, but due to the slow progress of the programme the 1:50 000 scale has now been adopted as standard. A new general soil map of the Federal Republic at 1 : 1 000 000 was published by BfGR in 1986.

Many other maps and map series in the environmental thematic area are published by those regional government and semi-government organizations responsible for forestry, water supply, conservation and other environmental issues. These include forest maps, maps of vegetation, maps of water quality and of environmental pollution. Addresses of many of these agencies have been included in the Address directory below.

Geomorphological mapping has been carried out by as many as forty university groups in the Federal Republic. In 1976, a Priority Programme on Geomorphological Mapping (GMK Schwerpunktprogramm) was set up by the Deutsche Forschungsgemeinschaft, Bonn. The aim was to map samples of a complete range of German landscapes to a standard specification using a legend which reached its final form in 1979 (4th edn). The cartography is carried out by IfAG and printed in eight colours using standard 1 : 25 000 scale base maps. The project was to be terminated in 1986 when 25 sheets should have been published in the 1 : 25 000 series (GMK25) and eight at 1 : 100 000 (GMK100). However, negotiations were under way to interest the geological services of the Länder in geomorphological mapping, and a group has also been formed to investigate digital production of these maps. These maps are distributed by **GeoCenter**, Stuttgart.

A major thematic mapping project is represented by the *Deutscher Planungsatlas* published by the **Akademie für Raumforschung und Landesplanung (AfRL)**, Hannover. This has been in progress since 1960 and a very wide range of thematic maps has been issued at scales usually of 1 : 500 000 or 1 : 1 000 000. There is a volume for each of the Länder, but many of the early complete volumes are out of print. However, numerous individual sheets are available and more are added from time to time: the atlas is an evolving concept. The last complete volume to be made available was *Band I—Nordrhein-Westfalen* (1982) in 102 sheets. AfRL publications are distributed by **Curt R. Vincentz Verlag**.

A major thematic mapping programme covering the whole Federal Republic at smaller scales is represented by the *Atlas zur Raumentwicklung*, essentially an atlas of regional planning. This has been prepared by the **Bundesforschungsanstalt für Landeskunde und Raumentwicklung (BfLR)**, Bonn-Bad Godesberg. It is in 11 volumes which combine texts, maps and graphics to analyse regional disparities in the Federal Republic in the 1970s. The themes of the volumes are administrative boundaries, labour, education, transportation, population, housing, leisure, land use, environment, disadvantaged areas and regional planning. The volumes are available singly or as a set. BfLR also collects information on administrative boundaries and publishes a number of administrative maps including a 1 : 300 000 series of the *Gemeindegrenzen*.

The **Deutscher Wetterdienst (DWD)**, Offenbach am Main, is publishing a climatic atlas of the Federal Republic, *Das Klima der Bundesrepublik Deutschland*, and two volumes (on rainfall data 1931–60) have been issued with a third (temperature) in preparation. DWD has also published a series of climatic atlases of the individual Länder, and a number of sheet maps.

The **Deutsche Bundesbahn (DBB)**, Mainz, publishes a number of general travel service maps and rail network maps of Germany and Europe.

Two important overseas mapping projects carried out in Germany are the *Afrika-Kartenwerk* programme sponsored by the Deutsche Forschungsgemeinschaft and published by **Gebrüder Borntraeger**, and the *Tübinger Atlas des Vorderen Orients*, prepared at the University of Tübingen and published by **Dr Ludwig Reichert Verlag**.

Commercial publishers

Of the exceptionally long list of commercial map publishers we discuss here only those which produce significant mapping within the frame of reference adopted for this book. We have not therefore included some well known firms, such as Karl Wenschow, Ernst Klett or Justus Perthes, whose production is almost exclusively of educational maps, atlases or globes.

The **Kartographisches Institut Bertelsmann** is a prolific commercial publisher, and also owns a number of other familiar names in cartography including **Reise- und Verkehrsverlag**. Among its products is the *Autoatlas Deutschland* 1 : 200 000, which also includes 37 town centre maps and a 35 000 entry gazetteer including postcodes. The RV maps are distributed by **GeoCenter Verlagsvertrieb**, München. They include series of walking and cycling maps, and a long list of tourist maps of foreign countries. Major urban areas in Germany have been covered by a new series of *RV Stadtatlanten*. These are described as being suitable for use in electronic navigation systems.

Bollmann Bildkarten Verlag publishes the isometric city centre maps of major cities both in Germany and abroad which are the work of Hermann Bollmann. A selection of these are available in three *Staedte* atlases.

The **Deutsche Alpenverein** works in association with the Austrian Alpine Club to produce the very high-quality alpine walking maps for which they are famous. Some 50 maps have been published, mostly at 1 : 25 000 with 20 m contours. Most sheets are of the alpine areas of Austria and south Germany alpine areas, but some sheets of the Himalayas, including Chomolongma (Mt Everest), are still available.

Falk-Verlag, Hamburg, is a major publisher of street maps and atlases of towns of Germany and of many other countries.

Heinz Fleischmann is the publisher of the well known *Kompass* series of walking maps which cover parts of Germany, Austria and Italy. There are also a number of *Kompass* street maps of towns in these countries.

Fritsch Landkartenverlag publishes a very interesting variety of outdoor recreation maps in the scale range 1 : 35 000 to 1 : 100 000, and covering areas of south Germany. A few street maps of north Bavarian towns are also published.

JRO Verlagsgesellschaft has a 1 : 300 000 series of road maps of Germany with postcode maps on the reverse, and also publishes several indexed administrative and business maps (*Organisationskarten*) giving administrative and postcode areas.

Karto+Grafik Verlag publishes a series of tourist maps which include some overseas areas for which maps are not usually easy to obtain. Initially published in German, many sheets are now also appearing in English.

Mairs Geographischer Verlag is famous for the *Generalkarte* 1 : 200 000, which covers the Federal Republic. Similar series cover the German Democratic Republic, Austria, Switzerland and Denmark. *Müllers Grosses Deutsches Ortsbuch* serves as an index to the *Generalkarte*. Mair also has a series of *Generalstadtpläne* of towns in the German Federal Republic and the German Democratic Republic.

Nelles Verlag distributes maps of north-east Nepal and the Kathmandu valley and also publishes its own series of APA maps of countries in Asia.

Ravenstein publishes a 1:250 000 German road map series and a series of international road maps at 1:800 000. The programme of this company has recently been extended to include maps of more distant areas overseas including Morocco, Tunisia, the USA, India and the Far East.

Städte Verlag, Stuttgart, is a long-established publisher of street maps of German towns and cities. About 580 such maps have been issued! In addition, there are series of *Kreis- und Freizeitkarten* at scales of 1:50 000–1:100 000. There are also *Stadt- und Wanderpläne* which combine large-scale street maps with smaller-scale leisure maps, the latter showing walking routes, cycle paths or Langlauf skiing routes. The maps have street and placename indexes.

VWK-Ryborsch Verlag, Obertshausen, has an active programme of publishing satellite image maps of regions of Germany and neighbouring countries. These image maps are in four colours, derived from Landsat imagery and accompanied by explanatory texts. The scale is 1:500 000. A number of conventional maps of the world and regions of the world are also published by VWK, and there is a map of airports in Europe, North Africa and the Near East.

Karl Wachholz Verlag, Neumünster, has published a number of topographical and air photo atlases of the Länder. The former are published in association with the Länder surveys and include extracts of the various official topographic and thematic maps to illustrate regional landscape descriptions.

Kartographischer Verlag Wagner, Berlin, publishes the so-called *Wagnerkarten* of continents, using a colour simulation of the vegetation cover.

Westermann, Braunschweig, is chiefly an educational publisher, responsible for the famous Diercke atlases. There is also a *Diercke Lexikon Deutschland* with descriptions of each city and Landkreis, and a series of A2 format *Handkarten* of regions of Germany which have satellite image and economic maps on the reverse.

Further information

Extremely comprehensive and well presented catalogues and indexes are published by all the main federal and provincial mapping authorities, and it is necessary to consult these to get a full appreciation of all the mapping available. Most commercial publishers also issue catalogues.

There is a large literature on modern German mapping. Of particular usefulness is the *Report on the Cartographic Activities of the Federal Republic of Germany in the period 1980–1984* prepared by U. Freitag for the **Deutsche Gesellschaft für Kartographie (DGfK)**, and the massive three-volume *Kartographie der Gegenwart in der Bundesrepublik Deutschland '84*, published by DGfK for the 3rd Dreiländertagung (this comprises a volume of text and two volumes of map examples).

The periodical *Kartographische Nachrichten* has numerous articles on new developments in German cartography, e.g. Grimm, W. (1983) Die Weiterentwicklung der Topographischen Karte 1:100 000, *Kartographische Nachrichten*, **33**(4), 126–128.

There is a large literature on geomorphological mapping. For English readers, a description of the GMK Schwerpunkt Programm is given by Barsch, D. and Liedtke, H. (1980) Principles, scientific value and practical applicability of the geomorphological map of the Federal Republic of Germany, *Zeitschrift für Geomorphologie NF*, Suppl. 36, 296–313; and a review of the literature itself is made by Moller, K. (1982) Verzeichnis der Literatur zum GMK-Schwerpunktprogramm, *Berliner Geographische Abhandlungen*, **35**, 127–130, 296–313.

Land resource mapping is described in Heide, G. and Muckenhausen, E. (1980) Land resource evaluation in the Federal Republic of Germany, in *Land Resource Evaluation*, edited by J. Lee and L. Van der Plas, Luxembourg: Commission for the European Communities, pp. 38–50.

Climatic maps of the Federal Republic are described and listed in: Kalb, M. (1979) Klimakarten de Deutschen Wetterdienstes für den Bereich der Bundesrepublik Deutschland, *Natur und Landschaft*, **54**, 250–252.

Addresses

Arbeitsgemeinschaft der Vermessungsverwaltungen der Länder der Bundesrepublik Deutschland (AdV)
Niedersächsische Minister des Innern, Lavesallee 6, Postfach 221, D-3000 HANNOVER 1.

Institut für Angewandte Geodäsie (IfAG)
Aussenstelle Berlin, Stauffenbergstrasse 11–13, D-1000 BERLIN 30; and Richard-Strauss-Allee 11, D-6000 FRANKFURT 70.

Deutsches Hydrographisches Institut (DHI)
Bernhard-Nocht-Strasse 78, Postfach 200, D-2000 HAMBURG 4.

Bundesanstalt für Geowissenschaften und Rohstoffe (BfGR)
Stilleweg 2, Postfach 51 01 53, D-3000 HANNOVER 51.

Akademie für Raumforschung und Landesplanung (AfRL)
Hohenzollernstrasse 11, D-3000 HANNOVER 1.

Bundesforschungsanstalt für Landeskunde und Raumordnung (BfLR)
Am Michaelshof 8, D-5300 BONN 2.

Deutscher Wetterdienst (DWD)
Abteilung Klimatologie, Frankfurter Strasse 135, Postfach 185, D-6050 OFFENBACH AM MAIN.

Deutsche Bundesbahn (DBB)
Zentrale Transportleitung, Kaiserstrasse 3, D-6500 MAINZ.

Bundesanstalt für Naturschutz und Landschaftsökologie (BFANL)
Konstantinstrasse 110, D-5300 BONN 2.

Deutsche Gesellschaft für Kartographie
Klüsenerskamp 10, D-4600 DORTMUND.

Deutscher Alpenverein
Praterinsel 5, D-8000 MÜNCHEN 22.

Kartographisches Institut Bertelsmann
Carl-Bertelsmann-Strasse 161, Postfach 5555, D-4830 GÜTERSLOH 1.

Bollmann Bildkarten Verlag GmbH
Lilienhalplatz 1, Postfach 1526, D-3300 BRAUNSCHWEIG.

Falk Verlag GmbH
Buchardstrasse 8, Postfach 10 21 22, D-2000 HAMBURG 1.

Heinz Fleischmann GmbH
Prinz-Karl-Strasse 47, D-8130 STARNBERG.

Fritsch Landkartenverlag
Hirschberger Strasse 7, Postfach 1144, D-8670 HOF/SAALE.

Gebrüder Borntraeger
Johannestrasse 3A, D-7000 STUTTGART.

GeoCenter ILH
Postfach 80 08 30, D-7000 STUTTGART.

JRO Kartographische Verlagsgesellschaft mbH
Kirschstrasse 12–16, D-8000 MÜNCHEN 50.

Karto+Grafik Verlagsgesellschaft mbH
Schonberger Weg 15, D-6000 FRANKFURT AM MAIN 90.

Mairs Geographischer Verlag
Marco Polo Strasse, D-7302 OSTFILDERN 4.

Post und Ortsbuchverlag Müller
Horather Strasse 172, D-5600 WUPPERTAL 1.

Nelles Verlag GmbH
Schleissheimer Strasse 371b, D-8000 MÜNCHEN 45.

Ravenstein Verlag GmbH
Wielandstrasse 31/35, D-6000 FRANKFURT AM.

Dr Ludwig Reichert Verlag
Reisstrasse 10, D-6200 WIESBADEN.

RV Reise- und Verkehrsverlag
Neumarkterstrasse 18, D-8000 MÜNCHEN 80.

E. Schweizerbart'sche Verlagsbuchhandlung
Johannestrasse 3A, D-7000 STUTTGART 1.

Städte-Verlag E.v. Wagner und J. Mitterhuber Abt. VE
Steinbeissstrasse 9, Postfach 2080, D-7012 FELLBACH b. STUTTGART.

Süddeutscher Verlag GmbH
Postfach 20 22 20, Sendlinger Strasse 80, D-8000 MÜNCHEN 2.

Curt R. Vincentz Verlag
Schiffgraben 41–43, Postfach 6247, D-3000 HANNOVER 1.

VWK-Ryborsch GmbH
Verlag für Wirtschafts- und Kartographie Publikationen, Postfach 2105, D-6053 OBERTSHAUSEN 2.

Karl Wachholtz Verlag GmbH
Gänsemarkt 1–3, Postfach 2769, D-2350 NEUMÜNSTER.

Kartographischer Verlag Wagner
Georg-Wilhelm Strasse 1, D-1000 BERLIN 31.

Westermann Verlag GmbH
Georg-Westermann-Allee 66, Postfach 5520, D-3300 BRAUNSCHWEIG.

Baden-Württemberg

Landesvermessungsamt Baden-Württemberg (LVAB-W)
Buchsenstrasse 54, Postfach 1115, D-7000 STUTTGART.

Geologisches Landesamt Baden-Württemberg (GLAB-W)
Albertstrasse 5, D-7800 FREIBURG i. Br.

Landesanstalt für Umweltschutz Baden-Württemberg
Institut für Ökologie und Naturschutz, Bannwaldallee 32, D-7500 KARLSRUHE 21.

Ministerium für Ernährung, Landwirtschaft, Umwelt und Forsten Baden-Württemberg
Marienstrasse 3, Postfach 491, D-7032 SINDELFINGEN.

Landessammlungen für Naturkunde
Botanische Abteilung, Erbprinzenstrasse 13, Postfach 3949, D-7500 KARLSRUHE 1.

Bayern

Bayerisches Landesvermessungsamt (BLVA)
Postfach, Alexandrastrasse 4, D-8000 MÜNCHEN 22.

Bayerische Geologisches Landesamt (GLAB)
Hesstrasse 128, D-8000 MÜNCHEN 40.

Bayerisches Landesamt für Wasserwirtschaft
Lazaretstr 67, D-8000 MÜNCHEN 19.

Berlin

Senator für Bau- und Wonungswesen (SfBWW)
Abteilung V, Vermessungswesen, Mansfelder Strasse 16, D-1000 BERLIN 31.

Senator für Stadtentwicklung und Umweltschutz (SfSU)
Lindestrasse 20–25, D-1000 BERLIN 61.

Bremen

Kataster- und Vermessungsverwaltung
Wilhelm-Kaisen-Brucke 4, D-2800 BREMEN 1.

Hamburg

Freie und Hansestadt Hamburg
Baubehörde-Vermessungsamt-Wexstrasse 7, Postfach 30 05 31, D-2000 HAMBURG 36.

Geologisches Landesamt Hamburg
Oberstrasse 88, D-2000 HAMBURG 13.

Hessen

Hessisches Landesvermessungsamt (HLVA)
Schaperstrasse 16, Postfach 32 49, D-6200 WIESBADEN 1.

Hessisches Landesamt für Bodenforschung (HLAB)
Leberberg 9, D-6200 WIESBADEN.

Der Hessische Minister für Landesentwicklung, Forsten und Naturschutz
Holderlinstrasse 1–3, D-6200 WIESBADEN.

Niedersachsen

Niedersächsisches Landesverwaltungsamt (NSLVA)
Landesvermessung, Warmbuchenkamp 2, D-3000 HANNOVER 1.

Niedersächsisches Landesamt für Bodenforschung (NSLAB)
Stilleweg 2, Postfach 51 01 53, D-3000 HANNOVER 51.

Niedersächs. Minister für Ernährung, Landwirtschaft und Forsten
Calenberger Strasse 2, D-3000 HANNOVER 1.

Niedersächs. Landesamt für Wasserwirtschaft
Langelinienwall 27, D-3200 HILDESHEIM.

Nordrhein-Westfalen

Landesvermessungsamt Nordrhein-Westfalen (LVAN-W)
Postfach 20 50 07, Muffendorfer Strasse 19–21, D-5300 BONN 2.

Geologisches Landesamt Nordrhein-Westfalen (GLAN-W)
De Greiff Strasse 195, D-4150 KREFELD.

Landesamt für Wasser und Abfall Nordrhein-Westfalen
Auf dem Draap 25, D-4000 DUSSELDORF 1.

Landesamt für Ökologie, Landschaftsentwicklung und Forstplanung (LOLF)
Nordrhein-Westfalen, Leibnizstrasse 10, D-4350 RECKLINGHAUSEN.

Rheinland-Pfalz

Landesvermessungsamt Rheinland-Pfalz (LVAR-P)
Ferdinand-Sauerbruch-Strasse 15, Postfach 1428, 5400 KOBLENZ.

Geologisches Landesamt Rheinland-Pfalz (GLR-P)
Emmeranstrasse 36, Postfach 2045, D-6500 MAINZ.

Landesamt für Wasserwirtschaft
Am Zollhafen 9, D-6500 MAINZ.

Saarland

Landesvermessungsamt des Saarlandes (LVAS)
Neugrabenweg 2, D-6600 SAARBRÜCKEN.

Geologisches Landesamt des Saarlandes (GLAS)
Am Tummelplatz 7, D-6600 SAARBRÜCKEN 1.

Landesamt für Umweltschutz (LAU)
Naturschutz und Wasserwirtschaft, Hellwigstrasse 14, D-6600
SAARBRÜCKEN 3.

Minister für Umwelt, Raumordnung und Bauwesen (MURB)
Staatliches Strassenbauamt Saarbrücken, Halbergstrasse 84,
D-6600 SAARBRÜCKEN.

Schleswig-Holstein

Landesvermessungsamt Schleswig-Holstein (LVAS-H)
Postfach 5070, D-2300 KIEL 1.

Geologisches Landesamt Schleswig-Holstein (GLAS-H)
Mercatorstrasse 7, Postfach 5049, D-2300 KIEL 21.

Minister für Ernahrung, Landwirtschaft und Forsten
Dunsternbrooker Weg 104, D-2300 KIEL.

Catalogue

Atlases and gazetteers

Atlas zur Raumentwicklung
Bonn: BfLR, 1976–.
11 volumes, 10 published.

Deutscher Planungsatlas
Hannover: AfRL, 1960–.
10 volumes, some out of print.

Geographisches Namenbuch Bundesrepublik Deutschland
Frankfurt AM: IfAG, 1981.
pp. 55 + pp. 738.

Müllers grosses deutsches Ortsbuch: Bundesrepublik Deutschland   21st
edn   Joachim Müller
Wuppertal: Post und Ortsbuchverlag Müller, 1982.
pp. 1194.

*Diercke Lexikon Deutschland. Bundesrepublik: Deutschland und
Berlin (West)*
Braunschweig: Westermann, 1985.
Descriptive gazetteer of cities and Landkreise.
pp. 480.

General

Bundesrepublik Deutschland   1 : 1 000 000
Frankfurt AM: IfAG, 1979.
Available in three versions: normal with shaded relief,
orohydrographic or administrative.

Bundesrepublik Deutschland/DDR Weltraumbildkarte   1 : 500 000
Braunschweig: Westermann, 1981.
2 sheets, both published.

Deutschland. Satellitenbildkarten   1 : 500 000
Obertshausen: VWK-Ryborsch GmbH, 1981–82.
4 sheets, all published.

Übersichtskarte   1 : 500 000   Grossblätter
Frankfurt AM: IfAG, 1978–.
4 sheets, all published.
Available in four versions: shaded relief, normal, orohydrographic
or administrative.

Topographic

Topographische Übersichtskarte   1 : 200 000   TÜK 200
Frankfurt AM: IfAG, 1974–.
44 sheets, all published. ■
Available in three versions: normal with shaded relief, normal
without shaded relief or road infills, or orohydrographic.
Some special sheets (*Umgebungskarten*) also available.

Topographische Karte   1 : 100 000   TK 100
Landesvermessungsämtern der Bundesländer, 1962–.
151 sheets, all published. ■
Several versions are usually available, i.e. normal, relief shaded
and orohydrographic.

Topographische Karte   1 : 50 000   TK 50
Landesvermessungsämtern der Bundesländer, 1956–.
558 sheets, all published. ■

Topographische Karte   1 : 25 000   TK 25
Landesvermessungsämtern der Bundesländer, 1953–.
2082 sheets, all published.

Geological and geophysical

Geochemischer Atlas Bundesrepublik Deutschland   H. Fauth,
R. Hindel, U. Siewers and J. Zinner
Hannover: BfGR, 1985.
pp. 79.

Geologische Karte der Bundesrepublik Deutschland   1 : 1 000 000
3rd edn
Hannover: BfGR, 1981.

*Bundesrepublik Deutschland. Gebiete mit oberflächennahen
mineralischen Rohstoffen*   1 : 1 000 000
Hannover: BfGR, 1982.
With text.

Schwerekarte der Bundesrepublik Deutschland   1 : 500 000
Hannover: BfGR, 1983–.
3 sheets.
Bouguer gravity anomaly map.

Aeromagnetische Karte der Bundesrepublik Deutschland   1 : 500 000
Hannover: BfGR, 1976.
2 sheets, both published.
With/without geology.

Geologische Übersichtskarte   1 : 200 000
Hannover: BfGR, 1973–.
42 sheets, 20 published. ■

Geomorphologische Karte der Bundesrepublik Deutschland   1 : 25 000
GMK 25
Berlin: IfAG/GMK Schwerpunktprogramm.

Environmental

Bundesrepublik Deutschland. Bodenkarte   1 : 1 000 000
Hannover: BfGR, 1986.
Soil map with 76 pp. legend and explanatory notes.
In German and English.

Naturräumliche Gliederung   1 : 200 000
Bonn: BfLR, 1959–.
74 sheets, 59 published.

Administrative

Gemeindegrenzenkarte   1 : 1 500 000   Gebietstand 1980
Bonn: BfLR.

Kreisgrenzenkarte   1 : 1 500 000   Gebietstand 1980
Bonn: BfLR.

Karte der Wahlkreise   1 : 1 500 000
Bonn: BfLR.
Constituency map, available for the 10th and 11th Bundestag
elections.

Gemeindegrenzenkarte   1 : 300 000   Gebietstand 1980
Bonn: BfLR.
25 sheets, all published.

GERMAN FEDERAL REPUBLIC

**1 : 200 000 topographic
1 : 200 000 geological**

Sheet Numbering System
Prefix CC

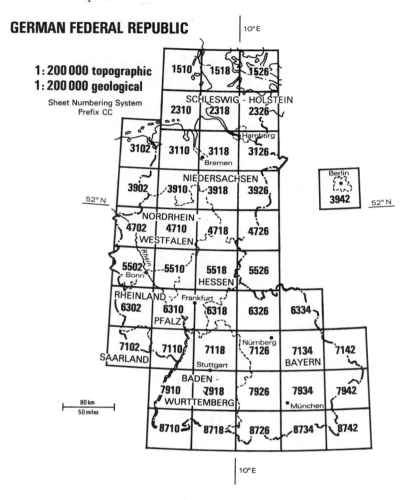

Human and economic

Streckenkarte der Deutschen Bundesbahn   1 : 750 000   2nd ed
Frankfurt AM: DBB, 1985.

Arbeitskarte Offentiche Wasserversorgung   1 : 200 000
Frankfurt AM: IfAG, 1977–.
44 sheets, all published. ■
Map of public water supply.

Town maps

For maps of Berlin, see regional section below. For street maps of
other towns, see commercial publishers cited in text.

BADEN-WÜRTTEMBERG

Atlases and gazetteers

Topographischer Atlas Baden-Württemberg
Stuttgart: LVAB-W/Neumünster: Karl Wachholtz, 1979.
pp. 259.

General

Baden-Württemberg   1 : 1 000 000   UK 1000
Stuttgart: LVAB-W, 1983.

Reliefkarte Baden-Württemberg   1 : 1 000 000   RK 1000
Stuttgart: LVAB-W, 1983.

Reliefkarte Baden-Württemberg   1 : 600 000   RK 600
Stuttgart: LVAB-W, 1983.
Also available at 1 : 300 000 (RK 300).

Übersichtskarte Baden-Württemberg   1 : 500 000   ÜK 500
Stuttgart: LVAB-W, 1985.

Topographische Übersichtskarte Baden-Württemberg   1 : 200 000
TÜK 200BW
Stuttgart: LVAB-W, 1984.
2 sheets, both published.

Strassenkarte Baden-Württemberg   1 : 200 000   SK 200
Stuttgart: LVAB-W, 1984.

Entfernungskarte Baden-Württemberg   1 : 200 000   EK 200
Stuttgart: LVAB-W, 1981.

Geological and geophysical

*Karte der Erdbebendenzonen für Baden-
Württemberg*   1 : 350 000   EBK 350
Stuttgart: LVAB-W, 1980.
Earthquake hazard map.

Geologische Übersichtskarte von Baden-Württemberg   1 : 200 000
GÜ 200
Stuttgart: GLAB-W, 1962.
4 sheets, all published.

Geologische Karte von Baden-Württemberg   1 : 100 000   GK 100
Stuttgart: GLAB-W, 1976–.
ca 20 sheets, 2 published.

Geologische Karte von Baden-Württemberg   1 : 25 000   GK 25
Stuttgart: GLAB-W.
286 sheets, *ca* 50% published.
Most with texts.

Environmental

Klima-Atlas von Baden-Württemberg
Bad Kissingen: DWD, 1953.
75 maps + 37 pp. text.

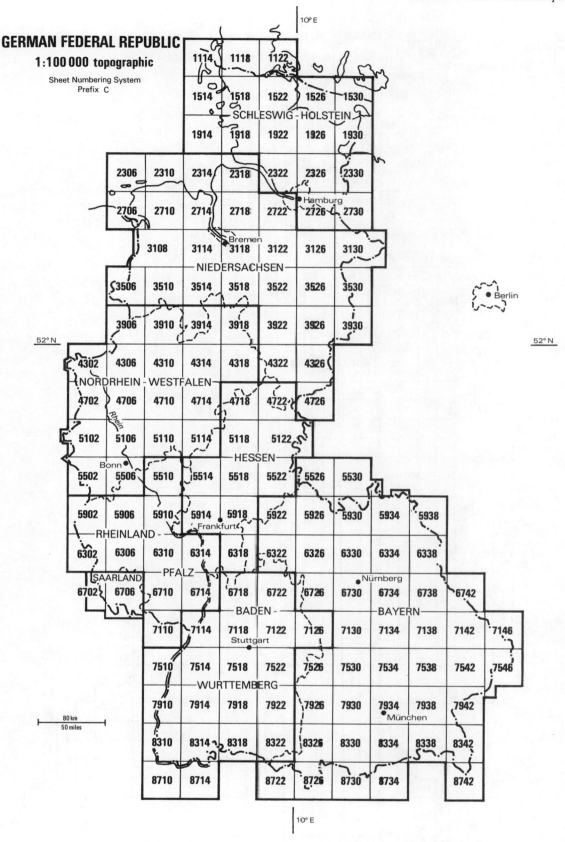

GERMAN FEDERAL REPUBLIC

1:100 000 topographic

Sheet Numbering System
Prefix C

Moorkarte von Baden-Württemberg 1:50 000 MK 50
Stuttgart: LVAB-W.
13 sheets published.

Administrative

Kreiskarte von Baden-Württemberg 1:1 000 000 KK 1000
Stuttgart: LVAB-W, 1985.
Also available at 1:250 000 (KK 250).

Gemeinde- und Kreiskarte von Baden-Württemberg 1:350 000
BW 350
Stuttgart: LVAB-W, 1985.
1984 situation.

Verwaltungs- und Verkehrskarte Baden-Württemberg 1:200 000
VVK 200
Stuttgart: LVAB-W, 1984.
2 sheets, both published.

GERMAN FEDERAL REPUBLIC

1:50 000 topographic

Sheet Numbering System
Prefix L
example:- L 3312
NB Northings before Eastings

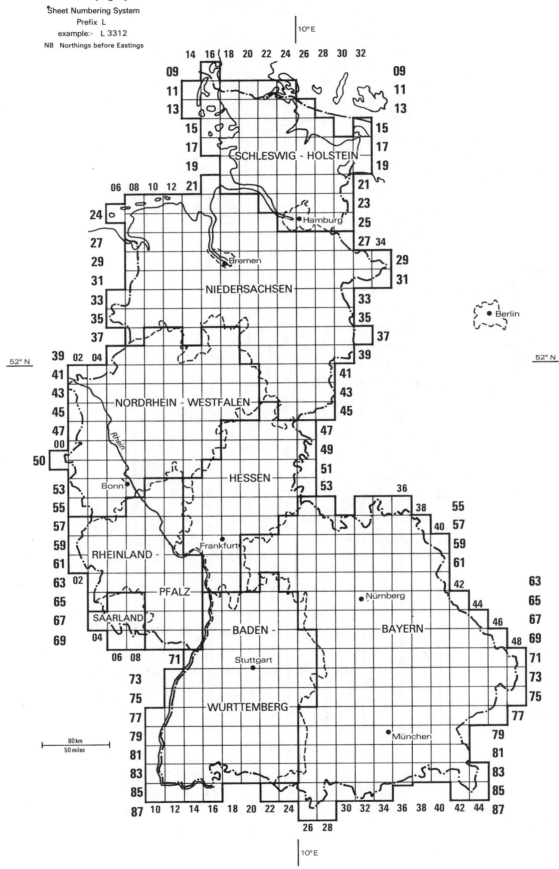

Verwaltungs- und Verkehrskarte Baden-Württemberg   1:100 000
VVK 100
Stuttgart: LVAB-W, 1981–.
18 sheets, all published.

Human and economic

Wandern und Radwandern in Baden-Württemberg   1:200 000
WRW 200BW
Stuttgart: LVAB-W, 1984.
2 sheets, both published.

BAYERN (BAVARIA)

General

Übersichtskarte von Bayern   1:500 000   UK 500
München: BLVA, 1983.
Available in normal, shaded relief with/without road distances,
and administrative versions.
Enlarged 1:300 000 edition (UK 300) also available.

Geological and geophysical

Geologische Karte von Bayern   1:500 000   3rd edn
München: GLAB, 1981.
With/without text.

Geologische Karte von Bayern   1:100 000
München: GLAB, 1951–.
43 sheets, 8 published.

Geologische Karte von Bayern   1:50 000
München: GLAB, 1978–.
157 sheets, 3 published.
With texts.

Hydrogeologische Karte von Bayern   1:50 000
München: GLAB, 1980–.
157 sheets, 2 published.
With texts.

Geologische Karte von Bayern   1:25 000
München: GLAB, 1955–.
558 sheets, *ca* 200 published.
With texts.

Environmental

Klima-Atlas von Bayern
Bad Kissingen: DWD, 1952.
79 maps + 23 pp. text.

Bodenkundliche Übersichtskarte von Bayern   1:500 000
München: GLAB, 1961.
With 168 pp. text.

Bodengüterkarte von Bayern   1:100 000
München: BLVA.
38 sheets, all published.

Bodenkarte von Bayern   1:50 000
München: GLAB.
157 sheets, 14 published.

Bodenkarte von Bayern   1:25 000
München: GLAB, 1957–.
558 sheets, 23 published.
With texts.

Administrative

Verwaltungskarte von Bayern   1:800 000
München: BLVA.

Amtsbezirksübersichtskarte von Bayern   1:100 000
München: BLVA.
38 sheets, all published.

BERLIN

General and topographic

Übersichtskarte Berlin (West)   1:50 000
Berlin: SfBWW, 1980.

Karte von Berlin   1:10 000
Berlin: SfBWW, 1982.
27 sheets, all published.

Geological and geophysical

Geologische Übersichtskarte von Berlin   1:50 000
Berlin: SfSU, 1971.

Environmental

Umweltatlas Berlin
Berlin: SfSU, 1985–87.
2 vols, both published.

HESSEN

General

Hessen   1:500 000   H 500
Wiesbaden: HLVA, 1981.
Available in normal or shaded relief version.
Also with Kreis boundaries, and as two-colour Arbeitsausgabe
with Kreis boundaries.

Hessen   1:200 000   H 200
Wiesbaden: HLVA, 1982.
Available in shaded relief version, normal version with Kreis
boundaries, an Arbeitsausgabe with Kreis or Gemeinde
boundaries.

Geological

Geologische Übersichtskarte von Hessen   1:1 000 000   2nd edn
Wiesbaden: HLAB, 1974.

Geologische Übersichtskarte von Hessen   1:300 000   4th edn
Wiesbaden: HLAB, 1986.

Karte der Bouguer–Schwere in Hessen   1:300 000
Wiesbaden: HLAB, 1984.

Übersichtskarte der Grundwasserbeschaffenheit in Hessen   1:300 000
Wiesbaden: HLAB, 1966.
+ text.

Mineral- und Heilwasservorkommen in Hessen   1:300 000
Wiesbaden: HLB, 1985.

Geologische Karte von Hessen   1:25 000
Wiesbaden: HLB, 1876–.
172 sheets, 160 published.
Text with each sheet.
Many early sheets out of print.

Environmental

Klima-Atlas von Hessen
Bad Kissingen: DWD, 1950.
75 maps + 20 pp. text.

Bodenübersichtskarte von Hessen   1:600 000
Wiesbaden: HLAB, 1958.

Bodenkundliche Übersichtskarte von Hessen   1:300 000
Wiesbaden: HLAB, 1951.

Bodenkarte von Hessen   1:25 000
Wiesbaden: HLAB, 1966–.
172 sheets, 28 published.

Administrative

Hessen 1:1 000 000 H 1000
Wiesbaden: HLVA.
Small monochrome map showing Kreise (H1000 K) or Kreis and
Gemeinde boundaries (H1000 G).

NIEDERSACHSEN

Atlases and gazetteers

Topographischer Atlas Niedersachsen und Bremen
Hannover: NSLVA/Neumunster: Karl Wachholtz, 1977.
pp. 289.

General

Übersichtskarte von Niedersachsen 1:500 000 UKN
Hannover: NSLVA, 1981.
Available in normal or administrative versions.

Entfernungskarten 1:200 000 EK
Hannover: NSLVA.
4 sheets published.
Special edition of *Bezirkskarten* with road distances.

Geological and geophysical

Geologische Karte von Niedersachsen 1:25 000
Hannover: NSLAB, 1924–.
435 sheets, 79 published.

Environmental

Klima-Atlas von Niedersachsen
Offenbach AM: DWD, 1964.
77 maps + 38 pp. text.

Bodenkundliche Standortkarte von Niedersachsen 1:200 000
Hannover: NSLAB, 1974–80.
7 sheets, all published.
Texts in preparation.

*Geowissenschaftliche Karte de Naturraumpotentials von Niedersachsen
und Bremen* 1:200 000
Hannover: NSLAB, 1979–.
8 × 12 thematic sheets, 4 published.
Themes include soils, groundwater, surface and deep mineral
resources.

Bodenkarte von Niedersachsen 1:25 000
Hannover: NSLAB, 1957–.
435 sheets, *ca* 90 published.

Administrative

Bezirkskarten 1:200 000 BK
Hannover: NSLVA.
4 sheets published.
Based on TÜK 200.
Edition with *Gemeinde* also available.

Kreiskarten 1:100 000 KK
Hannover, NSLVA.
38 sheets published.
Series of two colour on varying format based on TK 100.

NORDRHEIN-WESTFALEN

Atlases and gazetteers

Topographischer Atlas Nordrhein-Westfalen
Bonn: LVAN-W, 1968.
pp. 345.

General

Übersichtskarte Nordrhein-Westfalen 1:500 000 SK 500
Bonn: LVAN-W, 1980.
Available in normal, administrative, road and orohydrographic
versions.

Übersichtskarte Nordrhein-Westfalen 1:250 000 SK 250
Bonn: LVAN-W, 1976.
Available in normal, road distance and several administrative
versions.

Geological and geophysical

Geologische Karte von Nordrhein-Westfalen 1:100 000 GK 100
Krefeld: GLAN-W.
19 sheets, 12 published.

Hydrogeologische Karte von Nordrhein-Westfalen 1:100 000 HK
100
Krefeld: GLAN-W.
19 sheets, 4 published.

Hydrogeologische Karte von Nordrhein-Westfalen 1:50 000 HK
50
Krefeld: GLAN-W.
72 sheets, 6 published.

Geologische Karte von Nordrhein-Westfalen 1:25 000 GK 25
Krefeld: GLAN-W.
270 sheets, *ca* 200 published but many out of print.

Ingenieurgeologische Karte von Nordrhein-Westfalen 1:25 000
IK 25
Krefeld: GLAN-W.
270 sheets, 4 published.

Environmental

Klima-Atlas von Nordrhein-Westfalen
Offenbach AM: DWD, 1960.
77 maps + 38 pp. text.

Karte der Grundwasserlandschaften in Nordrhein-Westfalen
1:500 000 2nd edn
Krefeld: GLAN-W.

*Karte der Verschmutzungsgefahrdung de Grundwasservorkommen in
Nordrhein-Westfalen* 1:500 000 2nd edn
Krefeld: GLAN-W.

Bodenkarte von Nordrhein-Westfalen 1:100 000
Krefeld: GLAN-W.
19 sheets, 4 sheets published (1 out of print).

Bodenkarte von Nordrhein-Westfalen 1:50 000
Krefeld: GLAN-W.
72 sheets, 43 published.

Waldfunktionskarte Nordrhein-Westfalen 1:50 000
Recklinghausen: LÖLF, 1974–75.
72 sheets, all published.
Sheets are at 1:25 000 in Rhein and Ruhr area.

Bodenkarte von Nordrhein-Westfalen 1:25 000
Krefeld: GLAN-W, 1868–.
270 sheets, 20 published.

Administrative

Regierungsbezirkskarten 1:200 000 SK 200
Bonn: LVAN-W.
5 sheets, all published.

Kreiskarten 1:50 000 SK 50 K
Bonn: LVAN-W.
32 sheets, all published.

RHEINLAND-PFALZ

Atlases and gazetteers

Topographischer Atlas Rheinland-Pfalz
Koblenz: LVAR-P/Neumünster: Karl Wachholtz, 1973.
pp. 216.

General

Übersichtskarte von Rheinland-Pfalz   1 : 250 000   ÜK 250
Koblenz: LVAR-P, 1981.
Also available with UTM grid.

Rheinland-Pfalz Strassenkarte   1 : 200 000
Koblenz: LVAR-P, 1983.

Geological and geophysical

Geologische Übersichtskarte von Rheinland-Pfalz   1 : 500 000
Mainz: GLR-P, 1979.

Geologische Karte von Rheinland-Pfalz   1 : 25 000
Mainz: GLR-P, 1931–.
156 sheets, 13 published.
Text with each sheet.

Environmental

Klima-Atlas von Rheinland-Pfalz
Bad Kissingen: DWD, 1957.
77 maps + 37 pp. text.

Bodenübersichtskarte von Rheinland-Pfalz   1 : 500 000
Mainz: GLR-P, 1965.

*Übersichtskarte der Bodentypen-Gesellschaften von
Rheinland-Pfalz*   1 : 250 000
Mainz: GLR-P, 1968.

Naturschutz- und Landschaftsgebiete in Rheinland-Pfalz   1 : 200 000
Koblenz: LVAR-P.

Bodenkarte von Rheinland-Pfalz   1 : 25 000
Mainz: GLR-P, 1985–.
156 sheets, 5 published.

Administrative

Karte der Kreise und Verbandsgemeinden   1 : 500 000
Koblenz: LVAR-P.

Karte der Gemeindegrenzen Rheinland-Pfalz   1 : 200 000
Koblenz: LVAR-P.
Available with/without overprint of GGK 5.

SAARLAND

General

Karte des Saarlandes   1 : 100 000
Saarbrücken: LVAS.
Available with or without shaded relief.

Strassenkarte des Saarlandes   1 : 100 000
Saarbrücken: MURB.

Amtliche Entfernungskarte des Saarlandes   1 : 75 000   AE 75
Saarbrücken: LVAS.

Geological and geophysical

Geologische Karte des Saarlandes   1 : 50 000
Saarbrücken: GLAS, 1981.

Geologische Karte de Saarlandes   1 : 25 000
Saarbrücken: GLAS, 1965–.
24 sheets, 9 published.

Environmental

Gewasserkarte des Saarlandes   1 : 100 000
Saarbrücken: LAU.

Administrative

Verwaltungskarte des Saarlandes   1 : 100 000   VK100
Saarbrücken: LVAS

SCHLESWIG-HOLSTEIN

Atlases and gazetteers

Topographischer Atlas Schleswig-Holstein und Hamburg   4th edn
Kiel: LVAS-H/Neumünster: Karl Wachholtz, 1979.
pp. 235.

General

Schleswig-Holstein und Hamburg aus dem All   1 : 500 000
Frankfurt AM: VWK-Ryborsch, 1984.
Satellite image map.

Übersichtskarte von Schleswig-Holstein   1 : 250 000
Kiel: LVAS-H, 1979.

Strassenkarte von Schleswig-Holstein   1 : 250 000
Kiel: LVAS-H.

Geological and geophysical

Schleswig-Holstein. Geologische Karte   1 : 25 000
Kiel: GLAS-H.
176 sheets, 24 published.

Environmental

Klima-Atlas von Schleswig-Holstein
Offenbach AM: DWD, 1967.
63 maps + 43 pp. text.

Administrative

Gemeindegrenzen von Schleswig-Holstein   1 : 600 000
Kiel: LVAS-H.

Kreisübersichtskarte Schleswig-Holstein   1 : 250 000
Kiel: LVAS-H.

Amts- und Gemeindergrenzen von Schleswig-Holstein   1 : 250 000
Kiel: LVAS-H.

Gibraltar

Gibraltar has been mapped for the Gibraltar Government by the UK **Directorate of Overseas Surveys (DOS)** (now Overseas Surveys Directorate, Southampton, OSD) using air photographs flown in 1969. The **Public Works Department** is the local survey authority. Maps were compiled to the scale of 1:5000, 1:2500 and 1:1250. The maps are contoured (at 3 m up to and including 60 m, and 10 m thereafter). Heights are referred to the Alicante datum, and the sheets have a UTM grid. These maps were designed for planning purposes and are not generally available to the public. However, in 1986 a set of the 1:2500 series was available from stock at the map agents Edward Stanford Ltd, London.

Addresses

Public Works Department
23 John Mackintosh Square, GIBRALTAR.

Overseas Surveys Directorate (OSD/DOS)
Ordnance Survey, Romsey Rd, Maybush, SOUTHAMPTON SO9 4DH, UK.

For DMA, see United States; for Clyde Surveys, see Great Britain.

Catalogue

Atlases and gazetteers

Gibraltar. Names approved by the US Board on Geographic Names Washington, DC: DMA, 1984.

Topographic

Gibraltar 1:2500 Edition 1
Gibraltar Government, 1970.
5 sheets + reference sheet, all published.

Town maps

Gibraltar and the Costa del Sol 1:9000
Maidenhead: Clyde Surveys, 1987.
Two-sided map with ancillary maps and inset of town centre at 1:4500.

Gibraltar town plan 1:5000 Series M984 Edition 5
London: Directorate of Military Surveys, 1984.

Great Britain

Great Britain comprises the British mainland of England, Scotland and Wales. Northern Ireland is also part of the United Kingdom but is mapped by most agencies as a separate unit: we have therefore provided a separate section for Northern Ireland. Some maps cover the whole of the British Isles, i.e. Ireland and Great Britain. These are included in this section of the book.

The national mapping agency for Great Britain is the **Ordnance Survey (OSGB)**. It is currently responsible for the topographic, geodetic and photogrammetric surveying of the country and for the publication of official mapping. Though a civilian mapping agency OSGB had a military past and continues to cooperate closely with the **Ministry of Defence Mapping and Charting Establishment** and with the **Directorate of Military Survey** of the British Ministry of Defence in the establishment of mapping standards. OSGB mapped both Great Britain and the Irish Republic, but since the establishment of the Irish Republic it has only had a remit to survey and map Great Britain. However, from 1984 it absorbed the **Directorate of Overseas Surveys (DOS)** which became the **Overseas Surveys Directorate** of OSGB **(OSD)**, so OSGB is now involved in aid mapping in the third world, as well as being the official British mapping organization. OSD was established as the Directorate of Colonial Surveys (DCS) in 1946 and has since that date been active in the compilation of mapping funded through the British Overseas Development Administration in various third world countries. Most of this mapping has been topographic, but geological, soils and land use coverage has also been compiled in association with other GB public sector organizations.

The history of OSGB can be traced back to 1791, and mapping of Great Britain has been published at a variety of scales according to changing specifications since that date. By the middle of the nineteenth century England and Wales were mapped at 1:63 360 scale, though Scotland was not properly mapped until the compilation of 6 inch scale plans in the second half of the nineteenth century. In 1858 the decision was taken to adopt 1:2500 as the basic scale for all cultivated areas, to map the uncultivated areas at 1:10 560 and to map towns at 1:500 scale. This historical precedent is reflected in the current basic scales mapping of the country, though the 1:500 scale plans were abandoned in 1893. Until 1945 maps were published on the Cassini projection and large-scale maps were based not on one national projection but on a series of independent Cassini projections relating to a county or group of counties. Thus 'county series' 6 inch and 25 inch maps were published for the whole country, with independent sheet lines and numbering systems. The many different series, revisions and editions of the 1 inch map were published on different sheet lines, sheet sizes and specifications.

Only after World War II, and arising from the recommendations of the Davidson Committee, was it decided to introduce one unified National Grid referencing system, to be used for all scales, and to introduce a single Transverse Mercator projection. From the National Grid was derived the numbering system for all basic scale mapping and for the 1 : 25 000 first series maps started after the war. All OSGB maps have the National Grid overprinted at an appropriate interval. The basic scale (for uncultivated mountain and moorland areas) series for Great Britain is now the 1 : 10 560/1 : 10 000 scale: this is published as a derived map by generalization for the whole of the rest of the country. The 1 : 2500 scale is published for the whole country apart from moorland and mountain areas and is the basic scale for all minor towns and cultivated areas. Major towns are mapped at 1 : 1250 scale. A policy of continuous revision based on the number of units of change in the map is operated.

Reviews by Central Government have continued in the 1970s and 1980s. The Serpell Committee report of 1979 and subsequent official parliamentary committee reviews investigated both funding and future programmes of the survey, and made recommendations for the future, including a big increase in the rate of 1 : 1250 scale mapping. The later reviews encouraged the expansion of digital mapping activities by OSGB (see below).

The specification of the basic scale maps is as follows. 1 : 1250 and 1 : 2500 plans are single-coloured uncontoured line maps. The former cover 500 m quads, the latter 1000 or 2000 m quads. 51 694 basic scale 1 : 1250 plans and 170 163 basic scale 1 : 2500 plans are issued and a further 10 303 plans at 1 : 2500 were derived from the larger scale but are no longer published. These maps are available as paper copies, digital maps on magnetic tapes, or microfilm copycards as part of the *Survey Information on Microfilm* (SIM) service. Redrawn or up-dated hard copy printouts from SIM cards are available and unpublished up-dates derived from surveyors' drawings are also available for these series in the *Supply of Unpublished Survey Information* (SUSI) programme. OSGB has been using vector-based computer techniques in the production of these basic scale plans since 1972. To date about 25 000 of the required total of 225 000 large-scale plans have been digitized. The standard to which these digital maps are compiled has recently been under discussion, with a view to establishing a more structured database, rather than an automated map production system. It is planned that the large-scale digital programme will accelerate to give complete digital mapping of the UK for large-scale data by the year 2015.

The 1 : 10 000 series is a fully contoured two-colour map, with relief shown at 5 or 10 m contour intervals. It is compiled on the same sheet lines as the former 1 : 10 560 scale map and was started in 1969. There will be 1393 basic scale 1 : 10 000 scale plans and a further 9084 derived maps at this scale.

Derived series mapping of Great Britain is now published at 1 : 25 000, 1 : 50 000 and 1 : 250 000 scales. The 1 : 25 000 scale *Pathfinder series* will cover the country in 1371 sheets each covering 20 × 10 km. About 90 new sheets are being published each year and it is expected that complete coverage will be available by the end of the 1980s. This map contains all right of way information, unlike the first 1 : 25 000 series, which is being withdrawn as new Pathfinder sheets are published. Pathfinder sheets are not now being issued for areas where *Outdoor Leisure maps* are published: this series is intended to cover popular leisure and recreation areas of the country. These maps include additional tourist information on a Pathfinder base and are issued on irregular non-National Grid sheet lines. Thirty-one sheets have been published to date.

The 1 : 50 000 First Series was introduced as a photographic enlargement of the Seventh Series 1-inch map, in two instalments, for the south of the country in 1974 and for the north in 1976. The country is covered in 204 40 × 40 km quads. Sheets are being republished as full Second Series editions with metric contours at 10 m intervals and tourist information, and the series was redesignated 1 : 50 000 *Landranger* in 1980. It is intended to complete the Landranger Second Series in 1990. Experimental digital maps at this scale have been produced, but OSGB continues to use conventional methods in the production of this, its best selling series. The OSGB has published special tourist maps at 1 : 63 360 scale throughout the twentieth century. The areas at which maps have been published have changed as have the specifications of the series. The current range covers 11 popular holiday regions, with relief shown by a combination of contours, hill shading and hypsometric tints. Sheets are now printed using process colours. At 1 : 250 000 scale the OSGB publishes a nine-sheet *Routemaster series*, with relief shown by hypsometric shading. Insets on these maps show tourist information and distance tables. The colour editions of these maps are revised regularly.

The single-sheet 1 : 625 000 scale double-sided *Routeplanner* is revised annually and the 1986 edition was published for the first time as a full digital map, derived from the OSGB 1 : 625 000 scale database. This has also been available from 1986 as two separate datasets: coastline, hydrological features and administrative boundaries; and settlement and communications. This data is available at a national scale or for 100 km National Grid Squares.

Pathfinder, *Landranger*, *Routemaster* and *Routeplanner* maps are all available flat or folded, as either outline or full-colour editions.

OSGB has also compiled a range of administrative mapping. A 1 : 100 000 series is published on county sheet lines for England and Wales and uses reduced 1 : 50 000 outline topographic detail as a base and includes overprinted red local government and green parliamentary constituency information. A judicial boundary series on the same sheet lines and scale is also available and OSGB has begun to publish the outline bases without overprints. Wales is also covered in a 1 : 250 000 scale administrative map. Scotland is only mapped at 1 : 250 000 in these series. A two-sheet 1 : 625 000 scale edition is also published to give national coverage, showing European constituency and borough and county boundaries.

Archaeological and historical maps are published by OSGB: these are either small-scale general maps of Great Britain at different times or large-scale plans of places with particular historical or archaeological significance.

Following government pressure OSGB has begun to move towards the publication of more popular, revenue-generating products. It now produces a range of road atlases, tourist guides, street maps and specialist maps, often in association with commercial publishers. These include the useful large-format *Motoring Atlas*, revised on an annual basis, which contains the Routemaster series of maps enlarged to a scale of 1 : 190 090 and published with a gazetteer. A town map series at 1 : 10 000 scale using the two-colour 1 : 10 000 base with colour overprints and indexes has been started and now numbers about 20 maps, whilst town mapping of London has been issued in atlas format.

There are plans to increase the range of OSGB mapping that is available, including the 1986 publication of the *Ordnance Survey National Atlas of Great Britain*, which contains Routemaster mapping, a 33 000 name placename index and basic thematic coverage of 30 different topics. Despite its title this volume does not represent a true National Atlas; the thematic coverage is too small scale, and the volume resembles the Ordnance Survey Atlas issued in slightly smaller format in 1981.

Of rather more significance as a National Atlas is that covering the principality of Wales. The *National Atlas of Wales* is being issued in instalments in bilingual Welsh and English text in the first half of the 1980s by the **University of Wales Press**, and contains an excellent thematic overview of Wales. The third issue of this atlas is yet to be published.

In many ways fulfilling the same functions as a National Atlas is the *Domesday Project*. This major information storage and retrieval package is one of the first examples of the use of videodisk technology and was prepared to celebrate the 900th anniversary of the Domesday Survey. It is a collaborative project between the British Broadcasting Company, the Department of Trade and Industry, Phillips and the Ordnance Survey. A vast encyclopedia of information about the UK on a national scale, and about individual communities, is stored on two videodisks. It can be accessed using microcomputer, mouse and videodisk player. Included on the disks are over 22 000 Ordnance Survey maps, 40 000 photographs, gazetteer, most official statistics about Britain and land cover information, together with large amounts of textual information. Interactive colour mapping of over 20 000 variables for 33 data sets of geographical areas at ten different levels of resolution is possible using data on the National Disk. The Community Disk has been called a people's database and comprises 9000 4 × 3 km units, maps, photographs and associated free text information written by schools or community groups in the area.

The geological and geophysical mapping of Great Britain is the responsibility of the **British Geological Survey (BGS)** (formerly Institute of Geological Sciences, IGS). It recently moved to new headquarters in Keyworth, Nottingham. Most geological maps are published for the BGS by the OSGB in Southampton, but are available from either source (unless otherwise mentioned).

The history of the geological survey of England, Wales and Scotland can be traced back to the 1750s and the organization itself was founded in 1835. Maps have been published since the late nineteenth century on OSGB 3rd Series 1-inch sheet lines. The current basic scale geological mapping of GB is published at 1:50 000 scale, as either drift or solid or combined editions compiled from 1:10 000 scale survey information and borehole data. This *1:50 000 geological map* covers England and Wales in about 355 sheets, and Scotland in about 150 sheets: the imprecision applies particularly to Scotland since new 1:50 000 sheets for most areas are being published as a western and eastern section of the former 1-inch sheets. The earlier 1:63 360 scale series on the same sheet lines is still available for many areas where new sheets have not appeared. Monographs have been published to accompany each map, and are revised when new 1:50 000 coverage is published. There have been three types of revision: a simple photographic enlargement from 1 inch to 1:50 000 scale; a provisional recompilation to 1:50 000 involving no new geological survey but incorporating some non-survey information and some update of latest topography; and full new publication of a 1:50 000 map. Complete coverage of GB is not available at 1:63 360/1:50 000. Large areas of Wales have not been mapped and many 1-inch sheets are out of print in areas where no new 1:50 000 is on programme. There are few new sheets appearing in this series.

Some areas of the country have been published at 1:10 560 scale as either national grid or county series sheet lines. Coverage of these fully published maps is very

patchy, and mainly restricted to coalfield areas. Unpublished 1:10 000 scale and 1:10 560 scale maps are available for most areas as diazo copies from Keyworth.

BGS has since 1977 been publishing new full-colour UTM-based 1:250 000 scale geological maps of the whole of the United Kingdom and its continental shelf. These are issued in several different editions and include both land and seabed geology. The solid geology map will be completed in about 75 sheets: to date about 30 have been issued. Sea sediment editions are to be published for continental shelf areas: about 15 have so far been published. Quaternary editions are also published for some areas, often in combination with a seabed sediment map. In addition to these full-colour geological and sedimentological 1:250 000 maps BGS also holds complete aeromagnetic and Bouguer gravity anomaly maps on the same sheet lines, which are published with subdued topography and are available as diazo prints from Keyworth. An important series of regional geochemical atlases is being compiled by BGS at a scale of 1:250 000. Volumes for the northern highland areas of Scotland have : been published to date, and up to 30 elemental determinations at each of 10 000 sample sites a year are made. The data maps the distribution of trace elements in relation to surface features and geology. A national geochemical databank contains information for all mapped areas and for areas where the atlases have yet to be published. It is intended to extend this programme to cover eventually the whole of England, Scotland and Wales. Other smaller-scale maps published by BGS include 1:625 000 scale two-sheet geological, Quaternary sediments and aeromagnetic maps.

Another useful small-scale geological map of the country was compiled by the **Geological Society of London** in the early 1980s.

Soils mapping in Great Britain is carried out by two separate organizations. The **Soil Survey of England and Wales (SSEW)** is responsible for the compilation of soils and land use capability mapping to the south of the Scottish border. A six-sheet full-colour national soil map at 1:250 000 was published in 1983, and was followed by the publication of six accompanying Regional Monographs in 1984. This is available flat or folded, or as a boxed atlas set and is published with a National Soil Legend Booklet or legend sheet. There is no larger-scale complete coverage. 1:25 000 scale maps (on 1:25 000 topographic sheet lines) have been published for some areas, and 30 sheets of an old 1:63 360 series are still available. Some reconnaissance mapping on county sheet lines is also published. Financial cutbacks mean that a projected 1:50 000 soil map on the Landranger sheet lines is unlikely to be published for more than a very few areas. A 1:1 000 000 scale series is available as single sheets for a variety of environmental themes covering England and Wales.

Scotland's soils are mapped by the **Macaulay Institute for Soil Research (MISR)**. Seven-sheet 1:250 000 scale coverage is available as either a soils map or a land capability for agriculture map. It is published with seven accompanying monographs. Larger-scale mapping of Scottish soils is rather more advanced than south of the border. 1:63 360 scale mapping of the lowland areas of the country is available and the northern and western highlands are mapped at 1:50 000 as provisional and uncoloured diazo prints. Uncoloured 1:25 000 scale mapping on National Grid sheet lines are also available for lowland Scotland. There are plans for a full 1:50 000 scale land capability for agriculture series to be issued in 1987 to cover the main agricultural areas of the country in 31 sheets. MISR also compiles smaller-scale national coverage of the country to show climatic information, and land capability for agriculture. Perhaps reflecting the changing role of MISR, its name changed in mid-1987 to the **Macaulay Land Use Research Institute**.

The **Ministry of Agriculture, Fisheries and Food (MAFF)** was responsible for the compilation of a variety of agricultural mapping of England and Wales. A 1:63 360 scale series used 7th Series topographic map sheet lines to show the quality of agricultural land in a five-class division. Some explanatory booklets were issued with the 113 sheets that were published. From these maps was derived a seven-sheet 1:250 000 scale series. Type of farm classification maps were also compiled, and more recently MAFF has published small-scale land drainage and excess winter rainfall maps, and larger-scale agricultural mapping of some upland areas. MAFF maps were not published for Scotland and no equivalent coverage exists for areas north of the border. The Fisheries Directorate of MAFF in Lowestoft compiled a fisheries atlas of the seas around the UK which is still available. There are currently plans to compile an updated digital version of this atlas.

The **Department of the Environment** was responsible in the 1970s for the compilation of the important *Atlas of the Environment* and for its regional editions. These dealt with planning problems in specific areas, but they are no longer available. The Water Data Unit of the DOE compiled various hydrological mapping throughout the 1970s and 1980s, which are distributed by the DOE. These include 1:250 000 scale 10-sheet river quality mapping (to be revised early in 1987 with 1985 data) and, on the same water authority sheet lines, 1:250 000 scale 10-sheet river flow maps. Small-scale coverage of England and Wales was also published for both themes and is still available. The DOE also compiles small-scale administrative mapping of England and Wales, mapping of developed areas and of Green Belt distribution as defined by Structure Plans.

The **Scottish Development Department** carries out a wider range of functions in Scotland than those performed by the DOE south of the border. Its graphics group has compiled a range of thematic mapping, including a regularly revised oil and gas developments map but these maps are mainly prepared for internal use only.

The **Welsh Office (WO)** carries out a variety of mapping of Wales. The majority of published maps relate to boundary definition in the principality, but WO also compiles a wide variety of very useful thematic coverage of Wales, relating to all its areas of interest. Many of these full-colour maps are published as A4 format overviews and might in the future be issued as an atlas. We have listed only a selection of the more important WO maps.

Mapping of the Isle of Man is the responsibility of the **Isle of Man Local Government Board**: topographic maps are issued to OSGB National Grid specifications, at 1:10 560, 1:2500 and 1:1250, whilst a full-colour town map is published at 1:7500 for Douglas. Other thematic mapping of the island is carried out by mainland agencies.

The Channel Islands are not mapped by the Ordnance Survey. The **States of Jersey** has been responsible for the publication of official mapping of Jersey, which has been compiled on contract by **BKS Survey Technical Services Ltd**, who have also compiled mapping of other islands in the Group. General coverage of the islands is available in a Clyde Leisure map. The Channel Islands also have their own survey organizations, which operate on a similar basis to that of the Isle of Man. We have listed the Jersey and Guernsey Offices in our address lists.

The **Office of Population Censuses and Surveys (OPCS)** carries out mapping of the results of the 10-yearly Censuses of England and Wales. It participated in the compilation of a computer generated atlas of UK population, *People in Britain*, distributed through HMSO and still available, and publishes a variety of census maps in Census reports. A series of free small-scale population wallcharts and maps is distributed. Pre-Census mapping of enumeration districts is also compiled for OPCS on an OSGB 1:10 000 scale base to aid both in the interpretation of small area statistics and in the collection of Census data.

The **Civil Aviation Authority** compiles aeronautical charting of Great Britain, whilst hydrographic charting is the responsibility of the **Hydrographic Department**. The **Institute of Oceanographic Sciences**, funded by NERC, participates in the GEBCO programme and publishes bathymetric charting of the seas around the UK: we have listed these under Atlantic Ocean.

The **Institute of Terrestrial Ecology (ITE)** is engaged in a variety of resources mapping projects, many of which are computer based. It has published a small climatological atlas of the country and some larger-scale vegetation mapping of specific areas. ITE has developed an ecological digital cartographic database known as ECOBASE, using OSGB 1:250 000 and 1:50 000 scale data. ITE has also developed a number of applications for remote sensing of ecological resources.

Other resource map publishers include the **Forestry Commission** (forest mapping), the **Meteorological Office** (climatic mapping), the **Countryside Commission** (mapping of national parks and AONBs), and **British Gas, Britoil,** and **Oilfield Publications Limited** (hydrocarbon resource mapping).

The **National Remote Sensing Centre (NRSC)** acts as a national focal point for those using remotely sensed imagery, and publishes some image maps. Commercial agencies in this field include **ERSAC** and **Nigel Press Associates**.

There is a very active commercial sector in British map publishing. Firms concentrate on the publication of tourist maps, town maps and road maps and atlases.

Bartholomew in Edinburgh is one of the oldest and most prestigious commercial cartographic publishers in the world. It has been responsible for the compilation of the mapping that comprises the *Times Atlas of the World*, acknowledged as the standard large world reference atlas. Bartholomew also issues the *World Travel Series*, single-sheet maps from this same family of general relief-based small-scale coverage. We have cited these where published as being among the best small-scale general map series of individual countries. Other small-scale general worldwide products include several other atlases and a series of pocket maps. For Great Britain Bartholomew also publishes a wide range of mapping. The 1:100 000 scale full-colour *National Series* covers Great Britain in 62 sheets and shows relief by hypsometric tints at 50 m intervals for areas under 200 m and at 100 m intervals for higher areas. This map is not gridded but is regularly revised. An outline versions of this map is used as a base for the publication of postcode mapping at the same scale. Other postcode maps are published by Bartholomew at 1:250 000 (10 sheets) and for the major conurbations at larger scales. Bartholomew also publishes a range of road atlases and road maps, including a double-sided competitor to the OSGB routeplanner. All are frequently revised. A small range of city plans is also available. There has recently been a considerable investment by Bartholomew in digital mapping systems: world and Great Britain databases have been established and are to be used in the production of many Bartholomew products.

The George Philip Group can trace its history back to 1834, and comprises several different organizations. These are **George Philip and Sons** (publishers of general and educational atlases and globes), **Nicholson** (tourist map publishers), **Edward Stanford Ltd** (the World's largest map shop, with associated map publication activities), **Map Productions Ltd** (road and tourist maps), and **Intermap** (town maps). **Reader's Digest** has published road maps and atlases.

Wm. Collins and Son Ltd publish general and educational atlases and issue road atlases to the UK. **Macmillan Publishers Ltd** are also involved in the publication of atlases for the educational market in specific countries, often issued by associated companies in the Macmillan group in the relevant country. Among the more interesting recent Macmillan publications are a hard-copy version of the OS microfiche gazetteer. Macmillan is also investigating the possibility of producing a high-quality bound atlas version of the 1 : 50 000 Landranger series, upon completion of the second series of this map in the late 1980s. **Oxford Cartographers** is another commercial house with a long history of map and atlas compilation.

Road mapping of the UK is issued by the **Automobile Association** and by **RAC Publications**.

Quail Map Company is a specialist publisher and distributor of transport mapping, with the emphasis on simple black-and-white railway maps.

Chas E. Goad has for the last 20 years been publishing shopping centre plans at 1 : 1056 scale, showing retail uses, for over 1000 UK centres. The more important centres are revised on an annual basis.

Clyde Surveys, **Hunting Surveys** and **Geoprojects UK Ltd** are all mainly involved in the publication of mapping for other countries. Both Clyde and Huntings publish maps as a sideline to their main aerial surveying contracting activities, whilst Geoprojects has been particularly active in the Middle East and has published the *Arab Map Library series*.

Amongst other commercial map publishers are **Lascelles** and **Bradt Enterprises** (tourist mapping), **Imray, Laurie, Norie and Wilson** (yachting charts) and **Harvey Map Services** (recreation maps of mountain areas). Tourist maps are issued by **English**, **Welsh** and **Scottish Tourist Boards**.

Other town map publishers include **Geographers A–Z**, **Geographia**, **G. L. Barnett**, **E. J. Burrow**, **Estate Publications** and **Service Publications**: we have listed only a sample of these town maps for the capital cities of England, Scotland and Wales.

Further information

The best historical guide to the development of British official mapping is Harley, J. B. (1975) *Ordnance Survey Maps: a Descriptive Manual*, Southampton, OSGB.

For a review of current OSGB mapping see *Annual Reports* and *Trade Catalogue*, both revised annually, and *Information Leaflets* (23 free explanatory leaflets available, regularly updated).

Other recent articles on the OSGB include Farrow, J.E. (1986) National mapping in the UK and Republic of Ireland: Ordnance Survey GB, *The Photogrammetric Record*, **12**(68); and for a discussion of the Serpell Report and useful background see Smith, W.P. (1980) The Ordnance Survey: a look to the future, *Cartographic Journal*, **17**(2), 75–82.

The best overview of mapping of Great Britain as a whole is given in articles published in the *Cartographic Journal* for the Royal Society's British National Committee for Geography, which review on a regular basis cartographic activities in the country. The most recent available article is Cartographic activities in the United Kingdom 1980–83, *Cartographic Journal* (June 1984), **21**(1). Other useful articles about British mapping are regularly published in this journal.

Catalogues and annual reports are available for most of the organizations listed in the address lists. There are several useful articles in the *Natural Environment Research Council Newsletter* on geological and resources mapping of the UK.

Addresses

Ordnance Survey (OSGB)
Romsey Rd, SOUTHAMPTON SO9 4DH.

Ministry of Defence, Directorate of Military Survey
Elmwood Ave., FELTHAM, Middlesex TW13 7AF.

Ministry of Defence, Mapping and Charting Establishment
Block A Government Buildings, Hook Rise South, TOLWORTH, KT6 7NB.

Overseas Surveys Directorate (DOS/OSD)
Ordnance Survey, Romsey Rd, SOUTHAMPTON SO9 4DH.

University of Wales Press
6 Gwennyth St, Cathays, CARDIFF CF2 4YD.

British Geological Survey (BGS)
Keyworth, NOTTINGHAM NG12 5GG.

Geological Society of London
Burlington House, Piccadilly, LONDON W1V 0JU.

Soil Survey of England and Wales (SSEW)
Rothamstead Experimental Station, HARPENDEN AL5 2JQ.

Macaulay Institute for Soil Research (MISR)
Department of Soil Survey, Craigiebuckler, ABERDEEN AB9 2QJ.

Ministry of Agriculture, Fisheries and Food (MAFF)
Lion House, Willowburn Estate, ALNWICK NE66 2PF.

Department of the Environment (DOE)
Map and Air Photograph Library, 2 Marsham St, LONDON SW1P 3EB

Welsh Office (WO)
Cathays Park, CARDIFF CF1 3NQ

Scottish Development Office
New St Andrews House, EDINBURGH EH1 3SZ

Isle of Man Local Government Board, Department of Architecture and Planning
Bucks Road, DOUGLAS, Isle of Man.

States of Jersey
Royal Sq., ST HELIER, Jersey.

BKS Survey Technical Services
COLERAINE, Northern Ireland.

Office of Population Censuses and Surveys (OPCS)
Segensworth Rd, Titchfield, FAREHAM PO15 5RR.

Civil Aviation Authority (CAA)
Greville House, 37 Gratton Rd, CHELTENHAM GL50 2BN.

Hydrographic Department (HD)
Ministry of Defence, TAUNTON TA1 2DN.

Institute of Oceanographic Science (IOS)
Brook Rd, Wormley, GODALMING GU8 5UB.

Institute of Terrestrial Ecology (ITE)
68 Hills Rd, CAMBRIDGE CB2 1LA.

Forestry Commission
Publications Section, 231 Corstorphine Rd, EDINBURGH EH12 7AT.

Countryside Commission
19–23 Albert Road, MANCHESTER M19 2EQ.

Meteorological Office   Publications
London Rd, BRACKNELL RG12 2SZ.

British Gas
50 Bryonston St, Marble Arch, LONDON W1A 2AZ.

Britoil
150 Vincent St, GLASGOW G2 5LJ.

Oilfield Publications Ltd (OPL)
PO Box 11, LEDBURY HR8 1BN.

National Remote Sensing Centre
Space Department, Royal Aircraft Establishment, FARNBOROUGH GU14 6TD.

ERSAC Scientific Publications
Peel House, LADYWELL EH54 6AG.

Nigel Press Associates
EDENBRIDGE TN8 6HS.

John Bartholomew and Son Ltd
12 Duncan St, EDINBURGH EH9 1TA.

G. Philip/Edward Stanford/Nicholson/Intermap
12–14 Long Acre, LONDON WC2 9LP.

Map Productions Ltd
Olwen House, Quarry Hill Rd, TONBRIDGE TN9 2RH.

W. M. Collins
14 St James Place, LONDON SW1A 1PS.

Macmillan Publishing Ltd
Brunel Rd, Houndmills, BASINGSTOKE.

Oxford Cartographers
17 Beaumont St, OXFORD OX1 2NA.

Reader's Digest Association
25 Berkley Sq, LONDON W1X 6AB.

Automobile Association
Fanum House, BASINGSTOKE RG21 2EA.

RAC Publications
RAC House, Lansdowne Rd, CROYDON CR9 6HN.

Quail Map Co.
31 Lincoln Rd, EXETER EX4 2DZ.

Chas E. Goad
18 Salisbury Sq, OLD HATFIELD AL9 5BE.

Clyde Surveys Ltd
Reform Rd, MAIDENHEAD SL6 8BU

Hunting Surveys Ltd
6 Elstree Way, BOREHAMWOOD WD6 1SB

Geoprojects UK Ltd
Newton Rd, HENLEY UPON THAMES RG9 1HG

Roger Lascelles
47 York Rd, BRENTFORD Middlesex.

Bradt Enterprises
41 Nortoft Rd, CHALFONT ST PETER SL9 0NA.

Imray, Laurie, Norie and Wilson
Wych House, ST IVES PE17 4BT

Harvey Map Services
Mile End Rd, Main St, DOUNE, Scotland.

English Tourist Board
4 Grosvenor Gdns, LONDON SW1H 0DU.

Welsh Tourist Board
High St, Llandaff, CARDIFF CF5 2YZ.

Scottish Tourist Board
23 Ravelston Terrace, EDINBURGH EH4 3EU.

Geographers A–Z Map Co. Ltd
Vestry Road, SEVENOAKS TN14 5EP.

Geographia Ltd
63 Fleet St, LONDON EC4Y 1PE

G. L. Barnett
Graphia House, Rippleside Commercial Estate, BARKING, Essex.

E. J. Burrow and Co.
Publicity House, Streatham Hill, LONDON SW2 4TR.

Estate Publications
22A High St, TENTERDEN, Kent.

Service Publications
Caxton House, Ham Rd, SHOREHAM BY THE SEA, Sussex.

Catalogue

Atlases and gazetteers

Ordnance Survey National Atlas
Southampton and London: OSGB and Country Life Books, 1986.
pp. 256.

Ordnance Survey Motoring Atlas of Great Britain
Southampton: OSGB and Temple Books, 1986.
pp. 128.
Includes most of UK at 1:190 080
Annually up-dated.

Britain from Space: an atlas of Landsat images   R.K. Bullard and R.W. Dixon Gough
London: Taylor and Francis, 1985.
pp. 128.
Includes 32 1:500 000 scale satellite images.

Atlas Cenedlaethol Cymru/National Atlas of Wales   edited by H. Carter and H.M. Griffiths
Cardiff: 1980.
42 sheets in 9 parts.

Wales: A4 map series   ca 1:1 000 000
Cardiff: WO.
14 published sheets.
Thematic sheets available on individual basis.

Ordnance Survey Landranger gazetteer
Southampton: OSGB, 1986–.
33 fiches.
Contains all 1:50 000 scale names.
Updated annually. Also available in casebound edition published by Macmillan, June 1987.

Census 1981: index of placenames England and Wales
London: HMSO for OPCS, 1985.
2 vols.

Bartholomew gazetteer of places in Britain
Edinburgh: Bartholomew, 1986.
pp. 432 (pp. 112 of maps).

General

Britain and Ireland from space   ca 1:500 000
Farnborough: NRSC, 1984.
Simulated natural colour satellite mosaic.

Outline map of Great Britain   1:1 250 000
Southampton: OSGB, 1975.

Physical map of Great Britain   1:625 000
Southampton: OSGB, 1957.
2 sheets, both published.

Routeplanner of Great Britain   1:625 000
Southampton: OSGB, 1987.
Double-sided road map.
Revised annually.
Also available as outline edition (1980).

Bartholomew 1987 map of Britain   1:625 000
Edinburgh: Bartholomew, 1987.
Double-sided, with 18 small town map insets.

Maritime England   1:625 000
Southampton: OSGB, 1982.

Topographical map of Wales   1:500 000
Cardiff: WO.
Also available at 1:250 000 and 1:350 000

Topographic

Routemaster series   1:250 000
Southampton: OSGB, 1984–.
9 sheets, all published.
Also available as outline editions.

Bartholomew National Series   1:100 000
Edinburgh: Bartholomew, 1975–.
62 sheets, all published.
Ungridded.

Landranger series   1:50 000
Southampton: OSGB, 1980–.
204 sheets, all published. ■
Also available as outline editions.

Pathfinder series   1:25 000
Southampton: OSGB, 1981–.
1371 sheets, *ca* 1200 published. ■
Also available as outline editions.

Bathymetric

Atlas of the Seas around the British Isles   edited by A.J. Lee and
J.W. Ramster
Lowestoft: MAFF, 1981.
pp. 75.

British adjacent waters: fisheries chart   1:2 000 000
Taunton: HD, 1982 (Admiralty chart Q.6353).

Geological and geophysical

The Wolfson geochemical atlas of England and Wales   edited by
J. S. Webb
Oxford: OUP, 1978.
pp. 68.

Atlas of onshore sedimentary basins in England and Wales
Glasgow: Blackie, 1985.
pp. 96.

Geological structure of Great Britain, Ireland and the surrounding seas
London: Geological Society of London, 1985.

*Sub-Pleistocene geology of the British Isles and the adjacent
continental shelf*   1:2 500 000   2nd edn
Southampton: OSGB for BGS, 1979.

Geological map of the British Islands   1:1 584 000   5th edn
Southampton: OSGB for BGS, 1969.
Also available as uncol. edition.

Tectonic map of Great Britain and Northern Ireland   1:1 584 000
Southampton: OSGB for BGS, 1966.

Smoothed aeromagnetic map of Great Britain   1:1 584 000
Southampton: OSGB for BGS, 1970.

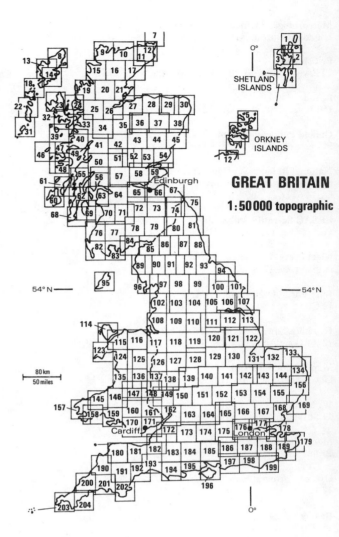

GREAT BRITAIN
1:50 000 topographic

Pre-Permian geology of the United Kingdom (south)   1:1 000 000
Keyworth: BGS, 1985.
With transparent overlay.

Geological map of Great Britain   1:625 000
Southampton: OSGB for BGS, 1979.
2 sheets, both published.
Available as solid (1979), Quaternary geology (1977) or
aeromagnetic (1965 and 1972) editions.

Hydrogeological map of England and Wales   1:625 000
London: BGS, 1977.

Geological series   1:250 000
Southampton: OSGB for BGS, 1977–.
ca 80 sheets, varying numbers published. ■
Published as 3 separate full-colour editions: solid geology; seabed
sediments and Quaternary.
Also available as diazo prints: Bouguer gravity anomaly;
aeromagnetic anomaly.

Geological Survey of Great Britain   1:63 360 and 1:50 000
Southampton: OSGB for BGS, 1923–.
ca 500 sheets, *ca* 250 published. ■
Published in two series: England and Wales, and Scotland.
Sheets published with accompanying memoirs.
Many 1:63 360 sheets now out of print.

Environmental

Atlas of drought in Britain 1975–76.
London: Institute of British Geographers, 1980.
pp. 87.

Protected areas in the UK   1:2 700 000
Manchester: Countryside Commission, 1984.

GREAT BRITAIN

**1:250 000 geological
UTM series**

80 km
50 miles

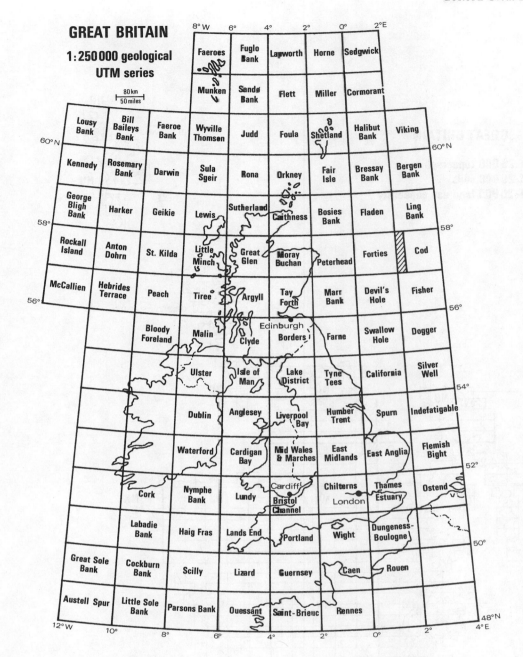

United Kingdom: climatic maps   1:2 000 000
Bracknell: Met. Office.
5 published sheets.
Maps cover: rainfall (3 sheets), potential evaporation, and water
authority boundaries.

England and Wales: generalised soil map   1:2 000 000
Harpenden: SSEW, 1974.

England and Wales: types of soil water regime   1:2 000 000
Harpenden: SSEW, 1974.

Soil Survey of England and Wales   1:1 000 000
Southampton: OSGB for SSEW, 1975–.
8 sheets, all published.
Single sheets cover the following themes: soils, land use
capability, winter rain acceptance potential, soil suitability for
grassland, accumulated temperature, average maximum potential
cumulative soil moisture deficit, wind exposure, and duration of
field capacity.

Forestry in Great Britain   1:1 000 000
Edinburgh: Forestry Commission, 1986.
Includes both private and public forests.

Land drainage in England and Wales   1:750 000
Alnwick: MAFF, 1983.

Soil Survey of England and Wales: bioclimatic
classification   1:625 000
Harpenden: SSEW, 1981.

Land capability classification for agriculture in Scotland   1:625 000
Aberdeen: MISR, 1982.
With accompanying explanatory text.

Mean annual excess winter rainfall   1:625 000
Pinner: MAFF, 1979.
Includes England and Wales.

Assessment of climatic conditions in Scotland   1:625 000
Aberdeen: MISR, 1971.
3 sheets, all published.
Sheets cover the following themes: accumulated temperature and
potential water deficit; exposure and accumulated frost;
bioclimatic subregions.

Soil Survey of Scotland: land capability for agriculture   1:625 000
Aberdeen: MISR.
With accompanying monograph.

Forestry in Great Britain   1:625 000
Edinburgh: Forestry Commission, 1976.
2 sheets, both published.

GREAT BRITAIN

1:25 000 topographic
1:25 000 soils
1:25 000 land use capability

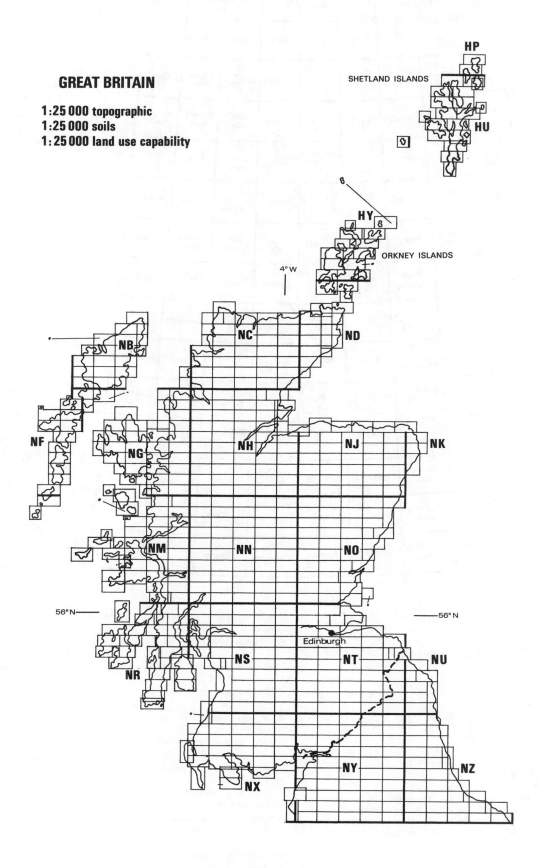

Sheet numbering example
SP 26-36

09-19	29-39	49-59	69-79	89-99
08-18	28-38	48-58	68-78	88-98
07-17	27-37	47-57	67-77	87-97
06-16	26-36	46-56	66-76	86-96
05-15	25-35	45-55 **SP** 44-54	65-75	85-95
04-14	24-34		64-74	84-94
03-13	23-33	43-53	63-73	83-93
02-12	22-3	42-52	62-72	82-92
01-11	21-31	41-51	61-71	81-91
00-10	20-30	40-50	60-70	80-90

From 1986 a sequential numbering
system will be introduced in parallel to
the National Grid based system for
1:25000 Pathfinder maps

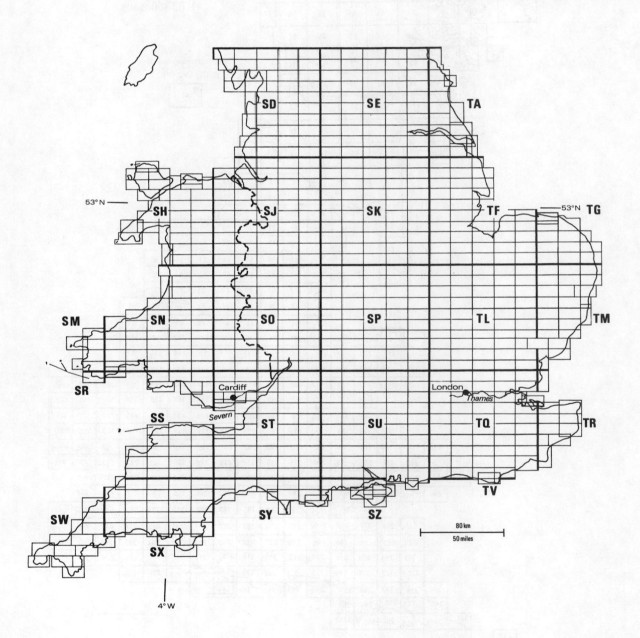

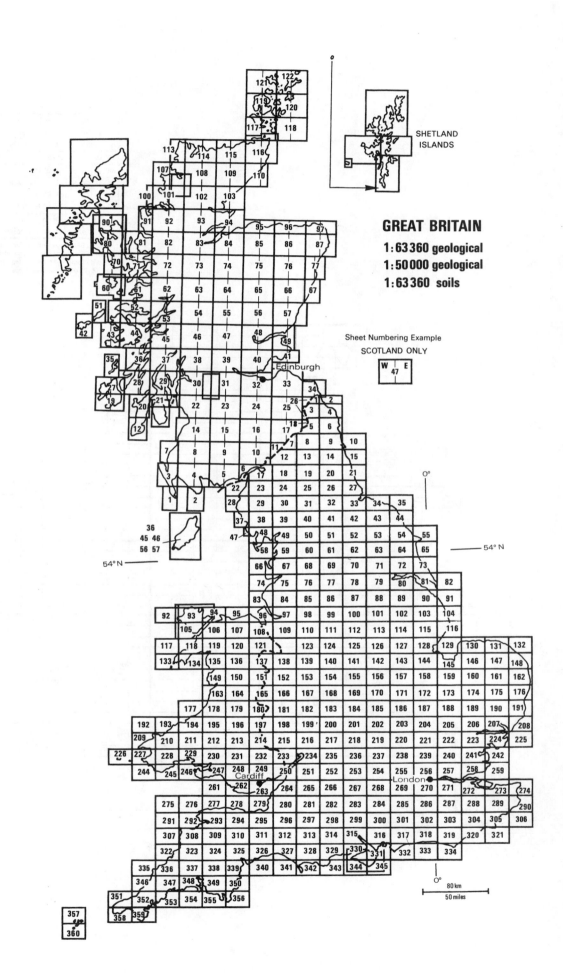

SHETLAND
ISLANDS

GREAT BRITAIN

1:63 360 geological
1:50 000 geological
1:63 360 soils

Sheet Numbering Example
SCOTLAND ONLY

Edinburgh

Cardiff

London

54° N

54° N

0°

0°

80 km
50 miles

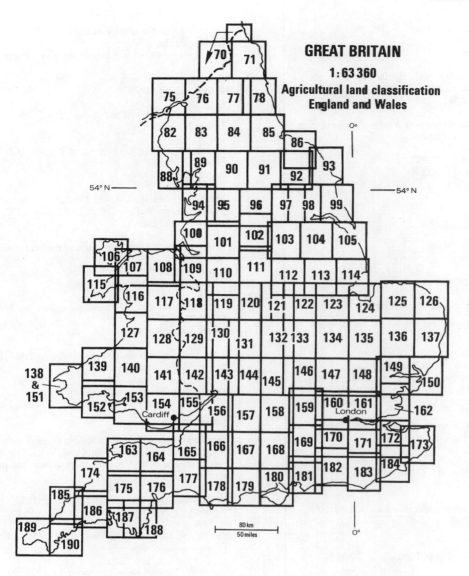

GREAT BRITAIN
1:63 360
Agricultural land classification
England and Wales

Map of average annual rainfall for Great Britain
1941–1970 1:625 000
Bracknell: Met. Office, 1977.
2 sheets, both published.

Soil map of England and Wales 1:250 000
Southampton: OSGB for SSEW, 1983.
6 sheets, all published.
With accompanying monographs, national legend sheet and soil legend booklet.

Soil and land capability map of Scotland 1:250 000
Aberdeen: MISR, 1982.
7 areas, all published.
2 sheets (soils and land capability) and an explanatory text published for each area.

River quality 1980 1:250 000
London: DOE, 1982.
10 sheets, all published.

Soil Survey of England and Wales 1:63 360
Southampton: OSGB for SSEW, 1955–.
360 sheets, 31 published. ■
Sheets published as either soils or land capability editions.
With accompanying memoirs.

Soil Survey of Scotland 1:63 360
Aberdeen: MISR, 1969.
131 sheets, 36 published. ■
4 sheets issued at 1:50 000.
With accompanying memoirs.

Soil Survey of England and Wales 1:25 000
Southampton: OSGB for SSEW, 1970–.
120 published quads. ■
Sheets published for quadrangles as either full-colour or outline editions showing soils, land use capability or soil drainage.

Administrative

New county map of the British Isles 1:1 250 000
Edinburgh: Bartholomew, 1984.
5-colour map showing counties and districts.

Great Britain: postcode key map 1:1 250 000
Edinburgh: Bartholomew, 1979.

Administrative areas 1:625 000
Southampton: OSGB, 1986.
2 sheets, both published.
Sheets show local government and European Constituency boundaries.

Scotland: local government and parliamentary constituency boundaries 1:250 000
Southampton: OSGB, 1984–.
5 sheets, all published.

Wales: parliamentary constituencies 1:350 000
Cardiff: WO.

Wales: local government boundaries 1:250 000
Southampton: OSGB, 1973.

Great Britain: postcode district maps 1:250 000
Edinburgh: Bartholomew.
10 sheets, all published.

Great Britain: postcode sector maps 1:100 000
Edinburgh: Bartholomew, 1982.
63 sheets, all published.■

England and Wales: local government areas and parliamentary constituency boundaries 1:100 000
Southampton: OSGB, 1975–.
54 sheets, all published.
Sheets arranged on county sheet lines.
Greater London published at 1:63 360 scale.

Human and economic

People in Britain: a census atlas
London: HMSO, 1981.
pp. 134.
Data from the 1971 census.

Atlas of cancer mortality in England and Wales M. Gardner
et al.
Chichester: Wiley, 1983.
pp. 116.

Atlas of mortality from selected diseases in England and Wales 1968–1978 M.J. Gardner, P.D. Winter and D.J.P. Barker
Chichester: Wiley, 1984.
pp. 164.

Rail atlas of Britain and Ireland 4th edn S.K. Baker
Poole: Oxford Publishing Co., 1984.
pp. 118.

The coalfields of Great Britain 1:2 000 000
London: NCB, 1979.

North West Europe: hydrocarbon exploration 1:2 000 000
London: British Gas, 1986.
Regularly revised.

UK onshore licence areas 1:1 000 000
London: British Gas.
Up-dated annually.
Also available as A3 or A4 size.

Oil and gas fields and the national gas transmission system
1:1 000 000
London: British Gas.
Up-dated annually.
Also available as A3 or A4 size.

Executive population density map of Great Britain 1:823 680
London: Philip, 1982.
1981 Census data, with enlarged insets of metropolitan counties.

Electricity supply industry map 1:625 000
London: CEGB, 1978.

Type of farm in England and Wales 1:625 000
Pinner: MAFF, 1978.

Agricultural land classification map of England and Wales 1:625 000
Pinner: MAFF, 1979.

British rail system map 1:475 200
London: Geographia for British Rail Board, 1977.
3 sheets, all published.

Agricultural land classification maps of England and Wales
1:250 000
Pinner: MAFF, 1976–77.
7 sheets, all published.

Type of farm maps for the regions of England and Wales 1:250 000
Pinner: MAFF, 1971–73.
9 sheets, all published.

Agricultural land classification map of England and Wales 1:63 360
Pinner: MAFF, 1966–75.
113 sheets, all published.■
With accompanying explanatory reports.

Town maps

Maps listed here cover only a sample of those published to cover the capitals of England, Scotland and Wales. For other maps of these cities and other UK town maps contact the publishers mentioned in the text.

ABC London street atlas 1:10 560 and 1:18 103
Southampton: OSGB and Newnes Books, 1984.
pp. 113.
Also available as hard or soft backed, 4-colour A4 sized car version.

New A–Z Premier London streetmap ca 1:21 500
Sevenoaks: Geographers A–Z, 1986.

City map of Cardiff 1:10 000
Southampton: OSGB, 1985.

Edinburgh street guide 1:15 000
Edinburgh: Bartholomew (Johnson and Bacon), 1985.
pp. 45.

Greece

The **Hellenic Military Geographical Service (HMGS)**, Athens, established in 1889, is responsible for the topographic mapping of Greece and has published complete map series at scales of 1:25 000 (1291 sheets), 1:50 000 (387 sheets) and 1:100 000 (128 sheets) together with town maps at 1:10 000 and 1:5000. Under present regulations, these series are not available outside Greece, access being limited to maps at 1:1 000 000, 1:500 000 and one or two sheets at 1:250 000. Therefore only these smaller scales have been listed in our catalogue.

HMGS has also published Greek alphabet gazetteers, and a roman-alphabet gazetteer is in preparation.

Much of the early topographic mapping of Greece was accomplished by foreign surveys, including British, French, Italian, Serbian and German military surveyors. Thus a *Carte de la Grèce* at 1:200 000 was produced in 1852 by the French Dépôt de la Guerre, while an Austrian Staff map was completed at 1:300 000 in 1880 and at 1:200 000 in 1916. Maps covering northern Greece in this latter series, the Austrian *Generalkarte von Mitteleuropa*, are still available (this series is described in the introductory section to Europe).

The best readily available cover of Greece is provided by the Greek alphabet *Karta Nomos* 1:200 000. This is published by the **National Statistical Service**. It is a provincial series, each map covering a single administrative area (*nomos*). Relief is shown by hypsometric tints and 200 m contours, there is a four-category classification of roads and tracks, and the maps also show census enumeration boundaries.

Greece has no National Atlas, but an *Economic and social atlas* of Greece was published in 1964 by the National Statistical Service, and is still in print. The text is in Greek, English and French.

Geological mapping of Greece is carried out by the **Institute of Geology and Mineral Exploration (IGME)**, Athens. The main series, in progress since 1956, is at 1:50 000. Over 50 per cent of the country is now covered in published form by this series, with the remaining sheets nearly all at the compilation or field mapping stage. The legends of these maps are in two languages—Greek and English or Greek and French.

A new geological map at 1:500 000 was published in 1983 in cooperation with the British Geological Survey. It is in Greek and English.

Soil mapping has been carried out by various institutes operating under the Greek Ministry of Agriculture. In 1977, a national soil mapping project was initiated with the goal of mapping the whole agricultural and forested area of Greece. A scale of 1:10 000 was adopted for the agricultural lowlands, with smaller scales (1:20 000 or 1:50 000) for the semi-mountainous and forest areas.

A parallel ecological land resource survey is also under way. It is carried out by the **Forest Research Institute**, Athens (also under the Ministry of Agriculture). This survey, published at 1:50 000, was intended to serve in the evaluation of the land capability of the more hilly areas of the country. Although this mapping began only in 1983, progress has been rapid, with about 30 per cent of the whole country published by 1987, and the survey has now been extended to cover the whole country. The land resource data collected by the Forest Research Institute is being incorporated into a computerized land evaluation information system.

There are several commercial map publishers in Greece. **B.G.D. Loukopoulis** publishes general, administrative and regional maps in Greek or Latin Script, and several town maps. The Greek islands are covered by a series of maps by **Toubis** at scales which vary (according to size of island) from 1:25 000 to 1:300 000. These maps lack contours and are rather generalized in specification. Tourist maps are also published by **D. and J. Mathioulákis**, Athens.

A new series of 1:50 000 walking maps of the Greek mountains is in progress, published by the **Greek National Mountaineering Association (EOS)** and based on the military topographic series.

Good general maps of Greece are also published by a number of foreign commercial publishers including **Bartholomew, Hallwag, Kümmerly + Frey, Ravenstein** and **Freytag and Berndt. Clyde Surveys** has published maps of several islands, as have **Nelles Verlag**, Munich and **Karto+Grafik**, Frankfurt AM.

Further information

Catalogues are issued by HMGS and IGME.
The land resource survey of Greece is described in Nakos, G. (1983), The land resource survey of Greece, *Journal of Environmental Management*, **17**, 153–169.

Addresses

Hellenic Military Geographical Service (HMGS)
Pedion Areos, GR-113 62 ATHINA.

National Statistical Service
14–16 Lycourgou, GR-105 52 ATHINA.

Institute of Geology and Mineral Exploration (IGME)
70 Messoghion Street, GR-115 27 ATHINA.

Forest Research Institute
Terma Alkmanos, GR-115 28 ATHINA.

B.G.D. Loukopoulis
10 Nikoloudi Arcade, GR-105 64 ATHINA.

Toubis
519 Vouliagmenis Avenue, GR-163 41 ATHINA.

D. & J. Mathioulákis
Andromedas 1, ATHINA.

Greek National Mountaineering Association (EOS)
Platia Agiou Vlasiou 16, Aharnes, GR-136 71 ATHINA.

Greek National Tourist Office
2 Amerikis Street, ATHINA.

For Kümmerly + Frey and Hallwag, see Switzerland; for Freytag and Berndt, see Austria; for Clyde Surveys and Bartholomew, see Great Britain; for Ravenstein, Nelles Verlag and Karto+Grafik, see German Federal Republic; for GUGK, see USSR.

Catalogue

Atlases and gazetteers

Oikonomikos kai koinonikos atlai tís Elládos/Economic and social atlas of Greece B. Kayser and K. Thompson
Athina: National Statistical Service, 1964.
pp. 508.

Lexikón tón dhímon, kinotíton ké ikismón tís Elládhos
Athina: HMGS, 1984.
pp. 341.

General

Road map of Greece 1:1 000 000
Athina: HMGS, 1985.

Hellas 1:1 000 000
Athina: HMGS.
5 sheets, all published.

Greece and the Aegean 1:1 000 000
Edinburgh: Bartholomew, 1983.

Hellas 1:500 000
Athina: HMGS, 1985.
10 sheets, all published.

Hellas general map 1:500 000
Athina: Loukopoulis.
4 sheets, all published.

Topographic

Hellas 1:250 000
Athina: HMGS.
32 sheets.

Chartis Nomon tís Elládhos 1:200 000
Athina: National Statistical Service, 1972–.
54 sheets, all published. ∎

GREECE
1:200 000 topographic provincial maps

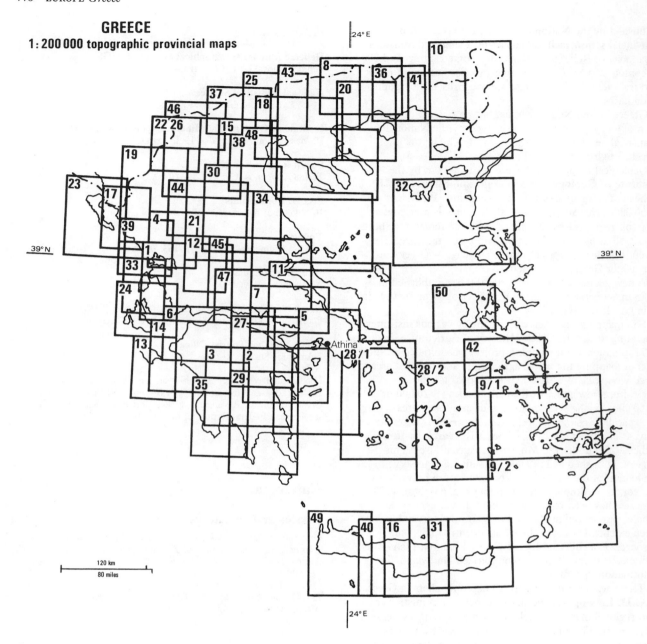

120 km
80 miles

Geological and geophysical

Geological map of Greece 1:500 000 2nd edn J. Bornovas and
Th. Randogianni-Tsiambaou
Athina: IGME, 1984.

Yeoloyikos chartis tís Elládhos 1:50 000
Athina: IGME, 1956–.
ca 420 sheets, 50% published. ■

Administrative

Hellas political map 1:1 000 000
Athina: Loukopoulos.

Hellas political map 1:600 000
Athina: Loukopoulos.
3 sheets, all published.

Town maps

Clyde leisure map of Athens and the Peloponnese 1:400 000
Maidenhead: Clyde Surveys, 1984.
Area map with ancillary 1:12 500 street map of central Athens.

Athina. Town plan in English 1:13 500
Athina: Loukopoulis.
With street index.

Greece, Athina, Attiki
Athina: Greek National Tourist Organization, 1983.
Main map is Athens city centre.
Legend in English, French and German.

Athens. City map with underground 1:8500
Bern: Hallwag, 1985.
Index in roman and Greek script.

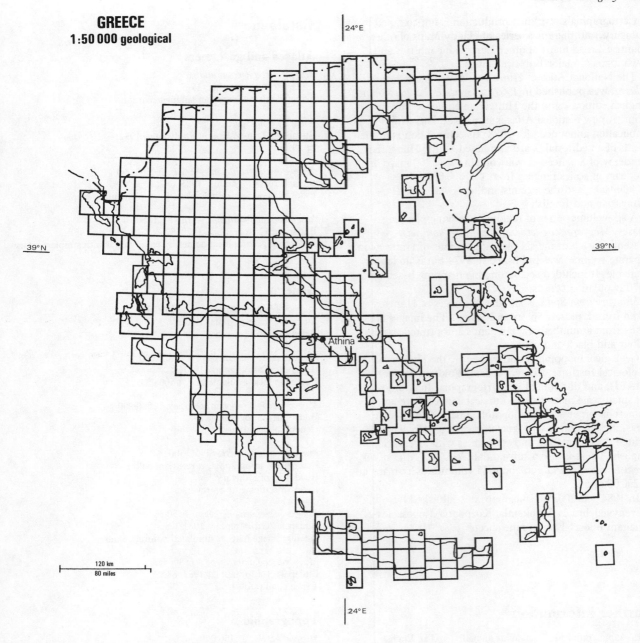

GREECE
1:50 000 geological

Hungary

Topographic mapping in Hungary is the responsibility of the National Office of Lands and Mapping (**Orszagos Foldúgyi es Terkepeszeti Hivatal**), Budapest, founded in 1954. Series are published at scales of 1:100 000, 1:25 000 and 1:10 000 (the last is not complete). A recessed cylindrical projection is used and the maps are in seven colours with a 10 m contour interval. These series are not available to the public.

Maps for public distribution and for educational use are published by the Hungarian Company for Surveying and Mapping, **Kartográfiai Vállalat (Cartographia)**, Budapest. This company was also founded in 1954. It publishes a wide range of maps, many in English and other language editions, and acts as the exporting agency both for its own maps and for the cartographic publications of other Hungarian institutions.

Most of the cartographic products of Cartographia are listed in our catalogue, and it can be seen that the most detailed public mapping covering the country at a consistent scale is the series of county maps, *Megye térképe*, at 1:150 000 scale. The content of these full-colour maps includes administrative boundaries, road network, forest areas and a contour interval of 100 m, with supplementary contours at 50 m. New editions were published in 1986. A new series of leisure maps of the counties has also been started recently. Many larger-scale tourist and water sport maps of particular areas are also published (scales of 1:80 000 to 1:20 000) and a symbol book is available to help foreign users decode the legends on these maps. Besides the street maps of Budapest, Cartographia also publishes street maps of about 100 other Hungarian cities, and of other major world cities.

Cartographia's own map production is not confined to Hungary, and there is a series of Handy Maps of other countries, road maps of other European countries and town maps of major European cities.

The National Atlas of Hungary, *Magyarország nemzeti atlasza*, was published in 1967 in separate Hungarian and English editions, and the Hungarian edition is still in print. A new National Atlas is in preparation, with publication announced for 1988. Its compilation is in the hands of an editorial board appointed by the Hungarian Academy of Sciences. It will contain some 400 maps of Hungary at scales ranging from 1:1 000 000 to 1:4 000 000, and the legends and texts will be in Hungarian and English.

A six-volume series of regional economic planning atlases (*Magyarország tervezesi-gazdasagi korzetei*), presenting each of the six planning regions of Hungary in a separate volume, was published in 1974 but is no longer available. It included socioeconomic mapping based on the 1970 population census.

Hungary from Space is an atlas of images of Hungary taken from Landsat, Soyuz and Salyut. The images are taken from a number of wave bands and printed in both colour and black and white.

Geological mapping is undertaken by the Hungarian Geological Institute, **Magyar Állami Földtani Intézet (MAFI)**, and distributed by Cartographia. In addition to the small-scale maps listed in the catalogue, there are numerous larger-scale monographic maps of particular areas, including many prognostic maps, usually with written documentation and sometimes with legends in English. There is also a substantial *Geological Atlas of the Great Hungarian Plain* in about 15 volumes, with maps at 1:200 000 scale.

In 1984 a 1:200 000 soils map was published by Novenyvedelmi és Agrokemiai Kozpontja Meliosacios es Talajtani Foosztalyán in nine sheets.

Further information

Cartographia publishes a catalogue annually, and an English language edition is available.

MAFI has published separate catalogues of maps and of written reports covering the period 1957–85.

Information about new developments in Hungarian cartography will be found in the periodical *Geodezia es Kartografia*.

Addresses

Orszàgos Földúgyi és Térképészeti Hivatal (National Office of Lands and Mapping)
H-1055 BUDAPEST, Kossuth Lajos ter 11.

Kartográfiai Vállalat (Cartographia)
H-1443 BUDAPEST, POB 132.

Magyar Tudomanyos Akademia (Hungarian Academy of Sciences)
Foldrajztudomanyi Kutato Intezet,
H-1388 BUDAPEST, VI. Nepkoztarsasag utja 62.

Magyar Állami Földtani Intézet (MAFI) (Hungarian Geological Institute)
H-1442 BUDAPEST, Népstadion út 14, POB 106.

Catalogue

Atlases and gazetteers

Magyarország nemzeti atlasza
Budapest: Cartographia, 1967.
National atlas: Hungarian language edition.

Hungary from Space
Budapest: Zrinyi Military Publishing, 1982.
pp. 95.

Magyarország földrajzinév-tára I   1:500 000
Budapest: Cartographia, 1985.
Index of geographical names with supplementary map.

Magyarország földrajzinév-tára II   1:500 000
Budapest: Cartographia, 1978–82.
Indexes of geographical names by county each with supplementary map.

General

Magyarország dombortérképe   1:650 000
Budapest: Cartographia, 1982.
Relief map of Hungary.

Magyarország munkatérképe   1:525 000
Budapest: Cartographia, 1981.
Base map (also available at 1:1 500 000)

Magyarország domborzati térképe   1:500 000
Budapest: Cartographia, 1982.
General physical map.

Magyarország autotérképe   1:500 000
Budapest: Cartographia, revised annually.
Road map of Hungary.
Index of settlements on reverse.

Magyarország ürfotótérképe   1:500 000
Budapest: Cartographia, 1981.
Satellite image map in simulated natural colour.

Magyarország alaptérképe   1:300 000
Budapest: Cartographia, 1987–88.
4 sheets, all published.

Topographic

Megyetérképek   1:150 000
Budapest: Cartographia, 1986.
19 sheets, all published. ■
Maps of counties.

Geological and geophysical

Magyarország geomorfologiai térképe   1:525 000
Budapest: Cartographia, 1972.
Geomorphological map of Hungary.

Magyarország földtani térképe   1:500 000
Budapest: MAFI, 1984.
Geological map of Hungary.
English language edition available.

Magyarország melyföldtani térképe   1:500 000
Budapest: MAFI, 1986.
Geological map without Cenozoic formations.

Magyarország paleozóos és mezozóos képzödményeinek fedetlan földtani térképe   1:500 000
Budapest: MAFI, 1967.
Geological map of Palaeozoic and Mesozoic basement.

Magyarország hasznosithato asvanyos anyagai. Nyersanyag elofordulasok es remenybeli teruletek   1:500 000
Budapest: MAFI, 1968.
Mineral resources. Mineral occurrences and prospective areas.

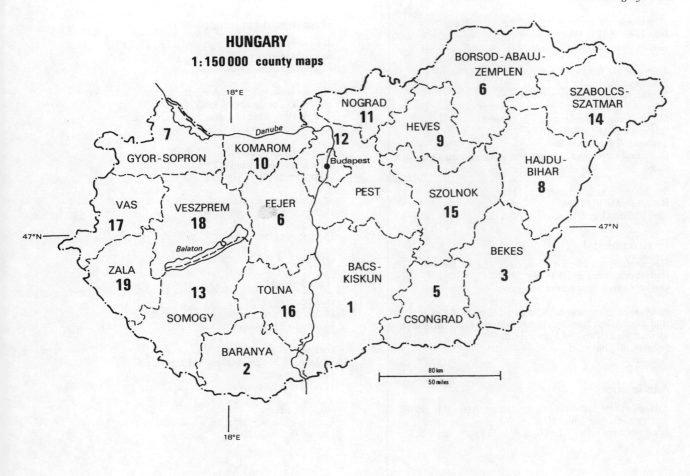

HUNGARY
1:150 000 county maps

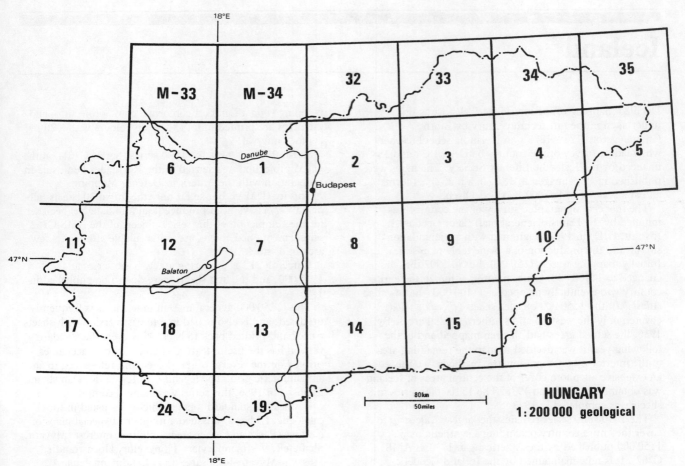

HUNGARY
1:200 000 geological

Magyarország talajvizforgalmi térképe   1:500 000
Budapest: MAFI, 1986.
Groundwater map of Hungary.

Magyarország mérnökgeologiai térképe   1:500 000
Budapest: MAFI, 1986.
Engineering geological map of Hungary.

Magyarország földtani térképe   1:200 000
Budapest: MAFI, 1966–.
25 sheets, 2 published.■
Geological map series, with several thematic variants of each
sheet.

Magyarország mélyfurási atlasza   1:150 000
Budapest: MAFI, 1984.
Deep-drilling location atlas.

Environmental

Magyarország felszini vizeinek minosege   ca 1:1 500 000
Budapest: Cartographia, 1979.
Quality of surface waters in Hungary.

Magyarország genetikus talajtérképe   1:200 000   I. Kovacs
Budapest: MEM Novényvédelmi és Agrokémiai Központja
Meliorácios és Talajtani Föosztályán, 1984.
9 sheets, all published.
Soils map.

Administrative

A Magyar Népköztársaság államigazgatási térképe   1:525 000
Budapest: Cartographia, 1985.
Administrative map of the Hungarian People's Republic.

Human and economic

Magyarország villamosenergia rendszere   1:800 000
Budapest: Cartographia.
Electricity network map.

Magyarország múszeumai és nemzeti parkjai   1:525 000
Budapest: Cartographia, 1982.
Text in Hungarian describing museums and parks with
accompanying map.

Magyarország vasúttérképe   1:500 000
Budapest: Cartographia, 1978.
Railway network of Hungary.

Magyarország villamosenergia rendszere   1:300 000
Budapest: Cartographia, 1983.
4 sheets, all published.
Electricity network map.

Town maps

Budapest térképe
Budapest: Cartographia, revised annually.
Legend in 5 languages.

Budapest belsö területe
Budapest: Cartographia, revised annually.
City centre map.
Legend in 5 languages.

Budapest atlasz
Budapest: Cartographia, revised annually.
Legend in 5 languages.

Iceland

Detailed mapping of the interior of Iceland was first
undertaken in the nineteenth century by Bjorn
Gunnlaugsson (1788–1876), an Icelandic school teacher,
who made a survey of the country between 1831 and 1843,
financed by the Icelandic Literary Society. The maps were
published in Copenhagen in 1844 in a four-sheet format
and at a scale of 1:480 000 by O.N. Olsen.

The modern topographic survey of the country was
initiated by the Danish General Staff (later Geodætisk
Institut, GID) in 1902, beginning with a triangulation
survey and the production of a plane table map series.
Initially mapping was at 1:50 000, but in 1907 this was
changed to 1:100 000. Consequently, although the entire
country was eventually mapped at 1:100 000 (the so-called
Atlasblöd), the 1:50 000 quarter sheets (*Fjórdungsblöd*)
covered only the western and southern coastal areas. By
1936, the settled areas had all been mapped and in the
following year it was decided to map the remaining areas
of the interior by photogrammetric methods. To this end
an extensive air photo cover of the central areas of Iceland
was obtained in 1937 and 1938. The 1:100 000 series in 87
sheets was completed in 1943.

The 1:100 000 series remains the largest-scale map to
cover the entire country (excluding a synthetic map at
1:50 000 published by the Americans in 1950 as AMS
C762). It is now maintained by the Iceland Geodetic
Survey (**Landmælingar Íslands, LI**), founded in 1946, but
the date of issue of individual sheets covers a considerable
time span, since many have not yet been revised. This

mapping is on a Lambert conformal conic projection, and
the maps are printed in four to seven colours with a 20 m
contour interval.

In the 1970s, a new 1:50 000 series (Series C761) on the
UTM projection was started in the Reykjanes Peninsula in
cooperation with the American Defense Mapping
Authority (DMA), but so far only the initial 10 sheets (all
dated 1977) have been produced, and no date has been set
for the completion of this series. Sheets of the AMS C762
series, mentioned above, are still available although a few
are out of stock.

Currently, LI is preparing a new basic series at
1:25 000, and the first sheets were due to be published in
1986.

The 1:250 000 general map in nine sheets is frequently
up-dated, and is also issued in a tourist version with sheets
paired and printed back to back. However, the standard
version has the useful feature of an index of placenames
printed on the reverse of the sheets, and referenced to an
alphanumeric grid. The legend is in Icelandic, Danish and
English. In 1986, this series was being recompiled.

Besides the standard topographic series listed in the
catalogue, LI has also issued a number of special maps of
National Parks and other areas of tourist interest (Mývatn,
Skaftafell, Vestmannaeyjar, Thingvellir, Hornstrandir,
Húsavík-Mývatn-Jokulsárgljufur, Landmannalauger-
Thórsmörk), a frequently revised street map of Reykjavik,
and several small-scale general maps.

The Survey's 1:250 000 series forms the base for the

standard geological series (*Jardfrædikort*), and for a series of aeromagnetic maps. The geological series—not yet complete—was begun in 1960 by Gudmundur Kjartánsson, and the sheets show both Quaternary drift and bedrock deposits. More recently, it has become a team effort and several sheets have appeared in much revised editions (sheets 3, 5, 6 and 7). Sheets 4 and 8 have yet to be published, but sheet 4 and a revised edition of sheet 2 are in preparation. *Supplementary notes to the legend of the Geological Map of Iceland* by Gudmundur Kjartánsson was published in 1962. The series is produced jointly by the Museum of Natural History (**Náttúrfrædistofnun Íslands, NI**) and the Iceland Geodetic Survey, and is obtainable from the latter. There is also a 1:40 000 scale geological map of the Reykjavik area, published in 1958.

A single-sheet geological and structural map of Iceland at 1:600 000 is due to be published in 1987.

Another important thematic mapping project in Iceland has been the vegetation mapping undertaken by the Agricultural Research Institute (**Rannsóknastofnun landbúnadarins, RL**), Reykjavik. This has been a continuous programme since 1957 and includes the production of a series of 1:40 000 scale land cover maps of the highland areas (mostly complete now but not all published), and a number of lowland sheets of the Borgarfjordur district at 1:20 000. The object of this mapping is to aid in the evaluation of land quality for agricultural purposes, to determine stock-carrying capacity of the grazing lands, and to provide an objective basis for land management decisions. Base maps for the 1:40 000 vegetation map are the American 1:50 000 series AMS C762. RL has also made a number of special vegetation maps at large scales of present or planned water reservoir areas. Future plans are to continue the highland maps at the 1:50 000 scale, while a new lowland series at 1:25 000 has been started which will use the new LI basic map at that scale.

The *Soil map of Iceland* was adapted from a survey by Iver J. Nygard, and prepared by the USDA Soil Conservation Service. There is a handbook available in Icelandic or English.

Iceland has no National Atlas, but a comprehensive gazetteer by Thorsteinn Jósepsson and Steindór Steindórsson has been reissued in an enlarged four-volume edition since 1980. The volumes are illustrated with colour photos and the entries include celebrated historical and physical features as well as settlements.

Further information

LI publishes index maps of the topographic and geological map series and in 1985 published an excellent descriptive brochure in Icelandic and English which includes extracts of the series.

A list of vegetation maps may be had from RL, and a description of the vegetation mapping of Iceland in English will be found in Gudbergsson, G.M. and Gísladóttir, G. (1984) Vegetation mapping in Iceland, *Bulletin of the Society of University Cartographers*, **18**, 93–98.

Addresses

Landmælingar Íslands (LI)
PO Box 5536, Laugavegi 178, 105 REYKJAVÍK.

Náttúrfrædistofnun Íslands (NI)
PO Box 5320, 125 REYKJAVÍK.

Rannsóknastofnun Landbúnadarins (RL)
Keldnaholt, 110 REYKJAVÍK.

United States Geological Survey (USGS)
Box 25286 Federal Center, DENVER, Colorado 80225, USA

Göteborgs Universitet Naturgeografiska Institutionen (GUNI)
Dicksonsgatan 4, S-412 56 GÖTEBORG, Sweden.

For DMA and USDA, see United States.

Catalogue

Atlases and gazetteers

Landid Thitt: Ísland Thorsteinn Jósepsson and Steindór Steindórsson
Reykjavík: Bókautgáfan Örn og Örlygur HF, 1980–84.
4 vols.

Iceland Gazetteer. United States Board on Geographic Names
Washington, DC: DMA, 1961.

General

Íslandskort: landgrunnskort 1:3 000 000
Reykjavík: LI, 1972.
Shows the continental shelf and 12-mile territorial limit.

Ísland: Litla ferdakort/The little touring map 1:1 000 000
Reykjavík: LI, 1985.
Two-sided map: relief on one side, roads and distances on the reverse.
Legend in Icelandic and English.

Ísland: Ferdakort/Touring map 1:500 000
Reykjavík: LI, 1986.
Two-sided map: half country printed on each side. Tourist symbols. Small thematic inset maps.
Legend in Icelandic, English, Norwegian and German.
Also available in book format.

Vatnajökull, Iceland: Fall scene. Satellite image map NASA Landsat-1 1973 1:500 000
Reston, VA: USGS, 1976.
False colour image.

Vatnajökull, Iceland: Winter scene. Satellite image map NASA Landsat-1 1973 1:500 000
Reston, VA: USGS, 1977.
Black-and-white image.

Topographic

Ísland: Adalkort 1:250 000
Reykjavík: LI, 1968–.
9 sheets, all published.

Ísland: Atlasblöd 1:100 000
Reykjavík: 1938–.
87 sheets, all published. ■

Ísland: Fjórdungsblöd 1:50 000
Reykjavík: LI, 1902–.
118 sheets covering western fjords and south coastal areas only. ■

Ísland: Stadfrædikort 1:50 000 Series C 761
Reykjavík: LI, 1977–.
209 sheets, 10 published. ■

Geological and geophysical

Geomorfologisk karta över Island 1:1 750 000 Jan Swantesson
Göteborg: GUNI, 1984.
2 sheets + text (*GUNI Rapport 18*).

Jardfrædikort 1:250 000
Reykjavík: LI/NI, 1969–.
9 sheets, 7 published.

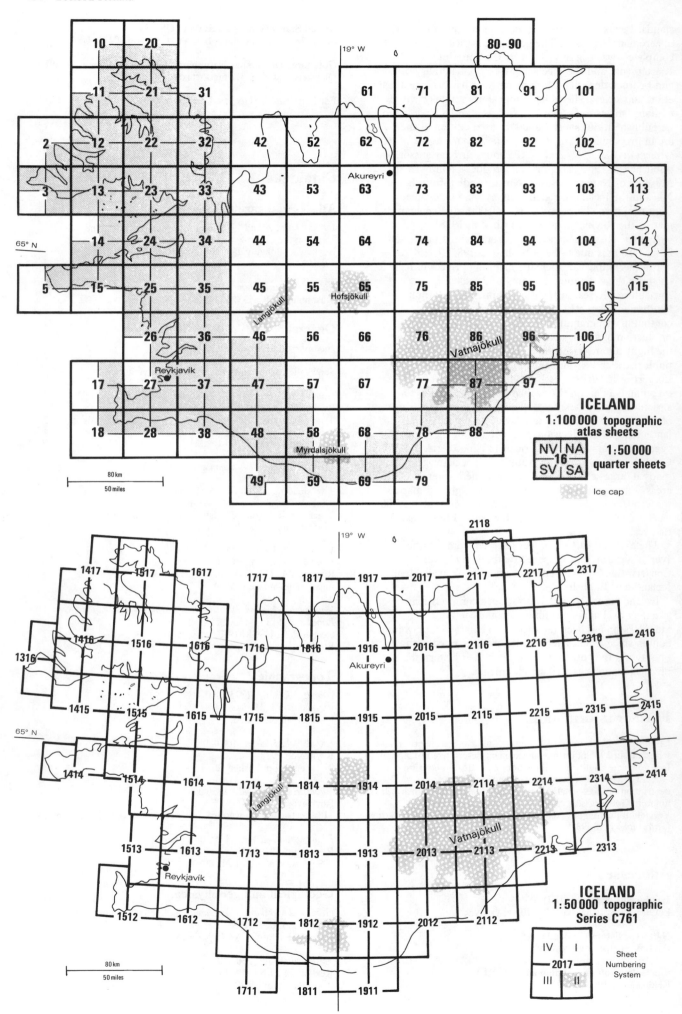

ICELAND
1:100 000 topographic
atlas sheets

1:50 000
quarter sheets

Ice cap

ICELAND
1:50 000 topographic
Series C761

Sheet
Numbering
System

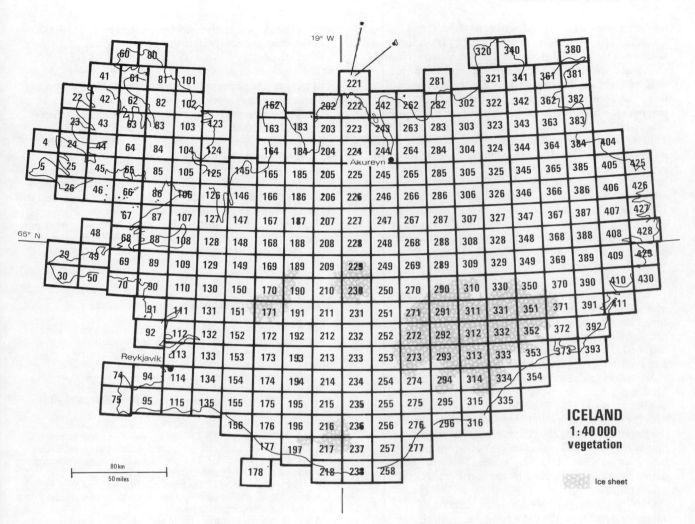

ICELAND
1:40 000
vegetation

Ice sheet

80 km
50 miles

19° W

65° N

Akureyri

Reykjavík

Segulkort af Íslandi 1:250 000
Reykjavík: Háskoli Islands/LI, 1970–.
9 sheets, all published.
Aeromagnetic profiles and total field intensity (partial cover).
Legend in Icelandic and English.

Environmental

Jardvegskort af Íslandi/Soil map of Iceland 1:750 000
Washington, DC: USDA Soil Conservation Service, 1959.
Legend in Icelandic and English.
Separate handbook published.

Gródurkort 1:40 000
Reykjavík: RL, 1957–.
Sheets covering highland areas only, 64 published. ■

Gródur og jardakort 1:25 000
Reykjavík: RL, 1985–.
9 sheets published.

Administrative

Sýslu- og hrepparkort 1:750 000
Reykjavík: LI, 1974.

Town maps

Reykjavík, Hafnarfjördur, Kópavogur, Seltjarnes, Gardabær,
Bessastadahreppur, og Théttbyli í Mosfellsveit 1:15 000
Reykjavík: LI, 1984.
Map of Reykjavik and adjacent communities.

Ireland, Northern

Topographic mapping in Northern Ireland is undertaken
by the **Ordnance Survey of Northern Ireland (OSNI)**,
Belfast, under the Department of the Environment. In
1985, OSNI moved into new headquarters at Colby House,
Stranmillis. The Survey has no responsibility for mapping
in the Republic of Ireland, although series share the same
projection and grid, and some conform with an all-Ireland
specification and sheet-numbering system.

OSNI was founded in 1922. Previously, the mapping of
the whole of Ireland had been carried out by the Ordnance
Survey of Great Britain and Ireland, originating with the 6-
inch survey, the first edition of which was completed in
1846.

As in Great Britain, a series of basic scales at 1:2500 and
1:10 560 with county sheet lines had evolved, and the
initial task of OSNI was to up-date and maintain these

series, together with derived series at 1:63 360 and 1:126 720. However, after World War II a programme similar to that adopted by the OSGB was introduced, and OSNI undertook a new triangulation from 1950 which was to serve as a basis for a new Irish Grid series of maps, using the Transverse Mercator projection. For postwar mapping the basic scales adopted were 1:1250 (for urban areas), 1:2500 (rural areas) and 1:10 560 or (after 1968) 1:10 000 (for mountains, moorlands and areas of low public demand). The 1:10 000 series also covers the whole of Northern Ireland, i.e. including the areas covered by larger basic scales. From 1960, photogrammetry was introduced for new survey, revision and contouring. By 1990, the revision and resurvey on to the Irish Grid is expected to be complete. Since 1981, regular print runs of the large-scale plans have been discontinued in favour of printing on demand using a high-quality photocopier.

The derived series at 1:63 360 (Third Series) has been superseded by a metric 1:50 000 series issued between 1978 and 1985. This was designed as an all-Ireland series in 89 sheets, of which 18 cover Northern Ireland (no sheets have yet been published for the Republic). Maps in this series have a 10 m contour interval with layer tints, and the Irish grid is printed at 1 km intervals. The series is also available in a special binder which includes the 1:500 000 map and a gazetteer (from **North-West Books**, Limavady).

The current 1:126 720 Second Series is not conformable with the original all-Ireland series which is still extant for the Republic. The 1:250 000 map is however one of a four-sheet all-Ireland series, and the other three sheets have been published by OSI, Dublin. This map includes tourist information supplied by the Northern Ireland Tourist Board.

OSNI is also responsible for the publication of administrative maps, the main series being the 1:63 360 Local Government District maps which show townlands and wards on a grey topographic base.

Several local recreational maps have been published by OSNI and a series of four-colour town maps with street indexes is in progress.

OSNI has plans to publish an official gazetteer of placenames, but at the time of writing no definite date is available.

In 1984, a full digital mapping system was introduced (Computer Mapping and Topographic Database, COMTOD) which will eventually develop into a complete topographic database for Northern Ireland. Initially, the 1:1250 urban sheets are being digitized, beginning with the Belfast area, and digital cover of all of Northern Ireland is to be achieved within about 10 years. This digital archive will be used for all map production, and is also intended for use by other public services in Northern Ireland for integration with their own data sets.

In 1986, a Regional Remote Sensing Centre was established within the Survey. SPOT, Thematic Mapper and imagery from airborne survey will be used for a number of project-orientated studies in the fields of soil, forestry, agriculture and pollution.

The **Geological Survey of Northern Ireland (GSNI)** was established in 1947 and is an agency of the Department of Economic Development. Although mapping had been completed in the nineteenth century by the Geological Survey of Ireland, the monochrome or hand-coloured maps had remained virtually unrevised and a major task after 1947 was a complete resurvey to be published in a series of 43 sheets at 1:63 360. These maps are all colour printed, are published in both solid and drift editions, and have the Irish Grid. From 1978 the

publication scale was changed from 1:63 360 to 1:50 000, but the sheet system remains that of the original nineteenth century all-Ireland survey, and does not therefore conform to the modern topographic series. About half the province has now been resurveyed, although only 25 per cent of the sheets have yet been published.

GSNI has also published a coloured solid geology map of Ulster at 1:250 000, and a companion drift edition is in preparation. Several small-scale maps covering Northern Ireland are also published by the British Geological Survey, including the 1:250 000 series of *Britain and the Continental Shelf*. Special maps published by GSNI include a 1:50 000 solid geology edition of *Mourne Mountains* and a 1:21 120 engineering geology map of Belfast.

Information about small-scale general and thematic maps covering the whole of Ireland will be found in the section on the **Republic of Ireland**.

Further information

The OSNI produces a map catalogue annually and a series of intermediate *Map Publication Reports* is also issued.

The development of the postwar topographic mapping programme has been described in Taylor, W.R. (1969) The Ordnance Survey of Ireland: an outline of its history and present mapping tasks, *Cartographic Journal*, **6**, 87–91; and Brand, M.J.D. (1981) Photogrammetry for national mapping in Northern Ireland, *The Photogrammetric Record*, **10**(58), 447–456.

The new developments in digital mapping and remote sensing are described in Brand, M.J.D. (1986) Towards a geographical information system for Northern Ireland, in *Mapping from Modern Imagery*, Proceedings of ISPRS Symposium held at the University of Edinburgh, 8–12 September 1986, pp. 473–479.

The work of the GSNI is described in *The Geological Survey of Northern Ireland*, published by the Survey, which also includes a list of publications (maps and textual).

A concise overview of mapping in Ireland (North and the Republic) is provided by Reeves-Smyth, T.J.C. (1983) Landscapes in paper: cartographic sources for Irish archaeology, *British Archaeological Reports British Series*, **116**, 119–177.

Addresses

Ordnance Survey of Northern Ireland (OSNI)
Colby House, Stranmillis Court, Stranmillis Road, BELFAST BT9 5BJ.

Geological Survey of Northern Ireland (GSNI)
20 College Gardens, BELFAST BT9 6BS.

North-West Books
23 Main Street, LIMAVADY BT49 0EP.

For Bartholomew, see Great Britain; for CNRS, see France.

Catalogue

General

Map of Northern Ireland   1:500 000
Belfast: OSNI.
Revised annually.
Coloured or outline edition.

The Ulster map   1:253 440
Edinburgh: Bartholomew.
Includes town map of Belfast.

Holiday Map: Ireland North   1:250 000
Belfast: OSNI, 1984.
Part of an all-Ireland series.

Topographic

The Half-Inch Map 1:126 720 Second Series
Belfast: OSNI, 1968–70.
4 sheets.

Ordnance Survey of Northern Ireland 1:50 000
Belfast: OSNI, 1978–.
18 sheets, all published. ■
Numbered as part of an all-Ireland series.

Geological and geophysical

Magnetic anomaly map of Northern Ireland 1:253 440
Belfast: GSNI, 1971.

Gravity anomaly map of Northern Ireland 1:253 440
Belfast: OSNI, 1967.

Carte géomorphologique du Nord d'Irlande 1:253 440
Paris: CNRS, 1972.

(Geological map) *Northern Ireland: solid edition* 1:250 000
Belfast: GSNI, 1977.

Geological Survey of Northern Ireland 1:63 360/1:50 000
Belfast: GSNI, 1961–.
43 sheets, 12 published. ■
First colour-printed editions, some with memoirs.
All available as separate solid or drift editions.
Copies of 19th century GSI monochrome maps also available.

Administrative

Northern Ireland Local Government Areas 1:250 000
Belfast: OSNI.

United Kingdom Parliamentary Constituency areas 1:250 000
Belfast: OSNI, 1983.
Shows 1981 boundaries.

Local Government District Electoral Areas (1985) 1:200 000
Belfast: OSNI, 1985.

Local Government District Maps (1984) 1:63 360
Belfast: OSNI, 1984.
14 sheets, plus 1 (Belfast City) at 1:25 000
Show townlands and wards.

Human and economic

Northern Ireland: a census atlas P.A. Compton
Dublin: Gill and Macmillan, 1978.
pp. 169.

Town maps

Greater Belfast street map 1:10 000
Belfast: OSNI, 1985.
2 sheets + index.

NB: for small-scale maps and atlases see also the Great Britain and
Republic of Ireland catalogue sections.

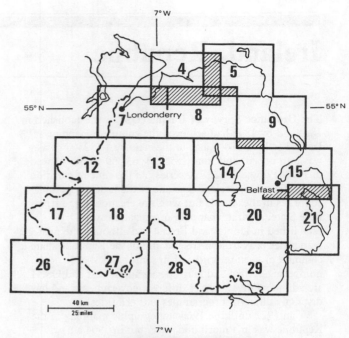

NORTHERN IRELAND
1:50 000 topographic

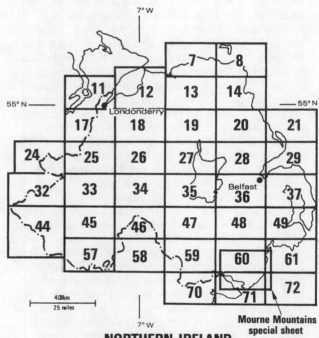

NORTHERN IRELAND
1 : 63 360 geological
1 : 50 000 geological

Ireland, Republic

The Ordnance Survey of Ireland (OSI) was established in 1824 and by 1846 had achieved the complete mapping of Ireland at 1:10 560. Later in the century a 1:2500 survey was introduced, and from 1855 a 1:63 360 map, followed in 1918–19 by a 1:126 720 series. At the time of partition (1922) therefore, Ireland was one of the best mapped countries in the world. But with the creation of the Irish Free State, separate mapping organizations were established in Dublin and Belfast and although the modern map series of Northern Ireland and of the Republic share a common projection, grid and in some cases sheet numbering system and map specification, OSI and OSNI are administratively under different governments and have distinctly different programmes and priorities.

From 1922 until the 1960s topographic mapping in the Republic was in a moribund state, but in 1964 a new national mapping policy was adopted, which included a new suite of basic scales (1:1000 for centres of more than 1000 population, 1:2500 as a basic scale for the whole Republic, and 1:5000 as a derived, and contoured, map for the whole Republic). These new Irish Grid maps were underpinned by a new first-order triangulation and levelling network completed in the mid-1960s. The task of new basic scale mapping was given priority, and it is necessary to appreciate the magnitude of this undertaking in order to understand why the smaller-scale derived mapping remains so out of date. The new basic survey has proceeded apace although the first sheets did not appear until 1977. Following a pilot study of computer assistance for this mapping programme carried out in 1979, a Kongsberg/MBB SysScan system has been acquired. This system handles the processing of photogrammetric and aerotriangulation information, and the digitizing and processing of contours for the 1:5000 map.

In 1978, OSI entered an agreement with OSNI to produce a joint series of 1:50 000 metric maps covering all Ireland in 89 sheets and drawn to a common specification. Although the 18 sheets covering Northern Ireland have all been published, production has not yet commenced in the Republic due to the higher priority of the large-scales programme. Hence the last revisions of the original 1:63 360 (1898–1914) remain in use. These sheets are a monochrome edition, usually with 100 foot contours, and all sheets covering the Republic (179 of the original 205 sheets) remain in print, but without further revision. There are also four special, layer coloured District maps covering Dublin, Wicklow, Cork and Killarney, and a single 1:50 000 recreational map of the *Wicklow Way*.

The two other small-scale series do incorporate recent revision. The 1:126 720 map, first introduced in 1911–18, is now represented in a layer coloured edition with 100 foot contours, and with the Irish Grid printed at 10 km intervals. The series began as an all-Ireland series in 25 sheets, but sheet 5 (Northern Ireland) is not now published, and OSNI has its own four-sheet series at this scale. The maps have up-to-date road classification and include some tourist information. The 1:250 000 series is a newly formatted and redesigned all-Ireland map, three of the four sheets being issued from Dublin. This map has tourist information supplied by Bord Failte, 10 km grid, and a 60 m contour interval. It is designed primarily as a tourist map.

OSI has published several modern, indexed street maps of major towns, including Dublin, Cork and Limerick.

In 1986, a new feasibility study for the production of the proposed 1:50 000 series was in progress. Complete air photo cover of the country at 1:30 000 has been flown for OSI by the Institut Géographique National, Paris. This may be used to set up a medium and small scale database using digital photogrammetric techniques of data capture. A pilot sheet has been started in the Killarney area.

The **Geological Survey of Ireland (GSI)** was founded in 1845, and achieved a complete published cover at 1:63 360 in hand-coloured solid and drift editions by 1890. Monochrome copies of these maps are still available from the OSI, Dublin. GSI moved in 1984 from 14 Hume Street, Dublin, its headquarters for 114 years, to new purpose-built premises at Beggars Bush.

A few colour-printed sheets appeared in 1901–03, but after that there was no new mapping until the 1970s. However, the Survey was revitalized in the late 1960s and early 1970s, following the discovery of major metallic mineral resources in Ireland. It was at first intended to produce a new series at 1:50 000 (using the sheet lines of the proposed all-Ireland 1:50 000 topographic series) together with 1:250 000 and 1:500 000 derivatives. Although much new and detailed fieldwork was completed in many parts of the country, publication policy has again had to change due to the non-appearance of the new topographic base map, and to renewed financial restrictions. Current effort is therefore being directed to the compilation of reconnaissance maps at the available 1:126 720 scale, and covering several earth science themes. The first trial sheets are in preparation.

Coloured solid geological maps of *Northwest and Central Donegal* (1972) and *Connemara* (1981) at 1:63 360 have been published by other organizations, but are available from GSI.

Soil mapping is undertaken by the **National Soil Survey (NSS)**, which was established in 1958 within the Agricultural Institute (An Foras Taluntais). Soils are classified by a modified version of the US 7th Approximation, and are published in colour at 1:126 720 on a county basis. Field sheets are compiled at 1:10 560 and are less generalized than the published sheets. They may be consulted at the NSS headquarters. The published maps accompany Soil Survey Bulletins which include textual description and supplementary maps of soil suitability and grazing capacity. A new edition of the *General Soil Map of Ireland* was published in 1980 and incorporated in an explanatory *Bulletin* (Gardiner, M.J. and Radford, T., Soil associations of Ireland and their land use potential, *Soil Survey Bulletin* 36). NSS has also published a *Peatland map of Ireland* (1978) with *Soil Survey Bulletin* 35.

The *Atlas of Ireland* was published in 1979. It was prepared under the direction of the Irish National Committee for Geography. It contains 97 plates and an index of placenames, and is in bound format. Most maps are thematic, at a scale of 1:1 000 000 or smaller, and cover the whole of Ireland, but there is also a six-page general map of Ireland.

Agriculture in Ireland: a census atlas has maps of agricultural statistics produced by automated cartography using the GIMMS mapping software.

A *Road Atlas of Ireland* has been published by Gill and Macmillan in association with OSI and OSNI and using the new 1:250 000 scale mapping. It includes simple street maps of major towns, a touring gazetteer and an index of approximately 3700 entries.

Further information

A brochure describing the maps of the Ordnance Survey is available from OSI, and the Geological Survey issues a list of recent publications. A list of soils bulletins is contained in *Bulletin 36* and may be up-dated by reference to the annual report of the NSS.

A concise review of Irish mapping is provided by Reeves-Smyth, T.J.C. (1983) Landscapes in paper: cartographic sources for Irish archaeology, *British Archaeological Reports British Series*, **116**, 119–177.

Further background information is provided in *Irish Geography*, **11** (1978), *viz.* Walsh, M.C., Current activities in the Ordnance Survey of Ireland (pp. 149–154) and Naylor, D., The Geological Survey of Ireland (pp. 155–160).

A brief historical review of the Geological Survey down to the present is also found in Archer, J.B. (1984) Adieu Hume Street: a retrospective view of the Geological Survey of Ireland, *Geological Survey of Ireland Bulletin*, **3**, 159–170.

Addresses

Ordnance Survey of Ireland (OSI)
Phoenix Park, DUBLIN.

Geological Survey of Ireland (GSI)
Department of Energy, Beggars Bush, Haddington Road, DUBLIN 4.

National Soil Survey (NSS)
An Foras Taluntais, Kinsealy Research Centre, Malahide Road, DUBLIN 5.

Department of Geography
University College Dublin, Belfield, DUBLIN 4.

Environmental Resources Analysis Ltd
187 Pearse Street, DUBLIN 2.

For Bartholomew, see Great Britain.

Catalogue

Atlases and gazetteers

Atlas of Ireland
Dublin: Royal Irish Academy, 1979.
pp. 104.

Ordnance Survey road atlas of Ireland
Dublin: Gill and Macmillan, 1985.
pp. 82.

General

Bartholomew Ireland Touring Map   1:760 320
Edinburgh: Bartholomew, 1979.
5 sheets, all published.
All Ireland series.

Map of Ireland   1:700 000
Dublin: OSI, 1982.
Town maps on reverse.

Ireland   1:575 000
Dublin: OSI, 1978.
An edition in Irish also available.

Topographic

Ireland travel map   1:253 440
Edinburgh: Bartholomew, 1977–.

Ireland   1:250 000   Edition 1
Dublin: OSI, 1981–82.
3 sheets, all published.
Part of all Ireland series in 4 sheets.

Ordnance Survey of Ireland   1:126 720
Dublin: OSI, 1978–.
25 sheets covering all Ireland, 24 (covering Republic) available. ■

Ordnance Survey of Ireland   1:63 360   2nd/3rd edn
Dublin: OSI, 1898–1914.
205 sheets (covering all Ireland), only 179 sheets covering the Republic available. ■

Geological and geophysical

Mineral deposits of Ireland   1:1 000 000
Dublin: Crowe Schaffalitzky and Associates Ltd/ Environmental Resources Analysis Ltd, 1984.

Geological map of Ireland   1:750 000   3rd edn
Dublin: GSI, 1962 (reprinted with minor corrections 1979).

Preliminary magnetic compilation map of Ireland
1:750 000   M.D. Max and T. McIntyre
Dublin: GSI, 1983.

A magnetic map of the continental margin west of Ireland, including part of the Rockall Trough and the Faeroe Plateau
1:625 000   R.P. Riddihough
Dublin: Dublin Institute for Advanced Studies, 1975.

Geological Survey of Ireland   1:63 360
Dublin: GSI, 1858–75.
205 sheets (covering all Ireland). ■

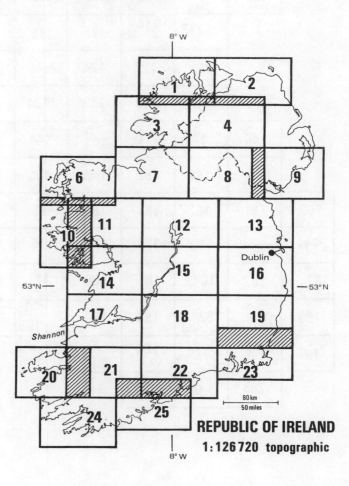

REPUBLIC OF IRELAND
1:126 720 topographic

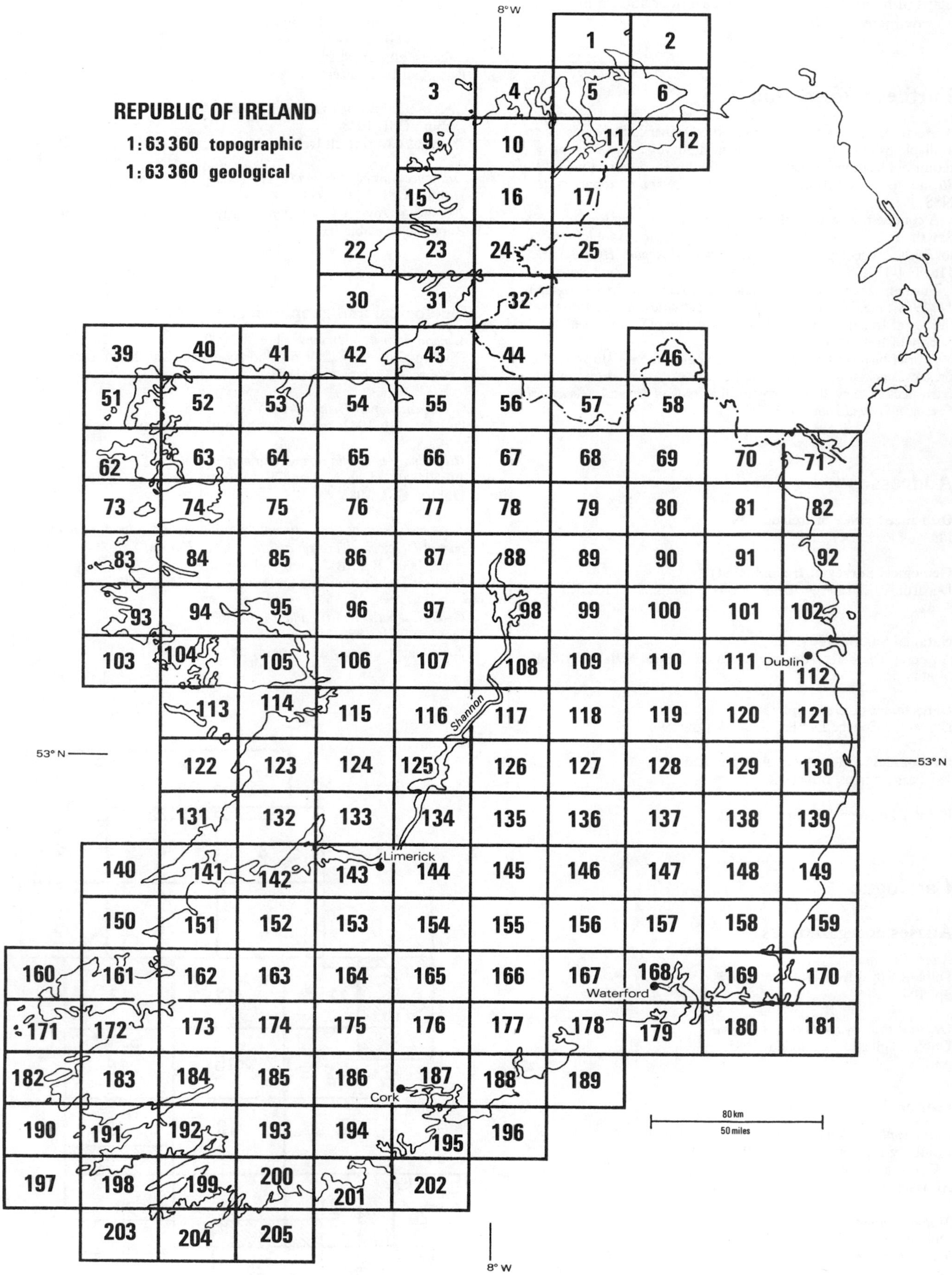

REPUBLIC OF IRELAND
1:63 360 topographic
1:63 360 geological

				1	2
3	4	5	6		

9 10 11 12
15 16 17
22 23 24 25
30 31 32
39 40 41 42 43 44 46
51 52 53 54 55 56 57 58
62 63 64 65 66 67 68 69 70 71
73 74 75 76 77 78 79 80 81 82
83 84 85 86 87 88 89 90 91 92
93 94 95 96 97 98 99 100 101 102
103 104 105 106 107 108 109 110 111 Dublin 112
113 114 115 116 117 118 119 120 121
Shannon
122 123 124 125 126 127 128 129 130
131 132 133 134 135 136 137 138 139
140 141 142 143 Limerick 144 145 146 147 148 149
150 151 152 153 154 155 156 157 158 159
160 161 162 163 164 165 166 167 168 169 170
Waterford
171 172 173 174 175 176 177 178 179 180 181
182 183 184 185 186 187 188 189
Cork
190 191 192 193 194 195 196
197 198 199 200 201 202
203 204 205

8° W
53° N
53° N

80 km
50 miles

8° W

Environmental

Energy and natural resources map—Ireland 1:1 500 000
Dublin: Natural Resources Group, 1980.

Ireland: general soil map 1:575 000 2nd edn
Dublin: NSS, 1980.
Published with explanatory bulletin.

Ireland: peatland map 1:575 000
Dublin: NSS, 1978.
Published with explanatory bulletin.

National Soil Survey of Ireland 1:126 720
Dublin: NSS, 1963–.
ca 50% counties published.
Sheets published by county.
Maps accompanied by bulletins.

Administrative

*Boundaries of administrative counties, county boroughs, urban and
dispensary districts and district electoral divisions* 1:250 000

Dublin: Department of Local Government and Public Health,
1962.
4 sheets, all published.

Human and economic

Agriculture in Ireland: a census atlas A.A. Horner, J.A. Walsh
and J.A. Williams
Dublin: Department of Geography, University College, 1984.
pp. 68 + 36 pp. maps.

Town maps

Dublin city 1:20 000 11th popular edition
Dublin: OSI, 1976.
East and west sheets printed back to back.
Street index.
Administrative edition available.

Dublin street guide 1:20 000
Dublin: OSI, 1985.
Pocket edition of the previous entry.

Italy

The topographic mapping authority in Italy is the **Istituto
Geografico Militare Italiano (IGMI)**, Florence. IGMI
originated in 1861 as the Ufficio Tecnico dello Stato
Maggiore Italiano, and assumed its present name in 1882.

The basic scale topographic map is the 1:25 000 series,
the earliest versions of which date from 1879 and cover the
entire country in 3545 sheets or *tavolette*, each covering 7′
30″ longitude and 5′ of latitude. The modern version is
postwar (Series M891) and has been compiled mainly from
aerial photogrammetric survey, but many areas are still
only covered by prewar sheets. The projection is
Gauss–Boaga. Some sheets are published in one, some in
three and some in five colours, with the UTM grid
overprinted in purple. The contour interval is usually 25 m.

Another long-established series is the 1:100 000,
composed of 278 sheets and published in editions with and
without shaded relief and also in a special administrative
edition in which the provinces and communes and their
administrative centres are shown in purple against a brown
background. This series is derived from the 1:25 000, with
each sheet covering 30′ longitude and 20′ latitude (the area
of 16 1:25 000 scale sheets). Contours are at 50 m
intervals. Current editions in this series are mainly of the
1960s and 1970s, and the series is not being up-dated,
although it is kept in print. There is also an archaeological
edition of some sheets in this series.

In 1964, the first sheets of a completely new series at
1:50 000 appeared. This high-quality map with elaborate
specification is printed in six colours, and there is also an
administrative edition in three colours. Projection is UTM,
and contours are at 25 m. Each sheet covers an area of 20′
longitude and 12′ latitude and the sheet system is not
conformable with the 1:100 000/1:25 000 mapping. This
series has been issued only slowly over the last 20 years.
Much of northern Italy, the whole of Sicily and Puglia are
covered. However, comparatively few new sheets have
been issued in the 1980s, and at the time of writing the
future mapping policy is under review, since it is now
considered that 1:50 000 is inadequate as a basic mapping
scale and that a newly designed 1:25 000 is required.

The former IGMI *Carta stradale d'Italia 1:200 000* has
been discontinued in favour of a series of regional road
maps at 1:250 000. The sheets are of variable size, and
printing is in 13 colours with shaded relief and 100 m
contours. Regional and provincial boundaries are shown,
and road distances are given in kilometres.

Over the last 5 years, IGMI has been investigating the
design of a national geographic information system and the
use of digital methods of map production. A system has
been developed for geodetic and geophysical data
management and another for topographic and cartographic
data. An experimental 1:25 000 sheet has been produced
by digital methods.

In 1984, the Italian Government established the
Commissione Nazionale di Geodesia, Topografia e
Cartografia, under the National Research Council, to take
responsibility for research and development in mapping
and to coordinate the mapping of regional bodies, since the
strongly regional administrative structure in Italy has
meant that much large-scale topographic (and thematic)
mapping has been undertaken by regional administrations
rather than IGMI.

Cadastral mapping in Italy is carried out by the
Direzione Generale de Catasto, and maps are published at
scales of 1:1000 and 1:2000. Marine charting and some
bathymetric mapping of the Mediterranean is carried out
by the **Istituto Idrografico della Marina**.

The national geological mapping agency is the **Servizio
Geologico d'Italia (SGI)**, Rome, and there is a complete
geological cover at 1:100 000, started in 1884. Although
many of the sheets and their monographs are out of print,
a number are available in a second edition begun in 1960
which adopts a new lithostratigraphic classification of the
rocks. A 1:50 000 geological series was initiated in 1970,
but only a few sheets have so far been published. The most
recent general synthesis of the geology of Italy is provided
by the series of five sheets at 1:500 000 published 1976–83,
each of which is accompanied by a monograph.

Much thematic mapping, often regional but sometimes
general, has been sponsored by the **Consiglio Nazionale**

delle Ricerche (CNR). This includes a number of geological and geophysical maps, among them the *Structural model of Italy* 1 : 1 000 000. Two pairs of complex maps cover the themes of lithostratigraphy, tectonics, bathymetry, total magnetic field intensity, gravity and seismicity. There is an accompanying 502-page monograph published (in English) as *Quaderni de La Ricerca Scientifica* 90. Currently, a series of volumes is being published as the results of the Progetto Finalizzato Geodinamica (*Quaderni de La Ricerca Scientifica* 114). Volumes published so far include an *Atlas of isoseismal maps of Italian earthquakes* (Volume 2A), and forthcoming volumes include a number of 1 : 500 000 scale maps of structural and tectonic geology and of mineralization.

Regional organizations have also carried out significant programmes of geological mapping, the most notable example being the *Carta geologica della Calabria* 1 : 25 000, complete in 182 sheets, published by the Cassa per il Mezzogiorno, in the period 1967–71 and accompanied by eight volumes of text. The Cassa per il Mezzogiorno was a state bank established to carry out state investment in southern Italy. In the late 1950s, it was also responsible for publication of a *Carta topografica della Calabria* 1 : 10 000 complete in 669 monochrome sheets with 10 m contours. The Cassa was abolished in 1984.

Much soil, land use, vegetation and natural hazard mapping has also been undertaken by regional authorities and by universities, and a number of monographic maps have been published. There is no national series of soil maps, but many provinces have been covered—Sardinia, Emilia-Romagna and Friuli-Venezia Giulia, for example—at various scales. At the University of Florence there is a **Centro di Studio per la Genesi, Classificazione e Cartografia del Suolo**, while the Istituto Sperimentale per lo Studio e la Difusa del Suolo, also at Florence, has been concerned with soil mapping and the methodology of land evaluation. The national series of land use maps at 1 : 200 000 produced by Touring Club Italiano and CNR in the 1960s is still available, while more recently a number of land capability maps have also been published of individual regions.

Many large-scale vegetation and other kinds of environmental maps were published as part of a CNR-funded project on promoting the quality of the environment (P.F. Promozione della qualità dell'ambiente—sottoprogetto 'Descrizione ecosistemi') which ran from 1976 to 1981. They are listed in the CNR catalogue.

Climatic maps have been published by CNR in association with the Instituto di Geografia dell'Universita di Pisa but may no longer be available.

Touring Club Italiano (TCI) published a National Atlas (*Atlante fisico-economico d'Italia*) in 1939, and it has been long out of print. A new national atlas has been in preparation since 1968 as a collaborative project between CNR and Touring Club Italiano. The first of four volumes of this *Atlante tematico d'Italia* was scheduled for publication in 1987, with the remaining volumes to follow in successive years until 1990. Italy's first regional atlas, a substantial *Atlante della Sardegna* is in progress, with two of the three projected volumes so far published.

There is a large number of commercial map publishers in Italy. **Touring Club Italiano (TCI)**, founded 1894, is well known for its tourist guides and maps, including the multi-volume *Guida d'Italia*, a three-volume *Atlante stradale d'Italia* at 1 : 200 000, numerous road maps, and the detailed alpine maps (based on IGM cartography) called *Carte delle zone turistiche*. A new set of 1 : 200 000 maps in 15 sheets entitled *Grande Carta Stradale d'Italia*

was issued in the period 1982–86. TCI also produces the *Annuario generale*, a comprehensive gazetteer of more than 32 000 places in Italy.

Litografia Artistica Cartografica (LAC), Florence, was founded in 1949 and publishes a wide range of mapping, including town maps, tourist maps, and provincial, regional and national maps.

The **Instituto Geografico de Agostini (IGA)** publishes provincial map-guides and maps of National Parks, while **Studio FMB** has a very long list of city street maps as well as provincial road maps. Detailed topographic maps tailored for the use of walkers and alpinists come from the publisher **Tabacco**, whose *Carta topografica per escursionisti* and *Carta dei sentieri e rifugi* at 1 : 25 000 and 1 : 50 000 cover most of the Italian alpine areas. **Verdesi** publishes street maps, atlases and directories including a map of Rome.

Many of the private publishing houses cited above also carry out thematic mapping projects for the various national and regional government agencies, and there are also many other surveying and cartographic companies, such as Cartografia D. Musielak, S.EL.CA. and Aquater SpA, which are primarily engaged in contract work for other agencies and do not have a significant list of their own.

Further information

IGMI publishes indexes, prices and descriptive material from time to time. SGI issues a catalogue of geological publications and maps, and CNR issues a catalogue of publications which includes some maps, as do most of the commercial publishers.

Italian map series are discussed *inter alia* in the *Bollettino della Associazione Italiana di Cartografia*, and this periodical also reports on new maps published by all the major mapmakers in the country.

Addresses

Istituto Geografico Militare (IGMI)
Via Cesare Battisti 10, I-50100 FIRENZE.

Istituto Geografico Militare (IGMI) For map sales
Sezione Vendite, Via Filippo Strozzi 14, I-50100 FIRENZE.

Istituto Idrografico Della Marina
Passo Osservatorio 4, I-16100 GENOVA.

Servizio Geologico d'Italia (SGI)
Largo S. Susanna N 13, I-00187 ROMA.

Consiglio Nazionale delle Ricerche (CNR)
Servizio Pubblicazione, Piazzale Aldo Moro 7, I-00185 ROMA.

Istituto Centrale di Statistica (ISTAT)
Via Cesare Balbo 16, I-00184 ROMA.

Centro di Studio per la Genesi, Classificazione e Cartografia del Suolo
Istituto de Geopedologia e Geologia Applicata dell'Università, Piazzale delle Cascine 18, I-50144 FIRENZE.

Touring Club Italiano (TCI)
Corso Italia 10, I-20122 MILANO.

Litografia Artistica Cartografia (LAC)
Via del Romito 11-13r, I-50134 FIRENZE.

Istituto Geografico de Agostini (IGA)
Servicio Cartografico, Corso della Vittoria 91, I-28100 NOVARA.

Studio FMB
Via Andrea, Costa 142/2. I-40067 RASTIGNANO DI PIANORO (BO).

Tabacco
Via Vardi 4, I-33100 UDINE.

Enrico Verdesi
Via degli Scialoia 19, I-00196 ROMA.

RIMIN SpA
Via Po 25/A, I-00198 ROMA.

For Bartholomew, see Great Britain; for Deutsche Bundesbahn, see German Federal Republic; for Hallwag, see Switzerland.

Catalogue

Atlases and gazetteers

Atlante tematico d'Italia
Roma: CNR/Milano: TCI, 1987–.
In 4 volumes.

Atlante economico-commerciale delle regioni d'Italia
Roma: SOMEA SpA, Istituto dell'encyclopedia Italiana, 1973.
2 vols, pp. 903.

Nuovo dizionario dei comuni e frazioni di comune con le circoscrizioni administrative 29th edn
Roma: ISTAT, 1984.

Annuario generale dei comuni e delle frazioni d'Italia Edizione 1980/1985
Milano: TCI, 1980.
pp. 1342 + 91 pp. maps.

Atlante della Sardegna
Cagliari: La Zattera Editrice, 1971 (Volume 1).
Roma: Edizioni Kappa, 1980 (Volume 2).
3 vols, 2 published.

General

Carta d'Italia 1:1 250 000
Firenze: IGMI, 1972.

Italy 1:1 250 000
Edinburgh: Bartholomew, 1982.
Inset maps of Rome and Milan.

Carta fisico-politico d'Italia 1:1 000 000
Milano: TCI, 1976.

Carta internazionalie 'Il Mondo' 1:500 000 Series 1404
Firenze: IGMI, 1966–.
14 sheets, 13 published.

Carta regionale d'Italia 1:250 000
Firenze: IGMI, 1970–.
16 sheets, 9 published.

Topographic

Carta topografica d'Italia 1:100 000 Series M691
Firenze: IGMI, 1881–.
278 sheets, all published. ■

Carta topografica d'Italia 1:50 000 Series M792
Firenze: IGMI, 1964–.
636 sheets, *ca* 230 published. ■

Carta topografica d'Italia 1:25 000 Series M891
Firenze: IGMI, 1879–.
3545 sheets, all published.

Geological and geophysical

Carta magnetica d'Italia della declinazione-isogone al 1973
1:2 000 000
Firenze: IGMI, 1974.

Carta magnetica d'Italia della componente orizzontale-isodinamiche al 1973 1:2 000 000
Firenze: IGMI, 1974.

Structural model of Italy 1:1 000 000 edited by L. Ogniban
Roma: CNR, 1973.
4 sheets, all published, plus text.
In English.

Carta geologica d'Italia 1:1 000 000 Edition 4
Roma: SGI, 1961.
2 sheets, both published.

Carta tectonica d'Italia, schema preliminare 1:1 000 000
Roma: CNR, 1981.

Carta mineraria d'Italia 1:1 000 000
Roma: RIMIN SpA, 1975.
2 sheets, both published, plus text.

Carta delle temperature sottoterranee in Italia
Roma: CNR, 1982.

Carta delle manifestazioni termali e dei complessi idrogeologici d'Italia 1:1 000 000
Roma: CNR, 1983.
2 sheets, both published.

Carta geologica d'Italia 1:500 000
Roma: SGI, 1976–83.
5 sheets, all published, plus text.

Carta magnetica d'Italia 1:500 000
Roma: AGIP SpA.
14 sheets, all published.

Carta geologica d'Italia 1:100 000
Roma: SGI, 1984–.
278 sheets, all published. ■
Many out of print.
Some available with a gravity overlay.

Carta geologica d'Italia 1:50 000
Roma: SGI, 1972–.
636 sheets, 15 published.

Carta gravimetrica d'Italia 1:50 000
Roma: SGI, 1973–.
636 sheets, 4 published.

Environmental

Carta dei Parchi Nazionali e delle aree protette d'Italia 1:1 250 000
Novara: IGA, 1983.

Carta della utilizzazione del suolo d'Italia 1:200 000
Roma: CNR/Milano: TCI, 1960–68.
26 sheets, all published.
Sheets are grouped together with illustrated regional monographs.

Administrative

Italia. Limiti administrativi 1:100 000
Firenze: IGMI, 1955–.
278 sheets, all published. ■
Special 'L' edition of topographic map.

Human and economic

Carta ferroviaria d'Italia/Map of Italian railways 1974. 1:1 000 000
Frankfurt: Deutsche Bundesbahn, 1974.
With index of stations.

ITALY

1:100 000 topographic
1:100 000 geological
1:100 000 administrative

SARDEGNA

SICILIA

160 km
100 miles

Town maps

Roma 1:15 000
Firenze: LAC.

Rome pocket plan 1:12 000
Roma: Verdesi.

Roma 1:13 500
Bern: Hallwag, 1985.

Vatican City
Washington, DC: CIA, 1984.

Roma 1:13 500
Bern: Hallwag, 1985.
Street map of Rome includes inset map of Vatican City.

Firenze 1:10 000
Firenze: IGMI, 1981.
2 sheets, both published.

Carta di Firenze Centro 1:5000
Firenze: IGMI.
With separate street index.
Maps of Florence also available at 1:10 000 and 1:25 000.

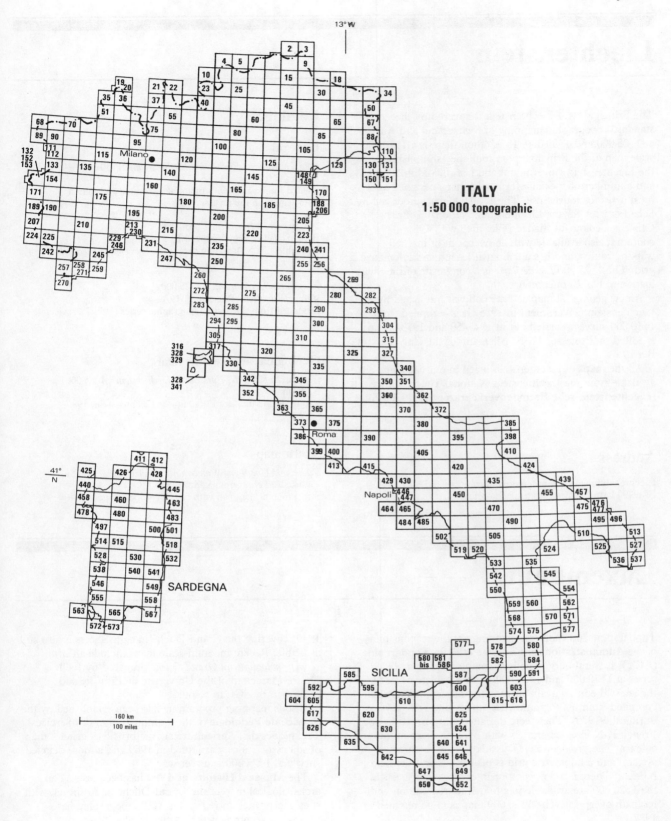

13° W

ITALY
1:50 000 topographic

41°
N

SARDEGNA

Napoli

Roma

Milano

SICILIA

160 km
100 miles

Liechtenstein

The Principality of Liechtenstein is covered by the standard topographic mapping of Switzerland and Austria at 1:25 000 and 1:50 000. In addition there is a 1:10 000 scale map of the Principality in four sheets published by the Liechtenstein Government and last revised in 1979, and a number of specially formatted maps designed in most cases for tourist use. The 1:50 000 map produced by Schad & Frey for the Liechtenstein National Tourist Office is a colourful shaded relief map with 100 m contours, also available with an overprint of marked walking routes, and has a trilingual legend and kilometric grid. The 1:25 000 *Landeskarte* is a composite of the Swiss maps and has 20 m contours.

A new geological map at 1:25 000 was published by the Liechtenstein Government in 1985. It is compiled from 1:10 000 surveys carried out in 1945–50 and 1973–83 and is edited by Professor Dr F. Allemann of the University of Bern.

All the maps of Liechtenstein listed in our catalogue are available from the Liechtenstein National Tourist Office (**Liechtensteinische Fremdenverkehrszentrale, LFVZ**), Vaduz.

Address

Liechtensteinische Fremdenverkehrszentrale (LFVZ)
FL 9490 VADUZ, Fürstentum Liechtenstein.

Catalogue

General

Agfa Tourenkarte Fürstentum Liechtenstein   1:50 000
Vaduz: LFVZ.
Shaded tourist map with pictorial panorama and tourist information on the reverse.

Fürstentum Liechtenstein   1:50 000
Bern: Schad & Frey/Vaduz: LFVZ, 1978.
Available also with walking routes as *Fürstentum Liechtenstein Wanderkarte*.

Fürstentum Liechtenstein   1:25 000
Vaduz: Fürstliche Regierung, 1976.
Produced from the Swiss topographic series (1971/72 revision).

Geological and geophysical

Geologische Karte des Fürstentum Liechtenstein   1:25 000
F. Alleman
Vaduz: Regierung des Fürstentums Liechtenstein, 1985.

Town maps

Vaduz und Umgebung Wanderkarte   1:10 000
Vaduz: Verkehrsverein Vaduz, 1975.
Street map of the town with circular walks in the vicinity.

Luxembourg

Topographic and cadastral mapping are the responsibility of the **Administration du Cadastre et de la Topographie (ACT)**, Luxembourg. The basic mapping is a parallel series at 1:20 000 and 1:25 000, with complete cover of the latest B edition available at both scales. The maps are compiled from photogrammetric survey and were revised during the 1970s. They were designed and printed by the French IGN in collaboration with ACT, and are in four colours. The projection is Gauss–Krüger (Luxembourg system) and a kilometric grid is printed around the map borders. There is a 5 m contour interval. Some sheets of the 1:20 000 are available in a folded tourist edition with footpath overprint. The 1:20 000 map is to be reissued in 1987.

A 1:50 000 series issued between 1964 and 1979 is no longer available. The 1:100 000 *Carte topographique et touristique* is issued as a single sheet in seven colours with 20 m contour interval and shaded relief. It has detailed road classification and includes road distances.

For cadastral and planning purposes, ACT issues a 1:10 000 enlargement of the *carte topographique* printed by diazo methods.

Geological mapping is carried out by the **Service Géologique de Luxembourg (SGL)**. A modern series at 1:25 000 in 13 sheets is in progress. An earlier series of the 1940s covering the Grand-Duché in eight sheets is also still available. Recent medium-scale maps include a *Carte géomorphologique du Grand-Duché* prepared by Joelle Desire-Marchand of the Université de Picardie and published by SGL in colour.

A soils mapping programme has been established by the **Service de Pédologie** of the Administration des Services Techniques de l'Agriculture at Ettelbruck. A general map of soil associations was issued in 1969 and a more detailed survey at 1:25 000 is under way.

The **Musée d'Histoire et d'Art** has been issuing an archaeological map of the Grand-Duché in 30 sheets, with sites overprinted in red on an ACT topographic base. Sheets are accompanied by monographs.

The **Ministère de l'Education Nationale** issued a national *Atlas du Luxembourg* from 1971 in three volumes. The first volume is out of print, but stocks of the second and third volume are still available. There are no plans to reissue this atlas.

Luxemburg in Karte und Luftbild uses sample maps, air photographs and satellite imagery with parallel texts in French and German to describe the geographical regions of Luxembourg. A pilot study has been undertaken for a proposed *Atlas Saar–Lorraine–Luxembourg*.

Further information

Both ATC and SGL issue lists and indexes of their publications and mapping.

Addresses

Administration du Cadastre et de la Topographie (ACT)
54, Avenue Gaston Diderich, 1420 LUXEMBOURG.

Service Géologique du Luxembourg (SGL)
43, bd G.D. Charlotte, 1331 LUXEMBOURG.

Service de Pédologie
Administration des Services Techniques de l'Agriculture, Avenue Salentiny, Boite Postale 75, 9001 ETTELBRUCK.

Musée d'Histoire et d'Art
Musée de l'Etat, Marche-aux-Poissons, 2345 LUXEMBOURG.

Ministère de l'Education Nationale
6, Boulevard Royal, 2910 LUXEMBOURG.

Catalogue

Atlases and gazetteers

Atlas du Luxembourg
Luxembourg: Ministère de l'Education Nationale, 1971–.
3 volumes, volume 1 out of print.

Luxembourg in Karte und Luftbild/Le Luxembourg en cartes et photos aeriennes G. Schmit and B. Wiese
Luxembourg: Editions Guy Schmit + Bernd Wiese, 1981.

General

Grand-Duché de Luxembourg 1:250 000
Luxembourg: ACT, 1984.

Grand-Duché de Luxembourg. Carte topographique, édition touristique 1:100 000
Luxembourg: ACT, 1983.
Available as 7-colour map or in black/blue/brown, natural environment and orohydrographic versions.

Carte des distances du Grand-Duché de Luxembourg 1:100 000
Luxembourg: ACT, 1980.
Map of road and rail distances.

Topographic

Grand-Duché de Luxembourg. Carte topographique 1:25 000/1:20 000 Edition B
Luxembourg: ACT, 1979.
30 sheets, all published. ∎
Sheets also available in black/blue edition.

Geological and geophysical

Carte hydrogéologique du Luxembourg 1:200 000
Luxembourg: SGL, 1981.

Carte géologique générale du Grand-Duché de Luxembourg 1:100 000 Edition 2
Luxembourg: SGL, 1974.

Carte géomorphologique du Grand-Duché de Luxembourg 1:100 000 J. Desire-Marchand
Luxembourg: SGL, 1984.
Legend in French and German.
Explanatory notice.

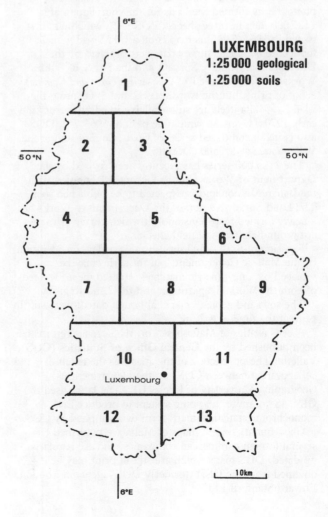

Luxembourg. Cartes géologiques detaillées   1 : 25 000
Luxembourg: SGL, 1971–.
13 sheets, 5 published. ■

Environmental

Carte des sols du Grand-Duché de Luxembourg   1 : 100 000
Ettelbruck: Service de Pédologie, 1969.

Grand-Duché de Luxembourg. Carte des sols   1 : 25 000
Ettelbruck: Service de Pédologie, 1971–.
13 sheets, 4 published. ■

Human and economic

Carte archeologique du Grand-Duché de Luxembourg   1 : 20 000
Luxembourg: Musée d'Histoire et d'Art, 1973–.
30 sheets, 11 published. ■

Town maps

Carte de la Ville de Luxembourg   1 : 5000
Luxembourg: ACT, 1954.
2 sheets, both published.
Monochrome map.

Malta

Formerly a British crown colony, Malta became fully independent in 1964 and in 1974 was constituted a Republic.

Modern mapping of the three islands of Malta, Gozo and Comino had previously been undertaken by the British Government in cooperation with the Office of Public Works, Valletta. Early in the present century, large-scale GSGS series at 1 : 2500 and 1 : 10 560 had been published, and there was also a general map at 1 : 31 680 (GSGS 3859) which forms the basis of some thematic maps which are still in print.

In the late 1950s, a new survey was initiated by the **Directorate of Overseas Surveys (DOS)** (now Overseas Surveys Directorate, Southampton, OSD). Air photo sorties were flown in 1955 and 1957 and in 1958 a complete cover at 1 : 8000 was obtained. This aerial photography formed the basis for the compilation of a 1 : 25 000 map in three sheets (DOS 352), which was first issued in 1962–63. This map is on a UTM grid with 25 foot contours. Field survey data was supplied by the Directorate of Military Surveys and the Office of Public Works, Valletta. Sheet 1 of this series (*Gozo and Comino*) is out of print, but the two sheets (2 and 3) covering the island of Malta itself are still available in editions from the early 1970s. Over the same time period (1963–75), DOS also issued a contoured series of Gozo, Comino and of the Valletta area at 1 : 2500 (DOS 152).

The 1 : 25 000 series has recently been revised by the **Department of Works**, Beltissebh, in collaboration with the Italian Mission and is expected to be published soon. The Land Survey Section of the Department of Works has a newly formed map-revision unit which currently carries out ground control and field checking.

Revision of the 1 : 2500 series is also under consideration by the Maltese Government, but the series may be replaced by a new metric map at 1 : 1000 to meet the needs of both the Works Department and the Land Registry.

The soils and geology maps, although dated, were still in print at the time of writing.

A census atlas of Malta based on 1985 census returns has been published by the **Central Office of Statistics (COS)**, Valletta. The production of the atlas was the result of collaboration between COS and the Department of Geography, University of Keele, UK. The latter used the GIMMS computer mapping system to produce the monochrome statistical maps from data supplied by COS.

Good-quality tourist maps of Malta are published by several foreign commercial companies, and are regularly up-dated. DOS maps which are still in print may be obtained from the OSD (formerly DOS) agents in London, Edward Stanford Ltd.

Addresses

Department of Works
Beltissebh, MALTA.

Central Office of Statistics (COS)
Auberge de Italie, VALLETTA.

Overseas Surveys Directorate (DOS/OSD)
Ordnance Survey, Romsey Rd, Maybush, SOUTHAMPTON SO9 4DH, UK.

For Hallwag, see Switzerland; for Clyde Surveys see Great Britain; for Karto+Grafik, see German Federal Republic.

Catalogue

General

Ferienkarte Malta   1 : 50 000
Bern: Hallwag, 1985.

Malta. Hildebrands Urlaubskarte   1 : 50 000
Frankfurt AM: Karto+Grafik, 1985.

Malta   1 : 40 000 and *Gozo*   1 : 50 000
Maidenhead: Clyde Surveys, 1986.
Includes tourist information and inset maps of Bugibba, Valletta and Sliema, Mdina and Rabat, Victoria.

Topographic

Malta   1 : 25 000   DOS 352 (M898)   Editions 2–3
Tolworth: DOS, 1969–72.
3 sheets, all published (Sheet 1 out of print).

Geological and geophysical

Malta geological   1 : 31 680
BP Exploration Co. Ltd, 1957.
2 sheets, both published.

Environmental

Malta and Gozo soils map   1 : 31 680   DOS (Misc) 258
Tolworth: DOS, 1960.
Accompanies report of D.M. Lang, *Soils of Malta and Gozo.*

Human and economic

Census '85 Volume III. A computer-drawn demographic atlas of Malta and Gozo   K.T. Mason and D.G. Lockhart
Valletta: Central Office of Statistics, 1987.

Monaco

Monaco is covered by standard French topographic mapping. A street map is provided by **Plans-Guides Blay**.

Addresses

Plans-Guides Blay
14, rue Favart, 75002 PARIS.

Catalogue

Principauté de Monaco et Beausoleil   1:8000
Paris: Blay.

The Netherlands

The Netherlands is a small but cartographically very active country, and maps are much used as instruments of planning so that a wide range of basic and derived mapping is produced by numerous government agencies. Our text and catalogue concentrate on those maps and map series which are of broad public interest and which are readily available for purchase.

The production of Dutch topographic maps is in the hands of the Topographic Service (**Topografische Dienst Nederland, TDN**), Emmen. The Service was founded in 1815 and is a branch of the Ministry of Defence. Originally called the Topografisch Bureau, it acquired its present name in 1931 when it was merged with its own lithographic press. Currently it is responsible for production of both civil and military topographic maps at scales 1:10 000, 1:25 000, 1:50 000 and 1:250 000. TDN moved to Emmen in 1984 having previously been at Delft.

In the nineteenth century the Topografisch Bureau quickly established a high reputation for the quality of its mapping—officers of the British Ordnance Survey visited it in 1826 for a whole year to study its methods—and an engraved map at 1:50 000 was completed in 1864, to be followed by the first full-colour map at 1:25 000.

Since 1932, topographic maps have been produced from vertical aerial photographs, and from these the basic scale mapping—a series of 658 monochrome sheets at 1:10 000—is compiled. Maps in this series are revised on a 10 year cycle, or more frequently if the rate of change requires it. The 1:25 000 scale map is produced by re-scribing and generalizing from a photoreduction of the 1:10 000 map.

The current 1:25 000 map is published in a civil edition only. It is a six-colour map on a stereographic projection, and a grid at 1 km intervals is printed on the face of the map. Contours are at 2.5 m intervals and are derived from a dense network of spot heights, compiled from spirit levelling and photogrammetry, and shown in this form (without contours) on a so-called 'altitude map' at 1:10 000. A new edition of the 1:25 000 has been appearing in the 1980s with a more public-user-orientated specification, and designed for use in a folded format. The legend is in Dutch and English. Although the series uses the numbering system shown in our index, the sheets have been reduced in number from 368 to 308 by amalgamation and rationalization of sheet lines.

The 1:50 000 map is published in civil and military editions, the former having the national grid system and the latter the UTM grid. This too is a six-colour map with 2.5 m contour interval. One of its more unusual features is the amount of land use information shown, including the differentiation of pasture and arable land.

A 1:100 000 series in 34 sheets was published in 1957–58, but for reasons of economy has not been revised, and will not be reprinted when existing stocks are exhausted. It is therefore not included in our catalogue. A 1:100 000 map in eight sheets produced by reduction from the 1:50 000 series (1970–84) has recently become available.

The 1:250 000 series is on the UTM projection and has a grid printed at 10 km intervals. It is printed in six colours with contours at 20 m, and is revised every 4 years.

In addition to the large-scale topographic series, TDN also publishes several regional maps (Walcheren, the Ommen district, Zuid-Limburg and the Wadden islands), and also the *Gemeentenkaart van Nederland* 1:400 000 (now a digital map), which is revised approximately every 3 years and shows provincial boundaries, provincial capitals and the parishes with more than 100 000 inhabitants. TDN has also introduced a series of *Military City Street Plans* beginning with The Hague and Rotterdam–Europoort.

TDN is engaged increasingly in digital methods of map production. A topologically structured database has been produced from the 1:250 000 series, and elevation data has been captured from the 1:50 000 map for the production of digital terrain models.

Statistical and administrative area maps are produced by the **Centraal Bureau voor de Statistiek**.

Larger-scale mapping in the Netherlands is carried out by the Cadastral Service (**Dienst van het Kadaster en de Openbare Registers**), by the Ministry of Public Works and by the municipal authorities of large towns (e.g. of Rotterdam, Amsterdam, the Hague and Utrecht, which publish fine multi-colour maps of these cities at 1:5000 and 1:10 000). Since 1973, the Cadastral Service has been incorporated in the Ministry of Housing and Physical Planning, and from 1975 the national government authorized the Cadastral Service to undertake a new Large Scale Base Map (the Dutch acronym is GBKN). This is a map designed for cadastral purposes and for use by utility services and it includes street names, house numbers and the national metric grid. Publication scales are 1:2000,

1:1000 and 1:500 according to degree of urbanization. Maps are also available in digital form. By 1982, 11 per cent of the country had been newly mapped in this programme.

Geological mapping is the responsibility of the **Rijks Geologische Dienst (RGD)**, Haarlem. The basic mapping is the 'new' 1:50 000 series begun in 1964 and replacing an older series published between 1925 and 1951. The map presents information about both surface and subsurface geology and is supported by sheet memoirs and supplementary single-value maps. Legends are in Dutch and English, and the memoirs include English summaries. Regional geological maps have been published of Zuid-Limburg (1:100 000) and of the Dutch Wadden area (in conjunction with the Rijksdienst voor de IJsselmeerpolders).

A new 1:250 000 geological map in about five sheets to cover the whole country has been planned for publication in the late 1980s, but no sheets have so far appeared.

In the last two decades there has been increasing interest in the geology of the North Sea, and the development of geophysical survey methods has made it possible to construct detailed maps of offshore geology. Currently a 1:250 000 scale series of 11 sheets is in preparation. This series links to the British Geological Survey series at the same scale, and the first sheet to appear *Flemish Bight* was published jointly with BGS. Triple sheets are planned, covering respectively Holocene lithology, Pleistocene formations, and the underlying Tertiary and older rocks. Survey is also in progress for a 1:100 000 series of 10 near-shore sheets. These maps are associated with a computer database storing borehole information which can be manipulated to produce computer drafted sections, contoured surfaces and fishnet diagrams.

Soil survey in The Netherlands is undertaken by the Soil Survey Institute (**Stichting voor Bodemkartering, STIBOKA**), Wageningen. The Institute was founded in 1945, and is a semi-governmental organization falling partly under the direction of the Ministry of Agriculture and Fisheries.

Soil maps are prepared by the Institute and distributed by the Centre for Agricultural Publication and Documentation (**Centrum voor Landbouwpublikaties en Landbouwdocumentatie, PUDOC**). The series at 1:200 000 was completed in 1960–61 and has also been incorporated into the *Atlas van Nederland* (1st edition). These sheets have also been issued separately, accompanied by monographs and supplementary maps (1965–75). Monographs include an evaluation of the soil suitability for agriculture. The 1:50 000 series is based on an adaptation of the 1:50 000 topographic sheet system. These full-colour maps now cover about 75 per cent of the country and the series is scheduled for completion within the next decade. The latest cartographic summary of Dutch soils is the 1:250 000 *Bodemkaart van Nederland* published in 1985 in four sheets with a separate legend book. Automated cartography has been introduced (the Computervision system) for making derived maps showing specific soil data, and for automated classification of landscape data and landscape simulation modelling. The Survey has also set up an earth science information system in cooperation with the Dutch Geological Survey, which will serve as a database for storage of surveys, analytical data and digital map information.

STIBOKA has published numerous monographic maps, usually accompanying regional reports, and these are all listed in the retrospective *Kaarten-catalogus* (see Further information).

The geomorphological map at 1:50 000 is a joint publication of STIBOKA and RGD. As with the geological and soil maps at this scale, it is based on the sheet system of the topographic map. But some sheets are issued in an extended format. An early series of 15 sheets was issued 1960–70, the current series beginning in 1975. The series is planned for completion in about 1990. A general legend has been issued for use with all maps in the series.

A uniquely Dutch map is the *Waterstaatskaart* 1:50 000 published by the **Rijkswaterstaat, Afdeling Waterstaatskartografie** and distributed by the **Staatsuitgeverij**, The Hague. This is a water control map, originating in 1865. It shows all the various features—drainage units, locks, bridges, pumps, water courses, water levels—relevant to water management registered on a 1:50 000 topographic base map. Sheets are revised on an 8 year cycle. Rijkswaterstaat also publishes road maps at 1:400 000 and road number and classification maps at 1:250 000.

The **Rijksplanologische Dienst** (National Planning Service) has developed a computer-based information system for physical planning and maintains a 1:25 000 scale 'KADRO' planning map for which data is captured by digital methods.

The **Rijksdienst voor de IJsselmeerpolders** carries out large-scale topographic and cadastral mapping and has published a number of small thematic and recreation maps of the polders and an excellent 1:50 000 map of Flevoland (*Flevoland in kaart*, 1981).

The first National Atlas of The Netherlands was issued in instalments from 1963 to 1977, and was up-dated with three supplements (comprising 16 further sheets), the last of which was issued in 1981. This major inventory of the geography of The Netherlands is now out of print, but a new edition in 20 bound parts is in progress (1984–88). This edition is radically different from its forerunner. It is less of a cartographic inventory of physical and economic resources, adopting a more human-centred approach and examining a variety of spatial/environmental problems, grouped around five main themes. There is a careful integration of text, maps and other graphics, and the aim is to attract a wide and non-specialist readership.

There have been discussions about publishing an English edition of this atlas (which unlike its predecessor is entirely in Dutch), but at the time of writing, no decision has been made. The atlas is produced at TDN by the Bureau of the Scientific Atlas of The Netherlands, and is published and distributed by Staatsuitgeverij.

Two higher educational institutions have a strong commitment to map production. The **Geografisch Instituut, Rijksuniversiteit Utrecht** has published a number of maps as part of its research and teaching programmes in cartography, including most notably the *Atlas Zuidoost Utrecht*. The **International Institute for Aerial Survey and Earth Sciences (ITC)** at Enschede, which specializes in training surveyors and cartographers for the developing countries, produces and distributes a number of useful monographic maps of overseas areas, as a by-product of these courses.

Among the maps produced for motorists and tourists, those of the motoring organization **ANWB Koninklijke Nederlandse Toeristenbond (ANWB)** are particularly outstanding. They include a 1:100 000 series in 13 sheets (*ANWB Toeristenkaarten*), updated every few years, and a series of *Waterkaarten* covering inland waters. The 1:100 000 maps are also included in the *ANWB Atlas voor Nederland*, published in 1985 in association with Zomer & Keuning Boeken BV, Ede. This volume also has index, descriptive gazetteer and a number of small street maps of towns.

A common specification for a series of tourist maps at 1:50 000 (*Kaart voor vakantie en vrije tijd*) was developed by the Dutch Association of Tourist Agencies, and individual sheets are commissioned by the regional tourist authorities and produced by various commercial cartographers.

The national forestry authority (**Staatsbosbeheer**) in association with ANWB has published a series of recreational *voetspoorkaarten* (footpath maps) of forest reserves, and also issues (free) small-format maps of nature reserves entitled *Uw natuurreservaat*.

The company **Omnium**, Waalwijk, publishes a series of 12 regional *Toeristenkaarten* (which includes a single-sheet road map of the whole country) under the Smulders Kompas imprint; also town maps of Amsterdam, The Hague, Breda and Leeuwarden, and a 1:400 000 administrative map of the country.

Wolters-Noordhoff NV is a major publisher of educational maps and atlases, including the famous *De grote bosatlas*, and also produces a number of tourist maps.

NV Europees Cartografisch Instituut produces a few educational and tourist maps under its own imprint, but operates mainly as a production department serving other publishers.

Born NV publishes maps of navigable waterways and other transport network maps.

All the Dutch towns and cities are covered by street maps, many of them excellent full-colour productions by major cartographic studios such as **Cito-Plan BV**, **NV Falkplan/CIB** and **Suurlands Vademecum**; others are simpler products issued by the local tourist offices or planning authorities. Suurland also has a useful atlas of town maps, entitled *101 Stedenboek*.

Further information

The Topografische Dienst issues a catalogue in Dutch which describes not only its own mapping but also that of other major mapping authorities. It includes map indexes and sample map extracts.

A Union list of maps published in The Netherlands is published annually (*Bibliografie van in Nederland verschenen kaarten*) by the Royal Library with the *Nederlandse Stichting Informatie en Documentatiecentrum*, and may be purchased from Koninklijke Bibliotheek, Lange Voorhout 34, Postbus 30469, 2500 GL 's-Gravenhage.

STIBOKA has issued a comprehensive catalogue of maps published since 1945 as *Kaarten-catalogus van de Stichting voor Bodemkartering* (Wageningen: STIBOKA, 2nd edn, 1982) with annual updates.

Articles in English on Dutch mapping include Piket, J.J.C. (1980) Modern mapping in the Netherlands, *Cartographic Journal*, **17**, 40–48; and Van Wely, G.A. (1982) Large scale mapping in the Netherlands, *Surveying and Mapping*, **42**, 347–350.

The geological mapping programme is described in Oele, E. *et al.* (1983) Surveying The Netherlands: sampling techniques, maps and their applications, *Geologie en Mijnbouw*, **62**, 355–372; while Dutch geomorphological mapping and its applications are discussed in Ten Cate, J.A.M. (1983) Detailed systematic geomorphological mapping in The Netherlands and its applications, *Geologie en Mijnbouw*, **62**, 611–620.

A good overview of modern Dutch mapping is provided by the National Report *Cartography in The Netherlands 1972–1980*, published by the Dutch Cartographic Society for the ICA General Assembly in Tokyo in 1980.

Addresses

Topografische Dienst Nederland (TDN)
Bendienplein 5, Postbus 115, 7800 AC EMMEN.

Centraal Bureau voor de Statistiek
Prinses Beatrixlaan 428, Postbus 959, 2270 AZ VOORBURG.

Dienst van het Kadaster en de Openbare Registers
Waltersingel 1, 7314 NK APELDOORN.

Rijks Geologische Dienst (RGD)
Spaarne 17, 2011 CD HAARLEM.

Stichting voor Bodemkartering (STIBOKA)
Postbus 98, 6700 AB WAGENINGEN.

Centrum voor Landbauwpublikaties en Landbouwdokumentatie (PUDOC)
PO Box 4, 6700 AA WAGENINGEN.

Rijkswaterstaat, Afdeling Waterstaatskartografie
Koningskade 4, Postbus 20906, 2500 EX 's-GRAVENHAGE.

Rijksplanologische Dienst
Rieteweg 25, Postbus 502, ZWOLLE.

Rijksdienst voor de IJsselmeerpolders
Zuiderwagenplein 2, Postbus 600, 8200 AP LELYSTAD.

Staatsuitgeverij
Christoffel Plantijnstraat 2, Postbus 20014, 2500 EA 's-GRAVENHAGE.

Atlas van Nederland
Topografische Dienst, Bendienplein 5, Postbus 115, 7800 AC EMMEN.

Rijksinstituut voor Natuurbeheer
Kasteel Broekhuizen, Postbus 46, 3956 ZR LEERSUM.

Geografisch Instituut van de Rijks Universiteit
Heidelberglaan 2, 3584 CS UTRECHT.

International Institute for Aerial Survey and Earth Sciences (ITC)
Postbus 6, 7500 AA ENSCHEDE.

ANWB Koninklijke Nederlandse Toeristenbond (ANWB)
Postbus 93200, 2509 BA 's-GRAVENHAGE.

Staatsbosbeheer
Griffioenlaan 2, Postbus 20020, 3502 LA UTRECHT.

Omnium
Prof. Lorentzweg 6, Postbus 162, 5140 AD WAALWIJK.

Wolters-Noordhoff NV
Oude Boteringestraat 24, Postbus 58, 9700 MB GRONINGEN.

NV Europees Cartographisch Instituut
ECI/Drukkerij Rijkswijk, Verrijn Stuartlaan 25, RIJSWIJK.

Born NV Uitgeversmij
Postbus 22, ASSEN.

Cito-Plan BV
Cartografisch Instituut, Groot Hertoginnelaan 156, 2517 EM 's-GRAVENHAGE.

NV Falkplan/CIB
Cartografisch Instituut, Zichtenburglaan 52, Postbus 43107, 2504 AC 's-GRAVENHAGE.

Suurlands Vademecum
Postbus 9510, 5602 LM EINDHOVEN.

NV Nederlandse Spoorwegen (NS)
Educatieve Voorlichting, Postbus 2025, 3500 HA UTRECHT.

Malmberg
Leeghwaterlaan 16, Postbus 233, 5201 AE DEN BOSCH.

Catalogue

Atlases and gazetteers

Atlas van Nederland
's-Gravenhage: Staatsuitgeverij, 1984–88.
20 parts.

General

Kaart van Nederland   1 : 500 000
Emmen: TDN.

Satellietbeeldkaart Nederland   1 : 275 000
Den Bosch: Malmberg, 1982.

Topographic

Overzichtskaart van Nederland   1 : 250 000   Series 1501
Emmen: TDN, 1985–.
6 sheets, all published.

Topografische kaart van Nederland   1 : 50 000
Emmen: TDN, 1970–.
110 sheets, all published. ■

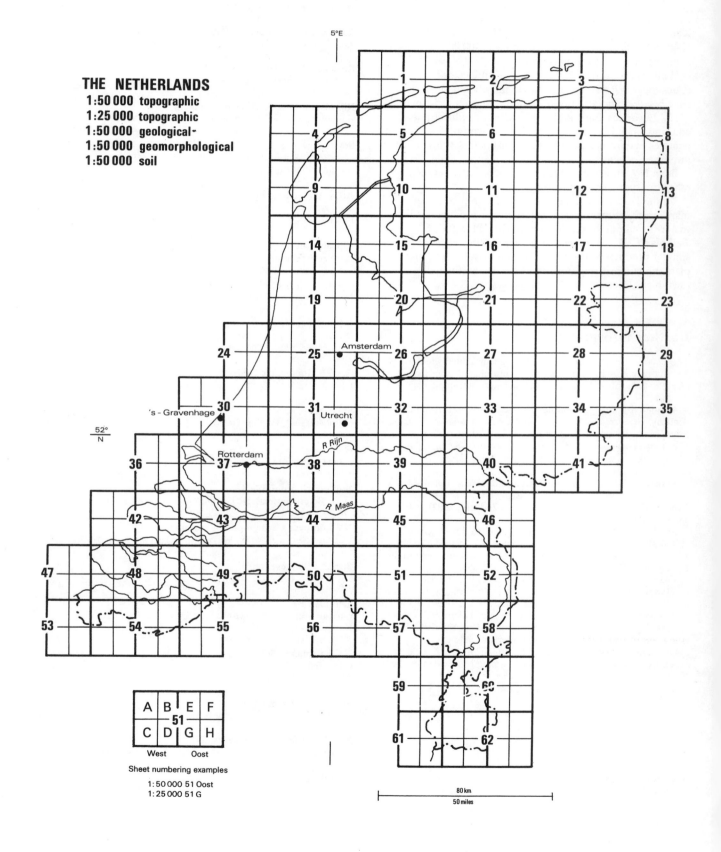

THE NETHERLANDS
1:50 000 topographic
1:25 000 topographic
1:50 000 geological-
1:50 000 geomorphological
1:50 000 soil

Sheet numbering examples
1 : 50 000 51 Oost
1 : 25 000 51 G

80 km
50 miles

Topografische kaart van Nederland 1:25 000
Emmen: TDN, 1960–.
308 sheets, all published. ■

Geological and geophysical

Hydrogeologische kaart van Nederland 1:1 500 000
Haarlem: RGD, 1972.
2 sheets: I, Superficial deposits; II, The main aquifers.

Geologische overzichtskaart van Nederland 1:600 000
Haarlem: RGD, 1975.

Overzichtskaart toegepaste geologie 1:600 000
Haarlem: RGD, 1975.
Map of applied geology.
This and the map above published as part of a set of maps with
134 pp. text, *Toelichting bij Geologische Overzichtskaarten van
Nederland.*

Geologische kaart van Nederland 1:50 000
Haarlem: RGD, 1964–.
ca 90 sheets, 18 published. ■

Geomorfologische kaart van Nederland 1:50 000
Haarlem: RGD/Wageningen: STIBOKA, 1975–.
49 sheets, 15 published. ■

Environmental

Klimaatatlas van Nederland
's-Gravenhage: Staatsuitgeverij, 1972.
24 double map sheets.

Globale bodemkaart van Nederland 1:600 000
Wageningen: STIBOKA, 1964.

Bodemkaart van Nederland 1:250 000
Wageningen: STIBOKA, 1985.
4 sheets + text, all published.

Bodemkaart van Nederland 1:200 000
Wageningen: STIBOKA, 1961.
9 sheets + legend, all published.

*Vegetatiekaart van Nederland en overzichtskaart van de ecologische
betekenis van het natuurlijk milieu in Nederland* 1:200 000
Leersum: Rijksinstituut voor Natuurbeheer, 1975.
4 sheets + text, all published.

Bodemkaart van Nederland 1:50 000
Wageningen: STIBOKA, 1964–.
ca 70 sheets, 36 published. ■

Administrative

De gemeentenkaart van Nederland 1:400 000
Emmen: TDN, 1987.

Nederlandse gemeenten 1:400 000
Waalwijk: Omnium.

Human and economic

Atlas van de kankerserfte in Nederland 1969–78
Voorburg: Centraal Bureau voor Statistiek, 1980.
pp. 173.

Spoorwegkaart van Nederland 1:300 000
Utrecht: NS, 1987.

Town maps

Amsterdam 1:13 500
Waalwijk: Omnium.

Amsterdam stadsplattegrond met centrum kaart 1:15 000
's-Gravenhage: Falkplan-Suurland BV, 1985.

Fietskaart van Amsterdam 1:12 500
Amsterdam: Gemeente Amsterdam, Afd. Verkeer en Vervoer/
ENFB, 1981.
Cyclists' map.

Kaart van 's-Gravenhage 1:15 000/1:7500
Emmen: TDN, 1979.

The Netherlands military city street plan Rotterdam–Europoort
1:35 000
Emmen: TDN, 1982.

Norway

The principal survey and mapping organization in Norway
is **Statens Kartverk (SKV)**, formerly Norges Geografiske
Oppmåling, Hønefoss, but established under its new name
in 1986 to be responsible for the country's land, sea and
resource mapping. The topographic survey was founded in
1773, but the first systematic mapping programme
commenced in 1867 with the establishment of a national
plan for mapping which included the provision of a
1:100 000 series, known as the *rektangelkart* and drawn on
the Cassini projection. Later, in north Norway, the
gradteigskart (or graticule map) was introduced, also at
1:100 000 but on the Polyhedric projection (later
transferred to the Gauss–Krüger). The four-colour maps of
these two series, drawn originally from plane table and
aneroid surveys, eventually covered most of Norway and
remained in production until 1966. At the time of writing,
many sheets of the *gradteigskart* are still available, but they
are no longer kept up to date and have been almost
entirely superseded by the modern 1:50 000
photogrammetric Series M711. They are not therefore
included in the catalogue.

The 1:50 000 Series M711 forms the basic scale modern
mapping of Norway. It has been in progress since 1955
and is scheduled for completion in 1988. The map is on a
UTM projection and is overprinted with a UTM
kilometric grid. It is in five colours with a 20 m contour
interval, and with a legend in Norwegian and English. The
map is a graticule map, with sheets each covering 15' of
latitude but varying in east–west extent from 22.5' of
longitude in the south to 36' in the north. Revision is on a
10 year cycle.

Since 1982, SKV has been digitizing hydrological and
contour details from the 1:50 000 series and this work was
to be completed by the end of 1986. Computer tapes
containing the digital height or hydrological data may be
purchased from SKV. A new 1:100 000 scale map is also
in preparation, derived from the 1:50 000, which will use
digital methods in its production. Sheets in this series will
have a similar specification to the 1:50 000 and include an
index of placenames on the reverse, referenced to the
UTM grid.

The 1:250 000 map series (Serie 1501) conforms to the

specification of the Joint Operations Graphic (JOG) charts and is available in air (contours in feet) and ground (contours in metres) editions. The topography of this map is somewhat dated, but a revision of the series to a new specification is in progress, and the first sheets were issued in 1984. The air edition is published by the Defence Mapping Service (Forsvarets karttjeneste) but is available from SKV. Both versions of this map have UTM grid and the GEOREF system. The 1:500 000 map (conforming to the Series 1404 World) is also a military series, although available for civilian use. There are plans to revise this series when the 1:250 000 revision has been completed.

SKV also publishes a large number of specially formatted tourist maps at scales ranging from 1:25 000 to 1:250 000 and covering the environs of major towns (*omlandskart*) and remoter recreational areas (*turkart*). Many of the latter have walking and skiing routes overprinted on them, and have been elaborated with the help of the Norwegian Tourist Association.

The **Statistisk Sentralbyrå** (Central Bureau of Statistics) collects and processes census data and in cooperation with SKV has produced population distribution maps (using the proportional circle technique) at 1:250 000 and 1:1 000 000 based on the 1970 and 1980 censuses. Their maps are published by SKV, and are also available in digital format.

The Norwegian Geological Survey (**Norges Geologiske Undersøkelse, NGU**) was founded in 1858, moving in 1957 from Oslo to Trondheim. It is the central institution for research and mapping of the geology of the country, and has completed detailed mapping of about 25 per cent of the land area. While some of this mapping is quite old, there are active modern series in progress at 1:50 000 and 1:250 000 (separate series for bedrock and Quaternary geology). New general maps of bedrock and Quaternary geology at 1:1 000 000 were published in 1983–84 and are available as part of the National Atlas series or separately. Since 1960, NGU has achieved almost complete aeromagnetic map cover of the land area, and a series of aeromagnetic maps of the continental shelf north of 62° is in progress. Other map series include the themes of geochemistry, hydrogeology, groundwater and mineral resources. Many of these latter are available only in transparency or diazo form, registered to the standard SKV topographic series.

The county map offices collaborate with the NGU in producing a 1:50 000 scale map of sand and gravel resources. Most sheets are in black and white. These maps are related to a computer database holding records of gravel deposits.

The Continental Shelf Institute (**Institutt for Kontinentalsokkelundersøkelse, IKU**), located at Trondheim, is an independent research and development institute under the Royal Norwegian Council for Scientific and Industrial Research. IKU has a regional mapping programme for the Norwegian continental shelf. The main series is at 1:500 000 with companion sheets covering the themes of bedrock geology, Quaternary geology and seabed sediments. There are also series of base maps at 1:100 000 and 1:250 000, seismic profile maps, and bathymetric maps, the latter made in cooperation with the Norwegian Hydrographic Office (**Norges Sjøkartverk, NSKV**), which is now a division of SKV. The NSKV *Fiskeriplottekart* (Fishery plotting charts) at 1:100 000 now incorporate detailed bathymetric information derived from recent geophysical survey.

The Economic Map (*Økonomisk kartverk*, ØK) is a large-scale series (1:5000 or 1:10 000) planned to cover over half the land area of Norway, mainly land under cultivation or of other economic importance. It is used for cadastral purposes and for the registration of prehistoric and other cultural monuments. This series was started in 1965, and is expected to be complete in 1990. SKV coordinates the preparation of this work and is responsible for the geodetic control, but production of the individual sheets is the responsibility of the county (*fylke*) authorities. The ØK 1:5000 maps (the majority) have 5 m contours, while those at 1:10 000 have a 10 m interval. There are also photographically reduced versions at 1:20 000 and sometimes at 1:10 000. The maps are produced in either printed or diazo form. For forestry purposes, orthophotos are combined with elements of the Økonomisk kart. The ØK is distributed from the *fylke* offices.

Soil, vegetation and land classification mapping are all undertaken by the Norwegian Institute of Land Inventory (**Jordregisterinstituttet, JRI**), Ås. Recent work has concentrated on the development of a 1:50 000 *arealressurskart*. This map, which is printed in colour using the M711 topographic series as a base, uses a land capability classification for cultivated land and plant association criteria for classifying mountain land. About 25 sheets have been published since 1982. JRI also undertakes large-scale soil and land capability mapping of selective areas using the *Økonomisk kart* as the base map. The *bonitetskart* 1:20 000 is also published by JRI. This is a photoreduced version of the ØK with land classification categories overprinted in colour (including the productivity of forest land). About 650 sheets have been published so far, with about 70 new sheets appearing each year. The *bonitetskart* is distributed by SKV.

The preparation of a Norwegian National Atlas began in 1979. The work is being carried out by SKV in close cooperation with other institutions, and the aim is to make it a 'decision-orientated information system for public administration, social and regional planning and educational work'. The maps are being published at scales of 1:1 000 000 or smaller and are grouped into 21 thematic sections. As each section is completed it is made available in a folded A4 format accompanied by descriptive texts. Data for the map production is being held digitally and this will facilitate rapid revision of the sheets when required. Individual sheets are available separately.

The Norwegian Highway Authority (**Vegdirektoratet**) issues several small-scale (1:1 700 000) maps, revised annually, showing maximum axle loads and maximum permissible lengths of vehicle on Norwegian roads, roads subject to snow clearance in winter, and recommended roads for caravans. In 1985, Vegdirektoratet began to issue a new series of 1:250 000 road maps in 21 sheets. The sheet lines are based on an old topographic series and do not correspond with either the Series 1501 or the population maps. This series is designed for the motoring public and is distributed by **Universitetsforlaget**. The sheets will be revised every 5 years.

The Water Resources Directorate (**Vassdrags-direktoratet**) prepares a number of thematic maps concerned with water resources. These include bathymetric maps of inland water bodies and an *Innsjøatlas* containing data and maps of 51 lakes. There is also a series of six hydrological maps covering the country at 1:500 000. Detailed glacier maps have also been prepared for large or especially important glaciers, and there is an inventory of all Norwegian (and Swedish) glaciers published in two volumes as *Atlas over Breer i Sør-Norge* (1969) and *Atlas over Breer i Nord-Skandinavia* (1973). Both are still available.

Norway has a long-established commercial mapping company in **J.W. Cappelen Forlag**, Oslo. Cappelen produces regularly updated road and tourist maps and in 1981 published a placename gazetteer which indexes all the 40 000 names appearing on their five sheet main series. Cappelen also publishes town maps of Oslo, Bergen and some smaller towns. A recent venture has been into the production of mountain recreation maps in cooperation with SKV. In 1983, the Readers' Digest Association (**Det Beste A/S**), Oslo, published an atlas of Norway *Det Bestes Store Norge Atlas* with maps at 1 : 300 000 compiled by Bartholomew and an index of 46 000 placenames.

Further information

Statens Kartverk produces an excellent catalogue annually which gives the status of all the topographic map series and the progress of the National Atlas, as well as brief information about other mapping (population mapping, the ØK, and maps from Vassdragsdirektoratet and JRI). This catalogue also includes full details of the mapping of the Norsk Polarinstitutt. Other catalogues are published by the NGU (new catalogue in 1986), the JRI, Cappelen and by the county map offices (whose addresses are listed in the SKV catalogue).

The history of topographic mapping in Norway is described in Whittington, I.F.G. (1980) A history of the survey and mapping of Norway, *Survey Review*, **25**(197), 291–312.

A discussion in English of the National Atlas is provided by Ouren, T. (1982) National atlas of Norway, *Geojournal*, **6**, 209–212.

Addresses

Statens Kartverk (SKV)
Monserudveien, N-3500 HØNEFOSS.

Statistisk Sentralbyrå
Oscars gate 1, Postboks 510 Stasj.S, N-2201 KONGSVINGER.

Norges Geologiske Underssøkelse (NGU)
Leiv Eirikssons vei 39, Postboks 3006, N-7001 TRONDHEIM.

Institutt for Kontinentalsokkelundersøkelser (IKU)
Håkon Magnussonsgt. 18, Postboks 1883 Jarlesletta, N-7001 TRONDHEIM.

Norges Sjøkartverk (NSKV)
Klubbgate 1, Postboks 60, N-4001 STAVANGER.

Jordregisterinstituttet (JRI)
Postboks 115, Drøbakveien 11, N-1430 Ås.

Vegdirektoratet
Kartkontoret, Gaustadalléen 25, N-0371 OSLO.

Universitetsforlaget
Postboks 2959 Tøyen, N-0608 OSLO 6.

Vassdragsdirektoratet
Postboks 5091, Majorstua, N-0301 OSLO 3.

J.W. Cappelen Forlag A/S
Postboks 350, Sentrum, N-OSLO 1.

Det Beste A/S
Postboks 1160, Sentrum, N-OSLO 1.

Fjellanger Widerøe A/S
Rolfstangveien 12, Postboks 190, N-1330 OSLO LUFTHAVN.

Catalogue

Atlases and gazetteers

Nationalatlas for Norge
Hønefoss: SKV, 1983–.
21 sections, in progress.

Det Bestes Store Norge Atlas, edited by J. Gjessing and T. Ouren
Oslo: Det Bestes Forlaget A/S, 1983.
pp. 232.

Register til Cappelens bil- og turistkart 1–10
Oslo: J.W. Cappelen Forlag, 1981.
pp. 224.
Index to Cappelen's road and tourist map series.

General

Norge fra satellitt 1 : 2 000 000
Oslo Lufthavn: Fjellanger Widerøe, 1985.

Kart over hele Norge 1 : 1 000 000
Oslo: Cappelen, 1984.

Konturkart (basiskart) over Norge med byer, jernbaner og riksveger 1 : 1 000 000
Hønefoss: SKV, 1976.
Outline map with towns, railways and main roads.
Also available showing only the drainage network.

Norge 1 : 500 000 Series 1404 World
Hønefoss: SKV, 1973.
14 sheets, all published.

Cappelen Norge: bil- og turistkart 1 : 325 000/1 : 400 000
Oslo: Cappelen, 1982–.
5 double sheets, all published.

Topographic

Topografiske kart 1 : 250 000 Series 1501
Hønefoss: SKV, 1969–.
46 sheets, all published. ■
New edition in progress.

Topografisk hovedkartserie 1 : 50 000 Serie M711
Hønefoss: SKV, 1955–.
727 sheets, almost complete. ■

Bathymetric

Norsk kontinentalsokkel. Geologisk temakart 1 : 500 000 B500T
Trondheim: IKU, 1983–.
Series of thematic maps of bedrock and Quaternary geology and seabed sediments.

Geological and geophysical

Glasialgeologisk kart over Norge 1 : 2 000 000 O. Holtedahl and B.G. Andersen
Trondheim: NGU, 1960.
Accompanies NGU 208, but available separately.

Malmforekomster 1 : 2 000 000
Trondheim: NGU, 1984.
Ore occurrences.

Berggrunnsgeologisk kart 1 : 1 000 000
Trondheim: NGU, 1983.
Also available in National Atlas.

Glasialgeologisk kart 1 : 1 000 000
Hønefoss: SKV, 1984.
Sheet 2.3.2 of National Atlas.

Grus- og sandtak i Syd-Norge 1 : 1 000 000 G. Holmsen
Trondheim: NGU, 1971.
Gravel and sand pits in South Norway.
Accompanies NGU 271, but available separately.

Jordartskart over Syd-Norge 1 : 1 000 000 G. Holmsen
Trondheim: NGU, 1971.
Surface deposits of South Norway.
Accompanies NGU 271, but available separately.

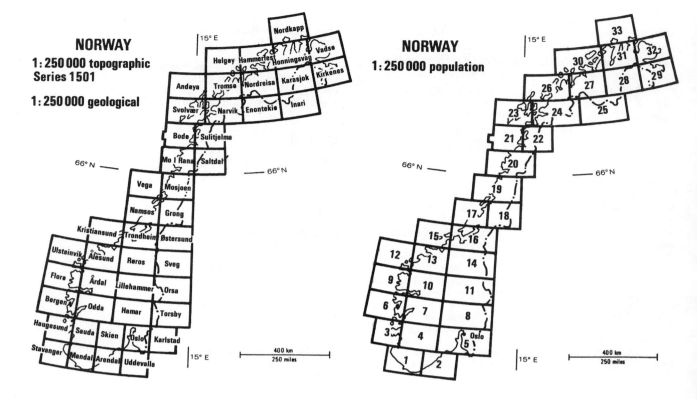

NORWAY
1:250 000 topographic
Series 1501

1:250 000 geological

NORWAY
1:250 000 population

Berggrunnsgeologisk kart   1:250 000
Trondheim: NGU, 1970–.
44 sheets, 16 published. ∎
Bedrock geology.
Further sheets available in provisional edition.

Vannressurskart   1:250 000
Trondheim: NGU.
44 sheets, 14 published. ∎

Aeromagnetisk kart over landarealene. Magnetisk totalfelt   1:250 000
Trondheim: NGU, 1971–.
44 sheets, 37 published. ∎

Berggrunnsgeologisk kart   1:50 000
Trondheim: NGU, 1972–.
727 sheets, 44 published. ∎
Some sheets with monograph.
Numerous further sheets in provisional, black-and-white edition.

Kvartægeologisk kart   1:50 000
Trondheim: NGU, 1972–.
727 sheets, 59 published. ∎
Some sheets with monograph.

Aeromagnetisk kart   1:50 000
Trondheim: NGU.
727 sheets, *ca* 468 published. ∎
Black-and-white maps.

Sand- og grus ressurskart   1:50 000
Trondheim: NGU.
Sand and gravel resources.
727 sheets, 258 published. ∎
Mostly black-and-white sheets.

Ressurskart for grunnvann i løsavsetninger   1:50 000
Trondheim: NGU, 1976–.
727 sheets, 42 published. ∎
Groundwater resources in surficial deposits.

Environmental

Skog og jordbruksområder   1:1 000 000
Hønefoss: SKV, 1984.
Forest and cultivated land.

Hydrografiske kart   1:500 000
Oslo: Vassdragsdirektoratet.
6 sheets, all published.
Hydrological map.

Vegetasjonskart   1:50 000
Ås: JRI, 1974–.
727 sheets, *ca* 75 published. ∎
Mostly printed in two colours.
Some further sheets available in manuscript.

Arealressurskart   1:50 000
Ås: JRI, 1982–.
727 sheets, *ca* 25 published. ∎
Land capability map.

Bonitetskart   1:20 000
Ås: JRI, *ca* 1980–.
About 650 sheets published.
Land classification map.
Maps mainly black and white.

Administrative

Offentlig grunn og bygdeallmenninger   1:2 000 000
Hønefoss: SKV, 1984.
Public lands administered by Directorate of State Forests and
Lands and parish common lands.

Kommunekart Norge   1:1 000 000
Hønefoss: SKV, 1978.
Commune boundaries.
Maps of police, judicial, civil defence and other administrative
divisions also available at this scale.

Human and economic

Kart over gods i norsk kysttrafikk
Oslo: Universitetsforlaget, 1981.
Map of coastal freight traffic.
Ad Novas 16.

Bosettingskart Folketelling 1970   1:1 000 000
Hønefoss: SKV, 1975.
Population distribution 1970 census.

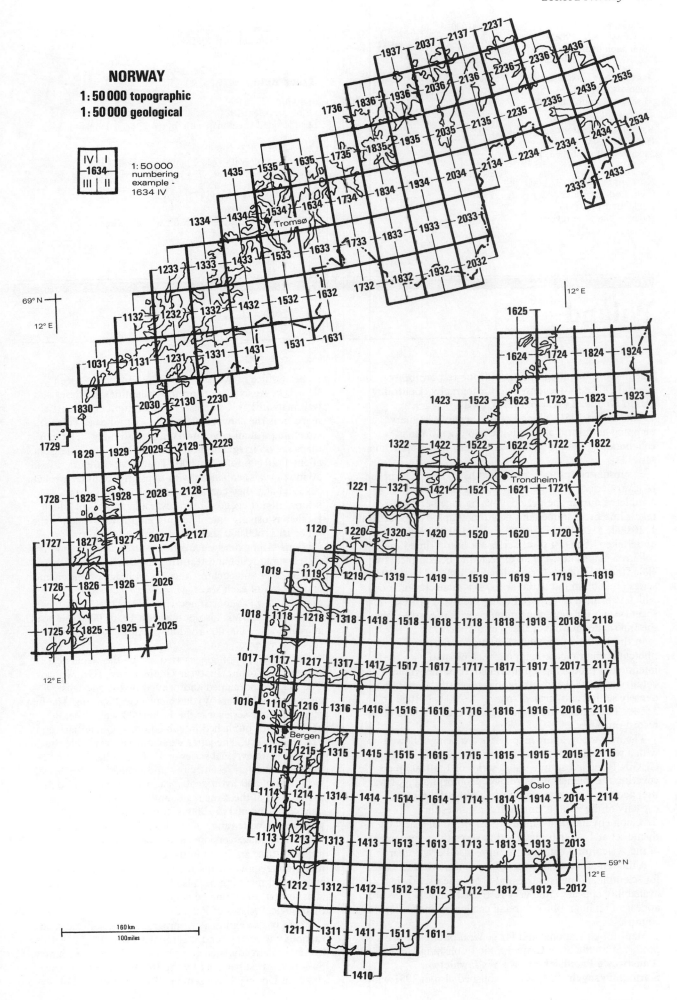

NORWAY

1 : 50 000 topographic
1 : 50 000 geological

1 : 50 000 numbering example - 1634 IV

Bosettingskart Folketelling 1980 1 : 1 000 000
Hønefoss: SKV, 1984.
Population distribution 1980 census.

Bosettingskart Folketelling 1970 1 : 250 000
Hønefoss: SKV, 1974–75.
33 sheets, all published. ■

Bosettingskart Folketelling 1980 1 : 250 000
Hønefoss: SKV, 1982–83.
33 sheets, all published. ■

Reisekart for Norge 1 : 1 000 000
Oslo: Forlaget Rutebok for Norge, 1980.
Public transport map.
Index of places on reverse.

Norge Vegkart 1 : 250 000
Oslo: Vegdirektoratet, 1985–87.
21 sheets, all published.

Town maps

Stor-Oslo 1 : 30 000
Oslo: Cappelen, 1984.
Double-sided street map with tourist information and index.

Oslo bykart 1 : 25 000
Oslo: Cappelen with Oslo Oppmålingsvesen, 1985.
City centre at 1 : 10 000.

Oslo kartboka 1 : 20 000/1 : 10 000
Oslo: Cappelen, 1986.
102 pp. maps and street index for whole of Greater Oslo.

Poland

The modern mapping of Poland is the responsibility of the Central Office for Geodesy and Cartography (**Centralny Urzad Geodezji i Kartografii, CUGiK**), Warsaw, established in 1951. There is also a separate military mapping authority, although the two organizations collaborate in the production of modern topographic map series.

A substantial resurvey of Poland has taken place in the postwar period. A new primary triangulation was completed during the period 1949–55 together with a relevelling programme, and new topographic map series at 1 : 10 000, 1 : 25 000, 1 : 50 000 and 1 : 100 000 were undertaken. The last three series are litho printed multi-coloured maps and were all expected to be complete by 1986 (the 1 : 50 000 was completed in 1983). The 1 : 10 000 series (1 : 5000 in industrial areas) is a monochrome map which is printed as required using the diazo process. The sheet system for these topographic series is derived from the International Map of the World system.

In the 1970s, a programme of large-scale photogrammetric base mapping was initiated with the intention of producing a uniform cover of the whole country at 1 : 5000 and substantial cover of the more densely settled and farmed areas at 1 : 2000, 1 : 1000 and 1 : 500. These maps are to be used for cadastral and other record purposes.

The new topographic map series are being followed up with a wide range of thematic mapping, and in 1980 CUGiK launched an integrated series of 1 : 50 000 maps covering the themes of hydrology, geomorphology, land use, geology, soils and agriculture, using the research capacity of the various university institutes for development and data capture. Test sheets of a new eight-theme set at 1 : 25 000, designed to serve the requirements of the agricultural community, have also been prepared.

Although all these various maps are described as being for scientific, administrative and economic use, their availability appears to be very restricted, and they do not appear in bibliographies or retail catalogues in Western countries.

Maps for educational and for general public use are produced by the State Cartographic Publishing House (**Panstwowe Przedsiebiorstwo Wydawnictw Kartograficznych, PPWK**), established also in 1951 with its head office in Warsaw and its production plant in Wroclaw. Apart from enormous quantities of atlases and wall maps, PPWK publishes a large number of tourist maps (it is the *only* producer of tourist maps in Poland), street maps of all large towns, and road maps. The tourist maps are on irregular sheet lines and range in scale from about 1 : 20 000 to 1 : 500 000. They have contours (usually 50 m intervals) enhanced with hypsometric tints or relief shading, and they contain an abundance of tourist information. Legends are in several languages of which English is usually one. The only map *series* as such is the set of 16 1 : 500 000 sheets which are issued in two versions. *Mapa przegladowa województw* is a general map with layered relief and featuring the boundaries of the provinces or voivodships: there is an index of placenames on the back of each sheet. *Mapa krajoznawczo—samochodowa* is a tourist and road users' version of the same series. PPWK also publishes a 1 : 500 000 scale road atlas and a 1 : 1 000 000 road map. All PPWK maps are regularly revised.

Geological mapping is carried out by the Geological Institute of Poland (**Instytut Geologiczny, IG**), Warsaw. The maps are designed and printed by the specialist geological publishers Wydawnictwa Geologiczne. The first major postwar series was the 1 : 300 000 scale general geology map, published in two editions, Quaternary and Pre-Quaternary. The latter version is still available, but since 1971 a new dual series at 1 : 200 000 has been in progress, and has recently become available in the West. Geophysical and hydrogeological map series are also in preparation at the same scale, and a detailed geological map of Poland at 1 : 50 000 is in progress. IG has also published a wide range of small-scale maps, many in English language versions, and most with multilingual legends (these are listed in the catalogue), as well as numerous regional maps at larger scales. Much geological and related mapping has also been incorporated into an atlas format, including geological, mineralogenic and hydrological atlases of Poland.

There is also an intensive programme of soil and land capability mapping, and it is reported that a 1 : 50 000 survey has already been completed for the whole country. Soil survey is supervised by the Institute of Soil Science (Instytut Uprawy Nawozenia i Gleboznawstwa), Pulawy.

Poland has a strong tradition of geomorphological mapping, and much detailed mapping at 1:25 000 has been carried out regionally since the 1950s by institutes of Physical Geography and of Quaternary Science in the universities. This work is summarized in a 1:500 000 general geomorphological map edited by L. Starkel and published in 1980 by the Polish Academy of Sciences (**Polska Akademia Nauk, PAN**). The map is printed in 16 colours and distinguishes 173 categories of landform and surficial deposit. A potential vegetation map of Poland is also in preparation by PAN.

Climatological data are collected by the **Instytut Meteorologii i Gospodarki Wodnej (IMGW)** and a climatic atlas of Poland was published in 1973 and a hydrological atlas in 1987.

The National Atlas of Poland was compiled by the Institute of Geography of the Polish Academy of Sciences (IG i PZ PAN) under the general editorship of S. Leszczycki. It was completed in 1978. A new, up-dated National Atlas is in preparation. This will have eight thematic units and is scheduled for completion by 1994. It will be published by PPWK in cooperation with CUGiK. A number of regional atlases have also been in preparation, chiefly by a team at the university at Cracow (an atlas of the City of Cracow voivodship was published in 1979, and others are soon to be published); and there is also a series of city atlases in progress, of which the first was the *Atlas of Warsaw* published in 1975.

The role of the academic institutions in Poland is thus important not only in their contribution to cartographic research but in the actual production of thematic maps and regional atlases.

Within CUGiK, there is also a research institute (IGiK) and a centre of remote sensing (OPOLiS) was established in 1976. The latter has explored the use of Landsat and Salyut imagery in thematic mapping, including the production of a land use map of Poland at 1:500 000, and in up-dating topographic maps from photogrammetric satellite images, while IGiK has recently been working on the development of geographical information systems.

In spite of the many exciting developments in Polish cartography, a rather limited range of this material is readily available on the international market. As with other East European countries, this reflects the lack of public availability of most detailed topographic and thematic mapping.

Further information

Recent developments in Polish mapping have been summarized by Ormeling, F.J. (1982) Achievements in Polish cartography, *ITC Journal*, **1982-2**, 174–177.

A detailed description of postwar mapping (including thematic mapping) is also found in Kretschmer, I. and Krupski, J. (1982) Die Entwicklung der kartographischen Darstellung Polens und der polnischen Kartographie, *International Yearbook of Cartography*, **22**, 105–146.

Geological mapping is included in the *List of Publications 1921–1/1979* and subsequent supplements issued (in an English version) by the Instytut Geologiczny.

Information about Polish mapping is also found in the quarterly Polish Cartographic Review (*Polski Przeglad Kartograficzny*) which includes summaries in Russian and English.

At the ICA International Cartographic Conference in Moscow in 1976, a special volume in English entitled *The Polish Cartography* was distributed to delegates. An up-dated volume was issued at Warsaw in 1982 and a shorter national report at Perth in 1984. All three volumes are in English and include bibliographic information about new maps as well as descriptions of modern mapping programmes.

Addresses

Centralny Urzad Geodezji i Kartografii (CUGiK)
Ulica Jasna 2–4, PL-00 950 WARSZAWA.

Panstwowe Przedsiebiorstwo Wydawnictw Kartograficznych (PPWK)
Ulica Solic 18–20, PL-00 410 WARSZAWA.

Instytut Geologiczny (IG)
Centralny Osrodek Informacji, Naukowej, Technicznej i Ekonomicznej, Rakowiecka 4, PL-00 975 WARSZAWA.

Centrala Handlu Zagranicznego Ars Polonica   Foreign orders for geological maps
Krakowskie Przedmiescie 7, 00-068 WARSZAWA.

Instytut Geografii i Przestrzennego Zagospodarowania PAN (IG i PZ PAN)
Ulica Krakowskie Przedmiescie 30, PL-00 325 WARSZAWA.

Instytut Meteorologii i Gospodarki Wodnej (IMGW)
Ulica Podlesna 61, PL-01 673 WARSZAWA.

Catalogue

Atlases and gazetteers

Narodowy atlas Polski
Warszawa: IG i PZ PAN, 1973–78.
127 pp. maps + 41 pp. text in 4 vols.
National atlas of Poland.
English text and legend available.

Samochodawy atlas Polski   1:500 000   8th edn
Warszawa: PPWK, 1983.
pp. 215.
Road atlas of Poland.

Wykaz Urzedowych Nazw Miejcowosci w Polsce
Warszawa: Wydawnictwa Akcydensowe, 1982.
3 vols, all published.
Gazetteer of geographical names of Poland.

Skorowidz miesjscowosci do mapy administracyjnej Polskiej rzeczpospolitej ludowek   1:500 000
Warszawa: PPWK, 1980.
pp. 230.
Placenames on the 1:500 000 administrative map of Poland.

General

Polska. Mapa Fizyczna   1:2 000 000
Warszawa: PPWK, annual.
Administrative map on the reverse.

Polska   1:1 000 000
Warszawa: PPWK, 1985.

Samochodowa mapa Polski   1:750 000
Warszawa: PPWK, 1984.
Road map.
Legend in Polish, Russian, English, German.

Polska, mapa fizyczna   1:500 000
Warszawa: PPWK, 1980.
4 sheets, all published.

Topographic

Województwa map przegladowa   1:500 000
Warszawa: PPWK.
16 sheets, all published.
General map with provincial boundaries.

Mapa krajoznawczo-samochodowa 1:500 000
Warszawa: PPWK.
16 sheets, all published.
Road and tourist map.

Geological and geophysical

Atlas geologiczny Polski 1:1 000 000
Warszawa: IG, 1955–61.
15 maps on geology-related themes, some out of print.

Geological map of Poland and adjoining countries without Cenozoic, Mesozoic and Permian formations 1:1 000 000 W. Pozaryski and Z. Dembowski
Warszawa: IG, 1984.
2 sheets, both published.
Polish, English and Russian versions.
With text.

Geological map of Poland and adjoining countries without Cainozoic formations 1:1 000 000
Warszawa: IG, 1979.
Explanation in Polish, Russian and English.

Photogeological map of Poland 1:1 000 000 J. Bazynski
Warszawa: IG, 1984.
Explanation in Polish, Russian and English.

Mapa wód mineralnych Polski 1:1 000 000 C. Colago *et al.*
Warszawa: IG, 1971.
Mineral waters of Poland.

Mapa geologiczna trzeciorzedu ladowego w Polsce 1:500 000
Warszawa: IG, 1965–68.
4 sheets, all published.
Geological map of the continental Tertiary in Poland.

Mapa sejsmiczna Polski . . . 1:500 000 J. Skorupa
Warszawa: IG, 1974.
Seismic map of Poland. Results of refraction survey conducted to recognize the deep basement.
27 pp. text in Polish.

Mapa kruszywa naturalnego w Polsce 1:500 000 J. Pawlowska and Z. Siliwonkzuk
Warszawa: IG, 1971.
Map of natural broken stone in Poland.
137 pp. text in Polish.

Map of mineral deposits in Poland 1:500 000 R. Osika
Warszawa: IG, 1971.

Map of mineral raw material deposits of Poland 1:500 000
R. Osika
Warszawa: IG, 1984.
Polish, English and Russian versions.
With text.

Geological map of crystalline basement of the East European platform in Poland 1:500 000
Warszawa: IG, 1975.
Explanation in Polish, Russian and English.

Geological map of Poland without Cainozoic, Cretaceous and Jurassic formations 1:500 000 E. Rühle *et al.*
Warszawa: IG, 1980.
Explanation in Polish, Russian and English.

Geological map of Poland without Cainozoic and Cretaceous formations 1:500 000 E. Rühle *et al.*
Warszawa: IG, 1978.
Explanation in Polish, Russian and English.

Geological map of Poland without Cainozoic formations 1:500 000 R. Osika *et al.*
Warszawa: IG, 1971.

Geological map of Poland without Quaternary formations 1:500 000 E. Rühle *et al.*
Warszawa: IG, 1977.
Explanation in Polish, Russian and English.

Map of natural aggregates in Poland 1:500 000 Z. Siliwonski *et al.*
Warszawa: IG, 1981.

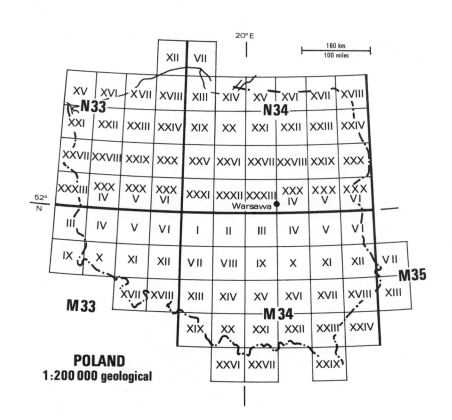

Map of the sea floor of the Southern Baltic 1 : 500 000
Z. Jurowska and W. Kroczka
Warszawa: IG, 1979.
Explanation in Polish, Russian and English.

Przegladowa mapa geomorfologiczna Polski 1 : 500 000 L. Starkel
Warszawa: PAN, 1980.
6 sheets, all published.
General geomorphological map.

*Przegladowa mapa geologiczna Polski. Wyd. B bez utworów
czwartorzedowych* 1 : 300 000
Warszawa: IG, 1946–55.
28 sheets + indexes and explanations.
General geological map Edition B without Quaternary deposits.
Explanation in Polish and Russian.

Przegladowa mapa surowców skalnych Polski 1 : 300 000
Warszawa: IG, 1965–70.
28 sheets + indexes and explanations.
General map of raw material deposits of Poland.

Mapa geologiczna Polski 1 : 200 000
Warszawa: IG, 1971–.
76 sheets, 64 published. ■
Available in A (Quaternary) and B (Pre-Quaternary) editions.
Texts.

Szczególowa mapa geologiczna Polski 1 : 50 000
Warszawa: IG, 1973–.
1064 sheets, *ca* 300 published.
Texts.

Environmental

Atlas klimatyczny Polski Waclaw Wiszniewski
Warszawa: IMGW/PPWK, 1973.
pp. 141.
Supplementary tables published 1973–79.

Atlas hydrologiczny Polski edited by Juliusz Stachy
Warszawa: Wydawnictwa Geologiczne, 1987.
2 vols, 78 maps + 735 pp. text.

Polska. Warunki glebowe i klimatyczne rolnictwa 1 : 1 000 000
Warszawa: PPWK, 1977.
Soil and climate conditions in agriculture.

Ochrona przyody w Polsce 1 : 750 000
Warszawa: PPWK, 1982.
Nature conservation in Poland.

Polska. Mapa uzytkowania ziemi 1 : 500 000
Warszawa: PPWK, 1980.
Land use map of Poland

Administrative

Polska Rzeczpospolita Ludowa. Mapa administracyjna 1 : 750 000
Warszawa: PPWK, 1984.

Polska Rzeczpospolita Ludowa. Mapa administracyjna 1 : 500 000
Warszawa: PPWK, 1980.
4 sheets, all published.

Human and economic

Polska. Przemiany spoleczno-gospodarcze 1970–1979 1 : 1 250 000
Warszawa: PPWK, 1980.
Socioeconomic changes.

Town maps

Plan miasta Warszawy *ca* 1 : 22 000
Warszawa: PPWK, annual.
Street map of Greater Warsaw.

Plan Warszawy Centrum
Warszawa: PPWK, annual.

Portugal

The central mapping agency for Portugal and its island
territories is the **Instituto Geográfico e Cadastral (IGCP)**,
Lisbon. IGCP undertakes geodetic, topographic and
cadastral survey, and is responsible for topographic map
series at scales larger than 1 : 500 000, providing also the
base maps and a printing facility for thematic maps
published by other government departments. Military
mapping is undertaken by the **Serviço Cartográfico do
Exército (SCEP)**, Lisbon, founded in 1932. The principal
series is the 1 : 25 000 (M888), which is on a Gauss–Krüger
projection and printed in six colours and with 10 m
contours. This is currently available to the public.

The basic scale civilian mapping is published at
1 : 50 000 (Series M7810). This is a six-colour map on the
Bonne Projection. UTM grid coordinates are given at the
corners of each sheet. Contours are at 25 m intervals and
each sheet covers an area of 32 × 20 km. The major road
network and settlements are shown in red, contours
brown, hydrography blue, forest and plantations in green
and toponyms and other cultural features in black. Many
sheets have appeared in new editions since their first
publication, and IGCP is currently considering production
of a new 1 : 50 000 series.

The 1 : 100 000 map (Series M684) was initiated in 1938
and is derived from the 1 : 50 000. It is on the same

projection and has contours at 25 m or (in mountain areas)
50 m. UTM grid is indicated at the neatlines, and similar
symbology and colours are used. Municipal administrative
divisions are shown. Each sheet covers the area of four
1 : 50 000 sheets.

A 1 : 200 000 map in seven colours with hill shading was
introduced in 1969. It is on the Transverse Mercator
projection. This series is not yet complete. A planning
version showing only planimetry and hydrography is also
available. IGCP has also published a number of smaller-
scale general and administrative maps, as indicated in the
catalogue.

Large-scale mapping is published by IGCP for planning
and cadastral purposes and includes a 1 : 10 000
topographic map begun in 1948 covering mainly the
Lisbon peninsula (this is now being superseded by an
orthophotomap series) and cadastral plans mainly at
1 : 2000 and 1 : 5000. Most of the latter are planimetric, but
some are available in a topographic version with 5 m
contours. All the large-scale mapping is used as source
material for the smaller-scale topographic series described
above.

SCEP has published a series of *Reportórios Toponímicos*
listing alphabetically all the toponyms on the 1 : 25 000 map
Series M888.

An *Atlas do Portugal* was first published in 1940, and a second edition issued in 1958 by the Instituto de Estudos Geográficos. This atlas is now out of print. However, a new *Atlas Nacional do Ambiente* has been in preparation since 1972. This atlas is being developed by a special **Comissão Nacional do Ambiente (CNA)** and is supported by government funding. More than half the projected total of more than 70 sheets have so far been published, covering mainly climatic, geological, biological and pedological themes. The maps are at a standard scale of 1 : 1 000 000 and are accompanied by texts in Portuguese with summaries in English and French.

Geological mapping is undertaken by the **Serviços Geológicos (SGP)**, Lisbon, and the principal series is at 1 : 50 000. Sheets have been published since 1937, each accompanied by a descriptive text, and it remains an active series with a number of new sheets appearing in the 1980s. One sheet of a new 1 : 200 000 series has also been issued. Although most of the geological mapping conforms with a standard series, there are also a few local geological maps, including a 1 : 20 000 map of Lisbon published in 1940.

Land use and land capability maps are published by the **Centro Nacional de Reconhecimento e Ordenamento Agrário (CNROA)** which is a branch of the Instituto Nacional de Investigação Agrária e de Extensão Rural. There is an extensive series of full-colour soil and land capability maps at 1 : 50 000 covering the southern half of the country, and a *Carta agrícola e florestal* 1 : 25 000 on the sheet lines of the military topographic map and covering all but the northern third of the country. This latter map gives detailed information about tree species and natural vegetation as well as agricultural land use. As well as the generalized maps at 1 : 250 000 and 1 : 500 000 listed below, there are maps of individual commercial tree species at the same scale. Currently, CNROA is working on maps which can be used for agricultural planning and management.

The **Instituto Nacional de Meteorologia e Geofísica (INMG)** has published a climatic atlas of mainland Portugal, and also maps of seismotectonics and of seismic activity. A geomagnetic map is in preparation.

Portugal also has an **Instituto de Investigação Científica Tropical (IICT)** which undertakes mapping projects overseas, principally in Angola and the Cape Verde Islands.

Maps of the *Regiões Autônomas* of the Azores and Madeira archipelagos are listed in the Atlantic Ocean section.

Further information

IGCP issues annually a descriptive catalogue of maps and publications entitled *Informação sobre documentação e elementos de estudo disponíveis*, and a catalogue with indexes is also issued by SCEP. SGP has issued a *Catálogo das publicações 1865–1981* with subsequent supplements, and this includes lists of geological maps and monographs. A list and index of soil and agricultural maps is available from CNROA.

Addresses

Instituto Geográfico e Cadastral (IGCP)
Praça da Estrela, 1200 LISBOA.

Serviço Cartográfico do Exército (SCEP)
Av. Dr A. Bensaude, Olivias-Norte, 1800 LISBOA.

Comissão Nacional do Ambiente (CNA)
Gabinete de Estudos e Planeamento, Praça Duque de Saldanha, 31, 5.o, 1096 LISBOA.

Centro de Estudos Geográficos de Lisboa
Alameda da Universidade, LISBOA.

Serviços Geológicos (SGP)
Rua da Academia das Ciências 19, 2.o, 1200 LISBOA.

Centro Nacional de Reconhecimento e Ordenamento Agrário (CNROA)
Rua Castilho 69, 1.o, 1200 LISBOA.

Instituto Nacional de Meteorologia e Geofísica (INMG)
Rua C do Aeroporto de Lisboa, 1200 LISBOA.

Instituto de Investigação Científica Tropical (IICT)
Rua da Junqueira 86, 1300 LISBOA.

Catalogue

Atlases and gazetteers

Atlas Nacional do Ambiente
Lisboa: CNA, 1972–.
41 sheets published.

Novo dicionário corográfico de Portugal   A.C. Amaral Frazao
Porto: Editorial Domingos Barreira, 1981.
pp. 1040.

General

Portugal continental e Regioes Autonomas   1 : 2 500 000
Lisboa: IGCP, 1985.

Carta de Portugal   1 : 1 000 000
Lisboa: IGCP, 1970.
IMW style.

Carta hipsométrica de Portugal   1 : 600 000
Lisboa: IGCP, 1984.
Relief and altitude map.

Carta de Portugal   1 : 500 000   2nd edn
Lisboa: IGCP, 1981.
2 sheets, both published.

Carta de Portugal   1 : 400 000
Lisboa: IGCP, 1968 and 1977.
3 sheets, all published.

Topographic

Carta militar de Portugal   1 : 250 000   Series M586
Lisboa: SCEP, 1965–68.
8 sheets, all published.

Carta de Portugal   1 : 200 000   Series M585
Lisboa: IGCP, 1972–.
8 sheets, 5 published.

Carta de Portugal   1 : 100 000   Series M684
Lisboa: IGCP, 1938–.
53 sheets, all published. ■

Carta corográfica de Portugal   1 : 50 000   Series M7810
Lisboa: IGCP, 1952–.
175 sheets, all published. ■

Carta militar de Portugal   1 : 25 000   Series M888
Lisboa: SCEP, 1941–.
639 sheets, all published. ■

Geological and geophysical

Portugal. Carta geológica da plataforma continental   1 : 1 000 000
Lisboa: SGP, 1978.

Carta geológica de Portugal   1 : 1 000 000   2nd edn
Lisboa: SGP, 1968.

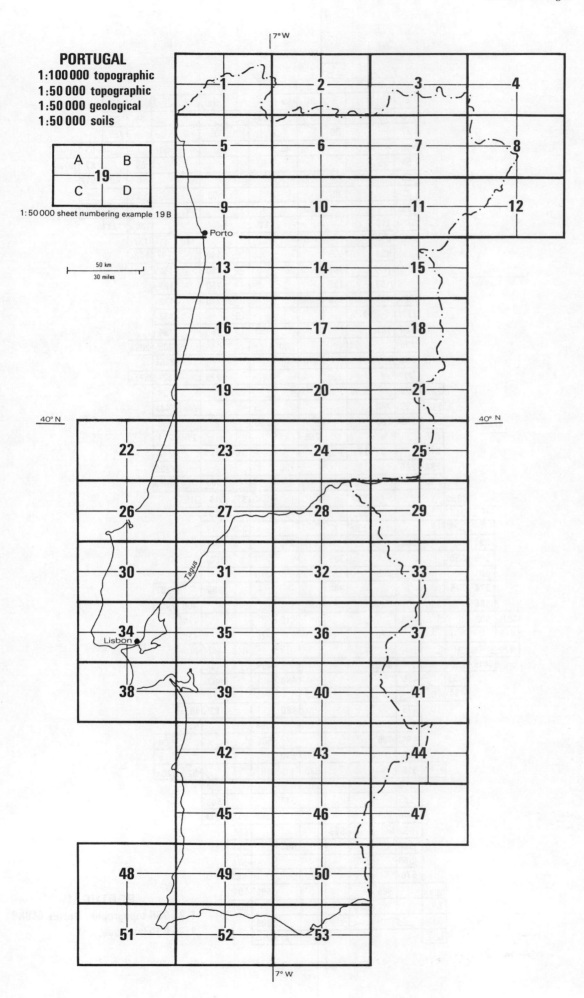

PORTUGAL
1:100 000 topographic
1:50 000 topographic
1:50 000 geological
1:50 000 soils

A	B
C	D

19

1:50 000 sheet numbering example 19 B

50 km
30 miles

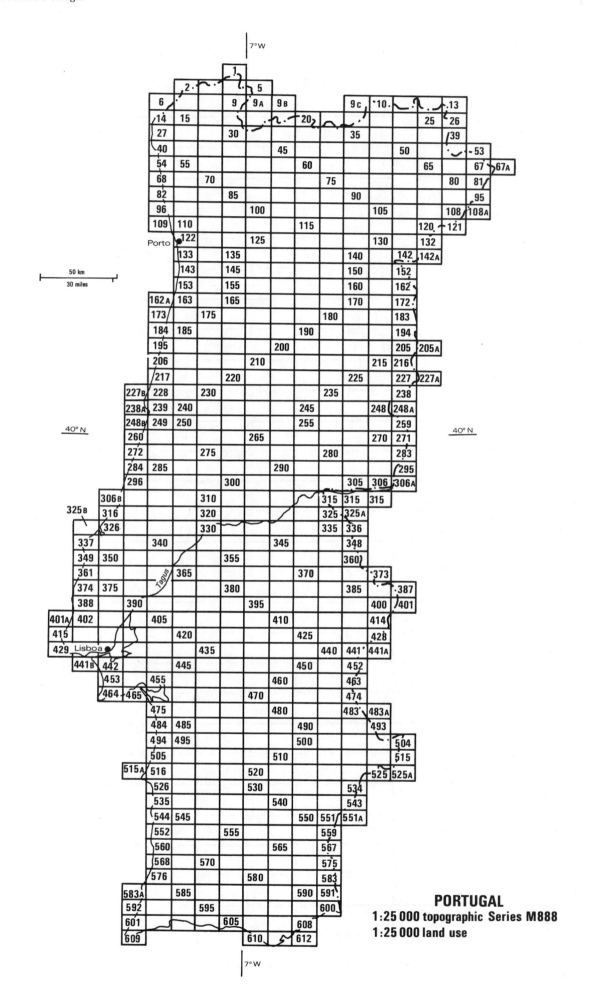

50 km

30 miles

7°W

40°N

40°N

7°W

Porto

Tagus

Lisboa

PORTUGAL

1:25 000 topographic Series M888

1:25 000 land use

Carta tectónica de Portugal 1:1 000 000
Lisboa: SGP, 1972.

Carta geológica do Quaternário de Portugal 1:1 000 000
Lisboa: SGP, 1969.
With 39 pp. text.

Carta hidrogeológica de Portugal 1:1 000 000
Lisboa: SGP, 1970.

Carta das nascentes minerais de Portugal 1:1 000 000
Lisboa: SGP, 1970.

Carta geológica de Portugal 1:500 000 4th edn
Lisboa: SGP, 1972.
2 sheets, both published.

Carta mineira de Portugal 1:500 000
Lisboa: SGP, 1965.
2 sheets, both published.
With 47 pp. text in French.

Mapa geomorfológico de Portugal 1:500 000 Denise de Brum Ferreira
Lisboa: Centro de Estudos Geográficos, 1981.
2 sheets.
Folded with descriptive text *Memorias do Centro de Estudos Geográficos* No. 6.

Carta geológica de Portugal 1:200 000
Lisboa: SGP, 1982–.
8 sheets, 1 published.

Carta geológica de Portugal 1:50 000
Lisboa: SGP, 1944–.
175 sheets, 88 published. ■

Environmental

Atlas climatológico de Portugal Continental
Lisboa: INMG, 1974.

Carta dos solos de Portugal 1:1 000 000
Lisboa: CNROA, 1971.

Carta agrícola e florestal de Portugal 1:1 000 000
Lisboa: CNROA.

Carta agrícola e florestal de Portugal (Grandes grupos de utilizacao do solo) 1:500 000
Lisboa: CNROA, 1969.
Reduced version of the next entry.

Carta agrícola e florestal de Portugal (Grandes grupos de utilização do solo) 1:250 000
Lisboa: CNROA, 1960–65.
3 sheets, all published.

Carta dos solos de Portugal 1:50 000
Lisboa: CNROA, 1950–.
175 sheets, *ca* 80 published. ■

Carta de capacidade de uso do solo de Portugal 1:50 000
Lisboa: CNROA, 1950–.
175 sheets, *ca* 80 published. ■

Carta agrícola e florestal de Portugal 1:25 000
Lisboa: CNROA, 1960–.
612 sheets, *ca* 400 published. ■

Administrative

Carta administrativa de Portugal 1:600 000
Lisboa: IGCP, 1984.

Carta administrativa de Portugal 1:250 000 Jose Correia da Cunha
Lisboa: CNA, 1979.
3 sheets, all published.

Human and economic

Rede elétrica de Portugal 1:1 000 000
Lisboa: Electricidade de Portugal, 1982.

Town maps

Carta do Concelho de Lisboa 1:10 000 Series M983
Lisboa: SCEP, 1975–.
4 sheets, all published.
Contours on reverse.

Cartas topográficas da zona da Grande Lisboa 1:10 000
Lisboa: IGCP, 1948–.
19 sheets published.

Lisbon and Alfama street map and tourist guide 1:6000
Lisboa: Portugalia Map Company.

Romania

The principal mapping authority in Romania is the **Institutul de Geodezie, Fotogrammetrie, Cartografie si Organizarea Teritoriului (IGFCOT)**, Bucharest. The basic series produced by IGFCOT is a 1:50 000 cadastral series which is nearing completion. There is also a 1:100 000 series of county (*judet*) maps comprising 41 sheets, although not all are published yet. These county maps include small thematic inset maps of soils, industries and features of cultural and tourist interest. There are also several 1:500 000 scale physical and administrative maps which are revised from time to time. Military topographic series have been prepared by the Direcția Topografică Militară, and by 1970 new series at standard scales ranging from 1:15 000 to 1:1 000 000 had all been completed, but are not available to the public.

Geological mapping is carried out by the **Institutul de Geologie si Geofizica (IGG)**, Bucharest. This institute was formed in 1974 by combining the previously separate institutes of geology and of applied geophysics. There is an active series at 1:50 000, started in 1969, and published in Romanian. About 90 sheets have been published, with more in press or preparation. It is aimed to complete this series by the year 2000. There is also a 1:1 000 000 scale *Geological Atlas* series, published in Romanian and English. The sheets are available individually and include a general geological map published in a second edition in 1978. The series also includes a soil map, while the latest publication is the second edition of the *Map of Mineral Resources, 1983–84*, which is accompanied by 33 supplementary illustrations and a 237 page text. These maps are available overseas on exchange with the IGG, and small stocks of some sheets have been held by GeoCenter in the German Federal Republic, and may therefore be obtainable from this source.

Soil surveying has been carried out by the **Institutul de Studii de Cercetari pentru Pedologie si Agrochimie**, and this includes 1:10 000 soil and land capability mapping carried out on a county (*judet*) basis. Smaller-scale soil maps at 1:200 000 have been published, and in 1976 a 1:500 000 scale soil erosion map *Harta eroziunii solurilov* was prepared. A general soil map at this scale was published by IGG in 1970–71.

Probably the best and most accessible cartographic material on Romania, at least outside the country, is the *National Atlas of the Socialist Republic of Romania* (*Atlasul Republicii Socialiste Romania*), prepared by the Geographical Institute of the Romanian Academy of Sciences (**Academia Republicii Socialiste Romania**), and issued during the period 1974–79. This is a substantial work with foldout maps at a principal scale of 1:1 000 000, grouped into 13 subject areas. Texts and legends are given in Romanian, French, English and Russian. The Academy has also published sheet maps, such as the four-sheet vegetation map at 1:500 000 (1961), and a series of geographical monographs.

Tourist maps are published in several languages by the **'Carpati' National Office of Tourism**.

A road atlas of the country, *România atlas rutier*, was issued in 1981 by **Editura Sport-Turism**. In addition to 73 pages of road maps at 1:350 000, it has 25 town maps and an index to about 14 000 places. Editura Sport-Turism also publishes a number of traveller's guides to Romanian cities, as well as mountain itinerary maps and district tourist maps.

The **Automobil Clubul Roman** publishes annually revised tourist/motoring maps at 1:250 000, and motoring itineraries which include street maps of towns and cities.

Romanian maps appear to be difficult to obtain outside the country, and consequently only a small amount of the material discussed has been included in the catalogue that follows.

Further information

A catalogue of geological maps including an index of the 1:50 000 series is issued in English by IGG.

A discussion of the Atlas of Romania placed in the context of postwar geography and cartography in Romania is found in Turnock, D. (1976) The National Atlas—a new achievement in Romanian cartography, *Bulletin of the Society of University Cartographers*, **10**, 24–28.

Addresses

Institutul de Geodezie, Fotogrammetrie, Cartografie si Organizarea Teritoriului (IGFCOT)
B-dul Expozitiei nr. 1 A, Sectoral 1, R-79662 BUCURESTI.

Institutul de Geologie si Geofizica (IGG)
Str. Caransebes 1, Sectoral 8, R-78344 BUCURESTI.

Institutul de Studii de Cercetari pentru Pedologie si Agrochimie
Md Marasti 61, Sectoral 1, R-71331 BUCURESTI.

Academia Republicii Socialiste Romania
Calea Victoria 125, R-BUCURESTI.

National Office of Tourism 'Carpati-Bucuresti'
Bd. Magheru, R-7 BUCURESTI.

Editura Sport-Turism
Str. Vasile Conta 16, Sectoral 2, R-BUCURESTI.

Automobil Clubul Roman
Str. Nikos Beloianis 27, 3 Bd. Poligrafiei, R-BUCURESTI.

For CIA, see United States; for Falk see German Federal Republic.

Catalogue

Atlases and gazetteers

Atlasul Republicii Socialiste Romania
Bucuresti: Academia RSR, Institutul de Geografie, 1974–78.
76 map sheets + introductory text.

Romania atlas rutier
Bucuresti: Editura Sport-Turism, 1981.

General

Romania   1:1 700 000
Washington, DC: CIA, 1970.
With 5 thematic inset maps.

Republica Socialista Romania. Harta fizica si administrativa   1:850 000
Bucuresti: IGFCOT, 1973.

Republica Socialista Romania. Harta fizico-geografico si administrativa   1:500 000
Bucuresti: IGFCOT, 1981.
4 sheets, all published.

Geological and geophysical

Geological Atlas of Romania   1:1 000 000
Bucuresti: IGG, 1969–.
15 sheets published.

Geological Atlas of Romania   1:50 000
Bucuresti: IGG, 1969–.
ca 600 sheets, 90 published.

Environmental

Harta pedologica   1:500 000
Bucuresti: IGG, 1970–71.
4 sheets, all published.
Soil map.

Administrative

Republica Socialista Romania. Harta administrativa   1:500 000
Bucuresti: IGFCOT, 1981.
4 sheets, all published.

Human and economic

Harta cáilor ferate   1:1 600 000
Bucuresti: Intreprinderea Poligrafica Brosov, 1975.
Map of railways.

Town maps

Bucharest. Plan of the town   *ca* 1:16 500
Bucuresti: 'Carpati' National Tourist Office.

Bucharest   1:18 000
Hamburg: Falk Verlag.

San Marino

San Marino, a tiny land-locked state within Italy, is covered by standard Italian map series. There is also a general map at 1:15 000 and a soils map at the same scale.

Address

Pitagora Editrice
Via Zamboni 57, BOLOGNA.

Catalogue

General

Repubblica di San Marino 1:15 000
Bologna: Gabicce Monte, 1973.

Environmental

Carta dei suoli della Repubblica di San Marino 1:15 000
Bologna: Pitagora Editrice, 1980.
+ 94 pp. text.

Spain

Spain has two centralized topographic mapping authorities, civil and military, the **Instituto Geográfico Nacional (IGN)** (until 1977, Instituto Geográfico y Catastral), founded in 1870, and the **Servicio Geográfico del Ejercito (SGE)**, both in Madrid. The two authorities work in cooperation and both produce topographic series which may be purchased by the public.

The principal topographic series is the 1:50 000, which exists in two versions, *Mapa topográfico nacional* (published by IGN) and *Mapa militar Serie L* (published by SGE). They share approximately the same sheet lines but use different numbering systems. The first sheet of the *Mapa topográfico* was issued in 1875 and the series was completed in 1968. It was on a Lambert polyconic projection. Since then a new edition has been in progress. The new edition is on a UTM projection and is printed in five colours with 20 m contours and the addition of hill shading on some sheets. Of the 1106 sheets in the series, 1036 cover the mainland, 26 the Baleares, and 42 the Canary Islands (more information on maps of the Canaries will be found in the Atlantic Ocean section of this book). The 1:50 000 Serie L was started in 1976 and now covers more than than 80 per cent of the country. This series is in seven colours with a 20 m contour interval. This too is on a UTM projection and has a UTM grid.

Following the completion of the 1:50 000 series in 1968, IGN initiated a new 1:25 000 scale basic series, and the first sheets were issued in 1975. The projection is UTM and the map is printed in five colours with contours at 10 m. A 1:25 000 sheet covers one-quarter of the area of the old 1:50 000 sheets (where 1:25 000 cover is published, no new 1:50 000 sheets are being issued). SGE has also published a few 1:25 000 sheets in the new *Serie 5V*, which replaces an older (and incomplete) military series, the *Plano Director 1:25 000*.

Smaller-scale multi-sheet series are published mainly by SGE. *Serie C* at 1:100 000 (40 m contours) is progressively replacing the earlier *Mapa mando*, and *Serie 2C* (1:200 000 with 100 m contours) and *4C* (1:400 000 with 100 m contours and hypsometric tints) have both totally replaced earlier series. IGN on the other hand has published two series of provincial maps (*Mapas provinciales*) at 1:200 000

and 1:500 000, and is currently producing a new series of *Mapas regionales* for the autonomous regions, at scales which vary between 1:200 000 and 1:500 000.

Cadastral mapping remains the responsibility of IGN, and is represented by a *Mapa topográfico parcelaria* at scales varying between 1:5000, 1:2000 and 1:1000. This map is uncontoured. More than 40 per cent of the country has been covered, and since 1977 orthophotography has been brought into use, which greatly increases the rapidity of survey.

The agency responsible for the mapping of geological resources is the **Instituto Geológico y Minero de España (IGME)**, Madrid, founded in 1849. The main series of geological maps is the *Mapa geológico nacional* 1:50 000. Sheets in the 2nd series (*Proyecto MAGNA*) are being issued very rapidly—200 in the first 4 years of the 1980s—with the cartography contracted to a number of private cartographic consultancies. The sheets conform to the 1:50 000 topographic sheet lines (with both the IGN and SGE numbering), are printed in colour and are issued in folded format in plastic wallets together with descriptive texts (*Memória*). Some sheets of the earlier first series are also still available. There is a new series of 1:50 000 hydrogeological maps, only a few sheets of which have been published. A few sheets of the old *Mapa geológico nacional* 1:400 000 are still in print, while the *Mapa sintesis geológica* 1:200 000 is being progressively replaced by a new series at the same scale.

Major series of land use and land classification maps have been in progress since 1974. The main series is the *Mapa de cultivos y aprovechamientos* published on the 1:50 000 topographic base map. These complex maps show about 20 land use types in colour with a surcharge of other information (shown by symbols) about aspects of land use and management. They are accompanied by booklets assessing the agricultural resources of the area. Some sheets are supported by companion land classification sheets. A provincial series of these maps at the scale of 1:200 000 was begun in 1982. This mapping is the responsibility of the **Dirección General de la Producción Agraria (DGPA)** which is a branch of the Ministerio de Agricultura, Pesca y Alimentación (MAPA), and forms part of an integral *Plan de evaluación de Recursos Agrarios*, which also includes the forthcoming production of a national *Atlas Agroclimático* at 1:500 000.

Maps of the Spanish National Parks have been published since 1981 by the **Instituto Nacional para la Conservación de la Naturaleza (ICONA)**. They are small-format topographic maps with tourist information on the reverse, and are distributed by MAPA.

The **Instituto de Edafológia** (under the Consejo Superior de Investigaciones Cientificas, CSIC) is publishing a series of provincial soil maps at 1:200 000 in cooperation with regional organizations. A 1:1 000 000 soil map of Spain was published in 1968.

CSIC also has an **Instituto de Geografía Aplicada** which has made a number of maps, including 1:1 000 000 maps of population density and change based upon the 1960 census, a land use map of Zaragosa, and various other regional projects carried out in collaboration with local research centres.

The **Ministerio de Obras Publicas y Urbanismo (MOPU)**, Madrid, publishes an annually revised road atlas at 1:400 000, as well as maps of traffic density and road conditions.

An important feature of Spanish cartographic activity during the 1980s has been associated with the transfer of many administrative and planning functions to the newly formed autonomous regions. This has led to a corresponding regionalization of certain mapping activities, and a demand for new maps, especially in the areas of environmental planning and large-scale topographic mapping. Further stimulus has come from the creation of new regional cartographic centres. In order to ensure a coordination of cartographic work done by these various bodies, the government has passed a Cartographic Regulation Act, operating through a National Cartographic Plan, and has established a central registry of mapping and standards for the different scales. Some examples of recent regional mapping activity are given in the following few paragraphs.

The Comunidad de Madrid is mapped by **CIDAMM (Centro de Informacion y Documentación del Area Metropolitana de Madrid)/CoplacO**, founded in 1971. Basic mapping of the whole province has been carried out at 1:5000, and of the urban areas at 1:2000, and there is a programme of frequent revision. A derived 1:25 000 map has recently been launched, and a number of thematic mapping projects are also undertaken including a programme of 1:25 000 land use mapping. An interactive computer mapping system is in operation.

In Catalonia, the **Institut Cartogràfic de Catalunya**, founded in 1982, is producing a wide range of new cartographic material of the region. This includes a 1:250 000 topographic map and a series of thematic maps at the same scale. A land use map, population density map and road map have already been published, and a climatic atlas of Catalonia is in preparation. There is also a false colour Landsat image map at 1:450 000.

In Andalucia, the **Agencia de Medio Ambiente** has set up an environmental database called SinambA (Sistema de informacion ambiental de Andalucia), which will integrate digital satellite imagery with information from maps, fieldwork and other sources. In addition a number of recent conventional maps have been published. These include a 1:300 000 *Mapa topográfico del la Comunidad Autonoma* (1984), a *Mapa geológico minero* 1:400 000 in two sheets with text (1985), a coloured Landsat 4 image map 1:400 000 (1986), and an *Atlas de evaluación de tierras de Andalucia* comprising four 1:400 000 scale resource maps and a 150-page text (1986).

In Galicia a 1:200 000 scale map *Capacidad productiva de los suelos de Galicia*, using FAO land evaluation methodology, was prepared by the **Universidad de Santiago de Compostela**, and published in a bound format in 1984. A two-volume *Atlas de Galicia* was published in 1982.

The *Atlas Nacional de España*, published 1965–68, is now out of print, although some individual folios are still available from IGN. However, a number of new regional atlases have begun to appear (of Galicia, Navarra and Aragon, for example). In Castellon, a recent atlas has made extensive use of Landsat Thematic mapper imagery (*Castellon desde el espacio*, published by **Editorial Confederación Española de Cajas de Ahorro**, 1986). Much of the work for these atlases has been carried out by the local universities, with support from CSIC and the Junta of the region.

Among Spanish commercial publishers, **Aguilar**, Madrid, has a large cartographic department, producing a range of atlases including a series of regional *Atlas gráficos*, the prestigious *Atlas gráfico de Espana* (of National Atlas quality), and a gazetteer keyed to the regional atlases, *Indice general de toponimos*. The *Atlas gráfico* has 214 pages of maps which include topographic cover at 1:250 000 as well as text, numerous colour photographs and statistical graphics. Aguilar has also published a *Gran atlas enciclopedico Aguilar* in 13 volumes, which includes a very large number of maps and cartograms and an index of 415 000 toponyms.

Editorial Alpina, Granollers, specializes in the publication of walking and climbing guides accompanied by contour maps, usually at the scale of 1:40 000 or 1:25 000. Principal areas covered are the coastal cordilleras of Catalonia and the Pyrenees. They also publish administrative maps of Catalonia.

Firestone Hispania, Bilbao, specializes in road and tourist maps, and produces a nine-sheet road map of Spain and Portugal at 1:500 000, a series of special maps covering the Balearics, Canaries and Bilbao at larger scales, and regional and tourist maps—*mapas turísticos*—mainly at 1:200 000. The latter series have detailed street maps of major cities on the reverse of each sheet.

A number of cartographic companies carry out map production on contract to national and local government agencies. These include **Cartografica Iberica** and **RHEA Consultares**. Projects range from detailed mapping of metropolitan areas to standard thematic mapping within the geological and land use series described above. Generally, these agencies do not publish their own maps, although some are beginning to publish a few tourist maps.

Further information

Both IGN and SGE issue catalogues and graphic indexes of current topographic map cover.

A *Catálogo de publicaciones del Ministerio de Agricultura, Pesca y Alimentación* is available, listing the land use and land classification maps published up to 1985 by Province.

Cátalogo por calabras clave del foldo documental generalado por el IGME (3rd edn, 1984) is a complete catalogue of IGME maps and publications arranged by keywords.

Addresses

Instituto Geográfico Nacional (IGN)
General Ibáñez de Ibero, no. 3, 28071 MADRID.

Servicio Geográfico del Ejercito (SGE)
Departamento de Distribución de Cartografía, C/Dario Gazapo
no. 8 (Ctel. Alfonso X), 28024 MADRID.

Instituto Geológico y Minero de Espana (IGME)
Rios Rosas, 23, 28003 MADRID.

Ministerio de Agricultura, Pesca y Alimentación (DGPA, ICONA)
Servicio de Publicaciones, Paseo de Infanta Isabel, 1, 28014
MADRID.

Instituto de Edafológia
C/Serrano, 115 Dpdo, 28006 MADRID.

Instituto de Geografía Aplicada
C/Serrano, 115 Dpdo, 28006 MADRID.

Ministerio de Obras Publicas y Urbanismo (MOPU)
Paseo de la Castellana, 67, 28046 MADRID.

CIDAMM/CoplacO
8a planta—Nuevos Ministérios, Plaza de San Juan de la Cruz, s/n,
MADRID 3.

Institut Cartogràfic de Catalunya
C/Balmes, 209, 08006 BARCELONA.

Servicio de Publicaciones y BOJA
Agencia de Medio Ambiente, Consejería de Presidencia, Junta de
Andalucia, Ac/Jesus, 16, 41002 SEVILLA.

Universidad de Santiago
Servicio de Publicacions e Intercambio Científico, Campus
Universitario, SANTIAGO DE COMPOSTELA.

Aguilar
Juan Bravo, 38, 28006 MADRID.

Editorial Alpina
Apartat Correus 3, GRANOLLERS (BARCELONA).

Firestone Hispania
Apartado 406, BILBAO.

Editorial Confederación Española de Cajas de Ahorro
C/Alcala 27, MADRID.

Cartografia y Diseno SA
Avda Alfonso XIII-75, MADRID 16.

Cartografica Iberica SA (CIBESA)
Conde de la Cimera, 4—Local 6, 28040 MADRID.

RHEA Consultores
Paseo La Habana, 206, 28036 MADRID.

For Hallwag, see Switzerland.

Catalogue

Atlases and gazetteers

Atlas gráfico de España
Madrid: Aguilar, 1984.
pp. 507.

Atlas gráfico de España (Indice general de toponimos)
Madrid: Aguilar, 1980.
pp. 153.

General

Mapa general de España 1:2 500 000
Madrid: IGN, 1976.

Península Iberica e Islas Baleares 1:2 000 000
Madrid: IGN.
Landsat Band 7 mosaic, monochrome.

Mapa de España 1:1 850 000
Madrid: SGE, 1983.

Península Iberica, Baleares y Canarias 1:1 000 000
Madrid: IGN, 1981.
In IMW style.

Mosaico fotográfico de la Peninsula Iberica e Islas Baleares 1:1 000 000
Madrid: IGME.

Península Iberica, Baleares y Canarias 1:800 000
Madrid: IGN, 1984.
2 sheets, both published.
Also available in edition of 9 sheets (1983).

Mapa de la Península Iberica, Baleares y Canarias 1:750 000
Madrid: IGN, 1974.
2 sheets, both published.

Atlas geográfico de Espana 1:500 000
Madrid: IGN, 1961.
15 sheets (including Canary Islands), all published.

Mapa Serie World 1:500 000 Serie 1404
Madrid: IGN.
12 sheets (including Canary Islands), 5 published.

Mapa oficial de carreteras 1:400 000
Madrid: MOPU, annual.
pp. 120.
Official road map.
Legend in Spanish, English and French.

Topographic

Mapa militar de España 1:500 000 Serie 4C
Madrid: SGE, 1972–.
30 sheets (including Canary Islands), all published.

Mapa militar de España 1:200 000 Serie 2C
Madrid: SGE, 1969–.
92 sheets, all published. ■

Mapa provincial 1:200 000
Madrid: IGN, 1970–.
47 sheets, all published. ■

Mapa militar de España 1:100 000 Serie C
Madrid: SGE, 1972–.
296 sheets, 50% published. ■

Mapa militar de España 1:50 000 Serie L
Madrid: SGE, 1968–.
1130 sheets, 90% published. ■

Mapa topográfico nacional 1:50 000
Madrid: IGM, 1968–.
1106 sheets, ca 25% published. ■
New edition on UTM projection.

Mapa topográfico militar 1:25 000 Serie 5V
Madrid: IGM, 1981–.
ca 4520 sheets, 908 published.

Mapa topográfico nacional 1:25 000
Madrid: IGN, 1975–.
4380 sheets, 536 published.
NB the two 1:25 000 series cover complementary areas.

Geological and geophysical

Mapa sismotectónico de la Peninsula Iberica 1:2 500 000
Madrid: IGME, 1960.

SPAIN
1:200 000 topographic
1:200 000 geological

Mapa metalogenético de España 1:1 500 000
Madrid: IGME, 1972.
17 sheets, each showing distribution of one mineral.

Mapa geológico de la Península Iberica, Baleares y Canarias 1:1 000 000 Edition 1
Madrid: IGME, 1980.

Mapa sismoestructural de la Península Iberica, Baleares y Canarias 1:1 000 000
Madrid: IGME, 1966.

Mapa tectónico de la Península Iberica y Baleares 1:1 000 000
Madrid: IGME, 1977.

Mapa de lineamientos deducidos de las imagenes Landsat 1:1 000 000
Madrid: IGME.
2 sheets, both published.

Mapa hidrogeólogico nacional 1:1 000 000
Madrid: IGME, 1972.

Mapa de vulnerabilidad a la contaminación de los mantos acuiferos 1:1 000 000
Madrid: IGME, 1973.

Mapa geológico nacional 1:400 000
Madrid: IGME, 1933–71.
64 sheets, 9 published.

Mapa geológico de España 1:200 000
Madrid: IGME, 1982–.
81 sheets, 4 published. ■

Mapa de sintesis geológica 1:200 000
Madrid: IGME, 1970–73.
81 sheets, all published. ■

Mapa hidrogeólogico de España 1:200 000
Madrid: IGME, 1982–.
81 sheets, 3 published. ■

Mapa metalogenético de España 1:200 000
Madrid: IGME, 1973–.
81 sheets, all published. ■

Mapa geotécnico general de España 1:200 000
Madrid: IGME, 1972–.
85 sheets, all published. ■

Mapa de rocas industriales de España 1:200 000
Madrid: IGME, 1973–.
85 sheets, all published. ■

Mapa geológico de España 1:50 000 Projecto Magna
Madrid: IGME, 1972–.
1130 sheets, 546 published. ■

Mapa hidrogeólogico de España 1:50 000 1a serie 1a edición
Madrid: IGME, 1982–.
1130 sheets, 3 published. ■

Mapa de orientación al vertido de residuos solidos urbanos 1:50 000
Madrid: IGME, 1977–.
1130 sheets, 46 published. ■

Environmental

Mapa de suelos de España 1:1 000 000
Madrid: Instituto de Edafologia, 1968.

Suelos naturales de la Provincia de . . . 1:200 000
Madrid: Instituto de Edafológia, 1982–.
Series of provincial soil maps.
2 published.

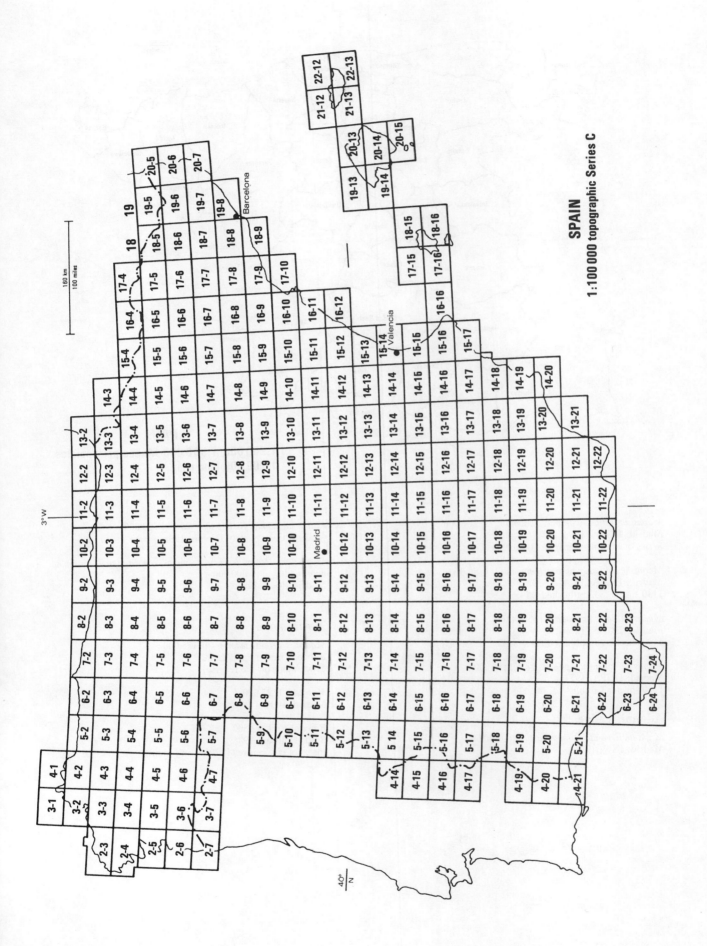

SPAIN

1:100 000 topographic Series C

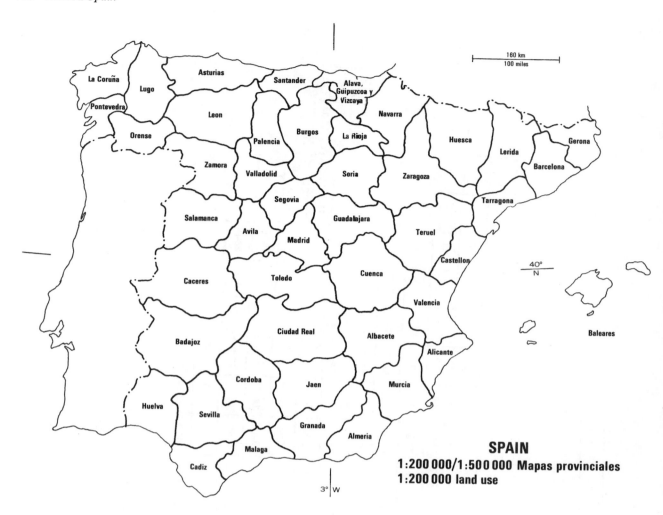

SPAIN

1:200 000/1:500 000 Mapas provinciales

1:200 000 land use

Mapas de cultivos y aprovechamientos 1:200 000
Madrid: DGPA, 1982–.
Series of Provincial maps, 23 published.

Mapas de cultivos y aprovechamientos 1:50 000
Madrid: DGPA, 1974–.
1130 sheets, 332 published. ■

Mapas de clases agricolas 1:50 000
Madrid: DGPA, 1975–.
1130 sheets, 78 published. ■

Mapas de ordenación productiva 1:50 000
Madrid: DGPA, 1974–.
1130 sheets, 1 published. ■

Administrative

Mapa de España (Autonomias) 1:2 000 000
Madrid: IGN, 1983.

Mapa autonómico de España ca 1:1 750 000
Madrid: Cartografia y Diseno, 1985.

*Mapa de la Peninsula Iberica (mudo), con division de terminos
municipales* 1:1 000 000
Madrid: IGN, 1983.

Town maps

Madrid y sus alrededores 1:175 000
Bilbao: Firestone Hispania.
Town maps on the reverse.

Madrid 1:25 000
Madrid: IGN.

Barcelona 1:12 500
Bern: Hallwag, 1985.
Insets of city centre (1:7000) and metro.

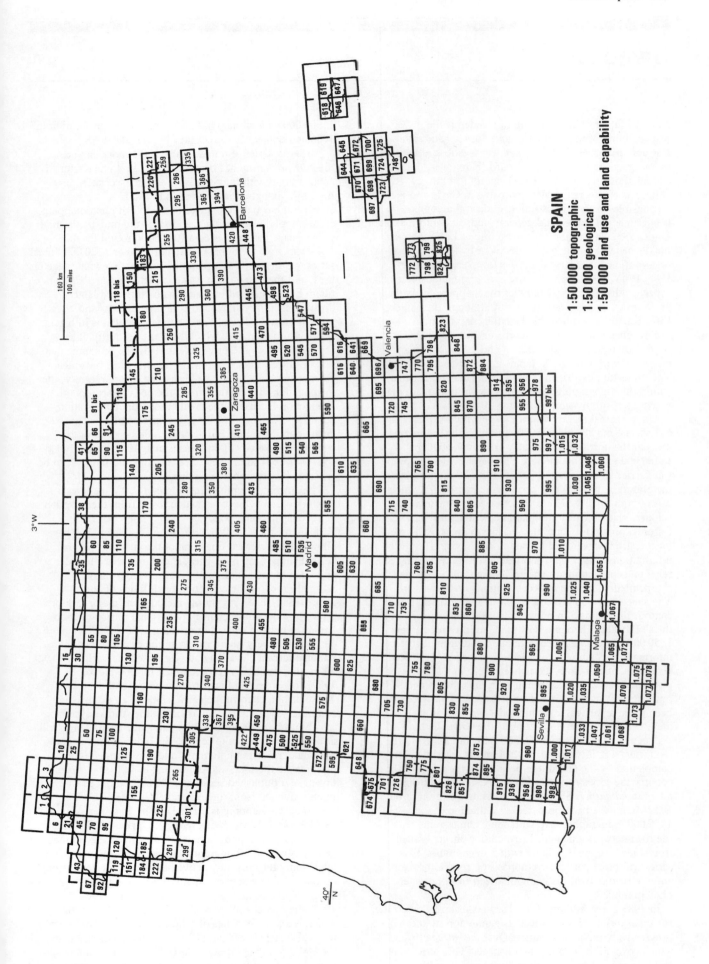

SPAIN

1:50 000 topographic
1:50 000 geological
1:50 000 land use and land capability

Sweden

National topographic mapping in Sweden is the responsibility of the National Land Survey (**Statens Lantmäteriverket, LMV**) which was established in 1974 by merging the Land Survey Board (Kungliga Lantmäteristyrelsen) with the Geographical Survey (Rikets Allmänna Kartverk). LMV is situated in Gavle and is also responsible for land registration, cadastral mapping, geodesy, aerial survey and general landscape information. Until 1985, elaboration and marketing of LMV mapping for the general public was undertaken by a commercial organization Liber Grafiska AB. However, in January 1985, LMV took over the entire cartographic business of Liber, so that LMV now handles the mapping for both professional and recreational users, although the name **LiberKartor** is retained for the latter map types.

The official mapping is undertaken by means of a contract between the government and LMV, and maps are published within the scale range 1 : 10 000 and 1 : 1 200 000. The projection used is the Gauss conformal, and sheets within the topographic series break down into quadrangles based upon subdivisions of a 50 × 50 km grid.

The principal map series, the *Topografiska kartan* is in 687 sheets, mostly at 1 : 50 000, although the mountain areas in the north of the country are at the smaller 1 : 100 000 scale. This modern photogrammetric mapping was carried out during the period 1953–79, and replaced the prewar 1 : 100 000 monochromatic map series which was of poor quality. The production is in four main colours: black for text and planimetry, blue for water features, brown for contours and marsh areas, and green for forest. However, red is used to add revision information to the maps, and yellow and grey for shaded relief on some mountain sheets. There is also a special overprinted telephone network edition (*telefältversionen*). The sheets have both Swedish and UTM grid and the contour interval is 5 m on the 1 : 50 000 maps and 20 m on the 1 : 100 000. There are now plans to introduce a new 1 : 50 000 containing more vegetation and ground surface information which will complement the new 1 : 100 000 series described below. Test sheets were due to be published in 1986. In common with the other general-purpose map series, the *Topografiska kartan* has recently been given a particular colour code—green—as an identifier, and is therefore dubbed the *Gröna kartan*.

A new 1 : 100 000 series (*Blå kartan*), orientated to the road user, was introduced in 1984 and is scheduled for completion by the end of the decade. In the north of the country the series merges with the existing 1 : 100 000 *topografiska kartan*. Sheets in a provisional edition of the *Blå kartan* have been prepared by reduction of the 1 : 50 000 topographic map with some names removed and the road network updated. Final edition sheets, issued from 1985, are in six colours with 5 m contours. This series will place particular emphasis on road information, and the road network is being digitized from new stereo air photography.

In 1983 a new edition of the 1 : 250 000 series (called the *Röda kartan*) in 23 large sheets began to appear and will progressively replace the earlier *Översiktskartan* (1972–80) which covered the country in 43 sheets. This series is in eight colours, with contours at 25 m, and carries the Swedish grid as well as UTM and Georef nets. A new

1 : 500 000 general map has also commenced publication in 1986 and will cover the country by the late 1980s.

A special series of tourist maps of the mountain areas, the *Nya fjällkartan* has been produced by LiberKartor in collaboration with the Swedish Tourist Association using the LMV topographic maps at 1 : 100 000.

Sweden has been a pioneer in the use of photomapping and orthophoto techniques, and since the mid-1930s an air photo image has been used as the background for its *Economiska kartan* published at 1 : 10 000 or 1 : 20 000. This series was designed as a planning map with a wide range of uses, and the first edition was completed about 1978. The photo base is printed in green, with cultivated land distinguished in yellow, and contours (at 5 m) and boggy areas shown in brown. The planimetric detail is in black, as are also placenames which have the authority of the *Ortnamnsarkivet* at Uppsala. Originally the economic map was produced from controlled photomosaics, but since 1966 orthophotomaps have been used. More recently, the introduction of computer aided methods in orthophoto production has led to the creation of a digital elevation database. The sheets of the *Economiska kartan* each cover an area of 5 × 5 km, and at present 80 per cent of the country is covered (11 600 sheets at 1 : 10 000 and 1100 at 1 : 20 000), but some sheets are very old and have yet to be replaced by orthophoto editions.

The current mapping progamme of LMV includes a number of new objectives which result from the adoption of 10-year mapping policy (1985–94) approved by parliament, the so-called *Kartpolitik 85*. This policy has allowed for increased financing of the survey through several supporting ministries, and introduces a broader approach to the handling of geographic data and a rationalization of the collection of all kinds of land information. The programme includes the major task of relevelling the entire country (a national retriangulation had already been carried out during 1968–82). *Kartpolitik 85* also gives very high priority to the establishment of a national terrain elevation database based on a grid of 50 × 50 m (this database is programmed for completion by 1989) and of several lower resolution general databases. These databases form the cornerstone of an extensive computer mapping system at LMV, the Autoka system, which has been under development since 1976. There will also be a major effort to replace the older economic maps with modern orthophotomaps, and it is planned to discontinue the four-colour printing of 1 : 10 000 sheets in favour of a published series at 1 : 20 000 produced by photoreduction from the larger scale.

Geological mapping is carried out by **Sveriges Geologiska Undersökning (SGU)**, Uppsala. SGU began mapping in 1862, and there is an extensive range of map series, as well as monographs accompanied by maps, from which only the currently active have been selected for our catalogue. These are principally the *Serie Ae*, which is a 1 : 50 000 Quaternary geology series begun in 1964 (until 1974, the sheets also showed solid geology), and *Serie Af*, which is a parallel series of solid geology and geophysical maps also at 1 : 50 000 and begun in 1967. The sheets use the LMV topographic series as base maps, but in northern Sweden they are issued in groups of four together with a joint monograph. The solid geology sheets are accompanied by aerial magnetic total intensity maps, and in central and south Sweden also by tectonic maps. Other modern series include the 1:50 000 hydrological *Serie Ag* started in 1971, and the 1:250 000 *Hydrogeologiska översiktskartor* (1981–). The latter is programmed to cover most of southern Sweden by 1989. The series *Ba* and *Ca* include a number of general and county (*Läns*) maps of solid and drift geology at scales of 1:200 000 to 1:400 000. SGU has also mapped the intensity of natural radioactivity in a series of *Geostrålningskartor* at 1:50 000.

Older geological series published by SGU have been discontinued although many sheets remain in print. They include solid and drift mapping of southern Sweden at scales of 1:50 000, 1:100 000 and 1:200 000.

SGU maps may be ordered from Liber Distribution, Stockholm.

There has been considerable interest in the mapping of vegetation in Sweden, and in 1975 a project to map the whole Swedish mountain region began at the Remote Sensing Laboratory in the Department of Physical Geography, University of Stockholm (**Naturgeografiska Institutionen, SUNI**). The mapping was carried out by interpretation of infrared colour air photographs in conjunction with ground truth observations. A series of 22 map sheets at 1:100 000 was published between 1978 and 1983. The maps are printed in five basic colours showing the vegetation units, with different shades indicating certain ecological factors and numerical codes defining vegetation communities. Further experimental mapping has been carried out in lowland areas, and there is group of 1:50 000 maps covering the Siljan area. These maps are available from SUNI in boxed sets grouped by county.

SUNI also produced a series of geomorphological maps of the mountain region in the 1970s. These maps, at 1:250 000 scale, were issued with monographs assessing the conservation value of the landscape and were produced as a contract for Statens Naturvårdsverk. They cover the whole of the mountain area and are available from SUNI. More recently, workers at the Department of Physical Geography at Uppsala University have produced an experimental series of 'topogeomorphological' maps which attempt to represent Swedish landscape types without classifying them genetically. There is considerable interest in the problems of incorporating more landform and vegetation information on the standard LMV 1:50 000 topographic maps.

A National Atlas of Sweden was edited by the **Svenska Sällskapet for Antropologi och Geografi** and published progressively from 1953 to 1971. It is now out of print, but a study has been undertaken by LMV with the Svenska Sällskapet and Statistics Sweden for a second edition. This will consist of 16–20 thematic books and the principal map scale will be 1:2 000 000. The production of this atlas awaits a decision of parliament on the question of financial support.

Census information is collected by the **Statistiska Centralbyrå (SCB)**—Statistics Sweden. A number of population maps have been published, mostly at 1:1 000 000 and based on previous censuses. The most recent was of 1970 census data by Hedbom and Norling. For the 1980 census, enumeration districts were identified by grid coordinates. Since the property registry is also currently being computerized by the Central Board for Real Estate Data, it has now become possible to produce large-scale census maps by combining census data with the coordinate index of the property register. Such maps can be produced on demand by SCB by outputting the thematic data on a line plotter and combining it with a base map. In 1984, SCB published a popular *Census Atlas*

of Sweden designed to give 'an animated and viewable presentation of the results from FoB 80'. The atlas has texts in Swedish and English.

The largest commercial map publisher in Sweden is **Esselte Map Service**, Stockholm. Esselte specializes in the production of maps and atlases for educational use, and their list also contains a variety of tourist, road, mountain and archipelago maps. The mountain maps cover the higher parts of Jämtland, Harjedal and Dalarna at a scale of 1:100 000, while the Archipelago maps cover the east coast to north and south of Stockholm. There is also a sectional tourist map of the whole country in eight sheets (*Sverigekartan*) incorporating a wide range of tourist information.

Further information

There is an extensive literature about Swedish mapping and developments in automation and in geographical information systems. A general description of the topographic series is given in Sandler, H. (1982) Amtliche Kartenwerke in Schweden, *Kartographische Nachrichten*, **32**, 179–184.

Kartpolitik 85 is discussed in Wickbom, S. (1983) En svensk kartpolitik, *Kart og Plan*, **75**, 169–175; and Wennström, H.F. (1984) A ten year programme for the official mapping of Sweden, *ICA 12th International Conference, Perth, Technical Papers*, **2**, 113–123.

LMV publishes a map catalogue *Kartor* and the latest edition appeared in 1986.

SGU also publishes a regularly updated catalogue.

Readers of Swedish will find that the 1984 issue of the periodical *Ymer*, devoted to the theme of maps in the community, has several papers on innovative thematic mapping, including developments in vegetation and topogeomorphological mapping.

Addresses

Lantmäteriverket (LMV)
S-801 12 GÄVLE.

LiberKartor
S-162 89 STOCKHOLM.

Sveriges Geologiska Undersökning (SGU)
S-751 28 UPPSALA.

Stockholms Universitet Naturgeografiska Institutionen (SUNI)
Box 6801, S-106 91 STOCKHOLM.

Svenska Sällskapet for Antropologi och Geografi
Kulturgeografiska Institutionen, Box 6801, S-11386 STOCKHOLM.

Statistiska Centralbyrån (SCB)
S-115 81 STOCKHOLM.

Esselte Map Service
PO Box 22069, S-104 22 STOCKHOLM.

Catalogue

Atlases and gazetteers

Svensk ortförteckning 1980
Stockholm: Postverket and Televerket, 1980.
pp. 754.

M.KAK bilatlas Sverige
Stockholm: Esselte.
Includes 46 000-entry placename index.
Legend in English, German, Swedish and Finnish.

General

Sverige 1:1 200 000
Stockholm: Liber.

Sverige 1:1 000 000
Gävle: LMV, 1985.

Internationella världskartan 1:1 000 000
Gävle: LMV, 1970.
3 specially formatted sheets published in IMW style.

M.KAK Sverige vägkarta 1:1 000 000
Stockholm: Esselte.

Generalkartan 1:500 000
Gävle: LMV, 1986–.
5 sheets, 2 published.

Sverigekartan 1:300 000
Stockholm: Esselte.
8 sheets, all published.
Motoring and tourist map.

Nya bil- och turistkartan 1:250 000/400 000
Stockholm: Liber.
9 sheets, all published.
Motoring and tourist map.
Northern sheets at 1:400 000.

Topographic

Röda kartan (Översiktskartan) 1:250 000
Gävle: LMV, 1983–.
23 sheets, all published. ∎

Blå kartan 1:100 000
Gävle: LMV, 1984–.
134 sheets, 44 published. (82 sheets in the new format shown on
the graphic index.) ∎

Topografiska kartan 1:100 000/1:50 000
Gävle: LMV, 1953–.
687 sheets, all published. ∎
Scale is 1:50 000 except for 41 sheets covering the mountain areas
of Norrland, which are at 1:100 000.

Geological and geophysical

Kart över Sveriges berggrund 1:1 000 000
Uppsala: SGU, 1958.
3 sheets, all published.
Pre-Quaternary rocks of Sweden.
English text available.

Kart over Sveriges jordarter 1:1 000 000
Uppsala: SGU, 1958.
3 sheets, all published.
Quaternary geology.
English text available.

*Karta över landisens avsmältning och högsta kustlinjen i
Sverige* 1:1 000 000
Uppsala: SGU, 1961.
Deglaciation and highest shoreline in Sweden.

Berggrundsgeologiska och geofysiska kart 1:50 000 Serie Af
Uppsala: SGU, 1967–.
ca 100 sheets published. ∎
In northern Sweden, issued in groups of four sheets with text.
Bedrock and geophysical (aeromagnetic) maps.

Jordartsgeologiska karta 1:50 000 Serie Ae
Uppsala: SGU, 1964–.
ca 96 sheets published. ∎
Quaternary geology.
Available with/without texts.

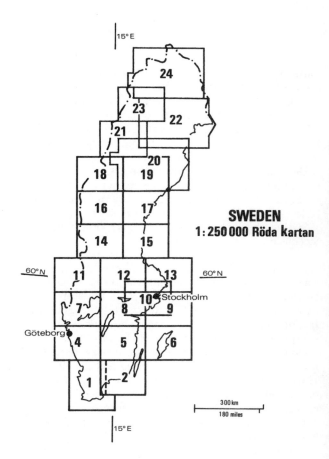

SWEDEN
1:250 000 Röda kartan

Hydrogeologiska karta 1:50 000 Serie Ag
Uppsala: SGU, 1971–.
ca 18 sheets published. ∎
Available with/without texts.

Geokemiska kartor: tungmineral 1:50 000
Uppsala: SGU, 1984–.
ca 40 sheets published. ∎

Geokemiska kartor: backtorv 1:50 000
Uppsala: SGU, 1984–.
ca 78 sheets published. ∎

Administrative

Sveriges Läns- och Kommunindelning 1:1 200 000
Stockholm: Liber, 1981.
Län and commune boundaries and centres of administration.

Human and economic

Atlas till Folk- och Bostadsräkningen/A Census Atlas of Sweden
J. Szegö
Stockholm: Statistiska Centralbyrån, 1984.

Town maps

Stockholm map and guide 1:12 000
Stockholm: Esselte.
Double-sided map with suburbs on the reverse and 20 pp. street
index and tourist guide.
Also available without the tourist guide as *Stockholm & fororter*.

Stockholm trafikkartan 1:40 000/1:10 000
Stockholm: Liber, 1985.
City centre 1:10 000, suburbs on reverse 1:40 000.
Parking and mass transit information.
Street index.

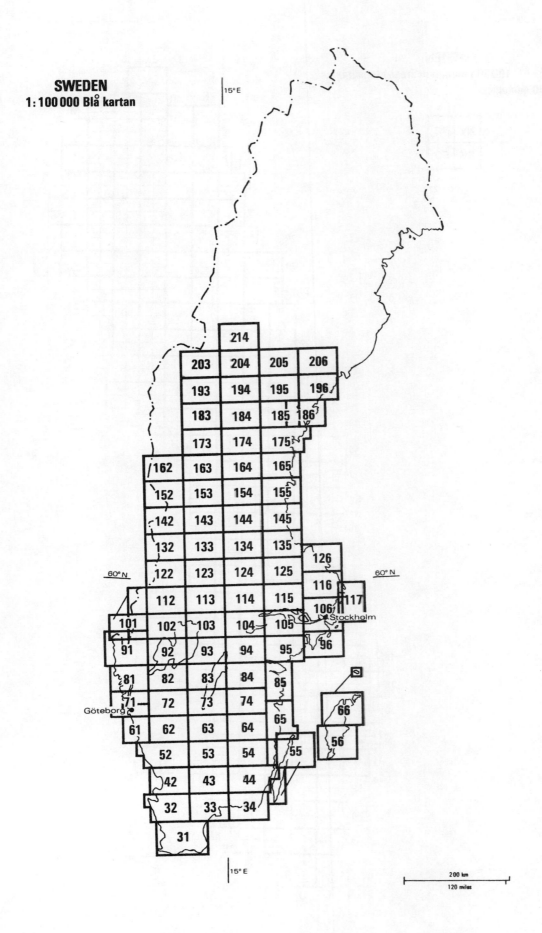

SWEDEN
1:100 000 Blå kartan

SWEDEN
1:50 000 (1:100 000 mountain areas) topographic
1:50 000 geological

NV	NO
SV	SO

Sheet numbering examples:-
1:100 000 22F
1:50 000 6C NV

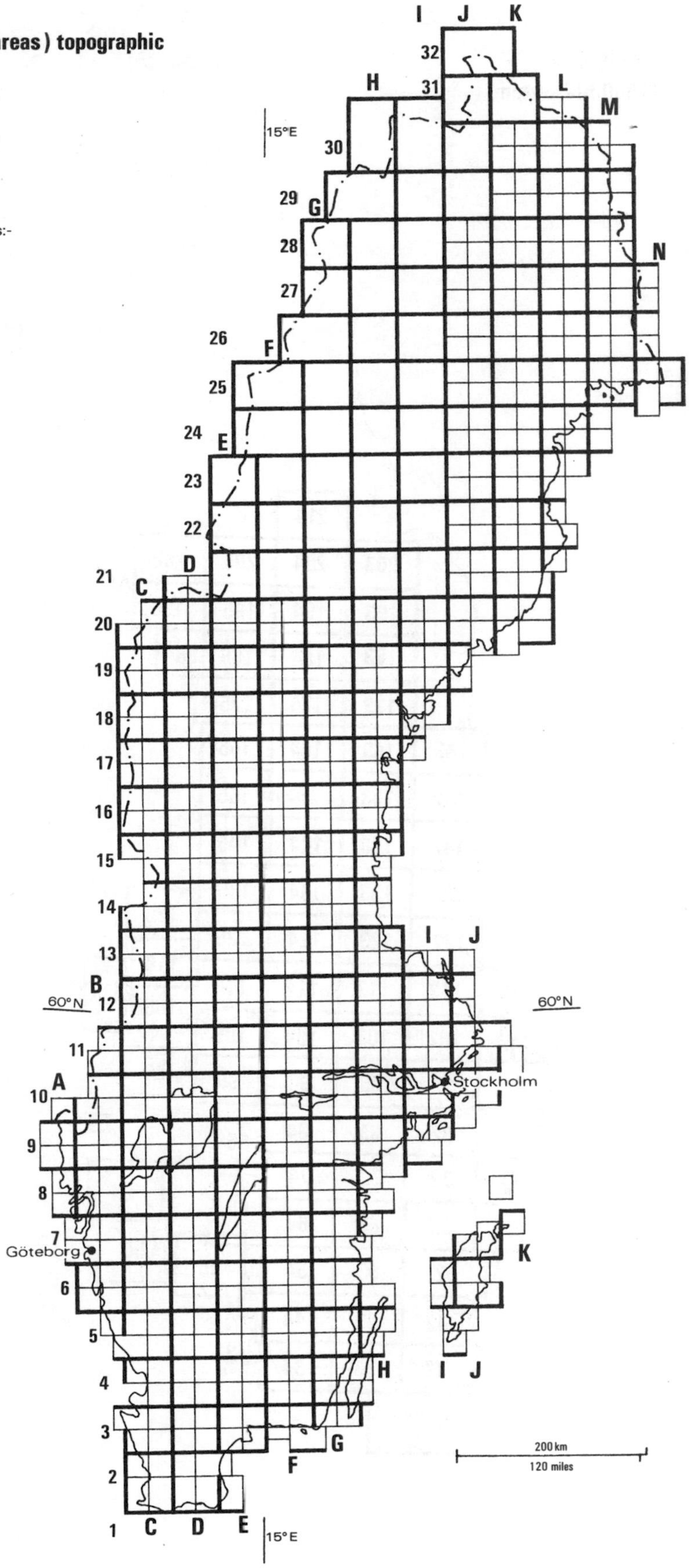

I J K
32
H 31
15°E
30
29 G
28
27
26 F
25
24 E
23
22
21 C D
20
19
18
17
16
15
14
13
B
60°N 12
11
A
10
9
8
7 Göteborg
6
5
4
3
2
1 C D E
15°E

L
M
N

Stockholm
60°N

I J

K

I J
H
G
F

200 km
120 miles

Switzerland

The Federal Office of Topography (**Bundesamt für Landestopographie, BLT**) is a division of the Swiss Ministry of Defence, and is responsible for the geodetic and topographic survey of Switzerland and the publication of the official map series. BLT was founded in 1838 at Geneva, but today operates from Wabern.

Antecedents of the modern Swiss topographic map were the famous 'Dufour' (1:100 000) and 'Siegfried' (1:25 000/ 1:50 000) maps of the nineteenth century. The *Siegfriedkarte* remained the official Swiss topographic map until 1935 when a Swiss federal law defined the publication of new official maps at scales of 1:25 000, 1:50 000, 1:100 000, 1:200 000, 1:500 000 and 1:1 000 000. These photogrammetric maps now form the present day mapping system, comprising some 350 sheets in all. The 1:50 000 series was completed in 1960, the 1:100 000 in 1964 and the 1:25 000 in 1979. These series now fall into a 6 year revision cycle.

The 1:25 000 eight-colour map forms the basic series. It has 10 m contour interval (20 m in mountain areas) and is on a conformal cylindrical projection. The 1:50 000 map is in six colours and the 1:100 000 in ten, and they employ a contour interval of 20 m and 50 m respectively. The 1:100 000 is available with or without a kilometric grid overprinted in violet. These three national series all feature hill shading very effectively. A separate legend is published explaining the symbols on all three maps. A number of composite sheets are published in all three scales covering popular tourist areas, and ski and walking route editions of some sheets are also available.

The 1:200 000 map covers Switzerland and surrounding territory in four sheets. Relief is shown by 100 m contours and hill shading, the road and rail network is also in colour, and the sheets include a legend and (on the reverse of folded editions) route access maps to major cities.

Large-scale cadastral mapping is the responsibility of cantonal and municipal survey authorities, with the Federal Directorate of Cadastral Surveys, Bern, as the coordinating body. Cadastral maps vary in scale from 1:250 in urban areas to 1:10 000 in the mountains. There is also a 'general cadastral plan' published at 1:5000 or 1:10 000 with 10 m contours, and derived from the more detailed cadasters. There are moves to revise the specification of the Swiss cadastral maps.

The *Atlas of Switzerland* was issued in its first edition from 1965 to 1978, as a series of loose folios. It was commissioned by the Swiss Federal Council, and the first edition had as its editor-in-chief Eduard Imhof. Since 1981, a second edition which both expands and up-dates the first edition has been in progress. The second edition, edited by Ernst Spiess, begins with the 10th issue, and the 12th issue was published in 1986. Maps are prepared with the aid of an interactive computer graphics system in the Department of Cartography, Swiss Federal Institute of Technology, Zürich. The *Atlas* is published by BLT.

Geological mapping is the responsibility of the Swiss Geological Commission (**Schweizerische Geologische Kommission, SGK**), Basel, which together with the **Schweizerische Geotechnische Kommission (SGtK)**, the **Schweizerische Geophysikalische Kommission (SGpK)** and the Hydrologische Kommission are organs of the Swiss Academy of Sciences. The publications of all four

bodies are distributed through **Kümmerly + Frey AG (K+F)**, Bern. The main geological series is the *Geological Atlas of Switzerland*, published as a progressive series since 1930. Maps conform to the *Landeskarte der Schweiz* sheet lines, although some are based on the earlier *Siegfriedkarte* base maps. A number of the early issues are now out of print.

A number of special topic maps for planning purposes have been published by the Federal Office for Spatial Planning (**Bundesamt für Raumplanung, BARP**), Bern. These include a series of environmental hazard maps at 1:100 000. The cantons and municipalities are active in preparing maps of land capability, landscape quality and environmental impact assessment in fulfilment of the federal law on spatial planning.

A new series of soils maps at 1:25 000 was initiated in 1981 by the **Eidgenössische Forschungsanstalt für landwirtschaft Pflanzenbau (FAP)**, Zurich-Reckenholz. This series is printed in colour on base maps of the national topographic series. By 1985, five sheets had been published with several more in preparation. Smaller-scale maps of land capability are also published, and these are distributed by **Eidgenössische Drucksachen- und Materialzentrale**, Bern.

The Swiss Meteorological Institute (**Schweizerische Meteorologische Anstalt, SMA**), Zurich, is engaged in a long-term project to investigate regional climates of Switzerland, and the results are being published as a series of climate maps forming the *Klimaatlas der Schweiz*, edited by Dr Walter Kirchhofer and published by BLT. The second folio of this atlas appeared in 1984, and the whole programme will take about 10 years to complete.

Several of the Swiss universities have active cartographic teaching and research units from which published maps sometimes emanate. They include the **Institut für Kartographie der Eidgenössische Technischen Hochschule**, Zurich, which, as mentioned above, has developed interactive computer mapping systems, and is currently developing a small-scale topographic database of the country, and the Departments of Geography at the University of Zurich and the University of Bern. The last named produces the series *Geographica Bernensia* (**Arbeitsgemeinschaft Geographica Bernensia**) which includes a range of environmental mapping of specific areas at home and abroad.

Many of the private cartographic companies in Switzerland are world famous for their small-scale tourist and educational mapping. They include **Hallwag AG**, Bern, Kümmerly + Frey (K+F), Bern, **Schad & Frey**, Bern, and **Orell Füssli**, Zurich. These do most of their own compilation and draughting, and their products include motoring and recreational maps of Switzerland, and street maps and atlases of Swiss towns. K+F has published large-scale hiking maps and cycling maps (*Velokarte*) based on BLT mapping.

Further information

BLT issues map indexes of the national map series, and leaflets describing their map products, revision policy and production methods.

SGK issues a *Verkaufskatalog der Publikationen* which lists the publications of the four commissions within the Swiss Academy of Sciences.

Indexes and leaflets are also issued by FAP and SMA.

The Swiss Society of Cartography has issued a report, available in English, entitled *Cartography in Switzerland 1980–1984*, which includes full-colour extracts of a range of Swiss mapping (SSC *Cartographic Publication Series* No. 7). It is available from Orell Füssli.

Addresses

Bundesamt für Landestopographie (BLT)
Seftigenstrasse 264, CH-3084 WABERN.

Schweizerische Geologische Kommission (SGK)
Birmannsgasse 8, CH-4055 BASEL.

Bundesamt für Raumplanung (BARP)
Bundesrain 20, CH-3003 BERN.

Eidgenössische Forschungsanstalt für landwirtschaft Pflanzenbau (FAP)
Reckenholzstrasse 191, CH-8046 ZURICH-RECKENHOLZ.

Schweizerische Meteorologische Anstalt (SMA)
Krahbuhlstrasse 58, CH-8044 ZURICH.

Eidgenössische Drucksachen- und Materialzentrale
CH-3003 BERN.

Institut für Kartographie, Eidgenössische Technische Hochschule
Winterthurerstrasse 190, CH-8057 ZURICH.

Arbeitsgemeinschaft Geographica Bernensia
Hallerstrasse 12, CH-2012 BERN.

Orell Füssli Graphische Betriebe AG
Dietzingerstrasse 3, Postfach 8036 Zürich, CH-8003 ZÜRICH.

Kümmerly + Frey (K+F)
Hallerstrasse 6–10, CH-3001 BERN.

Hallwag AG
Nordring 4, CH-3001 BERN.

Schad + Frey AG
Kartographisches Institut, CH-3018 BERN.

Ex Libris AG
CH-7178 ZÜRICH.

Catalogue

Atlases and gazetteers

Atlas der Schweiz
First edition (1965–78) edited by E. Imhof
Second edition (1981–) edited by E. Speiss
Wabern: BLT.

Schweiz: Strassenatlas mit Ortsverzeichnis
Bern: K+F, 1984.
pp. 90.

Neues Schweizerisches Ortslexikon
München/Lucerne: Verlag C.J. Bucher GmbH, 1983.
pp. 347.

General

Landeskarte der Schweiz 1:1 000 000
Wabern: BLT.

Schulkarte Schweiz 1:500 000
Zürich: Orell Füssli, 1983.
Reverse has satellite images and thematic maps.

Landeskarte der Schweiz 1:500 000
Wabern: BLT, 1979.
Also available enlarged to 1:300 000.

Schweiz—physische Karte 1:500 000
Wabern: K+F, 1984.

Relief der Schweiz 1:300 000 Kartengemalde von E. Imhof
Wabern: BLT, 1982.

Schweiz 1:250 000
Bern: K+F, revised annually.
Official road map of Swiss Automobile Club.

Topographic

Landeskarte der Schweiz 1:200 000
Wabern: BLT, 1976–.
4 sheets, all published.

Landeskarte der Schweiz 1:100 000
Wabern: BLT, 1954–.
23 sheets, all published. ■

Landeskarte der Schweiz 1:50 000
Wabern: BLT, 1938–.
78 sheets, all published. ■

Landeskarte der Schweiz 1:25 000
Wabern: BLT, 1952–.
249 sheets, all published. ■

Geological and geophysical

Atlas structurel de la Suisse 2nd edn
Zurich: Ex Libris AG, 1986.

Geologische Karte der Schweiz 1:500 000 2nd edn
Basel: SGK, 1980.

Tektonische Karte der Schweiz 1:500 000 2nd edn
Basel: SGK, 1980.

Deklinationskarte der Schweiz 1:500 000
Basel: SGpK, 1979.

Inklinationskarte der Schweiz 1:500 000
Basel: SGpK, 1979.

Totalintensitätskarte der Schweiz 1:500 000
Basel: SGpK, 1979.

Schwere-Karte der Schweiz (Bouguer-Anomalien) 1:500 000
Basel: SGpK, 1979.

Schwere-Karte der Schweiz (Isostatische Anomalien) 1:500 000
Basel: SGpK, 1979.

Das Geoid in der Schweiz 1:500 000
Basel: SGpK, 1980.

Erdbebengefährdung in der Schweiz 1:500 000
Basel: SGpK, 1980.
Earthquake risk map.

Seismizitätskarte der Schweiz 1:500 000
Basel: SGpK, 1981.

Aeromagnetische Karte der Schweiz 1:500 000
Basel: SGpK, 1982.

Geothermische Karte der Schweiz 1:500 000
Basel: SGpK, 1982.

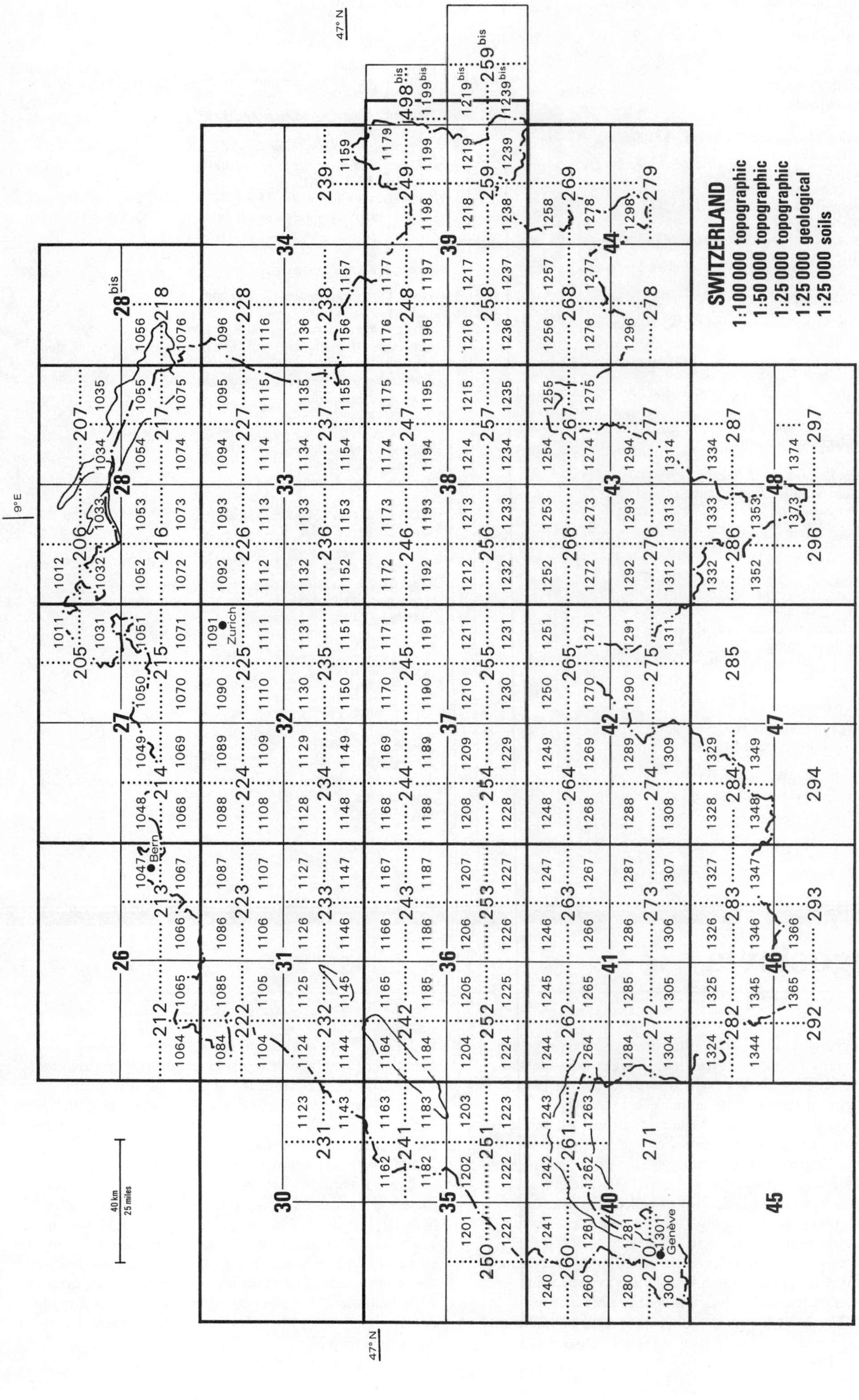

47° N

SWITZERLAND
1:100 000 topographic
1:50 000 topographic
1:25 000 topographic
1:25 000 geological
1:25 000 soils

40 km
25 miles

9° E

Zurich

Bern

Genève

47° N

9° E

Geologische Generalkarte der Schweiz 1:200 000
Basel: SGK, 1942–64.
Published in 8 sheets, 3 available.

Geotechnische Karte der Schweiz 1:200 000
Basel: SGtK, 1963–67.
4 sheets, all published.

Hydrogeologische Karte der Schweiz 1:100 000
Basel: SGtK, 1972–.
3 sheets published.

Geologischer Atlas der Schweiz 1:25 000
Basel: SGK, 1930–.
249 sheets, 83 published (some out of print). ■
Most with separate texts.

Environmental

Klimaatlas der Schweiz edited by W. Kirchhofer
Wabern: BLT, 1982–.

Karte der Lawinengefährdeten Gebiete 1:300 000
Zürich: Institut für Orts- Regional- und Landesplanung an der
ETHZ, 1970.
Avalanche risk map.
With explanatory brochure.

Bodeneignungskarte der Schweiz 1:200 000
Bern: Bundesamt für Raumplanung/Bundesamt für
Landwirtschaft/Bundesamt für Forstwesen, 1980.
4 sheets + 145 pp. text.
Package also includes 3 sheets at 1:50 000 and an extract of the
Bodenkarte at 1:50 000

*Klimaeignungskarten für die Landwirtschaft in der
Schweiz* 1:200 000
Bern: Geographica Bernensia, 1977.
4 sheets, all published.
Climate suitability for agriculture.

Gefahrenkarte der Schweiz 1:100 000
Bern: BARP, 1977.
22 sheets, all published. ■
Risk areas of Switzerland.

Schweiz—Bodenkarte 1:25 000
Zürich-Reckenholz: FAP, 1981–.
249 sheets, 5 published. ■

Administrative

Schweiz—politische Karte 1:500 000
Wabern: K+F, 1985.
Map of cantons.
Data on population, economy, trade and tourism on the reverse.

Schweiz. Kantone und Bezirke 1:375 000
Bern: Hallwag.

Die Gemeinden der Schweiz 1:200 000
Wabern: BLT, 1981.
4 sheets, all published.
Also available as single sheet reduced to 1:300 000 or 1:400 000.

Human and economic

Bahn-Karte der Schweiz 1:300 000
Bern: K+F, 1982.
Official map of Swiss Federal Railways.
On reverse: 8 town maps and map of international connections.

Strassenkarte der Schweiz 1:200 000
Wabern: BLT, 1982.
4 sheets, all published.

Karte der Kulturgüter 1:300 000
Wabern: BLT, 1970.
Map of cultural monuments.

Museumskarte der Schweiz 1:300 000
Wabern: BLT, 1982.

Burgenkarte der Schweiz 1:200 000
Wabern: BLT, 1978–.
4 sheets, all published.
Map of castles.

Town maps

Bern: Taschenplan
Zürich: Orell Füssli, 1985.

Zürich 1:15 500
Bern: Hallwag, 1985.
Inset: schematic map of mass transit system with index of
stopping places.
Street index.

Yugoslavia

The official topographic mapping agency in Yugoslavia is
the Military Geographical Institute (**Savezna Geodetska
Uprava, SGU**), Belgrade, which was founded in 1944.
The origins of the Institute go back to 1878, when a
Geographical Section of the General Staff was established
to carry out a topographic survey of Serbia. By the 1920s,
an extensive programme of triangulation and levelling was
in progress and topographic series at 1:50 000 and
1:100 000 were produced. This so-called 'Parisian'
mapping (because it used the Paris meridian) remained of
service until well after the war. However, soon after the
establishment of the new Institute in 1944, an entirely new
programme of topographic mapping was begun, using
photogrammetric techniques and tied to new survey
control. The basic mapping scale is now 1:25 000, and the
complete series of 3029 sheets was issued between 1959

and 1968. A second edition has subsequently been
completed, and further revision is currently in progress.
Derived series are published at 1:50 000 (the second
edition almost complete), 1:100 000 and 1:200 000. There
is also a 1:500 000 map in three sheets which is linked to
aeronautical and other thematic mapping, and three
1:1 000 000 sheets which conform to the IMW series.

Thus an entirely new system of mapping has replaced
the series available before, and immediately after, World
War II. The sheet lines are graticule based and refer to the
Greenwich Meridian. The 1:200 000 sheets are designated
by name, with numbers being added to identify sheets at
the larger scales. The maps are coloured, and are gridded,
with contours (20 m intervals at the larger scales), roads
colour classified and woodlands shown in green. The
1:500 000 map has hypsometric tints.

The maps of the Military Geographical Institute are designed primarily to meet the defence requirements of the Yugoslav military, and the majority of them are not available to the general public either within or outside the country. Their value for economic and social planning purposes is recognized, however, and the maps are distributed in large numbers to official non-military bodies. There are no restrictions on the distribution of maps at scales of 1:500 000 or smaller.

The Military Geographical Institute cooperates with the Federal Geological Bureau (**Zavod za geoloska i geofizicksa istrazivanja, GEOZAVOD**) in the production of geological maps. There is a basic series at 1:100 000, with some 163 sheets published, and a number of 1:500 000 scale earth science maps (geology, water resources, oil and gas deposits, metallogenic, mineral and thermal springs) most of the latter published in the 1980s. We have not been able to verify the general availability of these maps, though we have included them in the catalogue.

A 1:50 000 scale soil survey, *Pedoloska karta Jugoslavije*, is in progress.

Work is also in progress with the Federal Hydrometeorological Bureau (**Savezni hidrometeoloski zavod**) to produce a climatological atlas, of which several parts have already been issued, and vegetation and geomorphological maps are also in preparation.

The cartographic institute which produces maps for general public use is **Geokarta**, Belgrade. A number of general and thematic maps of the country as a whole are published as well as maps of the individual republics, and street maps of Belgrade.

The division of Yugoslavia into six republics and two autonomous provinces is reflected in the regionalization of much mapping. Most mapping at scales larger than 1:25 000 is carried out by civilian authorities on a Republic basis. A scale of 1:5000 has been adopted for populated areas and 1:10 000 for rural areas, and more than 30 per cent of the country has been covered at these scales. A series of Territorial Planning Atlases was published in the early 1980s by regional authorities, and some general tourist mapping is also published regionally.

Kartografija-Učila, Zagreb, was founded after World War II, and specializes in the production of educational maps and atlases. Some general maps of Yugoslavia (and of other countries) are also produced, and an atlas of SFR Yugoslavia is due to be published in 1987.

A series of 1:50 000 and 1:20 000 scale walking maps of parts of Slovenia has been published by **Planinska Zveza Slovenije**, Ljubljana/Maribor. These are detailed topographic maps, based on the military and regional civilian surveys, with 20 m contours and multilingual legends. They show marked walking routes, mountain huts, ski tows and other tourist information.

A number of foreign commercial publishing houses also produce good general tourist and motoring maps of Yugoslavia, including **Hallwag**, **Freytag und Berndt**, **Mair** and **Cartographia** (maps of Belgrade and Dubrovnik). Freytag and Berndt's series of *Wanderkarten* 1:100 000 also extend into the Julian Alps of northern Yugoslavia.

Addresses

Savezna Geodetska Uprava (SGU)
Dobrinjska ul 1D, YU-BEOGRAD.

Zavod za geoloska i geofizicka istrazivanja (GEOZAVOD)
48 Karadjordjeva, YU-11000 BEOGRAD.

Savezni hidrometeoloski zavod
6 Bircaninova, PO Box 604, YU-11000 BEOGRAD.

Geokarta Zavod za Kartografiju (Geokarta)
Bulevar Vojvode Misica 39, YU-11000 BEOGRAD.

Kartografija-Učila
Frankopanska 26, YU-41000 ZAGREB.

Graficki Zavod Hrvatske
Frankopanska 26, PO Box 227, YU-ZAGREB.

Planinska Zveza Slovenije
Planinska Za/ozba st. 24, Karto Izdelal IGF, YU-LJUBLJANA.

For Hallwag, see Switzerland; for Freytag and Berndt see Austria; for Mair, see German Federal Republic; for Cartographia, see Hungary; for CIA and DMA, see United States.

Catalogue

Atlases and gazetteers

Imenik mesta u Jugoslaviji
Beograd: Sluzbeni list SFRJ, 1973.
pp. 473.

Gazetteer of Yugoslavia: names approved by the United States Board on Geographic Names 2nd edn
Washington, DC: DMA, 1983.

General

Yugoslavia 1:1 860 000
Washington, DC: CIA, 1973.
Inset maps of population, land use and industry.

Jugoslavija. Geografska karta 1:1 000 000
Beograd: Geokarta, 1981.
Index of places on reverse.

SFR Jugoslavija 1:1 000 000
Zagreb: Kartografija Učila, 1984.
1:100 000 scale inset maps of major cities.

SFR Jugoslavija 1:1 000 000
Beograd: Geokarta.
1:100 000 scale inset maps of major cities.

Putna karta Jugoslavije razmjera 1:500 000 PK 500
Beograd: SGU, 1984.
3 sheets, all published.

Geological and geophysical

Geoloska karta Jugoslavije 1:500 000
Beograd: GEOZAVOD, 1970.

Hidrogeoloska karta SFR Jugoslavije 1:500 000
Beograd: GEOZAVOD, 1980.

Metalogenetska karta SFR Jugoslavije 1:500 000
Beograd: GEOZAVOD, 1983.

Karta mineralnih i termalnih voda SFR Jugoslavije 1:500 000
Beograd: GEOZAVOD, 1983.
Map of mineral and thermal waters.

Environmental

Atlas klime SFR Jugoslavije
Beograd: Savezni Hidrometeoroloski Zavod, 1971–.

Karta nalazista nafte i gasa SFR Jugoslavije 1:500 000
Beograd: GEOZAVOD, 1981.
Map of oil and gas deposits.

Administrative

Opstine u SFRJ 1 : 1 500 000
Beograd: 1980.

Human and economic

Jugoslavija industrija i rudarstvo 1 : 1 500 000
Beograd: Geokarta, 1978.
Map of mines and industry.

Karta Zeleznicke Mreze SFR Jugoslavija 1 : 800 000
Beograd: Geokarta, 1982.
Railway map.

Town maps

Plan Beograda 1 : 20 000
Beograd: Geokarta, 1980.

Beograd. Plan grada sa spiskom ulica 1 : 10 000
Beograd: Geokarta.
4 sheets, all published.

Zagreb. Plan grado po opcinama 1 : 15 000
Zagreb: Graficki zavod Hrvatske.

THE OCEANS

The oceans have been charted for navigational purposes by many nations, but such charts fall outside the range of this book, and in this and the following sections the maps discussed and catalogued are of two kinds: topographic and thematic maps of oceanic islands, and bathymetric, geological and environmental maps of the ocean basins themselves.

Interest in the production of maps of the form of the ocean floor—bathymetric maps—is international, and it was the co-sponsorship of two international bodies, the Intergovernmental Oceanographic Commission (IOC) and the **International Hydrographic Organization (IHO)**, that led to the preparation of the fifth edition of the *General Bathymetric Chart of the Oceans* (GEBCO). GEBCO was originally the brainchild of Prince Albert I of Monaco, who was responsible for publication of the first edition in 1903. The fifth edition consists of 16 sheets at 1:10 000 000 on the Mercator projection and two polar sheets at 1:6 000 000 on a Polar stereographic projection. There is also a single sheet of the whole world at 1:35 000 000. It was prepared from data supplied by 18 hydrographic offices compiled on 1:1 000 000 plotting sheets, and has been drawn up and published by the **Canadian Hydrographic Service (CHS)**. Digitization of the bathymetric contours of the GEBCO charts has been undertaken recently by the International Gravity Bureau, Toulouse, and the World Data Centre 'A', Marine Geology and Geophysics, at Boulder, Colorado has offered to distribute this data set. The hard copy GEBCO is available from IHO, Monaco, from the Department of Fisheries and Oceans, Canadian Hydrographic Service, or as a boxed set marketed by John Wiley and Sons. A sixth edition of GEBCO is planned for publication before the end of the century.

IHO is based at the International Hydrographic Bureau at Monaco and acts as a coordinating agency for the various national hydrographic offices. IHO provides a World Data Center for Bathymetry at Monaco which archives the unpublished 1:1 000 000 plotting sheets used for GEBCO.

In 1977, the Department of Navigation and Oceanography of the USSR Ministry of Defence also published a bathymetric chart of the oceans at the 1:10 000 000 scale. This is in eight sheets covering the world oceans between latitudes 82°N and 75°S on the Mercator projection. It was compiled from both Soviet and foreign cartographic data. The bottom configuration, shown by isobaths and relief shading, is said to be more detailed than on the GEBCO charts through use of a closer contour interval. However, the availability of this chart for export is in doubt.

Striking small-scale general maps of the oceans on a single sheet are provided by several publishers, including *The World Ocean Floor* by Bruce Heezen and Marie Tharp which is available in two versions of which the larger (1:23 230 300) has been painted by Heinrich Berann and Heinz Vielkind and gives a graphic wallmap-sized portrayal of seabed morphology.

The most comprehensive compilation of thematic ocean mapping is provided by the Russian *Atlas Oceanov*, prepared by the Naval Hydrographic Service of the USSR together with the Ministry of Defence and with the cooperation of institutes of the USSR Academy of Sciences and numerous other research institutions. This massive work is in three volumes, published 1975–80. It is distributed overseas by the Soviet foreign trade organization Mezhdunarodnaya Kniga, and may be obtained in the West from **Pergamon Press**.

At a less specialist level, the *Times Atlas of the Oceans*, published in 1983, presents a general introduction to the oceans through an integration of text and some 400 maps. There is a wide range of themes grouped broadly into sections on the ocean environment, ocean resources, trade and with a final section covering more disparate topics including pollution, strategic use of the sea, and the law of the sea. This atlas also includes some general bathymetric maps (scale not given) based on the GEBCO charts.

A stimulus to knowledge of the geology and tectonic structure of the oceans has come from the International Programme of Ocean Drilling (IPOD) in progress since 1975, and its predecessor, the Deep Sea Drilling Project (DSDP) begun in 1968 by the USA with the cooperation of several other countries. But, currently, the only sheet maps that synthesize the structural geology of the world's oceans as a whole are found in world or world series maps, rather than maps purely of the oceans. Thus the *Geological Atlas of the World* has three sheets (sheets 20–22), devoted to the Atlantic, Indian and Pacific Oceans, which draw on the findings of the DSDP, while the Russian *Tectonic Map of the World* 1:25 000 000 (1982) shows the structure of the oceans as well as the land masses, and the *Bedrock geological map of the world* 1:23 230 000 (1985) shows the ocean floor divided geochronologically. These maps are catalogued in the World section.

The *Atlas of the living resources of the sea*, published in a fourth edition in 1981 by the **Food and Agriculture**

Organization (FAO) of the United Nations, maps the distribution and migration patterns of the world's saltwater fish species of commercial importance. The atlas is available from FAO, Rome, or through the various national FAO agencies.

Currently there is a fast-growing interest and progress in ocean mapping stimulated both by the development of new mapping techniques (described in Chapter 2) and by the implications of the 1982 UNCLOS 3 convention, one outcome of which has been the declaration of Economic Exclusion Zones by more than 100 nations.

Many nations are now involved in mapping their new territorial waters and the continental shelf areas in considerable detail. For example, GLORIA II, the side-scan sonar system developed by the British Institute of Oceanographic Sciences, is currently engaged in a 6-year survey for the United States of the 5 million square miles (8 million km^2) of the US EEZ.

The USA was not party to the UNCLOS agreement but nevertheless declared a 200 mile Economic Exclusion Zone in 1983, and has set up an Ocean Assessments Division under the National Oceanic and Atmospheric Administration to carry out marine resources investigations. Publication of the findings will be in the form of a series of marine resource atlases issued by USGS and covering the entire EEZ of the USA.

Interest in coastal zone management has already resulted in the production of detailed bathymetric and resource mapping of coastal waters by a number of countries. We have included this kind of inshore mapping under the respective country headings. A number of resource and marine policy atlases have also been produced or are in preparation. The recent development of such multi-thematic mapping has coincided with the growth of computer-based Geographical Information Systems (GIS) for handling spatial data, and in future many marine environmental 'atlases' will be of the electronic kind, or will be produced from data held in digital form. It has been mooted that an FAO resource database should be developed to link these various national survey efforts.

Further information

The International Hydrographic Bureau at Monaco issues a list of publications and sells a *Catalogue of agents for sale of charts*, which includes details of the cartographic products of each of the IHO member states. A catalogue indexing the GEBCO bathymetric plotting sheets, defining the areas of responsibility accepted by each member state, and providing addresses from which the sheets may be obtained, may also be purchased from the IHB.

The GEBCO chart is discussed by Scott, D.P.D. (1981) Mapping the oceans' depth: the General Bathymetric Chart of the Oceans (GEBCO), *Nature and resources*, **17**, 6–9.

The Russian ocean atlas is described in the *First UN Regional Cartographic Conference for the Americas* (1979), **2**, 272–274.

The first 10 years of seabed survey using GLORIA I are described by Laughton, A.S. (1981) The first decade of GLORIA, *Journal of Geophysical Research*, **86**, 11511–11534. Subsequent survey by GLORIA II is described in Mapping the ocean floor (1982) *NERC Newsjournal*, **3**(1), 10–12.

An excellent overview of coastal zone mapping, with numerous map samples and an extensive bibliography, is provided by Perrotte, R. (1986) A review of coastal zone mapping, *Cartographica*, **23**(1 & 2), 3–72.

Addresses

International Hydrographic Organization (IHO)
7 Avenue Président J.F. Kennedy, BP 445, MC-98011 MONACO, Principauté de Monaco.

Canadian Hydrographic Service (CHS)
Chart Distribution Office, Department of Fisheries and Oceans, 1675 Russell Road, PO Box 8080, OTTAWA, Ontario K1G 3H6, Canada.

Pergamon Press Ltd
Headington Hill Hall, OXFORD OX3 08W, UK.

Pergamon Press Inc.
Fairview Park, ELMSFORD, New York 10523, USA.

Food and Agriculture Organization (FAO)
Distribution and Sales Section, Via delle Terme di Caracalla, I-00100 ROMA, Italy.

Celestial Products Inc.
Box 801, MIDDLEBURG, Virginia 22117, USA.

Celestial Arts
Box 7327, BERKELEY, California 94707, USA.

For Minsterstvo Oborony SSSR, see USSR; for AGS, GSA, NGS and DMA see United States; for CNR, see Italy.

Catalogue

Atlases and gazetteers

Atlas okeanov
Moskva: Ministerstvo Oborony SSSR, Vojenno-Morskoj Flot, 1975–80.
3 vols, pp. 302, 306 and 184.

The Times atlas of the oceans   edited by Alastair Couper
London: Times Books Ltd, 1983.
pp. 272.

Atlas of the living resources of the sea   4th edn
Roma: FAO, 1981.
24 pp. text + 69 pp. maps.

Gazetteer of undersea features   3rd edn
Washington, DC: DMA, 1981.
2 vols, each pp. 125.

General and bathymetric

Spilhaus whole ocean map   Athelstan Spilhaus
Middleburg, VA: Celestial Products, 1982.

Map of the ocean floor   Earl Bateman
Berkeley, CA: Celestial Arts, 1985.

The floor of the oceans   1 : 48 000 000   B.C. Heezen and M. Tharpe
New York: AGS/Washington, DC: Office of Naval Research, 1975.

World ocean floor   1 : 42 440 000
Washington, DC: NGS, 1981.

General bathymetric chart of the oceans   1 : 35 000 000
Ottawa: Canadian Hydrographic Service, 1984.
Reduced from the 1 : 10 000 000 GEBCO.

World ocean floor   1 : 23 230 300   B.C. Heezen and M. Tharpe
Washington, DC: Office of Naval Research, 1977.

General bathymetric chart of the oceans (GEBCO)   1 : 10 000 000
Ottawa: Canadian Hydrographic Service, Department of Fisheries and Oceans, 1975–82.
18 sheets, all published. ■
Arctic and Antarctic sheets scale 1 : 6 000 000.
Also available as boxed set with supporting text.

Geological and geophysical

Magnetic lineations of the oceans
Boulder, CO: GSA, 1974.

World seismicity map   1 : 48 000 000   A.F. Espinosa, W. Rinehart
and M. Tharpe
Washington DC: Office of Naval Research, 1981.

Administrative

Atlante dei confini sottomarini/Atlas of the seabed boundaries
Milano: Guiffrè Editore, 1979.
pp. 188.
Studi e documenti sul diritto internazionale del mare 5.

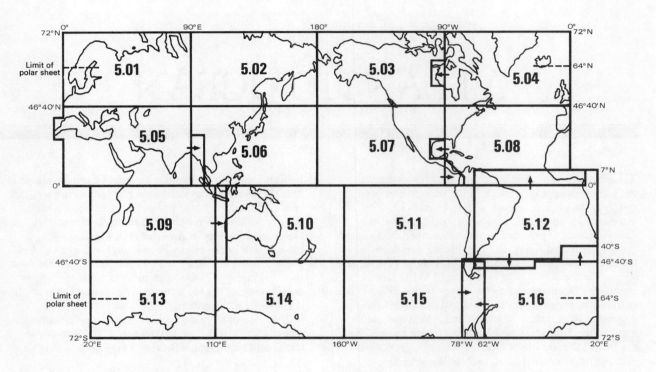

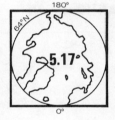

THE OCEANS
GEBCO 1 : 10 000 000
General Bathymetric Chart of the Oceans

ATLANTIC OCEAN

The islands of the Atlantic Ocean have been mapped by
the national mapping agencies of the various European
countries which have present or former colonial
associations with them. Thus topographic cover of the
Canary Islands is provided by an extension of the standard
Spanish mainland series, and in a similar way the maps of
the Azores and Madeira conform with standard Portuguese
series. The Faeroe Islands were mapped at the turn of the
century by Denmark and a new survey of these islands is
currently in progress. In the South Atlantic, St Helena,
Ascension Island, the Falkland Islands (Malvinas) and
many of the Sub-Antarctic islands have been mapped by
the British Directorate of Overseas Surveys (DOS/OSD).
Some of the remoter islands, such as Bouvet and
Inaccessible, have been surveyed only recently as part of
the research programmes of expeditions visiting them. The
small oceanic islands of Brazil (chiefly the Ilha Trindade/
Ilhas Martin Vaz and the Arquipélago de Fernando de
Noronha) fall within the Brazilian mapping and charting
programmes, while the islands of Bioko (formerly
Fernando Po) are within the mapping jurisdiction of
Equatorial Guinea. All Atlantic islands with significant
mapping are described and catalogued separately below.

The bathymetry and seabed geology of the Atlantic
Ocean itself are represented in World or World Ocean
series mapping. Thus four sheets of the fifth edition of
the *General Bathymetric Chart of the Oceans* (GEBCO)
1:10 000 000 provide bathymetric cover, while sheet 22 of
the *Geological Atlas of the World* 1:10 000 000 gives an
overview of the geology and tectonics of the whole ocean.

The most comprehensive mapping of the Atlantic in
atlas form is to be found in Volume 2 of the Russian *Atlas
okeanov*, which covers both the Atlantic and Indian
Oceans. The volume has six thematic sections devoted to
the history of exploration, ocean flow, climate, physical
properties of water, hydrochemistry and biogeography. A
seventh section consists of a series of general bathymetric
maps at a principal scale of 1:12 000 000. The atlas is
available with an English introductory text and gazetteer,
and with a legend for non-Russian-reading users.

The ICITA atlases published by UNESCO represent the
results of the International Co-operative Investigations of
the Tropical Atlantic.

Regional mapping of the Atlantic Ocean has been
carried out by several of the countries bordering it. An
extensive area of the north-east Atlantic is covered by a set
of four bathymetric charts prepared by the **Institute of
Oceanographic Sciences (IOS)** (UK) and published in
colour with isobaths at 200 m intervals. The French

**Institut Français de Recherche pour l'Exploitation de la
Mer (IFREMER)** has published geophysical maps of this
area. Sediment, tectonic and geophysical maps covering
large areas have been published *inter alia* by the
Geological Society of America (GSA) and the
Association of American Petroleum Geologists (AAPG).
More detailed mapping (1:1 000 000 and larger) has been
confined mainly to areas of continental shelf or to other
areas falling within the Economic Exclusion Zones recently
declared by many nations.

This kind of mapping includes the natural resource map
series (1:2 000 000, 1:1 000 000 and 1:250 000) of the
Canadian Hydrographic Service covering the waters

ATLANTIC OCEAN

marginal to Canada, the Brazilian continental shelf mapping (1 : 1 000 000) carried out under Project Ramec, the geological and geophysical mapping of the continental shelf adjacent to the British Isles (1 : 250 000) by the British Geological Survey and the bathymetric (1 : 250 000) and seabed (1 : 1 000 000) mapping carried out by the United States National Ocean Service. These series are described under the countries respectively responsible for them.

Addresses

For IOS and Hydrographic Department, see Great Britain; for NOAA, GSA, DMA, AAPG, NGS and US Naval Oceanographic Office, see United States; for UNESCO, see World; for GUGK, see USSR; for SHN, see Argentina; for IFREMER and BRGM, see France.

Catalogue

Atlases and gazetteers

World ocean atlas. Volume 2. Atlantic and Indian Oceans   1st edn
Moskva: Ministerstvo Oborony SSSR and Voenno-Morskoi Flot, 1978.
pp. 306.

ICITA oceanographic atlas. Equilant I and equilant II
Paris: UNESCO, 1973–76.
Volume 1: Physical oceanography, pp. 289.
Volume 2: Chemical and biological oceanography, pp. 358.

Oceanographic atlas of the North Atlantic Ocean
Washington, DC: US Naval Oceanographic Office, 1967–75.
6 vols.

South Atlantic Ocean. Official names approved by the US Board on Geographic Names
Washington, DC: DMA, 1957.
pp. 53.

General and bathymetric

Atlantic ocean   1 : 30 580 000
Washington, DC: NGS, 1973.

Atlanticeskij okean   1 : 20 000 000
Moskva: GUGK, 1981.

Carta batimétrica Atlántico Sur Occidental   1 : 5 000 000
Buenos Aires: SHN.

Bathymetry of the northeast Atlantic   1 : 2 400 000
Godalming: IOS, 1975–83.
4 sheets, all published.

Carte bathymétrique de l'Atlantique nord-est   1 : 2 400 000
Brest: IFREMER, 1985.

Geological and geophysical

South Atlantic Ocean and western part of the Indian Ocean. Magnetic variation 1985 and annual rates of change reduced to Epoch 1985.0   1 : 16 000 000   Chart 5376
Taunton: Hydrographic Department, 1985.

North Atlantic Ocean: bathymetry and plate tectonic evolution   1 : 8 753 909   MC-35
Boulder, CO: GSA, 1981.

Depth to basement/isopach map of sediments in the western North Atlantic Ocean   1 : 4 350 000
Tulsa, OK: AAPG, 1982.
2 sheets + text.

Brazil continental margin
Tulsa, OK: AAPG, 1977–79.
Set of 4 maps comprising bathymetry, free-air gravity anomaly, magnetic anomaly and sediment isopach map.

Argentine continental margin
Tulsa, OK: AAPG, 1977–78.
Set of 4 maps comprising bathymetry, free-air gravity anomaly, magnetic anomaly and sediment isopach maps.

Carte structurale du Golfe de Gascogne   1 : 2 400 000
Orléans: BRGM, 1982.
+ 32 pp. text.

Carte gravimétrique de l'Atlantique nord-est   1 : 2 400 000
Brest: IFREMER, 1981.

Carte magnétique de l'Atlantique nord-est   1 : 2 400 000
Brest: IFREMER, 1978.

Ascension Island

Ascension Island, in association with St Helena and the Tristan da Cunha group of islands, is a British Dependency, and has been mapped by the **Directorate of Overseas Surveys (DOS)** (now Overseas Surveys Directorate, Southampton, OSD). The principal map, at 1 : 25 000, was first published in 1964, and was compiled mainly from large-scale American maps of the island. A partially reconstructed second edition was issued in 1967. The map is on a Transverse Mercator projection with UTM grid. It is printed in three colours, and contours are given at 100 foot intervals.

Addresses

Overseas Surveys Directorate (DOS/OSD)
Ordnance Survey, Romsey Rd, Maybush, SOUTHAMPTON SO9 4DH, UK.

Catalogue

General

Ascension Island   1 : 200 000   DOS 977   Edition 3
Tolworth: DOS, 1972.

Topographic

Ascension Island   1 : 25 000   DOS 327 (Series G892)   Edition 2
Tolworth: DOS, 1967.

Azores

The Azores are an autonomous region of Portugal. Topographic mapping has been carried out by the **Instituto Geográfico e Cadastral (IGCP)**, Lisbon, and a complete cover at 1:50 000 in 10 sheets is available, the latest editions being from 1965 to 1971. These are photogrammetric maps and are printed in six colours. Projection is UTM and contours are at 25 m intervals. More recently, a cover of the archipelago in the 1:25 000 military series (M889) has been completed. This series is produced by the **Serviço Cartográfico de Exército (SCEP)**, Lisbon. Sheets are in six colours on a UTM projection and with a 10 m contour interval.

Topographic cadastral maps are available for parts of some islands, notably Ilha de S. Miguel and Ilha Santa Maria, and orthophoto mapping has commenced in Graciosa.

Geological maps of the archipelago were issued by the **Serviços Geológicos de Portugal** over the period 1958 to 1971. All islands are covered at either 1:50 000 or 1:25 000 and each by a single sheet, with the exception of São Jorge, São Miguel and Pico which each have two sheets.

Further information

Topographic and geological mapping of the Azores is listed in the catalogues issued by the IGCP, SCEP and SGP respectively.

Addresses

Instituto Geográfico e Cadastral (IGCP)
Praça da Estrela, 1200 LISBOA, Portugal.

Serviço Cartográfico do Exército (SCEP)
Av. Dr. A. Bensaude, 1800 LISBOA, Portugal.

Serviços Geológicos de Portugal (SGP)
Rua Academia das Ciências 19, 2, 1200 LISBOA, Portugal.

Catalogue

General

Arquipélago dos Açores   1:1 000 000
Lisboa: IGCP, 1965.
IMW special sheet.

Topographic

Arquipélago dos Açores   1:50 000   Series M7811
Lisboa: IGCP, 1965–71.
10 sheets, all published.

Arquipélago dos Açores   1:25 000   Series M889
Lisboa: SCEP, 1958–84.
35 sheets, all published. ■

Geological and geophysical

Carta geológica de Portugal   1:50 000/1:25 000
Lisboa: SGP, 1958–71.
12 sheets, all published.

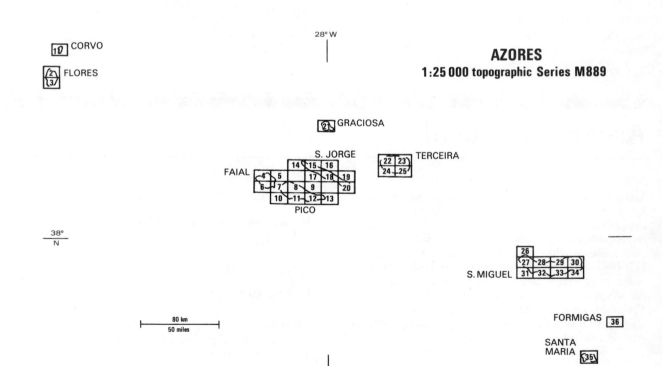

Bermuda

Bermuda, a British crown colony with its own constitution and internal government since 1968, was mapped in the early 1960s by the **Directorate of Overseas Surveys (DOS)** (now Overseas Surveys Directorate, Southampton, OSD) of the British Government.

The islands were mapped in a 74-sheet series at 1:2500 (DOS 111, now Bda 111) with 10 foot contours. A derived map at 1:10 560 (6 inches to the mile) in six sheets was also issued in the mid-1960s (DOS 311, now Bda 311). Projection is Transverse Mercator with UTM grid, and contours are at 10 foot intervals. Both map series were revised and published in second editions in the 1970s for the Bermuda Government, and a further revision took place in the 1980s, the contract for this work having been let to a British company in 1983 with arrangements for DOS to monitor the progress. The 1:10 560 series is being enlarged to 1:10 000 for this edition.

DOS has also issued a single-sheet tourist map at 1:31 680 (Bda 411), with contours at 50 feet. This includes inset maps of Hamilton and St George.

A gazetteer indexing the names on the 1975 1:2500 map was published by **G. J. Rushe** in 1978.

Addresses

Public Works Department
PO Box HM525, HAMILTON 5, Bermuda.

Overseas Surveys Directorate (OSD/DOS)
Ordnance Survey, Romsey Rd, Maybush, SOUTHAMPTON SO9 4DH, UK.

G.J. Rushe
PO Box 1271, HAMILTON 5.

Catalogue

Atlases and gazetteers

The Bermuda gazetteer
Hamilton: G.J. Rushe, 1978.

General

Bermuda 1:31 680 Bda 411
Tolworth: DOS for Bermuda Government, 1975.
Inset maps of Hamilton and St George.

Topographic

Bermuda 1:10 560 Bda 311 Edition 2
Tolworth: DOS for Bermuda Government, 1975.
6 sheets, all published.

Bouvet Island

Bouvet Island is a small island in the South Atlantic belonging to Norway and has been provisionally mapped by the **Norsk Polarinstitutt (NP)**, Oslo. The published map is in black and white with a 20 m contour interval. The map is distributed by **Statens Kartverk**, Hønefoss.

Further information

The mapping of Bouvet Island is discussed in Norsk Polarinstitutt *Skrifter* **175** (1981), and this publication includes two maps.

Addresses

Norsk Polarinstitutt
Rolfstangveien, Postboks 158, N-1330 OSLO LUFTHAVN, Norway.

Statens Kartverk
N-3500 HØNEFOSS, Norway.

Catalogue

Bouvetøya 1:20 000
Oslo: NP, 1981.
Provisional edition.

Canary Islands

The Canary Islands have been mapped by the two Spanish topographic mapping authorities, the **Instituto Geográfico Nacional (IGN)** and the **Servicio Geográfico del Ejercito (SGE)**, Madrid. The most detailed topographic mapping of the archipelago is at 1:50 000, cover being provided by the IGN in 42 sheets, and by SGE in 36. Most sheets currently available in the IGN series are of the first edition dating from the 1940s through to the 60s. These are on a polyhedric projection and are printed in five colours with 20 m contours. Since 1979, second edition sheets have begun to appear which conform to the new Spanish 1 : 50 000 specification and there is a modified sheet system for this edition. Projection is UTM and the sheets have relief shading. The SGE cover dates from the 1970s and is also on the new sheet lines. These sheets form part of the *Mapa Militar de España*, Series L, and are in six colours on the UTM projection.

IGN also publishes an *Edición para el Turismo* 1 : 50 000 of some individual islands. There are maps of Hierro, Gomera and La Palma in this series. All are on a UTM projection and have 20 m contours and shaded relief.

The Canaries are also covered in standard Spanish IGN and SGE series at smaller scales.

Geological mapping has been carried out by the **Instituto Geológico y Minero de Espana (IGME)**, Madrid. The original 1 : 50 000 *Mapa geológico* covered Tenerife, Lanzarote and Fuerteventura, but some sheets are no longer in print. This series uses the same sheet system as the older IGN topographic series. The Canaries are included in the new Spanish geological mapping programme *Proyecto MAGNA*. So far, a large part of Tenerife has been mapped by this project with publication at 1 : 25 000.

A series of 1 : 100 000 geological maps of the individual islands was prepared in the 1960s by IGME in cooperation with the Instituto Lucas Mallada de Investigaciones Geológicas CSIC. The four sheets published are still available from IGME. A series of hydrogeological maps of La Palma at 1 : 25 000 (*Proyecto Canarias SPA-15*) was published in 1974 by the Dirección General de Obras Hidraulicas, Madrid.

Sheets covering the Canaries in the series *Mapa geotecnico general* 1 : 200 000 and *Mapa de rocas industriales* 1 : 200 000 are also available.

Maps of National Parks on the Canaries have been published by the **Instituto Nacional para la Conservación de la Naturaleza**, under the Ministério de Agricultura, Pesca y Alimentación, Madrid. These contoured sheets are at scales varying from 1 : 20 000 to 1 : 30 000, and have tourist information on the reverse.

There are a number of good tourist maps of the Canaries published by external commercial publishers, including **Firestone Hispania, Clyde Surveys, Mair** and **Karto+Grafik. Aguilar** has published an *Atlas gráfico de las Islas Canarias*.

Addresses

Instituto Geográfico Nacional (IGN)
General Ibanez de Ibero 3, 28071 MADRID, Spain.

Servicio Geográfico del Ejercito (SGE)
C/Dario Gazapo 8, 28024 MADRID, Spain.

Instituto Geológico y Minero de Espana (IGME)
Rios Rosas 23, 28003 MADRID, Spain.

Instituto Nacional para la Conservación de la Naturaleza
Ministério de Agricultura, Pesca y Alimentación, Servicio de Publicaciones, Paseo de Infanta Isabel, 1, 28014 MADRID, Spain.

For Firestone Hispania and Aguilar, see Spain; for Clyde Surveys, see Great Britain; for Mair and Karto+Grafik, see German Federal Republic.

Catalogue

Atlases and gazetteers

Atlas gráfico de las Islas Canarias
Madrid: Aguilar, 1979.
pp. 63.

General

Canary Islands   scales vary
Maidenhead: Clyde Surveys Ltd. 1985.
Double-sided map.
Maps of each island individually and street maps of towns.
Legend in English, Spanish and German.

Generalkarte   1 : 150 000
Stuttgart: Mair.
2 sheets cover all the islands individually.
German text on reverse.

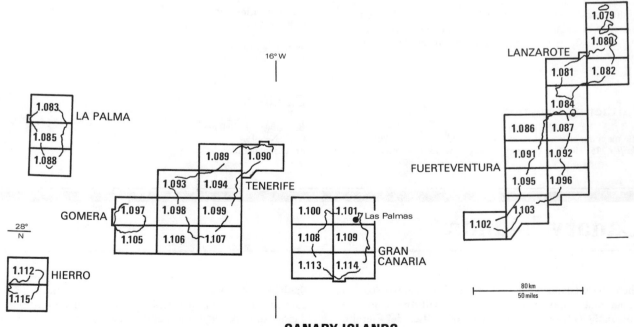

CANARY ISLANDS
1:50 000 topographic Series L

Canarias 1:500 000 Series 1404 World
Madrid: IGN.

Mapas provinciales 1:500 000
Madrid: IGN, 1978.
2 sheets, both published.

Mapa oficial de España—conjuntos provinciales 1:200 000
Madrid: IGN, 1972.
2 sheets, both published.

Topographic

Mapa militar de España 1:400 000 Series 4C
Madrid: SGE, 1972.
3 sheets cover the archipelago, all published.

Mapa militar de España 1:200 000 Series 2C
Madrid: SGE, 1971–.
6 sheets cover the archipelago, all published.

Mapa militar de España 1:1 000 000 Series C
Madrid: SGE, 1972–.
14 sheets cover the archipelago, all published.

Mapa topográfico nacional 1:50 000
Madrid: IGN, 1947–.
42 sheets, all published. ■
New edition in new format in progress since 1981.

Mapa militar de España 1:50 000 Series L
Madrid: SGE, 1975–.
36 sheets, all published. ■

Geological and geophysical

Mapa geotécnico general 1:200 000
Madrid: IGME, 1974–.
6 sheets, 4 published.

Mapa de rocas industriales 1:200 000
Madrid: IGME, 1973.
6 sheets, 4 published.

Mapa geológico nacional 1:50 000 1a Serie
Madrid: IGME, 1958–68.
42 sheets, 20 available. ■

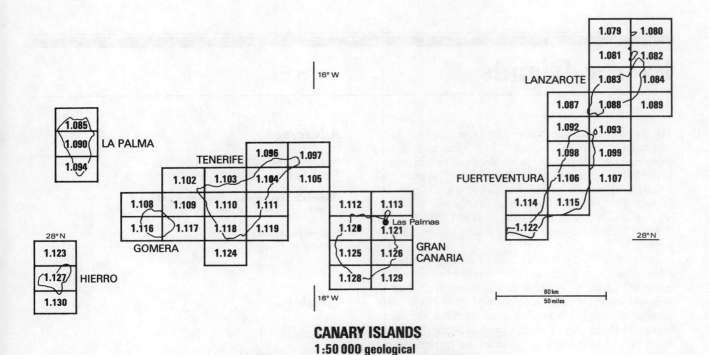

CANARY ISLANDS
1:50 000 geological
1:50 000 old 'Mapa topografico'

Cape Verde Islands

Formerly an overseas province of Portugal, the Cape Verde Islands became an independent republic in 1975. They have been mapped at 1:25 000 by the **Serviço Cartográfico do Exército (SCEP)**, Lisbon. The series began in 1968, and since 1976 has continued through a cooperative mapping agreement between Portugal and the República Popular de Cabo Verde. The maps are on UTM projection with 10 m contours. Several islands (Santo Antão, Boa Vista, Maio, Rogo, Brava) have still to be mapped in this series. At present these maps are not on sale to the public.

A general map of the archipelago is provided by the Tactical Pilotage Chart 1:500 000 published by the Defense Mapping Agency (DMA) of the USA.

The **Instituto de Investigação Científica Tropical (IICT)**, Lisbon, has undertaken thematic mapping in a number of the islands in collaboration with the Ministério do Desenvolvimento Rural da República de Cabo Verde (and partly with the participation of the Instituto para a Cooperação Económica, ICE).

Addresses

Serviço Cartográfico do Exército (SCEP)
Avenida Dr Alfredo Bensuade, 1800 LISBOA, Portugal.

Instituto de Investigação Científica Tropical (IICT)
Rua da Junqueira 86, 1300 LISBOA, Portugal.

Servicio de Topografia
Praia, SAO TIAGO.

For DMA see United States; for GUGK see USSR.

Catalogue

General

Ostrova Zelenogo Miysa 1:750 000
Moskva: GUGK.

Cape Verde Islands 1:500 000 TPC K-OA
Washington, DC: DMA, 1979.

Topographic

Carta hipsométrica da Ilha de Santiago 1:50 000
Lisboa: IICT/ICE, 1983.

Carta hipsométrica da Ilha do Fogo 1:50 000
Lisboa: IICT/ICE.

Geological and geophysical

Carta geológica da Ilha de S. Nicolau 1:50 000
Lisboa: IICT, 1983.

Environmental

Carta de zonagem agro-ecológica e da vegetação da Ilha de Santiago 1:50 000
Lisboa: IICT/ICE, 1986.

Carta de zonagem agro-ecológica e da vegetação da Ilha do Fogo 1:50 000
Lisboa: IICT/ICE, 1987.

Faeroe Islands

The Faeroe Islands were mapped by Danish surveyors, who began work in 1895 with a 1:20 000 series which was completed by 1901 and which was followed by a map at 1:100 000 completed in 1916. These plane table maps, in a revised form, are the basis of the current maps of the Faeroes. However, new surveying of the islands is in progress which will lead to a new series at 1:20 000 in 37 sheets. This new mapping is using aerial photogrammetric survey.

The present 1:20 000 series is in 53 sheets printed in six colours, with a contour interval of 10 m (or 5 m in cultivated areas). No graticule or grid is indicated. The style and content are similar to those of the old plane table series of Denmark. Placenames are in Faeroese with the towns and villages also spelt in Danish, and the legend is in Danish. Field revision of this series was carried out in 1971.

The current 1:100 000 map is available in two versions. There is a two-sheet edition with legend and glossary in Faeroese, Danish and English. This edition is available with or without UTM grid. There is also a single sheet edition, which is entirely in Faeroese but includes an index of places. The contour interval is 25 m. The 1:200 000 map, first published in 1932, is also in Faeroese, has a 50 m contour interval and is available with or without UTM grid. The topographic maps of the Faeroes are all published by the **Geodætisk Institut (GID)**, Copenhagen.

Geological mapping has been undertaken by **Danmarks Geologiske Undersøgelse (DGU)**, Copenhagen, and bedrock geology maps at 1:50 000 and 1:200 000 were published together with monographs as Nos 24 and 25 in the DGU *I. Række*. The Faeroes will also be covered by two sheets in the 1:250 000 UTM series of Britain and the continental shelf which is being published primarily by the British Geological Survey (for index, see Great Britain).

A bathymetric map with 10 m isobaths covering the Faeroe shelf has been published by **Føroya Jardfrødishavn**.

Addresses

Geodætisk Institut (GID)
Rigsdagsgården 7, DK-1918 KØBENHAVN K, Denmark.

Danmarks Geologiske Undersøgelse (DGU)
Thoravej 31, DK-2400 KØBENHAVN NV, Denmark.

Føroya Jardfrødisavn
TORSHAVN

Catalogue

General

Færøerne 1:1 000 000 ICAO 2106
København: GID.
Compiled to ICAO specifications but without aeronautical information.
Covers Faeroes and Shetlands.

Bathymetric

Føroyar. Bathymetric map of the Faroe shelf to 200 m depth based on echosoundings along Decca lines 1:243 200
Torshavn: Føroya Jardfrødisavn, 1977.
Land areas in outline only.

Topographic

Føroyar 1:200 000
København: GID, 1977.

Føroyar 1:100 000
København: GID, 1986.
Available as single sheet or two sheets (1983).

Færoerne 1:20 000
København: GID, 1972–74.
53 sheets, all published.

Geological and geophysical

Geology of the Faeroe Islands (Pre-Quaternary) 1:200 000
København: DGU, 1970.
Available with or without 142 pp. text.

Kortbladet Færoerne (Prækvarteret) 1:50 000
København: DGU, 1969.
6 sheets, all published.
Available with or without 370 pp. text.

Falkland Islands

Aerial photographic cover of the Falkland Islands was obtained in 1956 during the Falkland Islands and Dependencies Aerial Survey Expedition mounted by the British Government and carried out by Huntings Aerosurveys. This, in conjunction with a ground network established by triangulation and tellurometer survey during the following years, was used by the **Directorate of Overseas Surveys (DOS)** (now Overseas Surveys Directorate, OSD) to produce a 1:50 000 map in 29 sheets (DOS 453/Series H791). This map is on a Transverse Mercator projection with UTM grid and 50 foot contours, and was issued in 1962–63 with subsequent revision to some sheets. It is not at present on sale to the public.

The derived map at 1:250 000 (DOS 653) was first published in 1964 and a revised edition was issued in 1977. It has layered relief and a 500 foot contour interval.

The Falkland Islands (Malvinas) are also included in the publication programme of the **Instituto Geográfico Militar (IGM)**, Buenos Aires. Published maps available are sheet 5260 in the 1:500 000 series and an inset to sheet SN-20 in the Argentine 1:1 000 000 IMW series.

The 1:250 000 geological map was surveyed by the British Antarctic Survey under Dr R. J. Adie in 1969–72. Photogeological interpretation was carried out from the 1956 air photography.

A 1:2500 plan of Port Stanley, first produced by the War Office in 1943, was revised by DOS in the 1960s.

Addresses

Overseas Surveys Directorate (DOS)
Ordnance Survey, Romsey Road, Maybush, SOUTHAMPTON SO9 4DH, UK.

Instituto Geográfico Militar (IGM)
Av. Cabildo 381, BUENOS AIRES, Argentina.

Anthony Nelson
PO Box 9, OSWESTRY SY11 1BY, UK.

Ryder Geosystems
445 Union Street, Suite 304, DENVER, Colorado 80228, USA.

Catalogue

General

Satellite photomap of the Falkland Islands   ca 1:728 640
Denver, CO: Ryder Geosystems, 1982.

Falkland Islands   1:643 000   DOS 906   Edition 2
Tolworth: DOS, 1986.

Islas Malvinas   1:500 000
Buenos Aires: IGM, 1974.

The Falkland Islands   1:500 000
Oswestry, Shropshire: Anthony Nelson, 1981.

Topographic

Falkland Islands   1:250 000   DOS 653   Edition 2
Tolworth: DOS, 1977.
2 sheets, both published.

Geological and geophysical

Geological map of the Falkland Islands   1:250 000   DOS 1185A & B
Tolworth: DOS, 1972.
2 sheets, both published.

Madeira

The Archipelago of Madeira is an autonomous region of Portugal, and topographic mapping has been carried out by the **Instituto Geográfico e Cadastral (IGCP)**, Lisbon. A 1:50 000 map covers the islands in four sheets (two covering the Ilha da Madeira, one the Ilha do Porto Santo, and one the Ilhas Desertas e Ihas Selvagens). The projection is UTM and the map is in six colours with contours at 25 m. Recently, the 1:25 000 scale military series of the **Serviço Cartográfico do Exército (SCEP)** has become available. The islands are covered in 16 sheets (nine cover Madeira itself) on UTM projection in six colours and with 10 m contours.

Much of the main island is also covered by topographic–cadastral maps at scales ranging from 1:500 to 1:5000.

Geological maps of Madeira have been published by the **Serviços Geológicos de Portugal (SGP)**. The Ilha da Madeira (1:50 000) is in two sheets published in 1974, while Ilha Desertas (1:50 000) was published in 1972 and Ilha Selvagens (1:25 000) in 1978. There is at present no geological map of Ilha de Porto Santo.

Clyde Surveys publishes a tourist map of Madeira in English, German and Portuguese.

Addresses

Instituto Geográfico e Cadastral (IGCP)
Praça da Estrela, 1200 LISBOA, Portugal.

Serviço Cartográfico do Exército (SCEP)
Av. Dr. A. Bensaude, LISBOA, Portugal.

Serviços Geológicos de Portugal (SGP)
Rua Academia das Ciências 19, 2, 1200 LISBOA, Portugal.

For Clyde Surveys, see Great Britain.

Catalogue

General

A leisure map of Madeira   1 : 70 000
Maidenhead: Clyde Surveys Ltd, 1984.
Includes Porto Santo and street map of Funchal.

Topographic

Arquipélago da Madeira   1 : 50 000   Series P722
Lisboa: IGCP, 1970–71.
4 sheets, all published.

Arquipélago da Madeira   1 : 25 000   Series P821
Lisboa: SCEP, 1967–76.
16 sheets, all published.

Geological and geophysical

Carta geológica de Portugal   1 : 50 000/1 : 25 000
Lisboa: SGP, 1972–.
5 sheets to cover the *Arquipélago*, 4 published.

St Helena

St Helena, together with Ascension Island and Tristan da Cunha, forms a British Dependency and has been mapped by the Directorate of Overseas Surveys (DOS) (now **Overseas Surveys Directorate, OSD**) of the British Government. An original triangulation of the island had taken place in 1903 and was used as a basis for military survey, but a new triangulation was observed by DOS in 1971 using EDM methods to provide control for a 1 : 10 000 series compiled from helicopter photography flown in 1970. The 1 : 10 000 map (DOS 260) was published in 1974 in six sheets, on a Transverse Mercator projection. It is in five colours with contours at 1 : 10 000 (DOS 269). A derived map at 1 : 25 000 was also published in 1974 as Edition 4 of DOS 360. The map is in five colours with 40 m contour interval. An Edition 5 was issued with UTM grid, and Edition 6, published in 1983, has some amendments to contours.

The 1 : 10 000 series was used as a base map for a forestry and development project carried out by the **Land Resources Development Centre (LRDC)** and published as *Land Resource Report* 2 (the forestry maps accompanying the report are DOS 3025). In 1980, a new *Land Resource Study of St Helena* (LRS 32) was published by LRDC. This includes five 1 : 25 000 scale maps, available as separates, of geology, land units, soils, vegetation and re-afforestation as well as a three-sheet land use series at 1 : 10 000.

In 1982, a geomorphological map of St Helena was compiled by an expedition from the Department of Geography, University College London. The map, scale 1 : 10 000, was not published but is on open file at UCL.

Addresses

Legal and Lands Department
Government of St Helena, JAMESTOWN, St Helena.

Overseas Surveys Directorate (DOS/OSD)
Ordnance Survey, Romsey Rd, Maybush, SOUTHAMPTON SO9 4DH, UK.

Land Resources Development Centre (LRDC)
Tolworth Tower, SURBITON KT6 7DY, UK.

Catalogue

Topographic

St Helena   1 : 25 000   DOS 360   Edition 6
Tolworth: DOS, 1983.

Environmental

St Helena agriculture and forestry   1 : 50 000   DOS (Misc) 234
Tolworth: DOS, 1957.

The land resources of St Helena   1 : 25 000/1 : 10 000
Tolworth: LRDC/DOS, 1980.
9 sheets, all published.
Maps accompany *Land Resource Study of St Helena* (LRS 32).

St Pierre and Miquelon

The islands of St Pierre and Miquelon lie only a score of kilometres off the coast of Newfoundland but they form an overseas department of France. Triangulation of the islands was carried out in the early 1950s and linked to Newfoundland, and the **Institut Géographique National (IGN)**, Paris, published a four-colour 1 : 20 000 scale map of the islands in six sheets in 1955. The projection is UTM and the contours are at 5 m intervals. A derived single-sheet 1 : 50 000 map with 10 m contours was issued in 1957.

New air photo cover was obtained in 1985, and preparation of a new two-sheet map at 1 : 25 000 to the French *Série bleue* specification is in hand.

Address

Institut Géographique National (IGN)
136 bis, rue de Grenelle, F-75700 PARIS, France.

Catalogue

Topographic

Carte des Iles St Pierre et Miquelon 1 : 50 000
Paris: IGN, 1957.

Carte des Iles St Pierre et Miquelon 1 : 20 000
Paris: IGN, 1955.
6 sheets, all published.

São Tomé e Principe

São Tomé e Principe, two islands off the coast of West Africa in the Gulf of Guinea, have been an independent republic since 1975. Formerly they were an overseas province of Portugal.

Some topographic and soils mapping was done by the Portuguese prior to independence. Currently, a simple 1 : 75 000 general map of the two islands is available, published by the **Centro de Informação e Turismo de S. Tomé e Principe**. On the reverse are street maps of the Cidade de São Tomé and the Cidade de Santo Antonio.

Address

Centro de Informação e Turismo de S. Tomé e Principe
SÃO TOMÉ.

Catalogue

General

São Tomé e Principe 1 : 75 000
S. Tomé: Centro de Informação e Turismo de S. Tomé e Principe, 1971.

South Georgia

South Georgia was mapped from exploratory survey by the South Georgia Survey Expeditions between 1951 and 1957. The 1 : 200 000 scale map was first issued in 1958 to accompany a British Antarctic Survey Scientific Report *The History of Place-Names in the Falkland Islands Dependencies (South Georgia and the South Sandwich Islands)*. The 1980 monochrome edition includes minor amendments to geographical names. The projection is Lambert conic orthomorphic and contours are at 500 foot intervals. The island is also included in the 1 : 250 000 mapping programme of the British Antarctic Survey (BAS 250), but the sheet has not yet been published.

Partial mapping of South Georgia includes a 1 : 25 000

map of Royal Bay surveyed by the Combined Services Expedition in 1964–65 and printed in colour with 100 foot contours, and a 1 : 50 000 geomorphological map of the *St Andrews Bay—Royal Bay area*, based on field work by C. M. Clapperton and D. E. Sugden of Aberdeen University in 1975 and published in 1980 as BAS(Misc) Sheet 1.

There is a 1 : 500 000 scale map of the island published by the **Instituto Geográfico Militar (IGM)**, Buenos Aires.

Sheet 1 of the **British Antarctic Survey** 1 : 500 000 geological series (BAS 500G) covers South Georgia, together with the South Orkney and South Sandwich Islands. This series is listed under Antarctica.

Addresses

Overseas Surveys Directorate (DOS/OSD)
Ordnance Survey, Romsey Rd, Maybush, SOUTHAMPTON
SO9 4DH, UK.

British Antarctic Survey (BAS)
High Cross, Madingley Rd, CAMBRIDGE CB3 0ET, UK.

Instituto Geográfico Militar (IGM)
Av. Cabildo 381, BUENOS AIRES, Argentina.

Catalogue

General

Islas Georgias del Sur 1 : 500 000
Buenos Aires: IGM, 1971.

Falkland Island Dependencies. South Georgia 1 : 200 000 DOS
610A
Tolworth: DOS, 1980.
Monochrome map.

South Sandwich Islands

The South Sandwich Islands are a string of small volcanic
islands forming part of the Scotia Arc between South
Georgia and the Antarctic Peninsula. They fall within the
1 : 250 000 scale mapping programme of the **British
Antarctic Survey (BAS)**, but the sheets are not yet
published. A 1 : 500 000 scale map in the obsolete series
DCS 701 is out of print, but the islands are covered by two
sheets of the Argentine 1 : 500 000 series, published by the
Instituto Geográfico Militar (IGM), Buenos Aires.

The South Sandwich Islands are also included (together
with South Georgia and the South Orkneys) in Sheet 1 of
the British Antarctic Survey's 1 : 500 000 geological map
(BAS 500G), listed under Antarctica.

Addresses

Instituto Geografico Militar (IGM)
Av. Cabildo 381, BUENOS AIRES, Argentina.

British Antarctic Survey (BAS)
High Cross, Madingley Rd, CAMBRIDGE CB3 0ET, UK.

Catalogue

General

Islas Sandwich del Sur 1 : 500 000
Buenos Aires: IGM, 1971.
2 sheets (5727, 5927), both published.

Tristan da Cunha

The **Directorate of Overseas Surveys (DOS)** (now
Overseas Surveys Directorate, OSD) of the British
Government has published a small map of the main island
at 1 : 140 000 scale and with shaded relief. A 1 : 30 000 map
with 100 foot contours (DOS Misc 323) was prepared by
DOS in 1962 but not published.

The second largest island of the Tristan group,
Inaccessible Island, was surveyed accurately for the first
time in 1983 by an expedition from Denstone College,
UK. The island's position was fixed by satellite navigation
equipment on the ship which landed the party, and a map
at 1 : 10 000 was compiled by ground survey.

A geological map of Gough island was published by
DOS in 1958 at a scale of about 1 : 50 000. This
accompanies a monograph, *The geology of Gough Island,
South Atlantic*, by R. W. LeMaitre, published in the series
Overseas Geology and Mineral Resources, Volume 7. A
topographic survey of this island was carried out by a
research expedition which visited it in 1955, and a
1 : 40 000 formlined map accompanies the report of the
expedition given in the *Geographical Journal* in 1957, but is
not available separately.

Further information

The surveying of Inaccessible Island is described in Siddal, C.P.
(1985) Survey of Inaccessible Island, Tristan da Cunha Group,
Polar Record, **22**, 528–531.

The survey of Gough island mentioned above is described in
Heaney, J.B. and Holdgate, M.W. (1957) The Gough Island
scientific survey, *Geographical Journal*, **123**, 20–32.

Catalogue

General

Tristan da Cunha 1 : 140 000 Edition 2
Tolworth: DOS, 1972.

Geological and geophysical

Gough Island 1 : 50 000 DOS (Geol) 1095
Tolworth: DOS, 1958.
Geology map.

MEDITERRANEAN

There are relatively few maps which treat the
Mediterranean as a region, although the **Institut
Géographique National (IGN)**, Paris, has published two
such general maps, and the **National Geographic Society
(NGS)**, Washington DC, has a colourful physiographic
map of the seafloor. Of the two UNESCO maps of
bioclimatic regions and vegetation published in 1963 and
1970, the former is now out of print, but a new four-sheet
map of potential vegetation is in progress by the **Centre
National de Recherche Scientifique (CNRS)**, Paris; the
first sheet and monograph are available.

An important *International Bathymetric Chart of the
Mediterranean* at 1 : 1 000 000 has been published by the
Soviet Department of Navigation and Oceanography
(GUNO), Leningrad, under the authority of the
Intergovernmental Oceanographic Commission of
UNESCO. This has isobaths at 50, 100, 200, 300, 400, 500
and 1000 m. The original plotting sheets for this chart are
archived in the International Hydrographic Bureau at
Monaco (World Data Centre for Bathymetry). The IBCM
is to be used as a base for the production of geological and
geophysical maps which are currently in preparation.
There are also plans for a second edition of the IBCM.

A digital bathymetric map of the Mediterranean has
been prepared by **Petroconsultants**, Geneva, from data
digitized from the 1 : 1 000 000 maps and with isobaths at
mainly 200 m intervals. This product is available on
magnetic tape.

Considerable interest in the geology of the
Mediterranean seabed has been shown by oil prospecting
companies, particularly in the Tertiary deltaic deposits
lying offshore from major river mouths. Much of this *ad
hoc* mapping is geophysical, is held in digital format and is
not widely available. But this interest is also reflected in
some smaller-scale mapping such as the maps of structural
geology published collectively by the French **Institut
Français du Pétrole (IFP)**, **Centre National pour
l'Exploitation des Océans (CNEXO)** (now IFREMER)
and the Institut National d'Astronomie et de Géophysique
(INAG). Several French oil companies cooperated in the
publication in (1984 of a *Carte bathymétrique de la marge
continentale au large du delta du Rhône. Golfe du Lion*
1 : 200 000. The maps and publications of these
organizations are distributed by **Editions Technip**.
Magnetic maps of the western Mediterranean were
published by the Institut de Physique du Globe de Paris
(INAG), and are distributed by the **Bureau de Recherches
Géologiques et Minières (BRGM)**.

Countries bordering the Mediterranean have carried out
various bathymetric and thematic surveys relating to their
coasts and territorial waters. Notable among these is the
mapping of the French littoral discussed in the section on
France, and in the 1 : 100 000 BRGM series *Carte de la
nature des dépôts meubles*. Other examples will be found in
the publications of the Istituto Idrografico della Marina,
Italy, and the Geological Survey of Israel.

Addresses

Petroconsultants SA
8–10, rue Muzy, CH-1211 GENEVA 6, Switzerland.

Editions Technip
27 rue Ginoux, F-75737 PARIS Cedex 15, France.

Glavnoe Upravlenie Navigacii i Okeanografii (GUNO)
Ministetstva Oborony, 8111 Iliniya B34, LENINGRAD, USSR.

For IGN, BRGM, CNRS, IFP, INAG and IFREMER, see
France; for NGS see United States; for UNESCO see World; for
GUGK see USSR.

Catalogue

General and bathymetric

Sredizemnoje More 1 : 5 000 000
Moskva: GUGK, 1979.

The Mediterranean seafloor 1 : 4 371 000
Washington, DC: NGS, 1982.

Carte du Bassin Méditerranéen ca 1 : 4 000 000
Paris: IGN, 1978.
Map in shape of ellipse.

Bassin Méditerranéen 1 : 2 500 000
Paris: IGN, 1981.
2 sheets, both published.

Carte bathymétrique de la Méditerranée occidentale 1 : 1 500 000
Paris: IFREMER, 1981.

*Mezdunarodna batimetriceskaja karta Eredizemnogo
Morja* 1 : 1 000 000
Leningrad: GUNO, 1981.
10 sheets, all published.

Geological and geophysical

Carte géologique et structurale des bassins Tertiaires du domaine Méditerranéen 1 : 2 500 000
Paris: IFP/CNEXO/INAG, 1974.
2 sheets, both published.
+ 34 pp. text.

Esquisse photogéologique du domaine Méditerranéen. Grands traits structuraux a partir des images du satellite Landsat-1 1 : 2 500 000
Paris: IFP/CNEXO/INAG, 1976.
2 main sheets and 7 regional sheets.
+ 36 pp. text.

Carte magnétique de la Méditerranée occidentale 1 : 1 000 000
Paris: INAG, 1966–71.
Intensité du champ total.
Anomalies du champ total.
Each in 2 sheets (north and south).

Environmental

Vegetation map of the Mediterranean region 1 : 5 000 000
Paris: UNESCO, 1970.
2 sheets, both published.
+ text.

Carte de la végétation potentielle de la région méditerranéenne 1 : 2 500 000
Paris: CNRS, 1985–.
4 sheets, 1 published.
+ text.

INDIAN OCEAN

Mapping of the Indian Ocean and of its islands has been carried out by countries with an interest in the area. The French, British and Australians have produced mapping of most of the island groups, few of the newly independent island states have yet compiled any of their own mapping.

The ocean as a whole has been covered by American and Russian works, but there has not been nearly as much cartographic activity as in either the Atlantic or Pacific areas. The ocean is covered in Volume 2 of the major *Atlas okeanov*, prepared by the Naval Hydrographic Service in collaboration with the Ministry of Defence and many research institutes in the USSR Academy of Sciences. This is distributed through Pergamon Press in the west and

provides the best ocean-wide thematic coverage. Other Russian mapping of the Indian Ocean has been published by **Glavnoe Upravlenie Geodezii i Kartografii (GUGK)** and by the **Soviet Ministry of Geology (MGSSSR)**. A major investigation of the ocean resulted in the publication of several atlas volumes and map sheets by the International Indian Ocean Expedition in the 1970s. Most of these are no longer available. Several atlases have since been published, especially on meteorological and climatological themes. The other major map publisher for this area has been the **Indian National Atlas and Mapping Organization (NATMO)**.

The British **Directorate of Overseas Surveys (DOS)**

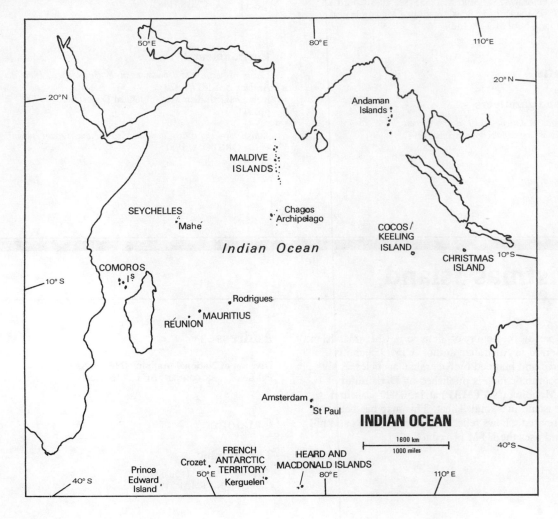

(now Overseas Surveys Directorate, Southampton, OSD) compiled orthophoto mapping for the Republic of Seychelles in the 1970s and a new edition of the 1 : 25 000 basic cover of Mauritius in the 1980s. It continues to aid these States. The Maldives have also been mapped with British aid.

The French **Institut Géographique National (IGN)** established topographic series for Réunion, the Comoros and for the various island territories of the French Antarctic territories. These maps use the IGN house style. Other French publishers of environmental mapping of the areas include **Office de la Recherche Scientifique et Technique Outre Mer (ORSTOM)**, **Centre National de la Recherche Scientifique (CNRS)**, **Bureau de la Recherche Géologique et Minières (BRGM)** and **Institut de la Recherche Agronomique Tropicale (IRAT)**. Australian maps have been compiled for Christmas Island, the Cocos/Keeling group and Heard Island, by the **Division of National Mapping (NATMAP)**.

Further information

There has been very little written that is specifically about the mapping of the Indian Ocean. See the Annual reports and catalogues issued by the organizations listed below for further information about their mapping.

Addresses

For DOS see Great Britain; for IGN, ORSTOM, CNRS, BRGM and IRAT see France; for NATMAP see Australia; for GUGK, Ministerstvo Oborony SSSR and MGSSSR see USSR; for CIA, NGS and USGS see United States; for Indian Meteorological Department and NATMO see India.

Catalogue

Atlases and gazetteers

Atlas okeanov: Atlanticheskij i Indijskij Okean
Moskva: Ministerstvo Oborony SSSR, 1977.
pp. 306.
In Russian.

Indian Ocean Atlas
Washington, DC: CIA, 1976.
pp. 80.

General and bathymetric

Indian Ocean floor   1 : 25 700 000
Washington, DC: NGS, 1975.

Indian Ocean   1 : 25 000 000
Calcutta: NATMO, 1978.

Indijskii okean   1 : 15 000 000
Moskva: GUGK, 1984.
In Russian.
With two small ancillary thematic maps showing geomorphology and currents.

Indijskij okean   1 : 10 000 000
Moskva: Institut Fiziki zemli and GUGK, 1977.
4 sheets, all published.

Bay of Bengal   1 : 6 000 000
Calcutta: NATMO, 1977.

Geological and geophysical

Geologo-geofizicheskij atlas Indiijskogo Okeana/Geological-geophysical atlas of the Indian Ocean
Moskva: Akademija Nauk, 1975.
pp. 167.
Maps compiled for the International Indian Ocean expedition.

The Indian Ocean: geology of its bordering lands and configuration of its floor   1 : 13 650 000 at equator
Reston, VA: USGS, 1963.

Environmental

Tracks of storms and depressions in the Bay of Bengal and the Arabian Sea 1877–1970
New Delhi: Indian Meteorological Department, 1979.
pp. 200.

Atlas hydrologique du Canal de Mozambique (Ocean Indien)
Paris: ORSTOM, 1981.
pp. 43.

Christmas Island

Formerly part of the colony of Singapore, Christmas Island was transferred to Australian authority in 1958 and is administered and mapped by Australian agencies. A six-colour topographic map is published by **Division of National Mapping (NATMAP)** at 1 : 50 000 scale and covers the island in a single sheet. This map has been regularly revised, shows relief with 20 m contours and hill shading and uses the UTM projection.

Addresses

Division of National Mapping (NATMAP)
PO Box 31, BELCONNEN, ACT 2616, Australia.

Catalogue

Topographic

Christmas island   1 : 50 000
Canberra: NATMAP, 1978.

Cocos/Keeling Islands

The Cocos or Keeling group is mapped by Australian agencies. The **Division of National Mapping (NATMAP)** has compiled topographic coverage of the group. A single-sheet 1 : 50 000 map was compiled in the early 1970s, and a two-sheet 1 : 25 000 six-colour map of Cocos Island was published by NATMAP in 1979. These maps are both compiled to full photogrammetric standards, use the UTM projection and show relief with contour intervals at 10 m.

Addresses

Division of National Mapping (NATMAP)
PO Box 31, BELCONNEN, ACT 2616, Australia.

Catalogue

Topographic

Cocos (Keeling) Islands 1 : 50 000
Canberra: NATMAP, 1973.

Cocos Island 1 : 25 000
Canberra: NATMAP, 1979.
2 sheets, both published.

Comoros

The island group of the Comoros became independent from France in 1975 and most of its mapping has been carried out by French agencies. The **Institut Géographique National (IGN)** compiled the basic scale series in the early 1970s, a five-colour map covering the group in five sheets, with relief shown by 20 m contours drawn on the UTM projection. This map is derived from aerial coverage. A 1 : 500 000 scale IGN map of the whole group is no longer available, but the American **Central Intelligence Agency (CIA)** map at 1 : 1 200 000 scale issued in 1977 provides a general overview of the group. Earth science mapping of the various islands has been carried out by different agencies. The main island Mayotte has been covered by 1 : 50 000 scale agricultural potential mapping compiled by **Institut de Recherches Agronomique Tropicale (IRAT)** in 1981. The **Office de la Recherche Scientifique et Technique Outre Mer (ORSTOM)** still distributes a soil map of Anjouan published at 1 : 100 000 and produced in the early 1950s, and vegetation mapping at 1 : 100 000 has been published for Grande Comoro by **Institut Français de Pondichéry (IFP)**.

Addresses

Institut Géographique National (IGN)
136 bis rue de Grenelle, 75700 PARIS, France.

Direction du Cadastre et de la Topographie
BP 486, MORONI.

For IRAT and ORSTOM see France; for IFP see India; for CIA see United States.

Catalogue

General

Comoros 1 : 1 200 000
Washington, DC: CIA, 1977.

Topographic

Carte générale des Comoros 1 : 50 000
Paris: IGN, 1973–79.
5 sheets, all published.

Environmental

Mayotte: cartes de l'inventoire des terres cultivables 1 : 50 000
Montpellier: IRAT, 1981.
3 sheets, all published.

La Grande Comoro climat et végétation 1 : 100 000
Pondichéry: IFP, 1969.
With accompanying explanatory text.

Carte pédologique d'Anjouan 1 : 100 000
Bondy: ORSTOM, 1952.

French Southern and Antarctic Territories

The Southern and Antarctic Territories were established in 1955 from former Dependencies of Réunion and comprise the Southern Indian Ocean islands of the Kerguelen and Crozet groups, New Amsterdam and St Paul. For Adelie land on the Antarctic continent see Antarctica. Mapping of these islands has been carried out by French agencies. **Institut Géographique National (IGN)** has flown a variety of aerial coverage which has been used to establish a modern map base for the islands. Sheets all use the UTM grid. Kerguelen is covered in six-colour 1:100 000 maps: the three sheets were published in 1967, and show relief with 50 m contours and hill shading. The basic scale for the Possession Island in the Crozet Group is 1:50 000, with 20 m contours, whilst work is under way to compile three sheets for the other islands in this group using 1982 flown aerial cover. Both Ile Amsterdam and Ile St Paul are covered by 1:25 000 three-colour maps, published in 1968 and 1973 which use 10 m contours. 1:200 000 six-colour general maps of Kerguelen and Crozet are derived from the larger-scale coverage and were published with 100 m contours and hill shading in the early 1970s.

A variety of earth science mapping of two areas of Kerguelen has also been published, including black-and-white geological and sedimentological and bathymetric maps of Morhiban Gulf and of Rallier du Baty Peninsula compiled for French Antarctic research organizations. This earth science mapping is no longer available.

Addresses

Institut Géographique National (IGN)
136 rue de Grenelle, 75700 PARIS, France.

Catalogue

General

Terres Australes et Antarctiques françaises: carte de reconnaissance   1:200 000
Paris: IGN, 1971–73.
2 sheets, both published.

Topographic

Terres Australes et Antarctiques françaises: carte de reconnaissance Ile Kerguelen   1:100 000
Paris: IGN, 1967.
3 sheets, all published.

Archipel Crozet carte de reconnaissance   1:50 000
Paris: IGN, 1965.

Terres Australes et Antarctiques françaises: carte de reconnaissance St Paul, Amsterdam   1:25 000
Paris: IGN, 1968–73.
2 sheets, both published.

Heard and Macdonald Islands Territory

These islands in the South Indian Ocean are administered by Australia. The only published mapping of the group has been carried out by the Australian **Division of National Mapping (NATMAP)**. A 1:50 000 scale four-colour topographic sheet covering Heard Island was revised to 1977, uses UTM projection and shows relief with 20 m contours and hill shading.

Addresses

Division of National Mapping (NATMAP)
PO Box 31, BELCONNEN, ACT 2616, Australia.

Catalogue

Topographic

Heard Island   1:50 000   2nd edn
Canberra: NATMAP, 1977.

Maldives

The Republic of the Maldives is a chain of over 1200 coral atolls which extend 600 miles in a north–south direction in the eastern Indian Ocean. No island in the chain reaches a height greater than 8 feet ASL, so conventional topographic mapping is a rather pointless activity. Full independence was granted from Britain in 1965, and British agencies have been responsible for most of the somewhat limited mapping activity which has taken place. Little mapping is available. The **Directorate of Overseas Surveys (DOS)** (now Overseas Surveys Directorate,

Southampton, OSD) compiled nine-sheet 1 : 1000 scale cover of the main island, Male, for the World Health Organization. These line maps were compiled from 1 : 8000 scale aerial cover flown in the early 1970s. The current Maldive agency responsible for surveying and mapping is the **Ministry of Planning and Development**.

Only small-scale general maps of the Republic are available, and are usually published in conjunction with mapping of Sri Lanka, for the tourist market: these include maps by **VWK Ryborsch** and **Astrolabe**.

Addresses

Ministry of Planning and Development
Ghazee Building, Orchid Magu, MALE.

For VWK Ryborsch see German Federal Republic; for Astrolabe see France.

Catalogue

General

Sri Lanka und die Maldiven 1 : 1 000 000
VWK Ryborsch, 1983.

Sri Lanka Maldives: carte routière 1 : 1 000 000
Paris: Astrolabe, 1981.

Mauritius

Mapping of the French Overseas Territory Mauritius is the responsibility of the **Survey Department** but has been carried out by French or British agencies. A 1 : 25 000 map in 14 sheets (now 13) has been revised by the British **Directorate of Overseas Surveys (DOS)** (now Overseas Surveys Directorate, Southampton, OSD). Fourth edition sheets in this series were begun in 1980 and completed in 1981. This five-colour map is drawn on the Lambert conical orthomorphic projection and shows contours at 10 m. Both British and French general maps of the island at 1 : 100 000 scale are available: the former was published by DOS in 1983 and includes an inset of Rodrigues Island; the latter dates to 1978 and was published by the French **Institut Géographique National (IGN)** as a tourist and road map, with relief shading. IGN published a 1 : 50 000 topographic map of Rodrigues Island in 1982, which uses the UTM projection and shows relief with 20 m contours. A special sheet of the International Map of the World is also published by IGN to cover the Réunion and Mauritius area.

Soils mapping of the island has been published by the **Mauritius Sugar Industry Research Institute (MSIRI)**. This 1 : 50 000 scale map covers the island in two sheets, is issued with a further legend sheet, and compiled by **Office de la Recherche Scientifique et Technique Outre Mer (ORSTOM)**. MSIRI, with the Food and Agriculture Organization (FAO), also compiled a two-sheet land resources and agricultural suitability map at 1 : 50 000 scale in the mid-1970s, which was published with an explanatory notice.

Karto+Grafik has issued a tourist map in its Hildebrands Urlaubskarte series.

Addresses

Survey Department (MSD)
Ministry of Housing, Lands and the Environment, Edith Cavell St, PORT LOUIS.

Overseas Surveys Directorate (OSD/DOS)
Ordnance Survey, Romsey Rd, Maybush, SOUTHAMPTON SO9 4DH, UK.

Institut Géographique National (IGN)
136 bis rue de Grenelle, 75700 PARIS, France.

Mauritius Sugar Industry Research Institute (MSIRI)
LE REDUIT.

For ORSTOM see France; for Karto+Grafik see German Federal Republic.

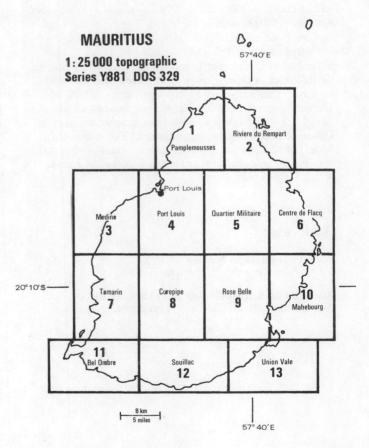

Catalogue

General

Mauritius: Hildebrands travel map   1 : 125 000
Frankfurt: Karto+Grafik, 1985.
With tourist notes on sheet.
Also available as German edition.

Ile Maurice: carte touristique et routière   1 : 100 000   2nd edn
Paris: IGN, 1978.

Mauritius   1 : 100 000   DOS 529 (Y 682)
Tolworth: DOS, 1983.
With inset of Rodrigues Island at same scale.

Topographic

Ile Rodrîgues: carte topographique   1 : 50 000
Paris: IGN, 1982.
With 1 : 10 000 inset of Port Mathurin.

Mauritius   1 : 25 000   DOS 329 (Series Y 881)
St Louis and Tolworth: MSD and DOS, 1980–84.
14 sheets, 13 published. ■

Environmental

*Land resources and agricultural suitability map of
Mauritius*   1 : 50 000
Le Reduit: MSIRI, 1975.
2 sheets, both published.
With accompanying explanatory text.

Carte pédologique   1 : 50 000
Le Reduit: MSIRI, 1983.
2 sheets, both published.
With 1 : 200 000 scale geological and hydrological map.

Réunion

Institut Géographique National (IGN) has carried out
mapping of the French Overseas Department of Réunion.
The basic scale series is the nine-sheet 1 : 25 000 scale *Série
bleu*, which uses UTM projection, is derived from
1 : 25 000 aerial coverage and shows relief in 10 m contours.
This map is revised to 1980 and published from 1983. It
replaced an earlier four-sheet 1 : 50 000 map (no longer
available) as the basic scale. A single-sheet 1 : 100 000
tourist map is published and IGN also issues a plastic relief
map at 1 : 100 000 scale of the island. The National Atlas
was compiled in the mid-1970s by IGN and includes nearly
200 maps mostly at 1 : 150 000 scale.

Geological mapping of Réunion is carried out by **Bureau
de Recherches Géologiques et Minières (BRGM)** who
have compiled four-sheet 1 : 50 000 (no longer available)
and single-sheet 1 : 100 000 maps. Soils mapping of the
island at 1 : 100 000 was compiled but is no longer in print.
**Office de la Recherche Scientifique et Technique Outre
Mer (ORSTOM)** has, however, carried out 1 : 40 000 scale
soils mapping of the areas around St Paul and St Pierre,
which is still available.

Addresses

Institut Géographique National (IGN)
136 bis rue de Grenelle, 75700 PARIS, France.

Bureau de Recherches Géologique et Minières (BRGM)
Ave de Concyr, BP6009, 45018 ORLEANS-CEDEX, France.

**Office de la Recherche Scientifique et Technique Outre Mer
(ORSTOM)**
70–74 route d'Aulnay, 93140 BONDY, France.

Catalogue

Atlases and gazetteers

Atlas des Départements Française d'Outre Mer   1. La Réunion
Paris: IGN and CNRS, 1975.
pp. 184.

General

La Réunion: carte touristique   1 : 100 000
Paris: IGN, 1984.

Topographic

Réunion   1 : 100 000
Paris: IGN, 1978.
Raised relief edition.

Île de la Réunion   1 : 25 000
Paris: IGN, 1983–84.
9 sheets, all published.

Geological and geophysical

Carte géologique du Département de la Réunion   1 : 100 000
Orléans: BRGM, 1967.
Also available as raised relief edition.

Seychelles

The Republic of the Seychelles comprises over 100 small granitic and coralline islands scattered over the middle of the western Indian Ocean. Mapping of this group has been carried out by the British **Directorate of Overseas Surveys (DOS)** (now Overseas Surveys Directorate, Southampton, OSD), under the Commonwealth African Assistance Plan. The **Department of Lands Planning and Survey** of the Seychelles Ministry of National Development is now responsible for the compilation of the Seychelles map base.

The main islands of Mahé, Praslin and La Digue are best mapped. These are covered by a six-colour 1:10 000 series with 10 m contours. From this series is derived the tourist map of Mahé issued at 1:50 000 with 50 m contours and a similar 1:30 000 map of Praslin and La Digue. A full-coloured photomap series at scales of between 1:5000 and 1:25 000 covers all the important islands in the group, the smaller and less populated islands being covered in detailed diazo maps. Diazo copies of larger-scale mapping of the main islands are also available from the Survey Division. DOS has issued small-scale maps of the archipelago. Other general maps are published by the American **Central Intelligence Agency (CIA)** and in the Hildebrands Urlaubskarte series by **Karto+Grafik**.

Geological coverage of Mahé and Praslin and the neighbouring islands was compiled by the **Geological Survey of Kenya** in the early 1960s and is available at 1:50 000 scale from the Survey Division. Soils mapping of the islands in four sheets has also been published. These maps were compiled by the DOS in the 1960s and are still available from the Survey Division.

Further information

Republic of the Seychelles map catalogue, Victoria: Survey Division, 1980. Up-dated at intervals with supplements.

Addresses

Department of Lands Planning and Survey
Ministry of National Development, Independence House, VICTORIA, Mahé.

Overseas Surveys Directorate (OSD/DOS)
Ordnance Survey, Romsey Rd, Maybush, SOUTHAMPTON SO9 4DH, UK.

For GSK see Kenya; for Karto+Grafik see German Federal Republic; for CIA see United States.

Catalogue

General

Seychelles 1:3 000 000 DOS 980 edn 8
Victoria and Tolworth: DOS, 1977.

Republic of the Seychelles 1:2 000 000 and 1:200 000 DOS 604
Victoria and Tolworth: DOS, 1984.
2 maps on 1 sheet.
Larger-scale map covers inner islands and Mahé.

Hildebrands Urlaubskarte: Seychellen various scales
Frankfurt AM: Karto+Grafik, 1984.
Tourist notes on sheet.
6 maps of the main islands in the group.

Seychelles: principal group 1:500 000
Washington, DC: CIA, 1977.

Topographic

Seychelles: island of Mahé 1:50 000 DOS 404 (Series Y762) 9th edn
Victoria and Tolworth: DOS, 1984.
Includes inset map of Victoria 1:12 500.

Praslin with La Digue and adjacent islands 1:30 000 2nd edn
Victoria and Tolworth: DOS, 1986.

Republic of the Seychelles 1:5000–1:25 000 DOS 104P, 204P and 304P
Victoria and Tolworth: DOS, 1978–80.
22 sheets, all published.
Photomap series.

Geological and geophysical

Geological maps of Mahé, Praslin and neighbouring islands 1:50 000
Nairobi: Geological Survey of Kenya, 1961–63.
2 sheets, both published.

Environmental

Soil survey of the Seychelles 1:50 000 DOS LR 3023
Tolworth: DOS, 1968.
4 sheets, all published.
With accompanying explanatory text.

PACIFIC OCEAN

Few of the islands in the Pacific have sufficient resources to devote to surveying and mapping. They have relied on their colonial administering power or on aid projects to establish mapping systems and standards. Many islands remain unmapped. Small-scale mapping and charting of the Ocean has also been carried out by the countries with an interest in the area. Thus the map base of this region has been created almost exclusively by agencies from New Zealand, the US, France, the UK and to a lesser extent Australia, Chile and Ecuador.

The New Zealand **Department of Lands and Survey (NZMS)** has carried out topographic mapping of the various New Zealand dependencies, of the Cook Islands and of Western Samoa. New Zealand aid established the Survey Offices in these territories. Small-scale mapping of the South-Western Pacific and of the Pacific as a whole has also been carried out by NZMS in the 1970s and 1980s and they recently published a second edition of the useful *Atlas of the South Pacific*. The **New Zealand Oceanographic Institute (NZOI)** has compiled full-colour bathymetric mapping of similar areas, including 1:1 000 000 (oceanic) and 1:200 000 (island) scale bathymetric charts. The **New Zealand Soil Bureau (NZSB)** has compiled soil mapping at large scale of several island groups.

British mapping of the Pacific has concentrated upon the establishment of topographic bases for Fiji, Kiribati, The Solomon Islands, Tonga and Tuvalu. Many of these series were published as 1:25 000 full-colour photomaps in the 1970s by the British **Directorate of Overseas Surveys (DOS)** (now Overseas Surveys Directorate, Southampton, OSD). DOS also compiled geological mapping for Vanuatu and produced some small-scale general maps of Vanuatu and Kiribati. Resources mapping was compiled by DOS for the Solomons and Fiji.

French interests in the Pacific have concentrated upon New Caledonia, Vanuatu and French Polynesia, and **Institut Géographique National (IGN)** established map series with French style specifications at 1:50 000, 1:100 000, 1:200 000 and 1:500 000 for these territories. The **Centre National pour Exploitation des Oceans (CNEXO)** carried out some bathymetric charting of French areas of interest. There has been little small-scale mapping of the basin as a whole produced by France.

American territorial interests in the Pacific include many of the Northern and Western islands, administered in the Pacific Trust Territory and the State of Hawaii. The **United States Geological Survey (USGS)** is responsible for the compilation of map bases in these areas, and the basic scale is the 1:24 000, or more recently 1:25 000 quad.

Little mapping has yet been published for the numerous Trust Territory islands. There has been a great variety of bathymetric and geophysical mapping at smaller scales produced for much of the Pacific by the **National Oceanographic and Atmospheric Administration, National Geophysical Data Center (NOAA)** and the **Geological Society of America (GSA)**. Other earth science mapping of the Pacific has also been carried out largely by American agencies. The most important current programme is the **Circum Pacific Map Project**. This is a cooperative international effort, established in 1973, which will involve the compilation of a total of 49 maps on eight different themes. The project is coordinated and final cartography is carried out by the United States Geological Survey, and the maps are published by the **American Association of Petroleum Geologists (AAPG)**. Circum Pacific maps are published as five overlapping 1:10 000 000 scale quadrants for each theme and use Lambert azimuthal equal area projection. A Pacific-wide map at 1:17 000 000 scale will also be published for most themes. Early basin-wide sheets were printed with a 1:20 000 000 bar scale but were in fact at the larger scale. A geographic map series was published in 1978, together with base maps. The latter are two colour and include overprinted 2° grids. Various thematic series have followed: plate tectonic, geological and geodynamic are now complete and work is continuing on the compilation and publication of mineral resources and energy resources maps. It is expected that the project will be complete by 1989 and that 47 different countries will have participated in some way.

The AAPG is also involved in the publication of other small-scale earth science mapping of the Pacific.

Other American agencies publishing mapping of the ocean include the **Hawaii Geographical Society** and the **National Geographic Society (NGS)**.

The **Geological Survey of Japan (GSJ)** and the Japanese **Maritime Safety Agency** have compiled small-scale geological coverage of the areas nearest to Japan, including a variety of coastal zone sediment, oceanographic and bathymetric mapping, but there has been little Japanese mapping of others parts of the Pacific basin.

The Russian Ocean Atlas, compiled by the Russian Naval Hydrographic Section of the **Ministry of Defence** and distributed by Pergamon Press, is perhaps the most important oceanographic work for the Pacific. It presents essential data in small-scale mapping of the basin and contains some larger-scale maps of major ports and islands. This volume is published in a Russian language edition

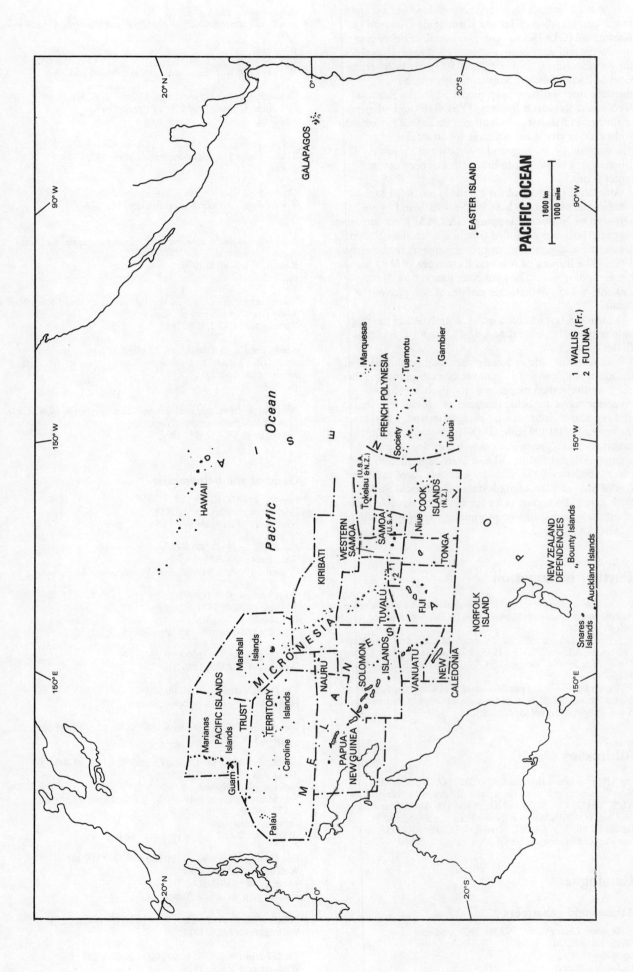

PACIFIC OCEAN

1600 km
1000 miles

GALAPAGOS

EASTER ISLAND

Marquesas

FRENCH POLYNESIA

Society

Tuamotu

Gambier

Tubuai

1   WALLIS (Fr.)
2   FUTUNA

Pacific   Ocean

HAWAII

Tokelau (U.S.A. & N.Z.)

WESTERN SAMOA

SAMOA (U.S.A.)

Niue

COOK ISLANDS (N.Z.)

TONGA

KIRIBATI

TUVALU

FIJI

NORFOLK ISLAND

NEW ZEALAND DEPENDENCIES

Bounty Islands

Snares Islands

Auckland Islands

MICRONESIA

Marshall Islands

NAURU

SOLOMON ISLANDS

VANUATU

NEW CALEDONIA

MELANESIA

PACIFIC ISLANDS TRUST TERRITORY

Marianas Islands

Caroline Islands

PAPUA-NEW GUINEA

Guam

Palau

only, but an English summary volume has been compiled to aid interpretation. **Glavnoe Upravlenie Geodezii i Kartografii (GUGK)** has compiled small-scale mapping of the Pacific and distributes a special 19-sheet extract from the Russian *Karta Mira* at 1:2 500 000 scale to cover the Central Pacific Islands. Numerous other small-scale thematic map sets have been produced by the Russian **Geological Research Institute (VSEGIE)** and other parts of the Soviet **Ministerov Geologii (MGSSSR)** at a Pacific-wide scale to give some coverage for structural geomorphology, geology and hydrocarbon resources. The Russian area of interest in the north has been covered at larger scales.

Australian mapping of the Pacific is restricted to the compilation of the map base for Norfolk island by the **Division of National Mapping (NATMAP)**. An important language atlas was compiled from the Australian National University and published in two instalments in the early 1980s. The **Bureau of Mineral Resources (BMR)** has also been responsible for the joint compilation with New Zealand of 1:5 000 000 scale geological coverage of the ocean.

Ecuador is responsible for the compilation of the map base of the Galapagos islands, and Chile for mapping of Easter Island.

The economic and technological changes referred to in the introduction to the Oceans are beginning to have their effect on the seabed mapping of the Pacific. Exclusive economic zones are being declared and resource mapping of these near-shore zones are being prepared, for instance the recent 1:500 000 scale GLORIA mosaics, to show bathymetry and preliminary detailed geological and sedimentalogical features. These were published in 1986 as part of the *Atlas of the Exclusive Economic Zone: western United States* by the **United States Geological Survey**. Such detailed mapping is for the first time possible because of the use of sidescan sonar technology employed in the GLORIA system.

Further information

For background information to the mapping of the Pacific see the Annual Reports and catalogues issued by the organizations listed below.

For more detailed information about the Circum Pacific Map Project see Addicott, W.O. (1985) *Scope and status of the Circum-Pacific Map Project*, Reston: USGS (USGS Open File Report 85–267).

A useful guide to the publication of geological mapping of the area is Thompson, B.N. (1984) *Geological maps of the islands of the South Pacific*, Wellington: NZGS.

Addresses

For DOS see Great Britain; for NZMS, NZSB and NZOI see New Zealand; for Hawaii Geographical Society see Hawaii; for NGS, AAPG, USGS, GSA, NOAA and DMA see United States; for CNEXO, BRGM and IGN see France; for GSJ and MSA see Japan; for GUNO, GUGK, NIIGA, VSEGIE, MGSSSR and Ministry of Defence see USSR.

Catalogue

Atlases and gazetteers

Atlas of the South Pacific   NZMS 295   2nd edn
Wellington: NZMS, 1986.
pp. 47.
Includes gazetteer.

Atlas okeanov: tikhii okean
Moskva: Ministerstvo Oborony SSSR and Voenno-Morskoi Flot, 1975.
pp. 340.
In Russian.
With English language translation to contents and titles.

Oceanographic atlas of the Pacific Ocean   R.A. Barkley
Honolulu: University of Hawaii Press, 1969.
pp. 176.

North-western Pacific Ocean: marine environmental atlas
Tokyo: Japan Hydrographic Association, 1975–78.
2 vols.

Oceanographic atlas of the North-western Pacific
Tokyo: Japan Meteorological Agency, 1975–76.
3 vols.

Atlas of the exclusive economic zone, western conterminous United States   1:500 000
Reston, VA: USGS, 1986.
pp. 152.

South Pacific Islands: official standard names approved by the United States Board on Geographic Names
Washington, DC: DMA, 1957.

South-west Pacific Islands: official standard names approved by the United States Board on Geographic Names
Washington, DC: DMA, 1957.
pp. 368.

West Pacific Islands: official standard names approved by the United States Board on Geographic Names
Washington, DC: DMA, 1957.

General and bathymetric

Pacific   1:40 000 000
Edinburgh: Bartholomew, 1981.
Numerous larger-scale insets of island groups.

Pacific ocean   1:37 700 000
Washington: NGS, 1975.
Printed on both sides; on verso: Pacific ocean floor.

Tikhij okean   1:25 000 000
Moskva: GUGK, 1984.
In Russian.
With 2 accompanying ancillary thematic insets.

The Pacific   1:20 000 000   NZMS 276
Wellington: NZMS, 1976.

The Pacific Islands   1:20 000 000
Honolulu: Hawaii Geographic Society, 1986.

Topography of the North and South Pacific   1:14 500 000
La Jolla, CA: NOAA, 1971–75.

Geographic map of the Circum-Pacific Basin   1:10 000 000
Tulsa, OK: AAPG, 1977–78.
6 sheets, all published.
5 quadrant sheets at 1:10 000 000, 1 total area map at 1:17 000 000.
Includes Antarctica.
Also available as 2-colour base map series.

The Pacific Hemisphere   1:10 000 000   NZMS 309
Wellington: NZMS, 1980.
4 sheets, all published.
Outline map including Antarctica.

Islands of the South Pacific   1:10 000 000   NZMS 275
Wellington: NZMS, 1974.

The Western Pacific   1:10 000 000   NZMS 285
Wellington: NZMS, 1979.

Bathymetry of the Pacific 1:6 442 194
Boulder, CO: GSA, 1977–.
4 published sheets.
Sheets published cover East and South East Asian seas, south-east
Pacific, north-central Pacific and north-east Pacific.

Bathymetry of the North and South Pacific 1:6 442 194
La Jolla, CA: NOAA, 1970 and 1974.
21 sheets, all published.

Bathymetric map of the north-east equatorial Pacific
ocean 1:5 000 000
Reston, VA: USGS, 1978.

Base map of the Aleutian–Bering Sea Region 1:2 500 000
Reston, VA: USGS, 1974.

Oceanic chart series: bathymetry 1:1 000 000
Wellington: NZOI, 1964–.
72 sheets, 19 published.
Also available as sediment and magnetic total force anomaly charts
(for New Zealand only).

Bathymetrie Pacifique sud 1:1 000 000
Paris: CNEXO, 1973–.
5 published sheets.

Island chart series 1:200 000
Wellington: NZOI, 1966–.
21 published sheets.

Area served by the South Pacific Commission 1:12 500 000
Canberra: NATMAP, 1984.

Topographic

Ostrova tikhogo okeana (Polinezija i micronezija) 1:2 500 000
Moskva: GUGK, 1980.
19 sheets, all published.
In Russian.
Special edition from Karta Mira with placename index and 5
language explanation.

Geological and geophysical

Metalliferous sediments, submarine volcanism, and submarine
geothermal activity in the South Pacific 1:25 000 000
Wellington: NZOI, 1974.

Preliminary tetrastratigraphic terrane map of the Circum-Pacific
region—Pacific Basin 1:17 000 000
Tulsa, OK: AAPG, 1985.
With accompanying explanatory text.

Structurno-geomorfologicheskaja karta dnja Tikhnogo okeana/
Structural geomorphology map of the Pacific Ocean 1:15 000 000
Moskva: MGSSSR, 1981.
Legend and title in Russian and English.

Geological map of the Circum-Pacific region—Pacific
Basin 1:10 000 000
Tulsa, OK: AAPG, 1983–86.
6 sheets, all published.
5 quadrant sheets at 1:10 000 000, 1 total area sheet at
1:17 000 000.

Plate tectonic map of the Circum-Pacific Region—Pacific
Basin 1:10 000 000
Tulsa, OK: AAPG, 1981–82.
6 sheets, all published.
5 quadrant sheets at 1:10 000 000, 1 total area sheet at
1:17 000 000.

Geodynamic map of the Circum-Pacific region—Pacific
Basin 1:10 000 000
Tulsa, OK: AAPG, 1984–86.
6 sheets, all published.
5 quadrant sheets at 1:10 000 000, 1 total area sheet at
1:17 000 000.

Geologicheskaja karta Tikhiokeanskogo podizhnogo pojasa i Tikhogo
Okean/Geological map of the Pacific Mobile Belt and Pacific
ocean 1:10 000 000
Moskva: MGSSSR, 1970.
10 sheets, all published.
Legend in Russian, English and Spanish.

Geological map of the world: Australia and Oceania 1:5 000 000
Canberra and Wellington: BMR and NZGS, 1965–71.
13 sheets, all published.

Free air gravity field of the south-west Pacific Ocean 1:4 400 000
Boulder, CA: NOAA, 1981.

Geologicheskaja karta Beringovomorskogo regiona/Geological map of
the Bering Sea area 1:2 500 000
Leningrad: NIIGA, 1982.
4 sheets, all published.
English and Russian text.

Free air gravity anomaly map of the Central Pacific 1:2 000 000
Yatabi-Machi: GSJ, 1982.
In English and Japanese.

Environmental

Manganese deposits in the South Pacific ocean 1:25 000 000
Wellington: NZOI, 1974–80.
7 sheets, all published.
Single sheets for different minerals.

Isopach map of sediments in the Pacific Ocean Basin and marginal sea
basins ca 1:14 000 000
Tulsa, OK: AAPG, 1979.
2 sheets, both published.

Mineral resources map of the Circum-Pacific Basin 1:10 000 000
Tulsa, OK: AAPG, 1985–.
6 sheets, 5 published.
5 quadrant sheets at 1:10 000 000, 1 total area sheet at
1:17 000 000.
Total area sheets published for manganese nodule/sediments.

Karta neftegazonosti i uglenosti Tikhiokeanskogo podvizhnogo pojasa/
Petroleum coal map of the Pacific mobile belt 1:10 000 000
Moskva: MGSSSR, 1978.
9 sheets, all published.
Title also in Spanish: legend and sheets in Russian and English.
With accompanying Russian language explanatory text.

South-west Pacific ocean 1:10 000 000
Wellington: NZOI, 1979.
7 sheets, all published.
Thematic set: one topic per sheet. Covers salinity, temperature,
water temperatures and dynamic height anomalies.

Surface sediments and topography of the North Pacific 1:6 442 194
La Jolla, CA: NOAA, 1972.
10 sheets, all published.

Manganese nodule distribution in the Central Pacific
Ocean 1:2 000 000
Yatabe-Machi: GSJ, 1983.
Japanese and English language.

Human and economic

Language atlas of the Pacific area
Canberra: Australian Academy of the Humanities, 1981–83.
47 sheets, all published.

Atlas of marine use in the North Pacific Region E. Miles et al.
Berkeley: University of California Press, 1982.
pp. 107.

Zones économiques des 200 milles nautiques du Pacifique Sud, Central
et Ouest/200 nautical mile exclusive economic zones in the South,
Central and Western Pacific 1:13 500 000
Paris: CNEXO, 1978.

200 Meilen Zonen im Pazifischen Ozean ca 1:50 000 000
Hamburg: Institut für Asienkunde, 1984.

American Samoa

American Samoa is a US territory administered through the Department of the Interior and is mapped by the national mapping agency of the United States, the **United States Geological Survey (USGS)**. Two 1 : 24 000 sheets cover the seven islands in the group. These four-colour maps use the Polyconic projection and show relief with 40 foot contours. The **University of Hawaii Press** has published a general tourist map covering both Western and American Samoa, which shows relief by hill shading. There is no indigenous mapping authority and no other published mapping of the group.

Addresses

United States Geological Survey (USGS)
National Center, 12201 Sunrise Valley Drive, RESTON, VA 22092, USA.

Department of Public Works
Pago Pago, TUTUILA 96799.

For University of Hawaii Press see Hawaii.

Catalogue

General

Samoa: reference map   1 : 200 000 and 1 : 80 000
Honolulu: University of Hawaii Press, 1980.
Double-sided: covers both American and Western Samoa, includes indexed town maps of Pago Pago and Apia.

Topographic

Topographic map of American Samoa   1 : 24 000
Reston, VA: USGS, 1963.
2 sheets, both published.

Cook Islands

The Cook Islands are administered by New Zealand and mapped by the various New Zealand survey authorities. Topographic coverage of the group has been undertaken by the New Zealand **Department of Lands and Surveys (NZMS)** which is producing four-colour mapping of the major islands of the group in a new metric 1 : 25 000 series that uses an orthophotobase and the Transverse Mercator projection. Sheets published to date cover Raratonga, Mitiaro and Aitutaki. Raratonga is also covered in a conventional 1 : 15 840 scale full-colour line map, which has been revised on a regular basis. Diazo preliminary editions at a variety of large scales are available to cover other islands in the Group, from the recently established **Cook Islands Department of Survey (CISD)**. Earth science mapping is carried out by various parts of the New Zealand Department of Scientific and Industrial Research (DSIR). At present geological coverage compiled by the **New Zealand Geological Survey (NZGS)** is available only for Raratonga. Soils have been mapped for the whole group by the **New Zealand Soil Bureau (NZSB)** at 1 : 15 000 scale. These six maps are published with accompanying textual explanations. **The New Zealand Oceanographic Institute (NZOI)** has published 1 : 200 000 oceanographic charts of the islands, which give a useful overview of the group.

Further information

Catalogues and annual reports of the below listed organizations.

Addresses

Cook Islands Survey Department (CISD)
PO Box 114, RARATONGA.

For addresses of NZMS, NZGS, NZSB and NZOI see New Zealand.

Catalogue

General

Island chart series   1 : 200 000
Wellington North: NZOI, 1966–69.
6 sheets, all published.

Topographic

Maps of the Cook Islands   1 : 25 000   NZMS 272
Wellington: NZMS, 1983–.
3 sheets, all published.

Map of Raratonga Cook Islands   1 : 15 840   3rd edn
Raratonga: CISD, 1982.

Geological and geophysical

Geology of the Cook Islands   1 : 25 000
Lower Hutt: NZGS, 1970.
2-colour photomap with explanatory text covering Raratonga.

Environmental

Soil map of the Cook Islands   1 : 15 000
Lower Hutt: NZSB, 1980–.
6 sheets, all published.
With accompanying explanatory texts.

Easter Island

Isolated Easter Island is situated in the Eastern Pacific ocean and is administered by Chile. Responsibility for the mapping of the island falls to the Chilean official mapping agency **Instituto Geografico Militar**, but no mapping of the island is available. The only published map of Easter Island was issued recently by **Motuiti** in Chile, in conjunction with the Spanish commercial house Editorial Alpina. This 1 : 30 000 scale full-colour map shows the island with 10 m contours and the typical Alpina style of 50 m hypsometric tinted relief and is compiled from a topographic base supplied by the Servicio Aerofotogrametrico in Santiago. Tourist and archaeological information is also added.

Addresses

Instituto Geográfico Militar
Castro 354, SANTIAGO, Chile.

Motuiti Ediciones del Pacifico Sur
Hernando de Aguirre, 720 Dep. 63, PROVIDENCIA, Chile.

Catalogue

General

Easter Island 1 : 30 000
Providencia: Motuiti.

Fiji

Fiji comprises over 300 islands in the South Pacific. Its survey organization is the **Department of Lands and Survey (FDLS)** but much of the mapping of the islands was carried out by the British **Directorate of Overseas Surveys (DOS)**. The basic scale is still the 1 : 50 000 map first issued by DOS in the 1950s. This series used the Cassini projection, and later UTM projection, is printed in three colours, with 100 foot contour interval, and covers the main islands of Viti and Vanua Levu, as well as the Yasawa and Lomaiviti groups, on irregular sheet lines. Revision of this map has been carried out by FDLS for areas in demand for planning and development purposes, but some areas are now only available as diazo copies. There are at present plans to upgrade the 1 : 50 000 map; it is intended to revise sheet lines, to use a Transverse Mercator projection and a metric grid and to introduce some larger-scale series using the metric system. Smaller-scale topographic coverage of the islands is also available from FDLS at 1 : 250 000 and 1 : 500 000 scales; this too was first compiled by DOS. In 1964 DOS compiled land use coverage at 1 : 250 000 for the two main islands: these may also still be obtained from FDLS. Other DOS compiled resources mapping is still available from the **Land Resources Development Centre (LRDC)** in the UK including 1 : 125 000 scale land capability mapping of the two main islands in eight sheets. Amongst the other mapping activities carried out by FDLS are cadastral surveys and the production of larger-scale series including a 1 : 10 000 town series and street directory of Suva.

Recent cartographic aid to Fiji has come from New Zealand agencies. A 1 : 500 000 layer coloured imperial topographic sheet was published in 1980 by the **New Zealand Department of Lands and Survey (NZMS)**. NZMS also compiled a report on the survey organization in Fiji, which recommended certain changes in the early 1980s.

Geological mapping of Fiji is carried out by the **Mineral Resources Department (MRD)** which has compiled 1 : 50 000 coverage using the same base as the topographic series for much of the group. This series is still in progress, though most sheets were issued in the 1960s with explanatory bulletins. Current activity is in the Lau group where sheets are being published at 1 : 25 000 scale. Other smaller-scale maps are also published by MRD including metallogenic coverage at 1 : 250 000 scale of the two main islands on a geological base (for Viti Levu no longer in print) and a 1 : 500 000 scale chronostratigraphic map. 1 : 250 000 scale full-colour geological mapping of Viti Levu is still available and MRD has recently published a 1 : 1 000 000 scale bathymetric map of the whole group.

Soils and other environmental mapping of Fiji was carried out in the 1960s by the **New Zealand Soil Bureau (NZSB)**. Eight-sheet 1 : 126 720 scale soil and land use maps of the islands were compiled, with smaller-scale single sheet coverage for soils, land use, geology, alienated land, land slope and population distribution. These maps are still available, as part of the soil resources of Fiji publication from NZSB, who are currently undertaking a major remapping of the resources of Fiji. This project includes the compilation of 38 1 : 50 000 soil maps of the major islands, from surveys carried out between 1981 and 1984. A 1 : 250 000 scale map of soil groups and a 1 : 750 000 soil moisture and temperature regime map are currently being prepared. Other larger-scale sheets of agricultural research stations have already been published.

Another important resources survey was published by the French **Office de la Recherche Scientifique et Technique Outre Mer (ORSTOM)** in 1983. This produced soils, geomorphological, topographic, land use and vegetation maps, mostly at 1 : 50 000 scale of the two Eastern Islands of Lakeba and Taveuni, and was published with a French and English language monograph.

Further information

Catalogues and annual reports from the below-mentioned organizations.

Addresses

Department of Lands and Survey (FDLS)
Government Buildings, PO Box 2222, SUVA.

Mineral Resources Department (MRD)
Ministry of Energy, Private Mail Bag, GPO, SUVA.

For DOS and LRDC see Great Britain; for NZMS and NZSB see New Zealand; for ORSTOM see France; for DMA see United States; for ITC see Netherlands.

Catalogue

Atlases and gazetteers

Fiji, Tonga and Nauru: official standard names approved by the United States Board on Geographic names
Washington, DC: DMA, 1974.
pp. 231.

General

Fiji 1:750 000
Enschede: ITC, 1981.

Fiji islands topographic map 1:500 000 NZMS 242
Wellington: NZMS, 1980.

Topographic

Fiji 1:250 000
Tolworth and Suva: DOS, 1966–.
7 sheets, 4 published. ■

Fiji 1:50 000 DOS 448 (Series X754 and X755)
Tolworth and Suva: DOS, 1966–.
48 sheets, all published. ■

Bathymetric

Bathymetric map of Fiji 1:1 000 000
Suva: MRD, 1984.

Geological and geophysical

Chronostratigraphic map of Fiji 1:500 000
Suva: MRD, 1983.

Geological map of Viti Levu 1:250 000
Suva: MRD, 1966.

Geological Survey of Fiji 1:50 000
Suva: MRD, 1960–.
ca 60 sheets, 33 published. ■

Environmental

Metallogenic map of Vanua Levu 1:250 000
Suva: MRD, 1976.
With accompanying explanatory text.

Fiji: land use 1:250 000
Tolworth and Suva: DOS, 1964.
2 sheets, both published.
Covers Vanua Levu and Viti Levu.

The soil resources of the Fiji Islands 1:126 720 and 1:760 320
Lower Hutt: NZSB, 1965.
23 sheets, all published.
Maps cover: Soils (8 sheets 1:126 720); land classification (8 sheets 1:126 720); soil, land use, geology, rainfall and climate, alienated land, landslope, population distribution (each at 1:760 320).
With accompanying explanatory text.

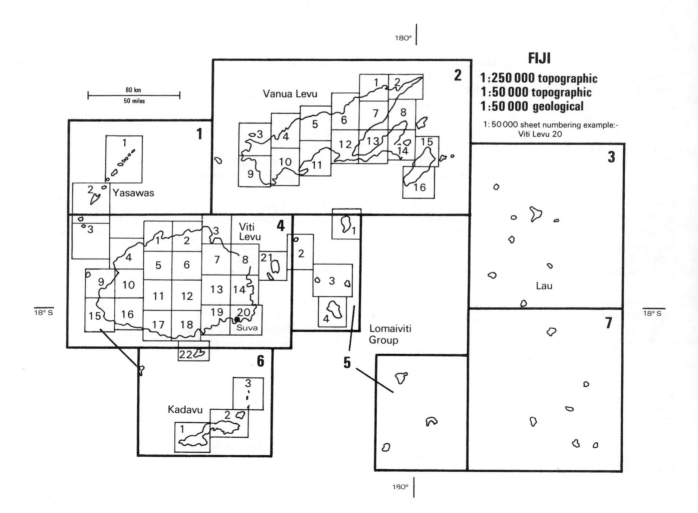

Land capability maps of Viti Levu and Vanua Levu 1:125 000
Tolworth: DOS for LRDC, 1964.
8 sheets, all published.

The Eastern islands of Fiji: a study of the natural environment, its use and man's influence on its evolution 1:50 000
Paris: ORSTOM, 1983.
4 sheets, all published.
In French and English.
With accompanying monograph.

Soil map of Rotuma 1:25 000
Lower Hutt: NZSB, 1983.

Town maps

Town map series: Suva 1:10 000
Suva: FDLS, 1975.

French Polynesia

French Polynesia comprises five archipelagos (Society, Tuamotu, Gambier, Tubuai and Marquesas islands) administered by France and mapped by the French **Institut Géographique National (IGN)**. There have been two main phases of IGN activity, from 1955 to 1967 and in the late 1970s and 1980s. The Society Islands and Tuamotu are covered in IGN base mapping dating from the 1950s and 1960s. These maps use a UTM projection and are compiled in three colours (four for Tahiti). They cover the major islands of Tuamotu in four sheets at 1:20 000 scale, and in 11 sheets at 1:50 000 scale. The Society Islands including Tahiti are mapped in a provisional series at 1:40 000 with 20 m contours issued in 1958. In the late 1970s sheets began to be published in a new 1:20 000 series, to cover Tahiti: this four-colour map also uses UTM projection, shows relief with 20 m contours and is derived from 1:25 000 aerial coverage flown in 1981. It was compiled by IGN for the **Service de l'Aménagement et de l'Urbanisme** in Papeete. IGN has issued a tourist map of Tahiti at 1:100 000 scale in 1977, also available as a relief edition. This is being revised and recast to include all the Society group in a map of the same scale but has not yet been published.

Other larger-scale mapping of the Tubuai group has recently been published to cover Tubuai and Rurutu, as single four-colour sheets for each island at 1:10 000 scale with 10 m contours. The Marquesas group has also recently been surveyed by IGN and maps are being published to cover the main islands. Two full-colour 1:50 000 sheets issued in 1985 cover Nuka Hiva and Hiva Oa, and photogrammetric plots at 1:50 000 have been published for three other islands in the group. A general 1:2 500 000 scale map of the whole of French Polynesia is currently being prepared by IGN.

Geological coverage of the Society Islands is provided by the French **Bureau des Recherches Géologiques et Minières (BRGM)**, whose six-sheet multi-colour series was issued in the mid-1960s.

The **Office de la Recherche Scientifique et Technique Outre Mer (ORSTOM)** has compiled a 1:20 000 soil map of Mangareva in the Gambier group, and 1:40 000 scale five sheet soils mapping of Tahiti.

Further information

Topographic mapping of French Territories of the Pacific, paper submitted by France to the 1977 UN Regional Cartographic Conference on Asia and the Pacific, in *UN Regional Cartographic Conference on Asia and the Pacific. Technical papers*, New York: UN, 1980.

Addresses

Institut Géographique National (IGN)
136 bis rue de Grenelle, PARIS 75700, France.

Service de l'Aménagement et de l'Urbanisme (SAU)
PAPEETE.

Bureau de Recherches Géologiques et Minières (BRGM)
Ave de Concyr, BP 6009, ORLEANS-CEDEX 45018, France.

Office de la Recherche Scientifique et Technique Outre Mer (ORSTOM)
70–74 route d'Aulnay, 93140 BONDY, France.

Service du Cadastre
rue Bruat, Paofai, PAPEETE.

Catalogue

General

Iles de la Societé/Iles Tubuai 1:2 000 000
Paris: IGN.

Iles Tuamotu/Iles Marquises 1:2 000 000
Paris: IGN.

Tahiti 1:1 000 000
Paris: IGN.
Special sheet from the International Map of the World series.

Topographic

Tahiti 1:100 000 3rd edn
Paris: IGN, 1977.
Also available as relief edition.

Polynésie Française 1:50 000 Prov. edn
Paris: IGN, 1967.
11 sheets, all published.
Covers main islands of Tuamotu Group.

Iles Marqueses 1:50 000
Paris: IGN, 1985.
2 maps, both published.

Polynésie Française 1:40 000 Prov. edn
Paris: IGN, 1958.
10 sheets, all published.
Covers main islands of Society Group.

Polynésie Française Archipel Tuamotu 1:20 000 Prov. edn
Paris: IGN, 1962.
4 sheets, all published.

Ile de Tahiti 1 : 20 000
Papeete and Paris: IGN, 1979–.
3 sheets, all published.

Geological and geophysical

Carte géologique des archipels polynésiens 1 : 40 000
Orléans: BRGM, 1965.
6 sheets, all published.
With accompanying explanatory texts.

Environmental

Tahiti: carte morpho-pédologique 1 : 40 000
Paris: ORSTOM, 1986.
5 sheets, all published.

Carte de situation et de répartition des sols de l'ile de Mangaréva 1 : 20 000
Paris: ORSTOM, 1972.

Galapagos Islands

The Galapagos islands are administered by Ecuador, and responsibility for the surveying and mapping of the group falls to **Instituto Geográfico Militar (IGME)** in Quito. None of the topographic series covering the mainland has yet been extended to the Galapagos: the only available mapping of the group is a tourist edition published by **Bradt Enterprises** at 1 : 500 000 scale, and some American hydrographic charting.

Addresses

Instituto Geográfico Militar (IGME)
Barrio El Dorado, Apartado 2435, QUITO.

For Bradt Enterprises see Great Britain.

Catalogue

General

Galapagos 1 : 500 000
Chalfont St Peter: Bradt Enterprises, 1985.
With tourist text on verso.

Hawaii

Hawaii has been administered by the United States since 1898 and became the 51st state in 1959. Its mapping is the responsibility of the **United States Geological Survey (USGS)**. The basic scale maps are 1 : 24 000 topographic quadrangles, which cover the five major islands in 122 sheets. These five-colour, $7\frac{1}{2}'$ quads use 20 foot contour intervals, are compiled to full photogrammetric standards using the polyconic projection, and are revised on a regular basis. A new seven-sheet 1 : 100 000 county map series to cover the islands was begun in 1980, and a six-colour 1 : 250 000 map with 200 foot contour intervals covers the group in four sheets. USGS also publishes 1 : 500 000 scale mapping of Hawaii, available in four versions: State, topographic (with 500 foot contours), relief (colour shaded) and hydrologic units. Other larger-scale hydrological maps have been issued by USGS in their series of Hydrologic Investigations Atlases. Most cover areas on Oahu island.

Geological mapping of Hawaii is also carried out by USGS. Few 24 000 quads have yet been published; most current mapping covers the volcanoes of Hawaii island. A 1 : 500 000 scale geological map of the whole group has appeared in the Geological Highway series produced by the **American Association of Petroleum Geologists (AAPG)**.

The **Land Study Bureau** in Honolulu compiled a 1 : 24 000 land capability for urban uses map series for the islands of Kauai and Mauai in the 1970s, and land classification maps using a soils overprint on to aerial photos for all the islands in the 1960s. None of these environmental maps is currently available.

Other useful general maps of the islands are published by the **University of Hawaii Press**. Compiled by James Bier these *Reference maps of the Islands of Hawaii* are published at a variety of scales according to the island size, are regularly revised, and include useful indexes to placenames and town maps. The University of Hawaii is also responsible for the publication of the State atlas, revised to 1983, and for a number of other general maps and atlases of the Pacific.

Rand McNally publishes small-scale mapping of the whole island group and a town plan of Honolulu.

Further information

Information about USGS mapping of Hawaii is given in the *Index to topographic maps of Hawaii, American Samoa and Guam* USGS, 1981; and in *List of Geological Survey Geologic and Water Supply Reports and Maps for Hawaii*, US Department of the Interior, 1979.

Addresses

United States Geological Survey (USGS)
National Center, 12201 Sunrise Valley Drive, RESTON, VA 22092, USA.

Hawaii Department of Land and Natural Resources
POB 621, HONOLULU H1, Hawaii 96809.

University of Hawaii Press
2840 Kolowalu St, HONOLULU, Hawaii 96822.

For AAPG, NGS and Rand McNally see United States.

Catalogue

Atlases and gazetteers

Atlas of Hawaii 2nd edn
Honolulu: University of Hawaii Department of Geography, 1983.

General

Standard reference map of Hawaii 1:950 400
Chicago: Rand McNally, 1986.
Includes gazetteer.

Hawaii 1:865 100
Washington, DC: NGS, 1983.

Hawaii 1:850 000
Chicago: Rand McNally, 1986.
Includes townplan of Honolulu.

State of Hawaii: principal islands 1:500 000
Reston, VA: USGS, 1971.
Topographic map, also available as 2-colour State edition and as Relief edition.

Reference maps of the islands of Hawaii
1:250 000–1:150 000 2nd edn
Honolulu: University of Hawaii Press, 1983.
5 sheets, all published.

Topographic

Hawaiian Islands 1:250 000
Reston, VA: USGS, 1961–.
4 sheets, all published.

Topographic quadrangles 1:24 000
Reston, VA: USGS, 1952–.
122 sheets, all published.

Administrative

State of Hawaii: county map series 1:100 000
Reston, VA: USGS, 1980–.
7 sheets, all published.
Sheets for Honolulu and Maui counties only available as monochrome planimetric editions.

Geological and geophysical

Hawaii 1:500 000
Tulsa, OK: AAPG, 1974.
Geological highway map series 8.

Geological quadrangles 1:24 000
Reston, VA: USGS, 1967–.
122 sheets, 4 published.

Environmental

Hawaii: hydrological units 1:500 000
Reston, VA: USGS, 1974.

Town maps

Honolulu 1:37 000
Chicago: Rand McNally.

Kiribati

The 30 islands of the Gilbert group became the independent state of Kiribati in 1979. Mapping of the group is the responsibility of the **Land and Survey Division (KLSD)**. Thirty-one dyeline atoll maps are available for most of the islands in the group, the scale depending on the size of the island. In addition South Tarawa is mapped at 1:2500 (19 sheets) and Betio the capital in four 1:1250 scale plans. Most of these maps were produced with the technical assistance of the British **Directorate of Overseas Surveys (DOS)** (now Overseas Surveys Directorate, Southampton, OSD). Full-colour maps of Kiribati are mostly published as orthophotomaps issued by DOS. A 1:25 000 scale series DOS 367P (X 042) covers most of the group. This 25-sheet series was completed in 1983 and uses the Transverse Mercator projection. A photomap is published at a scale of 1:12 500 for Marakei Island. A revised edition of the 1:50 000 sheet for Christmas Island (Kiritimati) was issued in 1984 with DOS aid and a new 1:50 000 scale photomap of the main island Tarawa was also published in 1985.

A general administrative map of the Republic was published by KLSD with DOS in 1984. A variety of thematic mapping of Christmas Island (Kiritimati) was compiled by DOS in the 1960s, including 1:50 000 topographic, soil and groundwater maps and 1:25 000 coverage of vegetation, coconut plantations and plantation development. None is any longer available.

Addresses

Overseas Surveys Directorate (OSD/DOS)
Ordnance Survey, Romsey Rd, Maybush, SOUTHAMPTON SO9 4DH, UK.

Land and Survey Division (KLSD)
Ministry of Home Affairs and Decentralization, PO Box 7, Bairiki, TARAWA.

Catalogue

General

Kiribati 1:2 000 000 DOS 867
Tolworth and Bairiki: DOS and KLSD, 1984.
2 sheets, both published.

Topographic

Kiritimati 1:50 000   DOS 436 (Series X782)   6th edn
Bairiki: KLSD, 1984.

Tarawa 1:50 000   DOS 467P
Bairiki: KLSD, 1985.

Republic of Kiribati 1:25 000   DOS 367P (X 042)
Tolworth: DOS, 1977–83.
25 sheets, all published. ■

Marakei Island 1:12 500   DOS 6007 (X884)
Tolworth: DOS, 1972.

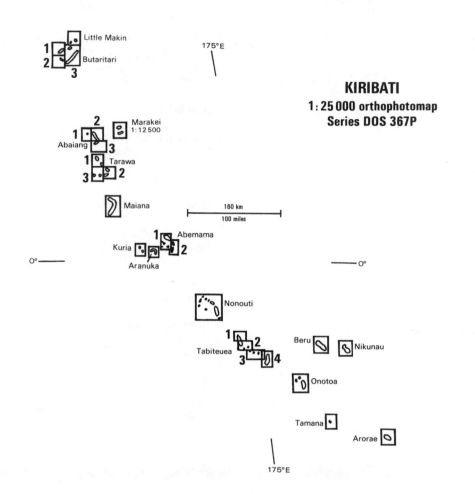

KIRIBATI
1:25 000 orthophotomap
Series DOS 367P

Nauru

The Republic of Nauru became independent in 1968 and is mapped by the **Directorate of Lands and Survey (NDLS)**, which has recently produced a new full-colour 1:10 000 scale map of the island. This small four-colour map shows relief by spot heights.

Addresses

Directorate of Lands and Survey (NDLS)
Lands Office, YAREN, Republic of Nauru.

For DMA see United States.

Catalogue

Atlases and gazetteers

Fiji, Tonga and Nauru: official standard names approved by the United States Board on Geographic Names
Washington, DC: DMA, 1974.
vi, 231 pp.

Topographic

Republic of Nauru 1:10 000
Yaren: NDLS, 1984.

New Caledonia

New Caledonia is a large island and French Dependent Territory in the Pacific. As such, mapping activities have been the responsibility of the various French survey authorities. Topographic coverage of the island has been carried out by the French **Institut Géographique National (IGN)**, and is now the responsibility of the **Service Topographique**. The basic map is the 46-sheet 1:50 000 scale series, which was completed in 1962. It was compiled from aerial photographic coverage using the standard French five-colour style of the time, and is drawn on a UTM projection. From this series was prepared an eight-colour, five-sheet derived map at 1:200 000, which shows both topography and communications. New aerial coverage at 1:40 000 scale was flown in 1982 and a revision programme for the 1:50 000 series is currently under way. Work is in progress on the sheets covering the southern coast and islands of Ouvéa and Lifou. Other recent mapping activity has been at smaller scales aimed at the tourist market, and the French are currently compiling a new two-sheet 1:200 000 scale map of the territory to replace the 1960s edition.

Geological mapping of the island has been carried out by the French **Bureau de Recherches Géologiques et Minières (BRGM)**, and its successor in New Caledonia, the **Service des Mines et de l'Energie (NCSME)**. A full-colour 1:50 000 series on mostly the same sheet lines as the topographic map is in progress and available now from NCSME. This series covers the main island only, and two sheets are issued to include double topographic sheets. The two-sheet 1:200 000 series is derived from the larger scale.

The French **Office de la Recherche Scientifique et Technique Outre Mer (ORSTOM)** has published a soil and land capability map and explanation at 1:1 000 000 scale, and is responsible for the compilation of the important *National Atlas*. A new 1:50 000 scale soil map is under way.

Further information

Topographic mapping of French Territories of the Pacific, paper submitted by France to the 1977 UN Regional Cartographic Conference on Asia and the Pacific, in *UN Regional Cartographic Conference on Asia and the Pacific*, New York: UN, 1980. Activities of the French Institut Géographique National in Asia and the Pacific: paper presented to the *11th UN Regional Cartographic Conference for Asia and the Pacific*, New York: UN, 1987.

For information about the ORSTOM atlas see Combroux, Jean (1985) L'Atlas de la Nouvelle Calédonie et dépendances: une activité cartographique de l'ORSTOM, *International Yearbook of Cartography*, **XXV**, 31–36.

Addresses

Institut Géographique National
136 rue de Grenelle, PARIS 75700, France.

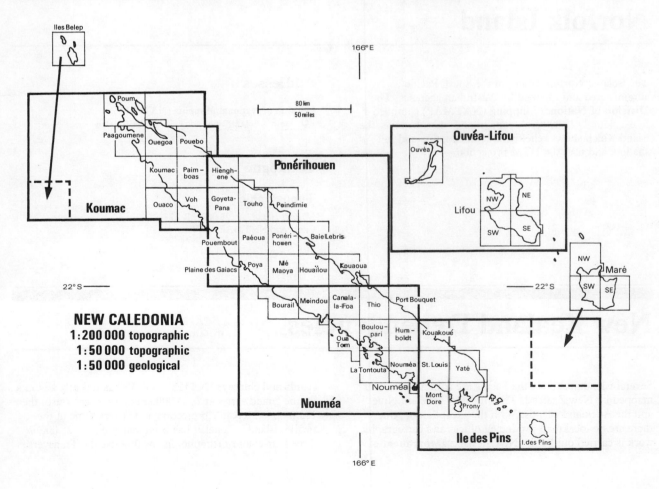

Service Topographique (NCST)
Nouvelle Calédonie et Dependances, Ave Paul Doumer, BP A2,
NOUMEA.

Service des Mines et de l'Energie (NCSME)
Nouvelle Calédonie et Dependances, BP 465, NOUMEA.

Bureau de Recherches Géologiques et Minières (BRGM)
Ave de Concyr, BP6009, ORLEANS-CEDEX 45018, France.

Office de la Recherche Scientifique et Technique Outre Mer (ORSTOM)
70–74 route d'Aulnay, BONDY 93140, France.

For DMA see United States.

Catalogue

Atlases and gazetteers

Atlas de la Nouvelle Calédonie et dependances
Bondy: ORSTOM, 1981.
53 sheets + explanations.

New Caledonia and Wallis and Futuna: official standard names approved by the United States Board on Geographic Names
Washington, DC: DMA, 1974.
pp. 100.

General

Carte internationale du monde: Nouvelle Calédonie 1:1 000 000
Paris: IGN, 1965.

Nouvelle Calédonie 1:500 000 Ed. 2
Paris: IGN, 1981.
Tourist map, also available as relief edition published in 1967.

Topographic

Carte de la Nouvelle Calédonie 1:200 000
Paris: IGN, 1965–66.
5 sheets, all published. ■

Carte de la Nouvelle Calédonie 1:50 000
Paris: IGN, 1956–62.
46 sheets, all published. ■

Geological and geophysical

Carte géologique de la Nouvelle Calédonie 1:200 000
Orléans: BRGM, 1981.
2 sheets, both published.
With accompanying explanatory text.

Territoire de la Nouvelle Calédonie: carte géologique 1:50 000
Paris: BRGM, 1956–.
38 sheets, 28 published. ■
Series excludes Loyalty Islands.

Environmental

Études des sols de la Nouvelle Calédonie 1:1 000 000
Bondy: ORSTOM, 1978.
2 sheets, both published.
Soil map and land use potential map.

Carte des sols et carte d'aptitude culturale et forestière de Nouvelle Calédonie 1:50 000
Bondy: ORSTOM, 1985–.
45 sheets, 1 published.

Norfolk Island

The isolated Norfolk Island in the South Pacific is
administered and mapped by Australian agencies. The
Division of National Mapping (NATMAP) compiled a
four-colour 1:25 000 scale topographic sheet to cover the
island which shows relief with 20 m contours and hill
shading and uses the UTM projection.

Addresses

Division of National Mapping (NATMAP)
PO Box 31, BELCONNEN, ACT 2616, Australia.

Catalogue

Topographic

Norfolk Island 1:25 000 R 873
Canberra: NATMAP, 1971.

New Zealand Dependencies

Several island groups in the Pacific are administered and
mapped by New Zealand. They include Tokelau, Niue
and the Auckland islands. With the exception of Niue
there are no local map-producing offices and cartographic
work is carried out by the New Zealand **Department of**

Lands and Surveys (NZMS). NZMS has recently begun a
Pacific Island series at 1:25 000 scale. This series uses the
Transverse Mercator projection and covers some of the
smaller islands. Tokelau has been mapped at 1:25 000 in
three four-colour orthophotomaps that use the Transverse

Mercator projection, and are published in the same series. Niue is mapped at 1:50 000 in a four-colour sheet with 20 m contours, with an inset map at 1:10 000 of the main settlement. The Tokelau group is covered in a general map at 1:750 000 with insets at 1:100 000 of the main islands. Little thematic mapping has been carried out, the exception being for Niue, where the **NZ Soil Bureau (NZSB)** has published a 1:63 360 scale soil map, and where a 1950s-produced **New Zealand Geological Survey (NZGS)** geology and hydrological map is still available.

Addresses

Department of Lands and Survey (NZMS)
Charles Fergusson Building, Bowen Street, Private Bag, WELLINGTON, New Zealand.

The Registrar and Director of Survey
Justice Department, PO Box 77, Alofi, NIUE.

For NZSB and NZGS see New Zealand.

Catalogue

General

General map of the Tokelau Islands 1:750 000 NZMS 254
Wellington: NZMS, 1969.
With inset maps of main islands at 1:100 000.

Auckland islands 1:95 000 NZMS 220
Wellington: NZMS, 1962.
Single colour map.

Map of Niue 1:50 000 NZMS 250
Wellington: NZMS, 1977.
With inset map of Alofi 1:12 500 scale.

Pacific islands series 1:25 000 NZMS 272
Wellington: NZMS, 1983–.
7 sheets published.
Series covers Snares and Bounty Islands, Antipodes Islands, Raoul Island, and Tokelau Group (Atafu, Fakaofo and Nukuonu).

Geological and geophysical

The geology and hydrology of Niue island South Pacific
Lower Hutt: NZGS, 1959.

Environmental

Soil map of Niue island 1:63 360
Lower Hutt: NZSB, 1958.
With accompanying explanatory booklet.

Pacific Islands Trust Territory

The Pacific Islands Trust Territory consists of the Micronesian island groups of the Mariana, Caroline and Marshall Islands and is administered and mapped by the United States. Guam is administered separately but geographically and cartographically falls into this grouping. Few topographic maps have been produced of these islands. The **United States Geological Survey (USGS)** issues a 1:24 000 nine-sheet series to cover Guam, which uses 25 foot contours and is drawn on a modified azimuthal equidistant projection. USGS has recently begun to publish 1:25 000 scale maps of the islands of Ponape (two sheets), Truk (six sheets), the Palau group (five sheets), and Yap (one sheet) in Micronesia and of certain islands in the Marianas group (Rota, Saipan and Tinian). These newer maps are issued with 10 m contours. A general map of the whole of the territory was published by the **Defense Mapping Agency Topographic Center (DMATC)** in 1985.

In Guam information about the mapping of the territory may be obtained from the **Department of Land Management**, the **Department of Public Works** and the **Bureau of Planning**.

Addresses

United States Geological Survey (USGS)
National Center, 12201 Sunrise Valley Drive, RESTON, VA 22092, USA.

Department of Land Management
Government of Guam, PO Box 2950, AGANA, Guam, USA 96910.

Department of Public Works
Government of Guam, PO Box 2950, AGANA, Guam, USA 96910.

Bureau of Planning
Government of Guam, PO Box 2950, AGANA, Guam, USA 96910.

Division of Lands and Surveys
Lands and Claims Office, SAIPAN, Mariana Islands, USA 96950.

For DMATC see United States.

Catalogue

Atlases and gazetteers

Guide to the placenames of the Trust Territories of the Pacific
E. Brian
Honolulu: BP Bishop Museum, 1971.

General

Trust territory of the Pacific Islands 1:4 000 000 DMATC 9203
Washington, DC: DMATC, 1985.

Topographic

Mariana Islands: Topographic map of Guam 1:50 000
Reston, VA: USGS, 1978.

Micronesia: Island of Truk   1 : 25 000
Reston, VA: USGS, 1985.
6 sheets, all published.

Micronesia: Island of Yap   1 : 25 000
Reston, VA: USGS, 1986.

Micronesia: Island of Ponape   1 : 25 000
Reston, VA: USGS, 1984.
2 sheets, both published.

Micronesia: Palau group   1 : 25 000
Reston, VA: USGS, 1983.
5 sheets, all published.

Marianas   1 : 25 000
Reston, VA: USGS, 1984–.
3 sheets, all published.
Sheets cover Rota, Saipan and Tinian.

Mariana Islands: Guam   1 : 24 000
Washington, DC: USGS, 1968.
9 sheets, all published.

Solomon Islands

The **Survey and Mapping Division (SISMD)** in the
Ministry of Agriculture and Lands maintains 1 : 50 000
base mapping of the Solomon Islands which was prepared
by the British **Directorate of Overseas Surveys (DOS)**
(now Overseas Surveys Directorate, Southampton, OSD)
in the 1960s and early 1970s. These four-colour maps have
40 m contour intervals, use UTM grid and were compiled
from British flown aerial coverage. Most sheets have not
been revised since compilation, but some new maps in this
series have recently appeared, produced by the Royal
Australian Survey Corps and by SISMD. In the 1970s the
Survey Division produced derived 1 : 150 000 scale 16-sheet
coverage, also on UTM projection, from the DOS
mapping. Effort is now being concentrated on 1 : 10 000
mapping of areas of economic importance, and 1 : 2500
scale town mapping projects. Large-scale aerial coverage of
many areas has recently been flown with New Zealand aid.
Cadastral surveying activities are also carried out by
SISMD and a gazetteer was produced to record names
appearing on the 1 : 150 000 scale maps.

Geological mapping is carried out by the **Solomon
Islands Geological Survey (SIGS)** within the Ministry of
Natural Resources. SIGS was established in 1950 but most
mapping has been produced through overseas aid and
technical assistance in the 1960s and 1970s. Small-scale
coverage is available at 1 : 1 000 000 to show either geology
or mineral occurrences, the latter with an explanatory
monograph. A full-colour 1 : 50 000 scale geology series has
been in progress since the early 1970s, and now covers
about 75 per cent of the group.

Resources mapping of the group was compiled for the
Land Resources Development Centre (LRDC) by the
British DOS in the early 1970s. The maps were published
in *Land Resource Study 18* and cover physiographic
regions, catchment areas, soil sample sites, soil
associations, land systems and land regions, land use,
agricultural opportunity areas, and forest types at
1 : 250 000 and 1 : 150 000 scale for each island. The eight
maps for each island were issued with a volume of
accompanying text per island. General 1 : 1 000 000 scale
coverage of the whole group was also published, covering
geology, landforms, population distribution and
agricultural opportunity areas. Single-sheet soil mapping of
the group at 1 : 1 000 000 scale was also compiled as part of
the same survey. These full-colour maps are still available
from LRDC.

Further information

Catalogues and indexes are published by SISMD and SIGS.

Background information is contained in Solomon Islands (1980)
Report of Cartographic activities in the Solomon Islands presented
to the *9th United Nations Regional Cartographic Conference for Asia
and the Pacific*, New York: UN.

Addresses

Survey and Mapping Division (SISMD)
Ministry of Agriculture and Lands, PO Box G24, HONIARA.

Overseas Surveys Directorate (DOS/OSD)
Ordnance Survey, Romsey Rd, Maybush, SOUTHAMPTON
SO9 4DH, UK.

Solomon Islands Geological Survey Division (SIGS)
Ministry of Natural Resources, PO Box G24, HONIARA.

For LRDC see Great Britain.

Catalogue

Atlases and gazetteers

*Solomon islands gazetteer: a list of place names in the British Solomon
Islands Protectorate*
Honiara: SISMD, 1969.
pp. 283.

General

Solomon islands   1 : 3 000 000
Honiara: SISMD, 1982.

Map of the Solomon islands   1 : 1 000 000
Honiara: SISMD, 1985.
2 sheets, both published.

Topographic

Solomon islands   1 : 150 000
Honiara: SISMD, 1970–.
16 sheets, all published. ■

Solomon islands   1 : 50 000   DOS 456 (X711)
Honiara and Tolworth: SISMD and DOS, 1965–.
83 sheets, all published. ■

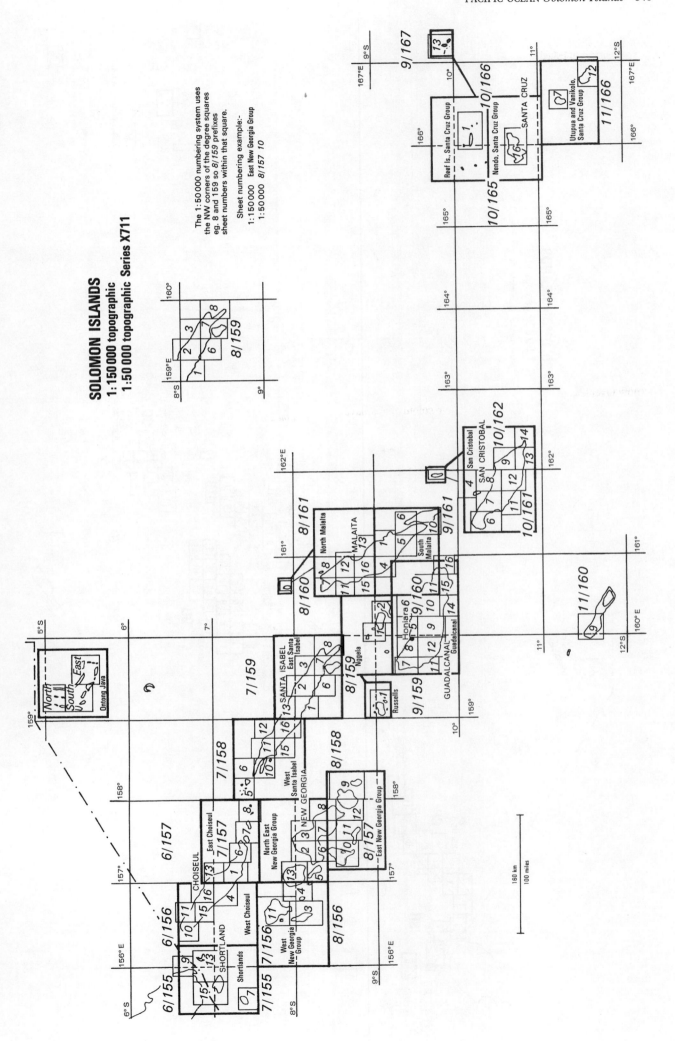

SOLOMON ISLANDS
1:150 000 topographic
1:50 000 topographic Series X711

The 1:50 000 numbering system uses the NW corners of the degree squares eg. 8 and 159 so 8/159 prefixes sheet numbers within that square.

Sheet numbering example:-
1:150 000 East New Georgia Group
1:50 000 8/157 10

SOLOMON ISLANDS
1:50 000 geological
Sheet numbering example SC 8

Geological and geophysical

Mineral occurrences map of the Solomon Islands   1 : 1 000 000
Honiara: SIGS, 1980.
With explanatory monograph.

Geological map of the British Solomon Islands   1 : 1 000 000   DOS
Geol 1145A   2nd edn
Honiara and Tolworth: SIGS and DOS, 1970.
1 sheet.

Solomon islands: geological map   1 : 50 000
Honiara: SIGS, 1969–.
ca 90 sheets, *ca* 60 sheets published. ■

Environmental

Land resources of the Solomon Islands   1 : 1 000 000, 1 : 250 000 and
1 : 100 000
Tolworth: DOS for LRDC, 1974.
60 sheets, all published.
With 8 accompanying explanatory volumes.

The soils of the Solomon islands   1 : 1 000 000
Tolworth: DOS for LRDC, 1974.
4 sheets, all published.
Maps cover sample sites, landforms, structure and lithology, and
soil associations.

Tonga

Mapping of Tonga is carried out by the **Ministry of
Lands, Survey and Natural Resources (TMLS)**. The
basic map series for this group is published jointly with the
British **Directorate of Overseas Surveys (DOS)** (now
Overseas Surveys Directorate, Southampton, OSD) and
was completed in 1976. This is the 23-sheet *1 : 25 000 Series
X872 (DOS 337)*, published on UTM projection in five
colours with 10 m contours. There are no plans to update
this map or to produce locally any other mapping. DOS
has published a 1 : 50 000 photomap of the main island
Tongatapu, and has published a small-scale general map of
the group, now very dated but still available. Tongatapu is
also covered by a single-sheet 1 : 100 000 scale full-colour
soil map prepared by the **New Zealand Soil Bureau
(NZSB)**. 1 : 25 000 scale soils mapping of some islands in
the group has also been compiled by NZSB.

Addresses

Ministry of Lands and Survey (TMLS)
PO Box 5, NUKU'ALOFA.

Overseas Surveys Directorate (DOS/OSD)
Ordnance Survey, Romsey Rd, Maybush, SOUTHAMPTON
SO9 4DH, UK.

For NZSB see New Zealand; for DMA see United States.

Catalogue

Atlases and gazetteers

*Fiji, Tonga and Nauru: official standard names approved by the
United States Board on Geographic Names*
Washington, DC: DMA, 1974.
pp. 231.

General

Tonga islands   1 : 2 000 000   DOS 990   Ed. 2
Tolworth: DOS, 1960.

Topographic

Tongatapu Island   1 : 50 000   Series X773 (DOS 6005)
Tolworth: DOS, 1971.

Kingdom of Tonga   1 : 25 000   Series X872 (DOS 337)
Nuku'alofa and Tolworth: TMLS and DOS, 1976–77.
23 sheets, all published. ■

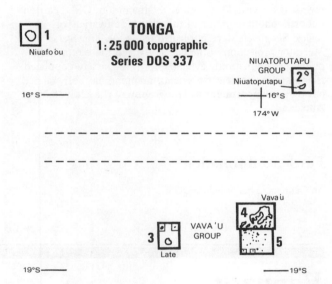

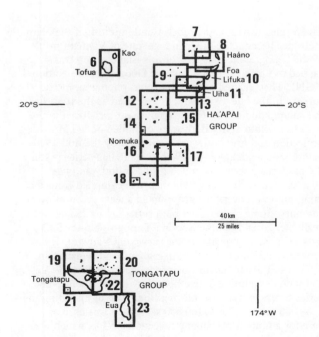

Environmental

Soil map of Tongatapu 1 : 100 000
Lower Hutt: NZSB, 1972.
With accompanying explanatory report published in 1976.

Soil map of the Kingdom of Tonga 1 : 25 000
Lower Hutt: NZSB, 1981–.
3 sheets, all published.
Sheets published cover island of Eua and parts of Vava'u and
Ha'apai groups.

Tuvalu

Tuvalu became independent in 1978, having formerly been
part of the Gilbert and Ellice Islands Colony. The **Lands
and Survey Division (TLSD)** of the Ministry of
Commerce and Natural Resources is responsible for
maintaining the map base for the territory, but mapping
has been carried out with overseas technical assistance.
The British **Directorate of Overseas Surveys (DOS)** (now
Overseas Surveys Directorate, Southampton, OSD) carried
out orthophoto mapping of the islands at a variety of
scales. Six sheets were published using the Transverse
Mercator projection in the late 1970s and these remain the
only accurate mapping available for this sparsely populated
island group. Some earth science mapping has been carried
out by the **Department of Geography of the University of
Auckland.**

Further information

Annual reports published by DOS.

Addresses

Lands and Surveys Division (TLSD)
Ministry of Commerce and Natural Resources, Vaiaku, FUNAFUTI.

Overseas Surveys Directorate (DOS/OSD)
Ordnance Survey, Romsey Rd, Maybush, SOUTHAMPTON
SO9 3DH, UK.

For University of Auckland see New Zealand.

Catalogue

Topographic

Tuvalu various scales Series X041
Vaiaku and Tolworth: TLSD and DOS, 1975–80.
6 sheets, all published.
Photomaps cover Vaitupu 1 : 10 000 (DOS 238P); Funafuti
1 : 50 000 (DOS 468P); and in Series DOS 368P at 1 : 12 500
Nanumea and Niutao; Nui and Nanumanga; and Nukulaelae and
Niulakita; and at 1 : 25 000 Nukufetau.

Vanuatu

The former British and French Condominium of the New
Hebrides became the independent state of Vanuatu in
1980. Topographic mapping of this island group was
carried out by the French **Institut Géographique National
(IGN)**. They carried out geodetic and photogrammetric
surveys in the late 1950s which were used as the basis for
the compilation of two topographic series produced in the
IGN five-colour style. Both were compiled on UTM
projection and grid and were completed by the mid-1970s
before independence. The 29-sheet 1 : 50 000 series is still
the basic scale series for much of the country and uses
20 m contours; the 1 : 100 000 map was compiled from the
same surveys, covers the group in 15 sheets and uses 40 m
contours. Some sheets from both series are no longer
available. After 1980 the **Service Topographique (STV)**
took over the responsibility for surveying Vanuatu. It has
concentrated on cadastral mapping and land registration. A
revised 1 : 1 000 000 map was produced in 1982 by STV,
and includes administrative boundary information. New
aerial coverage has been flown with a view to up-dating the
1 : 50 000 series. The 1 : 100 000 scale map was used as a
base for a limited placename revision by STV, which was
printed in black and white in 1982 by the Australian
Army. Amongst the other projects undertaken by ST with

foreign aid is the urban mapping of Port Vila and
Luganville. Full-colour 1 : 10 000 scale town maps were
published for both by IGN in the early 1980s. An official
gazetteer was compiled in 1979.

Geological coverage of the islands has been undertaken
by the British **Directorate of Overseas Surveys (DOS)**
(now Overseas Surveys Directorate, Southampton, OSD)
who began an 11-sheet 1 : 100 000 series in 1972. This full-
colour series is now complete and is distributed by the
Geological Survey in Port Vila. A 1 : 1 000 000 geological
map is currently being revised by this organization, and
some copies of the original edition are still available from
Edward Stanford in London.

The French organization **Office de la Recherche
Scientifique et Technique Outre Mer (ORSTOM)** has
compiled a major thematic atlas of the islands, which
covers all the group in a variety of medium-scale natural
resources maps. ORSTOM has also published 1 : 500 000
and 1 : 100 000 scale agronomic potential and land use
maps of the group, which were issued with a bilingual
explanatory volume, but these are no longer available. A
single-sheet bathymetric map compiled by ORSTOM also
covers the whole archipelago.

Further information

Topographic mapping of French Territories of the Pacific, paper submitted by France to the 1977 UN Regional Cartographic Conference on Asia and the Pacific, in *UN Regional Cartographic Conference on Asia and the Pacific, technical papers*, New York: UN, 1980.

Addresses

Institut Géographique National (IGN)
136 bis rue de Grenelle, 75700 PARIS, France.

Overseas Surveys Directorate (OSD/DOS)
Ordnance Survey, Romsey Rd, Maybush, SOUTHAMPTON SO9 4DH, UK.

Service Topographique (STV)
GPO, PORT VILA.

Geological Survey of Vanuatu
Department of Geology, Mines and Rural Water Supplies, GPO, PORT VILA.

Office de la Recherche Scientifique et Technique Outre Mer (ORSTOM)
70–74 route d'Aulnay, BONDY 93140, France.

Catalogue

Atlases and gazetteers

New Hebrides: official standard names approved by the United States Board on Geographic Names
Washington, DC: DMA, 1974.
pp. 76.

General

Republic of Vanuatu 1 : 1 000 000
Port Vila: STV, 1982.

Topographic

Carte de la Melanésie: archipel des Nouvelles Hebrides 1 : 100 000
Paris: IGN, 1965–71.
15 sheets, all published. ■

Carte de la Melanésie: archipel des Nouvelles Hebrides 1 : 50 000
Paris: IGN, 1961–76.
29 sheets, all published. ■

Bathymetric

Carte bathymetrique des parties centrale et méridionale de l'arc insulaire des Nouvelles Hebrides 1 : 1 036 358 (at 19 South)
Bondy: ORSTOM, 1984.

Geological and geophysical

Geological map of the Republic of Vanuatu 1 : 1 000 000 2nd edn
Vila: Geological Survey, 1975.

New Hebrides geological survey 1 : 100 000
Tolworth: DOS, 1972–81.
11 sheets, all published. ■

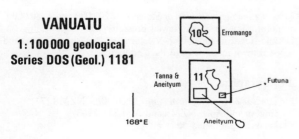

VANUATU
1 : 100 000 geological
Series DOS (Geol.) 1181

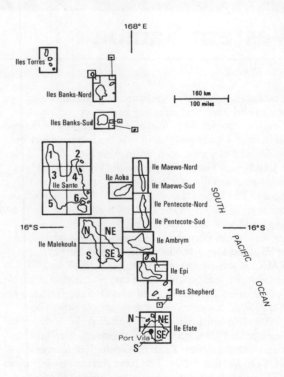

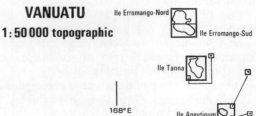

VANUATU
1 : 50 000 topographic

Environmental

Nouvelles Hebrides: atlas des sols et de quelques données du milieu naturel   P. Quantin
Bondy: ORSTOM, 1978.
7 vols of maps + texts.

Town maps

Plan de Port Vila   1 : 10 000
Paris: IGN, 1981.

Plan de Luganville   1 : 10 000
Paris: IGN, 1981.

Wallis and Futuna Islands

The French overseas territory of Wallis and Futuna is mapped by the French **Institut Géographique National (IGN)**. The first modern maps of the islands were prepared from 1982 flown 1 : 20 000 aerial coverage, and published as a two-sheet, four-colour 1 : 25 000 series in 1986. These maps use UTM projection and 20 m contour interval.

Addresses

Institut Géographique National (IGN)
136 bis rue de Grenelle, PARIS 75700, France.

Catalogue

Topographic

Wallis et Futuna   1 : 25 000
Paris: IGN, 1986.
2 sheets, both published.

Western Samoa

The island group of Western Samoa was administered by New Zealand until becoming the first independent Polynesian State in 1962. Much of the mapping of the two main islands, Upolu and Savaii, has been carried out by New Zealand mapping agencies or with New Zealand aid. The **Lands and Survey Department (WSLSD)** in Apia now has responsibility for topographic surveying and mapping. The basic scale series is a 28-sheet 1 : 20 000 six-colour map, which uses the Cassini–Soldner projection and a UTM grid and shows relief with 50 foot contours. This series was begun after independence with aid from the **New Zealand Department of Lands and Surveys (NZMS)**; it was completed by the early 1970s and has since been revised. Smaller-scale mapping is also available, including diazo 1 : 100 000 maps of each main island, a new 1 : 200 000 full-colour map of the whole country, and a 1 : 10 000 scale town map of the capital, Apia.

Geological, hydrological and bathymetric mapping was carried out by the **New Zealand Geological Survey (NZGS)** in the late 1950s. The **New Zealand Soil Bureau (NZSB)** carried out soil and land use surveys of Western Samoa in the 1950s, which were published with explanatory text in 1963. A 1 : 40 000 scale soil map of Upolu island in eight sheets was compiled for this survey using an orthophoto base. The 1 : 95 040 scale maps from this publication are also still available separately.

A two-sided tourist map of the islands was compiled by James Bier in the Reference Map Series and published by **University of Hawaii Press** in the early 1980s.

Addresses

Lands and Survey Department (WSLSD)
Main Beach Rd, PO Box 63, APIA.

Department of Lands and Surveys (NZMS)
Charles Fergusson Building, Bowen Street, Private Bag, WELLINGTON, New Zealand.

For NZGS and NZSB see New Zealand; for University of Hawaii Press see Hawaii.

Catalogue

General

Western Samoa   1 : 400 000   NZMS 279
Wellington and Apia: NZMS and WSLSD, 1979.

Western Samoa   1 : 200 000   NZMS 297
Wellington and Apia: NZMS and WSLSD, 1984.

Reference map of the Island of Samoa   1 : 200 000 and 1 : 80 000
Honolulu: University of Hawaii Press, 1980.
Double-sided, includes both Western and American Samoa, and indexed town maps of Pago Pago and Apia.

Topographic

Western Samoa topographical map   1 : 20 000   NZMS 174
Wellington and Apia: NZMS and WSLSD, 1963–.
28 sheets, all published. ■

WESTERN SAMOA
1:20 000 topographic Series NZMS 174

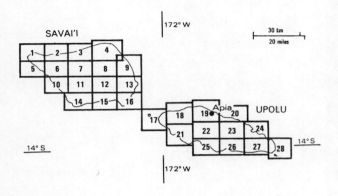

Geological and geophysical

The geology and hydrology of Western Samoa 1:100 000
Lower Hutt: NZGS, 1959.
5 sheets, all published.
2 geology, 2 hydrology and 1 bathymetric maps and accompanying explanatory text.

Environmental

Soils and land use of Western Samoa 1:95 040 and 1:40 000
Lower Hutt: NZSB, 1963.
12 sheets, all published.
2 soil maps and 2 land classification maps at 1:95 040; eight 1:40 000 scale soil maps of Upolu and accompanying explanatory text.

Town maps

Map of Apia city and environment 1:10 000
Apia: WSLSD, 1980.

THE POLAR REGIONS

The polar regions, although sharing certain common features and problems, are only rarely treated together cartographically. A number of world series maps cover the two regions to a consistent scale and specification but on separate sheets. Further information on these, and other, maps is given below under the respective sections on the Arctic and Antarctica.

Addresses

For CIA, PennWell and Woods Hole, see United States; for Admiralty Hydrographic Department, see Great Britain; for Ministry of Geology of the USSR, see USSR.

Atlases and gazetteers

Polar regions atlas
Washington, DC: CIA, 1978.
pp. 66.
Includes foldout maps of *Arctic region* and *Antarctic region*   1 : 17 250 000.

Geological and geophysical

The polar regions. Magnetic variation 1985 and annual rates of change reduced to the Epoch 1985.0   1 : 16 000 000
Taunton: Admiralty Hydrographic Department, 1985.

Polar oil and gas   1 : 10 000 000
Tulsa: PennWell, 1985.

Tektoniceskaja karta pol'arnych oblastej zemli   1 : 10 000 000
B. Ch. Egiazarov
Moskva: Ministry of Geology of the USSR.
Tectonic map of polar regions.
4 sheets, all published.

Administrative

Prospective maritime jurisdictions in the polar seas   1 : 20 000 000/ 1 : 25 000 000   M.E. Vaucher
Woods Hole, MA: Woods Hole Oceanographic Institution, 1983.

ARCTIC

In contrast to Antarctica, rather few maps have been published of the Arctic region, due presumably to the lack of a north polar land area, and to a dearth of detailed knowledge of the bathymetry and geology of the seabed. This situation is beginning to change, however.

The USSR has produced the most extensive range of arctic mapping, both in sheet and atlas form. A comprehensive oceanographic atlas of the Arctic Ocean was published in 1980 in the *World Ocean Atlas* series, but generally the best scale in this volume is only 1:15 000 000. However, in 1985 a major new synthesis of knowledge about the Arctic—*Atlas Arktiki* — was published by **Glavnoe Upravlenie Geodezii i Kartografii (GUGK)**, Moscow. This volume contains a large number of both regional and general maps covering a wide range of themes. There are also 140 pages of text (with English summaries). The contents are given in Russian and English. The maps include geographical maps of the arctic land areas at 1:500 000, and smaller-scale maps (generally 1:10 000 000 or 1:20 000 000) of geology, climate, hydrology, oceanography, seabed sediments, soils, vegetation, and population and administration. The atlas also includes some detailed glaciological and permafrost maps.

Another major piece of Arctic cartography is the bathymetric chart of the Arctic Ocean published by the **Geological Society of America (GSA)** in 1986. This was compiled by the Naval Research Laboratory from all data known to them to July 1986.

Good single-sheet cover of the Arctic is also provided by some world series, notably sheet 5.17 of GEBCO, 1:6 000 000, published 1979 and sheet 19 of the *World Geological Atlas* at 1:16 000 000, published 1983. (These series are more fully described in the World and Oceans sections.)

GUGK has published several general maps of the Arctic, while the **Ministry of Geology of the USSR (MGUSSR)** and its related research institutes have produced geological and tectonic sheet maps. Some of the small-scale geological maps we have listed may be difficult to obtain, although all are believed to be in print at the time of writing.

With the interest in possible petroleum discoveries in the Barents Sea, Norway has carried out geological and seismic investigations in this area and a number of maps have been published in association with these investigations. The Norsk Polarinstitutt (NPI) *Skrifter* No. 180 is devoted to 'Geoscientific investigations in the Barents and Greenland–Norwegian seas' and includes free-air gravity anomaly maps of these areas. NPI is developing a regional database of the entire Barents Sea, and a three-dimensional digital map at 1:500 000 has been generated from it.

Similar resource investigations in Arctic Alaska and Canada have included interest in the seafloors of Prudhoe Bay, Baffin Bay and the Beaufort Sea. The Canadian Arctic is also being covered by a series of 1:250 000 scale *Natural Resource Maps* prepared by the Canadian Hydrographic Service, but only a few sheets have so far been published (see under Canada).

Addresses

Glavnoe Upravlenie Geodezii i Kartografii (GUGK)
ul. Krahizhanovaskogo 14, korp. 2, V-218 MOSKVA, USSR.

Ministry of Geology USSR (MGUSSR)
Bolshaya Gruzinskaya 4, 12324, D-242 MOSKVA, USSR.

Chief Administration for Navigation and Oceanography
Glavnoe Upravlenie Navigatsii i Okeanografie (GUNO)
Ministetsva Oborony, 8111 Ininiya B34, LENINGRAD, USSR.

Research Institute of Arctic Geology (NIIGA)
Sredniy Prospekt 71, 199026, LENINGRAD.

Geological Society of America (GSA)
Box 9140, BOULDER, CO 80301, USA.

For NGS, AGS and CIA, see United States; for GSC, see Canada; for NPI see Svalbard.

Catalogue

Atlases and gazetteers

Atlas Arktiki
Moscow: GUGK, 1985.
pp. 204.

Atlas okeanov. Severnyy ledovityy okean
Leningrad: GUNO, 1980.
pp. 184.
Atlas of the oceans. Arctic Ocean.

Oceanographic atlas of the Bering Sea basin   M.A. Sayles *et al.*
Seattle, WA: University of Washington Press, 1980.
pp. 170.

General and bathymetric

Arctic region
Washington, DC: CIA, 1982.

Arctic Ocean floor 1 : 10 140 000
Washington, DC: NGS, 1976.
Painted by Heinrich Berann and Heinz Vielkind.
On the reverse: conventional line map.

Arktika. Fizicheskaya karta 1 : 8 000 000
Moskva: GUGK, 1981.

Map of the Arctic region 1 : 5 000 000
New York: AGS, 1975.
Sheet 14 of World Series.

Bathymetry of the Arctic Ocean 1 : 4 704 075 MC-56
R.K. Perry and H.S. Fleming
Boulder, CO: GSA, 1986.
Magnetic anomaly contours on the reverse.

Geological and geophysical

Geological map of the Bering Sea region (with Quaternary cover removed) 1 : 10 000 000
Moskva: MGUSSR, 1984.

Geological map of the Arctic 1 : 7 500 000
Calgary: Alberta Society of Petroleum Geologists, 1960.
Prepared for First International Symposium on Arctic geology.

Geological map of the northern polar regions of the earth 1 : 5 000 000
Moskva: MGUSSR, 1975.
6 sheets, all published.

Tectonic map of the northern polar regions of the earth 1 : 5 000 000
Moskva: MGUSSR, 1975.
6 sheets, all published.

Map of the Quaternary deposits of the Arctic and sub-Arctic 1 : 5 000 000
Moskva: MGUSSR, 1964.
4 sheets, all published.

Residual magnetic anomaly map chart of the Arctic Ocean region
Scales vary MC-53
Boulder, CO: GSA, 1985.

Geological map of the Bering Sea region (with the Quaternary cover removed) 1 : 2 500 000
Leningrad: NIIGA, 1984.
4 sheets + text, all published.

The physical environment Western Barents Sea 1 : 1 500 000
Oslo: NPI, 1983–84.
Sheet A: Surface sediment distribution.
Sheet B: Sediments above the upper regional unconformity.

Marine science atlas of the Beaufort Sea. Sediments edited by
B.R. Pelletier
Ottawa: GSC, 1984.

Administrative

Arktika. Politicheskaja karta 1 : 8 000 000
Moskva: GUGK.

Human and economic

Arktika. Ekonomicheskaja karta 1 : 8 000 000
Moskva: GUGK.

Greenland

The earliest genuine topographic survey was conducted in the north of Greenland from 1917 to 1923 during a series of scientific expeditions by Lauge Koch, and his map at 1 : 300 000 in 18 sheets is still available (*Map of North Greenland . . . surveyed by Lauge Koch in the years 1917–23*) from GID.

The Geodetic Institute of Denmark (**Geodætisk Institut, GID**) began surveying the settled coastal areas in 1927 which led to the production of a series of sheets at 1 : 250 000. Early sheets were based on plane table survey, but from 1932 to 1938 an extensive air survey was made of the coastal areas from which the photogrammetric compilation of subsequent sheets was made. Almost all the ice-free areas are now covered at this scale, and although no new sheets have been issued since 1972 it is planned to extend and up-date the series. Early sheets showed relief by hill shading and contours, with a brown tone for upland and green for lowland. Later sheets have a revised specification with no relief shading, although a green tone is retained for lowland, and ice-free upland is shown in a pale yellow. Contours are at 50 m intervals. Sheets in the series each cover 1° of latitude, with west coast sheets identified by a prefix V and east coast with a prefix Ø. A 1 : 50 000 series was also started in the 1930s, but was soon discontinued.

The GID also publishes several general single-sheet maps, including a 10-colour wall map at 1 : 2 000 000, and a map at 1 : 5 000 000 which shows towns, settlements and weather stations, and also the municipality and district boundaries. There are plans to revise these maps when new mapping from north Greenland is available.

Until recently, most of the inland ice of northern Greenland remained largely unsurveyed in detail, and, although there was an old American AMS series at 1 : 250 000, this was compiled without any reliable ground control and is known to have errors of up to 35 km in location, and scale errors of as much as 20 per cent. As recently as 1965, the chief of GID stated that the size of Greenland was imprecisely known because the island had not been accurately mapped. In 1977, GID obtained government approval to map all northern Greenland, and it was decided to produce orthophotomaps at a scale of 1 : 100 000 with overprinted contours. To this end, special high-altitude cover was flown in 1978 using a camera with superwide angle lens to give an image scale of 1 : 150 000. Colour infrared photos were taken at the same time to be used in support of future geological mapping. Publication of this series of approximately 100 sheets is in progress.

A programme of geological mapping has been carried out over many years by the Greenland Geological Survey (**Grønlands Geologiske Undersøgelse, GGU**). Initially this comprised a series of 1 : 100 000 bedrock geology sheets, but from 1976 it has been supplemented by a series at 1 : 500 000 which is intended eventually to cover all the ice-free areas in 14 sheets. Sheets of both bedrock and Quaternary geology are being issued in this series.

In 1975 and 1976, 52 000 line km of aeromagnetic data were acquired, and after editing and gridding this data, a

1 : 500 000 coloured contour map was produced in 1984 covering parts of southern and central West Greenland. This accompanies *GGU Rapport* 122 by Leif Thorning. In 1983, a new high-sensitivity aeromagnetic survey was successfully carried out over part of the Inland Ice in cooperation with the Geological Survey of Canada and the Canadian National Aeronautical Establishment.

In 1970 a tectonic/geological map was published showing the whole island in a single sheet at 1 : 2 500 000. The map uses 58 colour tones to differentiate the solid geology with structural data overprinted in black, and both surface and basal contours of the inland ice are given. A companion map of Quaternary geology was published the following year. A new edition of the tectonic/geological map is planned.

In 1979, Greenland was granted home rule from Denmark and full internal self-government 2 years later. With the rise of Greenlandic consciousness there has been a move to promote the use of Greenlandic placenames and to introduce a new orthography. This will be reflected in new and future maps, and is seen also in the *Atlas-håndbog over Grønland* which, following its comprehensive index, has a special Greenlandic index giving the old and new orthography. This small atlas also provides 48 pages of general maps of Greenland (mostly at 1 : 1 000 000) as well as 17 town maps and numerous small-scale thematic maps.

In spite of an active mapping programme by the Danish government, and many *ad hoc* surveys carried out by numerous expeditions from many countries, much of Greenland remains unsurveyed in detail.

Further information

The map catalogue of the Geodætisk Institut (Copenhagen) includes an index and brief descriptions of the topographic and general maps.

Grønlands Geologiske Undersøgelse issues a catalogue of geological publications and mapping. Mapping progress is also described in the annual 'Report of Activities' of the GGU, issued in the *GGU Rapport* series.

The orthophoto mapping of northern Greenland is described in Bengtsson, T. (1983) The mapping of north Greenland, *The Photogrammetric Record*, **11**(62), 135–150.

Addresses

Geodætisk Institut (GID)
Kortsalget, Rigsdagsgården 7, DK-1218 KØBENHAVN K, Denmark.

Grønlands Geologiske Undersøgelse (GGU)
Oster Voldgade 10, DK-1350 KØBENHAVN K, Denmark.

For DMA, see United States.

Catalogue

Atlases and gazetteers

Atlas-håndbog over Grønland—Kalaallit Nunaat Arne Gaarn Bak København: Nyt Nordisk Forlag Arnold Busk, 1978.

Gazetteer of Greenland: names approved by the United States Board on Geographic Names 2nd edn C.M. Heyda Washington, DC: DMA, 1983.
pp. 271.

General

Grønland 1 : 5 000 000
København: GID, 1973.
Enlarged edition at 1 : 2 500 000 also available.

Grønland 1 : 2 000 000
København: GID, 1974.
2 sheets.
Wall map. Available with/without names.

Topographic

Grønland 1 : 1 000 000 ICAO
København: GID, 1976–.
14 sheets, 8 published. ■
Drawn to ICAO specifications, but without aeronautical information.
2 additional sheets published by DEMR, Canada.

Grønland 1 : 250 000
København: GID, 1941–.
78 sheets, 71 published. ■
Series covers coastal (ice-free) areas only.

Grønland 1 : 100 000 ortofotokort
København: GID, 1985–.
ca 100 sheets planned.

Geological and geophysical

Tectonic/geological map of Greenland 1 : 2 500 000
København: GGU, 1970.

Quaternary map of Greenland 1 : 2 500 000
København: GGU, 1971.
Map described in *Report* No. 36: Short explanation to the Quaternary map of Greenland, A. Weidick.

Geologisk kort over Grønland 1 : 500 000
København: GGU, 1971–.
14 sheets, 5 published. ■
Series covers coastal, ice-free areas only.

Geologisk kort over Grønland 1 : 500 000
København: GGU, 1974–.
14 sheets, 2 published. ■
Series covers coastal, ice-free areas only.

Aeromagnetic anomaly map 1 : 500 000
København: GGU, 1984.
1 sheet 64°–68° 15′N.

Geologisk kort over Grønland 1 : 100 000
København: GGU, 1967–.
34 sheets published. ■
Series covers coastal, ice-free areas only.
Descriptive texts available for some sheets.

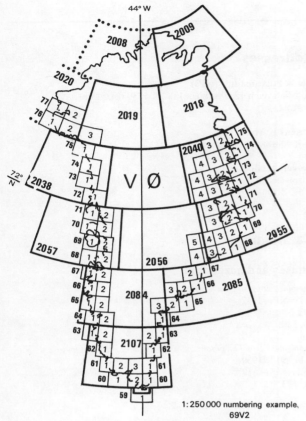

1:250 000 numbering example,
69V2

1:1 000 000 ICAO
1:250 000 topographic

GREENLAND

480 km
300 miles

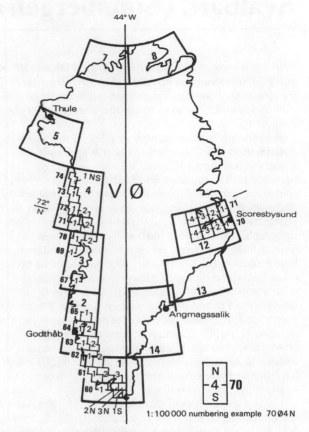

1:100 000 numbering example   70 04 N

1:500 000 geological
1:100 000 geological

Jan Mayen

Jan Mayen has been mapped by the **Norsk Polarinstitutt
(NPI)**, Oslo, at 1:50 000. The map was published in 1959.
The triangulation and fieldwork were carried out in
1949–50 and the compilation based on air photography
flown in 1949 and 1955. The map is contoured at 20 m and
glaciers are shown in white, ice-free areas in yellow.

The map is distributed by **Statens Kartverk** and its
overseas agents. NPI has also issued a gazetteer in its
Skrifter series.

Addresses

Norsk Polarinstitutt (NPI)
Rolfstangveien 12, Postboks 158, N-1330 OSLO LUFTHAVN,
Norway.

Statens Kartverk
N-3500 HØNEFOSS, Norway.

Catalogue

Atlases and gazetteers

The place names of Jan Mayen   A. K. Orvin
Oslo: NPI, 1960.
NPI *Skrifter* 120.
pp. 72.

General

Topografisk kart over Jan Mayen   1:50 000
Oslo: NPI, 1959.
2 sheets, both published.

Svalbard (Spitsbergen)

Svalbard is a group of islands lying between 76° and 81° N in the Arctic Ocean and belongs to Norway. The islands are commonly known as Spitsbergen in English-speaking countries, but the Norwegians use this name only for the main island.

Svalbard has been extensively mapped by the **Norsk Polarinstitutt (NPI)** (formerly Svalbard- og Ishavs-Undersøkelse) since 1947, and there is a continuing programme. Triangulation and field survey began in 1911 continuing into the 1920s and provided the control for the maps which are compiled photogrammetrically, from air photos flown in 1936 and subsequently. The main topographic series is at 1:100 000. The early sheets of south Spitsbergen, and a few of the most recent also, are litho printed in colour; the remainder are produced by diazo from the compilation plots, some with and some without toponyms. The contour interval is 50 m (25 m in low-lying areas), and the recent sheets have a UTM grid overprinted. A smaller-scale cover is provided in four sheets at 1:500 000. The first editions of these sheets (1964/1979) were described as provisional, and regular editions began to be issued in 1979. The 1:500 000 sheets have 100 m contours with glaciers shown in white and unglacierized areas in yellow.

There is also a series of diazo maps at 1:50 000 covering south Spitzbergen and a few sheets at 1:10 000. In 1979, NPI published a special tourist map of *Nordenskioldland* at 1:200 000 showing snow scooter routes.

Recently there have been extensive geophysical investigations in Svalbard carried out by the Scott Polar Research Institute, Cambridge, in association with NPI. Airborne radio echo sounding surveys of Nordaustlandet have produced detailed contouring of the ice sheet surface with a 10 m resolution, and this will be used by NPI in future maps of the area.

Bear Island (*Bjørnoya*) was surveyed in 1922–31 and a special map of the island was last published in 1955 but is currently out of print.

Geological and glaciological research is also undertaken by NPI, and a number of maps have been published with monographs in the *Skrifter* series. No. 154 in this series includes a four-sheet geological cover at 1:500 000. A geological map of Svalbard and Jan Mayen at 1:1 000 000 published in 1986 forms a plate of the *National Atlas of Norway*, and a glaciological map at 1:1 000 000 was also scheduled for publication that year, while a vegetation map is in preparation.

The NPI topographic maps are distributed by **Statens Kartverk** and its overseas agents. *NPI Skrifter* are distributed by **Universitetsforlaget**, Oslo. Other *Skrifter* volumes index placenames of the islands.

Further information

The NGO catalogue, published annually, lists the NPI maps and includes cover indexes. NPI also issues a catalogue of its publications.

Addresses

Norsk Polarinstitutt (NPI)
Rolfstangveien 12, Postboks 158, N-1330 OSLO LUFTHAVN, Norway.

Statens Kartverk
N-3500 HØNEFOSS, Norway.

Universitetsforlaget
Ensjøveien 10–12, Postboks 2977 Tøyen, N-0608 OSLO 6, Norway.

Catalogue

Atlases and gazetteers

The place name of Svalbard
Oslo: NPI, 1942.
NPI *Skrifter* 80.
pp. 539.

Supplement I to The place names of Svalbard A.K. Orvin
Oslo: NPI, 1959.
NPI *Skrifter* 112.
pp. 133.

General

Svalbard 1:2 000 000
Oslo: NPI, 1983.
Map also accompanies *Geography of Svalbard* (2nd edn, 1985) by V. Hisdal, published as *NPI Polarhandbok* No. 2, pp. 83.

Svalbard 1:1 000 000
Oslo: NPI, 1983.

Svalbard 1:500 000
Oslo: NPI, 1964–.
4 sheets, all published.
Sheet *Spitsbergen, sore del* also available as satellite image map.

Topographic

Svalbard 1:100 000
Oslo: NPI, 1947–.
60 sheets, all published. ■
12 sheets are colour litho printed, the rest diazo.

Svalbard 1:50 000
Oslo: NPI, 1949–.
14 sheets published.
Monochrome compilation sheets.

Geological and geophysical

Geological map of Spitsbergen 1:1 000 000
Oslo: NPI, 1969.
With NPI *Skrifter* 78 by A.K. Orvin.

Geological map of Svalbard 1:500 000
Oslo: NPI, 1970–84.
4 sheets, all published.
Accompanies the NPI *Skrifter* 154A-D.

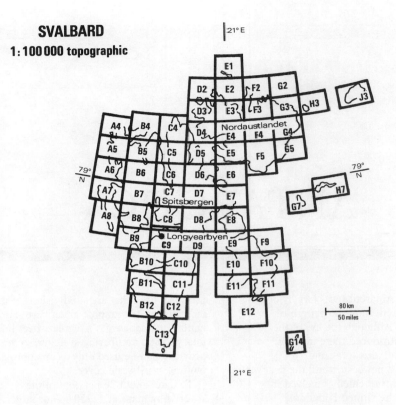

SVALBARD

1 : 100 000 topographic

ANTARCTICA

Until the 1950s, mapping in Antarctica was largely of a piecemeal exploratory kind, reflecting the pattern of expeditions to the continent. Although the use of the mapping camera had been introduced there in 1929, extensive and systematic use of air survey and photogrammetric methods did not occur until the postwar period. Before 1957, the countries chiefly involved in mapping the continent were the United Kingdom, Argentina, Chile, Australia and Norway and Sweden.

The structure of modern mapping in Antarctica has been strongly influenced by the establishment of a long-term programme of successful international cooperation in scientific research and data gathering following the International Geophysical Year in 1957. This cooperation was formalized in the creation of SCAR—the **Scientific** (originally Special) **Committee on Antarctic Research**—in 1958 by the 12 nations which had participated in the IGY. These countries also became the original signatories of the Antarctic Treaty of 1959. Membership of SCAR is restricted to nations with a strong commitment to research in Antarctica or on the sub-Antarctic islands. Currently 18 nations are members, of which 16 had all-year stations on the Antarctic continent or islands in 1985.

Among the groups established by SCAR is a Working Group on Geodesy and Cartography, which seeks to coordinate Antarctic mapping activities and has formulated a series of standing resolutions and recommendations to encourage standardization in the mapping of the continent and to ensure cooperation and the free exchange of information between the many national mapping authorities involved. The general aim is to foster the making and maintenance of up-to-date maps of Antarctica, using standard scales and symbols approved by SCAR. Its recommendations include adjustment of all mapping south of 60° S to the World Geodetic System (WGS72). Maps are to be metric, and there is a range of approved scales, namely 1 : 40 000 000, 1 : 20 000 000, 1 : 10 000 000, 1 : 3 000 000, 1 : 2 000 000, 1 : 1 000 000, 1 : 500 000, 1 : 250 000, 1 : 200 000, 1 : 100 000, 1 : 50 000 and 1 : 25 000. Thus, most Antarctic mapping conforms to these scales. Maps at scales less than 1 : 1 000 000 use a polar stereographic projection, while 1 : 1 000 000 mapping is drawn to ICAO specifications and projections. Larger scales are drawn on conformal projections such as the Lambert conformal conic. It is also recommended that IMW sheet lines are used along parallels of latitude, but with optional choice of meridians (with the convergence of meridians towards the pole, sheet lines will usually be amalgamated). The larger-scale mapping is supposed to use sheet lines that are subdivisions of the IMW system, as indicated on the graphic given here. However, many nations use somewhat irregular sheet lines and idiosyncratic numbering systems, so the graphic we have provided can be used only as an approximate tool for plotting small-scale cover.

To date, about 20 per cent of the continent has been covered by maps at 1 : 250 000 or larger. The recent trend has been to the use of Landsat image mapping in place of conventional line mapping, a policy which has been validated by the SCAR Working Group. Landsat cannot acquire images south of 80° S, however, and for this reason a joint NOAA/USGS Antarctic Mosaic Project was started in 1980, using AVHRR images from the NOAA satellites. USGS proposes to publish this mosaic.

Members of SCAR distribute their maps (both topographic and thematic) and gazetteers to the official Antarctic Mapping Centres in each participating country, so, theoretically, the entire Antarctic map production since 1958 should be available to researchers at these centres. Additionally, the maps of many SCAR nations are readily available for sale to the public, and are listed in the catalogues of the various mapping agencies.

A number of small-scale general maps and atlases of the continent have been prepared, usually by single national organizations, but drawing on the international pool of cartographic information. The most substantial atlas so far produced is the *Atlas Antarktiki* published in 1966 by the Soviet Academy of Sciences, Moscow. This is out of print, but a new edition is in preparation and expected at the end of 1989. Another substantial compilation, but again substantially out of print, is the *Antarctic Map Folio Series* published in 19 parts by the American Geographical Society over the period 1964–75. The best recent synthesis for the continent as a whole, although more restricted in theme, is the *Antarctica: glaciological and geophysical folio*, edited by D.J. Drewry and published by the **Scott Polar Research Institute (SPRI)**, University of Cambridge. This exceptionally fine work has large-format foldout maps of the Antarctic continent, with a maximum scale of 1 : 6 000 000. For these maps, the coastline was digitized from the best existing mapping and from Landsat MSS imagery, so that a computer-generated coastal outline could be combined with data from a geophysical database for the production *inter alia* of draft material for the maps in the folio. Much of the geophysical information comes from the radio echo sounding techniques pioneered by SPRI and used in Antarctica in a cooperative six-nation glaciological programme from the early 1970s.

In addition to the small-scale general and thematic maps listed in our catalogue, a number of world series also offer excellent single-sheet cover of the continent. These include the GEBCO chart at 1:6 000 000 for bathymetry (Sheet 5.18, 1980), the *Geological World Atlas* sheets 17 (covering the continent at 1:10 000 000, 1979) and 18 (covering the Southern Ocean at 1:20 000 000, 1980), and the Antarctica sheets of the Circum-Pacific Map Project (AAPG, Tulsa) which include a geographical map, a geodynamic map, and a plate tectonic map, all at 1:10 000 000.

Most SCAR members have undertaken some Antarctic mapping, but the most active nations have been the UK, the US, the USSR, Australia, New Zealand and Japan. Detailed topographic mapping has been mainly confined to the coastal zone, and to areas not submerged by ice, but geophysical survey using airborne radar techniques has been much more extensive. Most published mapping to date is either topographic, geological or geophysical in theme, but there has also been considerable interest in the biological and submarine resources of the Southern Ocean reflected for example in the BIOMASS (Biological Investigations of Marine Antarctic Systems and Stocks) programme started by SCAR in 1972. This has led to the establishment of a database of biological material held at the British Antarctic Survey at Cambridge.

A particularly important mapping activity has resulted from the long-range airborne echo-sounding (RES) and magnetometer survey which was initiated by SPRI with the British Antarctic Survey and in cooperation with the US National Science Foundation Division of Polar Programs (NSF-DPP) and the Electromagnetics Institute of the Technical University of Denmark. Studies using these methods included a survey of the surface form and thickness of the Antarctic ice and of the sub-ice geology, and about 50 per cent of the ice sheet has been covered by this survey.

Most Antarctic Treaty countries have geographic names committees, but there is no unified system for the whole of Antarctica, and only the US has attempted to publish a gazetteer for the whole continent. The USBGN Gazetteer lists about 12 000 names approved by the Board, as well as 3000 'unapproved variant names', and there is a useful discussion of the toponymic problems of Antarctica. Other countries have compiled gazetteers of their claim territories, and those published and believed to be available are included in our catalogue.

The following paragraphs summarize the post-IGY mapping of each of the SCAR member countries which have been most active in this area.

Australia

In Australia, Antarctic research is coordinated by ANARE (the Australian National Antarctic Research Expeditions) based at Kingston, Tasmania.

General and topographic maps are published by the **Australian Division of National Mapping (NATMAP)** and include a general map at 1:10 000 000 which has an index on the reverse, and sheets at 1:1 000 000 of the coastal areas of the Australian Antarctic Territory, two of which are drawn to IMW specification while the rest are in a preliminary two-colour edition without relief shading or bathymetry.

Detailed mapping by NATMAP is restricted to areas in the Australian Territory containing features not obscured by the ice sheet. There are two principal series, both at 1:250 000, one being a conventional four-colour line map showing rock and ice features, approved names and ice sheet contours at 200 m intervals, and the other a satellite image map produced from a mosaic of Band 5 Landsat imagery. Feature names and latitudes and longitudes have been added. These maps are in halftone, and dyeline copies are printed on demand. There is also a Landsat image series at 1:500 000.

Special large-scale maps of the Framnes Mountains and Vestfold Hills (1:100 000) and of Heard Island and Macquarie Island (1:50 000) have also been published.

Geological maps of the Australian Territory, mostly at 1:250 000, are produced by the **Bureau of Mineral Resources, Geology and Geophysics (BMR)**, Canberra. Some 1:500 000 maps of outcrop geology were published in 1985, with the outcrops shown in colour against a monochrome Band 7 Landsat mosaic.

Japan

Japan has carried out research in Antarctica since 1956. The programme is coordinated by the Japanese Antarctic Research Expedition (JARE). Several government agencies contribute to the programme. Thus topographic survey and mapping is the responsibility of the **Geographical Survey Institute (GSI)**, Tokyo, which carries out aerial photography, ground control and map compilation. There is an active series of 1:25 000 maps. Sheets are in colour with 10 m contours, and the projection is Lambert conformal conic. There are also some maps at 1:250 000 with 50 m contours, and some detailed mapping at 1:5000.

Studies in geomagnetism, geology, seismology and atmospherics are undertaken by the **National Institute of Polar Research**. There is an active *Antarctic Geological Map Series* at 1:25 000, with sheets published in English and covering the areas mapped topographically by GSI.

New Zealand

The New Zealand Antarctic programme is formulated by the Ross Dependency Research Committee, set up in 1958. Mapping, carried out by the Antarctic Division of the **Department of Lands and Surveys (NZMS)**, has been concentrated in the Ross Sea area and a number of 1:250 000 reconnaissance-type maps on IMW sheet lines were produced in the 1950s and 1960s using photo alidade. These maps were uncontoured and have not been subsequently revised or upgraded. In 1981, a number of contoured photomaps in dyeline form were issued. Since the 1970s NZMS has been experimenting with the production of Landsat image maps to the same scale and format. Computer Compatible Tapes (CCTs) have been subjected to image processing to optimize their value as a photobase. After ground truth work in 1980, it was concluded that such image maps showed more information than existing maps, and it is intended to produce such maps using rectified Landsat mosaics in a monochrome or two-colour style.

The **New Zealand Geological Survey (NZGS)** has also carried out geological mapping at the 1:250 000 scale (e.g. NZMS 166, 1961) while the **New Zealand Oceanographic Institute (NZOI)** published a *McMurdo Sound bathymetry map* 1:250 000 in 1985. Currently, New Zealand is cooperating with the USGS on the Dry Valley Seismic Project.

USSR

Soviet mapping in Antarctica began in 1955 when the first aerial photo sorties were flown. Cartography is the responsibility of the Central Administration of Geodesy and Cartography (**Glavnoe Upravlenie Geodezii i Kartografii, GUGK**), Moscow. General maps of the whole continent have been published at 1:5 000 000 and 1:3 000 000 (in nine sheets), and there is also a special set of the *Karta Mira* 1:2 500 000 covering the continent. 1:1 000 000 maps cover the coastal regions and areas where bedrock outcrops, while larger scales (1:200 000, 1:100 000 and 1:50 000) cover the coastal strip and other areas where research has been undertaken.

United Kingdom

Britain has published some 150 sheets of Antarctica since 1955. Mapping activity has focused on the British Antarctic Territory (south of 60° S) and the Falkland Islands Dependencies (north of 60° S). The original topographic series were at 1:200 000 (DOS 610/Series D501) and 1:500 000 (DOS 710/Series D401). These series were based on air photography and control obtained by the Falkland Islands and Dependencies Aerial Survey Expedition of 1955–57. They have now been largely superseded by a 1:250 000 series (BAS 250) started in 1973 as a three-colour line map with 100 m contours and shaded relief, and continued as a satellite image map (BAS 250P). This series is not yet complete. A number of islands in the British Antarctic Territory have been mapped at larger scales, including several sheets at 1:25 000 (DOS 310) and a few at 1:10 000 (DOS 210).

BAS is currently most active in publishing geological and geophysical maps. A major new series of 1:500 000 geological mapping in seven sheets covering mainly the Antarctic Peninsula was started in 1979 to summarize all the geological information currently available. A new topographic base was prepared for this series using satellite imagery and air photographs with Doppler satellite fixing for ground control. There is also a series of ice thickness maps at the same scale.

The maps have been published for the **British Antarctic Survey (BAS)** by the **Directorate of Overseas Surveys (DOS)** (now Overseas Surveys Directorate, Southampton, OSD).

United States

Mapping by the US is carried out by the **United States Geological Survey (USGS)** in cooperation with the National Science Foundation as part of the United States Antarctic Research Program. Topographic maps are published in four main series: 1:50 000 with 50 m contours (supplementary contours at 25 m) with sheets covering 15′ latitude by 60′ longitude (sheets cover the Dry Valleys area); 1:250 000 reconnaissance maps with 200 m contours and shaded relief; 1:500 000 reconnaissance maps with shaded relief of the coastal areas of Wilkes Land and Enderby land; and a number of 1:1 000 000 maps on IMW sheet lines. In 1975–76 the USGS published some experimental satellite image maps of McMurdo Sound and elsewhere and this has led to a proliferation of image maps in Antarctica. Future plans

include some further 1:1 000 000 satellite image maps, and a continental image map from AVHRR data, as mentioned above.

The most extensive series is the 1:250 000 topographic reconnaissance series with some 90 sheets published. Future sheets will be 1:250 000 scale satellite image maps which will replace the 1:500 000 sketch maps.

USGS has also undertaken a 1:250 000 geologic reconnaissance series with the collaboration of the Institute of Polar Studies, Ohio State University and under the auspices of the National Science Foundation. Eleven sheets have been published.

Other countries

Argentinian mapping covers the Argentine sector at a scale of 1:1 000 000 (covering the interior) and 1:500 000 (covering the Antarctic Peninsula). These maps represent extensions of the national mapping system at these scales, and were published in the period 1966–72 by the **Instituto Geográfico Militar**, Buenos Aires. Sheets in the 1:500 000 series also cover the South Orkneys, South Sandwich Islands and South Georgia (see under the Atlantic Ocean section).

Mapping by *Belgium* relates to a number of expeditions carried out during 1957–60 and is limited to a handful of sheets at various scales covering Prinses-Ragnhildkust, the Belgica Mountains and Queen Maud Land.

Chile has published some 1:500 000 scale maps of its claim area, but does not appear to have done much recent work.

French interest has been in the sub-Antarctic islands of Kerguelen and Crozet (see under Indian Ocean) and in Terre Adelie on the continent itself.

German mapping activity has included the production of satellite image maps of Neuschwabenland from AVHRR imagery captured by the NOAA-7 satellite. The **Institut für Angewandte Geodäsie (IfAG)**, Frankfurt AM has published several photomaps at 1:50 000 and larger relating to the 1983–84 Antarctic Expedition. The Alfred Wegener Institute for Polar and Marine Research has been established at Bremerhaven.

Norwegian interest in Antarctica is reflected in a series of 1:250 000 maps of Dronning Maud Land published by the **Norsk Polarinstitutt (NPI)**, Oslo. Twenty-three sheets were published between 1957 and 1975, and in 1984 a satellite image map of Sheet K5 *Filchnerfjella Nord* was issued. The Norwegian maps are distributed by **Statens Kartverk**, Hønefoss. Norway restarted regular summer expeditionary activities in Antarctica in 1985.

Limited local mapping has been carried out by a number of other countries. *Poland*, for example, has undertaken some 1:50 000 geological mapping, and has published large-scale maps of its polar base. It has also published an *Atlas of Polish Oceanographic Observations in Antarctic Waters 1981*.

South Africa published three 1:250 000 scale reconnaissance geological maps of Western Dronning Maud Land in 1981.

With the 30th anniversary of the Antarctic Treaty coming up in 1989, many countries are reviving their interest in the continent. The Dutch, after an absence of 20 years, are again active, although mapping was not specifically mentioned in their plan for 1987–91; and other recent signatories to the Treaty, such as China and Brazil, are developing research programmes which may include some mapping.

Further information

A complete catalogue of topographic maps and navigation charts produced by the Antarctic Treaty nations is compiled by SCAR, and new editions are issued from time to time. This *Catalogue of Antarctic maps and charts* is available on microfiche from NATMAP.

The activities of the SCAR Working group of Geodesy and Cartography are reported in the *SCAR Bulletin*, which is published in the periodical *Polar Record*. Several SCAR member countries issue indexes or catalogues of their Antarctic mapping.

A summary of *Antarctic maps and surveys 1900–1964* forms the first folio of the Antarctic Map Folio Series, published by the American Geographical Society, 1965.

A discussion of Soviet Antarctic mapping has appeared in English in Bazheeva, A. V. *et al.* (1978) Soviet mapping of Antarctica, *Polar Geography*, **2**.

The New Zealand mapping is described by Child, R.C. (1984) in Mapping in Antarctica, *New Zealand Cartographic Journal*, **14**(2), 12–14.

The early mapping of the British Antarctic Territory (up to 1964) was described by McHugo, M.B. (1965) in *Polar Record*, **12**, 395–402.

The radio echo sounding map is described by Drewry, D.J. (1975) in *Polar Record*, **17**, 359–374.

The problems association with placenames in Antarctica are discussed by Hattersley-Smith, G. (1980) Current sources of Antarctic and sub-Antarctic placenames, *Polar Record*, **20**, 72–78. This paper also lists the policies of each country, toponymic authorities and existing gazetteers.

Addresses

Scientific Committee on Antarctic Research
The Distribution Centre (for SCAR publications)
Blackhorse Road, LETCHWORTH SG6 1HN, UK.

The following addresses are in alphabetical order by country

Instituto Geográfico Militar (IGM)
Avenida Cabildo 381, 1426 BUENOS AIRES, Argentina.

Servicio de Hidrografía Naval (SHN)
Av. Montes de Oca 2124, 1271 BUENOS AIRES, Argentina.

Division of National Mapping (NATMAP)
PO Box 31, BELCONNEN, ACT 2616, Australia.

Australian National Antarctic Research Expeditions (ANARE)
Channel Highway, KINGSTON, Tasmania, Australia.

Bureau of Mineral Resources, Geology and Geophysics (BMR)
PO Box 378, CANBERRA CITY, ACT 2601, Australia.

Institut Français de Recherche pour l'Exploitation de la Mer (CNEXO/INFREMER)
Centre de Brest, BP 337, F-29273 BREST, France.

Institut für Angewandte Geodäsie (IfAG)
Richard-Strauss Allee 11, D-6000 FRANKFURT AM 70, German Federal Republic.

Alfred Wegener-Institut für Polarforschung
Columbus Center, D-2850 BREMERHAVEN, German Federal Republic.

Geographical Survey Institute (GSI)
Kitasato-1, Yatabe-Machi, Tsukuba-Gun, IBARAKI-KEN, 305 Japan.

National Institute for Polar Research (NIPR)
9–10 Kaga 1-chome, Itabashi-ku, TOKYO 173, Japan.

Department of Lands and Surveys (NZMS)
Charles Fergusson Building, Bowen Street, Private Bag, WELLINGTON 1, New Zealand.

New Zealand Geological Survey (NZGS)
DSIR, PO Box 30–368, LOWER HUTT, New Zealand.

New Zealand Oceanographic Institute (NZOI)
PO Box 12–346, WELLINGTON NORTH, New Zealand.

Norsk Polarinstitutt (NPI)
Rolfstangveien 12, Postboks 158, 1330 OSLO LUFTHAVN, Norway.

Statens Kartverk
N-3500 HØNEFOSS, Norway.

National Antarctic Programme
Cooperative Programmes, Council for Scientific and Industrial Research, PO Box 395, PRETORIA 0001, South Africa.

Scott Polar Research Institute (SPRI)
Lensfield Road, CAMBRIDGE CB2 1ER, UK.

British Antarctic Survey (BAS)
High Cross, Madingley Rd, CAMBRIDGE CB3 0ET, UK.

Overseas Surveys Directorate (DOS/OSD)
Ordnance Survey, Romsey Rd, Maybush, SOUTHAMPTON SO9 4DH, UK.

Her Majesty's Stationery Office (HMSO)
PO Box 276, LONDON SW8 5DT, UK.

United States Geological Survey (USGS)
Map Distribution, Federal Center, Building 41, Box 25286, DENVER, CO 80225, USA.

Naval Polar Oceanographic Center (NPOC)
4301 Suitland Road, WASHINGTON DC 20390–5180, USA.

Glavnoe Upravlenie Geodezii i Kartografii (GUGK)
ul. Krahizhanovaskogo 14, korp. 2, V-216 MOSKVA, USSR.

Ministry of Geology of USSR (MGUSSR)
Bolshaya Gruzinskaya 4, 123242, MOSKVA D-242, USSR.

For CIA, DMA and NGS, see United States; for University of Waterloo, see Canada.

Catalogue

Atlases and gazetteers

Antarctica: glaciological and geophysical folio edited by D.J. Drewry
Cambridge: University of Cambridge, Scott Polar Research Institute, 1983–.

Geographic names of the Antarctic Compiled and edited by Fred G. Alberts on behalf of USBGN
Washington, DC: DMA, 1981.
pp. 959.

Gazetteer of the Australian Antarctic Territory G.W. McKinnon
Melbourne: ANARE, 1965.
pp. 153.

Gazetteer of the British Antarctic Territory G. Hattersley-Smith
London: HMSO, 1977.
pp. 36.
Supplements published November, 1982.

Gazetteer of the Falkland Islands Dependencies
G. Hattersley-Smith
London: HMSO, 1977.

Provisional gazetteer of the Ross Dependency A.S. Helm
Wellington: Government Printer, 1958.
pp. 164.
With subsequent supplements.

Enciclopedico Antarctico Argentino
Buenos Aires: Direccion Nacional del Argentino, 1978.
pp. 130.

Toponimia del sector Antarctico Argentino
Buenos Aires: SHN, 1970.
pp. 746.

Diccionario de nombres geográficos de la costa de Chile V III.
Territorio Antárctico 1st edn
Valparaiso: Instituto Hidrográfico de la Armada, 1974.
pp. 312.

General and bathymetric

Antarctic regions and continents 1 : 26 345 000 NZMS
308 Edition 1
Wellington: NZMS, 1980.

Batimetricheskaya karta Antarktiki 1 : 15 000 000
Moskva: GUGK, 1974.
In Russian.

Antarktika. Fizicheskaja karta 1 : 15 000 000
Moskva: GUGK.

Antarctica/Antarktis 1 : 12 000 000
Karlsruhe: Geisler/Herrmann, 1985.

Antarctica 1 : 10 000 000 6th edn
Canberra: NATMAP, 1979.
Insets: Heard Island, Macquarie Island 1 : 250 000.

Antarctica 1 : 10 000 000
Tokyo: GSI, 1985.
In Japanese.

Territorio Chileno Antarctico 1 : 10 000 000
Santiago: IGM.

Antarctica 1 : 8 841 000
Washington, DC: NGS, 1975.

Antarctic regions 1 : 8 625 000
Washington, DC: CIA, 1984.

Antarktis (Konigin-Maud-Land) 1 : 6 000 000
Frankfurt AM: IfAG, 1984.
Satellite image map.

Antarktika 1 : 5 000 000
Moskva: GUGK.

Antarctique 1 : 5 000 000
Paris: IGN, 1969.

British Antarctic Territory (North of 82° S) with South Georgia and
South Sandwich Islands 1 : 3 000 000 BAS (Misc) 2 Edition 1
Tolworth: DOS, 1981.
Inset maps of South Sandwich Islands and South Georgia
1 : 1 000 000.

Neuschwabenland 1982 1 : 3 000 000
Frankfurt AM: IfAG, 1982.
Satellite image map.

West-Neuschwabenland 1981 1 : 3 000 000
Frankfurt AM: IfAG, 1982.
Satellite image map.

Ross Sea regions, Antarctica 1 : 3 000 000 NZMS 135
Wellington: DLS, 1970.

Antarktika 1 : 2 500 000
Moskva: GUGK.
17 sheets, all published.
Form part of the *Karta Mira* world series.

Antarctic Peninsula 1 : 2 000 000
Washington, DC: CIA, 1984.

Topographic

Australian Antarctic Territory 1 : 1 000 000
Canberra: NATMAP, 1967–.
19 sheets published.

Australian Antarctic Territory 1 : 500 000
Canberra: NATMAP.
33 sheets published.
Satellite image maps.

British Antarctic Territory 1 : 500 000 BAS 500P
Cambridge: BAS, 1981.
1 sheet published.

Reconnaissance series 1 : 500 000
Reston, VA: USGS, 1947.
11 sheets published.

Australian Antarctic Territory 1 : 250 000
Canberra: NATMAP, 1963–.
21 topographic and 59 satellite image sheets published.

British Antarctic Territory 1 : 250 000 BAS 250/250P
Cambridge: BAS, 1973–.
72 sheets, 20 published.

Ross Dependency (New Zealand) 1 : 250 000 NZMS 166
Wellington: NZMS, 1961–.
38 sheets published.

Dronning Maud Land 1 : 250 000
Oslo: NPI, 1957–.
23 sheets published.

Antarctica reconnaissance series 1 : 250 000
Reston, VA: USGS, 1959–.

Antarctica 1 : 50 000 Topographic series
Reston, VA: USGS, 1977.
8 sheets published.
Series covers Ross Island–Taylor Glacier area only.

Geological and geophysical

Magnetic maps of the Antarctic. Epoch 1975 1 : 15 000 000
Tokyo: GSI, 1978.
Set of 8 maps.

Free air gravity anomaly of Antarctic region 1 : 13 000 000
Tokyo: National Institute of Polar Research, 1984.
2 sheets, both published.

Tectonic map of Antarctica 1 : 10 000 000 G.E. Grikurov
Moskva: MGUSSR, 1978.
Legend in Russian and English.
81 pp. English text (1980).

Map of metamorphic facies of Antarctica 1 : 5 000 000
Moskva: MGUSSR, 1979.
4 sheets, all published.
43 pp. English text.

Geological map of Antarctica 1 : 5 000 000
Moskva: MGUSSR, 1976.
4 sheets, all published.

Metamorphic map of Antarctica 1 : 5 000 000
Moskva: MGUSSR, 1978.
4 sheets, all published.

Gravity map of the Antarctic 1 : 5 000 000
Moskva: MGUSSR, 1975.
10 sheets, all published.
(5 sheets Variant 1, Free air anomalies; 5 sheets Variant 2,
Anomalies Bouguer.)

Antarctica. Radio echo sounding map Series A 1 : 5 000 000
Cambridge: Scott Polar Research Institute, 1974.
3 sheets, all published.
Shows ice sheet surface and sub-ice relief.

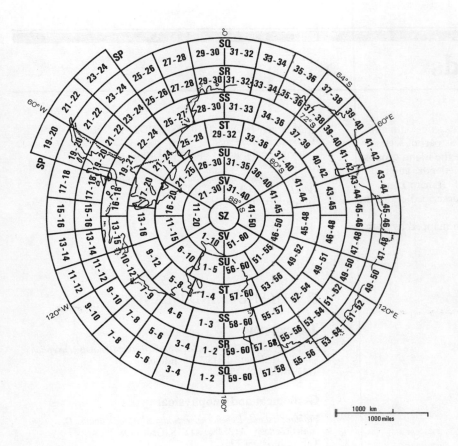

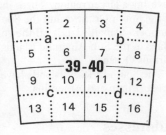

ANTARCTICA
1 : 1 000 000 topographic
1 : 500 000 topographic
1 : 250 000 topographic

Sheet numbering examples:-
1 : 1 000 000 SQ 39 - 40
1 : 500 000 SQ 39 - 40/a
1 : 250 000 SQ 39 - 40/13

1000 km
1000 miles

Tectonic map of the Scotia Arc 1 : 3 000 000 BAS (Misc)
3 Edition 1
Cambridge: BAS, 1985.

Carte géologique de l'Antarctique 1 : 2 500 000
Paris: IFP-CNEXO, 1977.
7 sheets + legend sheet + 1 general sheet at 1 : 10 000 000.
Text.

Bouguer anomaly map of the Antarctic Peninsula 1 : 1 500 000
BAS (Misc) 5 Edition 1
Cambridge: BAS, 1985.

Aeromagnetic anomaly map of the Antarctic Peninsula 1 : 1 500 000
BAS (Misc) 6 Edition 1
Cambridge: BAS, 1985.

British Antarctic Territory: geological map 1 : 500 000 BAS
500G Edition 1
Cambridge: BAS, 1979–.
7 sheets, 5 published.

British Antarctic Territory: ice thickness map 1 : 500 000 BAS
500R Edition 1
Cambridge: BAS, 1983–.
7 sheets, 1 published.

Reconnaissance geologic map of Antarctica 1 : 250 000
Reston, VA: USGS, 1970–.
11 sheets published.

Australian Antarctic Territory. Geological series 1 : 250 000
Canberra: BMR.
14 sheets published.

Administrative

Antarctica: stations and claims 1 : 17 250 000
Washington, DC: CIA, 1981.

Environmental

Southern Ocean Atlas
New York: Columbia University Press, 1982.

Sea ice climatic atlas: volume 1 Antarctica NAVAIR 50-IC-540
Washington, DC: Naval Polar Oceanographic Center, 1985.
pp. 131.

Atlas of Polish oceanographic observations in Antarctic waters 1981
Cambridge: SCAR and SCOR, 1985.
pp. 83.

Potential resources of Antarctica 1 : 60 000 000 D. Entwhistle
Waterloo, Ontario: University of Waterloo.

Antarctic Islands

A number of islands adjacent to the Antarctic continent have been mapped either separately or as part of series covering a larger region. These include the island groups of the South Orkneys and the South Shetlands, and a number of small islands off the coast of Graham Land. Maps of the Sub-Antarctic islands are described under the various Oceans.

The South Orkney Islands have been mapped by both Great Britain and Argentina.

Addresses

For IGM and DOS, see Antarctica.

Catalogue

SOUTH ORKNEY ISLANDS

Topographic

Destacamento Naval Orcades 1:500 000
Buenos Aires: IGM, 1972.
Sheet 6145 in Argentine series.

South Orkney Islands 1:100 000 DOS 510 (Series D601)
Tolworth: DOS, 1963.
2 sheets, 1 published (Coronation Island).

British Antarctic Territory. South Orkney Islands. Signy Island 1:10 000 DOS 210 Edition 2
Tolworth: DOS, 1975.

SOUTH SHETLAND ISLANDS

Topographic

Falkland Islands Dependencies. South Shetland Islands. Deception Island 1:25 000 DOS 310 Edition 1
Tolworth: DOS, 1960.

Geological and geophysical

Falkland Islands Dependencies. South Shetland Islands. Deception Island Geology 1:25 000 DOS (Geol 1108)
Tolworth: DOS, 1961.

GLOSSARY

aeromagnetic map   Map constructed from data acquired from airborne magnetometer survey showing variations in the earth's magnetic field; used in mineral prospecting.

aeronautical chart   A map designed for air navigation.

ancillary map   A small map provided as a supplement to a principal map, and printed on the same sheet.

azimuthal   Used of projections in which great circle arcs are shown as straight lines for all directions from the centre of the projection.

base map   A map showing selected topographic information (often in one colour) over which thematic information (often in multi-colours) may be printed.

basic scale   The largest scale used for publication of an original survey. The term is usually applied to the most detailed topographic series of a country, from which smaller-scale maps may also be derived (cf. *derived scale*).

bathymetric map   Map showing the relief of the floor of a lake, sea or ocean.

cadastral map   A map which delineates property boundaries and legal land divisions. It may also indicate ownership.

Cassini projection   A transverse equidistant cylindrical projection. Introduced in 1745, it was formerly much used for topographic and cadastral mapping, but is no longer popular.

compilation   The preparation of a map from original or derived data prior to the final draughting.

conformal   Used of a map projection in which the shape of small areas is shown without distortion.

controlled mosaic   A number of vertical air photographs assembled and joined together to minimize scale variations and fitted to a framework of surveyed ground control points.

database   A collection of information held in digital form and structured to facilitate automatic analysis and selective retrieval.

densification   The filling in of a survey control framework by higher-order triangulation or trilateration.

derived scale   Used of maps which have been produced by selection, simplification and generalization from larger-scale maps.

diazo (also **dyeline**)   A contact copy made from a transparency on to a paper coated with a compound which decomposes when exposed to light. The unexposed image is converted by developing to a coloured image formed by an azo dye.

digital map   Spatial information held in digital format and capable of being displayed graphically on a cathode ray tube, or plotted as a hard copy map.

digital elevation model/digital terrain model   A digital file of elevations stored with their coordinates and used to construct maps or three-dimensional representations of the land surface.

edition   Used of a map or map series to denote a particular issue representing a revision or resurvey of the information content. New editions do not usually incorporate any substantial change in specification (cf. *series*).

electronic map   Any map which is stored and which may be portrayed by electronic means. This may include maps stored digitally in a computer or as frames on a video disk.

equal-area/equivalent   Used of projections which maintain true area (in proportion to their scale). This attribute is achieved at the expense of correct portrayal of shape.

feature code   A number used in digital mapping to define the nature of a line (e.g. footpath), area (e.g. forest) or point (e.g. windmill).

formlines   Lines resembling contours, but which lack a control of accurately surveyed elevations, and which therefore give only a general impression of relief.

Gauss projections   A number of conformal projections named after the mathematician C. F. Gauss (1777–1855). The **Gauss–Krüger** projection is a term used in some European countries for the Transverse Mercator projection, having been elaborated by Gauss and Krüger, although the origin of the TM is usually ascribed to Lambert. The **Gauss–Boaga** projection is a modified form of the TM used for topographic mapping in Italy.

569

gazetteer A list of place and feature names and their locations, often referenced to a particular map series or atlas. Gazetteers sometimes include additional descriptive information about places.

geochemical map A map showing the distribution of the natural occurrences of particular chemical elements and compounds.

geodesy The science of measuring the size and shape of the earth, and the earth's external gravity field.

geodetic survey Survey which is used to define the framework of a map in terms of its relationship to the size and shape of the earth.

geographical information system (GIS) A computer system for the storage, analysis and display of spatially related data.

geophysical map A map giving information (indirectly) about the structure of the earth and produced by one of several geophysical survey methods, e.g. seismic, gravity or magnetic survey.

grad An angular measure of one hundreth part of a right angle (part of the centesimal system). It is used in place of degrees for defining latitude and longitude on some maps.

graphic index Used in this book to denote a sketch map to show the sheet lines of one or more map series, and which can be used for indexing information about coverage, editions, etc.

graticule A network of lines on a map which represent the meridians of longitude and parallels of latitude.

gravity anomaly map A map showing the variations in gravitational force caused by the varying density and thickness of crustal rocks. Such maps can be used for elucidating the structure of the earth's crust.

grid A network comprising two sets of uniformly spaced parallel lines drawn on the face of a map, intersecting at right angles and usually running north–south and east–west. Grids are often numbered eastwards (eastings) and northwards (northings) from the south-west corner of the map, and may be used to define position by rectangular coordinates.

ground control Accurate measurements obtained by ground survey of the horizontal and elevational position of features used for the adjustment and calibration of photogrammetric data.

hazard map A map showing the distribution of, or quantifying the spatial variation of the risk of, any stated natural or man-made hazard (e.g. earthquakes, floods, radioactive contamination).

hydrogeological map A geological map giving specific attention to the water-bearing attributes of rock strata.

hydrographic chart A map of sea, ocean or lake designed for navigation.

hypsometric relief/hypsometric tints A method of showing differences in elevation by a sequence of shades or colours applied to successive zones between contour lines (several other terms are also used, e.g. layer colouring, gradient tints).

image analysis Manipulation, usually by computer, of the pixels of a satellite image to extract information, remove distortion and error, or improve interpretability.

image map A map using for its base the tonal variations of satellite remote sensing or air photo imagery.

International Map of the World (IMW) A map series at 1 : 1 000 000 published by individual countries to a specification agreed internationally and last revised at a conference in Bonn in 1962. Currently IMW maps are drawn on the Lambert conformal conic or Polar stereographic projections.

isogonic map A map showing lines of equal magnetic declination.

Lambert conformal conic projection A conic projection usually with two standard parallels. Meridians are straight, parallels are curved and intersect the meridians at right angles. Area distortion near and between the standard parallels is small. The projection provides good directional and shape properties for areas of broad east–west extent.

Landsat The first civilian unmanned earth-orbiting series of satellites designed to investigate the earth's resources, and launched by the United States from 1972.

legend An explanation, usually given in the map margin, of the symbols used on a map.

levelling The method for providing a vertical survey control network to determine elevations.

multi-spectral scanner (MSS) A scanning device, usually mounted in a satellite, which collects imagery simultaneously in each of several spectral wave bands.

national atlas An atlas, usually published, funded or approved by a national government, providing a major cartographic synthesis of the physical, social and economic geography of a country.

neatline(s) The line(s) that surround the body of a map, separating it from the map margin.

orthomorphic See *conformal*.

orthophoto A photographic image made from a perspective photograph by differential rectification to remove the displacements caused by camera tilt and ground relief.

orthophotomap A map using the image of one or a number of orthophotos with the addition of some conventional cartographic enhancement such as placenames, elevations and grid.

photogrammetry The process of taking precise measurements from photographs (usually in order to construct maps).

photomap A general term for maps based on air photographs and retaining the photo image, usually with the addition of some conventional cartographic linework.

photomosaic A number of air photographs assembled and edge-matched to give a continuous representation of a larger area.

pictomap (Photographic Image Conversion by Tonal Masking Procedures)   A kind of photomap developed by the US Army Map Service, in which the continuous tones of the photo image are separated into components such as buildings or woodland and assigned masks for printing. Used mainly for the production of town maps.

pixel (picture element)   Area of ground represented by a single numerical value in a digital image. The size of this area determines the resolution of the image, e.g. Landsat MSS imagery has a pixel size and therefore a theoretical resolution of 80 m.

plane table   A board used in surveying on which maps of small areas are directly plotted by sighting points with an alidade from the ends of a measured base line. Much used for topographic survey before the advent of air photography.

planimetric map   A large-scale map accurately showing the horizontal position of features. Relief is not normally shown except sometimes by point elevations.

plat map   American term for a map that shows land divisions.

polyconic projection   A conic projection in which each parallel is treated as a standard parallel constructed on the tangent line of its own cone, and therefore the parallels are not concentric (cf. *Lambert's conformal conic*). Formerly used for the standard topographic mapping of the US.

polyhedric projection   A projection formerly popular in parts of Europe for topographic mapping. The parallels and meridians are rectilinear and their lengths correspond closely to the arc distances of the spheroid.

process colours   The printing inks used to produce polychromatic maps by the four-colour process. Percentage screens are assigned to the primary colours of magenta, yellow and cyan (together with black) which in combination produce the desired range of hues, while minimizing the required number of printing plates.

projection   The representation of the parallels and meridians as a graticule on a plane surface, according to a given mathematical formula.

radar   Acronym for 'radio detection and ranging'. A system of beamed, reflected and timed radiation, operating at wavelengths of 1 mm to 1 m.

raster data   Data organized into a grid of cells of a uniform size, and each representing a particular value.

rectification   The process of removing distortions caused by camera tilt from air photographs.

revision programme   The systematic up-dating of a map series. Revision may be continuous or cyclical.

satellite image map   A general term for a map which uses and retains the photo-like images obtained from satellite sensors.

series   A set of map sheets drawn to a common specification which when assembled in their correct relative positions cover a large area of land and which together may be regarded as constituting a single map.

series designation   A coded unique alphanumeric or numeric identifier applied to a map or map series by the publisher.

sheet lines   The layout of sheets forming a map series as defined by the pattern of sheet edges. Sheet lines may conform to the lines of a grid or graticule, or they may overlap.

shaded relief   A method of relief portrayal using shading and highlighting of slopes to simulate the light and shade which might result from an oblique (or sometimes vertical) illumination of the relief. With oblique shading, the supposed light source is from the north-west.

side looking airborne radar (SLAR)   A sensing system in which a narrow beam of pulsed microwave energy is transmitted from one side of an aircraft and the returning echoes recorded on film. This provides a continuous strip image of the terrain.

specification   The rules laid down for the content, compilation and design of a map or map series.

spot height/elevation   An altitude (in relation to some datum such as mean sea level) marked on a map by a point symbol and a numerical value.

stereographic projection   A conformal azimuthal projection often used for maps of polar areas.

synthetic aperture radar (SAR)   A radar system that uses a short antenna giving a wide beam width. The returning signals are then electronically processed and combined to produce a resolution in the azimuth direction otherwise obtainable only by using an excessively long antenna.

thematic map   A map which, in contrast to a topographic or general-purpose map, displays data on a specific subject such as geology, soils or population.

topographic map   A general-purpose map at medium or large scale which gives a systematic representation of the physical and cultural features of the land surface including relief (which is usually shown by contours).

town map   Term used in this book to include both detailed plaimetric urban maps and the usually less planimetrically correct street index maps.

Transverse Mercator projection   A cylindrical projection whose axis is rotated at right angles to that of the normal Mercator. The TM projection is conformal, but the scale factor increases away from the central meridian. The projection is much used, often in a modified form, for the topographic mapping of areas of small east–west extent.

triangulation   The method of providing a horizontal survey control network by the measurement of a base line and the construction from it of a series of triangles whose angles are measured by theodolite observations from triangulation stations.

trimetrogon photography   Photography taken by three cameras mounted in an aircraft. One takes vertical photographs while the other two simultaneously take high oblique photographs at right-angles to the line of flight.

Universal Transverse Mercator (UTM) grid   A plane coordinate system devised to cover the whole world between the latitudes of 84° N and 80° S. It is designed for use with specific applications of the Transverse Mercator projection (UTM projection) in which the world is divided into 60 zones each 6° longitude wide and each with its own standard meridian.

vector data   Digital data collected by capturing the coordinates of points on the component line segments of a map.

GEOGRAPHICAL INDEX

This index is not a list of all geographical names cited in the book. It lists alphabetically all the nation states and other 'mapping units' into which the main body of the text is divided. It also includes some alternative names, and the names of sub-national areas (e.g. Canadian Provinces) where these receive separate discussion in the texts.

PUBLISHERS INDEX

This index lists all organizations (survey authorities, publishers and suppliers) whose addresses are cited in the book. Pages where addresses will be found are given in **bold** type; other page numbers refer to the principal passages in the text which describe these organizations.